PHYSICS IN PERSPECTIVE

VOLUME I

Physics Survey Committee   •   National Research Council

NATIONAL ACADEMY OF SCIENCES   Washington, D.C.   1972

NOTICE: The study reported herein was undertaken under the aegis of the Committee on Science and Public Policy (COSPUP) of the National Academy of Sciences–National Research Council, with the express approval of the Governing Board of the National Research Council.

Responsibility for all aspects of this report rests with the Physics Survey Committee, to whom sincere appreciation is here expressed.

The report has not been submitted for approval to the Academy membership or to the Council but, in accordance with Academy procedures, has been reviewed and approved by the Committee on Science and Public Policy. It is being distributed, following this review, with the approval of the President of the Academy.

Available from

Printing and Publishing Office
National Academy of Sciences
2101 Constitution Avenue
Washington, D.C. 20418

ISBN 0-309-02037-9
Library of Congress Catalog Card Number 72-84752

Printed in the United States of America

June 1972

Dear Dr. Handler:

As outgoing chairman of the Committee on Science and Public Policy
I take pleasure in transmitting to you the report of the Physics Survey
Committee, chaired by Professor D. A. Bromley of Yale University. This
has been reviewed by COSPUP, and the final version reflects our comments.

This is the second report on the physics disciplines undertaken under the
auspices of COSPUP, the first, *Physics: Survey and Outlook,* having been
completed in 1966 under the chairmanship of Dr. George E. Pake. Both
physics itself and the environment in which it operates have changed
enormously since the first report. In a number of ways the present report
is considerably more ambitious than the earlier one, both in what it attempts
to accomplish and in terms of the effort and number of people involved in
the Survey Committee and its panels and reviewers. It may be worthwhile
to list a few aspects of this report that distinguish it from earlier efforts.

1. The report contains a much more extensive analysis of the manpower
and educational problems of physics. With the aid of the American
Institute of Physics and with tapes provided by the National Register, new
data on mobility of physicists, the sociological aspects of physics, and the
future supply and demand for physicists have been presented.

2. The report documents the rapid growth of several fields at the interfaces between physics and other disciplines, such as physics in biology,
in chemistry, and in earth and planetary physics, and gives much more
extensive coverage to some of the more "classical" areas of physics such as
acoustics, optics, and the physics of fluids. Thus physics is presented in a
larger scientific context than it was in the earlier report, with more emphasis
on the interaction between physics and other sciences as well as with
engineering.

3. Rather than recommending specific funding levels for physics and
for the various subdisciplines covered, the report attempts to delineate the
scientific, and to some extent the social, consequences of several different
rates of funding increase or decline for each subdiscipline. This proves
much easier to do in fields requiring expensive equipment such as

iii

elementary-particle physics, nuclear physics, and astrophysics than in fields consisting of many smaller, relatively independent projects such as condensed-matter physics or atomic, molecular, and electron physics. Nevertheless, the scientific opportunities foregone because of constricted funding are fairly clearly spelled out.

4. In Chapter 5 the report develops a rather elaborate scheme of criteria for assessing priorities among program elements within the subdisciplines. It presents one example of an attempt at a jury rating among program elements, which demonstrates that it is possible for a group of scientists from diverse branches of a broad field to reach a fairly good consensus on the relative importance of particular areas of research at the moment. It should not be inferred from this exercise, however, that such a set of priorities will have more than temporary validity, since the field as a whole and the interactions between its parts are constantly changing. This is by far the most sophisticated attempt at assessing scientific priorities within a broad discipline that we have yet seen, but the very magnitude of the effort shows how difficult the problem is.

Two general impressions emerge from the reading of the report. First, despite the growing funding difficulties of recent years, American physics has enormous intellectual vitality and has made considerably greater progress than was anticipated in the Pake report of six years ago. Although the rest of the world has been catching up rapidly, American physics is still strong, although there is increasing doubt in the physics community whether this relative strength can last very much longer in the current environment.

Second, the nature of physics is such that increasing concentration of effort in major facilities or "critical-size" research groups is necessary. Even if there were no restrictions on funding, the optimum development of the field would lead to the concentration of activities in fewer major laboratories, especially in elementary-particle physics, nuclear physics, and astrophysics but to some extent in all subfields. Restrictions on future available funding will greatly accelerate this trend and lead to increasingly severe sociological problems in the field. The "user-group" mode of operation will increasingly have to be recognized as a way of life if physics is to retain its vitality at the frontier while still maintaining necessary diversity, competition, and close involvement with education.

Finally, it is clear that physics is increasingly an international enterprise, and that it should be viewed not in terms of international competition but rather in terms of each nation making a contribution to a worldwide cooperative effort that is commensurate with its available talent and resources.

COSPUP is not in a position to endorse or criticize the specific priorities or programs recommended in the report, but we believe that the Survey Committee has made a well-documented case for its conclusions and recommendations.

Sincerely yours,

HARVEY BROOKS
for the
Committee on Science and Public Policy

This report, the latest in a series of comprehensive surveys of science disciplines, represents a major achievement. It offers an impressive and balanced picture of progress in physics and its subfields and relates the contribution of physics to progress in other scientific disciplines and the manner in which all of these have been utilized by our civilization. Further, the report suggests the future course of endeavor in the various subfields of physics and the extent to which these will require resources of manpower, facilities, and funding; it then assays the rational responses of the total research effort in physics to alternative funding levels. This constitutes the most successful effort, to date, to establish intradisciplinary priorities. Those who conducted this most impressive exercise in science policy formulation are much to be congratulated.

On behalf of the Academy, I wish to express our appreciation to the Committee on Science and Public Policy for its general guidance and extensive consultations with the Physics Survey Committee during the course of this work. For this prodigious accomplishment, I extend our gratitude to the Survey Committee and most especially to its energetic, imaginative, and dedicated chairman.

PHILIP HANDLER
President
National Academy of Sciences

Washington, D.C.
June 1972

vii

COMMITTEE ON SCIENCE AND PUBLIC POLICY

ix

PHYSICS
SURVEY
COMMITTEE

Ex officio

HARVEY BROOKS, Harvard University
E. L. GOLDWASSER, National Accelerator Laboratory
H. WILLIAM KOCH, American Institute of Physics
ROMAN SMOLUCHOWSKI, Princeton University

CONSULTANTS

MARVIN GOLDBERGER, Princeton University
ROBERT W. KEYES, International Business Machines Corporation

LIAISON REPRESENTATIVES *

C. NEIL AMMERMAN, Department of Defense
W. O. ARMSTRONG, National Aeronautics and Space Administration
MARCEL BARDON, National Science Foundation
DONALD E. CUNNINGHAM, National Science Foundation
PAUL F. DONOVAN, National Science Foundation
WAYNE R. GRUNER, National Science Foundation
GEORGE A. KOLSTAD, Atomic Energy Commission
JOHN J. McCAMBRIDGE, Department of Defense
WILLIAM METSCHER, Department of Defense
SIDNEY PASSMAN, National Science Foundation
JACK T. SANDERSON, National Science Foundation
A. W. SCHARDT, National Aeronautics and Space Administration
MILTON M. SLAWSKY, Department of Defense
HENRY SMITH, National Aeronautics and Space Administration

GEORGE W. WOOD, National Academy of Sciences, *Executive Secretary*

Staff Officers

BERTITA E. COMPTON
BRUCE N. GREGORY
CHARLES K. REED

* In general, no agency was represented on the Committee or any of its Panels by more than one person at a time.

xi

PREFACE

The Physics Survey Committee was appointed by the President of the National Academy of Sciences in mid-1969 and charged with an examination of the status, opportunities, and problems of physics in the United States. The Committee has interpreted this charge broadly and has attempted to place physics in perspective in U.S. society. It has evolved an approach to the establishment of priorities and program emphases that may have wider potential application; it has conducted detailed studies on education in physics and physics in education, on the production and utilization of physics manpower, on the dissemination and consolidation of physics information. It now presents a status report on both the core areas and interfaces of physics. Most striking among the findings of this Survey are the overall power and vitality of American physics—a power and vitality that are a tribute to generous support from the U.S. public over the past two decades.

But this strength is in danger. In this report we consider the sources of this danger and the changes that we believe are required to retain a leadership role for U.S. physics in international science.

Although we have dealt principally with problems of particular importance to the physics community, it is clear that many of these problems are much more general and affect all science. Our statistical data and discussions are necessarily rooted in physics, but it is our hope that our conclusions and recommendations may be of use in a broader context.

xiii

Throughout this Survey we have been reminded repeatedly of the great unity not only of physics but also of all science. This unity is all too often forgotten or ignored.

By its very nature a survey committee explores many alternatives and options in developing its report. We emphasize that the lack of explicit mention of one of these in the Report does not imply that it has not been examined or considered.

Our Survey has involved more than 200 active members of the U.S. physics community, listed in Appendix A; we are grateful to them for their most effective help and cooperation. Support for the Survey activity has been provided equally by the Atomic Energy Commission, the Department of Defense, the National Aeronautics and Space Administration, and the National Science Foundation. Additional assistance has been provided through grants from the American Physical Society and from the American Institute of Physics.

Members of all the federal agencies engaged in the support of physics have given generously of their time and effort in searching out and providing answers to innumerable questions. Those persons listed in Appendix A have participated also in long days of discussions as the Report developed. We are deeply grateful to all of them.

We cannot hope to acknowledge in detail all the assistance that we have received from many persons throughout the country during the course of this Survey. The Committee is greatly indebted to George W. Wood and Charles K. Reed and to Miss Bertita Compton, without whose continuing assistance this report would never have been completed. We would be remiss indeed, however, if we did not especially recognize the work, far beyond any call of duty, of Miss Beatrice Bretzfield, who has acted as Secretary to the Physics Survey at the National Academy of Sciences since 1970, and of Miss Mary Anne Thomson, my administrative assistant at Yale.

D. ALLAN BROMLEY, *Chairman*
Physics Survey Committee

CONTENTS

1
ORIGIN, OBJECTIVES, AND ORGANIZATION

BACKGROUND OF THE REPORT

The Survey described in this report began only months before Armstrong's first footprint on the lunar surface—a footprint marking the spectacularly successful completion of an age-long dream and the culmination of the greatest of man's technological adventures. Paradoxically, perhaps, the stimulus for the Survey was recognition of the rapidly growing problems affecting the continued progress and well-being of all science and technology.

This report deals principally with the opportunities that physics faces during the 1970's. These opportunities are of several kinds: opportunities for fundamental new insight into the nature and causes of natural phenomena; opportunities for the development of new devices and technologies; and opportunities for greater service, both direct and indirect, to U.S. society, of which physics forms an integral part. The problems addressed are largely those that we foresee in the realization of these opportunities.

Although it is clear that physics cannot speak for all science—much less technology—as one of the most basic natural sciences it has close ties and interactions throughout both science and technology. It underlies many of the new developments in these fields and shares many of the problems of the overall intellectual and practical enterprise.

In large measure these problems reflect significant changes that have occurred in recent years. Among these have been striking changes in the different subfields of physics, a marked change in the growth rate of federal

1

support for physics, changes in many aspects of contemporary U.S. society, changes in an ill-defined but nonetheless real ordering of priorities in the public mind, and changes in both the academic and federal communities that have particularly affected the motivation and basic philosophies of much of a student generation.

Why a New Report?

These changes, too, provide much of the answer to the very real question: Why yet another major survey report on a scientific field? In 1966, under the aegis of the Committee on Science and Public Policy of the National Academy of Sciences, a national committee chaired by George E. Pake published an extensive report on the status and opportunities of U.S. physics entitled *Physics: Survey and Outlook*. Why then the present Report?

The previous survey was completed at a time when the growth rate of U.S. science in general, and of physics in particular, was at an all-time high and the field was in a state of robust health. Under these circumstances, and within the framework of a burgeoning national economy, it was not surprising that physics was considered in a relatively narrow context that justified extrapolation of the needs and objectives of the field on the basis of internal considerations and the anticipated exploitation of most of the new opportunities that the field then presented (see Figure 1.1).

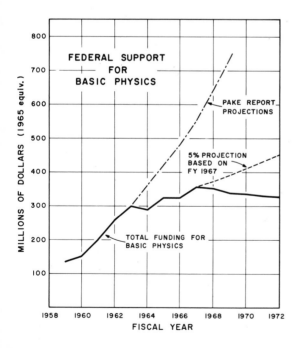

FIGURE 1.1 Federal support for basic research in physics.

Contingency Alternatives

These extrapolations have not been realized. Although it is extremely important for physicists to continue to emphasize these potentials, it has become necessary to address the difficult question of priorities among and within subfields of physics. Many of the frontier areas of physics, because of the scope and scale of the activity and instrumentation required, are inescapably dependent on federal support. And the competition for available federal support is increasing steadily as new attacks are mounted on major problems of national and social concern. Therefore, this Committee has attempted to define and develop contingency alternatives in an effort to ensure the most effective use of the available support throughout physics.

But there should be no misunderstanding: The opportunities and challenges for both internal growth and external service to society are still present; however, if current trends in the growth of support for U.S. physics continue, only few of these opportunities will be realized and the relative position of the United States in the international physics community will inevitably decline.

Physics and Society

Physics has contributed and continues to contribute to society in a great many ways through its concepts, its devices and instruments, and, more especially, the capabilities and activities of its people. Basic to the training of a physicist is the effort to reduce complex situations to their most fundamental aspects so that they can be subjected to mathematical tools and the philosophic rigor of natural laws. The research style and approach that characterize the physicist frequently constitute his major contribution to attacks on problems outside his own specialized field. In the past, the training of a physicist was characterized also by a breadth and flexibility that permitted him to range widely in his search for challenging problems, not only in physics but in other disciplines.

As the problems of society have multiplied—problems of population, poverty, and pollution, to name only three—largely coincidentally and unhappily, an erosion of these characteristics of breadth and flexibility has occurred, with a corresponding reduction in the effectiveness of physicists as partners in the solution of societal problems. This situation led the Committee to a much more intensive study of educational and manpower questions in physics than has been undertaken previously and to explicit recommendations to the scientific and academic communities for steps toward the resolution of problems in these areas.

Further, because our study has convinced us that a significant part of the apparent alienation of students and younger members of the scientific community in recent years and the growing tension between them and the academic, industrial, and federal institutions have reflected both lack of historical perspective and lack of awareness of current goals and objectives, we have included significantly more discussion of these aspects than has been the case in earlier surveys.

Interfaces with Other Sciences

Again in contrast to earlier surveys and in recognition of the rapidly growing importance of work at the interfaces between physics and other sciences and disciplines, we have devoted extensive effort to the achievement of a better understanding of what contribution physics might make in these interface areas and what implications such activity might, in turn, have for the structure of and activity and training in the physics enterprise. Work at the interfaces ranges from the most abstruse to the most practical. An example of the former is the astrophysicist's search for understanding of the mechanism for the total collapse of a star into inconceivably small dimensions, together with the study of its light, its magnetic fields, and all other evidence of its former presence. The search for better isotopes for diagnosis and treatment of cancer, or the search for understanding of the mechanisms of nerve action in biophysics, to alleviate ever more of the crippling afflictions that plague mankind are examples of the latter. In geophysics and planetary science, as a single additional example, the long-awaited understanding and prediction of major earthquakes may be at hand, with most profound practical and social consequences.

In short, in this Report the Committee has attempted to consider not only the internal logic of physics but also to put physics into a much broader context and perspective both in science and technology and in U.S. society. It has also addressed the crucial problems of priority involved in working toward most effective utilization of support under conditions in which that support may well be inadequate to permit pursuit and exploitation of many excellent new opportunities and ideas in the science.

ORIGIN OF THE SURVEY COMMITTEE AND PANELS

The Survey Committee

Early in 1969 and as a consequence of extended discussions within and among the Committee on Science and Public Policy of the National Academy of Sciences, the Division of Physical Sciences of the National

Research Council, the Office of Science and Technology, the President's Science Advisory Committee, and several of the major federal agencies that support physics, a decision was reached to initiate this Survey of the U.S. physics enterprise. Some of the Survey Committee members were appointed by then president of the National Academy of Sciences Frederick Seitz and others by Philip Handler, following his assumption of the presidency during the summer of 1969. At about the same time, a parallel committee, chaired by Jesse Greenstein, was charged with the responsibility of surveying the opportunities and needs of astronomy. Both committees have functioned under the auspices of the Committee on Science and Public Policy, chaired by Harvey Brooks.

The Subfield Panels

In developing its report, the Survey Committee early decided that it was essential that it obtain detailed input from panels of experts in each of a number of subfields. Several of these subfields have relatively well-defined and traditional boundaries in physics. These include acoustics; optics; condensed matter; plasmas and fluids; atomic, molecular, and electron physics; nuclear physics; and elementary-particle physics. In addition, there are several important interfaces between physics and other sciences. In the case of the interface with astronomy, which is a particularly active and overlapping one, the Physics and the Astronomy Survey Committees agreed to use a joint panel to report on astrophysics and relativity, an area of special interest to both. The broad area of overlap of physics with geology, oceanography, terrestrial and planetary atmospheric studies, and other environmental sciences was defined as earth and planetary physics, and a panel was established to survey it. In covering the physics–chemistry and physics–biology interfaces, the broader designations physics in chemistry and physics in biology were chosen to avoid restricting the study panels to the already traditional boundaries of these interdisciplinary fields.

Although each panel, particularly those in the core subfields, was asked to consider the interaction of its subfields with technology, the Committee anticipated that the emphasis would be on recent developments that advanced the state of the art and on what is generally known as *high technology*. Therefore, to include also the active instrumentation interface between physics and the more traditional manufacturing sectors of the economy—steel, drugs, chemicals, and consumer goods, to name only a few in which many old parameters are being measured and controlled in new and ingenious ways—a separate panel was established with the specific mission of examining the entire range of U.S. instrumentation activities.

For each of these subfields and interfaces a chairman was appointed by the Chairman of the Survey Committee, and groups of recognized experts, representing a cross section of activity in each, were brought together. Two additional panels were appointed to centralize the statistical data-collection activities of the Survey and to address the questions of physics in education and education in physics. An extended report on the dissemination and use of the information of physics was prepared by one of the Committee members. In addition, the Survey benefited enormously from assistance on a wide range of topics from a number of smaller working groups and individuals.

The Nuclear Physics Panel was established on an accelerated time scale, and in greater depth as compared with the remaining panels, in response to a specific request from the President's Science Advisory Committee (PSAC). The request was for a preliminary version of this panel's report for PSAC use in policy and planning discussions.

Appendix A provides a listing of the Committee and Panel members and their affiliations.

A number of subjects in classical physics, such as mechanics, heat, thermodynamics, and some elements of statistical physics were not considered explicitly in the Survey. This omission in no sense implies any lack of importance of these fields but merely indicates that they are mature ones in which relatively little research *per se* currently is being conducted.

Early in the Survey Committee activity, a lengthy charge, which appears as Appendix B, was developed and addressed to each panel. This charge was broad ranging and dealt with the structure and activity of a subfield, viewed not only internally but also in terms of its past, present, and potential contributions to other physics subfields, other sciences, technology, and society generally. Consonant with the overall survey objectives, each panel also was asked to develop several detailed budgetary projections ranging from one that would permit exploitation of all currently identified opportunities in a subfield to one that continued to decrease during the period under consideration.

Clearly, the charge was most directly relevant to the more traditional subfields; in the case of the interface panels, some questions were inevitably unanswerable without a survey of equivalent scope of the field or fields on the other side of the interface. In astronomy such a survey was available. The panels on statistical data, education, and instrumentation were, of course, special cases. As Volume II of this Report shows, the panels responded in depth to the questions asked.

Initial draft responses to the charge were presented to the Survey Committee by the panel chairmen during an extended working session in June 1970, and, following subsequent discussions and reviews, preliminary panel reports were submitted to the Committee during the summer of

1971. Whenever possible, each of these preliminary reports was forwarded also to a group of some ten readers, selected jointly in each case by the panel chairman and the chairman of the appropriate division of the American Physical Society or other Member Society of the American Institute of Physics. These readers were chosen, insofar as possible, to include very active scientists in each subfield, with particular emphasis placed on the inclusion of younger scientists who had not been previously involved with the Survey activity. The Committee received excellent cooperation from all these readers, who provided fresh insight and new viewpoints concerning many aspects of the panel reports. Their comments and those of the Survey Committee and other reviewers were carefully considered by the panels in the preparation of the final reports that appear in Volume II. Appendix C provides a listing of the readers who participated in this activity.

In the preparation of its Report the Survey Committee drew heavily on the panel reports for information on the various subfields and specialties. The panel chairmen also were most effective participants in many of the Survey Committee meetings and working sessions.

ORGANIZATION OF THE REPORT

The present volume of *Physics in Perspective* comprises Volume I of the complete Physics Survey Committee Report; it is divided into 14 chapters, including this one. It may be useful to provide here a brief summary of its contents.

Recommendations

Although recommendations occur throughout the entire report, together with appropriate background material and discussion, for convenience we have grouped major recommendations and basic premises in Chapter 2. In each instance we have indicated both the audience to which the recommendation is primarily directed and the specific chapter in this volume in which further supporting discussion may be found. More specialized recommendations and conclusions also appear in each of the panel reports in Volume II.

The Nature of Physics

Chapter 3 presents a discussion in some depth of the nature of physics as a science and as a part of Western culture. It addresses such basic questions as: What is physics? Why are physicists interested in it? Why should anyone else be interested in it?

The Subfields of Physics

Chapter 4 includes a summary for each subfield of physics and for the interface areas considered in the Survey. Each summary describes the present status, recent developments and achievements, and outstanding opportunities now identified, together with illustrations selected to demonstrate the vital unity and coherence of physics. This unity is the subject of the concluding section of Chapter 4, in which we illustrate first the remarkable similarity of concepts and theoretical techniques that are employed in condensed matter, atomic and molecular, nuclear, and elementary-particle physics, and then the even more remarkable extent to which almost every branch of classical and modern physics contributes to the understanding of the recent beautiful measurements on pulsars. In considering achievements and opportunities we have focused on those relating to the solution of major national problems and to other fields of science and technology as well as those internal to physics.

To facilitate a more detailed examination of each subfield, we have divided each into what we have defined as program elements. These are scientific subgroupings having reasonably identifiable and unambiguous boundaries with which it is possible, with reasonable accuracy, to associate certain fractions of the total manpower and federal funding in each subfield.

Priorities and Program Emphases in Physics

Chapter 5 is in many ways the heart of the report in that it attempts to address the very difficult questions of priority, program emphases, and levels of support. Because new developments and discoveries can change situations and priorities in physics, as indeed in any human enterprise, in rapid and unexpected fashion, the Committee thought it desirable to emphasize and develop the criteria that could be used as a basis for priority decisions rather than the specific decisions. Exceptions, of course, occur, as in the case of major facilities.

To this end we have examined in detail many of the approaches to priority determination in science that have been discussed in the literature in recent years, and we include in Chapter 5 our evaluation of the positive and negative aspects of each. From this broader examination we have evolved our own set of criteria—much modified in the course of trial application to the sets of program elements developed in Chapter 4. We have found it convenient and effective to divide these criteria into three classes: intrinsic, relating to the internal logic of a science and its fundamental bases; extrinsic, relating to its potential for application in other sciences; and structural, relating to available manpower, instrumentation, and institutions and to questions of opportunity and continuity.

We have refined these criteria by applying them, in a jury rating sense, to the program elements of the core subfields of physics. In Chapter 5 we also illustrate the application of the first two classes of criteria. Because the structural criteria depend to a much greater extent on detailed knowledge of the individual research project and investigator or group, we have not attempted any equivalent general illustration of their use. Instead, we have selected as illustrations certain situations in which structural considerations play a predominant role; for example, situations in which major facilities, approved and placed under construction in the mid-1960's, when quite different conditions and expectations in regard to the growth of federal support for science prevailed, are now becoming operational and require a step-function increase in funds for operation. If it should become necessary, under declining or even level support conditions, to obtain these incremental funds through selective termination of other ongoing programs, the much greater size and costs of the new facilities would require the elimination of much of the present high-quality activity in each of the subfields concerned.

Finally, we address the difficult questions involved in any attempt to establish a national funding level for physics—or indeed for any science or other activity that depends heavily on federal funds. We present our evaluation of these difficulties in relation to several recently proposed mechanisms.

We do not recommend an overall detailed national physics program level. As we emphasize later, this omission reflects the Committee's conclusion that it is impossible for any such group to develop, within the relatively limited time and level of activity that are possible, either adequate information or insight to make such a detailed attempt meaningful. Nor are we convinced that it is inherently desirable for any such small group to attempt to determine national priorities at this level of detail.

What we have attempted to suggest are criteria and a mechanism for their use that may enable intercomparison among subfields, or fields, of science on a more objective basis. We have elicited from the various panels detailed budgetary projections adequate to permit interpolation to match a wide range of possible funding levels and thus have attempted to provide contingency alternatives. Finally, using data from the panel reports, presentations by panel chairmen, and all other information available to us, we have selected some 15 program elements from the total of 69 that were subjected to our jury rating procedure that we believe to be particularly meritorious of increased support and attention.

The Consequences of Deteriorating Support

In Chapter 6 we attempt to describe the short- and long-range consequences for physics, science, the U.S. research enterprise, and the nation as a whole of continued deterioration of federal support. Beginning in 1967 with the abrupt change in the growth rate of support for physics, a wide variety of what were regarded as temporary or short-term mechanisms were developed to sustain individual research programs or groups until better days. Some of these measures had beneficial effects in forcing maximum efficiency in the use of available resources. But at the same time many of these have other effects—some of very subtle character—that in the long run can seriously weaken or destroy whole segments of the research community. Measures that are tolerable for a few years as stop-gap measures become frozen or institutionalized if too long continued—and often with very unfortunate consequences.

We do not claim any special position or consideration for physics, although for a variety of reasons that we discuss throughout the Report its problems are among the more serious in U.S. science. We recognize that other segments of the U.S. scientific, industrial, and technological sectors have suffered major disruption also. Our discussion focuses on physics because we know it intimately. However, we believe that the phenomena that we observe may have much greater generality and that it is extremely important that there be greatly increased public awareness of the consequences of a continuing deterioration in the support of U.S. science.

Physics and U.S. Society

Chapter 7 considers the role of physicists and physics in U.S. society. It illustrates through the discussion of selected examples the contributions that both have made and continue to make to society. It also highlights some problems that have developed in the interaction between physics and society in recent years. Here, too, we include a discussion of the connections between physics and the health of the national economy through the channel of high-technology industries, in particular, and a comparative discussion of research and development activities in a number of selected foreign countries.

International Aspects of Physics

The physics community has from its earliest days displayed a truly international character. Chapter 8 considers the implications of international interaction and cooperation for U.S. physics and the contributions that the physics community can make to better international understanding and communication.

The Institutions of Physics

In Chapter 9 we discuss the nature of physics and its institutions and show how the evolution of the science has led naturally to the development of three major foci—the universities, the industrial laboratories, and the national laboratories; we present a brief discussion of their historical evolution and of the career patterns of typical physicists in each.

In view of a continuing trend toward user-group activity in physics, in which experimental or even theoretical studies are necessarily conducted at a site remote from a scientist's home base to take advantage of a telescope, accelerator, computer, or magnet, we discuss in some detail the structural and organizational problems that can arise in such user-group situations and present some recommendations for their alleviation. We discuss, too, possible mechanisms for increasing the interaction among the various institutions and the changes in each that such increased interaction might require or imply.

The Support of U.S. Physics

Closely linked to the growth of U.S. physics has been the development of a complex federal support structure; indeed, the multiplicity of support channels has been one of the major sources of strength of U.S. physics. Because information concerning the way that this complex support structure functions has not been readily available in the past, and because we believe that much misunderstanding and misinformation are current throughout the physics community, we have attempted to provide a brief sketch of the historical origins of the present support mechanisms and of their current operation as viewed by a prospective recipient of federal support. Finally, we briefly discuss the structure and functioning of the many federal agencies that provide funds for research in physics.

Physics in Education and Education in Physics

Because education is so vital to the entire physics community and because, in turn, physics has much to contribute to education in its broadest interpretation, we have devoted extensive effort to an examination of the entire U.S. educational enterprise as it relates to both education in physics and physics in education. This examination encompasses the range from elementary school to graduate and midcareer education. At all levels we found significant problems that must be considered if physics is to realize its potential contributions to U.S. society, and we address recommendations to the institutions and agencies that we believe are best qualified to resolve these problems. We discovered in our study a wide variety of statistical information that was new and often surprising, at least to members of this Survey Committee. All this information, with discussion and recommendations, appears in Chapter 11.

An important aspect of education, broadly interpreted, is public awareness of science and of its possible contributions to a better life. In the United States, efforts to inform and educate the public about science are few and usually of poor quality. The scientific community must accept a large share of the responsibility for this situation. If the present antiscience trend is to be reversed, individual scientists and scientific organizations must be prepared to spend more of their time and money than has been the case previously in fulfilling their obligation to inform a much larger segment of the U.S. public.

Manpower in Physics: Patterns of Supply and Use

Using the extensive information that we have extracted from the National Register of Scientific and Technical Personnel data tapes as well as input from our own questionnaires and those of the Manpower Committees of the American Physical Society and the American Institute of Physics, we have assembled what we believe to be the most complete information concerning the training and use of U.S. physicists yet available. We include data on mobility, sociological aspects of physics, and a brief historical study, and we develop projections for physics manpower needs using several different models. Although serious problems clearly exist in regard to employment opportunities in the traditional types of jobs that physicists have held, there is also evidence to suggest that the scientific and academic communities are in some cases overreacting, and that unless some action is taken the oscillatory phenomena associated with the supply of U.S. physicists in the recent past will not be adequately damped. What emerges clearly from these studies is that the growth in tenured academic employment opportunities during the 1960's was indeed anomalous, and that it is essential to take steps to broaden the motivations and interests of young U.S. physicists beyond the academic sector—as was typical prior to the late 1950's, for example—or serious dissatisfaction and alienation will remain. We discuss these considerations in Chapter 12.

Dissemination and Use of the Information of Physics

Unfortunately, the available mechanisms for disseminating and consolidating the information of physics so that it is readily and conveniently available to all potential users have simply failed to keep pace in many areas with the rapidly increasing rate at which individual facts can be wrested from nature in modern research laboratories. Consequently, in many branches of U.S. physics a situation has developed in which it frequently is simpler, faster, and less expensive to remeasure a desired datum rather than to find its previously measured value in the scientific literature. This situation represents an unacceptable waste of the time and resources of the entire scientific community. In Chapter 13 we describe in detail the functioning of the various media for the communication of the information of physics. Again, we have developed an extensive body of new statistical information. We discuss these data in relation to current problems and develop specific recommendations directed toward different sectors of the physics enterprise. These recommendations suggest steps that, if implemented, can alleviate or resolve the present difficulties.

Policy Considerations: Conclusions and Findings

In addition to the major recommendations that we have collected in Chapter 2, the Committee activities during the Survey also led to a number of conclusions and findings that we believe are important for longer-range planning for U.S. physics and for the most effective use of the resources that become available to it. Many of these findings and conclusions may be found throughout the Report, with their supporting discussion and statistical information. However, for convenience, we have collected our conclusions and findings in Chapter 14 together with brief introductory remarks for each. We have grouped them according to the different audiences most directly concerned.

Consequently, Chapters 2 and 14 constitute a condensed presentation of the recommendations and conclusions that the Committee developed during the course of its Survey. The entire report (including both this volume and the one containing the reports of the various subfield panels) provides the detailed support and justification for these recommendations and conclusions.

2
RECOMMENDATIONS

INTRODUCTION

Science is knowing, and the most lasting and universal things that man knows about nature make up physics. As man gains more knowledge, what would have appeared complicated or capricious can be seen as essentially simple and, in a deep sense, orderly. To understand how things work is to see how, within environmental constraints and the limitations of wisdom, better to accommodate nature to man and man to nature.

For more than 25 years physics in the United States has set the style and pace of worldwide activity in this discipline. The major thrust of this Report is based on the belief that the interests of both the nation and this science will continue to be served best by the maintenance of physics as a vigorous enterprise at or near the frontiers of activity in each of its various branches. Such an objective is consistent with the vital role that physics plays in society, the unity of science, the importance that pre-eminence in science implies for the nation, and the expressed intent of both the executive and legislative branches of the U.S. Government.

The achievement of this objective in the face of changing national goals will require both making the most effective use of present resources and finding new sources of support for physics. Some of the difficult choices that may lie ahead, together with their probable consequences, are detailed in Chapter 5 of this Report and form the basis for the recommendations that follow. These recommendations are addressed to the community of physicists, which holds the responsibility for utilizing existing resources as wisely as possible, and to the federal government, from which additional support must be sought. They are augmented by the conclusions and findings 'presented in Chapter 14. Each has been cross-referenced to the chapter, or chapters, of the Report upon which it is based and to the specific audience to which it is addressed. More specific recommendations are to be found throughout the Report and in the panel reports in Volume II, with the discussion that supports them.

Throughout the Report, and in many of the recommendations that follow in this chapter, we necessarily have addressed problems that are not unique to physics but common to all the sciences. Our discussions and recommendations relate almost entirely to the physics aspects of these more general problems and must be understood in this context as a contribution from one of the sciences to what must be, in many cases, much broader considerations.

SOURCES OF SUPPORT

Let there be no misunderstanding. Even with the most judicious use of existing resources, this nation cannot continue as a leading contributor to world physics without support greater than is now available. This objective faces the hard facts of changing national goals. Three of the four federal agencies that currently are responsible for more than 90 percent of the federal support of U.S. physics—the Atomic Energy Commission (AEC), Department of Defense (DOD), and National Aeronautics and Space Administration (NASA)—continue to suffer reduced funding, especially for their fundamental research programs; only the National Science Foundation (NSF) has experienced budget increases, and these have been largely offset by the transfer of basic research projects from other agencies and by the diversion of funding to more technologically oriented projects. In general the federal component of support is by far the largest and consequently commands greatest attention. To satisfy the new national priorities and make possible the achievement of the stated objectives of many of the new federal agencies, it will be necessary both to maintain and expand the research programs of the agencies that presently support physics and to develop other appropriate sources of funding for physics within the federal government.

Recommendations

1. The federal agencies that have a long-term dependence on physics (the Atomic Energy Commission, Department of Defense, and National Aeronautics and Space Administration) should expand their support to a level more nearly commensurate with their stated needs. A strong reversal of the present downward trend in the support of basic science components of their programs is required to ensure the long-range capabilities of these agencies to fulfill their responsibilities.

2. All federal agencies with missions that rely to some extent on basic physics should accept the support of physics research as a direct responsibility. These agencies include, among others, the National Institutes of Health, Department of Transportation, Department of Housing and Urban Development, Department of the Interior, Environmental Protection Agency, and Department of Commerce. The Office of Science and Technology should work with these agencies to develop general guidelines for such support. A 100:10:1 ratio, corresponding to the value of the high-technology product, the support of the related development, and the support of the underlying basic research, has characterized some mission-oriented federal agency and industrial programs in the past. The new agencies should strive for such a ratio in the allocation of their resources. If the side effects as well as the major thrust of each new development are to be understood and steps taken to mitigate or avoid possible undesirable consequences, it will be necessary in the future to continue research as an integral part of the development process. This added requirement will inevitably increase the fractional research cost for new developments.

Recommendations (Continued)

FEDERAL GOVERNMENT
The Congress
Atomic Energy Commission
Chapter 10
Chapter 14

3. The recent addition to Section 31 of Paragraph (6) of the Atomic Energy Act as well as the revision of Sentence 1, Section 33, which now give the Atomic Energy Commission a general responsibility for research on energy, should be interpreted as encouraging support of those areas of basic research in physics that underlie this broadened responsibility of the Commission. To reflect the seriousness of the energy problems of the United States and the world, research appropriations of this agency should be increased substantially to permit vigorous attack on all aspects of research into energy generation and transmission. (See also page 50.)

FEDERAL GOVERNMENT
The Congress
Atomic Energy Commission
Chapter 10
Chapter 14

4. In view of the outstanding success with which the U.S. Atomic Energy Commission has developed and supported a broad program of fundamental physics research, both in the past and at present, the Atomic Energy Commission should seek the necessary support—and the appropriations to this agency should be correspondingly increased—to enable it to maintain and expand its support of basic physics research programs in *both* the universities and the national laboratories.

FEDERAL GOVERNMENT
Support Agencies
National Science Foundation
Chapter 10
Chapter 14

5. In accordance with its primary responsibility for federal support of basic science, the National Science Foundation should seek to maintain the integrity and balance of the national physics program through selective emphasis on those segments of basic physics that have less obvious relevance to the missions of other federal agencies supporting physics research. Balance in the national physics enterprise should take priority over balance of the National Science Foundation physics program itself. To function adequately in this role the

Recommendations (Continued)

national support in basic physics now provided by the National Science Foundation should roughly double, and appropriations to this agency should be increased for this purpose. This increase should not be at the expense of the ongoing programs of the mission-oriented agencies.

PHYSICS COMMUNITY
Nonacademic
Chapter 4
Chapter 7

6. In certain areas of basic physics, industrial support is comparable with that from federal sources. It is desirable that this industrial support be maintained or increased to reflect the probable increased relevance of such knowledge to industry in the future, in relation to both productivity and the increasing need for foresighted technology assessment prior to the marketing of new products. Productivity will be of rapidly growing importance in maintaining the international competitiveness of U.S. industry over a broad range of product sectors.

Stabilization of Support

As in any enterprise involving long-range goals, specialized facilities, and highly trained people, a degree of stabilization and continuity in the support of physics is essential to minimize the dislocation and waste of opportunity and training that all too frequently follow sharp changes and fluctuations in support. The consequences of the abrupt 1967 change in the rate of growth of U.S. physics support provide an instructive example and emphasize the need for developing and implementing long-range projections.

At the heart of the stability problem is the fact that the funding cycle has been an annual one, while the cycle for an experimental program, the completion of a graduate program of study, or the development of new research concepts to the stage at which they are widely used is more typically three to five years—if not longer. Many of the consequences of uncertainty could be removed if it were possible for funding agencies to provide investigators with reasonable assurances of support over this longer period. Certain agencies have attempted to establish their support of physics on such a forward-funding basis. Introduced some years ago during a period of increasing support, forward funding was rejected by most physicists because they felt that it would limit the flexibility and growth potential of their research programs. In the present period of more restricted funding, the advantages of forward funding have become much more widely appreciated and sought after. Unfortunately, however, this funding mechanism was an early casualty of the increasing pressure on agency budgets.

It should be emphasized that forward funding as such does not imply increased support levels. If the appropriations for a given fiscal year can be increased adequately to permit agencies to provide assurance of support for a portion of their programs for three to five years in advance, greatly increased stability can be obtained without increased treasury withdrawals for any given year. The annual cash flow would not be changed by such a procedural change, yet the return to science and the nation could be great. However, it will be important to avoid loss of flexibility, with the resultant inability to respond rapidly to new opportunities and needs; an appropriate balance between stability and flexibility might be achieved initially if one third of the program support were provided on a forward-funding basis.

Recommendations

FEDERAL GOVERNMENT
Support Agencies
Office of Management and
  Budget
Office of Science and
  Technology
Chapter 6
Chapter 14

7. The federal agencies supporting physics should seek an increase in their appropriations in fiscal year 1974 and any necessary authority such that approximately one third of their physics projects could be assured support for a three- to five-year period. This particular incremental appropriation should be carefully supervised by the Office of Management and Budget to assure that it is used only as needed for the purposes of stabilization and not for an increase in program expenditures. It would imply that for those projects supported under the forward-funding programs, planning would always cover a minimum of three years.

ALLOCATION OF SUPPORT

One of the difficult decisions that must be faced in allocating available support involves the balance between two important national goals: the maintenance and advancement of the most innovative and significant science and the distribution of the available support to as many promising individuals and institutions as possible. When support is level or decreasing, these goals frequently are competitive and the choices are especially difficult to make.

The support of the highest quality activity and most promising people has long been a feature of the U.S. funding pattern; under conditions of limited funding this feature takes on increasing importance.

Recommendations

FEDERAL GOVERNMENT
Support Agencies
Chapter 5

8. Under current and foreseeable economic constraints it is not possible to support adequately all those individuals and research groups identified as having excellent research ideas and high research potential as judged by federal agency review procedures and peer evaluations. In decisions on the allocation of funds, therefore, preference must increasingly be given to maintaining the position of individuals and groups who are at, or very near, the forefront of world activity in their subfields, consistent with maintenance of balance in the overall national program of physics.

FEDERAL GOVERNMENT
Support Agencies
Chapter 5

9. Under conditions of limited support, programs should be terminated and facilities should be closed in preference to continued operation of all under marginal conditions.

FEDERAL GOVERNMENT
Support Agencies
Chapter 5

10. The construction of new facilities and the initiation of new programs should be restricted to situations in which clearly defined new needs or opportunities exist; under conditions of limited funding, programs and facilities justified primarily on the basis of geographical or institutional equity should be deferred. At the same time it must be emphasized that failure to respond to new needs and opportunities, when a clear consensus regarding them exists in any

Recommendations (Continued)

scientific field, can have an unusually detrimental impact on the overall progress of that field.

11. While physics departments should continue to give students as wide a choice of fields of specialization as feasible, they should not support or initiate programs in areas of physics simply to have activity in all its major subfields. They should concentrate on those areas in which they can meet or exceed the critical level of activity required for high-quality work. Significant progress has already been made in the physics community in evolving regional cooperative arrangements to utilize most effectively particular strengths of the participating departments. These cooperative arrangements must be pursued with even greater vigor in the future.

PHYSICS AND NATIONAL GOALS

Limitations on man's ability to fulfill human needs often have technical components that can be removed only through research and development. Therefore, many industries and many federal agencies, such as the DOD, NASA, AEC, National Institutes of Health, and Department of Agriculture, invest heavily in research and development. However, some large industries and a number of federal agencies support little or no research and development.

Transportation, housing, and environmental quality recently have been designated national problem areas, and large federal agencies have been established to deal with them. To realize their potential for national service it is important for these new agencies to find ways to bring science fully to bear on the achievement of their missions.

Environmental monitoring provides an example of a problem area in which physics has an immediate and important role. The contaminants of greatest significance are frequently present in the natural environment at concentration levels so low that they elude detection with conventional monitoring instrumentation, yet the long-term consequences of their presence could be serious. Fundamental to any effective program of environmental improvement or control is the ability to detect these contaminants accurately with reliable and often portable instrumentation. The physics community has already responded effectively with a whole range of new ultrasensitive monitoring devices, but much remains to be done.

Recommendations

12. The Department of Transportation, Department of Housing and Urban Development, and Environmental Protection Agency, as well as other agencies, should be encouraged by the Congress, the Office of Science and Technology, the Office of Management and Budget, and the scientific community—through legislation, directed funding, and proposal pressure, respectively—to undertake and support substantial research and development programs in physics relevant to their missions, including the basic research that contributes to their technical capability.

By making physicists partners in the enterprise, several benefits will accrue to both the agencies and the national scientific and technical effort. First will be the advantages of a plurality of decision centers and a consequent diversity of criteria and viewpoints. This situation is healthy for science itself and also ensures an agency influence on the direction of evolution of the subdisciplines likely to be of particular significance to the agency in the future. Second, the association of both in-house and external scientists with the mission of an agency will assist it in the identification and appraisal of scientific discoveries made elsewhere in terms of their potential applicability to agency problems. Third, this association will help the agency in recruiting scientists who might later move into the more applied problem areas of the agency, or identify new significant areas of basic research deriving from the technology of particular concern to the agency. In this way agency support of basic and long-range applied research can serve to sensitize portions of the scientific community and the educational system to particular societal problems that are the responsibility of the agency. During the 1950's and 1960's the DOD played such a role. It was beneficial to both defense technology and the development of science, but, because defense support assumed a very large relative role, particularly in physics and engineering, some unbalance of effort may have resulted. Today national priorities are changing, and physics has much to contribute to the solution of the newly emerging problems, but a serious effort is required to discover and establish the appropriate links between physics and these new areas. Relevance is not always obvious at first, and its discovery requires serious intellectual effort from both scientists and potential users of science.

PHYSICS AND THE NATIONAL ECONOMY

The relationship of technology to a healthy economy and a high standard of living has received much study. Advanced technology is also widely regarded as playing a crucial role in maintaining U.S. leadership in international trade, as is discussed in Chapter 7.

Although a direct connection between the health of a nation's scientific enterprise and its economic strength may be difficult, if not impossible, to establish unambiguously, there is growing evidence to suggest that economic strength is linked not only to science itself but also to the scientifically trained manpower that flows into industry. In any steps taken by the federal government to improve the U.S. economy, certain measures relating particularly to the scientific aspects appear vitally important.

Recent developments have created serious demoralization in the scientific and technological community, largely because of the coincidence of three events. First, the financial crisis of the universities, cutbacks and policy changes in federal support of academic science, and rapid phase-out of student aid programs have combined to reduce abruptly the demand for new science faculty. Second, after 20 years of uninterrupted growth relative to manufacturing output, industrially financed research and development have been declining because of the national economic recession, and industrial basic research has been particularly seriously affected. Third, federal cutbacks of research and development programs in aerospace and electronics, which have been going on since 1965, have produced serious employment dislocations among technical people, accentuated in certain geographical areas. These three trends reinforcing each other after a long period of scientific and technological prosperity have had a truly devastating effect on the physics profession. Yet these trends have come at a time of growing realization that the nation faces serious challenges in the fields of energy, transportation, and environmental protection and management and a breakdown in the delivery of several important public services such as education, health care, social welfare, and other urban services. All of these challenges have important scientific and technological content.

Recommendations

---

13. The federal government should take immediate steps to develop new mechanisms and incentives for the support of substantially expanded industrial basic and applied research programs. The support by industry of the basic science that can contribute to its products and services—both in-house research and cooperative efforts with universities and governmental laboratories—should be strongly encouraged as one approach to strengthening the base of the nation's industrial economy. Means should be sought to stimulate support of basic research by associations of all member companies in an industry. The benefit to the nation's industry as a whole, resulting from any typical piece of industrial research, can usually be shown to be considerably greater than the benefit received by the particular industry that supported this research.

14. During the last decade there has been rapid growth in economic research on computing rates of return on investments in the innovation process in both industry and agriculture. Nevertheless, this field of research is still in a very rudimentary state, and very few definitive conclusions are possible. Furthermore, there is almost no adequate understanding of the interrelationships or relative contributions of the various components of the total innovation process ranging from basic research through development to production and marketing. Greater collaboration is strongly urged among natural scientists, economists, and sociologists in developing a more coherent and usable theory of the innovation process, and we urge federal agencies such as the Department of Commerce and the National Science Foundation to identify and support worthwhile projects in this area. At the same time great

Recommendations (Continued)

> caution should be exercised against drawing practical policy considerations from such studies prematurely.

Public Awareness of Science

Science is generally regarded as a vital element of western culture. Physicists, and indeed all scientists, owe it to themselves and to society to develop increased public awareness of this relationship. Yet, despite much lip service, little use has been made of public information media to fulfill this obligation to the public, and the potential of the professional and scientific societies for creative activity in this area also has been too little utilized. The time has come for individual physicists to demonstrate in more tangible fashion their support for oft-repeated statements of principle in this area. The BBC second channel, for example, presents an hour-long scientific documentary for each of 40 weeks each year; competing in prime time, these documentaries have an audited response that has reached five million persons, or some 11 percent of the British population. In contrast, U.S. television coverage of science is in a sorry state.

Recommendations

PHYSICS COMMUNITY
Chapter 11

15. All physicists, whatever the nature of their professional activity, should encourage those members of the physics profession with talent for such activity to devote a significant fraction of the time and resources available to them to introduce as many of their fellow citizens as possible—children and adults alike—to the pleasures and satisfaction that come from greater understanding of natural phenomena through application of the concepts and laws, as well as the style and approaches, of physics. A small but increasing number of physicists have written books and articles presenting the activities in their field at a level accessible to an interested public. Very much more remains to be done.

Recommendations (Continued)

PHYSICS COMMUNITY

Chapter 11

16. The member societies of the American Institute of Physics should assess each of their individual members not less than ten dollars per year to create a fund that would be used by the American Institute of Physics as seed money, and with matching assistance from foundations and other private sources, to further the use of mass media for informing the public in an understandable and interesting fashion concerning physics and its role in contemporary society.

PHYSICS COMMUNITY

Chapter 11

17. Recognizing the dominant position that television now enjoys in reaching the U.S. public and the future potential of cable television in particular, the American Institute of Physics, with support such as that recommended above, should actively explore the presentation of a continuing series of television programs concerning physics. A joint venture with other major scientific societies such as the American Chemical Society should be considered in order to reach a critical size at the earliest possible time. The National Academy of Sciences should take the lead in bringing together the interested parties and in coordinating this effort.

PHYSICS COMMUNITY

Nonacademic

Chapter 11

18. Industries and foundations should further develop channels and mechanisms for supporting the creative utilization of public communication media in areas of science. The U.S. Steel Foundation, for example, annually awards, through the American Institute of Physics, prizes for distinguished scientific journalism and scientific writing directed toward a broad public audience.

PHYSICS COMMUNITY

Chapter 11

19. Physicists should work actively to encourage and support science museums as ef-

Recommendations (Continued)

fective approaches to furthering public awareness of science. Washington, Boston, Chicago, New York, and San Francisco are examples of major U.S. cities with established science museums that are used as centralized teaching resources for the entire urban communities they serve. The San Francisco Exploratorium concept, which makes possible greater interaction between visitors and the museum displays, marks an important advance in the effectiveness of such institutions; other possibilities exist and should be developed.

PHYSICS AND PRECOLLEGE EDUCATION

In a viable democracy it is essential that each participating citizen appreciate the scientific and technological bases of his society. Unless the general public can understand something of the world of science and appreciate the nature and goals of scientific activity, it will not be able to fit science and technology properly into its perspective of life. As that life becomes increasingly conditioned by the products of science and technology, diffidence and even apathy grow, ultimately having adverse effects on the nation's capacity to maintain leadership, whether it be moral, intellectual, or economic. The science education of the general public beginning in the earliest school years should be a matter of grave concern to the physics community.

The typical U.S. teacher, at both elementary and secondary levels, is not well equipped to guide his pupils in learning that science is more than a collection of facts to be memorized or techniques to be mastered but is instead an inquiry carried on by people who raise questions for which answers are unknown and who have gained confidence in their ability to reach conclusions, albeit tentative ones, through experiment and careful thought sharpened by the open criticism of others. At the same time science is that body of established fact that is the common heritage of all men.

This inability of most teachers to impart some understanding of the nature of science results largely from a science education that fails to give the potential teacher an adequate appreciation of either of its above aspects.

Thus we recycle the attitudes that sustain widespread illiteracy about science and technology. There is a point of leverage in the cycle; all future teachers should receive increased exposure to science and appropriate mechanisms should be developed to this end. Physics departments and faculties in universities and colleges cannot afford to ignore the opportunity thus presented to initiate a long-term but sure approach to public understanding through education of the teachers who will provide the main point of contact between science and the average educated member of the public.

At the same time current indicators suggest that elementary and secondary school teaching is an oversubscribed employment market. These teachers typically acquire tenure after three years (as opposed to the seven years characteristic of college and university teachers). As a result, only a relatively small fraction of the total number of U.S. schoolteachers will be replaced in the near future. Consequently, changes in the education of new recruits to the teaching profession are not enough; the retraining and continuing education of the present teachers are essential and major components of any realistic effort to improve the quality of precollege education in the United States.

Recommendations

PHYSICS COMMUNITY
Academic
Chapter 11

20. Physics faculties in colleges and universities should acknowledge their clear responsibility for science education of the general public by developing and staffing courses, suitable for all the teachers of our young, that emphasize individual inquiry, contact with phenomena, and critical evaluation. Excellent models for these courses already can be found in the new elementary school science curricula to which physicists have made major contributions. Because such courses have merit for the nonscience student more generally, faculties of institutions not offering a degree program in education should involve themselves just as intensively in this effort as should those having direct responsibility for teacher preparation. Science is too vital a part of modern life to be taught only as a separate unit. More emphasis should be given to the importance of physics in other fields of endeavor.

Recommendations (Continued)

PHYSICS COMMUNITY
Academic
Chapter 11

21. U.S. colleges and universities should develop and make available to science majors, as well as recent graduates, the courses that would be required to enable them to meet state certification requirements for teaching in public schools. The justification, in principle, for appriate methodology requirements is recognized. However, many excellent candidates are presently precluded from precollege teaching because of certification requirements that frequently emphasize methodology at the expense of content.

FEDERAL GOVERNMENT
Office of Education
Chapter 11
Chapter 14

22. The Office of Education should encourage all state departments of education to work toward a uniform set of certification requirements with enhanced emphasis on content as opposed to methodology. Progress in this direction would serve to increase the pool from which scientifically talented and trained personnel might be attracted into precollege teaching.

PHYSICS COMMUNITY
Chapter 11
Chapter 14

23. The physics community should expand its current involvement in local educational activities and should actively support and encourage those of its members with talents for, and interest in, the improvement of science teacher training and of the science content in precollege curricula. Cable television, with its greatly increased capability for educational use, should not be overlooked by physicists as a tool in their attempts at improving physics education at all levels.

PHYSICS COMMUNITY
Academic
Chapter 11

24. Universities and colleges should make a particular effort to develop and make available to schoolteachers, in their local areas, courses designed to improve the level of their training in their various fields of academic specialization.

There is abundant evidence that practically all those who major in physics in college and go on to a PhD have taken at least one physics course in high school. Thus high school physics enrollment is a good indicator of the potential future supply of physics and, to some extent, engineering manpower. High school physics enrollments have been going sharply downward for several years, as discussed in Chapter 12, and the decline may well become even sharper in view of the present employment situation for physicists. As a practical matter there is approximately an 11-year lead time in the production of PhD physicists; therefore, it is important to keep careful watch on the pool of high school physics students. Both the scientific community and the federal government must take active measures to alleviate the possible consequences of overreaction to present employment problems insofar as this results in decreased scientific exposure in precollege education.

Recommendations

PHYSICS COMMUNITY

FEDERAL GOVERNMENT

Chapter 12

25. The physics community and the federal government should monitor the potential pool of high school graduates from which future physicists are drawn. There is an urgent need to develop more sophisticated dynamic models of the manpower flows in physics that take into account the influence of economic factors on the supply of potential talent at all levels and the demand for physicists at the BS and PhD levels. The government and the physics community should move to counteract present tendencies to escalate formal educational qualifications for employment and give more publicity to the career opportunities, including teaching below the college level, that are available to BS and MS level physicists.

PHYSICS AND UNDERGRADUATE EDUCATION

The goal of physics is a deep understanding of nature economically reduced to a few broad principles. Therefore, it provides a foundation on which other sciences and engineering can build. But physics is also a vital part of human cultures, as Chapter 3 indicates. During the past decade U.S. colleges and universities have emphasized the preprofessional aspects of undergraduate physics, frequently to the exclusion of courses directed toward nonscientists or nonphysicists. Frequently, too, this preprofessional bias has led to the exclusion from undergraduate curricula of courses in more applied subfields of physics such as optics, acoustics, and hydrodynamics.

Recommendations

PHYSICS COMMUNITY
Academic
Chapter 11

26. Physics departments should place renewed emphasis on the teaching of physics to nonphysicists. To do so requires the development of curricula parallel to those designed for preprofessional majors. Much progress has already been made in this direction in recent years.

PHYSICS COMMUNITY
Academic
Chapter 11

27. The physics community and physics departments should make a determined effort to regain the breadth and flexibility that have traditionally characterized education in physics. Accomplishing this objective will require a reemphasis on many branches of classical physics in undergraduate and graduate curricula, stimulation of broader interests among students, and exposure to the opportunities and challenges presented outside the core areas of contemporary physics and other than those that academic teaching and research can offer.

FEDERAL GOVERNMENT
PHYSICS COMMUNITY
Chapter 12

28. As noted above, the federal government and the physics community should act to maintain a level of BS physics production adequate to the nation's needs. The number of BS degrees granted in physics should not be allowed to fall much below the present 5000 per year if a supply of manpower adequate to fulfill future teaching and research needs is to be maintained. Current trends suggest that unless vigorous corrective action is taken now, the

Recommendations (Continued)

---

nation will experience yet another major oscillation wherein the supply of and need for trained physicists become grossly mismatched.

PHYSICS COMMUNITY

Chapter 11

29. The physics community, acting through the American Association of Physics Teachers, should prepare and publish objective descriptions of the curricula and the facilities available for the teaching of physics in all those institutions that now offer a stated undergraduate major in the field. This measure will not only act to provide much needed information to secondary school graduates embarking on a career but also will foster upgrading or elimination of marginal programs. The American Institute of Physics currently publishes an annual Directory of Physics Departments. This Directory, however, is all too rarely accessible to undergraduates and is not designed to provide information adequate for an informed choice of graduate programs. Too many physics students, having invested one or more years in graduate programs, find themselves trapped in schools or programs with faculties and facilities inadequate to provide them with an education of high enough quality to enable them to compete effectively in the employment market or to enable them to realize their inherent potential for career growth.

PHYSICS COMMUNITY

Academic

Chapter 11

30. Greater emphasis must be placed on regional cooperation wherein individual physics departments can specialize in their research activities to achieve "critical mass" while at the same time providing an adequately broad educational exposure to their students through cooperative utilization of faculty to teach advanced courses in their areas of specialization. A number of successful examples of such coop-

Recommendations (Continued)

eration already exist in the physics community. The physics community should face the fact that faculty requirements for teaching and research have very frequently been mismatched. All too often such mismatches have been effectively hidden by funding from external sources. In many physics departments the numbers of undergraduate physics majors and physics graduate students simply do not justify a faculty of adequate size to mount research programs of critical mass in all the areas in which they currently pursue research.

GRADUATE EDUCATION IN PHYSICS

Traditionally, graduate education in physics has involved a style of activity and a flexibility of approach that prepared students for a broad range of activities. Departments of physics in major institutions have a continuing responsibility to provide the best possible graduate education to those who want it and can profit from it. However, for at least the next few years it should be recognized that holders of advanced degrees in physics will less easily realize their aspirations, if these lie in the academic sphere, than in the 1960's. This rather abrupt shift in the market for physicists places an especially heavy responsibility on teaching faculties to provide realistic advice to their students concerning their projected career opportunities.

A number of current problems in graduate education in physics reflect the rapid growth of both faculties and facilities in the nation's physics departments during the 1960's. (See Chapter 12.) Although it is obvious that the quality of many of these departments improved markedly during this process, a fundamental instability was built into the academic system through the generally accepted practice of using federal grant and contract support to pay increasingly large fractions of the salaries of faculty members. Many departments used this mechanism to expand the size of their faculties, both tenured and nontenured, beyond the number that their institutions could afford to carry during periods of reduced federal funding of academic physics.

Recommendations

31. Physics department faculties should devote particular attention to the counseling and guidance of students who wish to undertake graduate education in physics to ensure that their career choice is a well-considered one. This responsibility involves an honest and realistic appraisal of the faculty and facilities of the institution involved and of how their contributions to the educational needs of the student compare with those obtainable elsewhere. It also involves supplying students with realistic assessments of the job market and of their abilities to compete in it. Physics departments have an obligation to offer an education that provides maximum flexibility in adapting to career options.

32. University administrations should recognize their local responsibility for the maintenance of the viability of their scientific departments. In many fields, and especially in physics, dependence on external support for research has become so great that fluctuations in that support can be disastrous, unless damped by local action. The present tendency to shift this responsibility to the federal government alone can have serious and unfortunate consequences, as the current situation all to clearly indicates. In particular, university administrations should exercise restraint in the establishment of new or expansion of old programs simply because of the apparent availability of federal or other external support. They should exercise equal restraint in terminating established and ongoing programs simply because of the disappearance of this same external support and, to a degree much greater than has recently been the case, should be prepared to devote institutional resources to the orderly readjustment of the many high-quality programs that have experienced

Recommendations (Continued)

drastic reductions in their external support. University departments and administrations should take steps to reduce their dependence on funding sources external to their institutions for the support of academic-year faculty salaries.

PHYSICS COMMUNITY
Academic
Chapter 9
Chapter 14

33. Physics departments should abandon the idea that to achieve or maintain greatness they must maintain research activity simultaneously in all the major subfields of physics. Rather, under conditions of limited funding they should select those areas in which they have the faculty and facility resources to achieve or maintain excellence and, if necessary, sacrifice marginal programs to permit the development of these selected areas. Many of the established physics departments, indeed, have long ago made such hard decisions; if many of the other departments do not follow this example, they risk mediocrity. Obviously, such actions, at the same time, place a high premium on the development of local or regional cooperative arrangements among institutions whereby the specialized training and research facilities of each are available to all. Otherwise there is a real danger that graduate education can become even more specialized and inflexible than it frequently is now.

INTERDISCIPLINARY ACTIVITIES

Physics and physicists have significant contributions to make to the solution of major national problems, many, if not all, of which require concerted attack by a wide variety of disciplines. The attention these national problems have received recently has fostered the rapid development of interdisciplinary programs, majors, and departments in various U.S. colleges and universities. Although many of these activities are of high quality, great care is needed to prevent the programs, the teachers, and the taught from becoming divorced from the underlying disciplinary roots. The problem of the second and later generation in interdisciplinary areas requires particular attention. Although the first generation comes, necessarily, from well-established fields with clearly defined requirements and intellectual standards, there is a tendency, in response to the often broader spectrum of activities related to an interdisciplinary field, to substitute, in the training of second-generation students, a somewhat cursory or introductory exposure to many subjects for fundamental mastery of a few basic subjects.

Recommendations

PHYSICS COMMUNITY
Academic
Chapter 11
Chapter 14

34. Since it is reasonable to assume that persons with a solid foundation in and open communication with a particular discipline can make the most significant contributions to interdisciplinary activities, colleges and universities should guard against proliferation of interdisciplinary degree-granting educational programs at the undergraduate level. Rather, they should concentrate on providing a strong grounding in the fundamental disciplines and use interdisciplinary programs to produce increased awareness of the application of these fundamentals.

PHYSICS COMMUNITY
Academic
Chapter 11
Chapter 14

35. Universities should facilitate the study of interdisciplinary and interdepartmental problems at the graduate level and should be prepared to recognize imaginative and innovative contributions to the solution of such problems as equivalent to the traditional departmentally oriented dissertations in satisfaction of degree requirements. At the same time the intellectual standards of the participating disciplines should be clearly maintained in the graduate course work components of these degree requirements.

THE ROLE OF PHYSICISTS IN THE EDUCATIONAL AGENCIES

Scientists in the United States in general, and physicists in particular, have had little direct contact with the activities of the Office of Education of the Department of Health, Education, and Welfare. Consequently, they have had relatively little influence in the community of professional educators or the development of national policies affecting education. This situation also exists at state and local levels. The major interaction between scientists and the professional education groups has occurred through the Education Directorate of the NSF.

Recommendations

FEDERAL GOVERNMENT *The Congress* Chapter 11	36. Legislation should be sought that would require the establishment, in the Office of Education, of advisory committees of scientists and educators for each of the major scientific fields, including physics.
FEDERAL GOVERNMENT *Office of Science and* *Technology* Chapter 11 Chapter 14	37. The Office of Science and Technology should take the initiative in ensuring that physicists and other scientists have access to the national educational policy levels through the above and other appropriate mechanisms.

ALLEVIATION OF SHORT-RANGE MANPOWER PROBLEMS

One of the most serious problems resulting from the abrupt change in the growth rate of physics in the United States is that of finding career opportunities for the young men and women who began their professional education during a period of expansion but now face a situation in which many of them will be unable to use more than a small part of their special training. Qualitative changes in physics department activities can alleviate some of the short-term aspects of this problem, albeit without really addressing its very serious long-range aspects. In the present difficulties the physics community may have to meet its responsibility to its recent graduates at the expense of at least a temporary reduction in the number of those entering the field. It is vastly preferable to exercise control at the beginning of the educational pipeline. (Chapters 6, 11, and 12 address the manpower problem in physics in greater detail.)

Recommendations

38. Admission of graduate students primarily on the basis of their anticipated contribution to the undergraduate teaching process should be abolished. Indeed, under the present employment conditions, departments that have traditionally used large numbers of graduate students as teaching assistants should consider replacing many of them with postdoctoral instructors. All too frequently, graduate education in physics, as indeed in all the sciences, has been looked on as a necessary by-product of the education of undergraduates, largely because graduate students have long provided a relatively inexpensive component of the undergraduate teaching staff. This problem has been a particularly severe one at some of the larger public universities under the pressures of burgeoning undergraduate enrollments.

39. Where feasible, physics departmental research groups should replace a significant fraction of their present graduate student complement with postdoctoral research associates. Evidence exists that such replacement is already in progress in many physics departments. Federal support agencies should also view postdoctoral positions supported by research grants with greater sympathy than has been the case in the past, even at the cost of supporting fewer graduate research assistants.

Both of these recommendations are intended to reduce temporarily the entering graduate student population and provide additional job opportunities for recent graduates. These measures offer no lasting solution to the current employment problem inasmuch as such positions are at best temporary, but they do provide a "holding period" for adjustment and for exploration of wider employment horizons.

Earlier, the support of excellence in allocating available funds for research was recommended. Yet, as discussed in this Report, the sharpest decrease (50 percent between 1969 and 1971) in the entering graduate student population in physics has occurred at the most distinguished physics departments in the nation. Thus, an increasing fraction of the young men and women of this country who are interested in physics are being educated in institutions with less than the best available faculties and facilities at a time when the highest quality institutions are operating far below capacity. This situation has been, in part, the result of a federal policy decision to phase out national fellowship and traineeship programs rapidly.

Recommendations

PHYSICS COMMUNITY
Academic
Chapter 11
Chapter 14

40. The nation's most able physics departments have a responsibility to make their faculties and facilities available to as many well-qualified candidates as can properly be accommodated. They should resist pressures toward reduction of their entering graduate student populations because of the general national reduction in employment opportunities for MS and PhD graduates. New student-support methods must be developed to prevent a deterioration in the average quality of U.S. graduate education that this, and other, present trends portend.

One of the fundamental problems here is the very natural desire of physics faculties in all colleges and universities to work with graduate research students. The problem is particularly severe in schools with marginal physics facilities and physics faculties of marginal quality and size that increasingly contribute a significant fraction of the total number of new PhD's each year. While attention is directed here to this problem as it affects physics, it is by no means unique to physics.

Recommendations

FEDERAL GOVERNMENT
Support Agencies
Chapter 11

41. To the extent that they can make funding available in competition with the needs of the nation's most able research groups (see Recommendation 1, page 16), federal agencies should develop, for at least the immediate future, research funding mechanisms and ap-

Recommendations (Continued)

priate criteria that would permit selected physics faculty members and colleges or smaller
universities to engage in research without training graduate students. This they might do either
locally or as members of user groups at regional
or national facilities or as part-time members of
research teams at the major universities. The
research support could include short-term postdoctoral staff. In recognition of the special
character of this support, it would be hoped that
colleges and universities involved would agree
not to initiate doctoral training programs (or
to admit new students to existing programs) in
the corresponding areas of physics.

INSTRUMENTATION

One of the major areas of contact between physics research and society
is through physics-derived instrumentation. Devices and concepts originating in physics research now play primary roles in medicine and industry
and throughout all of technology and other science. And, indeed, the
quality of physics research itself depends in no small measure on the
quality of its instrumentation. For decades the United States has been a
world leader in the commercialization of new instruments, and the instrument industry, though relatively small, has been an important source of
favorable trade balance. Among the early casualties of reduced funding
in major laboratories have been both instrumentation activities and instrumentation groups. Although the scientific groups in these laboratories, as
discussed throughout this Report, continue to originate new ideas and
concepts relating to instrumentation, their instrumentation activities have
been markedly reduced because of budgetary pressures. At the same time
the flow of new ideas and new devices from the dwindling instrumentation
groups themselves, which have long been a steady and continuing source
of innovation in instrumentation, has almost ceased.

Furthermore, with increasing pressures on budgets throughout the economy, the purchase rate for new instrumentation has decreased markedly,
with research, engineering, and medical groups living on their instrumentation capital. Continuance of this trend can strangle the U.S. instrumenta-

tion industry, which remains as the source for innovation following the effective withdrawal of the university and national laboratories from this activity.

The short-range instrumentation economies now apparent throughout physics, and science more generally, can have the most serious long-range consequences in significantly weakening the U.S. instrumentation industry both nationally and internationally.

Recommendations

SCIENTIFIC COMMUNITY
Chapter 4
Chapter 14

42. The National Academy of Sciences and the National Academy of Engineering should jointly establish a committee to assess the needs of the U.S. instrumentation activity and the ways these might most effectively be met. The recent NSF summer study on instrumentation provided a start in this direction; however, it was limited in scope and was also limited to instrument procurement and even then to procurement of relatively low-cost instrumentation. The questions of instrumentation development and the procurement of major instrumentation require comprehensive analysis and assessment.

FEDERAL GOVERNMENT
Support Agencies
Chapter 14

43. The National Science Foundation, Atomic Energy Commission, and other agencies supporting physics research and development should seek funding specifically for support of instrumentation development and instrumentation development groups. The committee recommended in 42 above, working with agency representatives, should develop the detailed criteria to be applied to those seeking support. In any program to support instrumentation development in universities and national laboratories, it is essential that mechanisms be included to ensure close and continuing interaction with active research groups; otherwise, the development activities can become sterile and instrumentation can be pursued for its own ends rather than as a support to research and de-

Recommendations (Continued)

velopment activities. It is recommended that instrumentation groups and activities be integral parts of the scientific departments or divisions of their host institutions.

CONSERVATION OF HELIUM RESOURCES

An adequate supply of helium is of crucial importance to any future implementation of superconducting technology—as in the transmission of electrical power or in new computer memory configuration, to give only two examples. Moreover, large components of contemporary physics research —in condensed matter, elementary particles, and nuclear physics—are entirely dependent on liquid helium in achieving adequately low working temperatures. While it may be argued that eventually superconducting systems may be found that can operate at the higher temperatures characteristic of liquid hydrogen, none has yet appeared.

In the meantime, although the United States holds a major fraction of the world helium supply in its natural gas wells, this irreplaceable resource is being squandered in alarming fashion. Current estimates suggest depletion of world reserves by the year 2000 or shortly thereafter.

During the past decade, pursuant to the Helium Act Amendments of 1960 (Public Law 86-777, 13 September 1960), the federal government has maintained a conservation program involving helium extraction from natural gas at the well heads and its underground storage against future national needs. This program was slated for termination by the fiscal year 1972 federal budget, but this termination has not been exercised because of a subsequent injunction against it, pending an environmental impact statement under the National Environmental Protection Policy Act. The Physics Survey Committee notes in passing the recommendation of the Committee on Resources and Man of the National Academy of Sciences–National Research Council (see *Resources and Man,* W. H. Freeman and Company, San Francisco, 1969) that deals with the importance of continuing such a conservation program.

Recommendations

FEDERAL GOVERNMENT
The Congress
PHYSICS COMMUNITY

44. The helium conservation program maintained during the period 1960–1970 should be continued pending new legislation to replace

Recommendations (Continued)

Chapter 4	the Helium Act Amendments of 1960. The new legislation should provide, on a viable and stable financial basis, for the maximum technically feasible and economic extraction of wasting helium, with storage of the excess over consumption. It should also provide for discouragement of wasteful consumption and of the development of critical uses likely to depend on a helium requirement incompatible with long-term supply.

A NATIONAL SCIENCE STATISTICAL DATA BASE

Fundamental to any long-range planning at a national level is the availability of reliable statistical and other information concerning the various scientific disciplines. The only current coherent program for the collection of such data in the United States comprises the various NSF survey series, such as *Federal Funds for Research, Development, and Other Scientific Activities* and the National Register of Scientific and Technical Personnel. These surveys do not give sufficient information broken down by disciplines to be very useful for long-range planning. The National Register, which is the most fruitful source of longitudinal data on scientific manpower and which has been maintained for over 20 years, is being terminated in fiscal year 1972.

In the present survey, as in all others previously, extensive effort has has been devoted to the development of the pertinent statistical data for the field under study. Unfortunately, no mechanism exists for maintaining these individual field data bases on the completion of the survey activities, although maintenance would be relatively simple compared to the completely new start that has been necessary in the past whenever a detailed examination of any scientific discipline was undertaken.

A major problem, too, is the lack of agreement concerning the types of statistical data collected by, or available from, different agencies, the definitions of such central items as scientific man-years and of the boundaries of different scientific disciplines and areas of specialization, and the allocation of funds and manpower among activities ranging from basic research to product development.

Recommendations

45. A coordinating committee should be established by the Office of Science and Technology and the National Academy of Sciences to develop, in collaboration with federal agencies, guidelines and definitions for use in the collection and reporting of manpower and funding data in major scientific, engineering, and other professional–technical disciplines. This committee should familiarize itself with previous and current efforts to clarify and coordinate federal reporting procedures and build on these efforts. It is essential that, to the greatest possible extent, intercomparability among the data from different disciplines be ensured.

46. The National Academy of Sciences, with funds provided by the National Science Foundation, should contract with the organizations representing major disciplines, for example, the American Chemical Society and the American Institute of Physics, to collect and continuously update manpower and funding data in these disciplines (drawing as necessary on existing compilations such as the National Science Foundation's series on *Federal Funds for Research, Development, and Other Scientific Activities*). Appropriate statistics on primary and secondary publications and other information-exchange media also should be compiled for each discipline. There should be a continuing program of operational research to assure that the implications of the data to be collected are developed and made available to all concerned audiences.

47. Recognizing that the National Science Foundation has the statutory responsibility to maintain a register of scientific and technical personnel and to make available such information, we strongly recommend that the National

Recommendations (Continued)

Register of Scientific and Technical Personnel, or an appropriate equivalent, be reinstated promptly. An important function of any such data compilation is to provide insight into systematic changes and trends. The Register data have only recently encompassed a time interval sufficient to be useful for this purpose. To permit an extended hiatus in statistical data collection would destroy the necessary continuity and detract from the utility of this resource.

DISSEMINATION AND CONSOLIDATION OF RESEARCH RESULTS

In many areas of science the problems involved in making the results of on-going research available to potential users have reached crisis proportions. The sheer volume of new research results, in the absence of effective consolidation and review, renders many of them inaccessible to most users. Several major activities are involved, including indexing, abstracting, current awareness services, and the preparation of critical reviews and data compilations in the different areas of specialization. Considerable progress has been made toward more effective abstracting services, although much remains to be accomplished; the situation in regard to compilation and consolidation becomes increasingly critical. Chapter 13 of this Report considers these problems in detail.

Recommendations

SCIENTIFIC COMMUNITY
Chapter 13

48. All groups involved in the conduct or support of basic research should pay greater attention to the extent of dissemination of the journals in which they publish. The physics community should work even more strongly toward a system in which prerun costs of publication are borne by the research itself, and primary publications are distributed at runoff cost. The same consideration also applies to

Recommendations (Continued)

many kinds of secondary services, such as abstracts, and to critical reviews.

FEDERAL GOVERNMENT
Support Agencies
Chapter 13

49. All federal agencies supporting physics research should allocate a specific small fraction of their resources for grants and contracts that would help to fund the abstracting services that are necessary to make the results of their work known and accessible and the data compilation and consolidation activities that will make it more easily applied. With rare exceptions, such services are performed most effectively by continuing groups assembled specifically for the purpose, which, in the absence of such specific allocations, are frequently early casualties of budgetary limitations.

PHYSICS COMMUNITY
Chapter 13

50. The physics community should strongly support and encourage those of its members with the talent for, and interest in, preparation of critical reviews. In particular, preparation of such reviews should be treated on an equal basis with original research in terms of logistic and other ancillary support provided. Specifically, this should include support for postdoctoral and student assistants and for various types of computer and information retrieval assistance.

INTERNATIONAL COMMUNICATIONS

During the past 25 years, because of international pre-eminence in almost all fields of science, U.S. institutions experienced a steady influx of foreign students, postdoctoral fellows, and scientists. Those foreign scientists who remained in the United States contributed in very significant fashion to the strength of the U.S. scientific enterprise; those who returned to their homelands provided a strong nucleus for the development of stronger national programs in science. They also played an important role in interpreting

U.S. aspirations to their countrymen and linking U.S. scientific activities with those in their countries.

As other national scientific programs increase in strength, dependence on the United States for leadership and training in science and the flow of foreign scientists to this country will decrease. As a result, the vital exchange of information between U.S. and foreign scientists will also decrease, unless measures are adopted to preserve and foster such communication. In addition, U.S. scientists increasingly will seek access to foreign facilities on a collaborative or user basis. Dollar for dollar at present levels, funds spent for these purposes probably yield a larger scientific return than additional funds for domestic research. Unfortunately, recent policy decisions concerning the use of federal agency funds for foreign travel or research at foreign centers and postdoctoral fellowship policies, particularly those of the NSF, have made such international activities increasingly difficult.

Recommendations

FEDERAL GOVERNMENT
Support Agencies
Chapter 8
Chapter 11

51. The National Science Foundation should reinstate and enlarge its program of postdoctoral and senior faculty fellowships. Other federal agencies should be encouraged to establish parallel programs.

FEDERAL GOVERNMENT
Support Agencies
Chapter 8
Chapter 14

52. The present budgetary ceilings for foreign travel and for collaborative research at foreign facilities that exist in some agencies should be removed. They are detrimental to international communication and advancement in science. However, requirements for prior justification and for posttravel reporting adequate to ensure, and to document, proper use of the funds involved must be retained. Within these limitations, the allocation of available funds among travel, other foreign activities, and other aspects of a research program should be the acknowledged responsibility of the principal investigator in optimizing the overall research return for a given level of support.

FEDERAL GOVERNMENT
Support Agencies
Chapter 14

53. In view of the particular importance to the nation of furthering scientific cooperation with its closest neighbors and the importance

Recommendations (Continued)

> of travel in such cooperation, Canada and
> Mexico should not be considered as foreign
> countries within the context of federal foreign
> travel regulations.

PROGRAM PRIORITIES AND EMPHASES

Chapter 5 describes a series of criteria—intrinsic, extrinsic, and structural
—for possible use in establishing program priorities and emphases. Intrinsic
criteria are those relating to the potential of a field for fundamental new
discoveries and insight into natural phenomena and are intimately related
to the internal logic of the field. Extrinsic criteria are related to the potential benefits from interaction of a field with other sciences, with technology,
and with society generally; they draw heavily on considerations external
to the field. Structural criteria relate to both internal and external considerations, to questions of continuity, return on scientific and economic investments, interdependence of different scientific fields, and the like. The application of the first two of these classes of criteria is illustrated through
detailed consideration of the program elements of each of the core subfields
of physics in a jury rating sense. Because the structural criteria frequently
require in each particular case a detailed knowledge of sociological, political, and other nonscientific factors for their evaluation, no equivalent
detailed jury rating has been attempted. A recommendation wherein the
structural criteria are of overriding importance will be found in the next
section. In the selection of certain program elements for special consideration herein, however, structural criteria have been included implicitly, if
not explicitly. In making this selection the Committee has considered the
panel reports in Volume II in detail and, working with panel chairmen, has
evolved the program elements for each subfield, as presented and discussed
in Chapter 4. The selection has been based on the Committee consensus
that in each case these were program elements wherein the gain in terms
of new scientific knowledge, new applications, new technology, and new
contributions to society would be especially large in proportion to the additional support required, provided that the specific projects and scientists
were selected on the basis of excellence adjudged by their peers. Chapter 5
includes a detailed listing of the selected program elements.

It must be stressed that the selection of particular program elements for
emphasis should in no way result in a compromise of the intellectual

standards in the selection of individual projects. It is the Committee's expectation that the proposals and people associated with these selected program elements will probably be a little more interesting and of a little higher quality than those that might be associated with program elements to which a lower jury rating has been given. Furthermore, the Committee recommends that somewhat more benefit of doubt be accorded to projects in the selected program elements when they appear risky or controversial. At the same time, it should be clearly recognized that if only the selected program elements were supported, the overall physics research program would be totally unbalanced.

Recommendations

PHYSICS COMMUNITY
FEDERAL GOVERNMENT
Support Agencies
Chapter 5

54. The selected program elements discussed in Chapter 5 represent the core subfields of physics meriting incrementally increased support in terms of their potential return to physics, to science, or to society. It should be emphasized that this increased support should not be at the expense of eliminating support of other program elements, although clearly some readjustment is not only necessary but healthy as the different program elements attain different levels of scientific maturity. The physics community is urged to consider whether readjustment of its activities to place more relative emphasis on these selected program elements might be in order. The federal support agencies are urged to encourage such examination and to support increased activity where possible in these selected areas.

FEDERAL GOVERNMENT
Support Agencies
Chapter 5

55. The criteria, the subfield program elements, and the procedures for applying the criteria to the program elements presented in Chapter 5 represent a first attempt at determining program emphases or priorities in a semiquantitative manner. Physicists, agency program officers, and review committees are urged, through study and application, to develop and refine this procedure further or to devise improved alternatives to the same end.

DEVELOPMENT OF FUSION POWER SOURCES

Occasionally, developments in a field of science or technology reach a stage at which the major impediment to substantial progress can be identified as the actual scale of activity in the field. In short, the fundamental scientific and technological questions have reached the point at which solutions and applications appear to depend in considerable measure on the level of effort, and a substantial increase in support might be expected to yield rapid and far-reaching returns. Before proceeding, the gamble must be balanced against the potential benefits.

The goal of fusion power, with its potential advantages in terms of cost and reduced environmental side effects, is of such significance to the nation and the world that the progress made toward its realization in recent years suggests that an enlarged national program directed toward achieving this goal is in order. Although we still cannot predict with confidence precisely when a self-sustaining fusion system will be demonstrated, there has been enough progress in the past several years to give distinct indication that this goal is attainable. Additional support seems to us an entirely worthwhile deployment of national resources.

In all these discussions the tendency to confuse fusion systems based on the deuterium–deuterium reactions with those based on deuterium–tritium should be avoided. There has been a tendency to combine the anticipated greater technical feasibility of the latter with the anticipated lower costs and lesser environmental problems of the former.

Recommendations

FEDERAL GOVERNMENT
Support Agencies
Chapter 4
Report of the Panel on
  Plasma Physics and the
  Physics of Fluids (in
  Volume II)

56. The federal government should announce a commitment to a full-fledged pursuit of fusion power with the immediate aim of achieving a self-sustaining reaction, provided that no scientific obstacles are found that would thwart this aim. The program to achieve fusion should be an orderly but vigorous one, and additional appropriation to support this program should be made. Significant industrial participation in the proposed program would be essential for most rapid development and utilization of this new energy source.

AREAS OF STRUCTURAL URGENCY

At the present time there are several areas in which the structural criteria, which we define above and in Chapter 5, assume an overriding importance. These are found in the subfields of astrophysics, elementary-particle physics, and, to a certain extent, nuclear physics, where work at the scientific frontiers requires major facilities or instrumentation such as satellites, telescopes, or accelerators. Because of their large size and large unit construction and operating costs, such facilities tend to dominate the funding but not necessarily the manpower or activity level in the respective subfields.

As discussed in some detail in Chapter 5, both the National Accelerator Laboratory (NAL) and the Los Alamos Meson Physics Facility (LAMPF) were approved and construction initiated during a period in the mid-1960's when support for physics was at an all-time high. In both cases there was a clear expectation that, while orderly termination or phasing down of some existing facilities was reasonable at the time of completion of the newer facilities, the support that could reasonably be diverted by the closing of these facilities would be much less than that required to operate and provide user funding for the newer ones, and that, while the new facilities would not be complete add-ons to the existing program, some incremental funding would be necessary and would be made available, if the overall program were to be scientifically viable.

These new facilities are now or will shortly be ready to begin operation. Yet, unless incremental operating funds are made available in fiscal year 1973, even fractional research utilization will be possible only at the expense of termination of significant segments of other research activities in their corresponding subfields. Despite the fact that a number of facilities have been closed since the mid-1960's, any flexibility that this might have introduced has been virtually eliminated by the leveling-off of support in physics and the increasing costs of doing research.

The investment in these facilities, both financial and in terms of scientific man-years, their potential for research at the frontiers of understanding, and their importance to the future of their subfields are so great that the Committee believes failure to exploit their potential would be unacceptable. At the same time, operation of these new facilities at the cost of terminating support for one third of the personnel and three fourths of the existing installations in the corresponding subfields is equally unacceptable.

Recommendations

FEDERAL GOVERNMENT
Support Agencies

57. The Atomic Energy Commission and the National Science Foundation should seek

Recommendations (Continued)

Chapter 5	and the Congress should provide incremental appropriations in fiscal years 1973 and 1974 sufficient to permit orderly and effective initiation of research operations—both in-house and through user-group activities—at new research facilities as they are ready to become operational.

NATIONAL SUPPORT LEVELS FOR PHYSICS

From the start of its deliberations this Committee has realized that it is unrealistic to consider a single level of support that physics "must have" over, say, the next five years. Rather, the proper level of support will necessarily be a compromise in which the benefits of work in physics are matched against national resources and against needs in other areas. Consequently, the charge to each panel asked for an assessment of consequences to the subfield, and to the nation, that would result from each of several conceivable levels of support. Specifically, details of program, funding, and manpower were requested for several different program levels, which, as they finally evolved, can be described as:

1. *An Exploitation Program* designed to exploit all the currently foreseen opportunities, both scientific and technological, in a subfield and to maintain a healthy development program directed toward long-range future facilities and approaches
2. *A Level Budget Program* designed to utilize a funding level held constant (after correction for inflation) in the most effective fashion
3. *An Intermediate Program* designed to exploit a moderate growth rate intermediate between the above two programs
4. *A Declining Budget Program* designed to explore the consequences of a funding level that, after correction for inflation, decreased at about 7.5 percent per year

The panels responded to these challenges. Consequences to the programs at various funding levels were much easier to predict in subfields centering on large facilities than in others. What emerged from the panel considerations, with reasonable consistency, was that an annual growth rate of 11 percent in fiscal year 1970 dollars would permit full exploitation of the opportunities presented by each subfield. Chapter 5 shows, quite independently and following a rather detailed examination of manpower figures and projections, that this growth rate would also permit most of the 1500 new PhD physicists who could be produced in each of the next five years to be absorbed into the general U.S. physics endeavor (university, government, and industry), with an approximate annual 3 percent escalation in the real cost of doing research. Even at a full 11 percent growth rate through fiscal year 1977, again as illustrated in Chapter 5, U.S. physics support would not regain the level that it would have reached had it been possible to maintain a steady 5 percent growth rate since fiscal year 1967 when the field was in a state of robust health.

Consideration of the effects of level funding tends to show that a wide variety of interim measures, introduced throughout a subfield to maintain viability during a hopefully brief funding pause, will necessarily be institutionalized and made permanent. This situation can result in major and serious consequences.

The Declining Budget Programs, almost without exception, demonstrate that whole areas of the different subfields would be abandoned; U.S. physics would no longer be, as it is at present, close to the forefront of progress in the great majority of areas. Contributions to the nation and to the national economy would be seriously eroded. In the face of burgeoning activity in other countries, the United States would find it necessary to accept a secondary role, attempting to retain a response capability such that important new discoveries, if not made in the United States, could nonetheless be exploited for U.S. society. The Committee believes that the U.S. public would not be willing to accept the consequences of such a situation. Development of declining budget programs, however, as is apparent in the panel reports, has forced a very salutary examination of the internal priorities in each subfield and has made more apparent the seriousness of the consequences of an extended period of deteriorating support, not only for science but for the nation.

Finally, the Intermediate Budget Program—typically involving a 6.5 percent annual growth rate—indicates the advances that can be made and the opportunities that can be followed up, as well as those that must be deferred or foregone. The individual panel reports of Volume II discuss all these programs in detail in terms of both support and manpower, in addition to their scientific consequences.

The fact that the Committee does not recommend a detailed national physics program appropriate to different possible levels of support in the growth range from 0 to 11 percent does not reflect an unwillingness to face the difficulties inherent in any such attempt. Rather it reflects the conclusion that it is impossible for such a group to develop either adequately complete information or insight to make such a detailed attempt meaningful. It is unrealistic to look upon the total support of U.S. physics as a reservoir from which funds for individual program elements may be distributed without cognizance of all the internal and external pressures and constraints within both the different funding agencies and the physics community. These, moveover, change rapidly with the magnitude of the overall funds available.

Furthermore, any detailed funding program for physics recommended by a single committee, no matter how wise, would tend to impart a rigidity to the effort that would soon become stultifying to further progress. Physics is a dynamic subject, which means that each major new discovery tangibly alters the priorities in the entire field. To recommend a funding plan that would inhibit responsiveness to such developments would be a serious disservice to the field. However, it is possible to provide a framework for evaluating the opportunities and needs of physics subfields according to various criteria. This the Committee has done and hopes that others will further refine and apply the procedure.

The pre-eminence of U.S. physics owes much to the complex process by which decisions determining the research to be supported by the nation are reached. It involves many working scientists, federal program administrators, economists, and legislators. In general, the science program is probably subjected to greater review than any other item in the federal budget. This Report represents only one small part of such a continuing review, but it has involved the efforts of several hundred physicists.

3

THE

NATURE

OF

PHYSICS

scire—to know
scientia—knowledge

Nam et ipsa scientia potestas est

FRANCIS BACON (1561–1626)
Of Heresies

INTRODUCTION

Science is knowing. What man knows about inanimate nature is physics, or, rather, the most lasting and universal things that he knows make up physics. Some aspects of nature are neither universal nor permanent— the shape of Cape Cod or even a spiral arm of a galaxy. But the forces that created both Cape Cod and the spiral arm of stars and dust obey universal laws. Discovering that has enabled man to understand more of what goes on in his universe. As he gains more knowledge, what would have appeared complicated or capricious can be seen as essentially simple and in a deep sense orderly. The explorations of physical science have brought this insight and are extending it—not only insight but power. For, to understand how things work is to see how, within environmental constraints and the limitations of wisdom, better to accommodate nature to man and man to nature.

These are familiar and obvious generalities, but we have to begin there

55

(*Above*) Hurricane Gladys was stalled west of Naples, Florida, when photographed from Apollo 7 on October 17, 1968. Its spiraling cumuliform-cloud bands sprawled over hundreds of square miles. A vigorous updraft hid the eye of the storm by flattening the cloudtops against the cold, stable air of the tropopause (then at 54,000 feet) and forming a pancake of cirrostratus 10 to 12 miles wide. Maximum winds near the center were then 65 knots. [From National Aeronautics and Space Administration, *This Island Earth*, O. W. Nicks, ed., NASA SP-250 (U.S. Government Printing Office, Washington, D.C., 1970).]

(*Below*) NGC 3031, spiral nebula in Ursa Major photographed with the 200-inch reflector of the Palomar Observatory. [Courtesy Hale Observatories]

if we want to discuss the value of physics in today's and tomorrow's world. Going beyond generalities evokes sharp questions from several sides. Will the knowledge physicists are now striving to acquire have intrinsic value to man, whether it has practical application or not? Is it possible to promise that material benefits will eventually accrue, at least indirectly, from most of the discoveries in physics? How does technology depend on further advances in physics (and vice versa)? How does physics influence other sciences? Is vigorous pursuit of new knowledge in physics still beneficial, as it demonstrably was in the past, to chemistry, astronomy, and the other sciences for which physics provided the base? Is physics perhaps approaching the end of its mission, without very much more to discover? Has the physicist himself an intrinsic value to human society? Must he justify his work by relating it to pressing social problems? Only one person in several thousand is a physicist; will it matter to the others what he does, or that he is there at all? Or is that the right test to apply, no matter how it comes out?

We speak briefly to such questions in this and the following chapter. The entire Report, including the reports of the various subfield panels, provides in copious detail answers to some of these questions or facts from which a reader can form his own judgment, for these are not questions that even all physicists would answer in just the same way.

FUNDAMENTAL KNOWLEDGE IN PHYSICS

Mathematics deals with questions that can be answered by thought and only by thought. A mathematical discovery has a permanent and universal validity; the worst fate that can overtake it is to be rendered uninteresting or trivial by enclosure within a more comprehensive structure. Mathematicians make up, or one could say discover, their own questions in the timeless universe of logical connection. In a science such as geology, on the other hand, the questions arise from local, more or less accidental features of nature. How was this mountain range formed? Where was Antarctica two billion years ago? To answer such questions one has to sift physical evidence. The answers are not universal truths. Geology, as its name attests, differs from planet to planet.

Physics, like geology, is concerned with questions that cannot be decided by thought alone. Answers have to be sought and ideas tested by experiment. In fact, the questions are often generated by experimental discovery. But there is every reason to believe that the answers, once found, have a permanent and universal validity. All the evidence indicates that physics is essentially the same everywhere in the visible universe. A physicist who asks, "Does the neutron have an electric dipole moment?" and turns to experiment to find out, could as well perform the experiment on any planet in any galaxy—it is just more convenient here at home. The question itself concerns a fact as general as (and perhaps even more basic than) the size of the universe. Physics is the only science that puts such fundamental questions to nature.

Take the question: Do all electromagnetic waves, including radio waves and light waves, travel through empty space at the same speed? Present theories assume so, but the contrary is at least conceivable. Perhaps there is a difference so slight that it has not been noticed. To decide, the physicist must turn to experiment and observation. In fact, this particular question has recently received renewed attention. As not infrequently happens, the most sensitive test was applied by asking what sounds like a different question but can be shown to be logically equivalent. Indeed, the experimental evidence shows that the speed of long and short electromagnetic waves is the same to extraordinarily high precision. The result implies that the light quantum, the photon, cannot have an intrinsic mass as great as 10^{-20} * of the mass of an electron. No one was astonished by the result. Most physicists have always assumed the photon rest mass to be exactly zero and can only be relieved that such peculiarly perfect simplicity has survived closer scrutiny. Those who examined the evidence may have been a little disappointed—but their time was not wasted.

This quite unsensational episode is characteristic of fundamental inquiries in several ways. Fundamental experiments in physics often—indeed usually—yield no surprises. However, had the result been otherwise, it would not have demolished electromagnetic theory. A generalization or enlargement of the theory would have been necessary. Finally, the test, sensitive as it was, could not settle the question once and for all, for no real experiment achieves infinite precision. So the question will doubtless be raised again, in one form or another, should a new experimental technique or a bright idea create the opportunity for a significantly more stringent test. An equally fundamental assumption, the proportionality of inertial and gravitational mass, was tested in experiments of successively higher precision by Newton (1686), Bessel (1823), Eötvös (1922), and Dicke (1964) —in the last case to an accuracy of 10^{-11}.

Thanks to such relentless probing of its foundations, even where they appear comfortably secure, physics has acquired a base far more solid than is popularly appreciated. When a physicist states that a proton carries

---

* Physicists commonly use powers of 10 as a convenient shorthand for expressing large numbers. Thus, for example, 1000 becomes 10^3 and 1,000,000 becomes 10^6. The numeral 1 followed by zeros equal in number to the value of the exponent is the rule of thumb. This notation is used throughout the Report.

a charge equal to that of the electron, he can point to an experiment that proved any inequality to be less than one part in 10^{20}. When he expresses confidence in the special theory of relativity, he can refer to a multitude of experiments under ultrarelativistic conditions in which even a slight failure of the theory would have been conspicuous. Electromagnetic interactions are today more completely accounted for—that is to say, better understood—than any other phenomena in physics. Quantum electrodynamics, the modern formulation of electromagnetic theory, has now been tested experimentally over a range of distance from 10^9 cm down to 10^{-14} cm, a range of 10^{23}. This theory was itself developed in response to experiments that revealed small discrepancies in the predictions of the much less complete theory that preceded it. No one will be much surprised if quantum electrodynamics in its present form fails to work for phenomena involving still smaller distances; that will not diminish its glory or its validity within the vast range over which it has been tested. Nor did the extension of electromagnetic theory into quantum electrodynamics deny the essential truth of Maxwell's equations for the electromagnetic field. Knowledge thus won is about as permanent an asset as mankind can acquire.

The most fundamental aspects of the physical universe are manifest in symmetries. In the history of modern physics, the concept of symmetry has steadily become more prominent. The beautiful geometrical symmetry of natural crystals was the first evidence of the orderliness of their internal structure. Exploration of the arrangement of atoms in crystals by x-ray diffraction, begun 60 years ago, has mapped the structure of thousands of substances and is now revealing in detail the architecture of the giant molecules involved in life. Meanwhile, physicists became concerned with more than just geometrical symmetry. Symmetry, in the broadest sense, involves perfect indifference. For example, if two particles are distinguishably different in some ways but show absolutely the same behavior with respect to some other property, a physicist speaks of symmetry. The notion of identity of particles is intimately related. All these ideas acquire their real importance in quantum physics, where an object, a molecule for instance, is completely characterized by a finite number of attributes.

In probing questions of symmetry in the domain of elementary particles, the physicist is again, like the first crystallographers, seeking a pattern of all-pervading order. A sobering lesson learned from modern particle physics, a lesson the Greek atomistic philosophers would have found unpalatable, is that man is not wise enough to deduce the underlying symmetries in nature from general principles. He has to discover them by experiment and be prepared for surprises. No one guessed before 1956 that left and right made a difference in the interaction of elementary particles. After it was found that the true and perfect indifference in the weak interactions is not left/right but left-electron/right-positron, it was again disconcerting to find even this rule of symmetry violated in certain other interactions. But it has by no means been all surprises. Symmetry rules guessed from scattered clues often have been amply corroborated by later experiments; and in particle physics, thinking about symmetries has been enormously fruitful. A grand pattern is emerging, largely describable in terms of symmetries, that makes satisfying sense.

The primary goal of research in fundamental physics is to understand the interactions of the very simplest things in nature. That is a basis, obviously necessary, for understanding larger and more complicated organizations of matter anywhere in the universe. The shape of a particular galaxy is not, from this rather narrow point of view, a fundamental aspect of nature, but the motion of an electron in a magnetic field is fundamental and has implications for many things, including life on earth and the shapes of galaxies.

However, physics has to be concerned with more than the elementary few-body interactions of particles and fields. It can be a gigantic step from an understanding of the parts to an understanding of the whole. To appreciate the total task of physics, a broader view of what is fundamental is necessary. Consider, for example, man's practically complete ignorance of the evolution of the flat, patchily spiral distribution of gas and stars that he calls his own (Milky Way) galaxy. (Only very recently some plausible theories have been developing; it is too early to say how much they can explain.) The interaction of molecules, atoms, ions, and fields is now well enough known for this problem, and Newtonian gravitation, on this scale, is unquestionably reliable. With these simple ingredients, why doesn't the problem reduce to a mere mathematical exercise? One good reason— perhaps not the only reason—is that a complete and general theory of turbulence is lacking. It is not just a lack of efficient methods of calculation. There is a gap in man's understanding of physical processes, which remains unclosed, even after the work of many mathematical physicists of great power. This gap is blocking progress on several fronts. When a general theory of turbulence is finally completed, which probably will depend on the work of many physicists and mathematicians, a significant permanent increase in man's understanding will have been achieved. That will be fundamental physics, using both fundamental and physics in a broad sense, as, to take an example from the recent past, was the explanation of the mystery of superconductivity by Bardeen, Cooper, and Schrieffer. Although pedantically classifiable as an application of well-known laws of quantum mechanics, it was truly a step up to a new level of understanding.

These two examples, turbulence and superconductivity, stand near opposite boundaries of a wide class of physical phenomena in which the behavior of a system of many parts, although unquestionably determined by the interaction of the elementary pieces, is not readily deducible from them. Physicists know how to deal with total chaos—disorganized complexity. Statistical mechanics can predict anything one might want to know about a cubic centimeter of hydrogen gas with its 10^{19} molecular parts. As for the hydrogen molecule itself, it presents the essence of ordered simplicity. Its structure is completely understood, its properties calculable by quantum mechanics to any desired precision. What gives physicists trouble, to continue the classification suggested by Warren Weaver, is organized complexity, which is already present in a mild form in so familiar a phenomenon as the freezing of a liquid—a change from a largely disordered to a highly ordered state. Here a general feature is that what one molecule prefers to do depends on what its neighbors are already doing. How drastically that feedback changes the problem is suggested by the lack of a theory that can predict accurately the freezing point of a simple liquid. In the physics of condensed matter, many such problems involving cooperative phenomena remain to challenge future physicists. A turbulent fluid, on the other hand, confronts the physicist with a system in which order and disorder are somehow blended. Complex it certainly is, but not wholly disorderly, admitting no clean division between the random flight of a molecule of the fluid and the organized motion of a row of eddies.

The solution of these major problems of organized (or partly organized) complexity is absolutely necessary for a full understanding of physical phenomena. Extraordinary insight and originality will surely be needed, as indeed they always have been. The intellectual challenge is as formidable as that faced by Boltzmann and Gibbs in the development of statistical mechanics. The consequences for science of eventual success could be as far-reaching.

Broadly speaking then, the unfinished search for fundamental knowledge in physics concerns questions of two kinds. There are the primary relations at the bottom of the whole structure. How many remain to be discovered and how small the number to which they can ultimately be reduced are not yet known. Then there is the knowledge needed to understand all the behavior of the aggregations of particles that make up matter in bulk. Here the mysteries are perhaps not so deep, although the remaining unsolved problems are of formidable and subtle difficulty. It is easier to imagine how this part of the development of fundamental physics could be concluded, if the even more difficult problem of organized complexity in living organisms is left for the physiologist, assisted by the biochemist and biophysicist, to solve.

What has been learned in physics stays learned. People talk about scientific revolutions. The social and political connotations of revolution evoke a picture of a body of doctrine being rejected, to be replaced by another equally vulnerable to refutation. It is not like that at all. The history of physics has seen profound changes indeed in the way that physicists have thought about fundamental questions. But each change was a widening of vision, an accession of insight and understanding. The introduction, one might say the recognition, by man (led by Einstein) of relativity in the first decade of this century and the formulation of quantum mechanics in the third decade are such landmarks. The only intellectual casualty attending the discovery of quantum mechanics was the unmourned demise of the patchwork quantum theory with which certain experimental facts had been stubbornly refusing to agree. As a scientist, or as any thinking person with curiosity about the basic workings of nature, the reaction to quantum mechanics would have to be: "Ah! So that's the way it really is!" There is no good analogy to the advent of quantum mechanics, but if a political–social analogy is to be made, it is not a revolution but the discovery of the New World.

The Question of Value

Most people will concede that fundamental scientific knowledge is worth its cost if it contributes to human welfare by, even indirectly, promoting the advance of technology or medicine. It is easy to support a claim for much of physics. But now that some frontiers of fundamental research have been pushed well beyond the domain of even nuclear engineering, that justification is not always plain to see. A connection between many-body theory and the latest semiconductor device is not much more difficult to trace than the connection between thermodynamics and a jet engine. But it is not easy to foresee practical applications of the fundamental knowledge gained from very-high-energy experiments or, say, tests of general relativity.

Two responses can be made, each of which has some validity. First, inability to foresee a specific practical application does not prove that there will be none. On the contrary, that there almost certainly will be one has become a tenet of conventional wisdom, bolstered by familiar examples such as Rutherford's denial of the possibility of using the energy of the nucleus. Applying this principle to strange-particle interactions will probably raise fewer doubts among laymen than among physicists. Even here the conventional wisdom may be sound after all. High-energy physics is uncovering a whole new class of phenomena, a "fourth spectroscopy" as it is termed elsewhere in this Report. In the present state of ignorance, it would be as presumptuous to dismiss the possibility of useful application as it would be irresponsible to guarantee it.

A secondary benefit that can be expected, as a return for supporting such research, is the innovation and improvement in scientific instrumentation that such advanced experiments stimulate. (This point is discussed in a subsequent section of this chapter on the contributions of physics to technology.) Other sciences also benefit from the development of experimental techniques in physics.

But these responses do not squarely face the question: What is fundamental knowledge itself worth to society? Elementary-particle physics provides an example. A permanent addition to physicists' knowledge of nature was the recognition, several years ago, that there are two kinds of neutrino. This fact, although compatible with then existing theory, was not predictable *a priori,* nor is the reason understood. The question was put to nature in a fairly elaborate high-energy experiment, at a total monetary cost that reasonable accounting might put at $400,000 (not including beam time on the Alternating Gradient Synchrotron Accelerator). The answer was unequivocal: The electron neutrino and the muon neutrino are not identical particles.

For physics this was a discovery of profound significance. Neutrinos are the massless neutral members of the light-particle or lepton family, of which the familiar electron and its heavier relative, the muon, are the only other known members. Just how these particles are related—even why there is a muon—is one of the central puzzles of fundamental physics, a puzzle that is as yet far from solution. Obviously, it was not about to be solved while physics remained ignorant of the fact that there are two kinds of neutrino, not just one.

Still, how does this bit of knowledge benefit the general public, interesting as it may be to the tiny fraction of scientists who know what "two kinds" means in this connection? The answer must be that the discovery was a step—a necessary step—toward making nature comprehensible to man. If man is going to understand nature, he has to find out how it really is. There is only one way to find out: Experiment and observe. If man does not fully understand the leptons, he cannot claim to understand nature.

On the other hand, the neutrino is a rather esoteric creature. It would be absurd to expect wide and instant appreciation of this fundamental discovery. Even of fundamental knowledge there is too much for most people to absorb. Many a physicist who could calculate on the back of an envelope the neutrino flux from the sun remains complacently ignorant of the location and function of the pituitary gland in his body. The point is that the value of new fundamental knowledge must not be measured by the number of people prepared to comprehend it. To say that man understands this or that aspect of nature usually means that some people do, and that they understand it sufficiently well to teach it to any who care to learn and to maintain a reliable base from which they or others can explore still further. The great thing about fundamental scientific knowledge is that it is an indestructible public resource, understandable and usable by anyone who makes the effort. When so used in its own domain, it is a thing of beauty and power.

The great American physicist Henry Rowland once replied to a student who had the temerity to ask him whether he understood the workings of the complicated electrostatic machine he had been using in a demonstration lecture: "No, but I could if I wanted to." Knowledge of fundamental scientific laws makes for economy of human thought. It is the great simplifier in a universe of otherwise bewildering complexity. It is not necessary to analyze every cogwheel in an alleged perpetual motion machine to know that it will not work or to keep tracking all the planets to be sure that they are not about to collide. The revelation that the electron and muon neutrinos are different, although it might appear to have complicated matters, was in fact a step toward ultimate simplicity, because it brought closer the essential truth about leptons.

Some of the fundamental ideas of physics have slowly become part of the mental furnishings of most educated people. The following statements probably would elicit general assent: All substances are composed of atoms and molecules; nothing travels faster than light; the universe is much larger than the solar system and much older than human life; energy cannot be obtained from nothing, but mass can be turned into energy; motions of planets and satellites obey laws of mechanics and gravity and can be predicted precisely. That is surely a rather meager assortment, but, even so, what an immense difference there is between knowing these few things and not knowing them—a difference in the relation of a person to his world. A child asks his father "What is a star?" or "How old is the world?" In this century he can be answered, thanks to hard-won fundamental knowledge. What is that worth? The answers can hardly contribute to anyone's material well-being, present or future. But they do enlarge the territory of the human mind.

Much of what modern physics has learned has not yet become common knowledge. Here is an example. Not only physicists but everyone who has studied quantum mechanics knows that all known particles, without exception, fall into one or the other of just two classes, called fermions and bosons, which differ from one another profoundly on a certain question of symmetry. The difference is as fundamental as any difference could be. Although usually expressed somewhat abstractly, the distinction is less recondite than some theological distinctions over which men have quarreled fiercely. Its concrete manifestations are vast, among them the astonishing properties of superfluid helium (a boson liquid), the electrical properties of metals, and, indeed, through the Pauli exclusion principle, the very existence of atoms and molecules, hence of life. Now it would seem that this profound, essentially simple truth about the physical universe ought to be known to most fairly well-educated persons, to as many, perhaps, as understand the difference between rational and irrational numbers. Yet, it is probably safe to say that a majority of college graduates have never heard of fermions and bosons, and that an even larger majority is not equipped to understand what the distinction means. Probably far less than 10 percent of current college graduates have had a course in physics or chemistry in which the exclusion principle was mentioned. Perhaps 10 percent will learn enough mathematics so that, if they are interested, they could be made to understand a statement such as "the wavefunction changes sign on exchange of particles."

But can anyone except physical scientists be interested in such a question? History suggests that it is possible. Long before the atomic bomb made mc^2 a catchword, the theory of relativity (both special and general) engaged the public interest more intensely than anything else in twentieth century physics. The fascination lay not only in the enigmatic figure of Einstein and the notion of a theory that, as the newspapers were fond of claiming (quite erroneously), only 12 men could understand. There was at the same time a sustained, genuine intellectual interest, at all levels of understanding commencing with zero, in the puzzling implications of new ideas about space and time. To this day, nothing beats the twin paradox for stirring up spirited argument in an elementary physics class. People who have any interest at all in ideas seem to be more interested, on the whole, in fundamental questions than in practical questions. It is usually easier to interest an intelligent layman in the uncertainty principle than in how the mass spectrograph works. Of course, that is true only if he or she can be given some idea, not wholly superficial, of the meaning of the uncertainty principle. This can be done; it has been accomplished many times, in different styles, by imaginative teachers and writers. Nor is it hard to convey to a thoughtful person the essential notion of antimatter, or the question of left- and right-handedness in nature, both ideas that intrigue many nonscientists. It may even turn out that nonphysicists of the next generation, many of whom were brought up with the new math and are on speaking terms with computers, will find the abstract rules of particle physics a more satisfying statement about nature than would an old-fashioned physicist.

Admittedly, there are difficulties in engaging the active interest of nonscientists in some of the most fundamental ideas of physics. They are illustrated in the example of the fermion–boson distinction. Unlike relativity, this subject makes no connection with familiar concepts such as time, space, and speed. No paradox or controversy stirs the imagination in first acquaintance. The intelligent layman can only listen to the explanation of fermions and bosons as if he were hearing a story about another world. There is nothing to argue about. It may serve as brief intellectual entertainment; most likely it will not impinge on or disturb the ideas he already has. He may not be eager to tell someone else about it. In that case it can hardly be claimed that the person has gained something of permanent value to him. And yet, when followed to a slightly deeper level, this idea has a direct bearing on a question that has engaged human throught for 2000 years—the ultimate nature of substance. It gives a most extraordinary answer to philosophical questions about identity of elementary particles, questions that were already implicit in the cosmology of Democritus but were never faced before quantum mechanics. Here, too, is a key to the wave–particle duality with which the quantum world confounded man's mechanistic preconceptions. The philosophical implications of the fermion–boson dichotomy are still, after 40 years, poorly understood by philosophers.

So the problem is one of teaching. Very many people who are not scientists are interested or can be interested in the basic questions that have always attracted human curiosity. The discoveries of physics, even those presently described in abstruse language, bear directly on some of these questions—so directly that when understood they can transform a person's conception of the atomic world or the cosmos. To promote that understanding is a task for the scientist as teacher, in the broadest sense of teacher. In the short run, drawing a potential audience from college graduates of the past 20 or 30 years, and perhaps the next 10, the physicist must apply his imagination and ingenuity to convey interesting and meaningful, and essentially true, accounts of some of the fundamental developments in physics. It is to be hoped that some day the educated layman he addresses will have had enough physical sciences and mathematics in his general education to turn a discussion of the symmetries of elementary particles into some sort of dialogue.

The audience need not be of a size that would impress a national advertiser but only a few million people—a few hundred, say, for every physicist. Of course, the distribution of potential interest and comprehension is a many-dimensional continuum. Everyone ought to be, and can be, given some glimpse of what fundamental physics is about. However, it is impossible to compare the value of a brief exposure of 10^8 people to news of a discovery in physics with the value of sustained and active interest on the part of 10^6 people. Both are valuable now, and both will help, in the long run, to make the fundamental knowledge that physics is securing meaningful and useful to all people.

The value of new fundamental scientific knowledge is not, after all, contingent on its appreciation by contemporary society. It really does not matter now whether Clerk Maxwell's ideas were widely appreciated in Victorian England. Their reception is interesting to the historian of science, but mainly as a reflection of the attitudes and structure of the society in which Maxwell worked. The full value of a scientific discovery is concealed in its future. But even as the future unfolds, the value that one may set on an isolated piece of fundamental knowledge often becomes uncertain because of the interconnection in the growing structure. In the end, one is forced to recognize that there is just one structure; understanding of the physical universe is all of one piece.

PHYSICS AND OTHER SCIENCES

Physics is in many ways the parent of the other physical sciences, but the relation is a continually changing one. Modern chemistry is permeated with ideas that came from physics, so thoroughly permeated, in fact, that the sudden demise of all physicists—with the exception of an important class calling themselves chemical physicists—would not immediately slow down the application of physical theory to chemical problems. The last great theoretical contribution of physics to chemistry was quantum mechanics. For another such contribution there is no room, almost by the definition of chemistry. From the point of view of the physicist, chemistry is the study of complex systems dominated by electrical forces. Strong interactions, weak interactions, gravitation—these are of no direct interest to the chemist. There is no reason to doubt, and voluminous evidence to show, that quantum mechanics and electromagnetic theory as now formulated provide a complete theoretical foundation for the understanding of the interactions between atoms and molecules.

An immense task remains for the theoretical chemist, a task that is in some part shared by the physicist interested in the same problems. One area of common interest is statistical mechanics, especially the theory of "cooperative" phenomena such as condensation and crystallization, where, although the forces that act between adjacent molecules are known, the behavior of the whole assembly presents a theoretical problem of singular subtlety. Other problems that attract both chemists and physicists include phenomena on surfaces, properties of polymers, and the fine details of the structure and spectra of simple molecules. A subject of very intense research in which physics and chemistry are thoroughly blended is the study, both experimental and theoretical, of reactions in rarefied partially ionized gases. This study has direct applications in plasma physics, the development of high-power lasers, the physics of the upper atmosphere, and astrophysics.

There is really no definable boundary between physics and chemistry.

There never has been. Approximately 5000 American scientists, on a rough estimate, are engaged in research that would not be out of place in either a physics or a chemistry department. Some call themselves physicists and their specialty chemical physics or just physics. Others are physical chemists. The label generally reflects the individual's graduate training and correlates with some differences of interest and style. These chemical physicists have illustrious predecessors, including Michael Faraday and Willard Gibbs. And those who, like them, have made a permanent mark on both sciences are likely to be thought of as physicists by physicists and chemists by chemists.

Physics serves chemistry in quite another way. It is the source of most of the sophisticated instruments that the modern chemist uses. This dependence on physics has been, if anything, increasing. Perhaps the infrared spectrograph and the x-ray diffraction apparatus should be credited to the physics of an earlier era; their present highly refined form is largely the result of commercial development stimulated by users. But mass spectrographs, magnetic resonance equipment, and microwave spectrometers, all of which originated in physics laboratories in relatively recent times, are found in profusion as well as are the more general electronic components for detecting photons and atoms—electron multipliers, low-noise amplifiers, frequency standards, and high-vacuum instrumentation. One might follow a research chemist around all day, from spectrograph to computer to electronic shop to vacuum chamber, without deducing from external evidence that he was not a physicist, unless, as might still happen today, the smell of his environment gave it away.

Radiochemistry is in a class by itself. The radiochemist and the nuclear physicist have been partners indispensable to one another since before either specialty had a name. The dependence of experimental nuclear physics on radiochemical operations is perhaps less conspicuous, seen against the whole enterprise of nuclear physics, than it was 10 or 20 years ago. On the other hand, advances in the use of labeled elements and compounds in chemical, biochemical, and medical research continue to be paced by improvements in detection methods. These came directly from physics. A spectacular recent example is the solid-state particle detector, with parentage in nuclear physics and solid-state physics.

Instead of viewing physics and chemistry as different though related sciences, it might make more sense to consider a science of substances, with its base in quantum physics and objects of study ranging from the crystalline semiconductor (now assigned to the solid-state physicist) to the alloys of the metallurgist, to the molecular chain of high-polymer physics and chemistry, to the elaborate molecular structures of the organic chemist. Through this whole range of inquiry one can discern a remarkable convergence in theoretical treatment, and also in experimental methods. The first comes about as fundamental understanding replaces phenomenology. When the properties of the complex system, be it a boron whisker or a protein molecule, can be systematically deduced from the arrangement of its elementary parts, which are nothing but atoms governed by quantum mechanics, a universal theory of ordinary substances will be at hand. Such a theory has not yet been achieved, but as theoretical methods become more powerful, they become, as a rule, more general, and there is steady progress in that direction. Already the language of theory in organic chemistry is much closer than it used to be to the language of theory in solid-state physics.

The convergence in experimental methods, which of course should never become complete, also reflects the tendency of more powerful analytic methods to be more general. The scanning electron microscope is equally precious to the biochemist and the metallurgist. The infrared spectrograph is almost as ubiquitous as the analytic balance. Radioactive labeling is practiced in nearly all the physical sciences.

Notwithstanding the staggering accumulation of detailed information in the materials sciences, a drastic simplification of scientific knowledge is occurring in these fields. As the facts multiply, the basic principles needed to understand them all are being consolidated. To be sure, the need for specialization by individuals is not declining; the quantity of information vastly exceeds what one mind can assimilate. But the specialist is no longer the custodian of esoteric doctrine and techniques peculiar to his class of substances. Quantum physics is replacing the cookbook, and the mass spectrograph is replacing the nose. The future organic chemist acquires a rough working knowledge of quantum mechanics very early—often earlier, the physicist must concede with chagrin, than his roommate who is majoring in physics. Soon, if it is not already so, any single section of this enormously rich and varied picture will be understood at a fundamental level by anyone equipped with a certain common set of intellectual tools.

Well under way here is nothing less than the unification of the physical sciences. This unification is surely one of the great scientific achievements of our time, seldom recognized or celebrated, perhaps, because, having progressed so gradually, it cannot be seen as an event. Nor can it be credited to one science alone. The influence of quantum physics on chemistry was clearly a central development, and, if one wishes to symbolize that development by one of its landmarks, there is Linus Pauling's *The Nature of the Chemical Bond*. In physics there are many landmarks in the theory of condensed matter, from the first application of quantum theory to crystals by Einstein and Debye to the solution of the riddle of superconductivity, among them the quantum theory of metals, the understanding of ferromagnetism, and the discovery of the significance of lattice imperfections in crystals. But the basic contribution of physics is the secure foundation on which all this knowledge is built—on understanding, confirmed by the most stringent experimental tests, of the interactions between elementary particles and the ways in which they determine the structure of atoms and molecules. The fruits of this immense achievement are only beginning to appear.

Biology obviously derives part of its nourishment from physics by way of chemistry. Biochemistry and molecular biology are equally dependent on physical instrumentation. X-ray diffraction, electron microscopy, and isotopic labeling are indispensable tools. Modern electronics is important in physiology, most conspicuously in neurophysiology, where spectacular progress has been made by observing events in single neurons, made accessible by microelectrodes and sophisticated amplifiers. Other examples are described in the Report of the Panel on Physics in Biology.

These are products of past physics. One might wonder whether future physics is likely to prove as fruitful a source of new experimental techniques for biology and medical science. There are two reasons for thinking that it will. First, there is no apparent slackening of the pace of innovation in experimental physics. In almost every observational dimension, short time, small distance, weak signal, and the like, the limits are being pushed beyond what might have been reasonably anticipated. If there is one thing experience teaches here, it is that quite unforeseen applications eventually develop from any major advance in experimental power. Through the Mössbauer effect, preposterous as it seems, motions as slow as that of the hand of a watch can be measured by the Doppler shift of nuclear gamma radiation. Even after this discovery, when Mössbauer experiments were going on in dozens of nuclear-physics laboratories, a physiological application would have seemed rather fanciful. In fact, the Mössbauer effect is being used today to study, in the living animal, the motion of the basilar membrane in the cochlea of the inner ear, perhaps the central problem in the physiology of hearing.

There is another reason to look forward to contributions to the life sciences from inventions not yet made. It is the existence of some obvious and rather general needs, the satisfaction of which would not violate fundamental physical laws, for example, an x-ray microscope with which material could be examined *in vivo* with a resolution of, say, 10 Å or a better way of seeing inside the body than the dim shadowgraphs, remarkably little better than the first efforts of Roentgen, that medical science has had to be content with for half a century. But the breakthroughs probably will again come in unexpected ways; one cannot guess what will play the role of Roentgen's Crookes tube. The physicist can only feel rather confident that an active, inventive period in experimental physics eventually will have important effects on the way research is done in the biological sciences.

The intellectual relations between physics and biology are changing, perhaps more because of what is happening in biology than what is happening in physics. Most physicists who have any acquaintance with biology, if only through semipopular accounts of the latest discoveries, find the ideas of current biology, especially molecular biology, intriguing and stimulating. No physicist could fail to be stirred by the elucidation of the genetic code or by the other glimpses into primary mechanisms of life. This wonderful apparatus works by physics and chemistry after all! But it is far more ingenious and subtle than any contrivance of wires, pulleys, and batteries. From the intricate engine of muscle fiber to the marvelous information processer in the eye, plainly there are hundreds of mechanisms in which physics, chemistry, and biological function are inextricably involved. Also, the evident universality of basic processes in the cell appeals strongly to a mind trained in the physicist's approach to structure and function. There is no doubt that biology is going to attract some students who would have made good physicists, which cannot be deplored. It is to be hoped that there soon will be a growing number of biologists who are not only well grounded in physics but who share, and possibly derive some encouragement from, the physicist's conviction that the behavior of matter can be understood in terms of the interactions of its elements; this behavior and these interactions are the goals of experimental study.

At the other end of the scale is astronomy. Physics began with astronomy, but after the foundations of Newtonian mechanics were secured, astronomical observations (not counting as such the observations of cosmic rays) did not directly generate new fundamental physics. However, astronomy did provide a rich field for the application of physics. Great advances in astronomy such as the elucidation of the structure and evolution of stars depended on an understanding of the structure of atoms. That came from the physics laboratory and from quantum theory as it developed. Then it was nuclear physics that supplied the keys to the generation of energy in the stars and to the production of the elements. These questions were highly interesting to physicists and inspired both theoretical and experimental work. But, broadly speaking, this work was merely physics applied to astronomical problems.

At a different level, though, astronomy has always had a powerful intellectual influence on physics. The heavens confronted man with tantalizing mysteries. His conceptions of what he saw there strongly influenced philosophical attitudes toward nature. Astronomy has given the physicist confidence that the universe at large is governed by beautifully simple laws of physics, discoverable from earth by man. That belief gives the explorations of physics a wider purpose and significance. It attracts the physicist's attention to cosmological questions, to the physics of gravitation, and to phenomena occurring under conditions utterly unattainable in a terrestrial laboratory.

Today the interaction of physics and astronomy is more vigorous than at any time since Newton. Astronomy has entered an astonishingly rich period of significant discovery. This is due, in part, to observing over a greatly widened spectrum, from the long waves of radio astronomy, which have in 25 years greatly increased the knowledge of the large-scale universe, to x rays and gamma rays, which are just beginning to produce interesting information. In part, too, it reflects the increased power and scope of astrophysical theory, working from a more complete base in atomic and nuclear physics. Also, nature has provided some incredibly marvelous, totally unexpected features for telescopes to discover, displaying on a grand scale phenomena that involve most of physics. Less than ten years after the maser was invented in a physics laboratory, the maser process was found to occur in clouds of interstellar gas. It is typical of the present intensive involvement of physicists in astronomy that this discovery was made by some of the same physicists, now turned radio astronomers, who had participated in the microwave spectroscopy that led to the invention of the maser.

Nuclear physicists and astrophysicists have been engaged for more than 20 years in a collaboration from which has come not only an understanding of the source of energy in stars but of the production of the chemical elements found in the universe. This knowledge bears directly on the history of the universe, providing much of the solid evidence against which cosmological theories can be tested. More surprising is the emerging importance to astronomy of elementary-particle physics. The opacity of matter to neutrinos turns out to be relevant not only to the reconstruction of a primordial big bang but to what is going on now at the centers of galaxies. Inside pulsars there is almost certainly "hyperonic" matter, composed of particles more massive than protons, known only in the laboratory as evanescent products of high-energy collisions. Perhaps a not negligible fraction of the matter in the universe is compressed into this state, a form of matter hardly speculated about before pulsars were discovered five years ago. It may be difficult to forecast commercial applications on planet Earth for high-energy physics, but its importance in the universe as a whole may have been greatly underestimated.

Of course, cosmic rays have been studied by physicists, not astronomers, for 50 years; and these particles, still the most energetic a physicist can hope to see, have been transcendentally important in the development of modern physics. It is hard to imagine how elementary-particle physics would have progressed if the earth had been shielded from cosmic rays. Although the source of cosmic rays was obviously astronomical, it is only rather recently that the importance of cosmic radiation as a constituent of the interstellar medium has been appreciated. Something like a merger of cosmic-ray and related high-energy physics with astrophysics has taken place; the new Division of Cosmic Physics in the American Physical Society is one indication. Magnetohydrodynamics and plasma physics are very lively subjects of common interest to members of both groups. Beyond these obvious cases of interest, even solid-state physicists have been drawn into astrophysics by the discovery of neutron stars.

In the same period, a resurgence of interest in gravitation has occurred among both astronomers and physicists—among astronomers because of the discovery of systems close to the theoretical conditions for gravitational collapse and among physicists because of experimental developments that bring some predictions of gravitational theory within the range of significant laboratory test.

All these developments are bringing again to physics and astronomy a wonderful unity of interest. Never before have so many parts of physics directly concerned astrophysicists; seldom before have astronomical phenomena so stirred the imagination of physicists.

The cosmos is still the place where man must look for answers to some of the deepest questions of physics. Were the fundamental ratios that characterize the structure of matter as found here and now truly precisely constant for all time? Observations of distant galaxies offer a view backward in time to an earlier stage of the universe. Is Einstein's general relativity an exact and complete description of gravitation? Is the visible universe a mixture of matter and antimatter in equal parts, or is what is called matter overwhelmingly more abundant throughout? Already astronomers have observations that bear on these questions. The conclusions are only tentative now, but it seems quite certain that the questions will be answered.

As for the earth sciences, a gap no longer exists between astronomy and geology. A look at the relation of physics to the earth sciences shows a network of interconnected problems, stretching from the center of the earth to the center of the galaxy. The earth's magnetic field provides a good example. How it is generated has always been a puzzle. Now it appears, although the explanation is not complete, that magnetohydrodynamic theory is about to produce a convincing picture of the electric dynamo that must be at work within the earth's fluid core. Furthermore, the same ideas may explain the generation of magnetic fields of stars and even, when applied on a very different scale, the magnetic fields that pervade the whole galaxy. These developments are the work of both geophysicists and astrophysicists, many of them people whose breadth of interest would justify both titles. In addition, the interplanetary magnetic fields in the solar system, which are dominated by the solar wind, are of interest to both the planetary physicist and the solar physicist.

The theoretical base for these interrelationships is the dynamics of highly conducting fluids, including ionized gases, which is also the base for such potentially important engineering developments as the magnetohydrodynamic generator. Not new fundamental physics but ingenious and insightful analysis and the development of more powerful theoretical tools are needed.

From the point of view of physics, the other sciences might be grouped into four very broad divisions: a science of substances, including chemistry and also a part of physics; life sciences; earth sciences plus astronomy (for which a good name that will comprehend the range from meteorology to cosmology is lacking); and engineering science. These divisions are, of course, multiply overlapping, with a topology that would defy a two-dimensional diagram. A category of current interest, environmental science, would overlap all four.

Engineering science is suggested as a fourth division, although it is not as extensive or as well recognized as the others, to emphasize a distinction between the products of technology and the growing body of knowledge—scientific knowledge—that constitutes the intellectual capital of engineering. To this knowledge both engineers and physicists contribute continually, with a mutual stimulation of ideas. To call different portions of this body of knowledge mere applied mathematics does not do justice to the imaginative work that goes on or to the potential influence on the other sciences of the ideas generated. A previously mentioned example is the important subject of fluid dynamics, with its ubiquitous problem of turbulence, in which engineering science naturally has a big stake. Consider, as another example, communication theory, developed in its many aspects by people calling themselves variously engineers, mathematicians, and physicists. Sophisticated treatments of fluctuation phenomena, including quantum effects, the relation of information to entropy, and the rich ramifications—including holography—of Fourier duality are just a few of the ideas it encompasses. Or consider the theory of automatic control with feedback, which was developed mainly within engineering science but is now an indispensable aid in most experimental sciences, including physics. The point is that between physics and engineering science there are strong intellectual, one might even say cultural, links. That is only one aspect of the relation of physics to technology, a topic explored more fully in the following section.

No one would question the importance of physics in the development of these fields of science. However, because chemistry needed physics does not necessarily imply that chemistry now needs help from physicists. Physicists have made fairly direct contributions to chemistry, even recently; and physics, at least as a source of new experimental tools and techniques for other sciences, may be as fruitful a source in the immediate future as it has been in the past. But can essential contributions from physics to the other sciences in the form of new and basic ideas be expected? Do the chemists or the earth scientists, who have fairly well assimilated the apparently relevant parts of physics, need the physicist for any service except to teach physics to their students? What is their interest in his hunt for quarks or gravity waves?

There are two ways to answer such questions from the physicist's point of view. One can meet example with example, explaining, for instance (as will be done in Chapter 4 when this question is addressed with specific reference to high-energy physics), how the isolation of the quark could have immense practical consequences. Or one can make a more general reply along the following lines. The increasing unity of the physical sciences at the basic level and the proliferation of interconnections among the fields, and especially with physics, make intellectual vigor widely contagious. New ideas tend to stimulate other new ideas. As long as physics has great questions to work on, its discoveries can hardly fail to excite resonances in neighboring fields.

TECHNOLOGY AND PHYSICS: THEIR MUTUAL DEPENDENCE

Everyone knows that today the main sources of new technology are research laboratories of physics and chemistry and not the legendary ingenious mechanic or Edisonian wizard. Actually, the relation of technology, that is, applied science, to basic science has been close for more than a century. Think of Faraday, Kelvin, Pasteur. It is true that Morse and Bell were amateurs in electricity, while Maxwell, it is said, found the newly invented telephone not interesting enough to serve as a subject of a scientific lecture. But the sweeping exploitation of electromagnetism that began in the latter half of the nineteenth century was based directly on the fundamental understanding achieved by Maxwell. No one would suggest that today's semiconductor technology could have been created solely by engineers ignorant of the relevant fundamental physics. Research in physics provided the base from which present technology is developing.

But physics research did, and is doing, more than that. Research is a powerful stimulator of fresh ideas. One reason is that in research, and especially in the most fundamental research, the scientist is often trying to break new ground. He may need to measure something at higher energy (remember Van de Graaff and the electrostatic generator) or closer to absolute zero (Kamerlingh Onnes discovering superconductivity) or in a previously inaccessible band of the spectrum. Years before World War II the magnetron was first exploited for the generation of 1-cm waves by the physicists Cleeton and Williams at the University of Michigan. They used it to make the first observation of the inversion resonance of the ammonia molecule.

Also, and this applies to both experimental and theoretical research, to be challenged by a puzzling phenomenon stimulates the imagination. One is likely to try looking at things from a new angle, questioning assumptions that had been taken for granted. P. W. Bridgman once described the scientific method as "the use of the mind with no holds barred." The uninhibited approach of the research scientist to a strange problem has even generated a whole discipline—operations research. Prominent among its creators were scientists like P. M. S. Blackett and E. G. Williams, who came from fundamental physics research, both experimental and theoretical, of the purest strain.

The research laboratory, including the theoretical physicist's blackboard or lunch table, provides the kind of freewheeling environment in which an idea can be followed for a time to see where it leads. Most new ideas are not good. In a lively research group these are quickly exposed and discarded, often having stimulated a fresh idea that may be more productive.

In such a setting, physicists are not generally intellectually constrained by the distinction between fundamental science and technology. For one thing, experimental physics heavily depends on some very advanced technology. The research physicist is not only at home with it, he has often helped to develop it, adapt it, and debug it. He is part engineer by necessity—and often by taste as well. An experimental physicist who is totally unmoved by a piece of excellent engineering has probably chosen the wrong career. One cannot make such a sweeping statement about theoretical physicists, but even they, as was spectacularly demonstrated long ago in the Manhattan Project, frequently can apply themselves both effectively and zestfully to technological problems. Currently, in fields such as plasma physics and thermonuclear research, there are many theoretical physicists, with a broad range of interest and expertise, some with a background in elementary-particle physics, intimately concerned with engineering questions.

Ongoing basic research is necessary for the translation of scientific discovery into useful technology, even after the discovery has been made. As a rule, the eventual value to technology of a discovery is seldom clearly evident at the time. It often emerges only after a considerable evolution within the context of fundamental research, sometimes as an unexpected by-product. Nuclear magnetic resonance (NMR) is now widely used in the chemical industry for molecular structure identification. This possibility was totally unforeseeable in the early years of NMR research. It came to light only after a major improvement in resolution had been achieved by physicists studying NMR for quite different purposes. However, a backlog of unapplied basic physics is not all that it takes to generate new technology; it may not even be the main ingredient.

Some of the most startling technological advances in our time are closely associated with basic research. As compared with 25 years ago, the highest vacuum readily achievable has improved more than a thousand-fold; materials can be manufactured that are 100 times purer; the submicroscopic world can be seen at 10 times higher magnification; the detection of trace impurities is hundreds of times more sensitive; the identification of molecular species (as in various forms of chromatography) is immeasurably advanced. These examples are only a small sample. All these developments have occurred since the introduction of the automatic transmission in automotive engineering!

On the other hand, fundamental research in physics is crucially dependent on advanced technology, and is becoming more so. Historical examples are overwhelmingly numerous. The postwar resurgence in low-temperature physics depended on the commercial production of the Collins liquefier, a technological achievement that also helped to launch an era of cryogenic engineering. And today, superconducting magnets for a giant bubble chamber are available only because of the strenuous industrial effort that followed the discovery of hard superconductors. In experimental nuclear physics, high-energy physics, and astronomy—in fact, wherever photons are counted, which includes much of fundamental physics—photomultiplier technology has often paced experimental progress. The multidirectional impact of semiconductor technology on experimental physics is obvious. In several branches of fundamental physics it extends from the particle detector through nanosecond circuitry to the computer output of analyzed data. Most critical experiments planned today, if they had to be constrained within the technology of even ten years ago, would be seriously compromised.

The symbiotic relation of physics and technology involves much more than the exchange of goods in the shape of advanced instruments traded for basic ideas. They share an atmosphere the invigorating quality of which depends on the liveliness of both. The mutual stimulation is most obvious in the large industrial laboratory in which new technology and new physics often come from the same building and, sometimes, from the same heads. In fact, physics and the most advanced technology are so closely coupled, as observed, for example, on a five- to ten-year time scale, that the sustained productivity of one is critically dependent on the vigor of the other.

EXPERIMENTAL PHYSICS

An experimental physicist is usually doing something that has not been done before or is preparing to do it, which may take longer than the actual doing. That is not to say that every worthwhile experiment is a risky venture into the unknown. Many fairly straightforward measurements have to be made. But only *fairly* straightforward! The easy and obvious, whether in basic or applied physics, has usually been done. The research physicist is continually being challenged by experimental problems to which no handbook provides a guide. Very often he is trying to extend the range of observation and measurement beyond previous experience.

A most spectacular example is the steady increase in energy of accelerated particles from the 200-keV protons, with which Cockcroft and Walton produced the first artificial nuclear disintegrations in 1930, to the 200-GeV protons of the National Accelerator Laboratory—in 40 years a factor of a million! This stupendous advance was achieved not in many small steps but in many large steps, each made possible by remarkable inventions and bold engineering innovations produced by physicists. Although numerical factors of increase do not have the same implications in different technologies, few branches of engineering have come close to that record. A possible exception is communication engineering. Transmission by modulated visible light, now feasible, represents an increase in carrier frequency of roughly a million over the highest radio frequency usable 40 years ago. What is really more significant, the information bandwidth achievable has increased by a comparable factor. This great advance was, of course, made possible by development in basic physics and at many stages was directly stimulated by the basic research of physicists. Even accelerator physics contributed at one stage by stimulating the development of klystrons. The accelerator physicists have a remarkable record of practical success as engineers. From the time of the early cyclotrons to the present, no major accelerator in this country, however novel, has failed to work; most of them exceeded their promised performance.

In other branches of experimental physics, too, people are doing things that would have appeared ridiculously impractical only 10 to 20 years ago. For example, a discovery in the physics of superconductivity (the Josephson effect) made it possible to measure precisely electric voltages and magnetic fields a thousand times weaker than could be measured previously. Electrons, also positive ions, can be electrically caged, almost at rest in space, for hours. By another technique, neutral atoms can be stored in an evacuated box for minutes without disturbing a natural internal oscillator that completes, in that time, about 10^{12} cycles of oscillation. One by-product is an atomic clock that is accurate to about a second in a million years. By recent laser techniques, light pulses of only 10^{-12} seconds' duration have been generated and observed. The local intensity of light that can be created with lasers is many million times greater than anything known in the laboratory ten years ago. Effects can be readily observed that in the past could be only the subjects of theoretical speculation, thus opening to investigation as entire field of basic research—nonlinear optics. Within the same decade, the highest magnetic-field strengths easily available in the laboratory, which had hardly changed in a century, were roughly quadrupled by superconducting magnets. In the same period, low-temperature physicists extended downward by a factor of 10 the temperature range usable for general experimentation.

These developments and others mentioned subsequently in this Report show that experimental physics is not running out of ideas or becoming a routine matter of data-gathering. In fact, experimental physics could be entering a new period, distinguishable (by criteria other than austerity of budgets!) from two preceding periods of conspicuous experimental advance in modern physics: the decade before World War II, which, in terms of tools and techniques, could be called the "cyclotron and vacuum tube" period, and the immediate postwar period, in which microwave electronics and nuclear technology, largely the fruits of wartime physics, made possible an enormous advance in experimental range. Within the past 10 to 15 years, several postwar developments have come of age that, considered as a group, promise a comparable advance in experimental capability. These include cryogenics in all its ramifications, semiconductor technology, laser–maser techniques, and the massive exploitation of computers. One trouble with any such historical formula, and the glory of physics as an adventure, is the existence of the important and unclassifiable exceptions. For example, the continuously spectacular progress in high-energy accelerators fits only very loosely into the scheme just outlined. A more modest exception is the simple proportional counter, one of the most elegant and sensitive devices of physics since the torsion pendulum, which has survived through all three periods, earning in each a new lease on life.

Whether it signifies a new era or not, the enormous advance in observational power that is occurring now will in all likelihood open still more fields of research in physics. It will certainly lead to applications yet unforeseen in other sciences and technology.

PHYSICS—A CONTINUING CHALLENGE

It is possible to think of fundamental physics as eventually becoming complete. There is only one universe to investigate, and physics, unlike mathematics, cannot be indefinitely spun out purely by inventions of the mind. The logical relation of physics to chemistry and the other sciences it underlies is such that physics should be the first chapter to be completed. No one can say exactly what completed should mean in that context, which may be sufficient evidence that the end is at least not imminent. But some sequence such as the following might be vaguely imagined: The nature of the elementary particles becomes known in self-evident totality, turning out by its very structure to preclude the existence of hidden features. Meanwhile, gravitation becomes well understood and its relation to the stronger forces elucidated. No mysteries remain in the hierarchy of forces, which stands revealed as the different aspects of one logically consistent pattern. In that imagined ideal state of knowledge, no conceivable experiment could give a surprising result. At least no experiment could that tested only fundamental physical laws. Some unsolved problems might remain in the domain earlier characterized as organized complexity, but these would become the responsibility of the biophysicist or the astrophysicist. Basic physics would be complete; not only that, it would be manifestly complete, rather like the present state of Euclidean geometry. Such an outcome might not be logically possible.

One might be more seriously concerned with the prospect of reaching a stage short of that, in which all the basic physics has been learned that is needed to predict the behavior of matter under all the conditions scientists find in the universe or have any reason to create. From chemistry to cosmology, let us suppose, all situations are covered, but one cannot predict with certainty the scattering cross section for e-neutrinos on μ-neutrinos at 10^{30} V. Suppose further that the experiments required to explore fully all the physics at 10^{30} V are inordinately costly and offer no prospect of significantly improving physics below 10^{20} V, which is already known to be sufficiently reliable. If some such state were reached, one might reasonably expect that research in fundamental physics would be at least brought to an indefinite halt if not closed out entirely as being in the state of perfection previously postulated. It would be said that all the physics that mattered had been learned.

Some physicists are supposed to have made this statement about physics at about the end of the nineteenth century. That they were wrong, spectacularly wrong, is a reminder that human vision is limited; it proves nothing more. For the state of knowledge of physics today is essentially different from that in 1890, just before the curtain was pulled back, so to speak, from the atomic world. Where the problems lie is evident. As far as the behavior of ordinary matter is concerned, it is hardly conceivable that the detailed picture of atomic structure, the product of quantum theory and exhaustive experimentation, should turn out to be misleading or that the main problem in nuclear physics should suddenly be revealed as one hitherto ignored. There are mysteries, but they lie deeper. If it were possible to fence off a particular range of application, for example, chemistry at temperatures below 10^5 deg, then a state in which all the relevant fundamental physics is essentially complete could be reasonably anticipated. Indeed, for a sufficiently restricted application, the day might already have arrived.

The trouble is that the range of interests continues to widen, and in unexpected ways. Because of pulsars, the structure of atoms exposed to a magnetic field of 10^{12} G (ten million times the strongest fields in laboratory magnets) becomes a question of some practical concern, as does the shear strength of iron squeezed to a billion times its ordinary density. Just now astronomy seems to be making the most new demands on fundamental physics; there the end is not in sight.

Even if the physicist could reliably and accurately describe any elementary interaction in which a chemist or an astronomer might be interested, the task of physics would not be finished. Man's curiosity would not be satisfied. Some of the most profound questions physics has faced would remain to be answered, if understanding of the pattern of order found in the universe is ever to be achieved. The extent of present ignorance still is great.

It is far from certain that in the presently recognized elementary particles the ultimate universal building blocks of matter have been identified. The laws that govern the behavior of the known particles under all circumstances are not known. It is even conceivable that the study of particle interactions at ever higher energy leads into an open domain of never-decreasing complexity. Probably most physicists would doubt that. Cosmic rays afford an occasional glimpse of matter interacting at energies very much greater than particle accelerators provide, and no bizarre consequences have yet been observed. It seems rather that physicists now face not mere complexity but subtlety, a strangeness of relationship among the identified particles that might render the question of which of them is truly elementary essentially meaningless.

Even if physicists could be sure that they had identified all the particles that can exist, some obviously fundamental questions would remain. Why, for instance, does a certain universal ratio in atomic physics have the particular value 137.036 and not some other value? This is an experimental result; the precision of the experiments extends today to these six figures. Among other things, this number relates the extent or size of the electron to the size of the atom, and that in turn to the wavelength of light emitted. From astronomical observation it is known that this fundamental ratio has the same numerical value for atoms a billion years away in space and time. As yet there is no reason to doubt that other fundamental ratios, such as the ratio of the mass of the proton to that of the electron, are as uniform throughout the universe as is the geometrical ratio $\pi = 3.14159$. Could it be that such physical ratios are really, like π, mathematical aspects of some underlying logical structure? If so, physicists are not much better off than people who must resort to wrapping a string around a cylinder to determine the value of π! For theoretical physics thus far sheds hardly a glimmer of light on this question.

The question was posed in even sharper form 40 years ago by Eddington, who argued that the structure perceived in nature can be nothing but a reflection of the methods of observation and description that must be employed. That view would reduce fundamental physics to metaphysics. But Eddington's own conception of the structure did not survive. Such evidence as he had adduced was soon washed away in a flood of discovery. The whole history of physics since then gives no sign that physics is about to become an exercise in deduction. Every attempt to close the theoretical structure to all changes except refinements has been confounded by an experimental discovery. This has happened so often that there has been some accession of intellectual humility along with the vast increase in knowledge of the underlying structure of matter. Surely the end of the story is yet far off.

The fundamental question survives, if not the attempts to answer it: Is there an irreducible base, or design, from which all physics logically follows? The history of modern physics warns that the answer to such a question will not be attained just by thinking about it. To be sure, brilliant theoretical ideas, probably many, will be needed, and some future Bohr or Einstein may become renowned for the flash of insight that eventually reveals a key to the puzzle (or the absence of a puzzle!). But without experimental exploration and discovery, new ideas are not generated. Physics will remain an experimental science at least until very much more is known about the fundamental nature of matter.

4

THE
SUBFIELDS
OF
PHYSICS

INTRODUCTION

This chapter provides a brief status report on physics and its subfields, including the historical background that has brought each subfield to its present status, current scientific activity in each subfield, and the distribution of activity in each.

Although each subfield is considered in some detail, to bring out its internal logic, emphasis is placed on the underlying unity of physics. Much has been said in recent years about the extent to which physics and, indeed, many of the other sciences were fragmenting with ever-increasing special-ization, so that effective communication within and among sciences is rapidly dwindling. We believe that, although there are obvious dangers that must be guarded against, the often discussed fragmentation of physics is exaggerated. For the entire field of physics is tightly linked by common techniques—both experimental and theoretical—but most of all by a common style or approach to problems. With increasing specialization, what tends to be forgotten is the remarkable extent to which techniques and concepts diffuse rapidly throughout physics and the degree of com-monality that exists. This underlying unity is illustrated in the closing section of this chapter in discussions of the four spectroscopies and of one of the most exciting new objects found in nature—the pulsar.

The basic principles of classical and quantum mechanics and of rela-tivity provide a unifying framework for all activity in physics. Modern physics encompasses a remarkable range in both space and time, as illustrated in Figure 4.1. The characteristic dimensions of the entities of physics span a range of some 10^{40}; strangely enough, the range of characteristic times—from the lifetime of the metagalaxy to a time characteristic of subnuclear phenomena, say the passage of a photon over a typical elementary particle—is again some 10^{40}. It is these enormous ranges that provide physics with much of its richness and challenge.

In moving from the macroscopic systems of classical physics and astronomy to the atomic domain, physicists found no necessity to add to their two natural forces—gravitation and electromagnetism. But they were

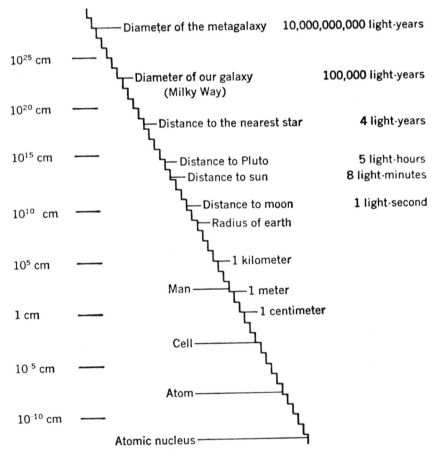

FIGURE 4.1   The range of sizes in physics. Each step corresponds to a factor of 10 increase in size.

Nature's Quartet of Forces

Gravitational force

Extends to infinity

Electromagnetic force

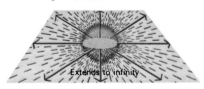

Extends to infinity

PLATE 4.I Nature's quartet of forces. We can detect a proton only through its forces. It is the sum of four effects. The gravitational force, surrounding all matter in all directions to infinity, controls the stars and galaxies. The much stronger electromagnetic force cancels out at long range, since there are equal numbers of positive and negative charges in the universe. It controls the world of atoms and molecules. The weak nuclear force is known to exist, but its carrier has not yet been detected. The strongest force—the strong nuclear force—controls most effects in the compacted nuclear and subnuclear world. [Source: *Science Year. The World Book Science Annual.* Copyright © 1968, Field Enterprises Educational Corporation.]

Weak nuclear force

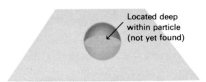

Located deep within particle (not yet found)

Strong nuclear force

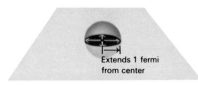

Extends 1 fermi from center

forced to develop a new arithmetic—quantum mechanics—whose assumptions reflect the indistinguishability of the microscopic constituents of matter and the abandonment of mechanical determinism. On moving deeper into the microcosm to the nuclear realm, physicists found that they had to double the number of kinds of natural forces, adding the weak and the strong nuclear interactions, but they were pleased to find that quantum mechanics, augmented with appropriate relativistic corrections required for high accuracy, was entirely adequate to describe the nuclear system, just as it had been in the case of the atom.

In moving on to elementary-particle physics, relativity has come to the forefront in all considerations. Fundamental interest has centered on whether additional natural forces, or modifications of quantum mechanics, or, more generally, quantum electrodynamics (QED) would be required to reproduce the new phenomena. Much effort has been devoted to these questions. At present there are hints that a superweak nuclear force may be required, but there is no evidence yet to suggest that QED is not entirely appropriate for the description of all phenomena in the range of energies and linear dimensions thus far accessible.

Ignoring for the moment the possible existence of a superweak nuclear force whose effects would be entirely negligible in all save the most esoteric situations, physicists have a mathematical framework—quantum electrodynamics—and four natural forces—gravitation, electromagnetism, and the two nuclear forces, strong and weak—within which they attempt to encompass and explain all natural phenomena (Plate 4.I). This economy of basic input and this focus on fundamental phenomena are the hallmarks of physics. But no less a part of physics is the development and application of new insights; in both areas, fundamental and applied, physics has played and continues to play an important role in education, in interaction with other sciences, in interaction with technology, and in addressing pressing problems facing man and society.

In developing this Report the Survey Committee early decided that it was essential to obtain detailed input from panels of experts in each of a number of subfields that it had identified. Several of these subfields have relatively well-defined and traditional boundaries within physics, for example, acoustics; optics; condensed-matter physics; plasma and fluid physics; atomic, molecular, and electron physics; nuclear physics; and elementary-particle physics. Even here, however, there is a considerable overlap of specific activities within subfields. For example, laser techniques appear in several of the subfields, as do those of colliding atomic and molecular beams. This situation reflects the close coupling within physics, as is emphasized in the section on The Unity of Physics at the end of this chapter. In addition to the subfields noted above, several important interfaces between physics and other sciences were identified.

There are several areas of classical physics, such as mechanics, heat, thermodynamics, and some elements of statistical physics, that have not been considered explicitly in this Survey. This omission in no way implies any lack of importance of these subjects but merely reflects the fact that they are mature areas of science and that relatively little research *per se* is currently taking place in them.

Early in the Survey, we, as a Committee, developed a lengthy charge, which was addressed to each of the panels. It is included as Appendix B to this volume of the Report. This charge was a broadly ranging one dealing with the structure and activity of each subfield, as viewed not only internally but also in terms of its past, present, and potential contributions to other physics subfields, other sciences, technology, and society generally.

Clearly the charge was most directly relevant to the traditional subfields; in the case of the interface panels some questions were inevitably unanswerable without a survey of equivalent scope of all fields working at the interface. In the case of astronomy, such a survey was available. The panels on data, education, and instrumentation clearly were special cases. As Volume II will show, the panels have responded in depth to the questions asked.

In Appendix 4.A to this chapter, activity in the subfields has been divided into program elements—components large enough to have some internal coherence and reasonable boundaries, and for which it might be possible to estimate present funding levels and PhD manpower involved. As discussed in Chapter 5, the purpose of this exercise was to divide the subfields into units of activity that the Committee could evaluate in terms of intrinsic, extrinsic, and structural criteria. In developing these program elements, the Committee worked with the panel chairmen; however, in some cases the elements used here are not identical with those suggested by the panel chairmen. In the subfields of elementary-particle physics and nuclear physics, it was possible to assign funding levels and manpower rather precisely. Similar assignments for some of the program elements in the other subfields may be in error by a factor of 2.

The principal objective in this chapter is to provide a summary of each of the panel reports, thus giving an overview of U.S. physics—its history, status, opportunities, and problems—as seen by a representative group of active physicists in 1970–1971.

General Activity in Physics

Although manpower, publication, and support data are presented in detail in Chapters 12, 13, and 5 and 10, respectively, to place the discussions of this chapter in perspective we include a number of figures showing some of these data for each subfield (with the exception of interfaces for which data are not fully available). (See Figures 4.2 through 4.8.)

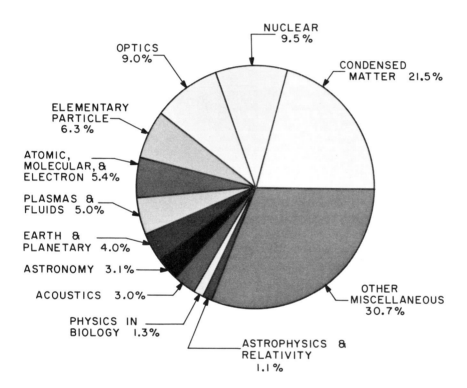

FIGURE 4.2   Distribution of physics manpower (PhD's and non-PhD's) by subfield.
[Source: National Register of Scientific and Technical Personnel, 1970.]

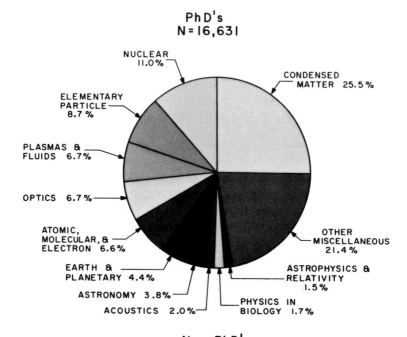

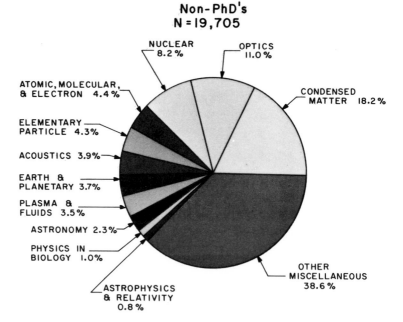

FIGURE 4.3   Distribution of physics PhD's and non-PhD's by subfield. [Source: National Register of Scientific and Technical Personnel, 1970.]

DISTRIBUTION OF PHYSICISTS BY EMPLOYING INSTITUTION

N = 36,336

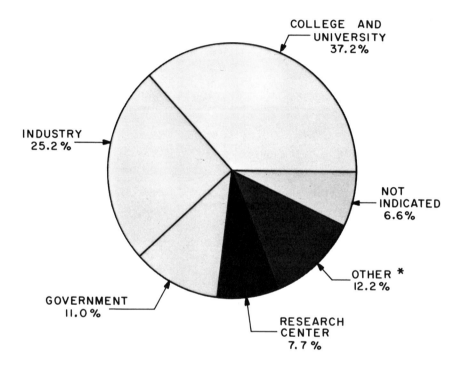

* Other includes medical school, secondary or elementary school, military service, self-employed, hospital, etc.

FIGURE 4.4 Distribution of physics manpower (PhD's and non-PhD's) by employing institution. [Source: National Register of Scientific and Technical Personnel, 1970.]

DISTRIBUTION OF PhD AND NON-PhD PHYSICISTS
BY EMPLOYING INSTITUTION

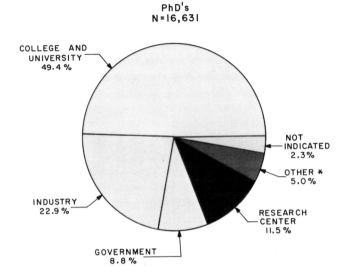

PhD's
N=16,631

COLLEGE AND
UNIVERSITY
49.4 %

NOT
INDICATED
2.3%

OTHER *
5.0%

INDUSTRY
22.9 %

RESEARCH
CENTER
11.5 %

GOVERNMENT
8.8 %

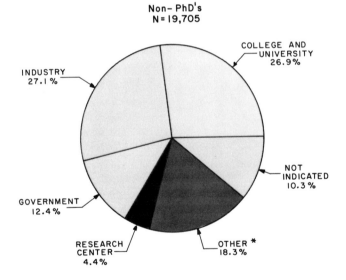

Non-PhD's
N=19,705

COLLEGE AND
UNIVERSITY
26.9%

INDUSTRY
27.1 %

NOT
INDICATED
10.3 %

GOVERNMENT
12.4%

RESEARCH
CENTER
4.4%

OTHER *
18.3%

* Other includes medical school, secondary or elementary school,
military service, self-employed, hospital, etc.

FIGURE 4.5 Distribution of physics PhD's and non-PhD's by employing institution.
[Source: National Register of Scientific and Technical Personnel, 1970.]

DISTRIBUTION OF PUBLICATIONS
AMONG PHYSICS SUBFIELDS *

N = 1,181

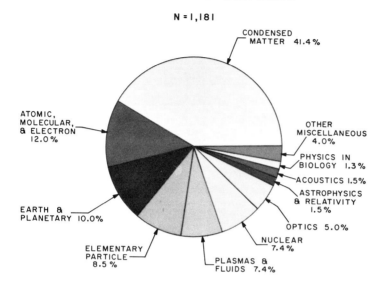

DISTRIBUTION OF PHYSICS PUBLICATIONS
BY ISSUING INSTITUTION *

N = 1,181

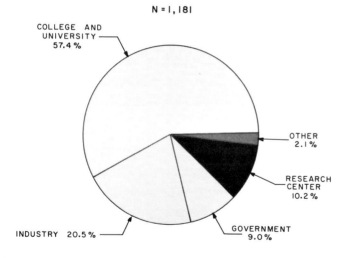

* Data are based on a sample of every tenth article listed in 1969
   issues of Physics Abstracts.

FIGURE 4.6   Distribution of publications in physics by subfield and by issuing
institution.

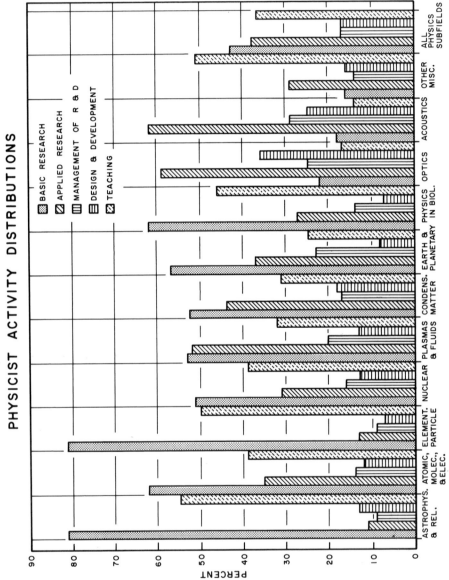

FIGURE 4.7 Principal work activities (those ranked first and second) of physicists (PhD's and non-PhD's taken together) in each subfield. [Source: National Register of Scientific and Technical Personnel, 1970.]

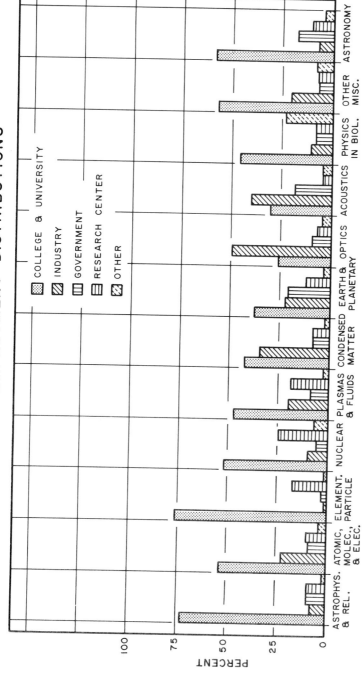

FIGURE 4.8 Distribution of PhD physicists in each subfield by employing institution. [Source: National Register of Scientific and Technical Personnel, 1970.]

The manpower data were derived from the National Register of Scientific and Technical Personnel for 1968 and 1970; the publication data were based on a sample of every tenth article listed in the 1969 issues of *Physics Abstracts;* and the information on support shown here was based largely on the reports of the panels. It should be noted that the totals here need not agree with those appearing in the National Science Foundation's publication, *Federal Funds for Research, Development, and Other Scientific Activities* (FFRDS) and elsewhere in this Report. Some of the panels carried out a more detailed examination of the overall funding pattern, and differences in categories and definitions in some cases have yielded numbers different from those in FFRDS.

Migration data for PhD physicists are displayed in Appendix 4.B to this chapter. As shown there, physics PhD manpower is much more mobile than has been commonly believed. During the period 1968–1970 about one third of the PhD's changed their subfields of major interest. This is a strong indicator of the unity of physics; the subfield interfaces are highly permeable.

> For we do not think that we know a thing until we are acquainted with its primary conditions or first principles and have carried our analysis as far as its simplest elements.
> ARISTOTLE (384–322 B.C.)
> *Physics,* Book I, 184

ELEMENTARY-PARTICLE PHYSICS

Introduction

Elementary-particle physics is concerned with the determination of the fundamental constituents of matter and energy, the behavior of matter under the most extreme conditions, the mathematical laws of nature governing this behavior, and the related underlying nature of microscopic space-time itself. The forms of elemental matter that occur naturally, or can be produced in the laboratory, include electrons, photons, protons, and neutrinos. They also include a great (and surprising) variety of unstable particles that are produced in very energetic collisions between the stable particles or between atomic nuclei.

Experimental observations range from measurements of fundamental properties of the particles, made with the greatest imaginable precision, to gross observations of the qualitative behavior or a particular form of matter under the extreme conditions of high-energy impact. Theoretical work includes interpretation of experimental observations based on generally accepted principles, suggestions for new directions of experimentation to answer well-defined questions, the invention of theoretical models and experiments to test them, speculations on the ultimate nature of matter and energy, speculations concerning the generalization of the laws of nature needed to overcome the inadequacies and paradoxes in the existing theory, suggestions of experiments to test these speculations, and deep mathematical investigations of the nature of the theory to reveal its strengths and weaknesses.

Theoretical research in this subfield requires a knowledge of the foundations and methods of most branches of physical theory, including both classical theory and relativistic quantum theory. The required mathematical techniques encompass a wide variety of fields and methods of applied mathematics. Experimental work requires access to a source of particle beams, usually an accelerator, although some types of investigations are carried out successfully with cosmic rays. The experimenter must also have access to very sophisticated instrumentation capable of identifying particles moving at speeds close to that of light or of measuring specific properties of the particles with great precision or of both. The large amount of information that must be sorted to make sense of the behavior of these particles under the extreme conditions to which they are subjected requires the use of very large computers. The development of accelerators (see Figures 4.9 and 4.10), detectors, and other equipment involves a major engineering effort. The people involved in the work include a large coterie of engineers, technicians, programmers, and scanners as well as physicists.

Historical Background

Elementary-particle physics has its roots in all aspects of physics that are historically concerned with the structure of matter, because the questions of structure and constituents of structure are closely related. The interpretation of the term elementary particle has a history going back to the original atomic theory. One need note only the use of the term chemical element, identifying an atomic species, to recognize the beginning of the subject.

The concept of basic building blocks from which the real world can be composed dates back to the ancient Greek philosopher Democritus, who coined the word, ἄτομος or atom, that is, indivisible particle. During the seventeenth and eighteenth centuries, the development of physics concentrated on macroscopic physical properties such as optics, heat, and mechanics, which did not emphasize the atomistic concept. With the development of chemistry and spectroscopy during the nineteenth century, however,

FIGURE 4.9   Aerial view of the Stanford Linear Accelerator Center. The 22-GeV electron accelerator at this center is the only controlled source of electrons in the world for research in the energy range above 12 GeV. In addition to electrons, this accelerator can provide many beams of secondary particles—pions, muons, kaons, and neutrinos. [Source: Stanford Linear Accelerator Center.]

FIGURE 4.10 The AGS magnet ring and linac junction. This 33-GeV accelerator at the Brookhaven National Laboratory is the principal source in the United States for research using protons in the energy range 12 to 33 GeV. The AGS has been undergoing a major conversion that is now nearing completion. The purpose of this conversion is to provide higher-intensity beams of protons and secondary particles and increased experimental capability, thereby making it possible to carry out research of a qualitatively different character than before. [Source: Brookhaven National Laboratory.]

the study of atoms, molecules, electrons, and ions became a most influential part of physics, culminating in the formulation of modern quantum theory in the 1920's. That the atom turned out to be a structured system, made up of electrons and nucleons, was the first in a series of experiences demonstrating the elusiveness of the ultimate elementary constituents of matter. A complete answer to this question is still lacking.

To determine its constituents, it is necessary to break matter apart until it is reduced to an unbreakable minimum. Thus, matter is subjected to extreme conditions by making the components collide with great vigor and break one another apart. This vigor is usually measured in terms of the energy of collision. The energy provided by a flame is great enough to break molecules apart into atoms (fractions of electron volts), but greater energy is needed to strip electrons from the atom (electron volts). After the electrons are stripped off, the bare nucleus of the atom might appear to be elementary unless even greater energy is brought to bear to shatter it.

The magnitude of energy needed to tear the nucleus apart is measured in millions of electron volts, that is, it is of the order of one million times greater than that required to tear electrons out of atoms. Sources of particles with such energy are naturally available in the emissions from radioactive nuclei, in cosmic radiations, and in the interiors of ordinary stars. Some of the earliest controlled experiments used radioactive sources, but an attack on the nuclear structure problem over a broad front was not possible until Cockcroft and Walton, Van de Graaff, and Lawrence developed machines (electrostatic generators and cyclotrons) capable of accelerating particles to energies of millions of electron volts.

At this stage, in the 1930's, physics appeared to have attained the goal of determining the ultimate constituents of matter and energy. The elementary particles of which atoms and ordinary matter are constituted were clearly the electron, proton, and neutron. The proton is the positively charged nucleus of the simplest atom—hydrogen. The neutron is its neutral counterpart, discovered by Chadwick in 1932. All other normal nuclei consist of neutrons and protons bound together.

The simple notion that all matter is made up of these three particles was augmented by the concept of the photon as the elementary particle or quantum of electromagnetic energy, a concept that came from the quantum theory of radiation. In this case, the particle is not a constituent of matter but a manifestation of energy released in the transition of matter from a state of higher energy (or mass) to a state of lower energy (or mass). This example shows the interchangeability of matter and energy, a well-known aspect of Einstein's special theory of relativity.

The theory of relativity combined with quantum theory also led to the suggestion that for every particle there should be a corresponding antiparticle, having the same mass but opposite electromagnetic properties, as, for example, the sign of its charge. This hypothesis was first suggested as a result of Dirac's relativistic version of the quantum mechanics of the electron. The positron, which is the antiparticle to the electron, and the first known antimatter, was discovered experimentally by Anderson in 1932.

Another elementary particle that appears only as a result of transitions between different states of matter is the neutrino. The existence of the neutrino (and of the antineutrino) was hypothesized by Pauli in that same era (about 1932) to account for what would otherwise have been gross violations of a fundamental law of nature—conservation of energy and of angular momentum—in the beta decay of radioactive nuclei. It is a particle having no charge or mass, and it has only recently—in very sophisticated experiments—been detected directly. The attempt to detect solar neutrinos is critically important for all of astrophysics. For example, their detection will provide the most direct test of the hypothesis that the sun is generating thermonuclear power (see Figure 4.11).

Although a rather complete picture of the basic constituents appeared to exist in the 1930's, little was known about the properties and structure of

FIGURE 4.11 The Brookhaven solar neutrino experiment located 4850 ft underground in the Homestake Gold Mine at Lead, South Dakota. Detection is based on observing the neutrino capture reaction $^{37}Cl + \nu \longrightarrow \, ^{37}Ar + e^-$ in 100,000 gallons of perchloroethylene, C_2Cl_4. The radioactive ^{37}Ar (35-day half-life) is removed from the liquid by a helium gas purge and placed in a small low-level proportional counter to observe the electron capture decay of ^{37}Ar.

matter at the nuclear level. Even the strong forces responsible for holding nuclei together (against the tendency of the positively charged protons to repel each other and push it apart) were a complete mystery. It was known that these forces were strong and acted over very short distances (short-range forces) of the order of nuclear dimensions—typically 100,000 times smaller than those of the atom. The shape of the potential, if indeed these forces could be described by a potential, and its other features were the subjects of intense experimental and theoretical investigation.

Investigation of the shape of such a short-range potential implied that measurements over very short distances were needed. Such measurements could be made only with particles of high energy, because the characteristic wavelength associated with a particle, in accordance with quantum mechanics, is inversely proportional to its momentum, and very short wavelengths are needed to see structure over very small distances. The energy corresponding to a proton wavelength comparable to the range of nuclear forces is about 10 million electron volts (MeV). Therefore, to study effects over much smaller distances requires proton energies of the order of at least 100 MeV (the energy goes inversely as the square of the distance scale).

For this reason, accelerators in the 100- to 600-MeV range were built in the late 1940's and early 1950's when methods to build such machines (synchrocyclotrons, linear accelerators) had been devised. However, there were many other reasons, too, not the least of them being simple curiosity regarding the open-ended question: How does matter behave under more and more extreme impact? From the viewpoint of particle physics the most important reason was given by Yukawa in 1935 when he proposed a field theory of the nuclear force analogous to electromagnetic theory.

Prior to the study of nuclear phenomena, the forces of nature were assumed to be two in number, gravitational and electromagnetic. Both were known to obey an inverse-square law, with the force, albeit rapidly weakening, extending to arbitrary distances from the force center. It was clear from the very earliest measurements, that the nuclear forces had a quite different character; within an exceedingly short range they were very much stronger than the electromagnetic forces, but they vanished abruptly beyond a distance characteristic of nuclear dimensions. Yukawa showed theoretically that the range of a natural force field was tied inversely to the mass of the characteristic particle or quantum of the field; this finding was in accord with the known zero rest mass of the photon and the effective infinite range of the electromagnetic field. It also suggested that the graviton, the quantum of the gravitational field, should have zero rest mass. To match the short range of the strong nuclear force, however, Yukawa was forced to postulate the existence of a nuclear field quantum of finite mass— roughly one tenth of that of the proton.

Because nuclear forces are capable of exchanging electric charge between neutrons and protons, the nuclear quanta had to appear in charged as well as neutral forms. Because the mass of this quantum corresponds, according to the relationship $E=mc^2$, to an energy between 100 and 200 MeV, energies in this range are required to produce them. They were first identified in cosmic rays in the late 1940's. They are produced in great abundance by machines giving protons with energies of 400 MeV and higher.* The strong nuclear field quanta are called pi-mesons or pions. Beams of these pions are now an essential tool for elementary-particle and nuclear physics research and are rapidly finding applications in medicine.

A major puzzle emerged after the initial postulation of the pion by Yukawa. By its very nature it was expected to interact strongly with neutrons and protons, hence with nuclei. But when particles of approximately the right mass and charge were first identified in the cosmic radiation, it appeared that, having passed through much of the earth's atmosphere before detection, they interacted very weakly with nuclei. It was finally realized that the difficulty arose because the free pion was itself unstable against decay. In a time of about one hundredth of a microsecond after being produced, the pion transforms spontaneously into a somewhat lighter particle called the mu-meson, or muon, and a neutrino. Because the muon interacts only weakly with nuclear matter, it can survive passage through the earth's atmosphere after being produced from the decay of pions, which result from the interaction of the primary atomic radiation with the matter in the upper fringes of the atmosphere. The muon was initially, and incorrectly, thought to be Yukawa's meson.

This muon is unstable and transforms spontaneously into an electron, a neutrino, and an antineutrino about 1 μsec after production (in essence, it beta-decays). Its observation at sea level simply shows that it is normally produced with very high kinetic energy following the initial cosmic-ray interaction, and the relativistic time dilation (rapidly moving clocks appear to run slow) is such that it survives passage through the atmosphere before decay. Some pions also survive passage through the atmosphere for similar reasons, despite their much shorter lifetime and the much higher interaction probability. Every observation made to date indicates that the muon behaves in every way like a heavy electron. It is not subject to strong interactions (of nuclear strength), it has identical electromagnetic interactions, and, as far as has been determined, it is subject to the same weak interactions that are responsible for beta-decay, except that the muon has

---

* The extra energy is needed in any collision process because the struck particles are not infinitely massive and recoil; in doing so they retain some of the available energy so that less than the total incident energy is available in the interaction itself. In the jargon of collision physics this is referred to as energy *in* the center-of-mass system as opposed to energy of the center of mass.

its own distinct neutrino (and antineutrino) associated with weak decay processes. It represents a major puzzle; physicists simply have no idea why there should be such a heavy electron or, indeed, whether still heavier electrons remain to be discovered.

Electrons, muons, the two kinds of neutrino, and all their associated antiparticles form a special class of particles called leptons. So far there is no evidence that they interact in a strong way with any form of matter or radiation. Possibly, for this reason, they appear to be truly elementary, that is, irreducible particles.

Particles like the proton, neutron, and pion that do interact strongly, that is, with interaction energies comparable with the potential energies between nucleons, are referred to as hadrons. There are also a great many other kinds of hadron with quite remarkable properties. The first hint of their existence came, again, from cosmic-ray observations, in which a few examples of what are now called hyperons and K-mesons were recognized as unusual events in cloud chambers and emulsions.

The characterization of these particles, their classification, and the relationships between them began to unfold only when the first accelerators in the billon-electron-volt (GeV)* range came into operation. The first such proton accelerator, the 3-GeV Cosmotron at Brookhaven, was planned before the existence of these strange particles was known, but their existence was clearly indicated by the time the Cosmotron was turned on. It was only shortly afterward, as a result of experiments at the Cosmotron, that the principle of associated production of strange particles, which is one of the fundamental concepts in the understanding of strangeness, was well established. Just as in moving from the atomic to the nucleon quantum system it was necessary to introduce a new quantum number, the isotopic spin, to characterize the different charge states of the nucleon—the neutron and proton—so also in going to these elementary-particle systems the addition of yet another label or quantum number, strangeness, was required.

The Cosmotron was built to study nuclear forces at a deeper level than had been possible before and to observe the behavior of matter under more energetic impact. Experiments showed that matter behaved in a totally unexpected manner, thus an entirely new field of investigation emerged. The Bevatron, at the University of California at Berkeley, planned for similar purposes, was completed shortly thereafter. In view of the theoretical reasons for believing that for every particle there is a corresponding antiparticle, the selected energy was sufficiently high (6 GeV) to produce antiprotons. (The first triumph of this theory had been the discovery of the positron.) The mass of the proton is large (about 1 GeV/c^2 in energy units), and antiprotons must be produced in proton–antiproton pairs (just

---

* For giga electron volt; this abbreviation has now replaced the formerly used BeV.

as positrons are produced in electron–positron pairs), requiring the conversion of 2 GeV of energy. This amount of useful energy (in the center-of-mass system) is produced in the collision of a 6-GeV proton with a proton at rest. Thus, the Bevatron produced antiprotons as well as other known particles.

At this point, it should be clear that the original objective of determining the constituents of matter is not distinguishable from the determination of the possible forms of energy. Not only are the photon and neutrino produced in transitions between energy states of matter, but electron–positron pairs, pions, proton–antiproton pairs, associated pairs of strange particles, and any other form of matter can be created in this way. The distinction between matter and energy has vanished.

Recent Developments in Elementary-Particle Physics

A continuous unfolding of surprises concerning the forms that elementary matter and energy can take characterizes the recent history of this subfield. The intrinsic properties, interactions, reactions, and all aspects of the behavior of these constituents, which are found under more and more extreme conditions, have led to entirely new concepts of the nature of the physical universe. Higher-energy machines, in the 10- to 70-GeV range, that became operative throughout the world during the past 15 years, and the great number of experiments conducted at them, produced many of these revelations. The nature of many of the most important discoveries could not have been anticipated when the machines at which they occurred were being planned.

Among these recent discoveries are the following:

1. Discovery of parity violation in weak interactions (that is, nature distinguishes between right- and left-handedness in these interactions);

2. Discovery of two kinds of neutrino;

3. Confirmation of the idea that the vector part of the weak nuclear interaction is generated in a manner remarkably similar to the generation of electromagnetic interactions;

4. Discovery of a difference between the world of particles and the world of antiparticles, even when the latter is viewed in a mirror (that is, violation of CP invariance), and the associated discovery that there is some aspect of the weak interactions that depends on the direction of flow of time;

5. Elucidation of the structure of neutron and proton in terms of their internal charge and current densities;

6. Discovery that protons appear to have an internal point particulate structure;

7. Realization that electromagnetic properties of particles not only relate

to the usual massless photons but also involve very massive vector mesons that are subject to strong (nuclear) interaction (reflecting the large masses, these interactions are of very short range);

8. Exploration of the limits of relativistic quantum electrodynamics to a distance of the order of 10^{-11} cm;

9. Opening of the field of hadron spectroscopy (Table 4.1) and discovery of the underlying $SU(3)$ symmetry relating the hadrons;

10. Realization of phenomenological theories for dealing with relativistic reactions between hadrons and interpreting them, especially at very high energy.

The report of the Panel on Elementary-Particle Physics in Volume II discusses a number of these findings in greater detail. What emerges is a

TABLE 4.1   Stable Hadrons

Name	Symbol	Electric Charge	Mag. of mc^2 (MeV)	Spin	Parity	Isotopic Spin	Strangeness
Pion	$\pi^\pm$	$\pm e$	140	0	neg.	1	0
	π^0	0	135				
K-meson	K^+	$+e$	494	0	neg.	½	$+1$
	K^0	0	498				
$\bar{K}$-meson	K^-	$-e$	a	a	a	a	-1
	$\bar{K}^0$	0					
Nucleon	p	$+e$	938.3	½	pos.	½	0
	n	0	939.6				
Antinucleon	$\bar{p}$	$-e$	a	a	a	a	0
	$\bar{n}$	0					
Lambda	Λ	0	1116	½	pos.	0	-1
Antilambda	$\bar{\Lambda}$	0	a	a	a	a	$+1$
Sigma	$\Sigma^\pm$	$\pm e$	1197	½	pos.	1	-1
	Σ^0	0	1192				
Antisigma	$\bar{\Sigma}^\pm$	$\pm e$	a	a	a	a	$+1$
	$\bar{\Sigma}^0$	0					
Cascade	Ξ^-	$-e$	1321	½	pos.	½	-2
	Ξ^0	0	1314				
Anticascade	$\bar{\Xi}^+$	$+e$	a	a	a	a	$+2$
	$\bar{\Xi}^0$	0					
Omega minus	Ω^-	$-e$	1672	½	pos.	0	-3
Antiomega minus	$\bar{\Omega}^+$	$+e$	a	a	a	a	$+3$

a Masses and indicated quantum numbers are the same for particle and antiparticle of opposite charge.

continuing progression of new discoveries and new surprises that represent salients into entirely virgin territory in man's understanding of the nature of space, time, and the basic laws that govern all natural phenomena. This push into the unknown is one of the greatest adventures of the human intellect; but it is much more. It is a continuing challenge to further exploration and fuller understanding of natural phenomena. It commands the devotion and attention of some of the most intellectually gifted scientists of this age.

Interactions with Technology

Successes in elementary-particle physics have depended heavily on the discoveries and developments made in almost all subfields of physics and in engineering. Many new technical developments were needed. The demands of this subfield presented a challenge to technology as great as any offered by either space research or military requirements. As the magnitude of the energy needed for exploration has increased, the size of the required apparatus has increased enormously and so has the degree of ingenuity required both to reduce costs and to provide methods for making the necessary measurements.

The development of new concepts in machine technology made the building of giant machines feasible. The particle beams of very high energy produced by these machines had to be manipulated in a way that required the development of intricate beam-transport devices, including huge bending magnets, magnetic lenses, electrostatic and radio-frequency velocity selectors, computer monitoring and control of beams, and the like. The detection and measurement of particles at high energy and intensity require special detection devices such as bubble chambers, Cerenkov and other counters, spark chambers, and wire chambers, all in combination with on-line computers (see Figure 4.12). Some of these devices must be capable of measuring time intervals of a few billionths of a second or of handling millions of particle events in seconds. Measurements on the photographic film on which information is stored has led to the development of automatic pattern-recognition devices that have many other applications. Computer and special methods had to be developed to perform the many intricate calculations that are needed before the physics can be recognized in the enormous mass of data emerging from an experiment. This task has strained the capabilities of the largest computers. Without such methods, the emergence of hadron spectroscopy, for example, would have been impossible.

Big superconducting magnets (see Figure 4.13) have been designed and built to provide the large magnetic fluxes needed to deflect high-energy particle beams at a reasonable cost, because the power requirements for some individual magnets would be as high as 10 MW. More recently,

FIGURE 4.12 The 12-ft-diameter bubble chamber that operates in conjunction with the 12-GeV ZGS proton accelerator at the Argonne National Laboratory. The chamber, which can be filled with hydrogen or deuterium, is the largest particle detector of its type in operation in the world today. The early work with the chamber has concentrated on neutrino physics experiments but will soon make use of more conventional beams, for example, pions, kaons, and antiprotons. [Source: Argonne National Laboratory.]

FIGURE 4.13. The 184-in. superconducting magnet (18 kG) built for the Argonne National Laboratory 12-ft bubble chamber. The 100-ton magnet consumes only 10 W of power, much less than the several megawatts required for conventional magnets of the same field strength. This is the largest superconducting magnet operating in the world. [Source: Argonne National Laboratory.]

alternating current and radio-frequency superconducting systems have become feasible and should soon make possible the building of accelerators at higher energy or with an improved duty cycle * without an increase in size or power requirements. Such concepts as the electron ring accelerator may also make it possible to go to much higher energy at reasonable cost.

This dependence of high-energy physics on technology and engineering frequently stretches the capabilities of existing technology to the utmost, requiring innovations and extrapolations that go well beyond any present state of the art. Because the resulting technological developments have implications much broader than their use in particle physics, all technology

---

* In lower-energy electrostatic accelerators, for example, the stream of particles in the beam current is constant in time, that is, it is a dc beam. In the larger accelerators, however, as a consequence of the accelerating mechanisms, the beam particles occur in bursts of greater or lesser length, with intervening periods in which there are no particles. The duty cycle is a measure of the extent to which a given accelerator approaches a dc beam character.

benefits from the opportunity to respond to this pressure. New technical developments occur sooner—sometimes much earlier—than they would in the absence of such pressure, and they often present new engineering opportunities, unrelated to high-energy physics, that can be exploited immediately. Examples of such engineering developments are manifold: large volume, very-high-vacuum systems; sources of enormous radiofrequency power; cryogenic systems; large-scale static superconducting magnets; variable-current superconducting systems; pattern-recognition devices; very fast electronic circuits; and on-line computer techniques. In this sense, elementary-particle physics has had a major impact on technology, but the effect has been an indirect one resulting from the urgency of the research requirements rather than the results of the research.

The history of physics offers many examples of the discovery of elementary particles, or properties of elementary particles, that have had extremely important applications. These examples are the basis of a belief that it will happen again. Examples of important innovations in technology based on particles whose existence is yet to be established have been suggested. Of course, physicists are on tenuous ground here, because such speculations could be based on false premises. That is why more research is needed—to determine the validity of the premises. If they are not correct, past experience suggests that the results could be completely unexpected and might have greater impact on technology than can be imagined at the present time.

It may be helpful to include an example of such a speculation suggested by recent discoveries in particle physics. The discovery of a multitude of particle resonances led to the development of hadron spectroscopy. This study has revealed many remarkable regularities in the relationships among the hadrons, which, in turn, have been classified in terms of several parameters that take on a very limited set of values and are called internal quantum numbers.

The situation is analogous to that which occurred in the nineteenth century in connection with the concept of chemical valence, which also can be regarded as an internal quantum number of the atom; the regularities exhibited a pattern that led Mendeleev to construct the periodic table of elements, which not only systematized existing information but also had predictive power. Every gap in the table was later filled by a chemical element whose existence was unknown at the time that the table was constructed.

In regard to the regularities of hadron spectroscopy, the internal quantum numbers have been associated with the symmetry patterns (representations) related to the group $SU(3)$ (long familiar in group theory) of simple three-dimensional unitary transformations. These patterns also can be used to predict the particles that should exist and some of their properties. The predictions thus far have been remarkably successful, although there are some contradictions and controversies that must be resolved.

The remarkable successes of the $SU(3)$ scheme, for which Gell-Mann was awarded the 1969 Nobel Prize in physics, have led to the type of situation of which physicists dream. In the case of the particle, then unknown, the theoretical prediction was remarkably specific. It included the mass, spin, charge, and all observable characteristics of the particle. Moreover, substantiation of the entire theory depended on its existence. Thus, discovery of the omega minus, with all the predicted properties, was a major vindication of this approach and the underlying understanding of fundamental phenomena on which it was based. Even more recently, the antiparticle, the antiomega minus, has been observed after strenuous effort by a group at Berkeley (Figure 4.14). Its discovery fills all the existing gaps in the $SU(3)$ ordering; however, this does not preclude the possible discovery of deeper symmetries or a broader ordering involving yet unknown species of particles.

The physics of the periodic table of the chemical elements was ultimately elucidated in terms of the electron structure of the atom, the quantum mechanics of the electron, the spin of the electron, the Pauli exclusion principle, and the like. To make an analogy with the situation in particle physics today, one could ask how the periodic table of elements might have been interpreted if the quantum mechanics of elementary particles had been known but the existence of the electron had not. It is not too difficult to imagine that someone could have come to the conclusion, on the basis of the periodic table of the elements, that the atom consisted of particles of spin one-half, satisfying Fermi-Dirac statistics (that is, the Pauli exclusion principle) and moving in a central field of force. This conclusion would not have been sufficient to characterize the electrons in atoms completely; but, in the context of the analogy, it would have been cause for subjecting atoms to severe collisions involving energies of kilovolts, which would have been high energy in the nineteenth century, to knock some of the particles (electrons) out of the atom for further study.

The corresponding problem in particle physics is the explanation of the observed $SU(3)$ symmetry of the hadrons. It has been suggested that $SU(3)$ may represent the structure of the hadrons as composites of ultimate particles of three distinct kinds, which have been given the name quarks. All mesons would be composed of appropriate pairs consisting of one quark and one antiquark each; the baryons, which are particles like the proton and neutron, would consist of three quarks.

To keep the model simple, that is, to limit it to just three kinds of quark and their associated antiquarks, it is necessary that the electric

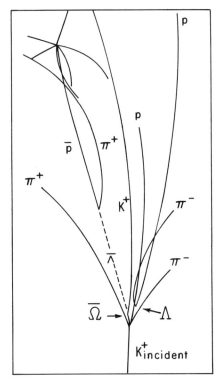

FIGURE 4.14. Almost all of the antibaryons have been observed in experiments, the most recent being the antiomega particle ($\bar{\Omega}$), in 1970, by a group from the University of California, Berkeley. The bubble chamber event is shown in the above figure. The experiment involved was a study of the K^+d interaction at 12 BeV/c carried out in the 82-in. Stanford Linear Accelerator Center bubble chamber. The production reaction is $K^+d \rightarrow \bar{\Omega} \Lambda \Lambda p \pi^+ \pi^-$, and the decay is $\Omega \rightarrow \Lambda K^+$. [Courtesy of Gerson Goldhaber.]

charges of these particles occur in units of one third of the elementary charge, *e*. There is, of course, no evidence whatsoever that this is actually possible. The three different kinds of quark would necessarily have charges of $\pm e/3$ and $2e/3$. The antiquarks would have charges of the opposite sign. The quark idea leads to simple interpretations of many of the known characteristics of hadrons, but the particles have other peculiarities and suggest many unanswered questions. That very massive quarks can combine, at least theoretically, to give much lighter resulting particles is simply a reflection of the correspondingly large binding energies that would be involved.

The key question is: Does the quark really exist? The energy of collisions between hadrons required to produce quarks depends on the as yet unknown quark mass, and it may be very high. They have not been detected at existing accelerators, either because the available energies were inadequate or because the quarks are a mathematical fiction, serving only as a simple and graphic representation of something even deeper in the theory. Despite this situation, one of the earliest experiments with the new National Accelerator will be a search for the quark. There is an elegance and an economy about this quark hypothesis that has great aesthetic appeal. (See Plate 4.II.)

The connection with technology is highly speculative and depends on the anticipated stability of the hypothesized quark. This stability implies that there may be ways to store quarks after producing them at an accelerator. The storage of quarks would correspond to the storage of enormous amounts of energy, and this energy could be released in a controlled fashion by allowing quarks and antiquarks to recombine at the desired rate.

Because electric charge is conserved in all reactions, there is no way for a quark to disappear in ordinary matter; therefore, it is stable not only in vacuum but also in the presence of matter. The making of exotic atoms by binding a proton to a negatively charged quark is conceivable. Since the quark is expected to be the more massive, the proton would play a role similar to that of the electron in an ordinary atom. Possibly, exotic atoms could form exotic molecules. One can immediately speculate that such exotic atoms and molecules might be used to catalyze thermonuclear reactions between protons and deuterons, because they would provide a mechanism to juxtapose the nuclei in spite of the repulsive forces caused by their positive electric charges. Relatively few quarks should be needed, since their release for reuse at each thermonuclear interaction would be expected. Thus, it is possible to imagine a controlled source of thermonuclear power consisting of a vessel of deuterons into which just enough quark impurity has been introduced to produce power at the desired rate. Although this speculation may seem farfetched, thermonuclear catalysis by muons in an analogous process already has been observed; only because muons are unstable are they useless as a practical source of energy.

It is quite possible that quarks do not exist. But the investigation of this question almost certainly will lead to the discovery of an even more fascinating world.

Impact on Other Sciences

Since elementary-particle physics has its origin in nuclear physics, there is a particularly close connection between these two subfields. They depend on one another in many ways. The research results of particle physics having the greatest direct bearing on nuclear physics are those that throw light on the nature of nuclear forces and those that provide methods for the study of nuclear structure. (See the following section on Nuclear Physics.)

Not only are there results of research that are of interest to both subfields, but there is a substantial overlap in the instrumentation, so that each learns from the other. There also has been a strong mutual impact in regard to methods for handling data, use of on-line computers, and the like.

High-energy physics also plays a special role in relation to astrophysics, because the astrophysicist is concerned with matter under the most extreme conditions. At sufficiently high temperatures the exotic and unstable particles produced by accelerators will exist in thermal equilibrium with matter and influence the equation of state of matter under these extreme conditions as well as determine the reactions that occur. Thus a knowledge of all the elementary particles and how they behave is essential to a full understanding of interstellar matter under some conditions. There is also an important relationship with the work in space-radiation physics. The cosmic rays were the first source of high-energy particles, and the study of these particles in space makes full use of both existing knowledge of their properties and instruments and techniques developed for their study.

In subfields other than nuclear physics and astrophysics, the connection is less direct. New instruments developed for particle physics are useful in many disciplines. For example, various pattern-recognition devices for scanning and measuring film from bubble chambers and spark chambers are valuable for a variety of applications in biological research, two of which are chromosome counting and measuring cross sections of nerve bundles. Image intensifiers, developed to photograph particle tracks in scintillators, have proved useful for quantitative observation of chemical luminescence in biological systems. Another important medical application has arisen in connection with the development of techniques for making and handling thin but very tough plastic films as support for multiwire particle detectors, plastic plumbing for very rugged thin-walled targets, and other plastic systems (Figure 4.15). Elementary particles could open new fields of research in medicine. For example, there is reason to believe that irradiation of tumors with pion beams may have therapeutic value because of the particular properties of the pion. To establish the validity of this idea requires extensive research on the biological effects of pion beams.

Theoretical methods of one branch of physics are usually applicable in others. The history of physics has been characterized by the unity of the theories (as discussed in the final section of this chapter). Elementary-particle theory is no exception. The methods of quantum field theory, developed to answer fundamental questions concerning particles, have

FIGURE 4.15   A preliminary version of a small, inexpensive ($15) artificial kidney, which has been developed at the High Energy Facilities Division of Argonne National Laboratory. It has been used successfully by nine patients at an Illinois Veterans Hospital. It has small enough volume ($6 \times 2 \times 2$ in.³) to be suitable for a child, and it is hoped that it can be made inexpensive enough to be used daily. The basic configurations of these dialyzer designs are also being explored for their potential as membrane oxygenators for heart–lung machines. [Source: Argonne National Laboratory.]

proved to be applicable to problems in many other subfields such as the nuclear many-body problem, many-body problems in condensed matter, superconductivity theory, and statistical mechanics. Methods for treating resonance reactions, a field that had its beginning in atomic physics and flowered in nuclear physics, have required deeper investigation to understand their meaning in the relativistic processes that occur at high energy. The result has been more general methods and insights into scattering processes in general, with applications whenever they occur.

The methods for dealing with the general problem of relativistic reactions have been the subject of intense investigation, with results that have widespread use and add to current understanding of all similar physical phenomena. However, not only the methods but also the content of these theories can be of great importance in other subfields of physics. Just as the full panoply of elementary particles and their excited states can exist in stellar interiors, so in atoms and molecules such particles exist in virtual form for minuscule instants of time and thus have a tiny influence on atomic and molecular properties that will eventually be detected when measurements become sufficiently precise. Therefore, it is possible that the limits of the known theory of electrons and electromagnetism will be passed by high-precision measurements of an atomic nature before they have been reached by high-energy experiments. The two approaches are complementary, and both have been pursued vigorously, but the only certainty is that a limit will eventually be reached, a limit such that even the theory of the electronic structure of atoms will require modification.

Although there are many specific examples of the impact of elementary-particle physics on other sciences, especially on other subfields of physics, it is not these details that represent its most significant contribution. This contribution is associated with the whole flavor and character of scientific exploration and discovery. The entire scientific organism flourishes when its major fields and subfields are flourishing and feeding the whole. In different fields, and in different subfields of physics, the nature and style of the work varies greatly, but all science gains strength when any subfield contributes in a substantial way to the overall store of knowledge or under-standing. Certainly all of physics gains from any significant new discovery, idea, information, or theory in any of its major subfields.

Of course, there is no way to guarantee the unexpected, and the pursuit of it is a gamble. But the gamble has been paying off handsomely, and there is no reason to believe that it will not continue to do so.

Distribution of Activity

Elementary-particle physics grew from academic research in atomic, nu-clear, and cosmic-ray physics, and, in spite of its enormous development in scale, its roots have remained in the universities. The early accelerators and the devices for studying the behavior of particle beams emerging from them were built at the universities. A few accelerators at universities continue to be used for elementary-particle physics experiments, but the emphasis has shifted to much greater energies and more complicated equipment than can be managed within the usual university framework. There-fore, the accelerators at these higher energies have been built at National Laboratories, some of them managed by single universities and others by consortia of universities (see Chapter 9). Although each of these labora-tories has its own in-house research activity, by far the larger part of the research in the National Laboratories is conducted by groups of professors, research associates, and students from universities throughout the country.

Table 4.2 lists the operating high-energy accelerators of the world. The Cosmotron, a 3-GeV proton accelerator at Brookhaven, was shut down several years ago, and, as noted in the table, the Princeton Particle Accelerator (PPA) has been shut down as an elementary-particle physics laboratory. These shutdowns resulted from funding limitations, not from any significant obsolescence of the scientific value of the equipment.

The 200–500-GeV proton accelerator at the National Accelerator Laboratory is nearing completion. It is possible that the first experiments will be carried out there in mid-1972. Already initial design work has begun

TABLE 4.2   Accelerators Operating above 1.5 GeV [a]

Accelerated Particle	United States		Western Europe		Soviet Union	
Proton	PPA [b]	3 GeV	Saturne	3 GeV		
Proton	Bevatron	6.2 GeV	Nimrod	7 GeV	ITEP	7.5 GeV
Proton	ZGS	12.7 GeV				10 GeV
Proton	AGS	33 GeV	CERN-PS	28 GeV	Serpukhov	76 GeV
Electron			Bonn	2.3 GeV	Kharkov	2 GeV
Electron	CEA [c]	6.3 GeV	NINA	5 GeV		
Electron	Cornell	10 GeV	DESY	6.2 GeV	Yerevan	6 GeV
Electron	SLAC	21 GeV				

[a] Abbreviations, names, and locations shown in table:

United States

PPA	Princeton Particle Accelerator, Princeton, N.J.
Bevatron	Lawrence Berkeley Laboratory, Berkeley, Calif.
ZGS	Zero Gradient Synchrotron, Argonne National Laboratory, Argonne, Ill.
AGS	Alternating Gradient Synchrotron, Brookhaven National Laboratory, Upton, Long Island, N.Y.
Cornell	Cornell University, Ithaca, N.Y.
CEA	Cambridge Electron Accelerator, Harvard University, Cambridge, Mass.
SLAC	Stanford Linear Accelerator Center, Stanford, Calif.

Western Europe

Saturne	Commissariat à l'Énergie Atomique, Saclay, France
Nimrod	Rutherford Laboratory, Chilton, Berkshire, England
CERN-PS	Proton Synchrotron, CERN, Geneva, Switzerland
Bonn	Physikalisches Institut, Bonn, Germany
DESY	Deutsches Elektronen-Synchrotron, Hamburg, Germany
NINA	Daresbury Nuclear Physics Laboratory, Daresbury, England

Soviet Union

ITEP	Institute of Theoretical and Experimental Physics, Moscow
Serpukhov	Institute of High Energy Physics, Serphkhov
Kharkov	Physical Technical Institute, Kharkov
Yerevan	Institute of Physics (GKAE), Yerevan, Armenian SSR

[b] Shut down as an elementary-particle physics facility at end of fiscal year 1971.
[c] Being used only as an intersecting-beam device.

that will extend the capability of this facility to 1000 GeV through the addition of a ring of superconducting magnets immediately above the present magnet ring (Figure 4.16).

At the present time, there are approximately 50 U.S. universities heavily involved in elementary-particle physics. In addition, the number participating to some degree, with hope of greater involvement, is about 125.

The sociology of high-energy physics has become rather complicated because the determining factors in the research and training activities in the universities are the decisions made concerning the experiments to be carried out or the ancillary equipment to be developed at the large accelerators at the National Laboratories. In order that the judgments and needs of the participating research community may be taken into account, an elaborate system of advisory committees and corporations formed by con-

FIGURE 4.16   Aerial view of main accelerator at the National Accelerator Laboratory. This 200–500-GeV proton accelerator will be for some years the only controlled source of protons in the world for research in the energy range above 80 GeV and the only one in the United States above 33 GeV. In addition to proton beams, the accelerator will provide many beams of secondary particles—neutrinos, pions, kaons, photons, and antiprotons. [Source: National Accelerator Laboratory.]

Combining Quarks

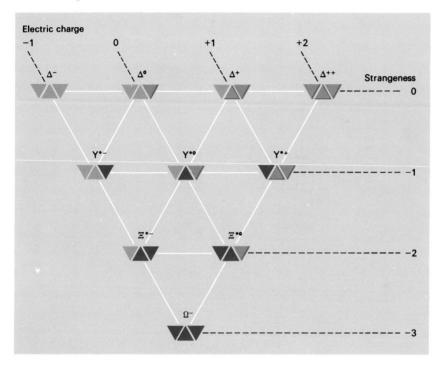

PLATE 4.II(a)   Combining quarks. Three kinds of object, shown as colored triangles, can be arranged into ten possible groups of three. In The Eightfold Way, each combination of three fictitious objects, called quarks, makes a different baryon. [Source: *Science Year. The World Book Science Annual.* Copyright © 1968, Field Enterprises Educational Corporation.]

One Baryon Family

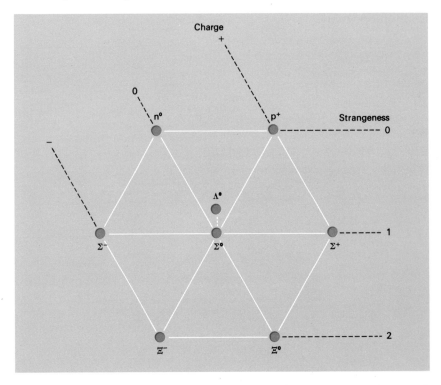

FIGURE 4.II(b) One baryon family. Complex combinations of the three quark species form other families of baryons, according to The Eightfold Way. The neutron and proton are in the top row. [Source: *Science Year. The World Book Science Annual.* Copyright © 1968, Field Enterprises Educational Corporation.]

sortia of universities has come into being. The arrangements are tailored to each laboratory and its special problems and needs. They seem to work rather well, at least as judged by the success of the research programs (see also Chapter 9).

There are approximately 1700 PhD physicists and engineers who can be identified as working on projects supported by federal particle-physics funds. Of these, about one third are theoretical physicists, the rest, experimental. This figure probably does not include a substantial number of theoretical particle physicists who do not receive federal support.

The number of graduate students at the thesis level working on programs supported by federal particle-physics funds is about 1100. A substantial number of students supported by other funds should probably be added to this number. Well over 300 PhD's annually are being granted in this subfield, and more than half of them are based on theses in theoretical physics.

About half of these particle physics PhD's have gone into other subfields after receiving their degrees. Those who remain usually spend from two to four years as research associates, or in equivalent temporary postdoctoral

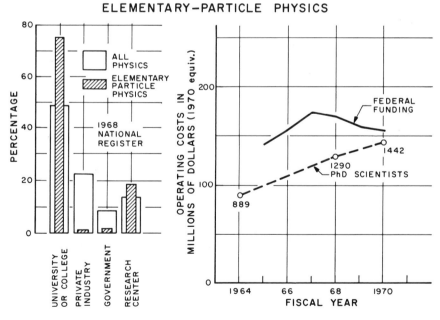

FIGURE 4.17 Manpower, funding, and employment data on elementary-particle physics, 1964–1970.

positions, in either a university or a National Laboratory. Because of the current limitations on permanent job opportunities, many research associates are extending their appointments beyond even the fourth year, and the number of PhD's leaving the subfield is increasing (see Appendix 4.B).

Elementary-particle physicists comprise about 10 percent of the physics PhD population, as reported in the 1970 National Register of Scientific and Technical Personnel. They work principally in academic institutions. One fifth were employed in research centers (for the most part, National Laboratories). The number of PhD's in the subfield has grown at an annual rate of 10 percent since 1964 (see Figure 4.17). Approximately 33 percent of the federal funds for basic research in the various subfields of physics in 1970 were allocated to the support of research in elementary-particle physics; there is essentially no direct industrial support (see Figure 4.18).

Problems in the Subfield

Funding   The funding of high-energy physics has developed serious inconsistencies in the past few years. Although the present capability, measured in terms of both equipment and manpower, probably provides the United States with the greatest potentiality for research in this subfield of any country in the world, the bleak funding pattern that has characterized federal budgets of recent years seems likely to lead to rapid dissipation of

ELEMENTARY-PARTICLE PHYSICS

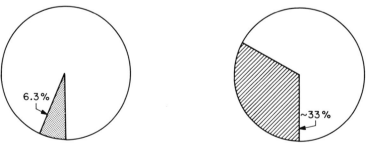

FIGURE 4.18   Manpower and federal funding in elementary-particle physics in 1970.

this strength. Accelerators are being utilized at less than 75 percent of capacity because of financial stringencies and are facing even further cuts. University groups are finding it increasingly difficult to obtain the funds needed to mount experiments that will take advantage of the available facilities.

The total funding of operations and equipment of the principal accelerator laboratories has been decreasing in absolute dollars at a time when one new major accelerator (SLAC) has come into operation, when a major improvement program increasing the capability of the AGS is being completed, and when university groups from all parts of the country are trying to prepare for experiments at the National Accelerator Laboratory (NAL).

As NAL operations get under way, it is essential that incremental funding be provided in the total operating and capital equipment budgets for all accelerators. Otherwise, the exploitation of NAL, which offers the most important opportunities in this subfield, will certainly force the demise of some of the lower-energy accelerators and a drastic reduction in the level of activity of others. Under these conditions much of the important work that remains to be done throughout the spectrum of energies below the 200 GeV available at NAL could not be carried out. This situation would seriously weaken elementary-particle physics research in this country.

Manpower The employment problem for theoretical particle physicists appears to be even more serious than it is for other physicists. The large number of such theorists produced in recent years and their high degree of specialization are often given as the causes of this difficulty. This narrow specialization is already an indication that the student of particle theory has been allowed to choose unwisely, because real success in any part of physics requires more breadth, and both great breadth and depth of perspective are required for a significant contribution, especially in theoretical particle physics.

It is imperative that university groups in elementary-particle theory act to discourage all but the most able potential students and that even these should be made fully aware of the employment problems and restricted career opportunities that now appear probable. University groups have a responsibility to expose their most brilliant and able students to the opportunities in all subfields of physics, particularly under present circumstances in which such students will be needed as effective partners in national attacks on major problems affecting society now and in the future. Because a sizable fraction of the most able students are attracted to elementary-particle physics, a correspondingly heavy responsibility rests on those who are now active in this subfield to provide objective and realistic advice concerning these career opportunities.

> First we must inquire whether the elements are
> eternal or subject to generation and destruction;
> for when this question has been answered their
> number and character will be manifest.
>
> ARISTOTLE (384–322 B.C.)
> *On The Heavens,* Book III, 304

NUCLEAR PHYSICS

Introduction

Nuclear physics includes the study of the structure of atomic nuclei and their interactions with each other, with their constituent particles, and with the whole spectrum of elementary particles that is now provided by the very large accelerators. The nuclear domain occupies a central position between the atomic range of forces and sizes and those of elementary-particle physics, characteristically within the nucleons themselves. As the only system in which all the known natural forces can be studied simultaneously, it provides a natural laboratory for the testing and extending of many of the fundamental symmetries and laws of nature.

Containing a reasonably large, yet manageable, number of strongly interacting components, the nucleus also occupies a central position in the universal many-body problem of physics, falling between the few-body problems, characteristic of elementary-particle interactions, and the extreme many-body situations of plasma physics and condensed matter, where statistical approaches dominate; it provides a rich range of phenomena and the hope of understanding these at a microscopic level.

Although still a relatively young science, nuclear physics has already had a profound effect on both peace and war and on man's view of his universe and himself. The release of nuclear energies in fission reactors holds high promise of providing the energies for civilization until fusion systems can be perfected or, perhaps, indefinitely; radioactive techniques have already had a major impact on technology and on both clinical and research medicine and biology.

Historical Background

The experiments of Rutherford and his colleagues at the beginning of this century established the basic structure of the atom and located the nucleus as a massive, positively charged, and very much smaller (by a factor of some 100,000) entity at the atomic center. Over the next four decades, the gross characteristics of nuclei were established—their sizes; their basic components, the neutron and proton; the characteristic energies by which they were bound together (some millions of times those in atomic systems);

and the nature of the radiations that were spontaneously emitted by some heavy nuclear species. Much of this information came from the study of collisions between nuclei and the nuclear projectiles as more of these became available at increasingly higher energies. In the beginning only the naturally occurring nuclear emanations—the alpha particles (helium nuclei) so dear to Rutherford—were available; the development of the early accelerators by Cockcroft and Walton, Van de Graaff, Lawrence, and many others led to a new era of experimentation in that, for the first time, the experimenter could choose and vary the nature and energy of his projectile, albeit within rather strict instrumental limits, instead of making do with what nature had provided.

Indeed, the development of accelerators of ever-increasing power and sophistication has been one of the main themes of nuclear technology; along with this has gone the development of nuclear radiation detectors of increasing sensitivity and resolution adequate to the challenge of sorting out the products and results of the accelerator projectile-induced nuclear interactions.

By 1949, a considerable body of knowledge had been accumulated on the simpler properties of the 300 or so stable nuclei found in nature and of the perhaps 500 additional radioactive isotopes that had been produced from these using either accelerators or the newly developed nuclear fission reactor. Apart from its energy-generation capabilities, the reactor for the first time made it possible to produce many of these radioactive species in relatively large quantities for study and application. While only a modest amount of information was available concerning any given nucleus, certain striking regularities already were becoming evident as the number of neutrons or protons was varied. These regularities led Mayer and Jensen to propose their remarkably successful shell model of the nucleus, quite analogous to the Bohr model of the atom in that the single nucleons were assumed to move almost independently in well-defined orbits in a central potential well.

This model proved to be a key idea in the understanding of nuclear structure and dynamics. But it was far from obvious! The success of the Bohr-Wheeler liquid-drop nuclear model in explaining the gross characteristics of nuclear fission in 1939 rested on the basic assumption that the nuclear constituents, the nucleons, interacted so strongly that the mean free path between interactions was much less than the diameter of the nucleus itself. This assumption was in apparent direct conflict with the well-defined orbital motion of the shell model, and it was not until 1956 that Weisskopf explained this apparent paradox as a consequence of the operation of the Pauli principle within the nucleus.

In the 1950's, a phenomenal increase in the amount of nuclear information available occurred as new accelerators, detectors, and instrumentation

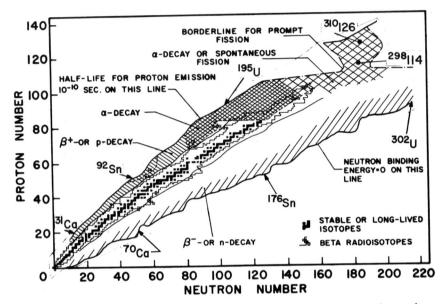

FIGURE 4.19 The nuclear stability diagram is obtained by plotting the nuclear binding energy as a surface deformed by the number of protons (Z) and the number of neutrons (N) present. For low N and Z the most stable species are those having $N=Z$; but with increasing Z, the electrostatic repulsion between the protons forces the stability toward nuclei where N is greater than Z. The black squares represent the nuclei that are stable in nature; this simply means that their lifetime against spontaneous decay is long compared with the lifetime of the solar system. There are some 300 in all. The region outlined with a light line is that which has been explored thus far in nuclear physics; it contains some 1600 different nuclear species (isotopes). The outer solid lines define the region in which it can be calculated that nuclear species should be stable against instantaneous decay via the strong nuclear forces. The lower line is the so-called neutron-drip line in that any species formed below it spontaneously and instantaneously emits neutrons and moves to the left until it reaches the line and the boundary of stability. Emission of tightly bound alpha particles (helium nuclei containing two protons and two neutrons each) competes favorably with direct proton emission at the upper boundary, and spontaneous fission limits the stability region for very heavy nuclei. $B_f=0$ indicates zero binding against such spontaneous fission as does $B_n=0$ against neutron emission. Two of the postulated islands of stability far beyond the natural range are indicated at $Z=114$ and $Z=126$. It is clear from such a figure that a vast range of nuclear species remains to be explored. While the most stable uranium isotopes have masses of 235 and 238, respectively, as shown here, uranium isotopes ranging in mass from 195 to 302 would be expected to be stable against instantaneous breakup.

derived from wartime technology became operational. Not only was much more known about each nucleus, but also many more nuclear species were produced and studied as the new facilities permitted forays from the valley of nuclear stability, which runs diagonally through any map of the nuclear domain wherein the number of neutrons is plotted against the number of protons. Such a map is shown in Figure 4.19. The total number of known nuclear species increased to about 1200, and, instead of

FIGURE 4.20 The first of the Emperor class of electrostatic Van de Graaff accelerators, installed in the Arthur Williams Wright Nuclear Structure Laboratory at Yale University. This accelerator is the largest of the tandem configurations in research use and is the first to make available all the nuclear species to precision study. Terminal potentials in excess of 11 MeV have been obtained on the machine. The large pressure vessel in the background contains the electrostatic accelerator structure; the magnetic systems in the foreground are part of the beam energy control and distribution system.

typically three or four quantum energy levels known for each, numbers such as 20 to 50 became typical.

Yet another wave of innovation and development in the instrumentation of nuclear physics took place in the 1950's. The large higher-energy electrostatic accelerators, for the first time, made all nuclei accessible to precision study with a very broad range of projectiles (see Figure 4.20); the semiconductor nuclear detectors improved the attainable energy resolution by factors between 10 and 100 (see Figure 4.21), and on-line computer techniques not only produced striking improvement in the quality of the available data but also made it possible to digest and analyze the flood of new results efficiently and effectively.

These instrumental innovations were reflected in a massive increase in the scope and quality of information about nuclei. Not only did the number of known nuclear species grow to some 1600, but the information on the quantum structure of each of these increased by orders of magnitude. In addition, entirely new insight into the nature of nuclear interactions and dynamics was gained—that is, how energy and linear and angular momen-

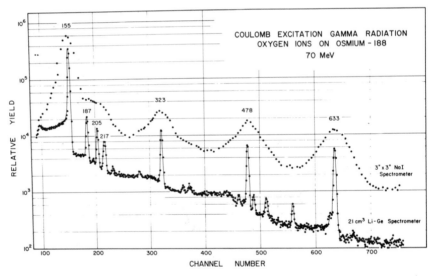

FIGURE 4.21 Comparison of the gamma radiation spectra, from Coulomb excitation of osmium-188 nuclei by a 70-MeV beam of oxygen ions, as measured with a sodium iodide spectrometer and with a lithium–germanium semiconductor spectrometer. In the region of channel number 200, for example, three isolated transitions appear in the latter spectrum that are entirely masked in the former. This is typical of the enormous improvement in resolution that has been attained with these new detectors.

tum are absorbed or emitted by nuclear systems; how nucleons or clusters deform, rotate, and vibrate; and how entirely new nuclear species might be created.

As a result of these studies, the fundamental modes of nuclear excitation are being uncovered and their microscopic structure analyzed. Crucial to this progress was the availability of a large number of complementary means of probing nuclear excitations, of many well-understood ways of building up an excitation, and of examining the way in which the fundamental modes vary systematically from nucleus to nucleus. It is perhaps particularly characteristic of nuclear physics that some of the most far-reaching discoveries came not from increasingly detailed study of a given nuclear system but rather from the recognition of certain underlying systematics traceable from nucleus to nucleus. A very broad effort—broad in kinds of facility and program—was required and has grown in response to this need.

Major progress has been made in understanding the general characteristics of the lower-lying nuclear excitations and nuclear dynamics. The shell model, with appropriate modifications, continues to provide the framework for a physically understandable picture of nuclear structure and motions (see Figure 4.22). To do so it had to encompass phenomena that at first seemed sharply at variance with the concept of single particles orbiting their separate ways. The collective excitations that were discovered and explored in the decades after its initial development clearly pointed to the cooperative motions of many nucleons. Sets of excitations that could only be interpreted as vibrations or rotations of large fractions of the nuclear charge or mass were especially striking testimony to this. The analysis of these collective modes in terms of the shell model appears qualitatively possible and, in some cases, also quantitatively possible, though complex. The new, unified shell model puts the single-particle and collective aspects together by including the residual interactions between the valence nucleons that tie their separate motions together and by recognizing that the assumed spherical central potential, or core, which defined the orbits in the initial shell model, is an oversimplification that, in large regions of the periodic table, must be replaced by spheroidal or ellipsoidal shapes. Besides forming a most useful physical model of the nucleus, this enlarged shell model is also a quantitative bridge between experimental findings and a deeper theory.

Development of a deeper theory has begun. The fundamental question of nuclear theory always has been how to connect the forces between the individual neutrons and protons of the nucleus and the observed nuclear processes. The work of the last 15 years already has achieved more than qualitative success in connecting these forces with both the

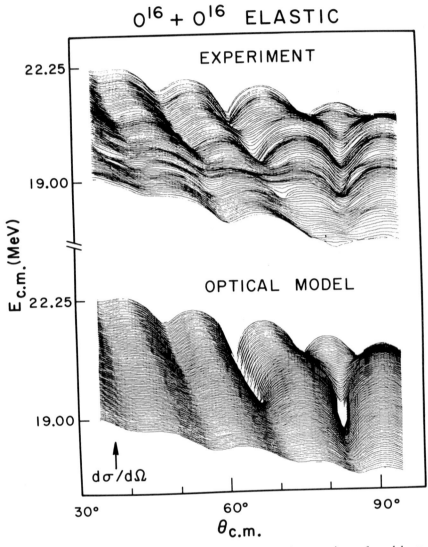

FIGURE 4.22   This figure illustrates the effectiveness of comparison of model pre-dictions with experimental data, which even simple on-line computer installations make possible in nuclear experimental work. Shown here are elastic scattering cross sections for ¹⁶O ions on ¹⁶O nuclei as functions of both energy and angle measured in the center-of-mass system. The upper figure shows actual experimental data obtained with an on-line computer-based data–acquisition system; the lower figure shows the corresponding model predictions. From examination of such comparisons of data surfaces rather than isolated angular distributions or excitation functions it is possible to determine at a glance whether the model correctly reproduces gross features of the experimental data without the confusion that otherwise arises from local detailed fluctuations.

gross properties of size and binding energies of nuclei and the chief ingredients of the shell model—the central force and the residual interaction between the valence nucleons. This fundamental problem is not yet wholly solved, for the agreement with observation is far from quantitative. But this work is well begun.

Nuclear physics owes much in terms of insight, concepts, and techniques of calculation used in advancing the theory of nuclear structure to molecular physics and the physics of condensed matter. The concepts of molecular physics have been extremely important in treating nuclear rotations and vibrations; the pairing concepts evolved to explain superconductivity in condensed matter also have played a crucial role in the nuclear physicist's understanding of the excitation spectra of nuclei and of the phenomenon of nuclear superconductivity. Much vital input concerning the nature of the nucleon–nucleon interaction has come, of course, from elementary-particle physics.

The early history of nuclear-reaction research was dominated by the phenomenon of the very narrow resonances discovered with low-energy neutrons. Such resonances correspond to the formation of a configuration of the nuclear system that lives a relatively long time before breaking up with the emission of a nucleon, a photon, or a more complex nuclear fragment. The postwar developments in experimental techniques made possible neutron beams of much greater energy range. A new phenomenon was uncovered, complementary in character to the nuclear resonance. A broad energy structure, which varied gradually and systematically with the atomic number of the bombarded nucleus, was found in neutron scattering. These regularities could be well described in terms of the scattering of neutrons by a smooth central force field of nuclear dimensions; in contrast to the longer-lived resonance phenomenon this is a prompt mechanism. In addition to the scattering, of course, some absorption of the incident neutron beam occurred, and it was recognized that the situation was analogous to the interaction of an incident light beam with a rather cloudy crystal ball. This optical model clearly has connections with the single-particle picture of the shell model. These concepts proved applicable to all nuclear projectiles.

The prompt mechanism applies not only to simple scattering but also to a great variety of transfer reactions in which one or a few nucleons are transferred between projectile and target. Because the process is simple and direct, these direct transfer reactions have provided a well-understood means for studying the way nuclear excitations are built up, and very much of the present detailed knowledge of nuclei comes about in just this way.

In the 1960's, nuclear physicists addressed themselves to the discernment and study of reaction cross-section structures intermediate between the broad ones of the optical model and the very narrow resonances. The underlying nuclear configurations are understood to be of a complexity intermediate between the individual orbits of the optical potential and the complications of the narrow resonances, which in many cases have been interpreted as shell-model excitations. These intermediate structures have implications for the unraveling of the nuclear dynamics and for nuclear structure.

An example of such intermediate resonance phenomena is the isobaric analog-state excitations in heavy nuclei that have been discovered and extensively investigated in recent years. The isobaric spin quantum number, discovered in the early days of nuclear physics, was long thought to be useful only in light nuclei, where the Coulomb forces resulting from the proton charges were still too small to destroy significantly the charge symmetries that the isobaric spin quantum number described. Recently, with the advent of precision techniques applicable to heavy nuclei, it was found that this symmetry has much wider validity than was anticipated; in retrospect, the reasons are clear. The isobaric analog of a state is one in which structure is the same except that a neutron has been changed into a proton, and therefore its level is higher by the extra Coulomb potential energy. In heavier nuclei, this extra energy lifts the analog excitation into the continuum. The importance of this finding for further nuclear-structure studies is enormous. It shows that states whose structure is known from the low-energy work on their analogs are available at high enough energies for many reactions to be possible. The reactions to final states carry, then, a great deal of information about the nature of these final states. A new rich technique resulted that already has proved highly effective. Further, shell-model concepts apply with textbook simplicity in these newly accessible heavy nuclei—much more so than in the lighter regions where they have been studied previously. (See Figure 4.23.)

So far only the relatively simple excitations and reactions have been mentioned. But perhaps the most important characteristic of nuclear physics is the diversity of phenomena that manifest themselves. The fission of nuclei into massive pieces was discovered more than three decades ago, quickly applied in a national emergency, and, subsequently, stimulated entirely new industries; it was studied intensively throughout this entire period. Discovery of a new and very interesting fission phenomenon, the interpretation of which has great import for the field as a whole and especially for its frontier extensions, occurred during the late 1960's in the Soviet Union. It had long been thought that once a nucleus had been stimulated to fission, it did so very rapidly. The Soviet scientists found, surprisingly, that many nuclei had excited states with a very much longer lifetime before fission than did the normal ground states—the

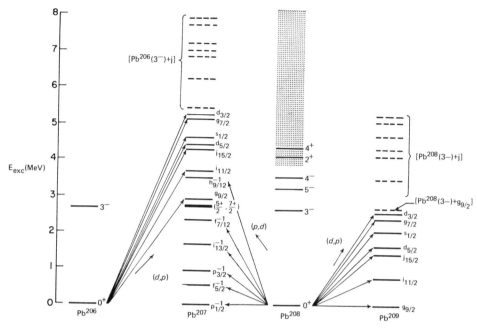

FIGURE 4.23 Experimental single-particle and single-hole levels in the lead region. ²⁰⁸Pb is doubly magic, having 82 protons and 126 neutrons. ²⁰⁸Pb states are formed by adding a single neutron to the ²⁰⁸Pb core; ²⁰⁷Pb states are formed by removing a single neutron from the ²⁰⁸Pb core *or* by adding a single neutron to the ²⁰⁶Pb core. As illustrated, all these states are very conveniently studied by deuteron stripping and pickup reactions. In addition to single-particle and single-hole states based on an unexcited ²⁰⁸Pb core, it has become possible to study equivalent states based on this core in any of its excited configurations; shown here are the states that are based on the octupole vibrational 3⁻ state at 2.6-MeV excitation in ²⁰⁸Pb. While the ground state of ²⁰⁷Pb is formed by the removal of a $p^{\frac{1}{2}}$ neutron from the ground state of ²⁰⁸Pb, the closely spaced doublet (5/2⁺ and 7/2⁺) at 2.6 MeV of excitation is formed by removing a $p^{\frac{1}{2}}$ neutron from the 3⁻ excited configuration of the core. Such systematic spectroscopic studies wherein the excited configurations are assembled, nucleon by nucleon, have provided extensive new nuclear structure information and have shown that the shell model concept of single-particle and single-hole orbits applies with classic simplicity in the lead region. Before the recent advent of the large electrostatic accelerators, this region of heavy nuclei was simply inaccessible to such high-precision studies.

so-called fission isomers. A massive international effort led to an understanding of these phenomena in terms of an outer balcony, which can develop in the nuclear potential barrier. The fission fragments, in essence, after penetrating the main barrier, can be trapped in the balcony, or

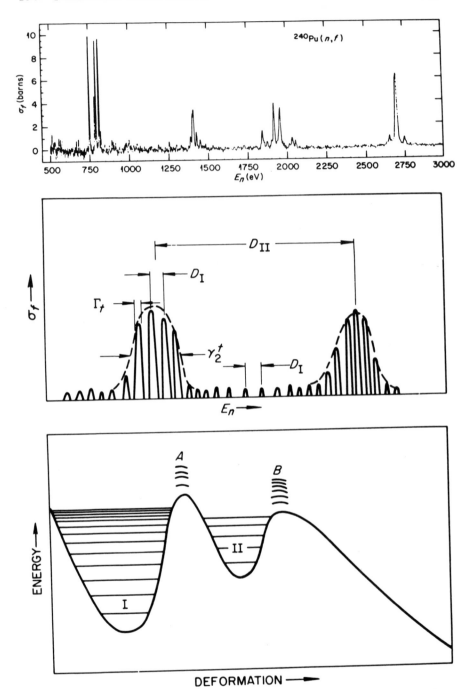

◄ FIGURE 4.24. The upper figure shows experimental data on the variation of the cross section for neutron-induced fission of plutonium-240 as a function of the neutron energy. Particularly striking are the periodic maxima in the cross section, with, in each case, a number of sharper resonances participating in the maximum. This phenomenon has led to important new insight into the mechanism of nuclear fission, as illustrated in the bottom figure, where the effective potential barrier is plotted against a parameter measuring the departure of the plutonium-241 compound nucleus from sphericity. In the past it was assumed that such a potential had only a single minimum (I) corresponding to the equilibrium shape of this nucleus and a single potential barrier through which the fission fragments tunneled before release. What the experimental data shown here have indicated, however, is that a second minimum (II) exists. How this reproduces the experimental results is shown schematically in the center figure. In some appropriate neutron energy range E_n, the compound, plutonium-241, is produced by neutron capture by plutonium-240 at a level of excitation such that the quantum states in the first potential well are separated by characteristic spacing D_I; because the second potential well is less deep, its quantum states at this same energy are more widely spaced (D_{II}). An enhanced correction occurs when the fission fragments tunneling through the first barrier find themselves at a quantum level energy in the second, where there is enhanced probability of tunneling through the outer barrier. These double potential shapes have now been recognized as being derivable from more microscopic nuclear shell models; they provide a textbook example of quantum-mechanical tunneling through barriers.

secondary minimum, for relatively long periods. Quite apart from some highly useful consequences, this finding constitutes a textbook example of penetration through a complex barrier. (See Figure 4.24.)

The existence of the second minimum had been predicted on the basis of shell-model calculations of the energy of nuclear deformation. That these calculations are closely related to those predicting the existence of as yet unseen islands of supertransuranic nuclei has provided no small part of the excitement (see Plate 4.III).

Future Developments

The frontiers of nuclear physics research are diverse. One of the main directions of general progress in the field has been toward the establishment of a coherent central model that reaches into all its phenomenological branches. As a result of this effort, the location of as yet uncharted regions has become clearer, and puzzles in the more definitely formulated areas have been identified. The future progress of nuclear physics will depend on a double-pronged attack against both kinds of problem.

Much of the progress in nuclear physics resulted from analyses of systematic studies of many types of systems and many nuclear properties. These studies must be extended with the newer techniques that have only recently become available. From 1900 to 1970, some 1600 nuclear species were identified and studied; reliable estimates suggest that collisions of 2-GeV uranium nuclei with uranium targets will produce at least 6000 different species. In addition, there is relatively little information as yet concerning the behavior of any nucleus as more excitation energy is pumped into it. Very recent evidence suggests that, in addition to heating and evaporating nucleons or fragments in analogy to a water drop, quasi-molecular complexes of long lifetime may form in which most of the available energy is bound in molecular fragments with the remaining available energy appearing as kinetic energy of simple relative motion of these fragments. This high-excitation energy region is still largely *terra incognita*.

Although much of this work can be done with existing techniques and probes, a quite different set of probes rapidly is coming into increasing use. New accelerator sources and detection devices now under construction will lead to precise nuclear investigations by high-energy protons and extend the capabilities of high-energy electrons. Examining the nucleus with these two different short-wavelength beams will correspond to putting the nucleus under high-resolution microscopes illuminated by complementary radiations. New copious sources of mesons will be used to probe the nucleus for different specific components of nuclear motions (see Figure 4.25). Mesonic atoms, formed by mesons orbiting the nucleus, sampling and reporting on the nuclear matter that they traverse, will provide still different complementary nuclear information. The intensive study of hypernuclei, formed by replacing one of the constituent neutrons or protons by a strange particle—a hyperon uninhibited by the Pauli Exclusion Principle—will provide yet another view from deep inside the nucleus. The possibility of the existence of a region of stable superheavy nuclei, lying well beyond the heaviest now known, has been much discussed. If current ideas on element production are correct, the superheavies are beyond nature's capabilities and can be created only by a leap over the instabilities that surround them. This leap can be achieved by the bombardment of massive nuclei with massive projectiles at energies sufficiently high for them to overcome their mutual repulsion and fuse. Such reactions appear also to be fruitful sources of the many other nuclear species that are sought. The new heavy-ion facilities now being planned will begin to open these possibilities to exploration with as yet unknown nuclei. If the superheavies do indeed exist, current calculations suggest that, among other things, they should emit about three times as many neutrons per fission as do the present fissile fuels. This capability could have major consequences in the development of a more convenient, perhaps portable, energy source.

Superheavy Nuclei

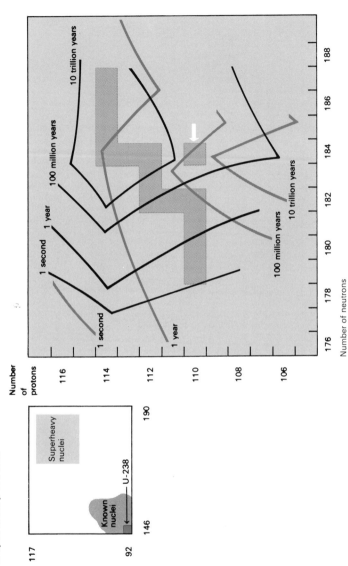

PLATE 4.III Superheavy nuclei. Theretical calculations predict great stability for several superheavy nuclei. Red lines show stability against alpha-particle decay, black lines against spontaneous fission, and colored squares against beta-particle decay. The most stable nucleus would thus be one having 110 protons (arrow) and several closed-shell nuclei having 114 protons. [Source: *Science Year. The World Book Science Annual.* Copyright © 1969, Field Enterprises Educational Corporation.]

FIGURE 4.25 Aerial view of the Los Alamos Meson Physics Facility. This facility features 1-mA proton beam at 800 MeV, plus simultaneously either 100-μA hydrogen ions at 800 MeV or 1-μA polarized hydrogen ions at 800 MeV. Beams of protons, neutrons, muons, and neutrinos will be available for simultaneous use in a multi-disciplinary program.

The use of heavy projectile beams is the most rapidly developing frontier in nuclear physics both here and abroad. These beams have become available under adequately controlled and precise conditions only in recent years and already have been widely exploited. They not only permit exploration of known nuclear phenomena in ranges of parameters —energy, angular momentum, and the like—well beyond any previously available, but they also give access to totally new phenomena—totally new configurations (the nuclear molecules and possibly superheavies are examples) and totally new dynamics (How do two heavy nuclei deform during a collision? What happens to this deformation energy?) Massive construction programs directed toward development of forefront facilities for such investigations are currently in progress in the Soviet Union, Germany, and elsewhere.

Applications of Nuclear Physics

The discovery and use of nuclear radioactivity, dating back to 1913, has provided research and clinical workers in the biological and medical sciences with a probe of unprecedented specificity and power and has revolutionized these sciences; nuclear radioactivity has provided a technological treasure trove that, even now, remains only little exploited. The discovery of nuclear fission in 1938 fundamentally changed the nature of both peace and war. At the same time, and much more important, it provided mankind with a wholly new resource of nuclear energy. In loosing the constraints imposed by Nature's caprice in locating her fast-dwindling energy resources and by civilization's prodigal use of them, nuclear energy stands as one of man's major weapons in his continuing struggle against poverty, hunger, and despair.

Less well known than radioactivity and nuclear fission but, in sum, perhaps equally important are the myriad small inventions, applications, and ideas that have been a part of nuclear physics since its inception. Even a cursory inspection of the intricate instrumentation in the intensive-care section of a modern hospital, the control section of a modern manufacturing plant, the nerve center of a major defense establishment, or the control rooms of Cape Kennedy or Houston reveals the debt owed to those pioneers in nuclear physics, who, often by intuition as much as logic, invented and devised the instruments that gave them a glimpse of the nuclear world. The commercial application of nuclear radiations is in its infancy but expanding rapidly; it ranges from the giant irradiation units performing vital but unseen service on production lines to the minuscule sources energizing, in virtually eternal and foolproof fashion, the emergency and warning signs now used around the globe in languages from Hindi to English. Medical applications of nuclear radiations run the gamut from the now familiar clinical use of diagnostic x rays and of these and harder radiations in the treatment of carcinomas to the newer uses of much more sophisticated and specific diagnostic probes such as the Anger camera (Figure 4.26), which probes lesions in the depths of the brain; neutron radiography units that make visible for the first time the soft tissues deep in the body; and new isotopes whose use permits unraveling of ever finer details of the intricate biochemical and biophysical bases of life.

At the present time, taking the world population at 3.2 billion and the total yearly consumption as approximately 3.5 million megawatt-years, the average yearly energy consumption per person corresponds to about one ton of coal. Harrison Brown *et al.** have estimated that by the year

* H. Brown, J. Bonner, and J. Wier, *The Next Hundred Years* (The Viking Press, New York, 1957).

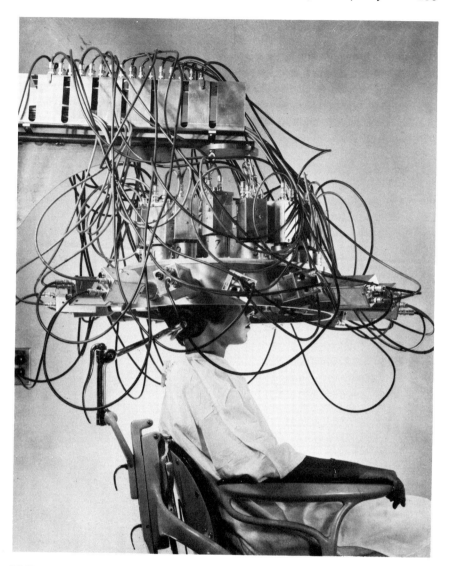

FIGURE 4.26 Brain tumor locator. The Anger camera forms an image of a portion of the brain. Gamma rays from radioisotopes pass through collimators to scintillators. Photomultiplier tubes detect signals, which are processed and displayed by a computer. About 10 million scans on patients are performed each year (three fourths of these in the United States). [Source: Brookhaven National Laboratory.]

2060, with a world population of 7 billion, this average equivalent per capita energy consumption will have increased to ten tons—or to a total of

70 billion tons of coal per year. At this rate, and, if anything, Brown's estimates seem conservative, for desalting of seawater is not taken into account, the present fossil fuel reserves of perhaps 2400 billion tons of coal equivalent would last only some 50 years. This provides the dimension of the problem that faces contemporary civilization; Table 4.3 gives the projected energy production pattern developed by Brown *et al.* for the year 2060. These estimates assume that nuclear sources will supply some 65 percent of the total world energy requirements. If desalting of seawater is included, this fraction would rise to about 75 percent.

Similar estimates project that by the year 2000 nuclear energy will be furnishing one half of the total electrical energy of the United States. Annual investments in nuclear power plants during the next ten years are expected to average some $3 billion. Nuclear power is fast becoming a major industry, and a major program directed toward the development of breeder systems has only recently been announced by the federal government. This action is in recognition of the fact that available resources of uranium-235 are limited; current reactors extract only 1–2 percent of the energy potentially available from their uranium fuel and are dependent on

TABLE 4.3   Projected Energy Production Pattern for the Year 2060 [a] (World Population 7×10^9)

Source	Equivalent Metric Tons of Coal (Billions)	Equivalent Heat Energy	
		10^{18} Btu [b]	MW-yr Heat [c]
Solar energy			
(2/3 of total space heating)	15.6	0.42	140×10^5
Hydroelectricity	4.2	0.10	38×10^5
Wood for lumber and paper	2.7	0.07	24×10^5
Wood for conversion to liquid fuels and chemicals	2.3	0.06	21×10^5
Liquid fuels and petrochemicals produced via nuclear energy	10.0	0.27	90×10^5
Nuclear electricity	35.2	0.96	320×10^5
TOTALS	70.0	1.88	633×10^5

[a] From H. Brown, J. Bonner, and J. Wier, *The Next Hundred Years* (The Viking Press, New York, 1957).
[b] Btu: British thermal unit = 252 calories.
[c] MW-yr: megawatt-year.

the availability of large amounts of low-price ores to keep their power costs economically competitive with those of fossil-fuel plants. Breeder-reactor power costs are expected to be relatively insensitive to the cost of uranium and thorium, thus making possible economic use of low-grade ores. Development of a successful, commercially competitive breeder may well be one of the most important scientific and technological tasks facing mankind at the present time.

That nuclear data for these systems is not yet known with adequate precision frequently is forgotten or ignored. Barschall, one of the pioneers in the precise measurment of neutron cross sections, has noted that some of the most basic cross sections used in the design of slow fission reactors are uncertain by as much as 10 percent; in the case of the cross sections needed in breeder-reactor work at higher neutron energies, uncertainties in excess of 50 percent are common. The nuclear engineer, faced with these uncertainties, is forced to more conservative and more costly designs to incorporate larger margins of safety. The absence of these data simply indicates that the necessary measurements are both difficult and sophisticated; only very recently have they become possible. For example, measurements on the ratio of the neutron capture to fission cross sections for ^{239}Pu, when carried to presently available precision, demonstrated that one particular breeder-reactor system, on which over $100 million in development capital was planned, was not feasible.

In the case of conventional reactors based on the use of thermal neutrons, the designs are not seriously limited by uncertainties or unknowns in the nuclear data. But even these limits involve very high stakes. Among the uncertainties are those associated with nuclear data affecting primarily the fuel costs. These uncertainties are estimated to be of the order of only 0.02–0.03 mil per kilowatt-hour of electric power. But applied to the entire projected nuclear power capacity of the United States, even this small uncertainty will amount to $20 million to $30 million in annual fuel costs by 1980 and $140 million by the end of the century. And in the case of breeder reactors, the uncertainties and possible cost savings, at present, are much larger.

Further off on the nuclear power horizon is the fusion reactor (see section on Plasmas and Fluids). While the main effort is in plasma physics, nuclear physics and physicists are involved in a number of essential ways with the primary nuclear phenomena. Also on the far horizon is an accelerator-based power system, contemplated by Lawrence in 1948, then shown to be much ahead of its time technologically but recently reinvestigated extensively by Canadian groups. This system takes advantage of the fact that neutrons produced by collisions of 1-GeV protons with matter cost roughly ten times less in energy than do those produced· in fission; since neutron economy is all-important in fission energy sources, this low-cost aspect is most attractive. In principle, this system appears less attractive than fusion, but it is necessary to maintain a level of activity in related research and technology such that it would be possible to move forward if major breakthroughs here—or major disappointments with fusion systems— should so dictate.

The many contributions of nuclear physics to the medical, industrial, agricultural, and general technological fields are described in detail in the report of the Panel on Nuclear Physics. Especially in technological fields, a close and fruitful symbiosis has evolved. Nuclear physics interacts with all these fields through five primary channels: radioisotopes as tracers, radioactive nuclei as energy and radiation sources, nuclear methods of materials analysis, direct utilization of electron and ion beams from accelerators, and, finally, the great diversity of instruments developed by nuclear physicists, or by others, in response to particular needs of nuclear research.

The availability of radioisotopes of all elements—and when desired, in large quantities—has worked revolutionary changes in many fields but perhaps most strikingly in medicine and the biological sciences. The specific example of the isotope ^{99}Tc is instructive. Technetium ($Z=43$) was early recognized as one of the two elements missing from the Mendeleev chart of the elements as far as terrestrial abundance was concerned. Identified in the 1930's as responsible for certain unassigned lines in the optical spectra of some unusual stars, technetium was first recovered in measurable quantities in the 1940's from an old and much-used molybdenum septum from the Illinois cyclotron, where it had been produced by the (p, n) reaction. Still it remained a laboratory curiosity.

By 1964 it had been recognized that ^{99}Tc had unique qualifications for use in brain scans. It had low-energy characteristic radiations that permitted its detection and identification with minimal extraneous or unnecessary patient exposure; it concentrated selectively and rapidly in diseased tissues; its radiations were readily collimated, permitting precise location of the isotope concentration; and its half-life (6 h) rapidly removed it from the body. In 1969, a substantial fraction of the entire Oak Ridge isotope separation effort was devoted to an attempt to meet the urgent demands of physicians throughout the country, and it has been estimated that ^{99}Tc production (through accelerator bombardment) and distribution will shortly form the basis for a multimillion-dollar-per-year industrial operation. Little did the nuclear scientists intent on separating micrograms of technetium from the defunct Illinois septum in the 1940's realize that kilograms of the element would be the production unit in the 1970's, or that in 1969 alone use of their element would be responsible for fending off death in literally thousands of cases. This example is perhaps extreme, but it illustrates the tangled and unpredictable linkage between discovery and ultimate use.

The growth of radioisotope utilization in industry has been phenomenal. In 1969, about half of the 500 largest manufacturing concerns used radioisotopes. About 4500 other firms also are licensed to use radio-isotopes. Virtually every type of industry is represented. The growth continues as new radioisotopes become available and as detector and instrumentation improvements continue to increase the sensitivity, selectivity, and reliability of these techniques. The estimated saving to U.S. industry resulting from the use of such methods in a myriad of flow-rate applications was $30 million to $50 million in 1963—a year for which statistics are available—and has increased greatly since then. Estimates in 1969 indicated that radioisotope gauges, together with their associated instrumentation, comprised a $35-million-per-year market that was expanding rapidly. The cost of nuclear oil-logging techniques amounts to some $25 million; the savings to the petroleum industry are many times that. Sterilization of medical supplies and materials by nuclear radiations has been put into routine production use and has led to new industrial ventures.

The technique of neutron radiography, long a workhorse in the field of nondestructive testing (see Figure 4.27) has been greatly enhanced by the availability of the new transuranic neutron-emitting radioisotope, californium-252.

Accelerators, developed as part of the nuclear-research program, have been applied to many other purposes. Over 1000 such machines are at work in medical, industrial, and technological operations. Some 200 accelerators around the country are furnishing the radiation therapy required by more than 300,000 patients yearly. New accelerator developments are quickly translated into improved radiation facilities. Thus, the high-energy machines, developed in the nuclear program, offer a narrow pencil of radiation that allows the delivery of more radiation to the lesion and less to surrounding healthy tissue. New advances in the technology of linear accelerators were quickly recognized as offering more efficient radiotherapy devices; many units are already in service, while others are being installed. Industry has built accelerators into many other functions: radiation processing, industrial radiography, and neutron radiography among others. The investment in such accelerators, over $130 million, affords some measure of these applications, independent of instrumentation or plant costs. Radiation-processed products manufactured annually amount to over $1 billion in sales.

The instrumentation of nuclear physics plays an important part in all the above applications and in many other ways. The scientific instrumentation industry is a key one, providing the tools on which much tech-

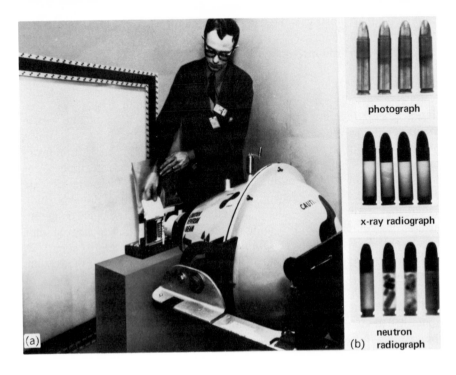

photograph

x-ray radiograph

(a)

(b) neutron radiograph

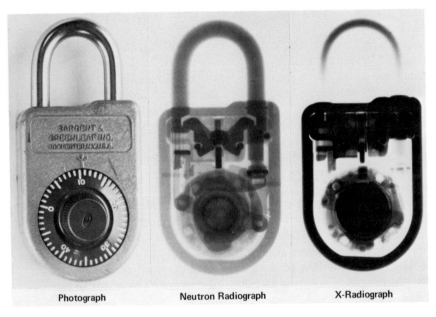

Photograph   Neutron Radiograph   X-Radiograph

◄ FIGURE 4.27 *Top:* Nondestructive testing. Neutron radiography. (a) Recently developed neutron-radiographic camera, based on californium-252, offers the significant advantage of portability (100-lb unit), so that the exposure may be conducted in a manufacturer's plant or at a field service facility, thus avoiding the necessity of shipping the item being neutron-radiographed to an atomic reactor facility. (b) Comparison of photograph, x-ray radiograph, and neutron radiograph of bullets. (Battelle Northwest.) [Source: John R. Zurbrick, *Yearbook of Science and Technology, 1971.* Copyright © 1971, McGraw-Hill Book Company, Inc. (Used with permission of McGraw-Hill Book Company.)]

Bottom: Comparative radiographs of a combination lock. [Source: This photograph was provided through the courtesy of Atomics International, a Division of North American Rockwell Corporation.]

nological progress depends. The nuclear-instrument section of this industry is large and rapidly growing beyond the $100 million sales volume estimated in 1970. Nuclear instrumentation sales abroad are also significant; in addition, they unite U.S. technology with that of the rest of the world.

Nuclear physics affects our technological society in other less tangible ways than those exemplified by its tools and products. The development of the computer industry provides an illustration. Much of the original logical circuitry was based on the electronics developed for nuclear-data purposes. Nuclear physicists were among the first to use a computer not only to analyze raw data instantly but also, acting on this analysis, to control an experiment. Such on-line processing is now common throughout industry, in high technology as well as the routine production of basic commodities. The sophisticated demands of nuclear applications have in themselves provided important stimuli for computer developments. Thus the need for analysis of nuclear explosions stimulated the development of the very large digital computer and its associated software. At the other extreme, the rapidly expanding minicomputer market owes part of its success to the pioneering use of these small processors in nuclear installations.

Much of the world's military arsenal is nuclear. Almost all of physical science is involved in describing a nuclear explosion or its effects; however, nuclear physics is central to this effort. Nuclear physics is equally relevant to establishing and then monitoring international test ban and disarmament agreements, the intention of which is to lessen or eliminate the chance that nuclear explosives will ever again be used in war. Nuclear physics and physicists continue to play important roles in every aspect of the U.S. national defense effort: the design of nuclear weapons, the testing program, the evaluation of weapons effects and a civil defense program, test ban monitoring and surveillance, and the arms-limitation efforts.

Interactions with Other Sciences

Nuclear physics is an integral part of the physical sciences. The mesonic nature of the force between the individual constituents of nuclei forms a close connection with particle physics. Many basic phenomena first seen in the nuclear domain apply in all of physics and the nucleus provides a well-explored laboratory. The weak interactions were first seen in nuclear radiations, and the violation of the fundamental parity law was demonstrated in a nuclear experiment. The concepts evolved in the course of developing nuclear physics have become part of the milieu of physical ideas in which physics as a whole grows.

Astrophysics is perhaps most closely and particularly related to nuclear physics, since nuclear processes produce much of the energy and achieve the elementary composition of the matter in the universe. A special subfield, nuclear astrophysics, concentrates on these problems. Both the problem of stellar evolution and the nucleosynthesis of the elements depend on knowing the dynamic features of reactions among a great variety of light and heavy nuclei at the energies of stellar processes. Although study of these problems constitutes a separate subfield, it is one that remains closely linked to the progress of nuclear physics. Thus the recent experiment on the measurement of the flux of solar neutrinos by the inverse of nuclear beta-decay has sharply thrown into question the picture of solar energy production that has been developed over the years. The exciting conjectures on the existence of neutron stars and other manifestations of superdense agglomerations of matter offer another and rather different example. It is a long extrapolation indeed from several hundred nucleons to 10^{56} nucleons, from normal nuclear densities to those so great that strange particles must be considered along with the nucleons, and from nuclear masses to those so enormous that nuclear and gravitational energies vie with one another. This domain is wholly new, one that tests the ideas and methods of nuclear physics, but one for which a solid conceptual foundation has been laid.

Solid-state physics and nuclear physics have interchanged ideas and tools for many years. Concepts based on many-body theory have proven of fundamental value in explaining nuclear phenomena; thus the formalism developed in understanding superconductivity has explained the pairing correlations in nuclear wavefunctions. The theory developed for nuclear matter has served for the many-body problems of condensed matter. The use of radioactive nuclei to probe their electric and magnetic environments in solids has been put to use in many problems of solid matter. The

Mössbauer effect is now a tool of the solid-state physicist. The use of neutrons to study the static and dynamic characteristics of solids is so well established that such work pre-empts the largest share of research time on the most modern reactors.

Chemistry, too, has formed two-way bonds with nuclear physics. A substantial fraction of the nation's nuclear research effort takes place in chemistry departments and laboratories. Hot-atom chemistry forms a classical region of overlap. Very-heavy-element chemistry and the discovery of new very heavy elements have always been directly related. Chemistry, like solid-state physics, has profited from the neutron capabilities of research reactors. An interesting new effort of chemical kinetics and structure is based on the study of elementary atom–atom, atom–molecule collisions, thus drawing directly on the methods and concepts of nuclear-reaction work—just as earlier nuclear physics took over the developments of atomic scattering methods.

From the time of the earliest studies of radioactivity, scientists realized that there was at hand a dating method for geological processes, and, by 1905, the first dating of rocks had been accomplished. Geochronology is now a fundamental part of the earth sciences and employs a great number of isotopes and isotope chains in the study of the time domain from one million to five billion years. Work on meteorites, lunar samples, and ocean-bottom-sediment cores is of wide public interest. Archeological studies based on the cosmic-ray-induced activity in carbon have provided a time scale commensurate with man's recent historical period. Radiocarbon dating is now a standard working method of the archeological laboratory.

Art history and forensic investigations also have profited from neutron activation analyses. Beyond this, nuclear systems provide the only effective clock, which has been running continuously since the formation of the solar system, and thus provide the possibility of testing cosmological speculations such as that concerning the fundamental physical constants and their possible variation with time over cosmological periods. The possibility that the elementary charge unit (that carried by the electron and proton) might vary in such a way that its square was proportional to time had been discussed for decades. Recently, examination of the stability systematics of heavy nuclei has demonstrated conclusively that, if this charge varies at all, its variation is less than 0.3 percent of that postulated by the cosmological arguments.

The biological and medical sciences have used the whole battery of radioactive isotopes to trace life processes. The materials as well as the detectors and instrumentation to measure their progress in biochemical reactions were the products of the nuclear laboratory.

Only recently, as part of the MAN Project at Oak Ridge, Anderson and his collaborators have begun to use centrifuges, developed in wartime for possible use in the separation of uranium isotopes, for biological research. For example, there have always been significant numbers of persons who were unable to receive protection from influenza because of marked allergy to the protein contaminants in the influenza vaccine. This inability to receive such protection was particularly serious in the case of older persons. During the past year alone over 50,000 of them were able to receive protection, which would otherwise have been inaccessible to them, because Anderson and his colleagues had discovered that ultra-centrifuging the influenza vaccine effectively removes all the protein contaminants. This work is only the beginning of a major instrumental change in the production and purification of biologicals.

Nuclear physics has also had a continuing interaction with the space-science effort. Nuclear instrumentation was a crucial part of space probes to measure the fluxes of high-energy electrons, photons, and nuclear fragments that form important parts of the environment of space. The radiation effects on spacecraft and the space traveler have been studied with nuclear accelerators. The first on-site studies of the lunar material were made by a nuclear scattering device; the instrumentation left behind to record lunar events is powered by nuclear isotope sources. These are, no doubt, mere preludes to man's future probing of more distant planets. The nuclear tools will grow in importance. Nuclear-powered rockets, already developed in prototype, may well be the essential means to move the huge payloads that currently are necessary. The connection will go much deeper than devices and instruments, for space science and nuclear physics share the problem of analyzing and understanding cosmic rays.

The Organizational Structure of Nuclear Physics

Nuclear-physics research is divided about equally between universities and government laboratories. A little over half of the scientific effort is university-based; most of the rest takes place in the AEC National Laboratories. If one includes both federal and nonfederal support, then the funds for operating these facilities are also about equally divided.

The essential nature of nuclear physics requires a broad effort with many kinds of programs. The present program utilizes more than 100 facilities—potential drop machines, cyclotrons (see Figure 4.28), and linear electron accelerators—ranging from those capable of being used by two or three scientists to major installations that require a large, trained crew to operate. The number of separate projects based on these facilities, or not requiring machines, is many times larger. There is a nearly continuous distribution of university-based projects from the very small to those of the same order as the large National Laboratory programs.

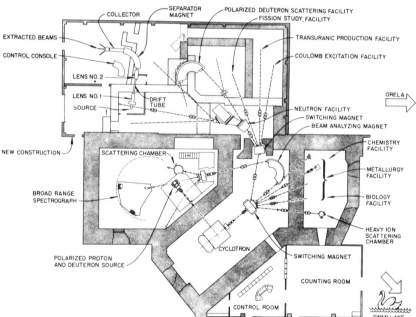

FIGURE 4.28 *Top:* The beam lines in the main cyclotron vault. The rf system of the cyclotron appears at the right and the beam preparation magnet at the left.

Bottom: A layout of the beam lines of the Oak Ridge Isochronous Cyclotron, showing the bending and analyzing magnets and the various shielded research rooms.

[Source: Oak Ridge National Laboratory.]

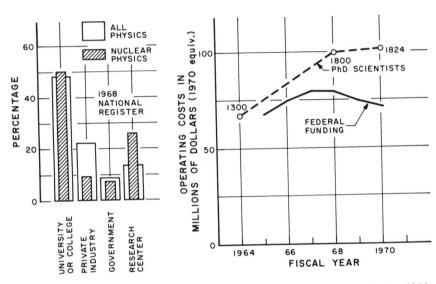

FIGURE 4.29   Manpower, funding, and employment data on nuclear physics, 1964–1970.

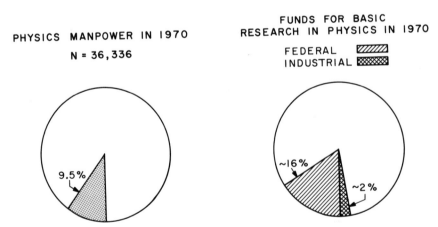

FIGURE 4.30   Manpower and funding in nuclear physics in 1970.

Nuclear physics has been largely an experimental program; only 15 percent of the scientific effort is devoted to theoretical research. The main experimental efforts, comprising about half of the total experimental program, have been based on Van de Graaff and cyclotron accelerators. This emphasis, of course, will shift as the facilities designed to open new phenomena to investigation come into operation, but a broad effort is still to be expected.

Nuclear physicists comprise about 13 percent of the physics PhD population, as reported in the 1970 National Register of Scientific and Technical Personnel. They are generally academically based; however, the research center is a close competitor for people. The number of PhD's working in the subfield grew at a rate of about 10 percent per year to 1968 but has since leveled off (Figure 4.29). In 1970, approximately 16 percent of federal funds and 2 percent of industrial funds for basic research in the physics subfields were allocated to this subfield. The federal entry includes $12 million in funding for basic nuclear-physics research supported under chemistry (Figure 4.30).

Problems in the Subfield

Funding trends in nuclear physics since 1967, if continued beyond the present, would appear to imply a decision to phase down activity in the subfield to a level at which it could no longer hope to remain at or near the frontier in major areas of current activity.

Three major areas must be considered. First is the broadly based activity involving a wide variety of techniques, facilities, and approaches that constitutes the central core of nuclear physics and the base on which all future activity in the science and its external utility must rest. Second is the group of major new facilities that received approval in the mid-1960's and are only now coming on-line, with corresponding large demands on dwindling operational funds; the Los Alamos Meson Physics Facility is the most visible example. Third are the major new starts that must be undertaken in the subfield in the next few years if it is not to stagnate or withdraw from the developing frontiers; a national heavy-ion physics facility is the outstanding current example. No one of these areas is viable over an extended period of time without the other two.

> And as regards the Atomists, it is not only clear
> what their explanation is; it is also obvious that
> it follows with tolerable consistency from the
> assumptions which they employ.
> ARISTOTLE (384–322 B.C.)
> *On Generation and Corruption,* Book I, 325

ATOMIC, MOLECULAR, AND ELECTRON PHYSICS

Introduction

At the beginning of the twentieth century it became clear that the atom was not indivisible but consisted of a small, heavy nucleus and a cloud of orbiting electrons. Although deeper probing of the nuclear structure led to the development of nuclear and high-energy physics, the detailed study of the properties of the electron clouds in atoms and molecules remained the subject matter of atomic, molecular, and electron physics.

It should be emphasized that these properties determine the structure of all chemical and biological compounds, and the forces between the electrons are also responsible for the cohesion of matter in liquids and solids. In fact, the electrical interactions between electrons and nuclei in atoms and molecules largely determine the physical phenomena of every-day life, for example, the emission of light in fluorescent bulbs, the boiling temperature of water, the course of electron beams and the lighting of a screen in television tubes, and the nature of chemical reactions such as the burning of coal or oil.

Clearly, atomic, molecular, and electron physics occupies a central position among the sciences and in the science curriculum. This subfield reached perhaps its apex as a research frontier in the first quarter of this century. The spectroscopic investigations of light emitted by atoms and molecules established the facts on which the central physical theory of quantum mechanics is based. In the late 1920's and early 1930's, the understanding of chemistry made great strides based on the quantum theory of electron orbits. Electronic technology was also rapidly developed during this same interval and spurred the revolution in communications that is exemplified by radio and television.

With the basic principles so well established, many physicists considered research in this subfield essentially complete; the most concentrated effort in physics since 1930 has been directed toward nuclear and high-energy physics. However, atomic, molecular, and electron physics continues to contribute substantially to the understanding of natural phenomena. Four Nobel Prizes in each of the past two decades were awarded in this subfield. The interaction of electromagnetic fields and electrons has

PLATE 4.IV Mode pattern of coherent light from an optical gas maser. To produce these mode patterns the normal operation of a helium–neon optical maser is
perturbed by placing a pair of wire cross hairs in the cavity. These wires interact with
the mode structure of the unperturbed cavity, suppressing some modes and, in certain
cases, coupling others together. By changing the angle between the cross hairs this
interaction can be altered and different mode patterns can be produced. [Source:
Bell Telephone Laboratories.]

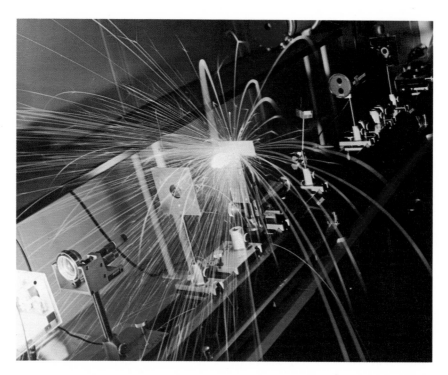

PLATE 4.V Experiments in laser machining techniques. [Source: General Electric Company.]

been pushed to new orders of magnitude in precision and intensity. Within one decade of its invention, originating in atomic and molecular physics, the laser has become a household word. Applications of this instrument appear not only in the physics laboratory but also in hospital operating rooms, high-quality machine tool lathes, military range finders— and even in James Bond movies!

Important Recent Developments and Current Activity

During the past ten years, the development of lasers and masers stands out as a critically important scientific achievement with broad technological relevance. The atomic hydrogen maser is the most accurate clock devised by man and makes possible timekeeping with an accuracy of 1 sec in 30,000 years. Improvements in navigational systems are an obvious by-product of this development. Lasers constitute a radically different and new type of light source, characterized by the extreme directionality, intensity, and color definition (wavelength) of the emitted light (Plate 4.IV). This new light source is used to measure distances with unprecedented accuracy for such diverse purposes as the precise control of machine-tool operations (Plate 4.V), the measurement of distances between points on earth (geodesy, continental drift, and crustal motion preceding earthquakes), and the measurement of the distance to moving objects (light radar, including ranging of low-flying aircraft and the distance to the moon). Measurement of the distance to the moon was accomplished with an uncertainty of only 6 in., which gives an idea of the high level of accuracy and great potential of this instrument.

The laser has rejuvenated the centuries-old science of optics and has created the entirely new subfield of nonlinear optics, the study of the necessary modifications of the well-known optical laws of refraction and reflection at extremely high light intensities. A new branch of technology, optoelectronics, is developing. The newly acquired physical knowledge of lasers and nonlinear optics is applied to optical communications, in which signals are transmitted and processed by light beams. Three-dimensional optical displays, the picturephone, and optical computer elements probably will be widely applied during the coming decade.

Another highly active field of investigation is the study of the interaction of individual molecules in specified states of vibration and rotation. The experimental technique makes use of molecular beams colliding in vacuum and gives much more detailed information about the causes and fundamental characteristics of basic chemical reactions than was previously attainable. This focus on the elementary chemical interactions holds high promise of an entirely new understanding of chemical processes. Use of merging beams, in which the relative velocities can be adjusted to permit study of collisions at relative energies measured in small fractions of electron volts, opens still further new vistas in basic collision phenomena.

Electron beams in high vacua are also used in electron microscopy, in which under favorable circumstances individual atoms have recently been made visible (see Figure 4.31), and in microprobe analysis of impurities and surfaces. The improved vacuum and beam-handling techniques used here hold high promise of increasing the present inadequate understanding of the physics of surfaces. The burgeoning low-energy electron

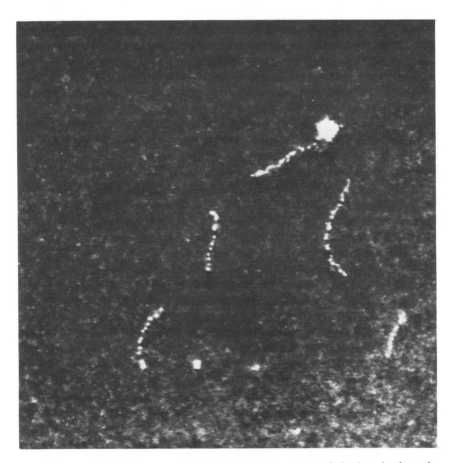

FIGURE 4.31   Single atoms and multiples of single atoms of thorium in the polymeric organic salt of 1,2,4,5-benzene tetracarboxylic acid are photographs on a 25-Å thick evaporated carbon film. The photograph is roughly 2100 Å to a side and was taken with the high-resolution scanning electron microscope in the laboratory of A. V. Crewe.

diffraction (LEED) field has opened up for study whole areas of surface physics for the first time.

The new field of beam-foil spectroscopy has rejuvenated the study of atomic spectra. In this work, high-velocity ions from accelerators are passed through solid foils in which very highly stripped and highly excited ionic species are prepared for spectroscopic examination in subsequent flight. Geometric shifting of the observation point permits detailed study of de-excitation and relaxation processes that previously were inaccessible. By this method higher states of ionization can be obtained in the laboratory than by any other method, and some transitions previously observed only in solar spectra have already been studied.*

Interaction with Other Subfields of Physics and Other Sciences

The nature of the collisions among electrons, atoms, and molecules has implications for a variety of other subfields and sciences. In fact, much atomic physics is supported and carried out in connection with other subfields, such as plasma physics, atmospheric physics, optics, and space physics. These elementary collision processes, which can now be studied in fine detail by colliding beam techniques, play a role in practical problems such as the re-entry of missiles and space vehicles into the atmosphere and the initiation and containment of plasmas necessary for controlled thermonuclear fusion; they are also extremely important in many problems in astrophysics, radioastronomy, chemistry, and biochemistry. (See Figure 4.32.)

Since the interactions between electrons and atoms determine most of the physical phenomena that man encounters, the central position of atomic and molecular physics is obvious. It contributes heavily to other disciplines and, in turn, benefits greatly from new developments in these other disciplines. For example, the development of microwave radar techniques during World War II contributed greatly to the rise of microwave molecular spectroscopy. New laser technology, born from atomic physics, provides better tools for spectroscopy and plasma diagnostics and has made Raman spectroscopy a practical method in chemical analysis.

The interaction with the physics of condensed matter is particularly strong, and the special field of quantum optics can be considered as a part of either subfield or of both. Laser phenomena in solids offer entirely new device possibilities, particularly when combined with some of the very

*Ad Hoc Panel on New Uses for Low-Energy Accelerators, NRC Committee on Nuclear Science, *New Uses for Low-Energy Accelerators* (National Academy of Sciences, Washington, D.C., 1968).

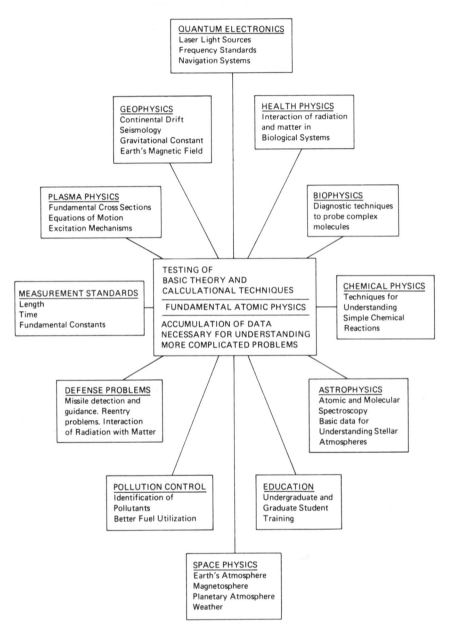

FIGURE 4.32   Relationship of atomic and molecular physics to other disciplines. [Source: Committee on Atomic and Molecular Physics, National Research Council, *Atomic and Molecular Physics* (National Academy of Sciences, Washington, D.C., 1971).]

recent tunable metal oxide surface situations developed in condensed-matter studies. LEED and other studies on clean surfaces provide not only the opportunity to study atomic and molecular interactions in two dimensions in monoatomic or monomolecular layers but also access to entirely new surface phenomena crucial to such applications as catalysis. With the rapidly increasing interest in the control of industrial pollutants, understanding of such catalysis phenomena holds promise of widespread and critically important societal applications.

Cold-field electron emission and field-ion microscopy have been developed to a point at which the diffusion of individual atoms along the surface and the influence of atomic configuration along different crystallographic planes can be followed visually. Spectroscopic relaxation techniques also are useful in the study of solid-surface interactions of atoms. Electron and ion-beam sputtering and ion implantation, and especially scanning-beam electron spectroscopy, provide other examples of the interaction of atomic, molecular, and electron physics with condensed-matter surface physics and high-vacuum technology. Another interface with the physics of condensed matter is the study of molecular fluids. Laser-beam light-scattering techniques are contributing very much more precise information about the mechanics and molecular arrangements in liquid crystal transitions and in mixtures of fluids. This is a research area on which atomic, molecular, and electron physics; condensed-matter physics; chemical physics; and biophysics all impinge.

Fundamental measurements of characteristically high precision (parts per million) in atomic and molecular physics have complemented those in elementary-particle physics in exploring the possible limits of validity of quantum electrodynamics. Such measurements provide crucial inputs to the precision determination of some of the fundamental physical constants and have other more esoteric physical uses. Atomic-beam measurements have been used to establish limits on possible mass anisotropies in the universe and to set stringent upper limits on possible differences in the electrical charge of various elementary particles. Atomic and molecular physics continues to stand at the forefront of ultrahigh-precision physical measurements. It has exported metrology and other precision techniques to all of science and engineering.

This subfield interacts also with nuclear and high-energy physics, as charged fundamental particles, such as positrons and mesons, antiprotons and hyperons, lead to the formation of artificial atoms, called, for example, positronium, muonium, and mu-mesic and pi-mesic atoms, in which one of the named particles substitutes for a normal atomic electron. Somewhat unexpectedly, the study of positronium annihilation in gaseous and condensed media paved the way for useful advances in chemistry and condensed-matter physics. The interaction of a polarized laser beam with an electron beam to obtain polarized high-energy gamma rays is another example of interaction with high-energy physics. Optical and radio-frequency pumping are also used to obtain polarized targets in nuclear and particle physics, an investigative area exciting much current interest.

Moreover, the experimental and theoretical analysis techniques perfected in nuclear physics for the study of elementary collision phenomena are now being exploited in atomic and molecular physics. Studies of resonance phenomena, originally exported to nuclear and particle physics, have been refined and extended and are once again being applied to atomic and molecular problems. It is not surprising, in view of the pervasiveness of electromagnetic interactions in determining the structure of all commonly known materials, that atomic physics also interacts strongly with technology.

The traditional physics subfields of acoustics, fluid mechanics, and optics have particularly close ties with atomic, molecular, and electron physics. Brillouin, Rayleigh, and concentration scattering have yielded new information about damping and kinetics in fluids, and the venerable science of optics has been rejuvenated by its contacts with this subfield. Holography, photon statistics, and the study of the concept of coherence have progressed rapidly in the past five years. Ultrahigh-resolution spectroscopy by means of correlations in photon arrival times has been developed. The field of nonlinear optics is approaching a peak of activity, and many new industrial applications appear likely.

This subfield also has strong ties with chemistry. Colliding-beam techniques have greatly advanced the study of low-energy atomic and molecular collisions. It is now possible to study a chemical reaction, not as a statistical thermodynamic average but with details about individual rotational and vibrational states. Experiments involving elastic, inelastic, and reactive scattering of atoms and simple molecules have led to an evaluation of interatomic forces and other phenomena, and even to the angular distribution of the products of elementary chemical reactions. The existence of relatively long-lived complexes or reaction intermediates in certain systems has been demonstrated. Evidence of the importance of the relative orientation of the colliding partners in a chemical reaction has been acquired. Obviously, atomic, molecular, and electron physics substantially overlaps chemical physics and physical chemistry.

In addition, atomic, molecular, and electron physics interacts strongly with plasma physics, astrophysics, and atmospheric physics. As mentioned earlier, atmospheric physics is determined by collisional rate processes involving electrons, atoms, ions, molecules, and electromagnetic radiation, as is also the study of the earth ionosphere, radiation belts, solar and stellar atmospheres, supersonic flight, shock waves, space vehicle entry, and the like. To draw a line between this subfield and plasma physics or space physics is difficult. In fact, a rigid definition of boundaries is neither useful nor meaningful. The rapidly expanding amount of basic data available on highly ionized atoms and collisions between more energetic particles obviously are of interest to all the subfields mentioned above. Beam-foil spectroscopy is proving extremely useful in this context.

Atomic, molecular, and electron physics has always been important to the interpretation of astrophysical phenomena. Recently developed observing techniques are opening new regions of the electromagnetic spectrum in which the universe can be observed; greatly improved precision in the basic data of atomic, molecular, and electron physics is required for the interpretation of these observations. In addition, deeper understanding of the enormous variety of processes that can occur in the universe is required. This subfield is of obvious relevance to cosmology. To cite one example, optical observation of the relative absorption by the two lowest rotational levels of cyanogen was the first measurement of what may be the temperature of the blackbody radiation remnant of the primordial explosion that is believed to be the origin of our present universe.

Interaction with Technology and Society

Some examples of the interaction of this subfield with technology and society were mentioned in connection with present communications capability. It should be noted that entirely new technologies are emerging in, for example, thermal imaging and communications (see Figure 4.33). Society's continuing need for improved channels of communication, with ever-increasing capacity, is reflected in the growth in capacity of long-distance telecommunication links (see Figure 4.34). Improved individual communication media could alleviate travel problems in commerce, industry, government, and other enterprises and also could enhance entertainment, education, and recreation. The picturephone has come into being and probably will be widely accepted, thus necessitating a large increase in communication channel capacity, which can be provided by laser beams. Three-dimensional television, using holographic techniques, is more distant. However, holography has important applications in information storage and is used industrially for the detection of small mechanical deformations of large objects, for example, automobile tires. (Several of these applications are discussed in greater detail in the section on Optics.)

More immediate applications of light sensing and optoelectronics are in short-range control and guidance. Light radar and range finding are already

FIGURE 4.33   Crystal of lead–tin telluride. Magnification ×60. Lead–tin telluride forms the basis of a new semiconductor device sensitive to infrared radiation of wavelength 8–14 μm, developed by Plessey. With a detectivity $D^{*}=2\times10^{10}$ cm Hz$^{\frac{1}{2}}$ W^{-1} and a response speed of 10 nsec, the device, which operates at $-77°$C, opens a new field of technology in thermal imaging and communications, where it can be used to produce more noise-free systems through heterodyning laser beams. [Photograph courtesy of Stereoscan Micrograph—Cambridge Scientific Instruments Limited, England.]

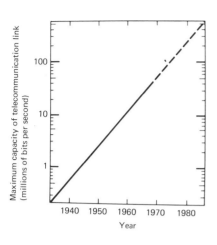
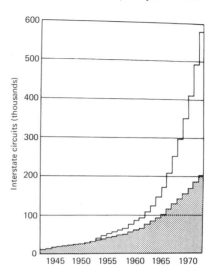

FIGURE 4.34 Increase in capacity of a single installed long-distance telecommunication link (left). Increase in number of long-line circuits in the Bell System (right). The upper curve shows thousands of interstate voice channels in the United States. The lower curve shows channels used for conventional telephone conversations. The difference between the lower and upper curves represents equivalent voice circuits for data transmission, facsimile, television, and special services. [Source: J. Martin and A. R. D. Norma, *The Computerized Society* (Prentice-Hall, New York, 1970).]

in use and may become important in air-traffic control. Public transportation, an increasingly important problem of the coming decades, also would benefit from better monitoring and control devices.

Although the conservation of resources and protection of the environment are primarily matters of economics and priorities in public policy, atomic, molecular, and electron physics can contribute to their solution through better instrumentation, which is essential for the development of the effective control devices that will inevitably be needed to implement an antipollution program. Laser-beam probing of the atmosphere and of smokestack exhausts is a new method for monitoring individual chemical constituents through high-resolution absorption and Raman spectroscopy and by Rayleigh scattering from dust particles. Measurement of temperature and water-vapor profiles in the atmosphere are essential to improved weather prediction and to possible future weather-modification programs. In the longer range, conservation of resources and preservation of the environment will require extensive and imaginative use of new technology.

This subfield also contributes, through better instrumentation, to the biomedical and health fields. Electron microscopy, holography, and various spectroscopic techniques are important analytic tools that require continuing improvement (see Figures 4.35–4.37). Fiber-optics cardioscopes, laser retina welding and cauterization, measurement of blood flow, automated spectroscopic analysis of body fluids—these all offer further possibilities for beneficial application.

It should not be forgotten that one of the most important diagnostic and clinical tools at the disposal of the medical profession is the x ray. Recent development of image intensifiers and computer techniques for image enhancement have vastly increased the potential for the use of x rays without dangerous radiological side effects. Imaginative and innovative use of x-ray diffraction also has contributed to the study and description of the structure of large biological molecules, as recounted in, for example, J.

FIGURE 4.35 A three-dimensional view of a cancer cell magnified 3000 times shows how these cells reach out to engulf neighboring cells. [Photograph courtesy of E. J. Ambrose, Chester Beatty Research Institute, London, England, and the Cambridge Instrument Company, which provided the stereoscan microscope.]

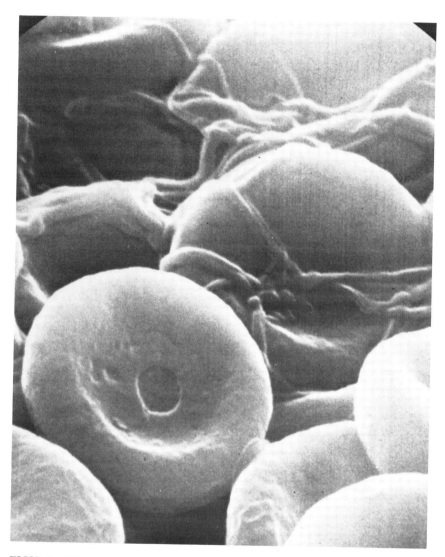

FIGURE 4.36 Human blood was allowed to clot in moist air, fixed in formalde-
hyde, and viewed in the scanning electron microscope. The sample is magnified
15,000 diameters. The disks are red blood cells, held in a meshwork of fibrin strands.
The cells are somewhat shrunken. [Photograph by L. McDonald provided through
the courtesy of T. L. Hayes, University of California at Berkeley.]

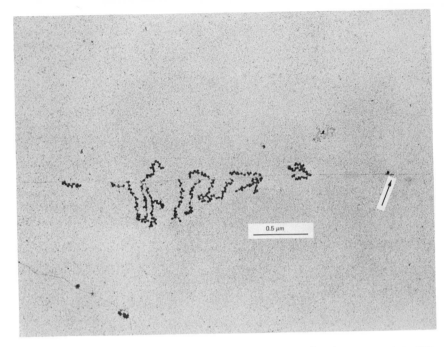

FIGURE 4.37 Visualization of bacterial genes in action by electron spectroscopy. Genetically active and inactive portions of *Escherichia coli* chromosomes are shown. The polyribosomes attached to the active segments exhibit imperfect gradients of increasing length. The shorter, most distal polyribosomes may have resulted from m-RNA degradation. The arrow indicates putative RNA polymerase molecules presumably on or very near the initiation site of these active loci. [Photograph courtesy of O. L. Miller, Jr., B. A. Hamkalo, and C. R. Thomas, Jr., Oak Ridge National Laboratory.]

Watson's *The Double Helix*. More generally, x-ray diffraction techniques continue to play a central role in crystallography and in much of the research on the physics of condensed matter.

Atomic, molecular, and electron physics also makes major contributions to national defense and space programs. Lasers and masers, for example, are used in radar defense, nighttime visual surveillance, bomb sighting, and satellite tracking (Figure 4.38). Molecular physics and infrared spectroscopy are employed in the detection and tracking of rockets and in studying space-vehicle and re-entry problems. The use of high-power laser beams for energy transmission in space, for antiballistic missile (ABM) defense purposes, and for possible ignition of thermonuclear plasmas also has received much attention. Realizing the relevance of atomic, molecular, and

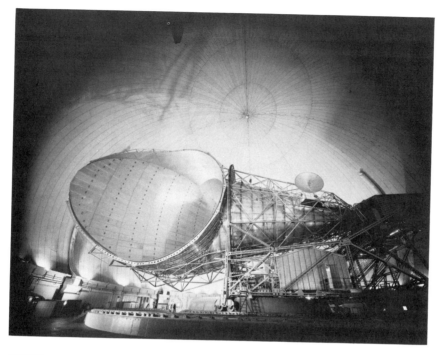

FIGURE 4.38 This huge horn antenna, weighing 380 tons, is used for satellite communication experiments at Andover, Maine. Small dish antenna in upper right of photograph was recently installed for special tests with the Communication Satellite Corporation experiments. [Photograph courtesy of American Telephone and Telegraph Corporation.]

electron physics research to questions of national security, the Department of Defense and the Atomic Energy Commission have supported a large fraction of the work in this subfield.

Future Activity

In the coming decade, much activity in a previously rather inaccessible region of the electromagnetic spectrum, the far infrared or submillimeter wavelength, will take place. Laser techniques have sparked a revolution in infrared spectroscopy that will have an impact on infrared molecular astronomy, thus continuing a historic trend as optical and microwave techniques finally converge and overlap. Extension of the new techniques of laser and quantum electronics deeper into the ultraviolet region will receive increasing emphasis in coming years. Chemical and biological laboratories are adopting physics-based methods and applying them in new ways. Nonlinear spectroscopy, made possible by these techniques, has just begun and promises exciting results. Its present state could be compared with that of radiowave and microwave spectroscopy in 1950. The huge development of these fields between 1950 and 1960 is well documented, and a similar development can be expected for nonlinear and quantum optics during the 1970's.

Work in beam-foil spectroscopy is increasingly important for the interpretation of observations of plasmas and astrophysical phenomena.

Continued progress in the making of precise measurements and the development of more accurate standards of length and time can be expected. These, in turn, will have applications in navigational systems and precision micromachining operations and will increase the sensitivity of seismic and other such monitoring devices. Direct time measurement recently has been extended to the region between 10^{-13} and 10^{-11} sec. Consequently, details of intermediate products in chemical and biochemical reactions may be studied directly; this capability can lead to extremely valuable information about the mechanisms involved.

The colliding atomic and molecular beam techniques have attained a level of refinement that permits the details of chemical reactions and atmospheric and astrophysical collision processes to be studied in the laboratory. An entire new line of research is now possible and is undergoing rapid growth throughout the United States.

These various opportunities are also widely recognized outside the United States, notably in the United Kingdom, West Germany, France, and the Soviet Union. These and other European nations have devoted a somewhat larger fraction of their total physics effort to atomic, molecular, and electron physics than has the United States. Their research output in this subfield relative to that of the United States reflects this emphasis. Whereas in many other subfields of physics, notably high-energy and nuclear physics, the United States holds a clear lead, it does not hold such a lead in this subfield. (Research effort in this subfield in the United States is about on a par at the present time with that in other countries.)

Distribution of Activity

The central position of this subfield in science and technology and its many direct interactions with other subfields of physics, other sciences, and technology are apparent in the distribution of its workers. Atomic, molecular, and electron physics accounts for approximately 6.5 percent of the physics PhD population, as reported in the 1970 National Register of Scientific and Technical Personnel. Approximately one half of these 1,964 physicists work at educational institutions, one fourth in industry, and one fifth in government laboratories and research centers (see Figure 4.39). A special characteristic of this subfield, in contrast to nuclear and elementary-particle physics, is the absence of large permanent installations; consequently, the cost of research per unit of scientific output is relatively low. Atomic, molecular, and electron physicists combine a variety of skills and, in contrast to some of the other subfields of physics in which greater specialization and division of labor are necessary, are often well versed in both theory and experiment. Not only is this subfield well suited to the training of future scientists, but it also acquaints them with problems and techniques relevant to the demands of a technological society.

In 1970, approximately 3 percent of federal funds and 7 percent of industrial funds used for basic research in physics were allocated to this subfield. A substantial amount of research in atomic, molecular, and electron physics is supported by sources outside the subfield. If included, these might double the above percentages (Figure 4.40).

Problems in the Subfield

Because of its relevance to defense questions and to technology, atomic, molecular, and electron physics has received a large share of its funding

ATOMIC, MOLECULAR & ELECTRON PHYSICS

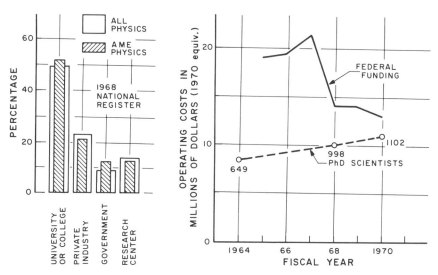

FIGURE 4.39 Manpower, funding, and employment data on atomic, molecular, and electron physics, 1964–1970.

ATOMIC, MOLECULAR & ELECTRON PHYSICS

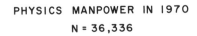

PHYSICS MANPOWER IN 1970

N = 36,336

FUNDS FOR BASIC
RESEARCH IN PHYSICS IN 1970

FEDERAL
INDUSTRIAL

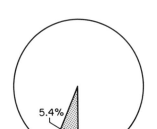

5.4%

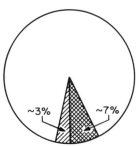

~3%   ~7%

FIGURE 4.40   Manpower and funding in atomic, molecular, and electron physics in 1970.

from the Department of Defense (DOD) and from industrial sources. Even though support of relevant research should be less susceptible to cuts, a general and rather rapid decline in DOD and private industrial funding of all types of physics research appears to be under way. This trend could greatly affect the conduct of atomic, molecular, and electron physics in university settings. Support from the National Science Foundation (NSF) in this subfield, especially at the larger academic institutions, has been low. Although some correction of the distribution of NSF funding among physics subfields has occurred in recent years, it is not sufficient to offset the drastic reductions now taking place in the support of this subfield by other agencies. Federal agencies concerned with pollution, health, and transportation, which would reap long-range benefits from a vigorous program in atomic, molecular, and electron physics, should assume greater responsibility for its support.

There has been an increasing need for sophisticated data-processing equipment. Molecular-beam apparatus now requires a mass spectrograph as a standard detection instrument, and optical spectrographs, costing $2000, have been replaced by integrated spectrometers that cost ten times as much. The need for updating equipment is felt most severely at the universities. Unless a substantial increase in support for research in this subfield at educational institutions is forthcoming, most of the equipment in these institutions will become obsolete within five years. If support were to remain at its present level, a choice between continuing most small-scale efforts or sacrificing many of them to allow the conduct of a few more-expensive major experiments would be necessary. No single research area would deteriorate entirely, but general progress would be sluggish. Equipment could not be upgraded, and a general erosion of experimental facilities would begin.

> Again matter is a relative term; to each form
> there corresponds a special matter.
> ARISTOTLE (384–322 B.C.)
> *Physics,* Book II, 194

PHYSICS OF CONDENSED MATTER

Introduction

Condensed matter consists of solid as well as liquid, glassy, and other amorphous substances. All these substances have in common atoms so closely packed that interactions among atoms play an important role in determining their properties. The largest class of such substances is that of the crystalline solids. Their mechanical, electrical, magnetic, optical, and thermal properties have been investigated and now are understood in a general way. Such properties have been of interest since man first concerned himself with the advantages of bronze over stone or the puzzle of why bits of iron cling to a lodestone. Condensed matter was the focus of physics and much of chemistry in the nineteenth century.

Historical Background

Modern physics of condensed matter has several roots: the discovery of the structure of the atom in the early part of this century; the discovery at about the same time of quantum mechanics, which made possible the quantitative understanding of the behavior of atoms, and their constituents, the nucleus and its electrons; and the discovery in 1912 of x-ray diffraction, which gave the first quantitative information on the ordered arrangements in other forms of condensed matter. In the late 1920's and during the 1930's, the new tool of quantum mechanics was applied with vigor to develop a more complete picture of condensed matter, and quantitative or semiquantitative understanding of the more prominent physical properties emerged rapidly. Today solids are classified according to the dominant binding forces that hold them together (molecular, metallic, ionic, or covalent bonding) and according to their electrical properties (conducting, semiconducting, or insulating). Some substances move from one part to another of the latter classification, with changes of temperature, pressure, or magnetic field, or with variations of purity. Indeed, the range spanned by the physical properties of condensed matter is probably the largest encountered in physics; as an example, the resistivity of common substances ranging from insulators to metals varies by more than 10^{10}.

Extremes of temperature and extremes of pressure have revealed astonishing variants on these usual states of matter. Thus, at very low temperatures some solids abruptly lose all resistance to the passage of electricity. This characteristic is called superconductivity. Although it was first discovered in 1911, at which time it was thought to be a property of only a small number of simple metals, it has been shown in the last decade that a remarkable variety of alloys and intermetallic compounds, and apparently some semiconductors as well, exhibit the property. A theoretical explanation of the phenomenon was not found until 1957, and many aspects of superconductivity are not yet properly understood.

A somewhat analogous loss of all resistance to flow (and other anomalous attributes) occurs in the liquid form of ordinary helium when this substance is cooled to within 2° of absolute zero. This condition is called superfluidity. Superconductivity and superfluidity depend on the quantum mechanical, as distinct from the Newtonian or classical, behavior of aggregates of atoms and electrons.

Mechanical properties of solids, such as strength, hardness, elasticity, plasticity, and the like, depend on cohesive qualities related to interatomic binding forces and also on certain characteristic imperfections in the ideal lattice structure of the solid. These imperfections, or lattice defects, have been the subject of intensive study in the last 25 years and now are generally understood; however, many parts of the picture, including most of its accurate quantitative features, still are lacking.

Similarly, electrical, magnetic, optical, and thermal properties of solids also are understood in general terms, and much quantitative knowledge has been achieved. The richness of the phenomena that matter can present, however, is enormous, and it is possible that today's understanding will one day be considered as crude and naive as that of half a century ago now appears.

Recent Developments

Many important developments have occurred in the study of condensed matter in recent years. Among them are advances in the understanding of the electronic structure and the elementary excitations of solids; the vast increase in knowledge of macroscopic quantum systems (super-conductors and superfluids) and their interplay with other systems; the improvement in the ability to produce many substances in states of great purity and crystalline perfection; experimentation with matter under extreme conditions and the resulting increased theoretical understanding of phenomena observed under such conditions; the achievement of broad understanding of lattice defects; the beginning of a quantitative science of amorphous materials; and the much deeper understanding of cooperative and many-body effects in condensed matter. In addition, there have been many applications of these and earlier developments to other sciences and to technology.

The electronic structures of a large number of simple crystalline solids have now been calculated from first principles in considerable detail. These calculations provide a good account of the binding and elastic properties of the substances. The motions of the carriers of electric charge have been studied, and transport properties are becoming rather well understood. Theoretical information has been checked with a variety of sophisticated experimental techniques—cyclotron resonance, de Haas-van Alphen experiments, magnetoacoustic experiments, optical absorption measurements, photoemission measurements, and many others. From this work a remarkably broad and detailed knowledge of electronic properties of a variety of pure substances resulted. This knowledge played a vital role in the design of solid-state electronic devices, solid-state lasers, and many other technological developments. This information also gives the necessary background for understanding solid surfaces (important for controlling corrosion, catalysis, adhesion, and other surface phenomena) and the effects of impurities, dopants, and alloying agents.

A solid can be regarded as possessing a state of lowest energy and many states of higher energy that correspond to various excitations. These are organized into elementary excitations of several kinds: the various modes of lattice vibration (phonons); excitation of individual electrons; excitations of the magnetic system, if any (magnons); collective excitations of conduction electrons (plasmons); and various combinations of these, which also can act as elementary excitations (excitons, polarons, and the like). Rapid progress in both theory and experiment has occurred in recent years in regard to the understanding of excited states of solids in these terms.

Although superconductivity and superfluidity have been known for decades, they continue to be subjects of great excitement, with important advances, both scientific and technological, taking place. The modern theoretical understanding of superconductivity dates from 1957, when Bardeen, Cooper, and Schrieffer proposed their now-famous theory. An outgrowth of this understanding was the proposal by Josephson in 1962 that a junction between two superconductors separated by a very thin insulating layer should possess anomalous properties. Josephson predicted that a current of limited magnitude could pass through such a junction with zero voltage between the two superconductors and that the magnitude of this current would be highly sensitive to magnetic fields, going periodically to zero as the field is changed. He also predicted that if a constant voltage bias were imposed across the insulating layer, an alternating current would flow with a frequency proportional to the voltage bias. These predictions were quickly confirmed by experiments and have led to the production of numerous practical laboratory instruments of unparalleled sensitivity and to better methods of measuring certain fundamental constants of nature. Superfluids also have exhibited a rich and astonishing range of phenomena in recent years, and a close parallel between properties of superconductors and superfluids has been found. The ideas involved have had important applications to theories of the atomic nucleus and the interior of certain kinds of stars.

The transistor and a number of other solid-state electronic devices require materials of unprecedented purity and perfection, for the electrical properties of semiconductors are influenced dramatically by impurities and crystalline imperfections. Techniques of producing many materials of exquisite purity, and in the form of nearly perfect single crystals, have been developed by necessity as a part of semiconductor technology. These methods have been useful in producing much better materials for research than were previously available. As a result, elucidation of the intrinsic properties of these substances and the extrinsic properties associated with impurities and imperfections became possible. These sensitive and far-reaching distinctions are a hallmark of modern experimentation in the physics of condensed matter.

An offshoot of the development of the high-purity germanium and silicon required for the device market has had revolutionary impact on nuclear and atomic physics. With adequate purity it was possible to construct semiconductor radiation detectors—in essence, solid ionization chambers—that provided energy resolutions typically one to two orders of magnitude better than those previously attainable; moreover, this paramount advantage is coupled to a wide variety of other inherent advantages including small size, elmination of high-voltage requirements, and insensitivity to magnetic fields and, when so designed, to ambient radiation backgrounds as well.

Continuing feedback into technology also occurred. Turbine blades for jet engines, for example, are now made in single crystals of metal, using well-developed techniques for crystal growth. These single-crystal blades have superior ductility and strength, because the boundaries between crystallites, present in ordinary metals, give rise to brittleness.

Interesting modifications occur in many substances when they are subjected to very high pressures. The material in the interior of the earth is subject to such pressures, which can now be attained in laboratory experiments. Static methods are used to reach pressures of several hundred thousand atmospheres. Shock waves, produced by chemical high explosives or other means, produce transient pressures into the range of millions of atmospheres. Theories of interatomic binding are tested by such experiments. Extension of the theoretical treatment to still higher pressures gives an account of matter in the interior of stars and is necessary in completing the understanding of astrophysics. Modern ideas of the mechanisms of crystal growth have guided the production of useful substances by means of high pressures combined with high temperatures. Thus diamonds have been made artificially, first of industrial quality, more recently of gem quality.

Very high magnetic fields have been developed for laboratory purposes. Static fields are produced in conventional coils, supplied with heavy currents, and in the coils made of recently discovered hard superconductors. Short-term transient fields of still higher strength are produced in various kinds of pulsed coils. These high magnetic fields subject the electrons in a solid to a perturbation that tests and elucidates the electronic structure of the solid. Such work has produced a vast amount of new information on electronic structures of solids.

Very high and very low temperatures have also been used to extend the knowledge of condensed matter. Ultralow temperatures are being achieved by sophisticated combinations of new methods, including the ^{3}He–^{4}He dilution refrigerator, and new kinds of cooperative phenomena are being sought. The quantization of magnetic flux in superconductors and of flow velocity in superfluids are two of the striking experimental findings relating to macroscopic quantum phenomena that have emerged recently from this work. Polarization, that is, spin alignment, of atomic nuclei has become a commonplace achievement. This is particularly interesting as a tool for the nuclear physicist. A number of new superconductors and new magnetic transitions have been found.

Lattice defects can be divided into two classes: point defects, which are of atomic scale, and extended defects, which are of macroscopic scale in one or two directions. Vacant sites on the lattice are point defects; lattice mismatches along an extended line, called dislocations, are extended defects. In many solids diffusion occurs by the hopping of vacancies from one lattice site to another, under the influence of the thermal agitation of the lattice. Catalytic effects often depend critically on the presence of lattice defects at a crystalline surface. Irradiation of metals by nuclear radiation, as in a reactor, causes important changes in properties, because of the point lattice defects produced by the radiation. Strength and plastic properties of solids depend markedly on the presence of dislocations and on point defects that impede their motions. All of these defects have been studied for almost three decades, but in the last decade the generally qualitative understanding is becoming quantitative. The variety of possible defects in different crystalline substances is enormous, and the quantitative knowledge that is in hand is but a small fraction of that which would be of immense practical value.

Glasses, liquids, and many other condensed substances are said to be amorphous; in other words, the arrangement of their atoms is only semi-regular. For many years electronic properties, such as electrical conduction, as well as lattice vibrations were treated by quantitative theories only in crystalline solids. Recently, these phenomena have been studied in amorphous substances, and a theory of electronic properties and lattice dynamics in such substances is beginning to emerge. Computer studies entailing the analysis of the behavior of models of amorphous substances have been of benefit as stimulants to intuition and guides to theory. Computer studies of the atomic dynamics in model liquids have been especially useful.

The sharp transition from a solid to a liquid, which occurs at the melting temperature, is an example of what is called a cooperative effect. Another example is the abrupt loss of long-range order in the magnetization of the atoms of an iron lattice when a certain characteristic temperature is reached. These phenomena depend on the cooperative interaction of many elements of a dynamical system and have been objects of study in physics for years. Recently, advances in theory have shown that many disparate cooperative phenomena are mathematically related. Many new examples of cooperative transition have been found and studied by dynamical as well as static methods. The scattering of laser light and thermal neutrons by such systems afforded new experimental information on the dynamical and static properties of these systems. Cooperative phenomena occur in what may be called many-body systems. Every system studied in the physics of condensed matter is actually a many-body system. At an earlier stage, physics usually progressed by reducing the many-body systems, conceptually, to an approximate replica in which one body at a time could be considered as moving. Although it is not yet possible to give a rigorous dynamical treatment of general many-body systems, much improvement has been made in methods of calculation so that good corrections for many-body effects can be made in an increasing array of problems. The understanding of superconductivity requires explicit allowance for such effects, and a great deal of the theoretical effort in the physics of condensed matter in the last decade has been devoted to understanding many-body corrections to earlier one-body models.

Other branches of physics have interacted recently in vital symbiosis with the physics of condensed matter. The Mössbauer effect is the phenomenon of emission of a nuclear gamma ray by an atom imbedded in a solid, without the recoil of that atom, that would ordinarily occur and somewhat reduce the energy of the gamma ray. This phenomenon combines the realms of nuclear physics (the emission of the gamma ray and its dependence on the state of the nucleus) with the freedom from recoil that is a consequence of the binding of the radiating atom into a crystalline lattice. The effect has been an important tool for learning about crystalline binding and the magnetic field (and the gradient of electric field) at the nucleus of the emitting atom. Some of this information complements that which can be obtained by the older phenomenon of nuclear magnetic resonance. Another measure of the magnetic field at the nucleus of an atom in a lattice results from observations of the angular correlation between two successive cascade gamma rays emitted by the nucleus. The phenomenon of radiation introduces lattice defects into a solid and constitutes a tool for studying lattice defects.

Still another example is the process of channeling of energetic charged particles by a crystal lattice (see Figure 4.41). In channeling, a nuclear particle, such as a proton or an alpha particle of a few million electron volts energy, when falling upon a single crystal at a direction nearly parallel to a prominent axis of the crystal, tends to be guided into a channel accurately parallel to this axis and to lose energy more slowly under these conditions than when going through the crystal in a random direction. Not only has this process added to the understanding of the process of energy loss by fast charged particles in solids (which is an old subject), it has also provided a new tool for investigating the perfection of crystals and learning the configuration of point defects in crystals. Even more recently, it has been used to measure nuclear lifetimes in the range of 10^{-15} sec, which previously have been completely inaccessible, through effective observation of flight times between lattice sites.

Relationship with Other Sciences

The physics of condensed matter has strong interactions with chemistry and metallurgy and with the broader field of materials science. These interactions are too numerous, and most are too obvious, to catalog here. Usually the solid-state physicist and the solid-state chemist are distinguishable only by their backgrounds or, occasionally, the emphasis in their work. The chemist is more often concerned with complex substances (although this difference is decreasing); methods of preparation, synthesis, crystal growth, and the like; and use of chemical methods of

channelling

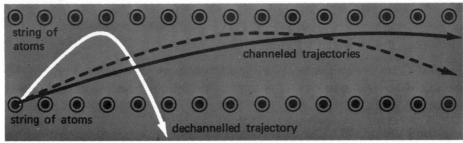

blocking

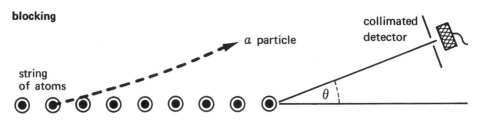

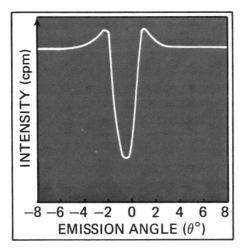

FIGURE 4.41a Channeling of charged particles through a crystal matrix can occur only if the incident angle is less than a certain critical value. Then the incident particles suffer a series of gentle repulsions from the string atoms and remain in the channel. When the angle is too large, the particle is strongly deflected and can no longer be channeled. The opposite process is known as blocking; particles originating in the center of the string cannot be channeled. Here alpha particles are shown being emitted from radioactive radon atoms situated on lattice sites in tungsten. Along the direction of the string, which corresponds to an emission angle of 0°, the intensity of the emission goes through a minimum value caused by the blocking phenomenon.

analysis. Metallurgy has been strongly influenced by concepts of electronic structure and binding originally developed by physicists. The large and productive field of defects in metals has become increasingly the province of the metallurgist.

A few selected examples will illustrate the widespread impact of the physics of condensed matter on other subfields of physics and other sciences. The Josephson effect, as noted above, consists of an oscillating

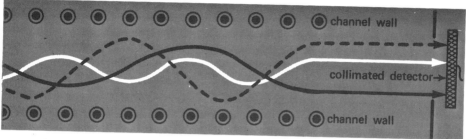

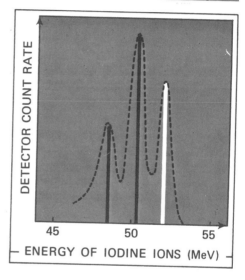

FIGURE 4.41b Position of the collimated detector imposes a simple geometrical boundary condition on the number of particles that are detected after they leave the channeling planes. Only those particles that leave the crystal in a direction very near to the plane can be detected if the detector is placed far enough away. This makes it easier to study the processes by which particles lose energy, by limiting quite drastically the number of oscillations that are observed. In the diagram above, the boundary condition allows only three oscillations to reach the detector, producing the three peaks in the energy spectrum (left). Studies of this kind, although difficult, make it possible to obtain information from the energy spectrum about the electron distribution and interatomic potentials in the individual atoms in the crystal lattice.

current between two superconductors in weak contact maintained at some potential difference V. The frequency of the oscillation is given by $v = 2eV/h$, where e is the charge on the electron and h is Planck's constant. From a measurement of v and of V, this relationship determines the important natural constant e/h. This method of obtaining e/h is several orders of magnitude more precise than any alternative experimental method. In conjunction with other measurements, this method has resolved a long-standing discrepancy in the quantum electrodynamic theory of the hydrogen atom and certain other elementary physical quantities. This finding reinforces belief in the validity of quantum electrodynamics, which is one of the most basic and far-reaching components of the current conception of the physical world. Thus, an understanding of the superconducting state in solids has had implications for the seemingly unrelated fields of quantum electrodynamics and atomic physics. The Josephson effect also promises to provide the basis for a very compact, high-efficiency computer memory system; such systems are being developed at the present time.

The impact of condensed-matter physics on nuclear physics is multidimensional. Nuclear magnetic resonance of atoms in solids, the Mössbauer effect, angular correlation of successive gamma rays, and blocking effects in crystals are examples of effects that, through sophisticated knowledge of solids, allow measurement of major parameters of nuclei. Perhaps the most dramatic impact of condensed-matter physics and associated technology on experimental nuclear physics, however, has been through semiconductor detectors of nuclear radiation.

In theoretical nuclear physics, major progress resulted from the application of the Bardeen-Cooper-Schrieffer theoretical breakthrough concerning superconductivity in condensed matter. This recognition of the joint dependence of these subfields on many-body phenomena greatly enriched both. The Josephson effect between superconducting solids, discussed above, is now being actively sought in the nuclear realm in the interface between colliding superconducting nuclei such as selenium and tin.

Physicists and chemists concerned with solids have devoted much effort to the study of surfaces. In spite of this work, heterogeneous catalysis remains a highly empirical field. Condensed-matter physics has contributed techniques that promise increased understanding. Thus it is now possible to cleave single crystals in very high vacuum to produce clean and regular surfaces that can be microscopically characterized. Then, by means of low-energy electron diffraction, two-dimensional order in a monatomic layer of adsorbed gas can be studied. Recently, a new technique, ion-neutralization spectroscopy, has been developed that is sensitive to processes going on within an atomic diameter of the metal surface by ions of noble gases, and the energy of the ejected electrons is measured. From these data information about the chemical bonds in molecules adsorbed on the surface can be deduced. Such molecules differ in interesting ways from free molecules, because they are constrained by the solid and can have structures that do not occur among free molecules but are likely to be of prime importance in both corrosion and catalytic reactions.

There are many ways in which condensed-matter physics is impinging on earth and planetary physics (an example is given in Figure 4.42). Our knowledge of the earth and its history is based to a large extent on solid-state physics and chemistry. The model for the composition of the mantle is based on the phase equilibria determined for different germanate systems that can be extrapolated to the silicates, which comprise the earth's mantle. The knowledge of the response of a solid to sound waves and studies of the equation of state of liquids allow one to hypothesize a liquid core; set

100μm

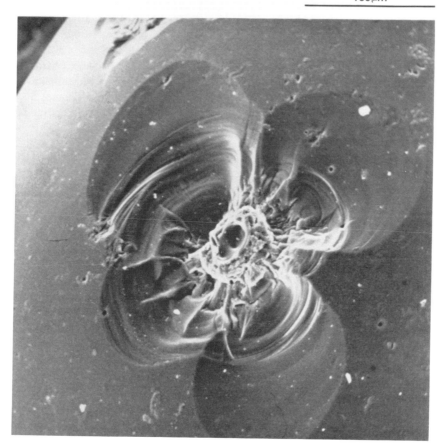

FIGURE 4.42 Apollo 11 moon rock sample. Micrometeorite impact crater on the surface of a glassy sphere showing a central crater (produced by melting) standing on a boss with irregular vertical fractures, surrounded by petal-like conchoidal (brittle) fractures. The small black areas are bubbles in the original glass which have been cut by the fracture or burst on the natural surface of the sphere. Magnification ×520. [Photograph courtesy of Stereoscan Micrograph—Cambridge Scientific Instruments Limited, England.]

a limit of the temperature and pressure in the mantle; and set some limits to the accuracy of the extrapolations that are made concerning the properties of materials in the earth.

Impact on Technology

Over the past several decades, the physics of condensed matter has had great impact on technology; its influence shows no sign of abating. Indeed, technological benefits from this subfield probably will emerge even more rapidly in the future. (Section 2 of the report of the Panel on Condensed Matter, in Volume II, presents an outline of recent developments in condensed-matter physics that have direct technological importance; Section 3 gives case histories of selected innovations; and Section 5 discusses the economic and social consequences of these technologies.)

Perhaps the single development with the most far-reaching consequences is the transistor, development of which began in 1947 with the discovery of the transistor effect by Shockley, Bardeen, and Brattain (see Plate 4. VI). An enormous number of solid-state electronic devices have evolved since. Today's large-scale integrated circuitry is even farther ahead of the original transistor than that device was ahead of the old vacuum tube (see Figure 4.43 and Plate 4. VII). Control circuits and computers of high speed, high reliability, low cost, and small size are only a few of the benefits

FIGURE 4.43   First commercial use of integrated circuits was made in 1966 RCA television receivers. [Source: *Science Year. The World Book Science Annual.* Copyright © 1966, Field Enterprises Educational Corporation.]

that these devices have made possible. Small chips of silicon are now made in routine fashion containing hundreds of thousands of electronic components. In the past five years, integrated circuits have come to account for sales of $500 million per year, which is 30 percent of the total semiconductor market. Recent innovations are solid-state microwave diodes and optoelectronic devices. The former, including Gunn-effect oscillators, probably will be widely used in communication and radar equipment. Optoelectronic devices include solar cells, which provide power for satellites, and light-emitting diodes, which make possible new indicator devices and display screens.

On the horizon are optoelectronic devices that include laser light sources, sophisticated large-scale computational elements, and very-large-scale memories, all fabricated within the same crystal wafer. Even more exciting is the possibility of overlapping multiple systems through the same optical channels, each operating on its own wavelength of radiation. Such devices, with ultraminiature size, microscopic power requirements, effectively infinite lifetime and reliability, and low intrinsic cost cannot fail to have at least as revolutionary an effect on society as did the introduction of computer techniques. (See Plate 4.VIII.)

Optoelectronic techniques also hold great promise for the development of whole new generations of visual displays in the all-important man–computer interface; these can be cheaper, safer, more compact, and more convenient by orders of magnitude than the typical cathode-ray tube band systems now in use.

Magnetic materials are widely used now and comprise many substances besides the steels of transformer cores and electric motors. Ferrites, which are in effect ferromagnetic insulators, are used in television transformers and computer memories. Other new magnetic substances appear in ferromagnetic memory devices, such as tapes, drums, and disks. Magnetic materials that can be stimulated by optical means, such as laser pulses, may also have an impact on display and information-storage devices.

Magnetic bubble technology has provided a radical new way to store information in magnetic bubbles in thin, transparent, magnetic crystals and to carry out logical operations by moving the bubbles over the crystal (see Figure 4.44). Magnetic bubble memories may replace both the core and disk file memories in computers and electronic central offices and interface directly with fast semiconductor devices. They may prove to be a very fast, compact, and inexpensive way to store and process data.

Superconducting materials have come into use for a variety of specialized needs, particularly for providing high magnetic fields for laboratory purposes. The greatest impact of these new materials, and of others under development, probably lies ahead, with power transmission lines (see Plate 4.IX), transformers, motors, mass-transport levitation and guide-

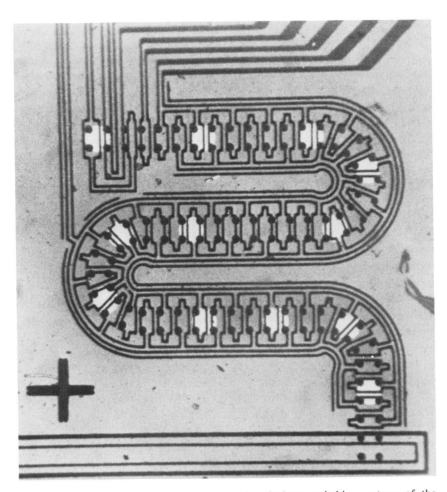

FIGURE 4.44 Tiny computers and electronic telephone switching systems of the future may accomplish counting, switching, memory, and logic functions all within one solid magnetic material, employing new technology now in exploratory development at Bell Telephone Laboratories. Looking more like a block diagram or a flow chart, this actual circuit, a photolithographic pattern on the surface of a sheet of thulium orthoferrite, can move magnetic bubbles (large white dots) through a shift register. The magnetic bubbles are 0.004 in. in diameter. [Source: This photograph was provided through the courtesy of Bell Telephone Laboratories.]

ways, and computer memories being potential areas for future application. Ultrasensitive electric and magnetic measuring devices, another application of superconductors, already have been developed.

Of crucial importance to any future implementation of superconducting technology, as, for example, in transmission of electrical power or in new computer memory configurations, is the maintenance of an adequate supply of helium. Helium is unique in its combination of unusual properties and critical uses. It is essential for cryogenics, superconductivity, some types of breeder reactors, and the space program. Moreover, large components of contemporary physics research—in condensed matter, elementary particles, and nuclear physics—are entirely dependent on liquid helium to achieve adequately low working temperatures. According to available estimates, this irreplaceable resource is in short supply, yet it is being wasted in alarming fashion.

Solid-state lasers of various kinds have had tremendous impact in many fields and promise further developments, including light-beam communication systems with a carrying capacity far greater than any system available today.

Secondary emission is a phenomenon from condensed-matter physics that has long found application in the electron multiplier. Image-intensifying devices based on this effect provide the starlightscope, a medical x-ray technique that permits much lower exposure of the patient to radiation and sensitive detectors that improve the seeing power of astronomical telescopes by large factors. New compounds have made possible sensitive detectors for the infrared. One example of their use is in the military snooperscope.

Fundamental discoveries in polymers, crystallization, morphology, radiation damage, point defects, dislocation, diffusion, and annealing have aided the development of a host of new materials: new rubbers and new composites, alloys for use in nuclear reactors, high-temperature and high-strength material, and new steels and alloys. Better materials are being developed for biological and medical uses. These include membranes for artificial organs and strong inert materials for replacement of joints and for heart-valve implants. Although progress has been made, the potential uses of biocompatible materials are just beginning to be realized.

Future Activity

Forecasting the trends of any active scientific field is difficult and uncertain. However, based on present activity, it appears reasonable to predict that the areas of vigorous activity and rapid progress in the physics of condensed matter in the next several years will include the following:

1. Surfaces and interfaces. New experimental techniques and increased theoretical attention should result in new breakthroughs.

2. Optical properties of solids. Lasers and new radiation sources, including synchrotrons for the far ultraviolet, have reinvigorated this field. Nonlinear optical properties of solids, exciton absorption and luminescence, and optical properties in the far ultraviolet and soft x-ray region offer a wealth of new opportunities.

3. Scattering studies on solids and liquids. New techniques and more intense sources of neutrons and photons have opened entirely new areas of research. This work has large potential for providing fundamental information on liquid and solid structure, and through that the potential for the development of both improved and entirely new devices.

4. Complex crystalline substances. Most of the substances that have been extensively investigated are relatively simple ones in which phenomena are not hopelessly complex and well-characterized specimens have been available. Silicon, germanium, the simple metals, and alkali halides have been among the most commonly studied solids. Not all useful solids are simple, however; the number of little-studied substances vastly exceeds the number of well-studied ones. Attention is turning now to a much wider and more complicated realm of substances, with three, four, or many atoms per unit cell.

5. Disordered condensed materials. The conceptual framework of the electron theory of solids and of lattice dynamics is being extended to include disordered materials. New materials in this category are being discovered and synthesized with new techniques. The development of useful electronic devices from such substances is likely.

6. Electrons and phonons as elementary excitations in solids. The motion of free charge carriers in solids and its elaborations offer rich ground for discovery and innovation. Bulk negative conductance in gallium arsenide, many-body effects in solid-state plasmas, propagation of helicon waves in high magnetic fields, and plasma instabilities arising from nonequilibrium conditions are examples. Investigation of the vibrations of crystal lattices (lattice dynamics) also continues to be an active area. Lattice vibrations play an important role in a wide variety of phenomena, for example, superconductivity, ferroelectricity, and antiferroelectricity. Moreover, significant advances have occurred in experimental techniques, including inelastic scattering of neutrons, Raman scattering of laser light, electron-tunneling spectroscopy, and ultrasonics.

7. Channeling, blocking, and related phenomena. These phenomena are finding important applications in studies of lattice structures and lattice defects. They provide insight into radiation damage and have important applications in nuclear physics, such as the measurement of very short lifetimes of nuclei. The related technique of ion implantation is becoming

PLATE 4.VI This is the first transistor ever assembled; the year was 1947. It was called a point contact transistor because amplification or transistor action occurred when two pointed metal contacts were presented onto the surface of the semiconductor material. The contacts, which are supported by a wedge-shaped piece of insulating material, are placed extremely close together so that they are separated by only a few thousandths of an inch. The contacts are made of gold, and the semiconductor is germanium. The semiconductor rests on a metal base. Public announcement of the transistor was made July 1, 1948, by Bell Telephone Laboratories. In 1956, John Bardeen, Walter Brattain, and William Shockley shared the Nobel Prize, the highest honor in science, for their discovery of the transistor effect. [Source: Bell Telephone Laboratories.]

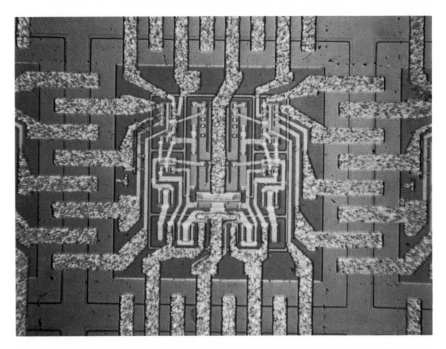

PLATE 4.VII   Close-up of a single microminiature beam-lead circuit. This circuit was developed by Bell Labs for possible use in electronic switching system equipment. A dual five-input NAND gate, it has 10 transistors, 18 diodes, and 12 resistors. Actual size of the circuit is 0.053 in. from beam tip to beam tip. [Source: This photograph was provided through the courtesy of Bell Telephone Laboratories.]

PLATE 4.VIII Weaving back and forth between two spherical mirrors, this laser beam travels nearly two miles. The beam forms an optical delay line on which information may be stored. In time, devices like this one at Holmdel might store telephone numbers, billing information, or other data for the telephone central offices. [Source: This photograph was provided through the courtesy of Bell Telephone Laboratories.]

PLATE 4.IX   A 40-ft section of high-voltage (450,000 V) cryogenic transmission cable undergoing testing at the General Electric Company.

increasingly important in the production of complex semiconductor devices.

8. Low-temperature phenomena and superconductivity. An important frontier of solid-state phenomena lies at the extreme low temperatures. New experimental techniques have pushed this frontier back, and new phase transitions and novel collective phenomena are sure to be found. Better understanding of superconductivity and superfluidity will follow.

9. Applications to astrophysics and extreme states of matter. Many of the principles in the theory of condensed matter at extremely low temperatures can be extrapolated to white dwarfs, neutron stars, and other problems in astrophysics. Along somewhat different lines, laboratory work in the extreme ultraviolet is called for in connection with the discovery of stellar x-ray sources. Matter under very high pressure, and also under very high magnetic fields, is being more widely investigated, both theoretically and experimentally.

All the foregoing areas appear ripe for vigorous exploitation, and unforeseen developments in this rich and rapidly moving subfield will undoubtedly play a prominent role in the years immediately ahead.

Distribution of Activity

The study of the physical properties of condensed matter and the search for understanding of these properties constitute, by a substantial margin, the largest subfield of physics. Condensed-matter physicists account for about 25 percent of the physics PhD population, as reported in the 1970 National Register of Scientific and Technical Personnel. Figure 4.45 shows the distribution of these physicists among employing institutions. About 40 percent of these work at educational institutions, 38 percent in industry, and about 18 percent in government laboratories and research centers.

The physics of condensed matter differs sharply from most of the other subfields because industrial support of the subfield exceeds government support. Much of the industrial effort is applied in character, but there is a continuous gradation between the application-oriented efforts and the basic efforts, and the industrial contributions to the basic side are highly significant. According to recent estimates, approximately 53 percent of the basic research in the subfield takes place in universities, 20 percent in government laboratories, and 25 percent in industry. About 75 percent of the more applied work is found in industrial laboratories. The industrial component of support cannot be assumed to act as a ballast, rising when the federal support decreases and decreasing after federal support rises. In fact, the trend over the past 20 years has been for the two to move synchronously.

The typical research project is relatively small in cost; and large devices

CONDENSED–MATTER PHYSICS

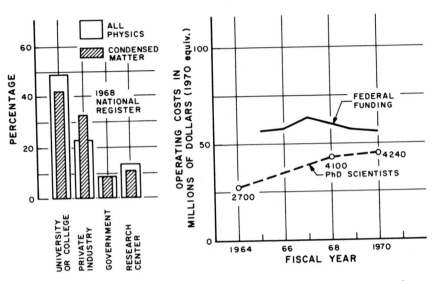

FIGURE 4.45   Manpower, funding, and employment data on the physics of condensed matter, 1964–1970.

do not consume a major portion of the funds. The contrast with elementary-particle physics and nuclear physics, in this regard, is particularly sharp. The diversity of the subfield manifests itself in the number and variety of research topics being pursued, as well as in the interplay of fundamental and applied research. The separation between basic discoveries and applications in condensed matter is less distinct than in many other areas of physics.

In 1970, approximately 12 percent of federal funds and 75 percent of industrial funds for basic research in physics were allocated to research in condensed-matter physics (Figure 4.46).

Problems in the Subfield

As noted before, the physics of condensed matter consists for the most part of many small to medium-sized research efforts widely dispersed in universities, industries, and governmental and national laboratories. Such a diverse effort, which has few individual projects with an immutable claim to survival, is at a disadvantage, in a time of retrenchment, in the compe-

CONDENSED MATTER

PHYSICS MANPOWER IN 1970

N = 36,336

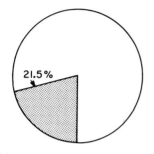

FUNDS FOR BASIC
RESEARCH IN PHYSICS IN 1970

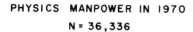

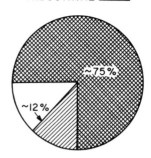

FIGURE 4.46   Manpower and funding in the physics of condensed matter in 1970.

tition with other subfields built around large projects that represent major items in the budget. In these circumstances, the small projects tend to become unduly eroded. At the same time, the condensed-matter subfield is developing an increasing need for large facilities: sources of intense magnetic fields, sources of high pressures and high temperatures, large computers, more-intense beams of thermal neutrons, improved electron microscopes, improved synchrotron sources of far-ultraviolet radiation, expanded facilities for preparation of pure materials, and special sources of particle radiation. These are not being provided in the United States at present, and, indeed, some existing facilities have recently been closed. Moreover, the increasing sophistication of the best laboratories in the subfield works to the increasing disadvantage of the smaller laboratories, which have always played a significant role.

To an increasing degree, research in condensed-matter physics, and other subfields of physics, is based on a wide array of experimental techniques that are applied to a single scientific question; the smaller laboratories cannot mount such broad attacks. Regional facilities that supplement the apparatus available to physicists in many small laboratories are needed, but new facilities of this character are not being set up. University departments with strong specialties in the subfield are now severely hampered by lack of funds for fellowships and for the research costs incidental to training.

All men by nature desire to know. An indication
of this is the delight we take in our senses; for
even apart from their usefulness they are loved
for themselves; and above all others the sense
of sight.

ARISTOTLE (384–322 B.C.)
Metaphysics, Book I, 980

OPTICS

Introduction

Optics is a basic and applied subfield of physics traditionally divided into physical optics, geometrical optics, and physiological optics. Although the major emphasis has always been on visible light, optics usually is generalized to include the techniques and phenomena of electromagnetic radiation extending from the far ultraviolet to the far infrared and occasionally to x rays, microwaves, and even electron optics.

Classical nineteenth-century optics emphasized optical instruments such as the microscope and telescope; visual phenomena such as the sensation of color; and physical optics of interference, diffraction, spectroscopy, polarized light, and crystals. Modern optics has added a tremendous number of intricate new phenomena produced by the interaction of light with matter. Holography, photon counting, the biochemistry of vision, photography, photoconduction, light-emitting diodes, photoemission, and the laser are examples. Modern optical instruments include luminescent display panels, image intensifiers, electrooptic light modulators, tunable lasers, guided waves, amplifying fibers, pulse compressors, image-enhancement systems, tracking devices, and combinations of these and other new tools.

Throughout the nineteenth century, optics was a central subject of physics and commanded the attention of the greatest physicists of the time, but the skill and insight of these great physicists seemingly exhausted the topics to which they addressed themselves. For example, the Abbé theory of the microscope established a clear-cut limit to the resolving power of the instrument and, when commercial instruments reached that limit, the subject was seemingly closed. The major exception in optics—the subject that was not exhausted—was the field of atomic and molecular spectroscopy. This subject, which still challenges many physicists, now constitutes a separate body of knowledge in atomic, molecular, and electron physics, though it still is closely related to optics.

Much of the stimulus to optics in the last two decades developed as a deeper understanding of the limiting factors in many experiments and devices was achieved. Many problems throughout science were found to be "optics-limited." That is, the speed or accuracy with which a measurement could be made, a device controlled, an object detected, or a chemical analysis completed often was limited by fundamental optical problems of intensity, resolving power, stability, or photon statistics. The search for ways to overcome these limitations led to extensive programs of applied optical physics. In addition, the astonishing attributes of the laser made it productive to re-examine all these optics-limited situations. For example, it is now possible to determine with unprecedented precision the Raman spectrum of a few milligrams of a water-soluble biological compound using a laser as a light source. Without the laser, only massive samples can be analyzed. A laser beam shining on a beam of high-energy electrons suffers Compton scattering and is shifted in wavelength to the gamma-ray region without loss of polarization; the scattering cross section is low, but the intensity of the laser beam is high. This technique has already provided a powerful new tool for elementary-particle physicists.

There appears to be no shortage of frontier areas in optics at this time, partly because so many situations are still optics-limited, so that each new fundamental development has a rapid and direct effect on applied optics. The need for re-examination of long-established assumptions has had many important consequences in optics. Abbé and Rayleigh had assumed that the purpose of the microscope was to see small opaque or self-luminous objects, thus they defined the figure of merit according to this purpose. Zernike chose a different figure of merit—the ability to see small transparent living objects. He discovered the phase microscope, a simple method of great subtlety, and won a Nobel Prize.

The pattern continues today. The majority of optical scientists (and there are some 6000 members in the Optical Society of America) currently work on practical problems in optical engineering or applied optics. But questions arise frequently that challenge present understanding of the basic physics, demonstrating that there are not only unanswered questions but that the classical questions often are not the correct ones.

Spectographic instruments provide an important contemporary example. Every physics student is taught something about the wavelength resolving power of a spectrometer and shown how it depends on the properties of the dispersing element, but the arguments that he hears do not deal with the time devoted to the observation or the signal-to-noise ratio. The modern Fourier-transform spectrometer, built by Pierre and Janine Connes, has produced planetary spectra with 100 times greater resolving power and 10 times better signal-to-noise ratio than the best prism spectrometers. Much of their success arises, of course, from their technical skill, but the underlying physics is beautiful and profound. Their new atlas of planetary spectra shows more detail in the absorption spectra of the atmosphere of

Venus than is shown in corresponding spectra of the earth's atmosphere that have been obtained with prism spectrometers (so-called solar atlases).

New Developments in Optical Physics

Although optics is now predominantly an applied science, advances in basic optical physics continue to occur. By far the most important recent advances in optical physics are the laser and the new devices and techniques that it has made possible. The invention of the laser can be credited to atomic and molecular physicists, but optical specialists played a basic role from the beginning. The availability of copious coherent light has made possible a wide variety of new optical techniques such as the hologram and related image-processing methodology (see Figure 4.47), and the high intensity of lasers has led to the development of the exciting new work in nonlinear optics.

Currently there is no apparent slowing of the pace of development associated with the laser and laser devices. Laser techniques are being extended into new wavelength regions; lasers are being made tunable to different wavelengths; and new techniques are being developed for deflecting and modulating laser beams and producing and controlling extremely short pulses of light. (See Plate 4.X).

A wide variety of special situations has resulted from the use of this new tool. The ability to produce ultrashort pulses of light, for example, has opened the time domain in chemical kinetics to investigation in a way that was impossible with more conventional sources; now the sequence of events can be unraveled and followed. Very short pulses can also produce fantastically high-power densities and field strengths—so high that even nuclear forces may be affected. A completely new tool thus becomes available for studies of nuclear fusion as well as for other purposes. Eventually, these high photon densities may allow the direct experimental observation of the scattering of light by light; however, a vacuum more perfect than any now obtainable will be necessary.

The frequency stability of even a simple laser gives light of quite remarkable purity, but stabilized lasers can be built whose frequency is fixed to one part in 10^{11}. The application of this stability to phenomena now thought to be well understood will certainly result in surprises. Meanwhile, even much simpler lasers in unequal path interferometers allow measurement of strain, tilt, and shear in the earth so that earthquakes can be studied and perhaps predicted with new accuracy.

A particularly significant development arising from the laser is the invention of the modern hologram in 1962 by Leith and Upatnieks. Gabor had formulated the conceptual base for the hologram in 1949, but it was

Deblurring Photographs by Holography

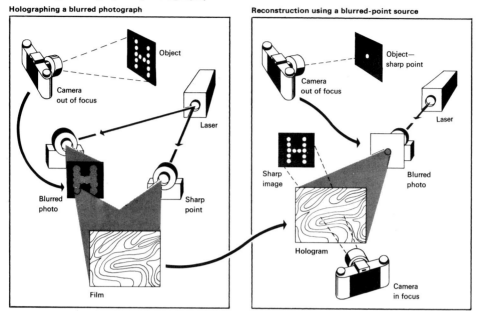

FIGURE 4.47   A photograph taken out of focus can be sharpened with the use of holography. First, a conventional hologram is made, above left, as laser light from a sharp source simultaneously interferes with light from each blurred point in the scene. If the hologram were then illuminated with a sharp light source, the same blurred scene would be reconstructed. In a technique created by George Stroke of the State University of New York at Stony Brook, the image is instead reconstructed, above right, using a point of light made out of focus by the same camera. Each point on the reconstructed scene then appears as sharp as the sharp light source originally used to make the hologram. [Source: *Science Year. The World Book Science Annual.* Copyright © 1968, Field Enterprises Educational Corporation.]

not until coherent light was available in copious supply that large-scale, intense holographic displays became possible. The theory of the nature of optical images has been challenged by the hologram, and a wide variety of new experimental techniques have become necessary. Throughout this decade of development of holography, optics has been stimulated constantly by the prospect of useful devices, and so substantial investments of money and manpower have been made in this subfield.

The hologram can preserve and reproduce an image, and, within certain limits, a three-dimensional image of an extended object is possible. The image need not arise from a real object; a hologram can be produced from a computer printout. Thus, it is possible to generate an image of an object that did not exist or to tabulate the sound field around an object illuminated with coherent ultrasound or to recreate an image of the object for visual examination.

In principle, holography also is useful as a means of storing data for use in a computer or other information-retrieval system. Because the information in the hologram is not localized, scratches, dirt, or other blemishes in it do not introduce errors or destroy information. In the same way that communications engineers have learned to analyze signals in either the frequency domain or the time domain, the hologram makes it possible to transform images between object and aperture spaces. It is not always clear which type of space has practical advantages over the other, but the analysis of the hologram has increased understanding of information storage systems and offered new possibilities for extending their scope.

The scope of image storage can be increased in another respect as photochromic materials with improved properties become available. With such materials, it is possible to generate and record an image using one wavelength of light, to read it or extract information from it by using a second wavelength, and, finally, to erase it with a third wavelength or the application of heat.

Holographic interferometry has begun to have substantial application in engineering. If a hologram of an object is made and subsequently compared with the real object, any change in the object can be seen as an interference pattern spread over the surface. Thus the elastic modes of vibration of complex objects can be determined readily. This technique is already in routine use, for example, as a production-line test method in the fabrication of automobile tires (see Figure 4.48). The number of applications of lasers and laser technology is legion and growing rapidly. At the moment, the only limitations appear to be the imagination of the user.

It would be wrong, however, to suggest that contemporary optics is concerned only or even primarily with lasers. Optical physicists are still very much concerned with atomic and molecular spectroscopy and solid-state optical phenomena. In much of this activity, the useful devices that are anticipated motivate and justify financing the work, but the research involved is frequently of the most fundamental character.

One of the older but still highly productive branches of modern optics is that concerned with thin films (Plate 4.XI). It is now possible to design and construct multilayered thin films that will reflect chosen wavelengths and transmit others in complete analogy to the much more familiar electrical filter systems. These may find very important applications in the large-scale utilization of solar power in which they can be designed to

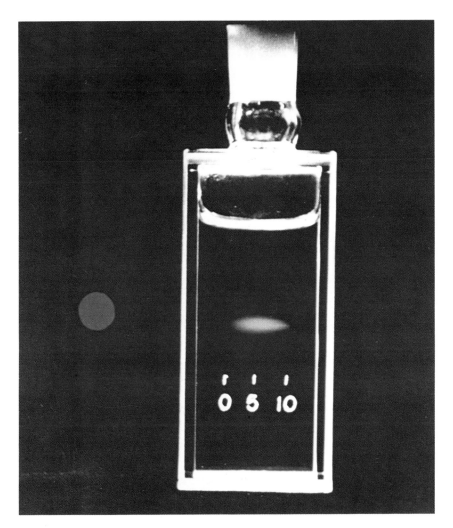

PLATE 4.X Stop-motion photograph of ultrashort green pulse in flight through a water cell. The green pulse was moving from right to left. The scale is in millimeters. The camera shutter opening time was about 10 psec. The red spot is the impression left on the high-speed Ektachrome film by an infrared laser pulse used to activate the ultrafast shutter. A neodymium glass laser generates the infrared pulse, which in turn gives rise to the green pulse by passing through a nonlinear optic crystal (not shown). This photograph is thus simultaneously an illustration of second harmonic generation from the infrared (red spot) to the green. [Source: M. A. Duguay, Bell Telephone Laboratories.]

PLATE 4.XI A thin-film prism. A ZnS film was deposited on a glass substrate. The film in the triangular area is thicker than that in the rest of the area. A helium–neon laser light propagating in the film was deflected after passing through the triangular area that acted as a prism. [Source: P. K. Tien and R. Ulrich, *Journal of the Optical Society of America, 60,* 1325 (1970).]

NONDESTRUCTIVE TESTING

hologram recording

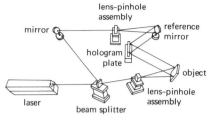

(a)

hologram viewing

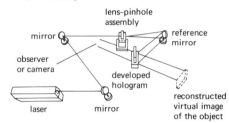

(b)

a

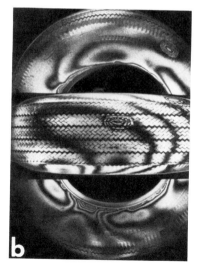

b

FIGURE 4.48 Nondestructive testing. *Upper left:* Beam-path scheme for the laser hologram process. (a) Construction of a hologram diffraction pattern in a photographic emulsion. (b) Reconstruction of the three-dimensional image in space.

Right: Testing of automobile tires. (a) Holographic tire analyzer reveals (b) hidden tire defects, such as tread- and shoulder-area ply separations, when the tire is inflated or subjected to various temperatures, captured here by double-exposure interferometry. Real-time viewing of a defect may be likened to observing the telltale moving ocean-wave patterns in the vicinity of a sunken rock. (GC Optronics, Inc.)

[Source: J. R. Zurbrick, *Yearbook of Science and Technology 1971.* Copyright © 1971, McGraw-Hill Book Company, Inc. Used with permission of McGraw-Hill Book Company.]

inhibit reradiation of the absorbed energy in the form of infrared radiation as described earlier in this section. High-pass, low-pass, bandpass, or low-band filters can be built economically and reliably. An interesting class of such filters transmits only a narrow band of wavelengths; when made with one or more layers tapered in thickness, the filter will transmit different wavelengths in different regions. Such wedge filters can be used to build a particularly simple and compact spectrometer for use in space vehicles or in inexpensive commercial instruments.

Another important development of thin-film technology is the high-efficiency mirror, reflecting 99.6 percent to 99.8 percent of the incident light in a chosen wavelength region. The availability of such mirrors has been one of the indispensable tools of the technical development of the laser. Still another by-product is the well-known low-reflection coating. Without such coating, complex microscope or camera lenses produce images having only low contrast because of the veiling glare of multiple reflected light inside the barrel of the lens. With such coating, complex high-performance lenses become possible.

Optical thin films have also played an important role in military infrared devices by making them wavelength-selective so that military targets can be distinguished from background sunlight. Progress in the design of very-high-precision optical systems, using large-scale computer techniques, has reached its highest point in the development of such remarkable systems as those used in the U-2 high-altitude planes and surveillance satellites. The possibility of such surveillance is fundamental to current discussion of, and hope for, international disarmament activities.

One of the most striking new developments in optics is the use of large-scale digital computer techniques for image enhancement in such fields as clinical x-ray fluoroscopy and the analysis of the Mariner photographs of Mars (see Figure 4.49). Not only can substantial improvement be achieved in apparent signal-to-noise ratios in such photographs; but also, the final images can be corrected for either deficiencies in the original optical system or lack of adequate focusing during the original exposures. The apparent improvements are frequently startling. The development of practical, mass-produced image-intensifier systems has dramatically reduced the radiation exposure required in medical fluoroscopy and has pushed back the observable frontiers of our universe when used with large optical telescopes. Paralleling this conquest of the ultralarge is that of the ultrasmall. Within the past year Crewe and his associates have succeeded in photographing the outlines of the double-helix structure of the DNA molecule through judicious tagging with thorium atoms.

Indeed, the scanning electron microscope, in itself, has given biologists, metallurgists, and all who deal with microscopic phenomena a perspective

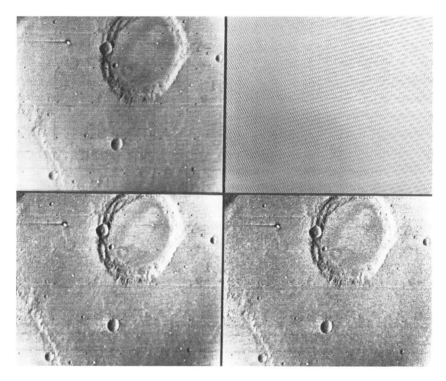

FIGURE 4.49 Example of preliminary computer enhancement of Mariner 6 near-encounter picture 18 of Mars. A portion of this frame, as originally recorded, is shown at the upper left. Apparent in it is a faint basket-weave pattern due to electronic pickup in the sensitive preamplifier of the camera system, as well as a general softness due to the limited resolution of the vidicon image tube. Computer analysis reveals the pickup pattern shown at the upper right. When the appropriate numerical value, determined from this pattern, is subtracted from each of the 658,240 elements of the picture, the result is as shown at the lower left. Two further computer programs may then be used to compensate for the smearing due to the vidicon tube, with the result shown at the lower right. The final processing procedure will involve more refined versions of these steps, as well as programs designed to remove the numerous small blemishes, correct for electronic and optical distortions of the image, and correct for the photometric sensitivity of each picture element of the vidicon tube. The computer will also be used to combine the digital and analog video data into a single, photometrically accurate picture. [Source: *Yearbook of Science and Technology 1971*. Copyright © 1971, by McGraw-Hill Book Company, Inc. Used with permission of McGraw-Hill Book Company.]

and depth of focus that, in effect, offer an entirely new window on their subjects (see Figure 4.50). The contrast between the visual impact and the immediate information content of a transmission electron micrograph

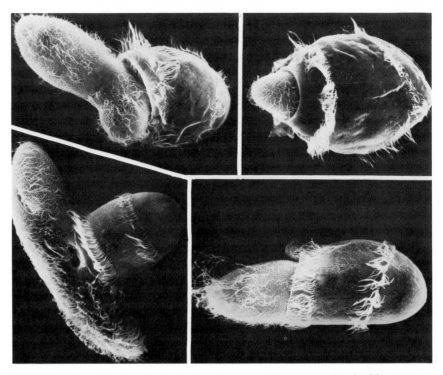

FIGURE 4.50   A single-celled animal captures and devours another in this sequence of pictures taken with the aid of the electron microscope. Exceptional details in the drama indicate the potential of this instrument. [Source: This photograph was provided through the courtesy of G. Antipa and H. Wessenberg, Argonne National Laboratory, and the *Journal of Protozoology*.]

as recently as a few years ago and what is routinely available now from even relatively inexpensive scanning electron microscopes is remarkable. It is worth noting, too, that a significant fraction of the new technology underlying these new instruments—ultraprecise magnetic lenses, superconducting lenses, and the like—has been derived from frontier projects in subfields as diverse as elementary-particle and condensed-matter physics.

Physiological Optics

Physiological optics has always been an important part of optical science, and physicists still have a role in it, although it is secondary to the work of electrical engineers and psychologists. Observations of the eyes of animals and humans enhance the understanding of the scientific basis of pattern recognition. Mechanisms have been found in the eye of the frog that respond to certain moving patterns and not to others. The human eye loses its response if an image is held stationary on the retina, showing that tremor is a necessary condition, not a defect, of vision. New adaptive mechanisms have been found in the eye to assist in the perception of color and contrast. That some parts of the pattern-recognition process are visual instinct in nature and some computed in the brain is now recognized. The human eye remains one of the truly remarkable optical instruments in its sensitivity, range, and efficiency. The processes of human vision retain large areas of mystery. Research in these areas is of direct physiological value and affords possibilities of guiding certain optics technology.

Applications of Optics in Other Sciences

Because observations are at the heart of all experimental science, optics plays a crucial part in research in nearly all disciplines. The range of optical instruments and techniques involved in physics beggars description. They are so widespread and commonplace that their origin in optics is frequently entirely forgotten. However, there are many specific and specialized applications.

The use of optical techniques in astrophysics and astronomy is obvious and requires little comment. New techniques of image enhancement, image digitization, aperture synthesis, and long-baseline interferometry have only very recently revolutionized broad areas of research in these fields.

When light from a laser beam is scattered by high-energy electrons from an accelerator, the scattered light has a higher frequency because of the Doppler effect. If the electrons have very high energies, the scattered photons become nearly monochromatic gamma rays, with energy in the range of several billion electron volts, and retain the polarization of the original light. The polarized gamma-ray beam obtained in this way is nearly free of low-energy background radiation and is ideal for many studies in particle physics.

Optical monitoring of the earth's atmosphere already has become an important tool for weather prediction with now-familiar global cloud cover satellite pictures—pictures that even a few years ago would have been considered little short of miraculous. In addition, and perhaps more important, the properties of the upper atmosphere can be measured and the effects of pollutants on the lower atmosphere can be determined from the satellite and rocketborne spectral observations of various atoms and molecules. Satellite observational techniques, using ultraviolet, visible, and infrared spectroscopy, make it possible to monitor continuously the global distribution and vertical profiles of natural and man-made variations of specific chemicals in the earth's atmosphere and to have rapid knowledge of changes. These techniques provide an immediate means by which a physicist can apply his specialized knowledge to problems of pollution and ecology. They also have provided a new observational basis for earth and planetary science.

Recent developments in frequency synthesis from the microwave to the optical region make it feasible to consider defining length and time with the same molecular transition. These achievements, coupled with the stabilities obtained through coherent laser stabilization by saturated absorption of molecules, probably will lead to the development of new and more-precise standards. Frequency synthesis to a CO_2 laser transition near 10.6 μm and stabilization of this transition by saturated absorption in SF_6 have already been demonstrated. The application of similar techniques to the near infrared is anticipated in the next few months, and following that, their extension to visible frequencies.

The newly developed saturation absorption (Lamb dip) spectroscopy can attain spectral resolution exceeding one part in 10^{10}. Thus an additional powerful tool is available for studying spectral line shapes and level splittings to within natural line widths of atoms and molecules. The new super-Kerr cell systems (see Figure 4.51), developed by Duguay and his collaborators, with effective shutter times of picoseconds, will make possible the detailed following of atomic and molecular interactions in a totally new fashion. These phenomena will be widely exploited in atomic and molecular physics and in chemistry.

There are many other applications of optics. Of particular interest recently are the laser measurements that will show variations in the earth–moon distance during the next few years and should offer a new and highly accurate check of gravitational theory. These data also will improve substantially present knowledge of: (a) lunar size and moment of inertia, (b) the earth's rotation rate and polar rotation, and (c) the present rate of continental drift. Uncertainties of less than 6 in. are involved. Further, satellite-based optical navigational systems hold promise of unprecedented precision, measured in inches anywhere on the earth.

Future Opportunities in Optics

Optical scientists and optical techniques can make major contributions in a variety of fields in addition to those already described. Consider, for example, the wretched state of technology in the Postal Service compared to that in color television or the telephone system. Or consider the fragmented leadership and lack of innovation in crime prevention and detection. A rather trivial optical system, introduced a few years ago to keep

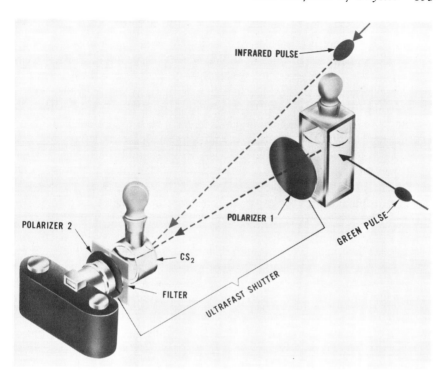

FIGURE 4.51   Setup used to photograph a green laser pulse in flight through a water cell. A drop of milk was added to the water to enhance light scattering and thus render the pulse brightly visible from the side. The ultrafast shutter is essentially an electrodeless Kerr cell. A powerful ultrashort infrared laser pulse is used to induce Kerr birefringence in the liquid, thus opening the shutter for ~10 psec. The filter greatly attenuates the infrared pulse and prevents it from damaging the camera. Both pulses are generated simultaneously by a Nd:glass laser. [Photograph courtesy of M. A. Duguay, Bell Telephone Laboratories.]

track of at least some of the nation's freight cars, excited wide newspaper coverage and attention. A great stimulus to research and development, and thus to the effective use of talented manpower, will arise as federal agencies establish programs and laboratories of their own commensurate with the complexity of their tasks and the importance to society of their missions.

The development of the picturephone is progressing and will require substantial optical support. Optical communication through glass fibers is beginning to be taken seriously, and useful links probably will be installed experimentally within two years. These two efforts could stimulate one another. Further, the picturephone will call for a wide variety of peripheral devices to introduce signals into it and distribute the output.

The importance of the picturephone concept can scarcely be over-emphasized in the context of the congested traffic and travel systems. Picturephone linkages of high quality, and with appropriate privacy controls where required, can reduce substantially the amount of business travel now considered essential. A linkage such as that now in experimental use between the Bell Telephone Laboratories at Murray Hill and at Holmdel permits effective large or small conferences, with participants at both laboratories fully involved both technically and psychologically. This mechanism may offer a partial solution to the national air and surface traffic problems.

Optical memories for computers are still in the developmental stage, and their utility is not yet fully demonstrated; but the need is so great and there are so many possibilities in optical techniques that substantial additional work will undoubtedly take place. In particular, the costs and the relationship between capacity and real-time appear favorable.

Further in the future lies integrated optics; this combines a laser beam trapped in a surface film, which is modulated or deflected by electrical or acoustical signals. In such devices, it is now possible to construct literally thousands of gate and switching elements in optical paths of fractional millimeter length and to operate them simultaneously in parallel with beams of different frequencies from integrally constructed lasers. The possibilities for miniaturization of digital instrumentation through this approach are enormous.

Much emphasis today is placed on optical methods of detecting and measuring atmospheric contaminants. Undoubtedly, there will be new problems and jobs in this field, but it seems unlikely that they will be sufficient to justify more effort than already is being allocated to this branch of physics.

Fiber optics, which for a few years was both challenging and lucrative, is now a mature and highly competitive industry with many groups ready to take advantage of any new applications that are identified.

Pattern recognition seems to hold great promise. It is possible that pattern recognition in the next ten years could change society in fundamental ways at the same time that it changes the understanding of the ways man gains knowledge and interacts with his environment. Much of the drudgery in which man engages in a highly industrialized society consists of optical pattern recognition of some kind, from the checkout counter in the supermarket to the reading of electrocardiograms. Physicists have an opportunity to make a fundamental contribution to this growing field.

It is clear that optics will continue to be of great value in military problems. Aerial reconnaissance and the image-intensifier telescope have proven invaluable to American troops in Vietnam. Reconnaissance is not only a tool for waging war more effectively but a tool for maintaining peace. It is one of the few techniques available for verifying compliance with international agreements such as arms limitations.

Structure of U.S. Optics Activity

The applied science character of optics led to a steady decrease in emphasis on optics in the physics departments of U.S. universities until about five years ago when the influence of the laser began to be felt. Through the same period, however, the Optical Society of America continued to grow, as more and more people found challenging opportunities in optics. There were some 6000 members of this Society in 1970. In 1962, the Society establishd a new journal, *Applied Optics,* which now publishes 50 percent more pages annually than does the *Journal of the Optical Society of America,* the traditional journal of this subfield. Optics thrives today primarily through its applications, though it also has strong roots in fundamental physics.

Who are the optical scientists of today? The 1970 National Register of Scientific and Technical Personnel shows 3280 physicists who indicated optics as the specialty in which they were employed. Of these 3280, one third (1111) held a PhD degree and accounted for 6.7 percent of the total PhD population in physics. The number of PhD's in this subfield shows a substantial increase from the 743 included in the 1968 Register survey, a growth rate of about 25 percent per year at a time when the rate of increase of PhD's for physics as a whole was about 7 percent per year. (See Figure 4.52).

Of the physics participants in the 1970 Register survey, 2494 indicated membership in the Optical Society of America. Other disciplines with substantial representation in the Society are psychology, chemistry, and electrical engineering. In addition, many of the members are technicians and manufacturers of optical products.

Because many practical devices are optics-limited, strong optics groups engaged in applied research and design and development work have been established in many industrial laboratories. Half of the optics PhD's work in industry. This pattern is well illustrated by the development of the ruby laser at Hughes Aircraft, the gas laser at the Bell Telephone Laboratories, and the glass laser at American Optical Company. And this pattern continues. Exceptional work in ultrashort light pulses is taking place at United Aircraft, IBM's Thomas J. Watson Research Center, and Bell Telephone Laboratories. Liquid crystals were developed at Westinghouse, and stabilized-frequency lasers at Perkin-Elmer.

One fifth of the optical scientists are working for the U.S. Government

OPTICS

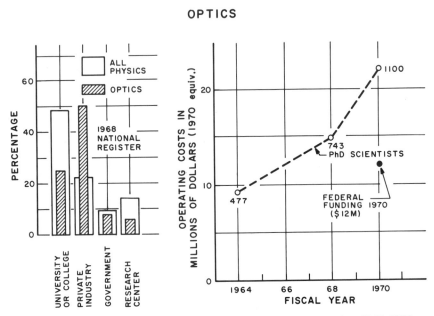

FIGURE 4.52   Manpower, funding, and employment data on optics, 1964–1970.

or in federally funded research laboratories. Since many of the laboratories were established only recently, they probably represent a substantial fraction of the new jobs in optics in the last decade. In addition, the equipment needs and extramural research programs of these laboratories have stimulated industry to employ many optical specialists (70 percent of all optical scientists work in industry).

A remarkable development of optical skill has occurred in government-sponsored nonprofit laboratories, such as the Cornell Aeronautical Laboratory, University of Michigan Institute of Science and Technology at Willow Run, and Lincoln Laboratories at MIT. Most of these groups are concerned principally with the development and operation of large special-purpose optical equipment, but many of them conduct significant fundamental studies of the underlying physics. Some have made major contributions, such as the development of the hologram by Leith and Upatnieks at Michigan.

Finally, a smaller part of modern optics has developed in university laboratories. For example, nonlinear optics was discovered at Michigan and has been studied with notable skill by Bloembergen at Harvard. But, although a very large fraction of contemporary optical scientists were trained in physics, optics has received relatively little emphasis in most physics departments. Most scientists entering this subfield have been trained primarily in some other branch of physics. However, this situation is changing rapidly. One fourth of the optics PhD's now work in academic institutions.

The hidden character of optics support makes it very difficult to establish its magnitude with any degree of certainty. Basic research in optics as a branch of physics has had very little *direct* government support in the past and seems unlikely to obtain it in the future. The estimate of 3 percent of the total federal funds for basic research in physics, in 1970, shown in Figure 4.53 may be in error.

Problems in the Subfield

Optics now draws its support primarily from industry and from government mission-oriented agencies. Both sources have invested substantial sums in basic research in optics. This pattern of support should be continued and widened. There are optical needs in transportation, the Postal Service, crime control, environmental control, printing and publishing, training devices, mapping and surveying, earthquake prediction, and information

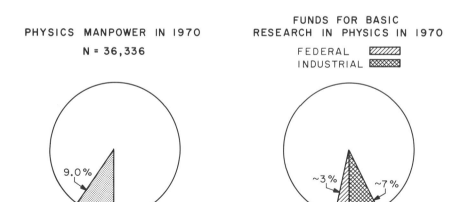

FIGURE 4.53   Manpower and funding in optics in 1970.

storage and retrieval that await exploitation. Many of these needs can be funded only at the national level. Federal agencies should recognize their opportunities and responsibilities for the support of research in optics.

> Hearing may be devided into Direct, Refracted
> and Reflex'd, which are yet nameless unless we
> call them Acousticks, Diacousticks, Catacousticks,
> or (in another sense, but with as good purpose),
> Phonicks, Diaphonicks, Cataphonicks.
>
> RT. REV. NARCISSUS, Lord Bishop of Ferns
> and Leighton
> *Philosophical Transactions of the Royal
> Society* (*London*), *14,* 473 (1683)

ACOUSTICS

Introduction

To most, the term, acoustics covers the process of hearing as well as those devices, methods, and materials that assist the hearing process, for example, hearing aids, microphones, loudspeakers, acoustical materials, and room acoustics, but no more. The average physicist might add the study of speech, high-frequency sound (ultrasound), and noise. However, a perusal of the pages of the *Journal of the Acoustical Society of America* will convince the reader that all elastic and inelastic waves in matter come within the realm of modern acoustics. Therefore, music, architectural acoustics, shock waves and sonic booms, vibrational phenomena, high-intensity sound, and seismic waves are all part of acoustics. The problem in this brief review is to group these manifold phenomena into a brief presentation that indicates how such a wealth of subject matter has developed.

Acoustics can be defined most succinctly as the study and use of mechanical waves in aggregate matter. The waves need not be elastic; they can be inelastic, with attenuation and dispersion, and can have large amplitudes, with consequent nonlinear effects. Acoustics deals in large part with techniques and devices for the generation, transmission, and detection of these waves. The study of the transmission of acoustic waves in matter is one of three main techniques (acoustic, electromagnetic, and the interaction of matter with particles) used to investigate the properties of matter.

Because acoustics is so old a subject, early having been associated with a basic sensory and perceptual mechanism, it sometimes is viewed only as a part of classical physics and as a subject fully explored in the nineteenth century, which is not true. Lord Rayleigh's famous *Theory of Sound,* first published in 1877, did not represent the end of acoustics. On the contrary, the theory and Rayleigh's subsequent research contained the elements of a new beginning in which new concepts and techniques were generated. These new approaches and methods not only caused but resulted from new applications in science and technology. Consequently, acoustics has today both a classical and a quantum character.

As with optics, and more generally with electromagnetic radiation, acoustics has a wide spectrum ranging from approximately 10^{-4} to 10^{14} Hz, as indicated in Table 4.4. Generally, the techniques differ greatly in the different ranges, although the physical principles are similar.

Acoustic waves often penetrate media (for example, oceans or solids) that electromagnetic waves do not enter. For example, all our knowledge about the interior of the earth (liquid core, temperatures, and pressures) comes from acoustical studies together with equations of state, which are determined largely with acoustical measurements. This example is but one of many in which acoustics provides the only means available for direct study of the properties of matter.

Defining the scope of acoustics is difficult, for the subject matter of this subfield cuts across the definitions of the subfields of physics selected by the Physics Survey Committee. In fact, acoustics has extensive ramifications not only in the various subfields of physics but in many other scientific disciplines and in technology. Consequently, people using the concepts and techniques of acoustics can be found in many physics subfields, related sciences, and engineering disciplines. However, in spite of the variety of topics dealt with in acoustics and the strong association with other sciences, engineering disciplines, and technologies, acoustics remains fundamentally a part of physics. It includes all material media. It requires the mathematics of theoretical physics. Its methods play a primary role in exploring the characteristics of the various states of matter. In addition, it provides

TABLE 4.4   Frequency Ranges in Acoustics

Frequency (Hz)	Name of Frequency Range
10^{-4}–20	Infrasonic
$20–2 \times 10^4$	Audio
$2 \times 10^4–5 \times 10^8$	Ultrasonic
$5 \times 10^8–10^{12}$	Hypersonic or praetersonic
$10^{12}–10^{14}$	Thermal vibrations

many techniques, devices, and inventions for other subfields. And, in education, there is a unity about the subject of acoustics and its manifold applications that tends to be lost if it is taught in a fragmented way in a number of different courses.

The constantly changing nature of acoustics as new discoveries occur also contributes to the problem of developing a practical definition of this subfield. Subjects within acoustics grow and gradually separate from it as understanding progresses through the development of the geometry of transmission to the response of materials to acoustic waves. Thus, acoustics stimulates the development of new areas of investigation that later become either independent or subdivisions of other physics subfields that deal with the properties of matter. However, the techniques of generating and receiving waves, as well as the geometrical aspects (as in geometrical optics) of transmission, remain a fundamental part of acoustics.

The currently changing emphases in physics also have an impact on the definition of acoustics. Emphasis on more applied work relevant to critical national problems necessarily means increased attention to acoustical studies. Examples of disciplines that are undergoing major development with acoustical methods are biophysics and geophysics. The large programs planned for oceanography and ocean engineering will be limited by the available understanding of acoustics.

In addition, the shift in emphasis of federal programs from defense to environmental and social problems (pollution, public transportation, housing, health and safety, and communications) will have an impact on the relative place of acoustics in the hierarchy of physics subfields.

In public transportation, there are critical problems of noise, shock waves, materials, communications, and control. Physics, and acoustics in particular, has much to contribute to their solution.

Recent developments in acoustical holography promise to make striking contributions to problems of health. One example is shown in Figure 4.54. Further, safety devices necessarily involve electromagnetic and acoustical signals. Electromagnetic–acoustic devices now exist that will locate astronauts to within 1 ft of a spacecraft at all times. The potential applications of such devices have not yet been fully explored.

The recent developments of ultrasonic surface-wave physics will probably exert a strong impact on the communications and computer industries, which by now pervade all aspects of modern life with examples familiar to everyone.

In this brief summary, acoustics is discussed under the headings of instrumentation and the relation of acoustics to the medium, to man, and to society.

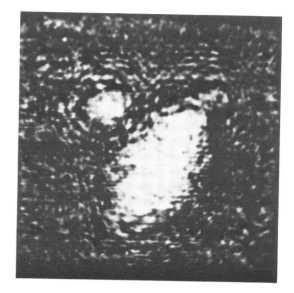

FIGURE 4.54 Reconstruction with visible light of acoustic hologram of a living human eye with a retinal detachment. (P. Greguss, Jr., Budapest Medical University.) [Source: *Yearbook of Science and Technology 1970*. Copyright © 1970, by McGraw-Hill Book Company, Inc. Used with permission of McGraw-Hill Book Company.]

Instrumentation

Seventy-five years ago the sound sources available to man were limited largely to the human voice, musical instruments, and a few whistles and other mechanical devices, all producing sound energy almost exclusively in the range audible to the human ear—20 to 20,000 Hz. The discovery of the piezoelectric effect by Jacques and Pierre Curie in 1880, the effect in which mechanical stresses on many crystalline solids can produce electrical voltages and *vice versa,* subsequently was used widely to produce ultrasonic waves (ultrasound) whose frequencies can be as much as a million times greater than the highest frequency audible to the human ear. At the other extreme of frequency, the use of geophysical receivers (geophones) and of low-frequency (infrasound) detectors has led to an entirely new range of measurements. With a modern infrasound detector, for example, it is possible for a detector in New York City to hear an Apollo launch at Cape Kennedy. The range of acoustic measurements now extends from below 1 cycle per hour to well over 5×10^9 Hz.

The individual techniques for the production of ultrasound are myriad. The discovery, during World War II, that certain ceramic materials, such as barium titanate, could be electrically polarized and used in a fashion similar to piezoelectric crystals made available electromechanical transducers that could be molded into virtually any shape or size. Another technique involves the generation of ultrasound by means of semiconducting films, such as those of cadmium sulfide, painted onto the test material. Most recently, the direct production of ultrasound through the impinging of electromagnetic radiation on a metal surface has become a practical method for the generation of high-frequency sound in solids.

Acoustics and the Medium

Although the mathematical analyses of the processes by which sound is absorbed in its passage through air and other media were well developed in the nineteenth century, accurate experimental measurements were lacking even 50 years ago. The reason is not difficult to find. In the range of audible frequencies, the absorption is so small that enormous path lengths would be needed for measurement—distances that were simply too great to be experimentally feasible. Furthermore, the wavelengths of such sounds are so long that diffraction effects enormously complicate experimental measurements. It was not until substantially collimated beams of ultrasound became available that reliable measurements on acoustic propagation were obtained—between 1920 and 1940 for gases, 1930 and 1950 for liquids, and more recently for solids.

These measurements produced a major surprise. In air, the absorption coefficient, the measure of energy loss in the medium, was considerably greater than that predicted through consideration of the viscosity and thermodynamic conditions developed by Stokes and Kirchhoff in the mid-nineteenth century. The corresponding measurements in liquids yielded values for the absorption coefficients that were often 1000 or even 10,000 times the predicted ones.

These findings quite naturally led to vigorous activity, both theoretical and experimental. The result was the identification of relaxation processes, involving the excitation of internal rotational and vibrational states of the molecules involved, as the cause of this extra absorption in gases. In liquids, the relaxations frequently could be associated with structural re-arrangements inside the molecule or aggregates of the molecules involved. These studies resulted in a steady accumulation of acoustical information about liquids and the successful use of this information as a probe to study the structure of the substances involved. (See Figure 4.55)

Because sound absorption generally decreases as the medium becomes more condensed, absorption measurements in solids at low frequency are even more difficult than in liquids; therefore, it is not surprising that absorption measurements in solids are the most recent. Measurement of sound velocities and absorption in different directions in single crystals and

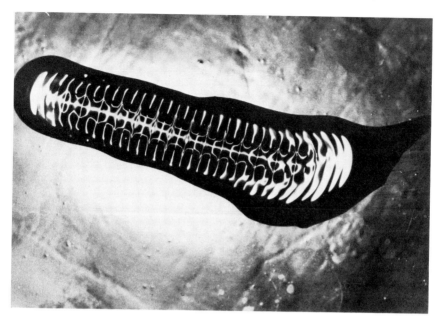

FIGURE 4.55 Visible effect of sound on a layer of glycerin at 30 Hz and weak amplitude. (Photo by J. C. Stuten, in H. Jenny, *Cymatics*, Basilius Presse, 1967.) [Source: *Yearbook of Science and Technology 1971*. Copyright © 1971, by McGraw-Hill Book Company, Inc. Used with permission of McGraw-Hill Book Company.]

in crystals under various external stimuli, such as magnetic fields, varying pressures, or varying temperatures, have contributed significantly to the understanding of crystalline imperfections, the behavior of superconductors, and the nature of Fermi surface metals.

A number of areas in physical acoustics merit special mention because of their recent rapid growth, for example, surface waves, acoustic holography, high-accuracy velocity-change measurements, nonlinear acoustics, acoustic emission and heat pulses, and infrasonics.

As a group, these research areas typify the steady evolution of investigations in physical acoustics from basic physical research studies to applications in other sciences and in technology. Two examples are given in Figures 4.56 and 4.57.

Surface waves, known from the days of Lord Rayleigh, only recently have been exploited for their applications to electronic systems. They have marked advantages over bulk waves in information storage and in signal filtering. They require only a single surface and are better suited to piezo-

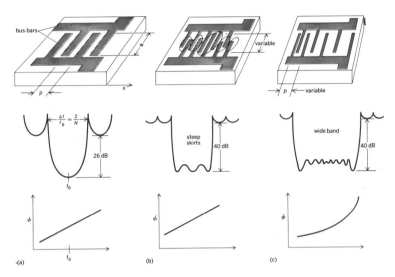

FIGURE 4.56 Transducer configurations and corresponding insertion loss and phase characteristics of electric filters: (a) interdigital construction; (b) amplitude weighting technique; (c) graded periodicity technique. [Source: *Yearbook of Science and Technology 1971*. Copyright © 1971, by McGraw-Hill Book Company, Inc. Used with permission of McGraw-Hill Book Company.]

electric amplification because of their low surface-to-volume ratio. As filters, they allow the designer to prescribe the phase characteristic independent of the amplitude characteristic.

Acoustic holography, through development of a water–air interface method of acoustical imaging, allows the transfer of the spatial modulation of a sound field onto a light field, which can then be used to produce optical images. At the water–air interface, the surface of the water is deformed by the incident sound pressures. A light beam is then reflected from this deformed water surface to obtain the required spatial modulation. Acoustic holography offers new possibilities for more accurate imaging of objects within the human body and for nondestructive-testing applications generally.

The use of the ultrasonic imaging technique in nondestructive testing is increasing rapidly. This technique is useful, generally, for detecting the same types of flaw as those recommended for conventional ultrasonics tests. In the medical field it may find use in studying the movement of fluids in the body. An ultrasonic image of a fingertip is shown in Figure 4.58.

High-accuracy measurement of velocity change has proved useful in the study of solids at low temperatures. With recently achieved increased

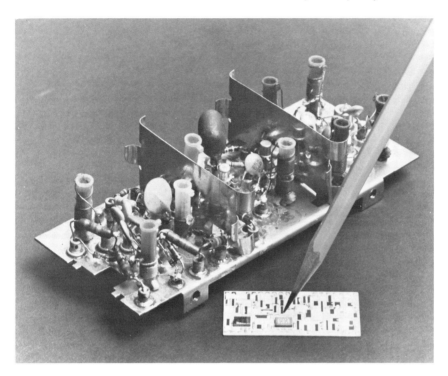

FIGURE 4.57 A Zenith-engineered thick-film circuit (foreground) is designed to replace the present-day large television intermediate-frequency amplifier in background. Its tiny ultrasonic filters (at pencil tip and left) select certain frequencies and reject others. [Photograph courtesy of Zenith Radio Corporation.]

sensitivity, these measurements are now used in the determination of third-order elastic constants, especially of soft materials. These constants have application in many aspects of the solid state, for example, establishing of equations of state, providing connections between measured mechanical properties and thermal effects (through measurement of the Grüneisen constant), and developing an acoustic thermometer.

The work on third-order elastic constants in solids was preceded by a burst of corresponding nonlinear acoustical research in liquids. This led to the determination of higher-order coefficients in the adiabatic equation of state for liquids, and also to the realization and study of the sound-with-sound interaction and the practical application of high-resolution parametric sonar.

The discovery of emission from a single dislocation wall in a solid has

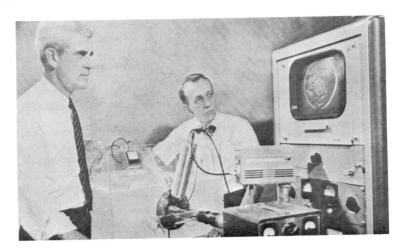

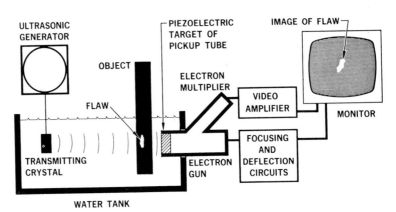

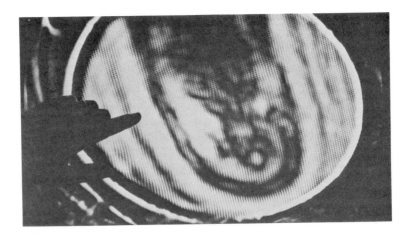

◄ FIGURE 4.58 *Top:* Photograph and diagram of an ultrasonic inspection system that yields a televised picture of the ultrasonic radiation reaching the pickup tube detector. Water is used to transmit the ultrasonic vibrations in a flashlight-type beam from the transmitting crystal (left) to the object and then to the detector. This detector is a camera in a closed-circuit television system. (Courtesy James Electronics, Inc.)

Bottom: Television picture of a human finger obtained with the ultrasonic system shown above. The image of the finger is the oval in the center of the round white area, which is the image of the sensitive part of the television camera tube. The tip of the bone in the finger is easily seen. The hand at left points to the television image.

[Source: Atomic Energy Commission, Division of Technical Information.]

suggested the possible identification of characteristic signatures of various solid defects, and thus the creation of yet another new nondestructive-testing tool. The use of heat pulses as sources of acoustic waves in the 10^{11}- to 10^{12}-Hz range has also become possible recently, leading to new investigations in the physics of condensed matter.

Infrasound studies deal primarily with the pressure fluctuations in the atmosphere, with periods from 1 sec to several hours. Causes of such pressure fluctuations are mountain-induced vortex shedding by the jet stream in the upper atmosphere, severe storm systems, aerodynamic turbulence, and volcanic eruptions. Infrasound has defense applications in the detection and identification of major rocket launches. It also offers the possibility for major advances in the knowledge of the large-scale behavior of the atmosphere, with applications to clear-air turbulence detection, the tracking of severe weather systems, and the capability to provide advance warnings of destructive open-sea earthquake waves or tsunamis.

When research in plasma physics bloomed, it was early noted that the ionized medium was sensitive to the passage of sound waves. In addition to the normal, longitudinal sound wave, an additional transverse wave appears, which is known as the Alfvén wave, after its discoverer, the 1970 Nobel laureate in physics.

Study of the propagation of sound in liquid helium also has been productive. In the 1940's, Landau predicted the existence of a periodic temperature wave in liquid helium below the λ point. This wave, known as second sound, was later detected experimentally by his colleague, Peshkov. The study of first and second sound in liquid helium resulted in the identification of two other forms of sound. Third sound is the longitudinal oscillatory motion of the superfluid component of a thin superfluid helium film. The motion is parallel to the surface of the film and involves only the superfluid component. Fourth sound, a compressional wave of the superfluid component, moves through the pores of a finely dispersed solid when the pores have been filled with liquid helium. The study of all these phenomena provides an important method for probing the physical character and quantum behavior of liquid helium.

In geophysics and ocean science, the respective media, the earth and the ocean, provide gigantic laboratories in which ultrasonic, sonic, and infrasonic waves are used routinely to provide data that frequently are inaccessible by other means.

Study of the ocean requires a detailed knowledge of ocean depths and the nature of the ocean floor. Information on both can be obtained from a study of sound transmission and scattering. The problems of underwater communication and sonar development have led to sophisticated use of computer, correlation, and filter techniques to detect minute signals in a world of noise.

In recent years, the observed periods and amplitudes of the fundamental modes of vibration of the earth have enabled scientists to derive a more accurate model of the earth's interior. The implanting of a geophone on the surface of the moon has produced similar data regarding the moon's interior. By the use of suitable systems of recording instruments, relative changes in the earth's internal displacements can also be monitored, so that we now have a realistic possibility of achieving earthquake predictions—a feat that could save untold thousands of human lives, quite apart from the material advantages to be gained from such predictions.

Acoustics and Man

Speech It has long been a goal of speech scientists to produce synthetic speech. The development of sound recording and reproduction techniques already has been combined with the computer storage and retrieval of information to such an extent that a breakthrough appears imminent. A major problem has been the modifying effect that adjacent sounds (phonemes) and near-adjacent ones have on any given speech sound. The basic question is: How does one program a computer so that it will select, modify, and present the sequence of sound that carries the desired meaning?

Hearing The process of hearing involves both psychological and physiological acoustics. Two recent, fundamental discoveries have been made in psychological acoustics. The first is that man apparently possesses an internal psychological scale that enables him to express the experienced auditory sensation magnitudes quantitatively, with reasonable consistency, even when intersensory comparisons such as the loudness of a sound with the brightness of a light are involved. The second discovery is that the problem of auditory signal recognition reduces to a discrimination between noise-plus-signal and noise-alone events. These discoveries have led to increased social and industrial applications and to increased rigor of psycho-acoustic experimentation. For example, the instrumentation and methodology developed for acoustic impedance measurements at the eardrum have recently been accepted for diagnosis of the highly prevalent middle-ear disorders. An example of such instrumentation is given in Figure 4.59.

An exact knowledge of the working of the human auditory system still

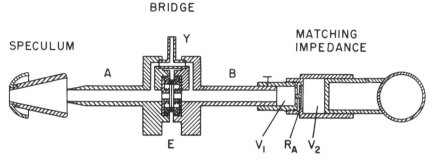

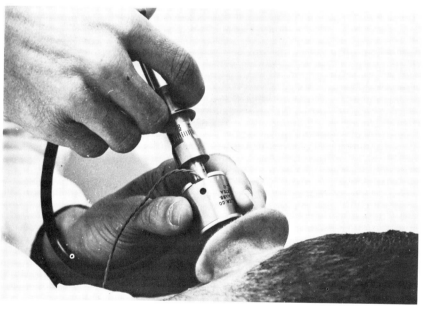

FIGURE 4.59 Acoustic bridge. *Top:* For measurement of acoustic impedance at the eardrum. Air volume V_1 is adjusted to match the residual air volume in the ear canal. R_A and V_2 are variable impedance elements. [Source: J. Speech Hearing Res., 6, 304–314 (1963).]

Bottom: Acoustic bridge held in the ear by hand. [Source: American Speech and Hearing Association, ASHA Monograph 15, Washington, D.C. (1970).]

[Photographs courtesy of J. Zwislocki, Syracuse University.]

is lacking. Extensive research is required. Among the goals of this research are the determination of the way in which sound parameters are encoded in the nervous system and the explanation of auditory characteristics—that is, the nature of the sensory brain function.

Acoustic physicists have rapidly adapted techniques from other subfields of physics to their specific purposes. As an example, the nuclear Mössbauer effect has recently been used to measure precisely the normal motions of the human eardrum and the components of the auditory mechanism through appropriate deposition of a radioactive emitter on the drum or bones. These measurements of velocities of the order of 0.1 mm/sec, or equivalently, one mile per year, confirm the remarkable fact that the threshold of human hearing corresponds to a drum motion of the order of the diameter of a hydrogen atom. Still not understood is the mechanism by which man can distinguish, with some reliability, tones differing by only a few cycles in tens of thousands!

In these areas of speech and hearing, the application of ultraminiaturized electronic components (implanted and otherwise), which are just now coming into use as a consequence of continued research in both basic and applied physics of condensed matter, undoubtedly will have revolutionary consequences in the alleviation of defects that afflict a significant fraction of mankind.

Medicine   In addition to providing hearing and speech aids, tests, and the like, acoustics has a special role to play in medicine. Most of the possibilities for medical applications so far are underdeveloped or undeveloped. The use of ultrasound in therapy is an example. Many ultrasonic generators are in active use, but accurate knowledge of how ultrasonic therapy helps is still lacking. Evidence exists of acceleration of wound healing by low-intensity ultrasound and of enhanced cancer radiation therapy effectiveness through simultaneous administration of ultrasound and of x rays, but general procedures have not yet been developed. Whether it is the local heating or the modification of flow of body fluids that is the effective characteristic is not yet known.

Ultrasound has been tested widely as a probe for the visualization of internal body organs (see Figure 4.60). In this, it is similar to the x ray, although lacking many of the latter's side effects, but its use has been far more limited. Its greatest success has been in providing information on acoustic impedance discontinuities in soft tissues and in situations such as examination of the brain, abdomen, and heart, where x radiation provides little or no discrimination. The use of acoustic holography in such nondestructive testing of the human body is still in its infancy but holds promise of success.

Other techniques that have reached the clinical stage are the application of Doppler shift and returning echoes to the measurement of blood-flow

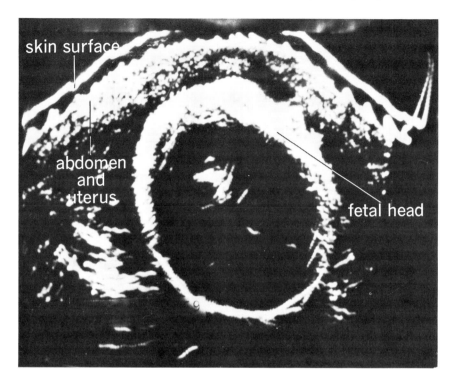

skin surface

abdomen
and
uterus

fetal head

FIGURE 4.60   A typical ultrasonic scan of the pregnant abdomen at level of fetal
head. Echo pattern at left may be an extremity. (University of Colorado Medical
Center.) [Source: *Yearbook of Science and Technology 1970.* Copyright © 1970,
by McGraw-Hill Book Company, Inc. Used with permission of McGraw-Hill Book
Company.]

rates, focused ultrasonic beams in neurosurgery to cause highly localized
damage deep in the brain without injuring surrounding tissues, and ultrasonic radiation of the ear to treat disorders of the balancing mechanisms.

Perhaps the medical ultrasonic application most familiar to the citizen
is in modern dentistry. Use of ultrasonic drills, with their ability to cut
hard dental structure without simultaneous damage to soft support tissue
and without generation of excessive heat, has done much to reduce the traditional, and often irrational, horrors of the dental chair.

Acoustics and Society

Music   The impact of acoustics on society through audio recording and
reproduction requires little comment. The production and reproduction of
music, speech, and other sounds by means of tape and disk recorders,

pocket radios, and electronically amplified musical instruments are commonplace.

The advent of electronic and computer music, however, has opened new vistas for both composers and performers. After an initial period of bizarre experimentation, the musical content of which may well have been debatable, a new stage has been reached. The music produced by a Moog synthesizer, for example, has attracted serious interest. This instrument, typical of others as well, is capable of producing virtually any desired musical sounds, some of which resemble those of existing musical instruments and some of which are entirely new. Soon a new version of musical form and sound can be expected to evolve, and recreational composing could very well occupy the leisure time of many individuals.

The on-line computer also will play its part by allowing traditional composers to perfect their compositions with an entire orchestra effectively at their fingertips through the medium of computer storage.

Perhaps the greatest progress will be made by those trained from youth in both the musical arts and in physics, so that the novel ideas of both disciplines can be combined to produce results inconceivable to the traditional composer. Early stages of this form of education already are evident.

Architectural Acoustics   Serious musical performances depend on the nature of the room in which they are given, so that architectural acoustics must receive attention. Composers have always had some specific hall-reverberation characteristics in mind, whether consciously or not, for each of their works. Some modern composers now see the exciting possibility of the expansion of artificial reverberation to permit reverberation times that change for different parts of a composition and could be different for different musical frequencies. Here again an entirely new musical experience becomes possible.

Recent research has made it clear that reverberation time is only one of the factors contributing to acoustical quality in concert halls. Of greater importance, probably, is the detailed signature of the hall reverberation—the response during the first 200 msec after the direct sound from the orchestra is heard.

There are many other subjective attributes to musical acoustical quality besides the liveness (reverberation time), including richness of bass, loudness, brilliance, tonal blend, and related binaural spatial effects. Although the computer simulation of real halls could lead to the separation of a number of variables, it is probable that model experiments, as well as tests on full-scale halls, will be necessary to improve the basic understanding of the relative importance of the many factors involved.

Only recently have significant efforts been made to control the acoustical environment of dwellings. Building codes on all government levels will be required before major achievements can be realized. Essentially all the fundamental acoustical knowledge exists to control the transmission of sound from one apartment to another; however, new techniques of application will only develop as standards for acoustical privacy become an accepted part of building specifications. With growing encroachments on the privacy of the individual as the population burgeons, such standards are already underdeveloped.

Noise   Perhaps the principal impact of acoustics on society in the years to come will be in relation to the problem of noise pollution. In addition to the specific damage to the hearing process that can occur from excessively loud or prolonged sounds, noise that is well below the physical damage level is characteristic of contemporary urban civilization and interferes increasingly with speech, relaxation, and sleep. Control of noise will not be achieved, however, until society demands it and is prepared to pay the price in both money and possible inconvenience.

The most intense sources of noise that an average person encounters are those involving aerodynamic phenomena. The exhaust streams from jet or rocket engines, the rotating blades of aircraft propellers, and, of course, the sonic boom are all examples. It has been traditional in acoustics to remark on the small amount of power dissipated in the sounds produced by the loudest voices or musical instruments. (The analogy customarily used is that all the shouting of the fans in a large football stadium would serve only to heat one cup of tea!) Therefore, it may come as a surprise to many to learn that the power radiated as *noise* by the engines of a large commercial jet airliner would be sufficient to operate the average automobile, and that the noise power radiated by the Saturn V rocket is of the order of that required to propel a large aircraft carrier. Noise remains a big problem. It is today one of the major roadblocks to the continued orderly development of air transportation systems. Noise-control considerations and noise-reduction technology are vital in the future development of aviation (and, indeed, of mankind). Scientists from many different disciplines have made and will continue to make important contributions toward both the understanding and the solution of aerospace noise problems.

Distribution of Activity

Because of the interdisciplinary character of acoustics, research is found not only in physics departments but increasingly in engineering, geophysics, oceanography, and life science as well as music departments. In regard to biology, most acoustics-related research is performed at universities and medical schools. In addition to university activities, government laboratories are very active in acoustics research and development. The most easily identified facilities are those operated by the U.S. Navy. These laboratories are supplemented by other national security installations, the National Bureau of Standards, and laboratories of the Environmental Science Services Administration. Industry supports wide-ranging research in this subfield.

Physicists who designated acoustics as their subfield of employment in the 1970 National Register of Scientific and Technical Personnel represented 3 percent of the 33,336 participants in this survey. The PhD physics population in acoustics represented 2 percent of the total number of PhD physicists. These data indicate that acoustics is among the least populous physics subfields and that a gradual decrease in the relative number of physicists has occurred in recent years. For example, in 1964, acoustics accounted for 6 percent of the total physics registrants.

Physicists identified with acoustics are found principally in industry; 44 percent reported employment in industry in 1970. More than one fourth (29 percent) worked in government laboratories, and 17 percent in colleges or universities (Figure 4.61). Academic employment of PhD's

ACOUSTICS

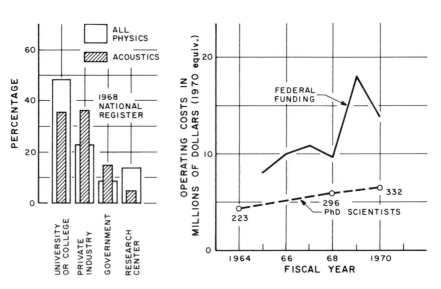

FIGURE 4.61   Manpower, funding, and employment in acoustics, 1964–1970.

was nearly four times greater than was the case for non-PhD's, with roughly equivalent percentages of doctorates indicating academic and industrial employment. Substantially higher percentages of PhD's in acoustics indicated industrial and government employment than was true of all physicists. Basic research and teaching received more emphasis among the PhD's in acoustics than the non-PhD's, but involvement in these activities was far less in acoustics than in all physics subfields taken together.

Data on the support of work in acoustics are difficult to obtain. The DOD is the major source of support for basic research in acoustics. Some relatively small-scale support comes from NASA and the NSF. The heavy concentration of acoustics personnel in industry suggests that this subfield derives much of its support from the private sector. A very rough estimate is that 3 percent of the federal funds and 1 percent of industrial funds allocated to basic research in physics in 1970 were applied to basic research in acoustics (Figure 4.62).

Problems in the Subfield

For many years, an outstanding problem facing acoustics as a subfield of physics has been the preservation of its identity. Some subject matter has been taken into other disciplines and subfields because acoustics provided a successful tool (for example, ultrasonic measurements in condensed-matter physics), some has always spread across other classifications (noise *vis à vis* turbulence), and some has been so largely outside physics that it

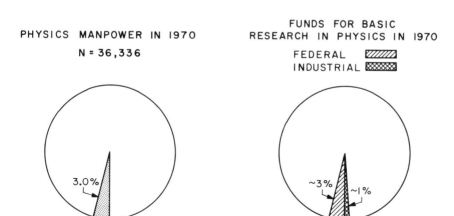

FIGURE 4.62   Manpower and funding in acoustics in 1970.

has virtually been forgotten (speech, hearing, and bioacoustics). It matters little what label attaches to the subfield, but it is important that it be maintained.

A corollary of the problem above has been instruction in acoustics and fields related to acoustics—mechanics, fluid dynamics, classical wave theory. Whether or not it is called acoustics, the study of the whole area in physics departments should be encouraged.

Finally, physical acoustics has had a long tradition of heavy support by the Office of Naval Research. The amount of this support to universities is now in the process of severe curtailment, and means to induce other agencies to support a larger fraction of basic acoustics research that is relevant to their needs must be found.

> In the next place we have to consider that there are diverse kinds of flames.
>
> PLATO (c. 428–c. 348 B.C.)
> *Timaeus* 58

> The wind goeth toward the south and turneth about into the north; it whirleth about continually and the wind returneth again according to his circuits.
>
> *Ecclesiastes* I:6

PLASMA AND FLUID PHYSICS

Introduction

The physics of fluids deals with the study of liquids and gases, and plasma physics, with the study of ionized gases. Human beings spend their lives in intimate contact with air and water, both inside and outside their bodies. The science that studies the forces and motions of liquids and gases is called fluid dynamics (the physics of fluids or fluid mechanics). Fluid dynamics covers an enormous range of topics from the most basic to the most applied. It is one of the oldest sciences in its hydraulics aspects, yet one of the newest in fluidics, the basis for a new family of computational devices. Turbulent fluids pose one of the outstanding challenges in theoretical physics; and success in the understanding and control of turbulence will have far-reaching consequences not only in all aspects of fluid dynamics but also in other sciences and technology.

When a liquid is heated, it becomes a gas. When a gas is further heated, the negative electrons are thermally stripped from the positive ions, giving

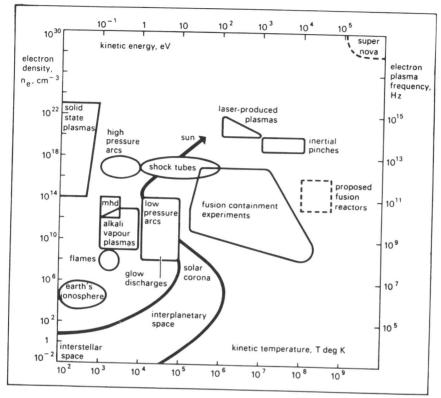

FIGURE 4.63   The plasma state—density and temperature (by I. G. Spalding).

rise to an ionized gas or plasma. This so-called fourth state of matter *
exists with a very large range of physical parameters as illustrated in Figure 4.63. Particle densities range from 1–100 cm^{-3} in interstellar gas,
through 10^8–10^{20} cm^{-3} in laboratory plasmas, to 10^{22}–10^{25} cm^{-3} in stellar
interiors and nuclear explosions. Plasma temperatures range from fractions
of an electron volt in low-current discharges to 10^5 eV in fusion plasmas
and reach relativistic energies in cosmological plasmas, such as in the Crab
nebula.

In its simplest form, plasma consists of fully ionized hydrogen, and at
high temperatures and low densities collisional effects become negligible.
This is the realm of astrophysical and thermonuclear plasmas, which has
been established as an experimental science only in the last 15 years or so
and which is often called the physics of fully ionized gases. In lower-
temperature plasmas, the interactions among many different particle species,

* The designation, fourth state of matter, was used by Crookes in 1879; Langmuir introduced the term plasma in 1928.

positive and negative ions, molecular ions, and neutral atoms become important. In many laboratory plasmas, the interaction of these species with surfaces of condensed matter must also be taken into account. Plasma physics also incorporates many of the disciplines of magnetohydrodynamics, since in many cases plasma behaves as a conducting compressible fluid dominated by interactions with magnetic fields. In addition, studies of hypersonics—shocks and detonation waves at high Mach numbers—and high-temperature chemical flames are a part of plasma physics.

Almost all matter in the universe exists in the ionized state except for the relatively small but important fraction in planets and gas clouds. Man spends his life immersed in fluids but is surrounded on the cosmic scale by plasma.

Since so much of man's environment comprises either fluids or plasmas, it is small wonder that so much intellectual effort has been spent attempting to understand them. The most obvious and dramatic part of the environment, the weather, is the subtlest and most complicated problem in the physics of fluids. There is only one other comparable problem in fluids—comparable because there is not the slightest indication of its origin—and that is how the first giant aggregates of matter, clusters of galaxies, were formed during the early stages of an expanding universe. Between these two applications, one of great social concern and impact and the other a spectacular edifice of human reason, lies the science of fluids. An activity in plasma physics that holds high promise of great success and concomitant social impact is fusion research.

Plasma Physics and Fusion Research

Plasmas do not impinge on man's immediate senses in any fashion so direct as the weather. They exhibit a far more complex and intricate behavior than do classical fluids. This very complexity allows modes of behavior that are far removed from those normal to terrestrial matter. In particular, the extreme of ultrahigh temperature makes feasible the release of controlled fusion power on earth. This particular aspect of the behavior of plasmas—the quest for fusion power—has by and large dominated plasma physics for the last 20 years. The knowledge that has flowed from this quest has led to scientific understanding of such diverse topics in plasma physics as the behavior of the magnetosphere around the earth, of magnetic fields in the galaxy, and of the plasmas surrounding stars; the propagation of radio waves in galactic and intergalactic space; and the embryonic considerations concerning quasars and pulsars. This knowledge offers a dramatic example of how man, in the pursuit of a very practical goal, must face the overriding requirement of developing fundamental scientific knowledge.

The controlled fusion problem was not recognized initially as one requiring a great extension of the existing knowledge of plasma physics. Many technical approaches were pursued in parallel; however, grave difficulties of a plasma-physical nature seemed to frustrate them all. It gradually became clear that much more basic research would be required before any practical, large-scale device or fusion reactor would be possible. Theoretical effort directed toward the fundamental understanding of plasmas received high priority. This understanding was reinforced by many experiments in the fusion program directed more toward the development of plasma physics than toward the immediate objective of fusion power. The results were, first, steady progress toward understanding plasma physics in detail and, second, steadily accelerating progress toward controlled fusion power. A variety of experimental approaches are being pursued in the United States and abroad (see Figures 4.64 and 4.65), and technology is advancing on several fronts (see, for example, Figure 4.66). Progress has now reached the point at which realistic reactor designs and development time scales are being discussed.

At the inception of the controlled fusion project, scientists knew that they had to achieve a millionfold increase in both temperature and a somewhat

FIGURE 4.64 Controlled fusion research device of the Tokamak type (in toroidal geometry) located at Princeton University Plasma Physics Laboratory. The principal goal of this program is to measure particle and energy confinement times under many conditions and to predict how these relationships may change as the size of the torus is increased. [Courtesy Plasma Physics Laboratory, Princeton University.]

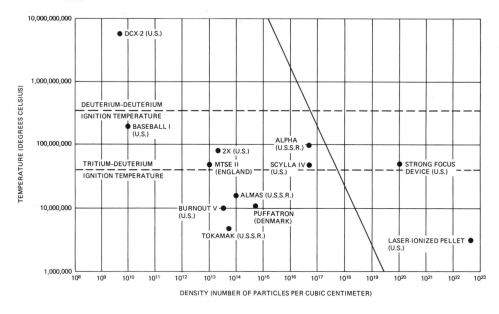

FIGURE 4.65 Plasma experiments that have achieved temperatures near or above the fusion ignition temperatures of a deuterium–tritium fuel (bottom horizontal line) and a deuterium–deuterium fuel (top horizontal line) are identified by the name of the experimental device and the country in which the experiment took place. The diagonal line represents the limit beyond which the materials used to construct the magnet coils can no longer withstand the magnetic-field pressure required to confine the plasma (assumed to be 300,000 G in this case). Beyond this limit, only fast-pulsed systems (in which the magnetic fields are generated by intense currents inside the plasma itself) or systems operating on entirely different principles (such as laser-produced, inertially confined plasmas) are possible. The record of 6×10^{9}°C was achieved with the aid of a high-energy ion-injection system associated with the DCX-2 device at the Oak Ridge National Laboratory. [Source: W. C. Gough and B. J. Eastland, "The Prospects of Fusion Power," Scientific American, *224*, 55 (Feb. 1971).]

abstract quantity called confinement. Confinement implies maintenance of the reacting nuclei in a restricted volume to permit the reaction to continue. At the required temperatures, no material container was conceivable, and so development of the magnetic bottle began. (A more prosaic example of confinement is the preservation of thermal energy in a thermos bottle with a tight cork.) To achieve fusion power, these two quantities, temperature and confinement, had to be increased simultaneously. The goal of past work was to obtain a 10^{12} increase in the conditions dealt with in ordinary furnaces. Remarkable progress has been achieved. At present, this factor has been reduced to about 10^{3}, and it is still decreasing rapidly.

Practical fusion reactors will involve either the deuterium–tritium (D–T) or the deuterium–deuterium (D–D) nuclear fusion reactions. The former is technically more feasible in the early stages of the program inasmuch as the ignition temperature for the D–T plasma is only 45 million deg, whereas that of the D–D plasma is some 400 million deg. But the former has certain disadvantages. The neutrons produced are of relatively high energy ($\sim$14 MeV), and these must be recycled in a lithium blanket to breed the tritium fuel in a $^6Li(n,T)^4He$ reaction. Not only will this create severe shielding problems, but also the world supply of lithium is not greatly different from that of uranium, hence ultimately limited. In the D–D case, on the other hand, an essentially limitless supply of stable deuterium is available in the oceans. Furthermore, the D–D neutrons are much lower in energy ($\sim$2.5 MeV) and consequently are much more readily handled. It seems clear that

FIGURE 4.66 Photograph of the 2X II pulsed magnet set prior to construction of the external glass and epoxy reinforcement. The completed structure, which contains approximately seven tons of plastic reinforced with glass cloth, is the latest and largest experimental pulsed magnet to be built using techniques that were pioneered at the Lawrence Livermore Laboratory and that are now commonplace in industrial application. [Courtesy Lawrence Livermore Laboratory.]

practical fusion reactors will be achieved via the D–T system, followed by an eventual conversion to D–D systems.

Present progress in the U.S. program suggests the possibility of a self-sustaining fusion system by 1980 and economically competitive fusion-based power sources by 2000, assuming adequate support for the program.

As noted earlier in this chapter, too, the availability of very-high-power lasers has stimulated great interest in the possibility of using such lasers to stimulate fusion reactions in a small pellet of a frozen D–T mixture. The target goal here is perhaps a tenfold increase beyond the Lawson criterion of $n\tau = 10^{14}$ needed to achieve a self-sustaining system. Here, n is the plasma density of the reacting species in particles per cm^3 and τ is the containment time in seconds. In the magnetically contained plasmas, the densities are low, hence, substantial containment times of the order of seconds will be required. By starting with a solid-fuel D–T pellet, a very high n value is obtained, and only a very short τ, comparable with that inherent in the inertia of the interaction fuel following ignition, is required. Here again the reaction neutrons would be used to breed tritium fuel in a lithium blanket.

It is still too early to be able to estimate with any degree of certainty exactly which approach to economic fusion power holds the highest promise of early, or of ultimate, success. However, progress in recent years has been so promising that significantly increased activity and support seem imperative in view of the national and social benefits that success in this field will bring.

The first and most important step in achieving this degree of success was the expansion of basic knowledge of plasma physics. Without this knowledge, either of two catastrophes could have taken place: (a) there would have been inadequate clues to the most productive lines of research, with a consequent delay in progress; and (b) had fusion power proved scientifically impossible on this planet, resources of time, manpower, and money would have been committed to countless costly trial-and-error experiments, with no hope of success. There is an element of trial and error in every experiment, but it is fundamental knowledge—mathematical theory and physical understanding—that guides experiments toward what is called a rational approach. In this way, the multidimensional infinity of possible experimental attempts is reduced to a finite set of logically related steps.

The validity of the scientific method for the solution of man's practical problems is so well documented that the further example of fusion research might seem redundant. Nevertheless, throughout the current pattern of support for science, there is recurring overemphasis on immediate practical relevance. The enthusiastic and optimistic attempts of the early 1950's to forge ahead to the development of a fusion reactor, without consideration of the gaping holes in the knowledge of plasma physics (many then totally unrecognized), were doomed from the outset. A major fraction of the knowledge that has guided the work in controlled fusion has been fundamental in nature; that is, the original creative thought did not arise in response to a specific short-range goal but, rather, stemmed from a desire to understand the nature of things. This pattern has implications for the development of any national policy for science, either basic or applied.

Fluid Dynamics

Thousands of problems of the physics of fluids have been and are being solved for the benefit of man. An enumeration of even part of these would constitute a telephone directory of scientific achievement. Testifying to these achievements are such stupendously successful engineering results as airplanes, ships, rockets, oil pipelines, the functioning of the Weather Service, and the understanding of the motion of galactic gas clouds.

Of principal concern in this brief review are the unknown and the structure of a science that continually pushes back man's frontiers to this unknown. Flow of fluids separates into two broad classes, laminar and turbulent. By and large, laminar-flow patterns, a flow that is smooth and steady, are now understood.

But when flow stream lines become distorted, convoluted, and mixed in that wondrous random pattern called turbulence (for an example see Figure 4.67), the fundamental knowledge of fluids breaks down almost completely. In leaning over the fantail of a ship and watching the complicated ever-changing patterns of the wake in the ocean, the mind is teased by the almost total unpredictability. The need to develop an adequate description of turbulence motivates the largest fundamental effort in the physics of fluids. To a certain extent, the lack of predictability of turbulence can be predicted, but the problem of quantifying such a notion still remains. Some examples of the inadequate predictability of unpredictability follow.

The problem of the motion of stirred tea leaves is familiar from allusions to the tempest in the teapot, but the general convective patterns in a rotating fluid are still far from solution. Rotating fluids with a turbulent boundary layer are involved in many forms of pollution separators, and the short-circuiting of the separation work by turbulent-flow patterns is a significant and unresolved problem. The most important of these separators is, of course, the atmosphere of this rotating planet, which encompasses the extraordinarily complex phenomena of cyclonic storms, thunderstorms, tornadoes, and smog. Each of these components may be many orders of magnitude removed from full understanding, but in total they constitute an overwhelming impact on our environment. Understanding of the electrifica-

FIGURE 4.67   Air flowing past a cylinder rolls up into a regular succession of vortices. A cross section of these vortices is made visible by injecting a sheet of smoke
in the center of a wind tunnel. (Photograph taken by Gary Koopmann at the Naval
Research Laboratory, Washington, D.C.) [Source: *Yearbook of Science and Technology 1971*. Copyright © 1971, by McGraw-Hill Book Company, Inc. Used with
permission of McGraw-Hill Book Company.]

tion process and subsequent precipitation in a thunderstorm still is lacking.
Arguments focus on whether a tornado requires electrical energy input to
release its awesome power; understanding is insufficient to reach an answer.

Two-dimensional turbulence describes many of the large-scale atmospheric phenomena on this planet as well as on other planets. Two-dimensional turbulence also describes much of the surface circulation on the sun
and the stars. Yet little is known about how to quantify and predict even
the behavior of such two-dimensional situations. An important phenomenon resulting from two-dimensional turbulence is the jet stream that me-
anders across the continent at high altitude and affects so dramatically
many transcontinental airplane flights, quite apart from weather effects.
Another familiar example is the motion of a fish; in many cases, the
fish uses far less energy to move than would be predicted from normal
turbulence theory. Perhaps some particular waggle, developed in a long
evolutionary process, somehow sheds a turbulent boundary layer more
rapidly than otherwise would be the case. It is also possible that the skin
of some efficient swimmers, such as the porpoise, can be neurologically and

physiologically active instead of passive, so that it can damp incipient eddies in the boundary layer before they develop.

These, and many similar problems in turbulent flow, stimulate the imagination of many scientists, but the creativity and effort of the theoretical hydrodynamicist is challenged most of all by the need to understand the fundamental aspect of all these problems—namely, the behavior of the turbulent fluid. An increase in the knowledge of such a fundamental aspect of fluids could have wide impact and applications.

The current lack of understanding of thunderstorms, tornadoes, jet streams, clear-air turbulence, and other atmospheric phenomena affects the everyday life of society. With fuller knowledge, prevention of tornadoes in some cases might be possible. Or scientists might be able to establish a continuously operating thunderstorm in the Los Angeles basin and similar areas during times of temperature inversion through the imaginative use of the waste heat from a very large power plant. One such large thunderstorm is enough to wash and purify the air of the Los Angeles basin every 12 hours.

The addition of traces of long-chain organic molecules to fluid streams is already reducing turbulent drag by as much as 80 percent in certain cases; with fuller understanding of the way that these additives work, further improvements might be feasible, with a reduction in costs of pumping fluids and of transportation through and on the surface of fluids.

Bistable fluid jet devices, which can assume either of two states and which may be switched by a relatively small signal, may be used either in logic circuits or as power relays (see Plate 4.XII). As compared with electronic or electrical devices, they are not nearly so fast, but they have advantages in cost and in freedom from damage owing to heat (as in proximity to jet engines) or to radiation fluxes (as in the interior of nuclear reactors).

Plasma Physics Other Than Fusion Research

There are clearly important applied goals in plasma physics other than fusion. New fundamental knowledge in plasma physics has been applied also to the problem of direct conversion of electrical energy from hydrodynamic flow [the magnetohydrodynamic (MHD) generator] (see Figure 4.68). Progress has been made in MHD generation, using fossil fuels, to the point at which it appears practical to consider power plants considerably more efficient than the current ones and emitting significantly less pollution in the atmosphere. Very large MHD stations are now under construction in the Soviet Union for inclusion in that nation's national electrical grids. In the United States, MHD has been largely an industrial research problem,

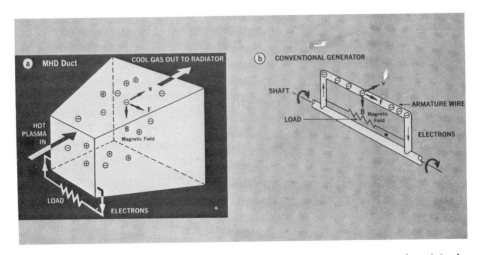

FIGURE 4.68 Magnetohydrodynamic (MHD) generator. In the MHD duct (a), the electrons in the hot plasma move to the right under the influence of force F in the magnetic field B. The electrons collected by the right-hand side of the duct are carried to the load. In a wire in the armature of a conventional generator (b) the electrons are forced to the right by the magnetic field. [Source: USAEC/Division of Technical Information.]

with federal funding sought on a matching basis. Unfortunately, the matching federal funds for the presently proposed pilot plants have not been made available, and the utility industry has been reluctant to invest in the development of a new power source in the face of the immediacy of the present power shortage. It would seem a wise federal policy to assure that new power sources, with a potential for dramatic improvement in efficiency and in the quality of the environment, are vigorously pursued.

The fundamental understanding of plasma physics will lead to a better knowledge of conditions beyond the earth—the magnetosphere, solar wind, solar corona, cosmic rays, and ionized clouds in galactic space. Most matter outside the earth's atmosphere is in the plasma state.

There is also a comparatively large research area of great potential impact—the study of the dynamics of highly ionized plasmas—in which the laboratory experimental work is beginning to expand in both the United States and the Soviet Union. It includes wave motions in plasma, turbulence (both MHD turbulence and electrostatic turbulence), shock waves, instabilities, the generation and emission of high-energy particles and nonthermal radiation, wave echo phenomena, and properties of relativistic plasmas. These phenomena, which are increasingly invoked in the inter-

WALL ATTACHMENT SWITCH

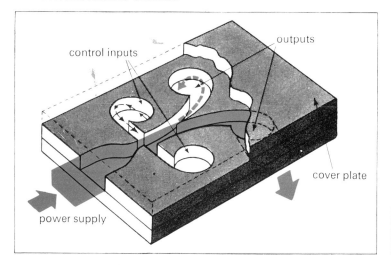

control inputs

outputs

cover plate

power supply

TURBULENCE AMPLIFIER

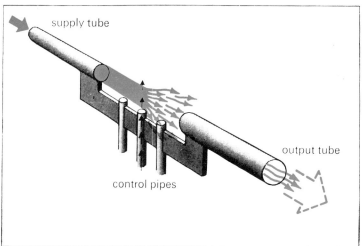

supply tube

output tube

control pipes

PROPORTIONAL AMPLIFIER

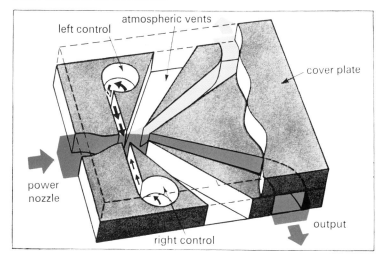

left control

atmospheric vents

cover plate

power
nozzle

right control

output

VORTEX AMPLIFIER

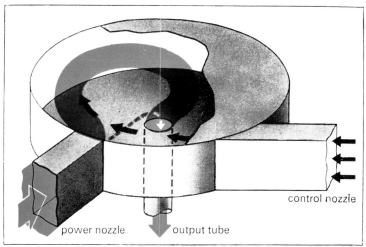

control nozzle

power nozzle

output tube

PLATE 4.XII Fluidic devices are frequently based on one of four basic concepts. The wall attachment switch will stabilize with the output wholly along one channel, depending on microscopic asymmetry in manufacture and other variables. But if the control nozzle on that side is turned on, fluid is entrained (black arrows) and the asymmetry at once switches the flow to the other output. Reblocking the control nozzle will not cause the flow to revert (broken colored line); it will stay attached to the new output. The turbulence amplifier has a supply tube and output tube in alignment. Flow from any control pipe at once converts the laminar main jet to turbulent flow, causing sudden and drastic reduction in pressure in the output (original, full output shown as broken arrow). The beam deflection proportional amplifier divides its jet according to relative flows through left and right control nozzles. In condition shown, the left control is greater, giving higher pressure in the right output. The vortex amplifier offers little resistance when the control flow is switched off (main flow then following line of broken color). With control pressure applied, the main jet is made to swirl round the vortex chamber so that resistance rises and flow is reduced. High control pressure reduces main flow almost to zero.

pretation of astrophysical and space observations, are now being studied under controlled laboratory conditions.

The study of the dynamics of plasmas in magnetic fields has broad significance. Unlike most media studied by physicists in the laboratory, plasmas cannot yet be fully controlled and are highly sensitive to the exact details of the apparatus used to study and generate them, and especially to impurities. Although magnetic control of plasmas has been the main aim of thermonuclear research for many years, improvements in the understanding of plasma dynamics leading to better control of plasmas could have an impact in many other fields. For example, applying existing knowledge might lead to important developments in ion sources for ion propulsion and heavy-ion sources for nuclear research. Many suggestions have been made in the last ten years for accelerators incorporating plasma in one form or another; recently, groups in the Soviet Union announced the successful operation of the first electron ring accelerator (ERA) embodying some of these principles.

In the general area of low temperatures and partially ionized plasma, basic work is required that will improve the understanding of ionospheric physics and reinforce various technological applications of plasmas. Three areas are worthy of special attention. First, the discovery of gas lasers provides powerful sources of radiation over an enormous range of the electromagnetic spectrum, particularly in the three decades between 1-μm and 1-mm wavelength. Although the potential applications in this range are many, the understanding of the physical processes in many gas lasers is comparatively poor.

A further example is the interaction between plasmas and surfaces of condensed matter. Here technological devices have been developed and used for many years with comparatively little understanding of the basic processes. For example, there is much dispute and uncertainty about the basic physical principle of an arc spot, and spark erosion machines are used with little understanding (other, perhaps, than crude energetics) of how the metal is removed.

An essential feature of ionization-physics studies is the provision of information on elementary processes such as collision cross sections, reaction rates, and details of atomic and molecular interactions with charged particles and radiation. At low gas pressures, sophisticated studies with electron, atomic, and molecular colliding beams, which can be readily carried out in academic laboratories, will lead to new cross-section data that are urgently required in many technical applications of plasma physics.

The recent development of high-intensity coherent light sources will lead to advances in the understanding of the interaction of radiation with matter and of mechanisms of plasma production in gases and solids. Improved determinations of molecular structure will follow from investigations of the scattering of laser radiation in gases and liquids. Further high-power laser development and extensions from the red end of the spectrum toward shorter wavelengths in the ultraviolet are likely to make possible, with the aid of nonlinear optical media, intense sources of ultraviolet radiation that are urgently needed in other plasma studies.

Studies of the mechanism of the electric spark extended to high-gas pressures and large-electrode separations will provide information on electron–atomic ionization and electron–ion recombination. All these investigations will be associated with the development of sensitive electronic detectors, fast optical and electronic techniques, and computing methods. Techniques involved in the production, separation, and detection of fundamental charged particles produced by high-energy nuclear accelerators are also closely linked with developments in ionization physics.

The technological application of the results of basic studies in plasma physics are widespread. Applications include electrical communications, space-vehicle design, and the control and transmission of electrical power. To a large extent the technical long-term future of ionization and plasma physics will be concerned with the objectives of fusion power, plasma and ion propulsion, and the development of MHD power generation and direct conversion systems.

In the short term, improvements can be expected in a wide variety of devices employing gaseous electronic phenomena, ranging from optical frequency and microwave communication systems to ultrahigh vacuum pumps and gauges, arc rectifiers, and lighting equipment. The possibility of the full control of the basic physical processes that lead to an electrical discharge opens the prospect of suppressing discharges for times sufficient to effect electrical switching operations free from danger and unreliability. Studies of the microplasmas at separating electrical contacts can lead to improved switching technology in low-power applications. Associated studies of the mechanism of electrical breakdown in gases at high pressures could facilitate new developments in large circuit breakers and to improvements in high-voltage power cables using high-pressure gas insulation.

Applications in high-power technology are perhaps more obvious but no less important. Extension to yet higher power transmission voltages (greater than 1 MV) will require basic studies of the physical mechanism of the electrical breakdown of insulators of a wide range of media, including gaseous, liquid, and solid insulators, as well as of vacuum.

Use of Computers in Plasma and Fluid Physics

Plasma physics is the most complex of the classical sciences, and the mathematics needed to describe the phenomena are correspondingly

FIGURE 4.69 A computer's view of galactic evolution. Astronomers at the Goddard Institute for Space Studies, New York City, used a computer to simulate the growth of a galaxy. From top to bottom, the frames show the initial shapeless gas clouds, two growing spiral arms and two condensations, the start of a spiral pattern, and, finally, a well-developed spiral. [Source: *Science Year. The World Book Science Annual.* Copyright © 1971, Field Enterprises Educational Corporation.]

complex. For this reason, computers have played an especially important part in the development of deeper understanding of plasma and fluid dynamics phenomena. For an example, see Figure 4.69.

Since von Neumann's pioneering efforts in the use of computers in weather prediction, the problems of the fluid physics of the atmosphere and oceans along with nuclear weapons requirements, have excited the greatest pressure for the development of ever larger and faster computers. And the computer industry has responded. Experimental tests involve great costs, and in some cases they are physically impossible to carry out in complete fashion so that even partial replacement of them by numerical simulation requires large funding and high priority. This close coupling between plasma and fluid physics and computer technology will continue and even grow.

International Cooperation

Both plasma physics and the physics of fluids have benefited from, and added greatly to, the scientific knowledge of the world. They have been among the subfields of most vigorous international interchange and collaboration.

In the period from 1950 through 1958, controlled fusion research was conducted secretly in the United States, the United Kingdom, and the Soviet Union and to a much lesser extent in other countries, with major emphasis on the production of a fusion energy source. The increasingly evident long-range nature of the research task encouraged declassification in 1958. It was then found that very similar theoretical approaches and minor experimental achievements had characterized the different national programs. Soviet–U.S. relations at this time were personally amiable but distant and highly competitive.

The Geneva II Fusion Exhibits, particularly that of the United States, stimulated worldwide participation in controlled fusion research in 1958. The first International Atomic Energy Agency-sponsored conference on Plasma Physics and Controlled Nuclear Fusion Research was held in Salzburg and attended by 29 national delegations. The conference was marked by highly productive scientific exchanges, as well as by vigorous controversies along national lines.

Since 1961, the worldwide fusion research effort has moved toward real integration. There is a large-scale exchange of visitors and visitors-in-residence among participating nations, including the Soviet Union. Personal relations have become extremely cordial, and controversies, though still lively, form along scientific rather than national lines. The scientific value of this interaction has been so important for the U.S. fusion program that it is difficult to imagine what alternate course it might have followed in the absence of declassification. Certainly the cost of research to achieve comparable progress would have been much higher.

A similar impact on East–West relations can be claimed for meteorological research. During the original development of computer analysis of global circulation, the weather reporting of Soviet stations was crucial. The initial cooperation—a direct TWX line into the Lawrence Livermore Laboratory—has now been extended to the worldwide effort known as the Global Atmospheric Research Program, in which almost all the nations of the world are cooperating in an effort to understand weather on a worldwide basis. The impact of this and other international programs of cooperative meteorological research are of major importance to East–West relations.

Distribution of Activity

Work in plasma and fluid physics is widely distributed throughout the scientific and engineering communities. The major fraction of the PhD scientists working in these areas—6.7 percent of the physics PhD population as reported in the 1970 National Register of Scientific and Technical Personnel—is concentrated in the universities (48 percent), with substantial percentages found also in industrial laboratories (22 percent) and government and federally funded research laboratories (29 percent) (see Figure 4.70). A large part of plasma and fluids research is performed by people who consider themselves engineers or applied mathematicians. The combined physics research in plasma and fluids produces roughly 7 percent of the world's physics publications and 8 percent of the U.S. publications. Of the physics theses produced in 1965 and 1969, roughly 4 to 5 percent were produced in plasma physics and only 1 percent in the physics of fluids. On the other hand, 30 percent of those engineering theses (in *Dissertation Abstracts*) that might equally have been classified as physics were produced in fluid dynamics, so that the combined plasma and fluid theses constitute 17 percent of the total number dealing with the subject matter of physics, the contribution from fluids being about three times that of plasma physics.

The total annual funding of physics of fluids by the federal government amounts to approximately $37 million. By far the largest fraction of this support is motivated by military objectives; ships, planes, missiles, fluidic computers, and the like account for some $20 million a year. Much of this research is applied, but it also contributes to the store of fundamental knowledge. The DOD funds some work in universities, though much less than in previous years. The AEC, NASA, NSF, and NOAA each support the physics of fluids at a rate of about $4 million to $5 million a year.

The current funding for plasma physics is about $30 million. Nearly 85

PLASMA & FLUID PHYSICS

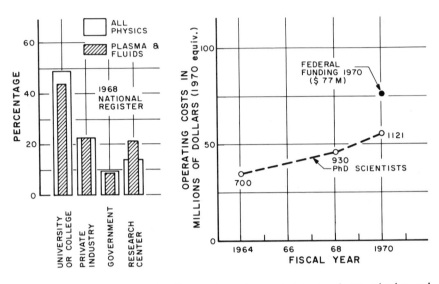

FIGURE 4.70 Manpower, funding, and employment data in plasma physics and the physics of fluids, 1964–1970.

percent of these funds are provided under the controlled thermonuclear research program of the AEC. The support of controlled fusion research in the United States is concentrated on large confinement experiments in a number of major laboratories. The universities have pursued smaller, but important, experiments, have trained young scientists for the major laboratories, and in turn have tended to recruit their professors from major laboratory personnel. Support of plasma physics as a basic science is largely the responsibility of the NSF. In 1970, approximately 17 percent of federal funds and 9 percent of industrial funds for basic research in physics were allocated to the support of basic research in fluids and plasmas (Figure 4.71).

Problems in the Subfield

In view of the great potential importance of controlled fusion research to the world, and of the steady progress being made toward the goal, the lack of reinforcement by increased support is surprising. By way of comparison, the ratio of U.S. to U.S.S.R. manpower in this area dropped from

PLASMAS AND FLUIDS

PHYSICS MANPOWER IN 1970

N = 36,336

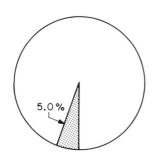

FUNDS FOR BASIC
RESEARCH IN PHYSICS IN 1970

FEDERAL
INDUSTRIAL

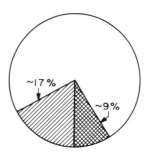

FIGURE 4.71   Manpower and funding data in plasma physics and the physics of fluids in 1970.

1 in 1960 to about 0.3 in 1970. Federal support over the past five years has been essentially constant, with a consequent significant reduction of manpower in the major fusion research efforts, particularly in the AEC laboratories. This reduction has the effect of reducing experimental capability at a time when the possibility of major advances has been demonstrated by other nations and when progress in the U.S. program has been marked. The nation cannot afford continuation of these downward trends, as one major recommendation in Chapter 2 for a stated national commitment in this area implies.

The current funding of plasma physics as a basic science is such as to reduce drastically the support of new proposals. The dollar investment by the NSF and the AEC declined in 1971; yet the motivation for federal support of research and new knowledge in this subfield is vastly greater now than before. The present funding of MHD is indeed dismal. Considering the very large application and relevance of plasma physics, there is need for supporting basic work at a much higher level than is presently the case.

The overwhelming relevance of fluids has so far maintained the level of funding in the physics of fluids. On the other hand, the near-constant funding for the last four years implies a fractional decrease of 20 to 25 percent in the net effort expended. Serious dislocations of the subfield will take place if present funding trends continue much longer. In view of the fact that the foremost phenomenological aspect of fluids, namely turbulence, is not yet subject to a quantitative analytical description, such a reduction of effort seems indeed shortsighted. Compounding the gradual inflationary reduction of effort is the current political expediency of relevance. Application of this criterion to proposals dealing with a basic understanding of turbulence does a disservice to all concerned. First and foremost, it discourages work on the most important problem of fluids.

> We have next to speak of the stars, as they are called, of their composition, shape and movements.
> ARISTOTLE (384–322 B.C.)
> *On the Heavens,* Book II, 289

> Hence it is incumbent on the person who specializes in physics to discuss the infinite and to inquire whether there *is* such a thing or not, and if there is, *what* is it.
> ARISTOTLE (384–322 B.C.)
> *Physics,* Book II, 202

> When I, behold upon the night's starr'd face,
> Huge cloudy symbols of a high romance
> KEATS (1795–1821)
> *When I Have Fears*

ASTROPHYSICS AND RELATIVITY

Introduction

The interface between physics and astronomy is one of the most active in the physical sciences, and indeed the division between these fields is increasingly blurred and artificial. All of physics is required for any coherent understanding of some of the most recent and most spectacular discoveries in the heavens. Nuclear astrophysics has evolved into almost a separate area of specialization in nuclear physics, atomic physics has long played a central role in both observational astronomy and the study of stellar atmospheres and the interstellar medium, elementary-particle physics is now being called on to contribute to the understanding of the totally new high-energy phenomena at the hearts of stellar objects, and condensed-matter physics is needed to elucidate the structure of their cooler regions. The importance of plasma physics and optics in astrophysics is obvious.

Because a special relationship exists between astronomy and physics in regard to relativistic and gravitational phenomena and cosmology, and because developments in recent years have dramatically changed the nature of the physics in these areas from inspired and intuitive theoretical prediction to a more balanced situation in which these predictions are accessible to experimental test, astrophysics and relativity receive special consideration and discussion in this review.

One of the classic writings on astrophysics and relativity begins with the subtitle, *The Bigger They Are, The Harder They Fall.* It could well serve as a motto of this subfield, which deals with very large bodies such as stars and galaxies, the strong gravitational forces they create, and the way they behave under the influence of such forces. The mysterious force of gravitation was described accurately by Newton in the seventeenth century, and his classic work on the subject has been the basis for virtually all calculations about the way bodies move under the influence of gravitation, from the orbits of planets and, more recently, space probes, to the slow turning of the vast systems of stars, known as galaxies, in space.

But Einstein, in 1915, proved that gravitation can be reinterpreted as a curvature of space–time. The power of this concept lies in the finding, noted by Galileo and Newton, that in ordinary empty space bodies not acted on by forces travel in straight lines; in Einstein's space–time, curved by gravitation, bodies still travel the shortest path between two points, but this path is now curved. Technically, the curvature of space–time refers to the distortion of geometrical figures formed by light rays between various points. For example, in a curved space, the angles of triangles no longer sum to 180°, just as is the case for triangles drawn on the curved surface of the earth. But the concept of shortest distance—like the great circle route between cities on the earth—is still valid, and, according to Einstein, this is the type of path actually pursued by bodies under the gravitational forces exerted by a star, planet, or galaxy.

Although Einstein enunciated his theory early in this century, it has had little impact on the thinking of most physicists and astronomers, let alone nonscientists. The reason is not hard to find. Newton's conception that space itself is flat (described by normal Euclidean geometry), and that bodies in the solar system travel on curved paths within this space, yields highly accurate calculations. Indeed, the three classical tests of Einstein's theory of general relativity present such extremely small deviations from Newton's classical theory that they are still the subject of continued discussion and measurement.

A sequence of discoveries in astronomy, starting in the post-World War II era, using a variety of new techniques such as radio astronomy, have brought to light a variety of extraordinary phenomena—radio galaxies, quasars, cosmic background radiation, and pulsars. In each of these cases, in contrast to previous experience, scientists could be dealing with systems in which the differences between the theories of Einstein and Newton are profound. Relativistic astrophysics brings together astronomers, who need a theory such as Einstein's to interpret what they see, and physicists, who are eager to test the validity of Einstein's theory and to show how it applies in cases where space is very much curved.

Application to Cosmology

Because the effects of gravitation are important only for large amounts of material such as are found in a star (whose typical mass is one million times that of the planet earth), one might expect major effects to be observed on the largest possible scale in the universe. Indeed, Edwin Hubble in the 1920's formulated the remarkable law of the red shift, according to which galaxies seem to be receding from the earth at enormous speeds that are greater as the distance of the galaxy from earth increases. Friedmann showed that this phenomenon is a natural consequence of Einstein's theory; thus the science of relativistic cosmology—the application of relativity to the universe—came into being. Hubble's observations had indicated velocities up to about 20 percent of the velocity of light. Relativistic cosmology predicts that at even greater distances material would be seen to be moving away from earth at much higher speeds. Only recently, a quasar was found (a strange astronomical object described later in this review) that appears to be receding from earth with a speed nearly 90 percent that of light. Moreover, a variant of Friedmann's calculations, one frequently identified with Gamow and known as the big-bang model of the universe, predicts that radiation should be reaching earth from the very limits of the universe, where matter is receding from earth at a speed differing from that of light by only one part in a million. (In connection with this phenomenon, see Plate 4.XIII.) Recently, this prediction was partially confirmed by measurements at radio wavelengths. Radiation emitted as visible light from the limits of the universe arrives at the earth as radio waves (this is simply a consequence of the familiar Doppler effect).

Pulsars, Supernovae, and Black Holes

There are two ways to obtain a strong gravitational field. One is to have a large mass, which, of course, is the case with the universe. Another is to have a smaller mass, such as that of a star, and to come extremely close to it, which can happen only if the mass is compressed into a very small volume. Some stars have extraordinarily small diameters compared with those of normal stars such as the sun. White dwarfs, which are a rather common type of star but too feeble to be seen without a large telescope, are only 1 percent of the size of these normal stars. Strong gravitational fields have been found for these white dwarfs; however, extremely strong gravitation was expected only in the case of what was a purely theoretical concept—the neutron star—now believed to be the cause of the phenomenon known as the pulsar. In 1967, pulsars were found quite by accident as radio sources in the sky whose intensity flickers rapidly but with extreme precision. The initial reaction to the observed precision was that it represented a signal from extraterrestial intelligences. The most famous of these sources, the pulsar in the Crab nebula, pulses every 1/30 sec with extreme regularity (see Figure 4.72). The basic phenomenon is apparently a rotating star that has one or more bright spots on it. The radiation from the star then appears to vary as the bright spots pass overhead. To fit this interpretation, the star must have a diameter of only 1/100,000 that of an ordinary star but a comparable mass. Under such conditions, as Oppenheimer and Volkov had shown 30 years before, ordinary matter is crushed into nuclear matter, with a density 10^{14} times that of water. Space–time in the neighborhood of a neutron star must be extremely curved if Einstein is right, and calculations are now under way to predict what phenomena might be observed as a consequence.

A neutron star could be formed as a result of a nuclear explosion occurring inside an ordinary star. The core of the star collapses to the fantastic density of nuclear matter, and the outer layers leave the scene at tremendous speeds, causing the phenomenon that astronomers have long known as a supernova. In fact, the Crab nebula, first observed by court astronomers of the Ming Dynasty in China on July 4, 1054, is probably the remains of the supernova explosion that left behind the pulsar that is spinning rapidly in the center of the nebula today. (The concluding section of this chapter describes the neutron star and its structure.)

There is evidence that the Crab nebula is filled with fast electrons traveling at nearly the speed of light. Probably they were accelerated to their enormous speeds by a whiplash effect of the spinning neutron star. Since such fast electrons are observed throughout the Milky Way galaxy, it is quite possible that they originated in other supernova events similar to those in the Crab nebula. This hypothesis offers a solution to the riddle of the cosmic rays—fast protons and electrons that have long been known to bombard the earth and were man's first source of very-high-energy particles.

If one takes seriously the predictions of general relativity, and the example of the red shift is convincing, an extraordinary dilemma arises in regard to the fate of certain types of star. Generally stars are suspended in a delicate balance between their radiation energy, which continually escapes to space, and the compensating energy generated deep inside them

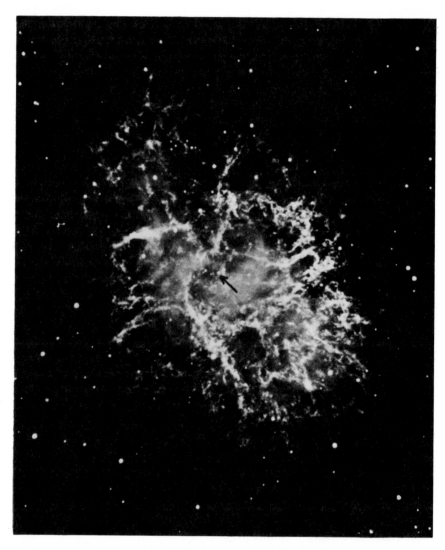

FIGURE 4.72 The Crab nebula. Arrow marks the pulsar NP 0531. (Lick Observatory photograph.) [Source: *Yearbook of Science and Technology 1971*. Copyright © 1971, by McGraw-Hill Book Company, Inc. Used with permission of McGraw-Hill Book Company.]

by thermonuclear reactions. Ultimately, a star must exhaust its nuclear fuel; when it does so, it attempts to compensate its radiation loss by releasing gravitational energy in a slow contraction. The dwindling radiation pressure from the interior no longer balances the gravitational forces, and the star contracts. In the case of moderately massive stars, this process may lead to the formation of a supernova and a pulsar. But very massive stars, when they reach the stage at which a neutron star might form, have such strong gravitational fields that ordinary matter is crushed completely. The star goes into what is called a gravitational collapse, which, at least theoretically, brings all the matter in the star to infinite density at a geometrical point of vanishing spatial dimensions. The gravitational field of such an object is so strong that light waves either originating inside it or attempting to get past it to distant observers are sucked in and disappear forever, with the result that the object appears to be a black hole. Although no black holes have yet been detected unambiguously, theorists fully expect them to exist. Some possible candidates have been located in certain binary systems. To discover them requires techniques capable of detecting an absolute black void a few miles across, located trillions of miles away in space—this truly is a challenge!

It should be emphasized that this prediction of infinite density faces a most important paradox that can have far-reaching consequences for all physics. When a physical theory predicts infinite parameters in this sense, it simply indicates that the theory is inadequate and that major new physical insight must be brought to bear on the problem.

Some 60 years ago, physicists faced a similar paradox in the atom. Accelerated electrons were known to radiate electromagnetic energy, but still atoms, with electrons in closed orbits—and therefore subject to acceleration—were stable. Resolution of this paradox required the discovery and development of quantum mechanics to replace the classical arithmetic; this discovery revolutionized physics and much of philosophy.

It may well be that a new discovery of comparable importance awaits physics in the phenomena of gravitational collapse.

It has recently been suggested that in this process of gravitational collapse a star that is not perfectly spherical will emit a special kind of wave predicted by Einstein—a gravitational wave. Such waves, invisible and extremely hard to detect, should have the property that, as they pass the earth, all material objects are simultaneously squeezed in one direction and stretched in the other. Weber and others seeking to detect such waves have set up apparatus capable of recording displacements as little as 10^{-14} in. It appears to be recording such waves about three times a day, but this discovery is so difficult to explain that confirmation is urgently needed. These waves are much more frequent than had been anticipated; if this finding is correct, it may mean that gravitational collapse to a black hole is a rather common occurrence in the universe.

The discovery of the pulsars and the convincing identification of a neutron star as their source have given great impetus to what previously had been inspired speculation concerning gravitational collapse phenomena. Nothing is yet known about the behavior of matter at the inconceivable densities that would be achieved in the latter stages of collapse nor about the validity of natural laws as the dimensions of this remarkable system shrink without limit. And what happens to the star after it collapses to a point? Physics currently has no answer. This situation challenges some of the ablest minds. It is truly at the brink of the unknown.

Galaxies and Quasars

Radio galaxies and quasars were first detected as a result of their radio emission; subsequent observations have shown them to be extremely bright emitters of visible light as well as of x rays and infrared radiation. The total amounts of energy radiated are stupendous—for example, in one case, at least 10^{14} times the radiation from the sun. At first these objects were related to normal galaxies, which are collections of individual stars. Normal galaxies emit up to 10^{11} times as much energy as the sun, but still only 1/1000 as much as a bright quasar. The extraordinary aspect is that quasars are much smaller than ordinary galaxies, and it seems impossible to explain their radiation in terms of many individual stars. Particles moving in a sufficiently strong gravitational field can emit a huge amount of energy. Up to 50 percent of the theoretical absolute maximum energy ($E=mc^2$) should be possible, but nuclear reactions in stars yield only 0.1 percent or less. Perhaps quasars represent the first known energy source powered by general relativistic collapse. But it has also been suggested that at least some of them correspond to collisions between matter and antimatter components of our universe, with the observed radiation originating through total annihilation processes.

Eventually the enormous energy production by quasars may be attributed to gravitational collapse; this is perhaps the key significance of the discovery of quasars for physics. However, there are awkward problems, even if it can be shown that the basic energy source is gravitation. Much of the electromagnetic emission in all wavelength regions, including radio, infrared, optical, and x-ray wavelengths, is believed to be synchrotron emission, a type of radiation that was first predicted and found in nuclear electron accelerators called synchrotrons. When an electron, a tiny electrical particle, moves at speeds near that of light in a magnetic field, so that it is accelerated into a circular orbit, it emits a peculiar glow, which, in the synchrotrons where this effect was first noted, is concentrated in the ultraviolet part of the spectrum. This emission of radiation by an accelerated charge is a classical phenomenon predictable on the basis of Maxwell's equations.

Because of various clues, astrophysicists are certain that the same emission process operates in quasars and radio galaxies (as well as in the Crab nebula). The radiation comes from a huge number of very rapidly moving, accelerated electrons. The question is, just how can these electrons be accelerated to the required speeds, even granting that a powerful energy source is available? In the laboratory, it is necessary to build an elaborate machine, such as a synchrotron, to accelerate electrons to the required energies. It is difficult to imagine how even a collapsing star, or possibly a collapsing galaxy, would produce great clouds of fast electrons. This problem falls within the scope of plasma physics, or its specialized offspring, plasma astrophysics—the subfields that study charged particles and their interactions. Solar flares—explosions on the surface of the sun—which are known to emit copious quantities of fast electrons, provide some clues. This phenomenon is known to be associated with magnetic fields in the solar atmosphere; since magnetic fields are also present in quasars, it could well be that these phenomena are related. But to show how the gravitational energy of collapse can be converted into the electromagnetic energy of synchrotron radiation remains a most challenging problem.

It is not only the general relativist and the plasma physicist who are interested in quasars but also the astronomer studying the evolution of galaxies. Quasars and radio galaxies may form from part of a class of objects in distant space * of which the normal galaxies are but another part. According to one interpretation of quasars, they are associated with the birth of a normal galaxy, the enormous explosive energy found in them coming from the massive stars that are expected to form and die explosively early in the life of a galaxy. A contrasting but equally arguable interpretation suggests that quasars are associated with the final throes of a dying galaxy, in which stars, drawn by the relentless force of gravitation to the center of their galaxy, are packed so closely that they begin to collide with one another, with disastrous results for them and a correspondingly high energy output. Whether the outburst of a quasar is the birth pang or the death rattle of a galaxy remains obscure, but astronomers and astrophysicists are trying to find out which, if either, is correct.

---

* If the red shift turns out to be gravitational rather than cosmological in origin, these objects are much closer to the earth than is now commonly believed.

Relation to Man's Fate

What have such exotic phenomena as quasars and pulsars to do with man's understanding of his own history and significance? The key idea in understanding this relationship is evolution. Anthropology discusses the cultural evolution of man from the time he first emerged from a primitive state. Biology discusses the evolution of plants and animals over a much greater time scale, culminating in man. Thus, natural selection, operating over millions, and even billions, of years, has led simple one-celled organisms to evolve into the extraordinarily complex phenomenon we know as man. Recent studies have shown that living matter was present on earth as long as 3 billion years ago. It is believed that the earth is only 4.7 billion years old; and since older life forms are constantly being found, it is possible that life may have originated on earth soon after its creation. The very recent discovery of materials that are essential for life—the amino acids—embedded deep inside a meteorite suggests that prebiological material of the sort that would be needed for early life on earth may already have been present in the solar system at the time the earth was formed.

It is believed that life formed in a natural and predictable way on the earth, and it seems likely that a similar phenomenon has occurred throughout our galaxy, in which there are billions of stars that are similar to the sun in light and heat and composed of the same chemical elements found on the earth, which are necessary for life. This being the case, life could well be ubiquitous throughout our galaxy and in many distant galaxies as well. Literally billions of life systems may exist independently throughout the universe. Some of them may have evolved to a state of intelligence, such that one can contemplate communication with them. Although no such communication has yet occurred, in the minds of at least some dreamers it seems a reasonable possibility. As Condon has discussed, however, unless we grant these intelligences the wisdom to prolong their civilization for periods much longer than 100,000 years—a wisdom that we seem as yet to lack—the probability of our actually achieving this communication appears vanishingly small simply because of the vastness of even our own galaxy. Intrinsic in any such statement is a trace of human arrogance; for it assumes that velocities and channels of communication that are now inconceivable to human physics do not, and cannot, exist.

The point of this discussion is that evolution toward higher forms of life appears to be an inevitable consequence of the uniformity of nature as we now understand it from astronomy. Not only are individual stars very similar to each other, with the chemical elements from which they are built in almost exactly the same proportion from one star to the next, but other whole galaxies seem to be very similar to the one in which we live. This

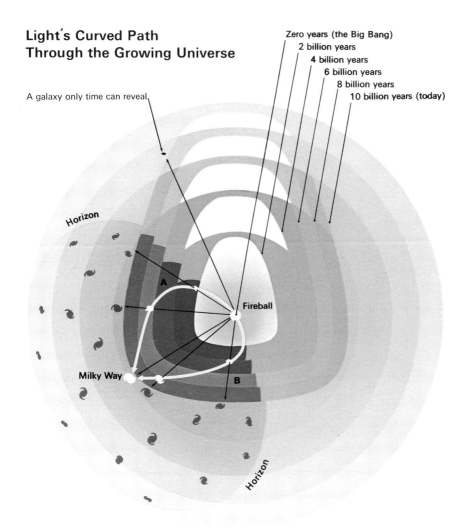

Light's Curved Path Through the Growing Universe

Zero years (the Big Bang)
2 billion years
4 billion years
6 billion years
8 billion years
10 billion years (today)

A galaxy only time can reveal

Horizon

A

Fireball

Milky Way

B

Horizon

PLATE 4.XIII Light's curved path through the growing universe. The universe has swelled since its genesis in the primordial fireball. The two-dimensional spherical shell is shown at 2,000,000,000-year jumps. The light from only relatively nearby galaxies has had time to move across the shell and reach us. In fact, the oldest light we can see, that of the fireball itself, has reached us from only as far as the horizon. As we look out to the horizon, then, we see galaxies at earlier times, when the universe was smaller and they were closer to us. Their light must have taken the same curved path as the runner on an expanding race track. The light from every one of the billions upon billions of galaxies we see today must have come to us along curved lines like A and B. [Source: *Science Year. The World Book Science Annual.* Copyright © 1967, Field Enterprises Educational Corporation.]

situation would seem to call for a special explanation, until one realizes that the physical processes operating in the formation of stars or planets and of chemical elements appear to be the same throughout the universe. The extraordinary fact is that the atoms that can be detected by sensitive means, even in radio galaxies so far away that they are receding from earth at half the speed of light, appear to be identical to those found in a sample of dirt from a nearby lot. That this is far from coincidence is explained by the physicists who study the nuclei of atoms and the elementary particles from which they are made. They have shown in countless terrestrial experiments that all atoms on the earth belonging to a given chemical element are interchangeable, and that the same is true of the individual particles of which these elements are made. Furthermore, if Einstein was right, the gravitational forces that govern the evolution of stars, planets, and galaxies, and the collapse of interstellar clouds to form stars and planets from the dust floating in the interstellar space inside of galaxies, are also of uniform character throughout the universe. Relativistic cosmology, according to which material billions of light years from the earth moves under the same laws of gravitation as that on earth, rests entirely on this principle of uniformity. Possibly, astrophysicists will be able to surmise the entire evolutionary history of the sun, the earth, the biosphere (living matter on the surface of the earth), and man from observations of distant but similar systems that could be in earlier phases of evolution. This procedure is used in stellar astronomy, in which stars both younger and older than the sun, but otherwise having the same characteristics, are observed. In the case of galaxies, the study of quasars as a possible birth or death stage of galaxies may provide a clue to the fate of the Milky Way system. Consequently, astrophysicists continue to probe the past—although not in the time–travel fashion of science fiction—for clues to the future.

The Origin of Things

How did this whole evolutionary cycle begin? Currently there is no definitive answer; perhaps there never will be. A simple, yet ultimately obscure, theory apparently agrees with presently known facts in most respects. The big-bang or Friedmann model of the universe suggests that all matter from which galaxies and stars are made originated in a catastrophic explosion at the origin of time as we measure it. The density of matter was infinite at that original moment, but the matter was expanding outward so rapidly that its density quickly became low and normal galaxies could form. Two lines of reasoning, both associated with the fact that the explosion must have been very hot, provide evidence for this theory. At the high temperatures involved in the first second or so of the life of the universe, many reactions between atomic nuclei would necessarily have taken place and certain of the chemical elements would have been formed. Calculations indicate that the two main elements produced would have been hydrogen and helium, light gases familiar on the earth in their application to lighter-than-air flight. About 90 percent of the atoms would have been hydrogen, and about 10 percent would have been converted, according to this theory, into helium in the very early expansion phases. Study of the stars and interstellar matter confirms this hypothesis.

Another consequence of the very high temperature of the early universe would be cosmic blackbody radiation. Astrophysicists assume that this radiation originates at a surface far from the earth that was of high density and had not yet formed into galaxies at the time the light left it. In essence it is seen as it was just after the birth of the universe. This surface has a temperature about that of the surface of the sun, but, because the radiation it emits is red-shifted by the Doppler effect to 1000 times longer wavelengths, it can be perceived only by radio telescope and not by the human eye, as in the case of the sun. Again, observation generally confirms prediction in this respect. Einstein's theory of relativity is capable of describing the expansion of matter and the radiation in the early universe, once they had been created and sent on their way. However, it cannot explain the explosion or the initial presence of the matter or energy, and therein lies an extraordinarily frustrating puzzle.

The assumed uniformity of physical processes in the universe is based on the observation that the elementary particles involved are identical throughout. Apparently these elementary particles were created in the initial instant of the expansion; this event cannot be explained by elementary-particle physics at the present time. Relativistic cosmology and elementary-particle physics come together in the study of this initial catastrophe. An ultimate goal of relativistic astrophysics is the unraveling of the phenomena that led to the early explosion. (What happened before it is, at present, in the realm of metaphysics.)

A less ambitious goal is to understand the phenomena in the universe that occurred after the explosion, including formation of the chemical elements, formation of galaxies, and the relationship of quasars and radio sources to distant galaxies. There is good evidence that the elements heavier than hydrogen and helium, including the carbon, oxygen, and nitrogen of which living matter is composed, were formed inside stars. When certain stars such as supernovae explode, they eject these elements into interstellar space, where they are available to form new stars and planets. Thus, the study of supernovae and pulsars is linked to the origin of the elements. One can even go so far as to speculate that all mankind is composed of atoms that were once inside an exploding star such as that which gave rise to the Crab nebula.

In summary, the picture of cosmic evolution that fits much of the present data is this: Following an initial and inexplicable big bang in which the elementary particles were formed, the universe expanded and cooled until hydrogen and helium formed. Later these elements condensed into galaxies in which stars formed that were capable of further nuclear "cooking" to produce heavier elements. These elements, ejected into space by supernovae explosions (see Figure 4.73), formed more stars and planets, which are the habitats of widespread life forms throughout the universe. It is a

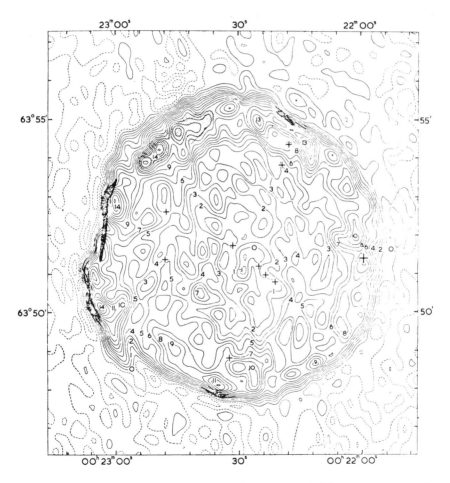

FIGURE 4.73 Tycho Brahe's supernova. The remnant of this supernova was observed by aperture synthesis at 1407 MHz with the 1-mile interferometer at the Mullard Radio Astronomy Observatory, Cambridge. [Source: J. E. Baldwin, "Radioastronomy and the Galactic System," IAO Symposium 313 (1967).]

major triumph of the human intellect that, on the basis of a knowledge of physics, the results of laboratory measurements, and mathematical skills, it is possible to reproduce, with remarkable precision and detail, the life cycle of the universe, assuming, of course, the initial explosion.

Tests of General Relativity

In addition to the question of the origin of the universe, astrophysics and relativity seek to work out the fundamental laws that govern the universe, even though many of them may not be of immediate application in the astronomical world as currently perceived. Thus, the theory of general relativity predicts that already in the solar system there should be minute deviations from Newtonian physics. So far no one has shown that these deviations are of any practical importance, but as a matter of principle it is important to search for the small deviations that are predicted to find out whether Einstein was right, for the correct gravitational theory has implications for the evolution of the universe. For many years attempts have been made to do this, but as yet with mixed success. Now new methods using radio and radar astronomy are being brought into play, with the hope that the precision of measurements can be made vastly greater than was previously possible and that a decision between the Newton and Einstein theories can be made once and for all. Present precision is already great enough to be confident that Newton was wrong in principle, but measurements still are not accurate enough to distinguish between two versions of the theory of general relativity—Einstein's and a variant, called the scalar–tensor theory, developed more recently by Dicke and Brans. It is hoped that further measurements in the solar system will give a final answer to this problem and thereby give physicists confidence in applying one or the other of these theories to distant matter in the universe.

Ways of Testing General Relativity

The number of scientists who are actively engaged in studying relativity *per se* is rather small, less than a few hundred. On the other hand, the equipment required for their studies is expensive, with the result that a substantial fraction, perhaps one fourth, of the budget devoted to the study of astronomy and astrophysics is allocated to research in astrophysics and relativity. For example, the United States has radio telescopes of varying sizes, ranging from 10 to 1000 ft in diameter (see Figure 4.74). The larger installations are expensive to operate and rely heavily on the federal government for support. As telescopes are multipurpose instruments, capa-

FIGURE 4.74 The radio–radar Arecibo Ionospheric Observatory in Puerto Rico. This giant has many uses, ranging from studies of the earth's ionosphere to the most distant radio sources. [Source: *Science Year. The World Book Science Annual.* Copyright © 1966, Field Enterprises Educational Corporation.]

ble of receiving radiation from any part of the sky and at a variety of wavelengths, they are used to study a wide variety of objects, ranging from the atmosphere of Venus to the radiation of distant quasars. However, because of the great interest in objects such as quasars and pulsars, a very significant fraction of their time, and consequently of the associated support budgets, may be considered as devoted to the study of relativistic objects. If one looks only at the expenditures of the scientists who are studying general relativity, one finds a total expenditure by the federal government of a few million dollars per year. On the other hand, if the various astronomical investigations that are related to this subfield are included, the expenditures may be ten times higher.

The nature of the enterprise involves theoretical physicists and astronomers, working with paper, pencil, and computer to delineate the implications of Einstein's theory; it also involves a much larger number of physics or astronomy PhD's who work with a variety of instruments to detect radiation from distant sources. Before World War II, the only such information available was from optical telescopes in the western part of the United States, many of them built with private funds, such as the 100-in. telescope on Mt. Wilson. Since the war, federal funds have been used to build a variety of optical telescopes in this country, with two new 150-in. optical telescopes scheduled for completion in 1973. Their power to penetrate space will rival that of the justly famous 200-in. telescope on Mount Palomar (see Figures 4.75 and 4.76.) In addition, smaller instruments have been built and used for the exploration of infrared emission by distant quasars. Because the atmosphere inhibits such observations, the telescopes are often mounted in airplanes or balloons. National observatories funded by the federal government account for a sizable part of the effort in this subfield; they provide larger instruments than any one institute or university can support. These instruments are then used by university personnel, as well as by the staff of the observatory, to study the faint signals from space on which the science depends.

In recent years, rockets and satellites above the earth's atmosphere have been used to study those wavelengths that cannot penetrate the earth's atmosphere, particularly x and gamma radiation. Such studies have shown

FIGURE 4.75   200-in. Hale telescope pointing to zenith; seen from the east. (Photograph courtesy of the Hale Observatories.)

FIGURE 4.76 NGC 5194, spiral nebula in Canes Venatici. Photograph taken with the 200-in. telescope. (Photograph courtesy of the Hale Observatories.)

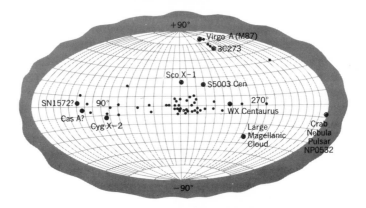

FIGURE 4.77 X-ray astronomy. *Top:* Distribution of the cosmic x-ray sources plotted in the galactic ($/^{11}$, b^{11}) coordinate system. Most of the sources lie close to the equator and are in the Milky Way galaxy. Two sources, Virgo A and the Large Magellanic Cloud, have been positively identified with other galaxies. Encircled sources have been optically identified.

Bottom: X-ray telescope to study solar x-ray phenomena from Skylab. The x rays enter the narrow annulus in the front and are focused by glancing-incidence fused-silica mirrors. (NASA photograph.)

[Source: *Yearbook of Science and Technology 1971.* Copyright © 1971, by McGraw-Hill Book Company, Inc. Used with permission of McGraw-Hill Book Company.]

that a nearby galaxy noted first for its radio emission emits even more energy in the form of x rays. All the instruments used are expensive. A large radio telescope costs approximately $10 million, and a large optical telescope costs a comparable amount. Space instrumentation can be even more expensive, because of the requirement of operating the facility by automated means at a great distance from the earth and because of the rocket power that is necessary to put such devices into orbit. The detailed instrumentation needs for relativistic astrophysics have been examined in the Report of the Panel on Astrophysics and Relativity (see Volume II) and are the subject of further remarks in Chapter 5 as well as many of the recommendations of the Astronomy Survey Committee of the National Academy of Sciences.

Present Status and Future Hopes

Progress in this subfield has been little short of astonishing, given the instruments that are available for detecting faint radiation. Optical astronomers have used the large West Coast telescopes to study the physics of quasars and galaxies. Infrared astronomers have made very significant measurements with their still rather modest instruments. X-ray astronomers, using rockets and orbiting observatories, have surveyed the sky and pinpointed interesting x-ray sources (see Figure 4.77). Radio astronomers have utilized radio telescopes in various parts of the United States to survey a large number of radio sources.

In the years ahead optical astronomers will bring into operation the two new 150-in. telescopes already mentioned, which will yield a new stream of information, particularly if this information can be processed at the focus of the telescope with the most recent electronic devices. Infrared astronomers also plan a significant increase in aperture capabilities. X-ray and gamma-ray astronomers are placing their hopes on a series of satellites, particularly a high-energy astronomical observatory, which will embrace a whole series of x-ray and gamma-ray experiments in one satellite. Since this satellite will operate for a year or more and survey the entire sky instead of being confined to a few minutes of observation, as are rockets, an enormous amount of new information on this subject should result. Radio astronomers for some time have planned a series of new instruments that will permit them to form a clearer picture of the distant radio sources that have been detected previously. So far it has not been possible to obtain funds for the largest of these, an array of radio telescopes capable of forming a fine beam for probing distant space, but it is hoped that this lack can be remedied in the next few years.

Distribution of Activity

Astrophysics and relativity, defined as the application of general relativity to astronomical systems, is actively pursued at comparatively few institutions in the United States and abroad. With a broader definition, one finds that much of contemporary astronomy and astrophysics, a significant amount of theoretical and experimental physics, and a considerable fraction of the space program are related in one way or another to this subfield. Several hundred physicists and astronomers are members of special divisions of the American Astronomical Society and the American Physical Society that deal with subjects in the scope of this subfield. Work is proceeding in dozens of astronomy and physics departments across the country, as well as at several research centers maintained by federal funds.

Einstein's theory is extraordinarily complicated mathematically. Because, until recently, there were few problems subject to observations for which it seemed necessary to use this theory, its study was primarily carried on by mathematicians, although a few theoretical and experimental physicists scattered about the country maintained an interest in it. But with the discovery of the extraordinary astronomical phenomena mentioned above, greater effort has gone into applying the theory to realistic models of stars and galaxies. As a result, relativity is no longer a mathematical plaything but a practical tool for finding out what is taking place in the universe.

Astronomers and physicists engaged in work in this subfield find it exciting, for it touches on fundamental questions of the evolution of matter and life. To obtain fundamental proof that some version of general relativity is correct would provide a new philosophical framework preferable to that developed by Newton in the sixteenth century, in which events in space–time, related by inexorable laws, give rise to the extraordinary structure and beauty of the astronomical universe. Increasing numbers of physicists engaged in other subfields, such as nuclear and plasma physics, are attracted to astrophysics and relativity by the opportunity to use their specialized knowledge in solving such enigmas as the pulsar. Substantial funding is required for such research. The support of the general public must be gained if this work is to progress.

Unlike most subfields of physics and astronomy, astrophysics and relativity is not widely recognized as a distinct discipline. The Panel attempted to identify a core group whose research is primarily in this subfield and to identify those federal funds that clearly are supporting the efforts of this group. It also attempted to estimate the additional funds allocated to the support of observational programs that are used less directly by this group, even though these funds have not been specified by the federal agencies as being for this purpose. The estimated percentage of federal funds for research in this subfield (Figure 4.78) includes about one fourth of the total ground-based astronomy program ($12 million per year). When space-based research is included, total support is about $60 million per year. This estimate is quite uncertain. The cost per PhD, of course, reflects the high cost of space experiments. Without the space component, which provides crucial observations of distant relativistic objects, the cost per PhD drops to a figure corresponding more closely to the level of support in most physics subfields.

Problems in the Subfield

Astrophysics is continuously rocked by a stream of astonishing discoveries —quasars, x-ray sources, cosmic radiation, pulsars, and infrared galaxies. Hints of bizarre phenomena such as black holes are on the horizon. These dynamic phenomena, interpreted in terms of Einstein's theory of curved space–time, are leading to a new conception of the universe.

Recent discoveries, following on the opening of the whole electromagnetic spectrum by new techniques, were made by pioneer telescopes, which can be enormously improved by a new generation of instruments now coming off the drawing boards.

ASTROPHYSICS AND RELATIVITY

PHYSICS MANPOWER IN 1970

N = 36,336

FUNDS FOR BASIC
RESEARCH IN PHYSICS IN 1970

FEDERAL

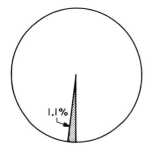

FIGURE 4.78   Manpower and funding in astrophysics and relativity in 1970.

The goal now is the construction of these major new facilities, including a large radio array, a variety of infrared telescopes, digitized imaging devices for major optical telescopes, and a high-energy astronomy observatory in earth orbit. The manpower to perfect and apply these instruments is already available.

That exciting discoveries about the universe will continue to be made cannot be doubted. Other nations have responded by committing major resources to this subfield. The United States must now decide whether it will grasp the opportunity it now has to participate in an all-out effort on the physics of the universe.

> They take place in the regions nearest to the motion of the stars.—It studies also all the affections which we may call common to air and water and kinds and parts of the earth.
> ARISTOTLE (384–322 B.C.)
> *Meteorology*, Book I, 338

EARTH AND PLANETARY PHYSICS

The Nature of the Subfield

Earth and planetary physics is a rather general title that encompasses a wide variety of scientific activities of ever-increasing importance in which physics and physicists play a significant role. The principal concern in this review is the use of the methods and concepts of physics in a number of research areas that deal with large, natural systems that generally cannot be controlled or altered by the observer.* Instead of applying the familiar techniques of laboratory investigation, the physicist studying such systems can only observe and record natural events (see Figures 4.79–4.81). His prime objectives are to understand the observed events and to predict their possible future occurrence.

Two other aspects of this subfield distinguish it from the rest of physics. First, money and effort are invested in studies of the earth and in certain aspects of planetary physics because they relate directly or indirectly to the relationship of man to his physical environment. The questions asked and the investigations pursued, although basic, have strongly applied overtones. For example, in regard to the earth, the prediction of hail, understanding of earthquake damage, analysis of storm surges, reliable

---

* However, many laboratory experiments are performed to study microscale phenomena, for example, water-droplet formation.

FIGURE 4.79 A submarine volcano erupts near Iceland. [Photograph courtesy of Photoreporters, Inc.]

FIGURE 4.80   *Top:* Barnard Glacier and Mt. Natazhat and Mt. Bear from the
south southwest.
   Bottom: Mt. St. Elias looking northwest from over Malaspina Glacier.
   [Photographs courtesy of Bradford Washburn.]

radio communications, discovery of material resources, and better weather predictions are major objectives. There are, of course, both long- and short-term approaches to these problems. As an illustration, almost all research in weather modification thus far has been directed toward understanding the physical bases for weather phenomena rather than toward attempting weather-modification operations, which, in the absence of such understanding, could well have disastrous consequences.

There are many reasons for studying the physics of earth, and although there is a strong aesthetic appeal in understanding phenomena that surround us, the roots of the subfield and the ultimate justification for the high level of activity that it enjoys are securely tied to the practical needs of society. This is in contrast to the physics of planets (planetology), which is concerned with the interiors, surfaces, atmospherics, and satellites of planets and is motivated primarily by scientific curiosity and interest arising from opportunities in the nation's space program. The search for extraterrestrial life belongs to this category, because so much depends on physical techniques and physical phenomena.

Another distinguishing aspect of the subfield is its interdisciplinary character. It is not a closed subfield of physics, although physicists can contribute to much of the activity—much more, indeed, than they have in the past. Its problem-oriented programs and its large-scale projects have brought together a loosely allied community of physicists, chemists, engineers, and technicians with a wide variety of specializations. Each group is dependent on the others and must learn the discipline of working with colleagues whose attitudes and backgrounds differ and with problems whose hallmark is complexity.

The subfield falls into several separable components. Meteorology and atmospheric physics have to do with the lower regions of the atmosphere, with an emphasis on the study of weather and climate. Aeronomy is concerned with the chemical processes and electrical effects in the upper atmosphere, out to the magnetopause where the earth's atmosphere merges with the solar wind.

Oceanography embraces a wide range of disciplines that include marine biology. The subdiscipline of physical oceanography is concerned with global ocean circulation, currents, waves, temperature, salinity, and other physical features.

Solid-earth geophysics embraces a variety of disciplines that relate closely to geology, mineralogy, petrology, and the like. Some of the main concerns here are such areas as: geomagnetism (study of the magnitude, fluctuations, and origins of the earth's magnetic field), seismology (study of the propagation of elastic waves in the earth and of earthquakes), tectonics

FIGURE 4.81 Deep ocean swell. [Photograph courtesy of Jan Hahn, Woods Hole, Massachusetts.]

(the secular motions of the earth's crust and of the underlying mantle), and the physical and chemical properties of natural minerals at high temperatures and pressures.

All branches of atmospheric and solid-earth geophysics are reflected in planetary studies. Although remote from this environment, planetary investigations provide a different look at problems of vital interest to man and, as in the case of the recent studies on the moon, can effectively give a view of the earth at both earlier and later phases of its history. Comparative studies of the earth and the planets complement each other and lead to new approaches and to fresh ideas about fundamental processes.

Finally, as a product of the space age, an entirely new field of study—that of the interplanetary plasma—has appeared. This is related directly to the established physics subfields of cosmic rays and plasma physics; therefore, departments of physics have shown more of an interest in it than in most of the other components of this subfield. The term space physics often is used to describe this subject, as well as the upper atmospheric (aeronomy) studies requiring rockets and satellites.

Historical Perspective

The history of the subfield starts in the distant past, with the ancient interest in terrestrial phenomena and the solar system, but its character has continually changed in response to the demands of society and technical development and opportunity.

Aristotle divided the physical sciences into physics and meteorology, the latter embracing all beneath the orbit of the moon and corresponding roughly to what we now call physics of earth. He understood that physics would embrace the more orderly branches of nature but that meteorology would have to deal with complicated phenomena in which descriptive methods, as opposed to the more abstract processes of induction and deduction, would play an essential role.

In the last two centuries, conventional physics and studies of the sea, atmosphere, and earth have drawn apart. New experimental methods led physics rapidly toward goals concerned with the fundamental nature of matter and radiation, while industrial needs coupled to new technological advances, such as the telegraph and deep-drilling rigs, led studies of the earth to increasingly technical and descriptive methods, with little emphasis on fundamental understanding. These trends are now beginning to reverse, and the proper role of physics in earth and planetary investigations is becoming clearer. Deep understanding of the natural phenomena has become recognized as a necessary step to prediction, understanding, and control, just as it has in other subfields such as plasma physics. One of the primary means of reaching this understanding is through physical research.

A few milestones in the recent history of the various disciplines that contribute to the study of earth and planetary physics are the following:

Meteorology began to move rapidly with the invention of the telegraph and accelerated with the demands of the aircraft industry during and after World War I. Theories of fronts and air masses were the mainstay of the subject through World War II, although the technologies of collection and dissemination of meteorological data improved continuously. Since 1950, the outstanding advance, spurred by von Neumann's insight, was the introduction of the computer, coupled with an increased understanding of the physics of atmospheric processes. Now meteorologists are beginning to exploit a combination of satellite (see Plate 4.XIV) and computer technology and are perhaps on the threshold of modifying climate and weather.

Aeronomy began with the discovery and study of the ionosphere, by Appleton and by Breit and Tuve, in the 1920's. Advance was slow until rockets became available after World War II. Only in recent years, however, was the systematic use of satellites for aeronomy research proposed. The separation between meteorology and aeronomy now appears artificial, and the distinction may disappear in the future.

Physical oceanography is a relatively young science. Studies of ocean tides are not new, but most other aspects of the subject are related to modern dynamical meteorology. In many schools the two subjects are taught together, and research workers cross readily between fields. This subject, too, appears to be on the threshold of rapid advance as more oceanographic research ships become available, new instruments are developed, and satellite technology is used.

Seismological stations have existed for a long time, but solid-earth geophysics did not become a major area of study until the years between the two World Wars when the needs of the petroleum and mining industries gave it impetus. This increased activity was paralleled by the gradual evolution of the belief that the present and past state of the earth could be understood in physical terms. The work of Bridgman, Birch, and others on the physics and behavior of matter at high pressures was a cornerstone in this respect. Since World War II, the origin and behavior of the earth's magnetic field on a geological time scale have emerged as major research topics. The Vela program for detection of underground explosions gave major impetus to seismology, and the gradual increase of observing facilities on ships together with more sophisticated instrumentation and methods for coring the deep-ocean sediments led, in the past few years, to the concept of sea-floor spreading. There is now a completely new picture of the earth's crust, in terms of large mobile plates floating on the underlying mantle, and through their impinging motions producing continental drift, sea-floor spreading, mountain building, and zones of seismic and volcanic activity. Even more important, experiments here and abroad suggest that with better instrumentation of geological faults and a better understanding of earthquake mechanisms it may be possible to predict the occurrence of earthquakes. Russian scientists believe that they can make ten-day predictions of earthquake magnitudes and locations in the Tadzhikistan Province. There is also some hope for preventing earthquakes (when their foci are shallow) by injecting fluid to relieve the strain along rock fractures.

Physics of the interplanetary medium is a direct product of space research. Its beginning can be placed a little earlier, but advance was not rapid until Van Allen's discovery of the radiation belts in one of the original International Geophysical Year (IGY) satellite flights in 1958.

A similar statement can be made about planetary studies generally. They have had a long history as part of astronomy, but advances were slow, and most astronomers professed little interest in earthlike objects. With the Mariner and Venera explorations of Mars and Venus and the Surveyor and Apollo missions to the moon, the subject changed rapidly and is now very actively studied, especially planetary surfaces, interiors, and atmospheres and the whole problem of the evolution of the solar system.

How Research Is Conducted

In its approach to observations, earth and planetary physics differs importantly from conventional physics; this difference is a crucial part of the character of the subfield. First, there is the problem of field observations. Since control of the atmosphere, the oceans, or the earth is not possible, it is necessary to measure them in three dimensions and follow the changes with time. Many problems arise. The observations can be difficult, even dangerous, and almost invariably they involve the discomfort and inconvenience of grappling with the environment. But this situation has its converse, for many scientists relish this closer contact with their environment.

Observations on a large scale are invariably expensive. Few research workers in this subfield would complain that the resources needed for research have been inadequate. However, the need to seek and justify large funds and to handle them responsibly are immensely time-consuming occupations, often involving political aspects. As in many other subfields that require large facilities and support, an effective scientist must be prepared to devote a significant fraction of his time to these essentially administrative matters.

The physical phenomena of the earth know no national boundaries, and, since the earliest times, investigations in this subfield have been international. A network of scientific organizations (principally under the International Union of Geodesy and Geophysics and the Committee on Space Research) and intergovernmental organizations (for example, the World Meteorological Organization) exists to initiate and coordinate research programs, to disseminate information, to aid less-developed countries scientifically, and to perform many other essential functions (see Plate 4.XV). Research programs have occasionally been such remarkable international successes that they have attracted the explicit attention and support of both diplomats and politicians. The IGY and the continuing cooperation in the Antarctic are two of the most striking examples of such international ventures.

Organizational and logistical problems associated with earth and planetary research, and the obvious interest of the military in some of its aspects, led to large in-house federal research efforts. For example, meteorological research is principally, but not exclusively, carried out by the Weather Bureau, now a part of the National Oceanic and Atmospheric Administration (NOAA) of the Department of Commerce. Partly to enable the universities to participate in big science, a number of national activities have developed: national research laboratories, research facilities, national programs, and data centers for dissemination of information. National programs are frequently organized by government agencies, sometimes also with international collaboration. The Arctic Ice Deformation Joint Experiment is an example of a largely national program involving universities, government agencies, and industry.

The National Center for Atmospheric Research (NCAR) is an example of a laboratory operated by a consortium of universities using government funds. The huge radio telescope at Arecibo, Puerto Rico,* is an example of a research facility, open to all competent researchers, operated by a single university—Cornell. The Greenbank radio-telescope installation is operated by Associated Universities Incorporated (AUI), the same organization that operates the elementary-particle, nuclear, and other facilities and programs of the Brookhaven National Laboratory.

Many of the observational data in this subfield are obtained for operational purposes. Meteorological information will be collected from ground stations, balloons, and aircraft for purely practical reasons regardless of any research requirements. The same is true for much ionospheric, seismographic, and oceanographic data. The existence of these data in reasonably processed form is one of the most important requirements for research in the subfield, and additional data centers are needed for their dissemination.

Similarly, studies pertaining to earth, studies of other planets, and satellites require the use of major astronomical observatories with optical, radar, and radio instrumentation, as well as major projects and programs such as Mariner, Pioneer, TOPS, the Grand Tour, and Stratoscope. Many of these studies require international collaboration, but usually at a lesser scale than the terrestrial research programs discussed above.

* Now the National Astronomy and Ionospheric Center (NAIC).

Recent Achievements

Much more than conventional physics, this subfield is characterized by a gradual evolution of new concepts reflecting the combined thinking and contributions of a number of scientists. Discoveries of totally new phenomena are rare, as is the sudden emergence of new ideas. But it is also rare to look back over a decade without finding impressive advances.

In the past 20 years in physical oceanography, for example, major advances in both observing methods and theory have occurred. From the observational point of view the development of new sensors, or remote observing systems attached to buoys, and the use of numbers of ships to collect synoptic data have laid the foundation for a three-dimensional picture of temperatures, chemical composition, and currents in the sea. However, surprises still occur; after centuries of ocean travel and observation, not until the 1950's and early 1960's was the Equatorial Countercurrent discovered and described. This current is a remarkable narrow and shallow flow, with a maximum intensity almost exactly on the equator, moving in a direction *opposite* to that of the easterly trade winds.

On the theoretical side, methods developed by dynamical meteorologists have proved remarkably successful when applied to oceanic phenomena. There are now rather elaborate theories of the Gulf Stream, the Kuroshio, and the equatorial undercurrents and their relationship to the slower circulations in the main part of the oceans.

In the geophysics of the solid earth, the outstanding discovery of recent years is sea-floor spreading and its spectacular confirmation of the idea that continents are drifting apart, having once been joined (see Figure 4.82). The crust of the earth is now regarded as in a continual state of motion, with huge plates driven by upwelling currents from the interior.

Also of major importance has been the gradual evolution of ideas about the generation of the earth's magnetic field. That the field must be related to a dynamo motion in the earth's conducting, liquid interior is now known, although the precise mechanism has yet to be fully understood. Remarkably precise evidence shows that the direction of the earth's magnetic field has reversed many times, suddenly and for periods of varying duration, throughout geologic history (see Figure 4.83). Indeed, the record of these reversals written in the solidifying magma from midocean upwelling, can be read with as much confidence as tree rings in establishing a much older and more far-reaching chronology of the earth. Physical techniques have played a crucial role here in evolving magnetometers of ever-increasing sensitivity and convenience.

Meteorology since World War II has changed in almost all of its major

120 million years ago

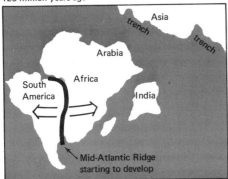

60 million years ago

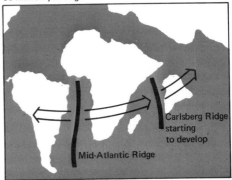

30 million years ago

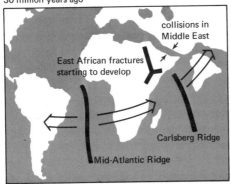

Present

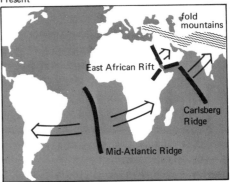

FIGURE 4.82 African and American continents probably started their drift apart some 120 million years ago. The initial separation of the combined African and Indian land masses from South America seems to have been caused by ocean-floor material spreading out of the mid-Atlantic Ridge. By about some 60 million years ago, the Carlsberg Ridge had begun to develop between Africa and India and helped to drive India toward Asia; one dramatic consequence was the formation of the Himalayas. About 30 million years ago, another ridge seems to have formed in the Gulf of Aden and, separated from the Carlsberg Ridge by the Owen Fracture Zone of transform faults, this has continued to develop independently. [Source: *Science Year. The World Book Science Annual.* Copyright © 1968, Field Enterprises Educational Corporation.]

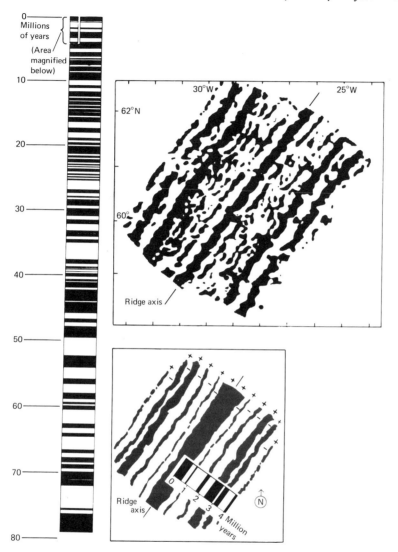

FIGURE 4.83 History of undersea magnetic reversals. The geomagnetic time scale
(left) of 80 million years of magnetic reversals along one side of the mid-Atlantic
Ridge matches an identical pattern on the opposite side of the Ridge. This time scale
represents belts of rock of alternating polarity such as revealed in a map of Reykjanes
Ridge, near Iceland (below). Each belt reflects the direction of the earth's magnetism
while molten rock welled up through the Ridge rift. The magnetic map (top) has
been clarified in the diagram (bottom). Note the close comparison between ridge
reversals on the map and the geomagnetic time scale. [Source: *Science Year. The
World Book Science Annual.* Copyright © 1968, Field Enterprises Educational
Corporation.]

aspects. The mapping of upper air currents (notably the midlatitude jet stream) has given rise to studies of the stability of atmospheric motions and the upgradient transport of momentum. The outline of a theory of the general circulation is now available. Weather forecasting has been revolutionized by the advent of the computer, but the realization of this tool required the solution of many theoretical problems since the introduction of the idea by Richardson—prematurely, soon after World War I—and by von Neumann, following World War II. The last 20–30 years have brought about a rather complete understanding of the processes of precipitation, so that some forms of weather control now appear feasible. Fundamental questions of uncertainty of prediction in fluid problems are receiving attention. Theoretical climatology is imminent. Air chemistry is an established science and has provided the scientific basis for much of the recent concern about atmospheric pollution.

New laser-based techniques that permit extremely rapid and precise vertical density and composition profiles in the atmosphere, like those developed in oceanography, are beginning to make the construction of a realistic three-dimensional picture of the earth's atmosphere a reality. Much more extensive implementation of these new techniques is essential to further progress.

In aeronomy, a rather complete phenomenological description of the ionosphere has been developed. Chemical processes are studied in increasing complexity, so that the subject is now ready to come to grips with problems of upper-atmosphere pollution—questions of ever-increasing urgency as the density of supersonic high-altitude aircraft traffic increases. As dynamic transport processes by winds are included in the calculations, the subject slowly becomes more like lower-atmosphere meteorology.

Almost all recent planetary discoveries have come from space probes. Lunar samples have extended the history of the solar system back in time by a billion years. The surface of Mars (see Figure 4.84), its meteorology, and its aeronomy have become familiar. We are now poised for an attempt to detect life forms on that planet. Knowledge of Venus, with its surface at dull red heat, is evolving more slowly because of the ubiquitous cloud cover, but the Soviet and U.S. probes have reported back from this inhospitable surface (see Figure 4.85). The outer planets are less well understood because of their strangeness and remoteness, but here, too, ideas of composition and structure are developing at an increased pace.

Finally, the interplanetary plasma or the solar wind is mapped in detail near the earth, revealing a complicated system of shock waves and a hydromagnetic tail. The interaction of the solar wind with the moon and with Mars and Venus is also partially understood, and scientists now look toward the limits of the heliosphere, where the planetary system blends with the galactic system.

PLATE 4.XIV   Earth as seen from space. [Source: National Aeronautics and Space Administration.]

Weather Observation: Present and Planned

- (grey shape) Existing: observations considered adequate
- ◆ Existing: more equipment and/or observations needed
- ⊡ Proposed: upper air
- + Proposed: fixed ship

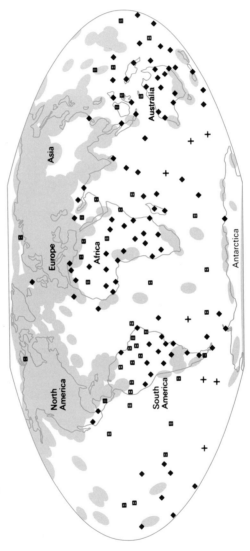

PLATE 4.XV Weather observation—present and planned. Deficiencies in weather observation stations are most noticeable in the Southern Hemisphere. The World Meteorological Organization has called for more equipment and increased observations at existing stations and for more surface and upper-air observatories. [Source: *Science Year. The World Book Science Annual.* Copyright © 1967, Field Enterprises Educational Corporation.]

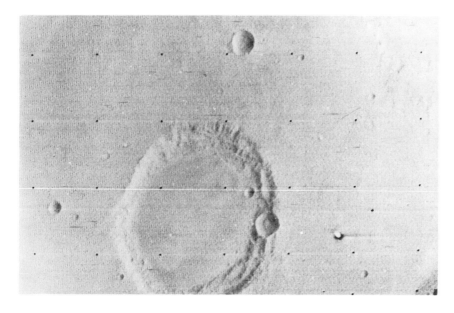

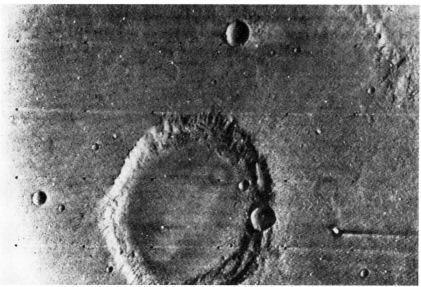

FIGURE 4.84   An illustration of computer enhancement of a Mariner probe photograph of the Martian surface. This is photograph 6N18 taken on July 31, 1969. The
upper photograph is as received from the probe, and the lower is that obtained after
computer enhancement. Enhancement is accomplished using the techniques developed
at the California Institute of Technology Jet Propulsion Laboratory under NASA sponsorship and described by T. C. Rindfleish *et al.* in the *Journal of Geophysical Research, 76,* 394 (1971).

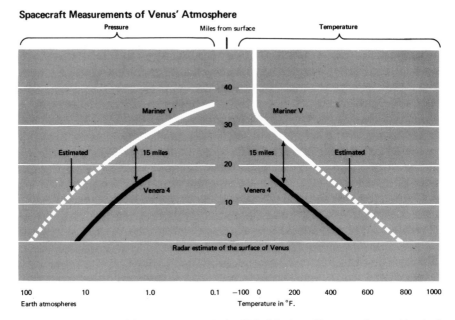

FIGURE 4.85   Russia's Venera 4 and the U.S. Mariner V returned near-identical data on the temperature and pressure of the atmosphere of Venus. The unexplained 15-mile gap could be from errors in earth-based radar measurements of the surface of the planet. [Source: *Science Year. The World Book Science Annual.* Copyright © 1968, Field Enterprises Educational Corporation.]

All of these developments have depended heavily on techniques and devices of physics, and the phenomena involved are those of classical physics, modern theory, condensed matter, hydrodynamics, optics, and electromagnetic theory, as well as those of the more modern subfields such as plasma physics and magnetohydrodynamics.

The Future of Earth and Planetary Physics

It is difficult to see limits to the development of this subfield. The pace of advance accelerates year by year, while the range of questions to be answered increases as applications become ever more apparent.

In physical oceanography, a better understanding of the transport of heat, mass, and momentum on both large and small scales should emerge. The large-scale transport couples with transport by the atmosphere and may be related to climactic changes. Transport by small-scale motions is important for the dispersal of pollutants. But the frontier of knowledge lies in the deep abyssal circulations. These have to be recorded and explained; advances in this area can be expected as the technology for abyssal study improves.

Meteorology should advance rapidly in the next decade, with the completion of the Global Atmospheric Research Program (GARP). The ten-day numerical forecast could well become a reality. And, importantly for many underdeveloped parts of the world, tropical meteorology should begin to catch up with midlatitude studies. The first major steps in weather modification could come in the next decade. Theoretical climatology and an explanation of climatic change on many time scales, including the ice ages, may be just over the horizon.

Solid-earth studies will change greatly as the ideas of plate tectonics and sea-floor spreading are exploited. The beginnings of earthquake forecasting have emerged. In this period an acceptable theory of the earth's magnetic field could be developed, as well as a clearer understanding of the seismological tool for studying the earth's interior. As increasing attention is paid to laboratory study of the properties of matter at high temperatures and pressures, studies of planetary interiors and of the early events in the solar system should become less speculative.

In space physics, measurements of solar and stellar particles in the solar ecliptic and beyond the limits of the solar system are anticipated. The purpose of the atmosphere Explorer missions will be to establish quantitatively the effects of specific processes that produce the F- and upper E-regions of the earth's ionosphere. Planetary exploration is likely to proceed at a rapid pace. Mercury will be reached, and its surface photographed. Probes and orbiters, tailored to a relatively modest budget, are likely to be launched to Venus. Mars will be the subject of extensive exploration, including the elaborate Viking lander that seeks to detect a new form of life. The first steps toward the exploration of the outer planets are now under way. Lunar studies may benefit from new unmanned techniques, although the Apollo program, after a remarkable record of achievements, is drawing to an end. All of these investigations are exploratory, and each of them should bring surprises and completely new insights.

Relationship to Societal Needs

Despite many new fundamental developments, the bases for support of this subfield rest, and will continue to rest, on the extent to which it can answer questions of more direct concern to mankind. Thus, in oceanography the main thrusts probably will relate to the production of food from the sea, the interaction of sea and coastline, the global distribution of water pollutants, and exploration and exploitation of the resources of the ocean floor. But the whole subject is in an early stage of development, and relatively undi-rected exploration of the motions and physical properties of the ocean at all depths must also be a feature of the studies conducted in coming decades. New insights emerging from fundamental studies will continue to generate opportunities for more applied investigations, but applications alone are neither a sufficient nor an efficient guide to the choice of problems for study.

In meteorology, the key problems will be, as they always have been, the prediction of weather and the possible control of weather and climate. Again, these questions can be answered only through deep exploration and understanding of fundamental questions. Problems of air pollution, which have long concerned meteorologists, will receive increasing study as their importance is brought to the attention of the general public.

In solid-earth geophysics, the traditional application has been petroleum and mineral exploration. With the rapid increase in the rate of utilization of these resources, locating additional reserves—often in less concentrated and accessible forms—assumes new and growing importance. Here the geophysicist, with his new instruments—magnetometer, earth conductivity probes, portable ^{252}Cf neutron probes, radiation-measuring equipment, and the like—plays a major role. To some extent emphasis will shift from local exploration to a more global geochemical census of mineral provinces. In addition to the vitally important goal of mineral exploration (scientific prospecting), the prediction of at least some features of the occurrence of earthquakes is a possibility.

Planetary exploration and space physics do not have direct applications to man's needs, except in certain specific cases. Radio transmission is still an important topic and was the original reason for the development of upper-atmospheric physics. There are obvious military applications of space research, and satellite techniques will continue to yield powerful new techniques for many kinds of terrestrial observation and monitoring, includ-ing resources. Further, planetary studies support, enlarge, and illuminate the findings of terrestrial studies. Although this is an indirect reason for support of planetary research, it remains a substantial consideration.

Manpower

A discussion of the role of physicists in the earth and space sciences presents a number of aspects that are not encountered in the traditional core subfields of physics. There are two major sources of difficulty. First, earth and planetary physics is not a separate, clear-cut field. Physics is one of the disciplines involved in the earth and space sciences. To be effective, the physicist working in these areas must combine his efforts with those of engineers, chemists, life scientists, meteorologists, geologists, and others. In doing so, his view of himself and the research process can change significantly from that engendered by his early training and experience as a physicist. Possibly, he will cease to identify himself as a physicist.

The second difficulty relates to the inhomogeneity of the earth sciences. Although all have some similarity in approach, the differences are real and important. Thus the geological sciences employ by far the largest number of scientists and differ from other earth and space sciences in having a large industrially employed segment and a large proportion of teachers. The geological sciences constitute a mature field, transfers to and from which are slow. Meteorology also is a mature field with low turnover of personnel, but industrial involvement is small and that in government-supported activities, high. In meteorology, research and development tend to be conducted by non-PhD's. On the other hand, both aeronomy and interplanetary research have always been populated largely by physicists. Data based on earth and space scientists in the societies composing the American Institute of Physics (AIP) tend to be weighted toward this group. Oceanography only recently has become an important field for physicists. It has a large interface with the life sciences. Both oceanographic and space sciences employ small numbers of scientists compared with the atmospheric sciences, which, in turn, are dwarfed in comparison with the geological sciences.

The main concern here is the relationship between the discipline of physics on the one hand and the earth and space sciences on the other. The interface defines the earth and planetary subfield. The relationship to physics usually is established by the nature of the highest academic degree, AIP membership, and self-identification. All three methods of identification are fallible; in particular, all can fail to identify a person whose research methods are those of physics but whose graduate department was not physics. Thus estimates of the number of physicists working in the interface are only approximate.

The 1968 edition of the NSF publication, *American Science Manpower,* provides data on those portions of the National Register of Scientific and Technical Personnel identifying the members of the American Meteorological Society and the American Geological Union. These data, together with 1968 and 1970 data on physicists identified with earth and planetary physics in the National Register survey, afford a partial description of the characteristics of scientists working in the subfield.

American Science Manpower 1968 shows that approximately 10 percent of the 298,000 participants in the 1968 National Register survey were identified with atmospheric, geological, and oceanographic disciplines, and that about 11 percent were identified with physics. About 1600 of the physicists work in the three former disciplines; however, very few of those in the earth and space science groups work in physics. A higher percentage of PhD's, composed largely of physicists, is found in the space sciences than in atmospheric, geological, and oceanographic disciplines taken together. In the space sciences, the percentage of PhD's (49 percent) is, as might be anticipated, close to that for physics (44 percent). Only 16 percent of the 23,160 scientists identified with atmospheric, earth, and ocean sciences in the 1968 Register survey held PhD's.

Scientists who indicated earth and planetary physics as their subfield of employment in 1968 numbered 1193, or 3.6 percent of the 32,491 physics respondents to the Register. In 1970, 1448—4 percent—of the 36,336 physics respondents to the Register were identified with earth and planetary physics. Exactly half of the physicists in this subfield held PhD degrees and comprised 4.4 percent of all PhD's in physics. The 724 nondoctorates in this subfield accounted for 3.7 percent of the total number of doctorates in physics in 1970.

The largest fraction of the PhD physicists working in earth and planetary physics is concentrated in the universities (38 percent); industry and government and federally funded research laboratories also account for substantial percentages (23 percent and 35 percent, respectively).

A high proportion of PhD's and non-PhD's in earth and planetary physics is engaged in research (67 percent and 64 percent, respectively). The comparable figures for all physics are 55 percent for PhD's and 47 percent for non-PhD's. In the earth sciences as a whole, the figures are 32 percent (PhD) and 18 percent (non-PhD). Thus patterns of activity and employment in earth and planetary physics differ from those that characterize earth sciences and physics.

Funding Patterns

In recognition of the importance of earth and planetary physics to societal needs, funding from federal sources has been relatively generous. Approximately $150 million per year is spent on research and research support, not including the NASA program. This sum divides in roughly equal proportions among meteorology, oceanography, and solid-earth geophysics; the major part of the expenditure is for work in government laboratories. Associated with this sum spent on research are far greater expenditures on operations and applications. The NASA program varies greatly from year to year, and the costs of hardware and development are difficult to evaluate.

These expenditures have generally been on a sufficient scale to support major demands, at least outside the space program. Promising new lines of research have usually been funded when the need arises. However, it is characteristic of most problems related to societal needs that the end is never in sight. Each improvement in weather forecasting leads to further demands: Man is rarely satisfied with existing capabilities until the point is reached at which the cost of advance is greater than the probable return. This last problem could be closer than scientists care to admit and should receive constant scrutiny.

The rate of advance is closely related to the level of support. The level of funding of research has not significantly changed in the last five years. However, the real value of this money has decreased. For example, progress in weather and climate control during this interval has been relatively slow. It is true that this is a very difficult area that contains complicated legal and political elements. However, the slow rate of advance should be a matter of greater concern than it has been.

In addition, solid-earth geophysics is entering a period of discovery and potential progress. Understanding of the dynamics of the earth's crust and the nature of earthquakes is growing. So far, federal agencies have not responded adequately to this situation. Perhaps the time has been too short, but adequate support for research is essential in the near future.

The final major question in funding patterns concerns the NASA program, particularly the planetary program. There is no area of knowledge that has been and will be more radically changed by the space program than that of the planetary system. The transition from the remote methods of astronomy to direct contact has created completely new fields of study for which the space probe is the unique tool of investigation. These fields are concerned with problems of historical importance to mankind that also have an impact on human needs.

Problems in the Subfield

The major problems in earth and planetary physics differ from one branch to another of this complex subfield. In most areas of the earth sciences the critically important task is to upgrade the research personnel so that they can undertake the extremely difficult physical problems that are now being revealed. The key to this problem lies in the universities. It is essential that earth and planetary physics be integrated in the physics teaching structure. By this means some of the excellent students entering physics programs will be able to appreciate the opportunities for research outside the laboratory and outside the core subfields of physics. The Report of the Panel on Earth and Planetary Physics (Volume II) offers recommendations on this and other subjects related to the educational problem and makes suggestions to foster the employment of more physicists in the earth sciences.

In the space sciences, the problems are fiscal and organizational. Again the Panel Report offers a variety of recommendations about the level of funding in space physics, the balance between large and small missions, NASA's advisory structure, the rationale for a space-science program, and other matters related to these questions.

Chemistry is undergoing a renaissance. This is a new science. It is, perhaps, oldest of the laboratory sciences but it had fallen into the doldrums about the time of World War II; even the chemists found chemistry rather dull. Just after World War II the situation wasn't much improved. What has changed chemistry again has been the availability of a new set of tools. They have names which you may have read or heard of in one place or another: Nuclear magnetic resonance, spectrophotometry, very careful spectroscopy, electron diffraction, electron paramagnetic resonance spectrometry, instruments for measuring circular dictroism, mass spectrometry, and a few others combine to give the chemists a completely new set of handles on that aspect of the world for which they are responsible. This has permitted the chemists to understand the structure of molecules in a way no chemist understood them before. Instead of vague, two-dimensional chicken tracks written on paper, chemists now have a very clear understanding of the three-dimensional structure and electronic configuration of a large number of molecules as well as of chemical reaction on mechanisms.

Statement by Philip Handler, testimony before Daddario Committee, 1970 NSF Authorization Hearings, p. 10, Vol. 1 of transcript.

PHYSICS IN CHEMISTRY

Introduction

Throughout the long history of the physical sciences, physics and chemistry have maintained a very active and fruitful interchange of ideas and approaches. Progress has been complementary in the sense that major advances in one have opened new horizons or spurred new developments in the other. Interaction at the chemistry–physics interface is increasing rapidly. This discussion describes developments at this interface, which traditionally has been defined as *chemical physics* but includes other closely related subjects.

It is neither simple nor necessary to formulate precise definitions to distinguish physics from chemistry. Even so, it is useful to recognize certain characteristics that differentiate most of physics from most of chemistry. These are best expressed in terms of goals and attitudes, rather than explicit content. Physicists aspire to establish the *general* laws of behavior common to all of nature, while chemists establish the laws that distinguish one substance from another. Physicists tend to be more concerned with the common features of phenomena of matter; chemists, with those features that differentiate one species of matter from another.

Other distinctions between these disciplines have been suggested. Some people would, for example, restrict the definition of chemistry to studies concerned with extranuclear matter, so that nuclear structure would fall outside chemistry but chemical shifts in Mössbauer spectra would lie within chemistry. Much of what has been called the physics of condensed matter falls into the intermediate area between physics and chemistry because of its concern with materials and its attention to the distinctions among these substances. Yet even here, for example, condensed-matter physics tends to be concerned with the generic properties of crystals, while chemistry is more concerned with their classification and differentiation. Quite obviously, the interface moves with time; it is markedly different from what it was a few decades ago.

Subjects usually considered part of chemical physics include theory pertaining to wave properties of atoms and molecules; spectroscopy of atoms and molecules, from the radio-frequency range to some aspects of gamma-ray spectroscopy; chemical kinetics and collision processes, including hot-atom chemistry, radiation damage, and atomic and electronic collision processes at energies up to kilovolts or, perhaps, tens of kilovolts; the entire field of the structure of liquids (see Figure 4.86); polymeric molecules; statistical mechanics of both equilibrium and transport processes; molecular crystals, but only certain highly selected properties of other types of crystals, such as metals or semiconductors; constitutive thermodynamic properties of all phases of matter except perhaps superfluids, superconductors, and plasmas; and the use of lasers to study properties of matter (for example, nonlinearities). In the United States, these topics are frequently included within the scope of chemical physics; however, several of these subjects are considered pure physics in Europe and Japan.

Among the subjects that are included in the interface but are frequently excluded from chemical physics are properties of ionic, metallic, and covalent crystals and other periodic structure; amorphous materials such as glasses; semiconductors; the origin of the elements and several other aspects of nuclear physics, including many aspects of nuclear spectroscopy of radioactive species and some aspects of nuclear reactions; and superconductivity,

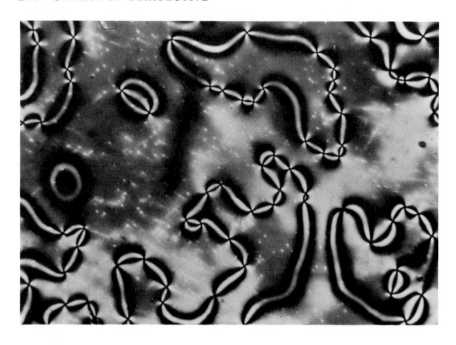

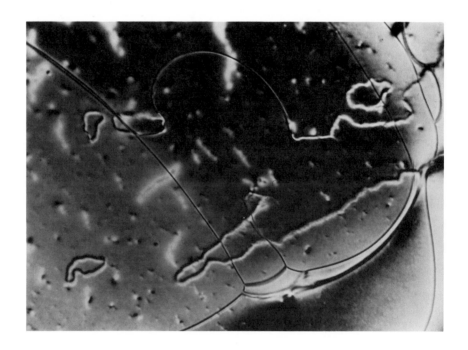

◄ FIGURE 4.86   Liquid crystals. *Top:* All the molecules along any black curve are aligned in the same direction in this polarized light view of the nematic crystal MBBA.

Bottom: This nematic liquid crystal seen in polarized light reveals the internal threadlike structures that are the result of spontaneous orientation of the molecules along their long axes. In this relatively thick specimen (about 100 μm) of MBBA, the thicker threads are on the surface, and the thinner ones are connected to the surface only at their ends.

[Photographs courtesy of Glenn Brown, Kent State University and the National Science Foundation.]

particularly with regard to its microscopic level of interpretation and the development of superconducting materials having specific properties.

Just as some subjects clearly relate chemistry and physics and develop quickly because of this relationship, some appear to suffer unduly slow development because of inadequate communication between these fields. The development of materials having specifically desired properties is one such area—for example, crystalline and amorphous solids having particular optical or electrical properties, liquids with particular viscous properties, or gas mixtures with special capabilities for energy transfer. As later examples will indicate, the communication between physicists and chemists dealing with properties of materials appears to be much more frequent and effective in industrial laboratories than it is in most universities, where too often there is relatively little communication between the two sciences.

The Influence of Physics and Chemistry on One Another

In some respects the relationships between physics and chemistry are asymmetrical. One illustration is the flow of manpower from subjects traditionally in one discipline to subjects associated with the other. A second illustration relates to the influence of one discipline on the other.

Although people trained as physicists in some instances practice what they call chemistry, movement from one discipline to the other generally is from chemistry, especially physical chemistry, into chemical physics or physics. Often these physical chemists and chemical physicists move into subjects in physics that pose challenging and important problems but currently receive little attention from the physics community. Possibly, the tendency for persons trained in chemistry to work on problems formerly regarded as belonging to traditional physics reflects the changing nature of chemistry and the delay between the time an approach is first used to solve chemical problems and the time when it is well assimilated into the chemical literature.

In regard to the influence of the content of one discipline on the other,

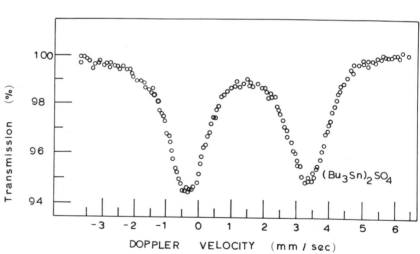

FIGURE 4.87 *Top:* Source–absorber–detector arrangement in a typical Mössbauer experiment. The radioactive source is attached to the velocity transducer on the right, which provides the energy modulation necessary in Mössbauer spectroscopy. The gamma radiation passes through the experimental sample, which is held at cryogenic temperature in the Dewar whose tail section is located in the center of the op-

we are clearly in a period in which most of the basic theoretical framework for chemistry and most of the chemist's physical methods for experimental research trace their origins to physics. Quantum mechanics, statistical mechanics, and thermodynamics—the entire theory of the extranuclear structure of matter—are essentially physical laws that underlie any theoretical interpretation of chemistry. Many (but not all) chemists now feel that the theory of the chemical bond has been reasonably well explained in terms of quantum mechanics. Fundamental and productive insights continue to occur in chemistry as a result of the application of quantum-mechanical methods to chemical problems. One particularly powerful tool, recently developed, is a set of interpretive and qualitative rules governing the spatial arrangements preferred by reacting species. These rules often permit the synthetic chemist to select from several possibilities a specific path by which he can control in detail the geometric arrangement of atoms within a molecule. He can, for example, find methods for synthesizing only the active form of a chemotherapeutic drug, instead of having the greater part of his product produced in inactive forms. Formerly, only natural biosynthetic processes exhibited this kind of efficiency. Only recently has it been recognized that the human olfactory system has a remarkable sensitivity to molecular structure such that the stereoisomers of the same molecule, except in rare circumstances, have strikingly different smells. There may be many more situations wherein the ability to control molecular architecture in detail can have important and practical consequences.

Experimental chemistry has been deeply influenced by physical methods. Spectroscopy, from the radio-frequency region of nuclear magnetic resonance through microwave, infrared, visible, ultraviolet, x-ray, and even some aspects of gamma-ray spectroscopy (see Figure 4.87), is vital to the chemist. Nuclear magnetic resonance (see Figure 4.88) has become an exceedingly powerful tool for analyzing chemical structures, because in most cases the response of a nucleus to this probe is highly characteristic of its immediate environment. For example, one can count the numbers of

---

tical path and impinges on the radiation detector on the left. A multichannel analyzer (not shown) stores the counting rate versus source–absorber velocity data for each cycle of the velocity transducer.

Bottom: Typical Mössbauer spectrum of the ^{119}Sn resonance (23.8 keV) in an organometallic compound. The presence of two resonance maxima (at ~ -0.3 mm/sec and at $\sim +3.3$ mm/sec) is due to the fact that the tin atom occupies a noncubic symmetry lattice site in the experimental compound. These data were recorded with the absorber held at liquid nitrogen temperature (78 K) in the Dewar tail section shown in the above figure.

[Courtesy of R. H. Herber, Rutgers University.]

FIGURE 4.88 Bruker HFX-90 Multinuclear cw and pulsed FT NMR spectrometer consisting of (left to right) a Nicolet 1083 computer, magnet, and transmitter–receiver console. Variable-temperature and double-resonance capabilities are also present. [Photograph courtesy of the Department of Chemistry, Florida State University.]

hydrogen atoms in each of many kinds of site, even in rather complex molecules. The method is being developed and used now for the study of the structures of protein molecules. Electron spin resonance, in the microwave region, permits scientists to probe the location, hence the reactive site [as well as the structure (see Plate 4.XVI)] in the large class of highly active intermediates known as free radicals, the species responsible for the formation of many polymers. Microwave and infrared spectroscopy have enabled the chemist to study the sizes and shapes of molecules by letting him see how they rotate and vibrate. They also make possible the recognition of characteristic reactive parts of molecules for purposes of identification and monitoring. Ultraviolet and x-ray spectroscopy are tools for studying the electrons within the molecule; they permit the chemist to study, for example, the chemical bonds and the colors of substances. Chemists have applied lasers in much of their work, including photochemistry, the study of energy conversion, and the probing of the energy levels of bound electrons.

Other techniques from physics have equally broad use. Mass spectrometry and x-ray crystallography are now necessary tools for the chemist.

One of the most striking recent transfers from physics to chemistry pertains to atomic and molecular beams. In this burgeoning chemical field the elementary atomic and molecular interactions, the electronic transfers and rearrangements involved, and the structural and energy dependence of the interactions are all subjected to detailed study. These studies for the first time reduce chemical kinetics to its elementary interactions. The entire arsenal of techniques, both experimental and theoretical, that have been developed in atomic and nuclear physics are thus being assimilated rapidly by chemistry. Most of the major chemistry departments now have chemical accelerators roughly parallel to the nuclear accelerators found in physics departments in the 1950's (see Figure 4.89). Through the use of crossed beams it becomes possible to extend the study of the chemical collisions to molecular species in highly reactive or metastable states. With merging beams it becomes possible to study these collisions at extremely low effective energies, as the relative velocity of the two reaction beam species is reduced toward zero. Adoption of this technique marks a major transformation of experimental chemistry from a macroscopic to a microscopic science.

Complementary to this activity is that of hot-atom chemistry. Here the time sequences of chemical reactions can be studied by following radioactively labeled atoms of carbon, nitrogen, fluorine, phosphorus, and the like as they move from one to another molecular site during the reaction process. Typically these labeled atoms, because of their characteristically short lifetime, are produced and studied in the environs of a large nuclear accelerator.

Many aspects of nuclear chemistry, as in the Lawrence Berkeley Laboratory, are not distinguishable from similar activities called nuclear physics. (In this survey all the nuclear chemistry activity *per se* is included in the discussions of nuclear physics.) Some 15 percent ($12 million) of the total current operating funds in nuclear science is designated for the support of nuclear chemistry.

It is extremely important to recognize a characteristic pattern associated with the assimiliation of any physical method into chemistry. When the method is first discovered, a few chemists, most often physical chemists, become aware of chemical applications of the method, construct their own homemade devices, or work with a physicist in an informal fashion to

FIGURE 4.89   Apparatus for crossed-molecular-beam experiments examining the reactions of alkyl and aryl halides with alkali metal ions at low kinetic energies. [Source: Photograph courtesy of the Department of Chemistry, Florida State University.]

demonstrate the utility of the new tool. At some point, according to the pattern, commercial models of the device are put onto the market. These are sometimes superior, sometimes inferior, to the homemade machines in terms of their ultimate capabilities to provide information. However, the commercial instruments generally are easier to use and far more reliable than the homemade devices. The impact of the commercial instruments is rapidly felt and often very far reaching; it sometimes virtually revolutionizes a field. Chemists with little interest in learning elaborate new techniques from physics can now apply a physical tool to answer questions of direct interest to them without an unreasonable expenditure of time and learning effort.

The chemical accelerators noted above, with energies measured in tens to hundreds of electron volts rather than MeV, are excellent examples. So also are the nuclear magnetic resonance and laser spectroscopy systems.

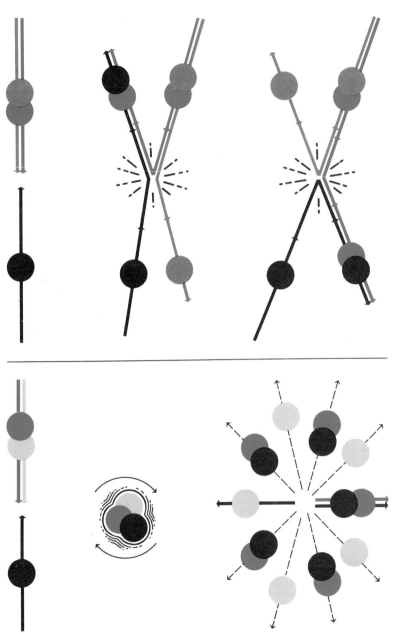

PLATE 4.XVI Determining complex intermediates from direct reactions. *Top left:* Colliding reactions may form a short-lived complex intermediate that rotates and pulsates with internal energy. *Bottom left:* The initial orientation lost, the complex decays into products that emerge in random directions. *Top right:* In a direct reaction between approaching molecules, one reactant may strip off atoms from the other in passing. *Bottom right:* Or it may collide head-on with the other and rebound attached to the atoms. In both cases, the directions of the emerging products are not random. [Source: *Science Year. The World Book Science Annual.* Copyright © 1968, Field Enterprises Educational Corporation.]

Chemists with the new instruments typically have not been concerned with the principles or the development of the device; they have concentrated their efforts on extracting the chemical information that the device can provide. In optical, infrared, and radio-frequency spectroscopy; mass spectroscopy; and x-ray crystallography this pattern has been followed. In the past 20 years, the chemistry curriculum has added the subjects of nuclear resonance; electron resonance; rotational, vibrational, and electronic spectroscopy; and x-ray crystallography. Any reasonably trained chemistry student has learned enough of each of these techniques to be able to recognize their capabilities and to apply them to chemical problems for which they are appropriate.

Other experimental methods that have great impact on chemistry include radioactive tracer and activation-analysis techniques, which are employed in both qualitative and quantitative analysis. These methods use the properties of the nucleus in a way analogous to that in which infrared and ultraviolet spectroscopy use the properties of electrons and atoms. Some of the most dramatic applications of these methods are in the interfaces between chemistry and other disciplines. For example, radioactive dating methods (initially developed by Libby, a chemist) have become an integral part of archaeology. Neutron activation and x-ray fluorescence are also used in archaeology, as well as in the study of fine arts and in criminology. The first analyses of the composition of the moon were accomplished by a simple backscattering version of Rutherford's initial experiment in which alpha particles emitted by a radioactive source are scattered from a target material, in this case the lunar surface (see Figure 4.90). This experiment, flown on NASA's Surveyor series, was developed by Turkevich, also a chemist.

Naturally enough, when a physical method has proven its ability to give accurate and reliable chemical data, it usually finds a place as a monitoring probe, often in industrial process control. The oil industry could hardly exist as we know it now without real-time mass analysis of the contents of its cracking towers by means of mass spectrometers.

The flow of new physical methods for the chemist continues. Electron spectroscopy—the analysis of electrons knocked from the molecules of a sample, either by light, electrons, or other means of excitation—is a very new and powerful tool for studying the way electrons hold a molecule together. In this area, the first commercial instruments are just appearing, and one can watch the rapid growth of a new kind of chemical information. That electron spectroscopy is a very powerful probe and monitor for environmental pollutants already has been demonstrated; it combines extremely high sensitivity with reliability and relatively low cost (see Figure 4.91).

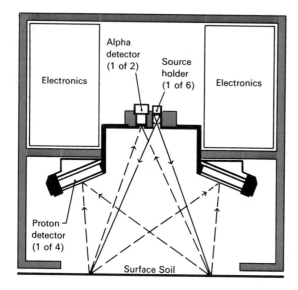

FIGURE 4.90 This alpha scattering instrument was designed to bombard a section of the lunar surface with alpha particles from radioactive curium-242. The energies of the back-scattered particles provided information on the chemical makeup of the lunar surface. [Source: *Science Year. The World Book Science Annual.* Copyright © 1968, Field Enterprises Educational Corporation.]

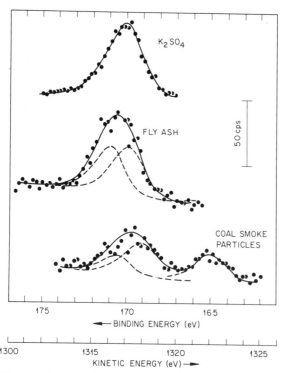

Photoelectron Spectra of Sulfur on Various Substrates.

FIGURE 4.91 Photoelectron spectrum of sulfur in fly ash and coal-smoke particles. The sulfate spectrum is included for comparison. These data are indicative of the presence of sulfur in various chemical forms, the higher energy peak being due to the presence of sulfate on the surface of the particle. [Photograph courtesy of T. Carlson, Oak Ridge National Laboratory.]

Another current example is the use of low-energy electron diffraction for the study of surfaces. This technique has important potential for increased understanding of the role of surfaces in catalysis; even small improvements here can be reflected in enormous industrial economies.

A third example is slightly less developed, but its promise is great; this is the use of very-long-wave infrared spectroscopy, actually interferometry, to study surface properties such as epitaxial growth. Chemists have used this method less than have physicists studying properties of solid materials; it illustrates the delay that can result from weak interaction between fields. Still another example is the application of Mössbauer spectroscopy to the study of dynamical properties of ions dissolved in water and other solvents.

Theoretical concepts, as well as hardware, continue to flow from physics to chemistry. Modern methods for dealing with interactions among electrons—problems of the electronic structure of atoms and molecules and of scattering of electrons by atoms and molecules (as well as molecule–molecule interaction)—have strong ties to the many-body methods developed in the context of nuclear and condensed-matter physics. In some cases, very closely related methods were developed independently by scientists, some of whom identified themselves as chemists and others as physicists.

Only recently have modern computer techniques been brought to bear on the enormously complex calculational problems involved in the microscopic quantum-mechanical descriptions of chemical reactions. Nonetheless, these calculations have already evolved to a point at which they are among the most demanding in all of science in terms of computer speed, size, and sheer number-crunching ability. Clementi and his associates at the Large Scale Computation Laboratory of IBM have played a leading role in such computer applications (see Figure 4.92). This clearly is one of the most rapidly developing fields at the physics–chemistry interface.

Chemical information, other than theory, also feeds back into physics, often in the form of specific information about substances having some particular desired properties. The ruby and neodymium lasers, the dye lasers, and liquid lasers in general were made possible in part because of the availability of the kind of information that the chemist now collects routinely. The subject of properties of materials also includes problems of preparation and purity. The development of solids for microcircuitry required new chemical research into methods of purification and analysis. Properly functioning microcircuit elements of one large computer were developed only when the physical and chemical problems of ion migration under conditions of high current flow and methods to achieve an exceedingly high degree of purification had all been studied in a quite fundamental way. (See section on condensed-matter physics.)

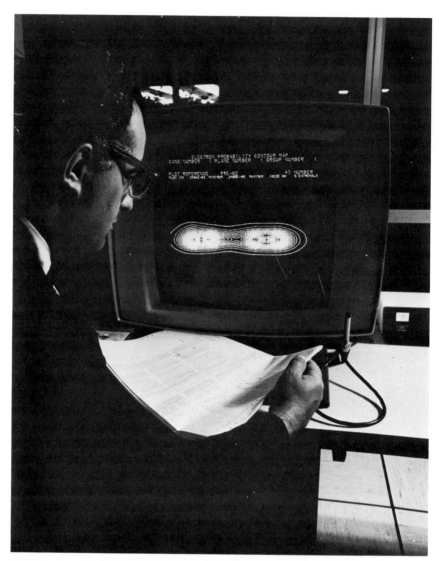

FIGURE 4.92 Computer-generated electron density map of the reaction $NH_3 + HCl$ → NH_4Cl, taken in a plane along the NCl axis, is projected on a display tube. [Courtesy J. D. Swalen, IBM Research Laboratory, San Jose, California.]

An example of an active area of materials preparation, in which chemistry and physics interact strongly, is the new field of bubble memory devices. It is hoped that these devices, based on the behavior of orthoferrite materials, will lead to a new generation of computer memories with far fewer failures and faster access time than are presently available.

One type of interaction between chemistry and physics falls outside the categories that we have mentioned. It relates to data compilation. The American Petroleum Institute has produced extensive tables of physical properties that, in effect, correlate properties of substances with their molecular structure. The National Bureau of Standards and the Joint Army, Navy, NASA, Air Force Interagency Propulsion Committee (JANAF) prepare similar compilations with other kinds of data. The National Bureau of Standards also has a system to provide standard samples for those who need them.

Fields Deriving Information from the Physics–Chemistry Interface

Many areas outside physics and chemistry receive benefit from a combination of ideas generated in these fields. In some cases the contributions take the form of very specific information; in others the contributions are more in the nature of general approaches and techniques of common currency in physics and chemistry that are just now being recognized as useful in other fields.

Engineering and technology are the most obvious fields in which the knowledge obtained from basic studies in chemistry and physics finds early application. Studies of corrosion, the synthesis of new materials such as crystalline polymers, the nature of surfaces, heterogeneous catalysis including enzyme reactions, and thin films require concepts and methods from both physics and chemistry. The techniques employed, such as low-energy electron diffraction, electron microscopy, electron microprobes, and electron spectroscopy are being assimilated into chemistry and physics. An example will show the interaction between chemistry and physics and the close relationship to technological developments. A manufacturer recently developed a material for photocopying; the material is an organic charge-transfer photoconductor. The development of this material can be traced from the basic concepts of the quantum theory of the chemical bond through the chemical idea of a charge-transfer complex to the recognition of photoconductivity. Current knowledge in basic physics and chemistry permitted the research for this material to advance rapidly to the point at which it was apparent that an electron acceptor was desired, thus reducing the need for expensive trial-and-error research.

The study and management of the oceans, the earth, and the atmosphere rely heavily on both physics and chemistry. In oceanography, for example, the turnover of water in the oceans can be understood only through a study of both hydrodynamic flow (traditionally a subject of physics) and molecular flow (as much a part of chemistry as of physics). The carbon dioxide balance in the air and the absorption of carbon dioxide by the oceans are critically important and difficult basic problems in physics and chemistry, as well as in biology, meteorology, and environmental management. The management of fisheries and water resources generally requires inputs from physics, chemistry, and biology. Air resource management is perhaps one of the clearest examples in which chemistry and physics play an integral role. Whether one is concerned with control of dangerously acute pollution levels or with long-range planning and legislation, air-quality models are essential to the analysis. Such models require a knowledge of the behavior of masses of gas (fluid dynamics), the reactions within the gas (aerochemistry), and the influence of the earth's shape, motion, and position on the way this gas behaves (meteorology and geodesy). Only within the past two years have the first models begun to appear that incorporate all these components at a level of accuracy high enough to be useful for an air resource management program. One of the major drawbacks has been the limited knowledge of the basic physics and especially of the basic chemistry of polluted air.

In social or economic modeling, a particular kind of contribution that the physicist or chemist can make merits attention. The constraints that one puts into a model are, in fact, of two kinds, although models now in use do not make this distinction. One type of constraint is essential and unavoidable, depending only on the laws of thermodynamics and other equally general laws. The other type of constraint is basically a limitation due to the state of current technology. By sorting the first type of constraint from the second, one can sometimes determine the ultimate limiting behavior of a system—the condition it could approach (but never reach) if its technology were optimum. Such an analysis could be made, for example, to determine the optimum long-term choices for supplying sufficient water to an industrial nation. A similar kind of analysis, using population biology instead of thermodynamics, could be used to determine the amount and pattern of insecticide to apply in an area to maintain the long-term agricultural productivity and the health of the animal species in the associated food chain.

Some other fields of research or development to which physics and chemistry are contributing include: the structure of biomolecules, the process of photosynthesis, the structure of cell walls, and the action of nerves; new methods of materials preparation; discharge methods for petroleum cracking; surface curing by electron bombardment; geographics, including the origins of vulcanism, the chemical origin of rocks, and minerals, and the composition of the planets; and the upper atmosphere and astrophysics, including the entire complex set of problems associated with the earth's atmosphere, as well as the problem of the formation of molecules in interstellar space.

Manpower

Because the definition of the physics–chemistry interface is not precise, estimates of the number of persons working at the interface are only approximate. The total interface contains between 4000 and 8000 scientists at the PhD level. (The larger figure includes condensed-matter physics.) This estimate is based on figures from the 1970 National Register of Scientific and Technical Personnel and on an independent estimate derived from the number of subscribers to the *Journal of Chemical Physics*. In chemical physics alone, the number of scientists is some 3000 to 4000. Of these, approximately 250 hold academic positions in chemical physics, and approximately 500 hold academic positions in some other subfield included in the interface.

The 1968 National Register recorded 8466 scientists working in physical chemistry, 4836 of whom had PhD's. The Register category, physical chemistry, is composed of an aggregation of detailed specialties. A scientist is classified as working in a field on the basis of his own statement of a detailed specialty that most nearly corresponds to the subject matter with which he is principally concerned in his work.

Many chemical physicists, of course, are doing work in subfields outside the scope of the Register definition of physical chemistry and would be included in, for example, the condensed-matter or fluid-physics subfields.

According to the National Research Council Office of Scientific Personnel, there are about 500 PhD's produced each year in "Chemistry, Physical" (459 in 1968 and 554 in 1970). (These physical chemists are produced by chemistry departments; no record is kept of "Physics, Chemical" as a subcategory of physics department PhD production.) For purposes of comparison, it is worth noting that in 1970 physics departments produced 400 condensed-matter doctorates, 470 elementary-particle- and nuclear-physics doctorates, and 730 doctorates in 11 other subfields of physics, including 160 in "Physics, Other." The total number of physics PhD degrees reported in 1970 was 1600.

The throughput—the replacement rate required to keep existing positions filled—currently is estimated to be about 6 ± 1 percent per year in the interface. This figure is based on several estimates, such as the normal throughput in a typical industrial laboratory. The figure appears to be fairly firm, except in years such as 1970, when the rate of hiring was held at a lower level. Consequently, the total number of replacement personnel required at the PhD level is approximately 240–280 per year for the over-all interface and about 180–240 per year for the more strictly defined field of chemical physics.

Based on a detailed analysis of manpower needs and production of degrees in subjects at the physics–chemistry interface, the Panel on Physics in Chemistry concluded in its report (see Volume II) that the rate of production of trained manpower in chemical physics is much higher than can be assimilated through the traditional past employment patterns of chemical physicists.

Problems at the Interface

The effects of reductions in financial support for this interface are less obvious, perhaps, than in areas in which a few large facilities dominate the field. The interface is characterized by a large number of small groups; most of the academic groups consist of one senior scientist, perhaps one or two young postdoctoral fellows, and a few graduate students. In government and industrial laboratories, the groups constituting research units consist of one, two, or perhaps three PhD scientists with, at most, the same number of technicians.

Financial cutbacks have thus far apparently taken the forms of non-replacement of staff lost by natural attrition, releasing staff, and, in some cases, loss of contract or grant support. Insofar as the elimination of support could be used to maintain excellence in the interface, this method would perhaps be the least unpalatable means of absorbing cutbacks. As the Report of the Panel on Physics in Chemistry states in its recommendations, supporting a smaller amount of excellent work at an adequate level is preferable to supporting a considerably larger amount of excellent, mediocre, and low-quality work at a level too low for any group to do an adequate job.

It is not clear, however, that the means of reducing funds are actually preserving excellence. The continued reduction in research support from the DOD and AEC has reached the point at which some of the best research is now threatened with loss of funds, and it appears that the NSF will not be able to assume this support at the level necessary to continue the work, at least within the current NSF budget.

Several national laboratories have experienced difficult problems; it has become necessary to terminate the employment of many able scientists.

So large a number has been released—people respected highly by their peers—that morale in the physics–chemistry interface at a number of national laboratories is at a low level.

> Biology has become a mature science as it has become precise and quantifiable. The biologist is no less dependent upon his apparatus than the physicist. Yet the biologist does not use distinctively biological tools—he is always grateful to the physicists, chemists, and engineers who have provided the tools he has adapted to his trade.
>
> Until the laws of physics and chemistry had been elucidated it was not possible even to formulate the important penetrating questions concerning the nature of life.
>
> PHILIP HANDLER
>
> *Biology and the Future of Man* (1970)
> (Chapter I, pages 3 and 6)

PHYSICS IN BIOLOGY

Nature of the Interface

An especially active scientific interface, and one that is attracting an increasing number of physicists, is that between physics and biology. The problems posed relate to fundamental questions of life and hold the promise of substantial contributions to the alleviation of human ills and misery. This interface combines fundamental science and both immediate and long-range social goals in a close and balanced relationship. Growing application of the methods, devices, and concepts of physics to some of the central biological problems should result in increasingly rapid progress. The scientific returns from such applications are already impressive.

The title of this section, "Physics in Biology," rather than the more usual biophysics, is indicative of the Committee's approach to the interface. Rather than a survey of all biophysics, this report focuses on the role of physics and physicists in attacking some of the major problems of modern biology. The implications of such a role for physics education and the overall physics enterprise also receive attention.

To describe the interaction of physics with biology requires study of the flow of manpower, ideas, and procedures through the interface. The title, "Physics in Biology," implies a flow from left to right. It is convenient to divide the subject into three parts: (a) the flow of physicists into biology;

(b) the flow of physics into biology, which includes the flow of ideas, techniques, and equipment, often through the intermediaries of engineering and chemistry; and (c) the interface called biophysics, in which the powers of physics are merging with the problems of biology in new ways that require nonstandard combinations of skills from both disciplines.

The goal of biophysics is to be biologically useful. When it is successful, it merges with other branches of biology, such as biochemistry and molecular biology, a major goal of which is to understand, in molecular terms, how genetic information is transmitted. The following example illustrates the contributions of biophysics to molecular biology. In 1944, it was shown that DNA, rather than proteins, contained the genetic information with which molecular biology is principally concerned. In 1953, J. Watson and F. Crick, working in the Cavendish Laboratory, developed from the accumulated chemical studies of DNA and a consideration of its biological function an interpretation of the x-ray studies of M. Wilkins in terms of the now famous double helix. Since then, molecular biology has developed rapidly, and biological function is now so interwoven with the structure of DNA that the term "structure and function" has become a platitude.

At the same time, and in the same laboratory in Cambridge, the crystal structures of the proteins were first being determined by x-ray crystallography, with a parallel influence on enzymology and biochemistry. In all these cases the isolation of the molecules and the definition of their biological functions were accomplished after roughly a century of chemical research. The physicists who determined the crystal structures were attacking a biologically important problem. Their boldest and most original step was their starting assumption that these large biological molecules had unique structures that could be determined by x-ray crystallography. Physicists are accustomed to this kind of simplicity in science; assuming such simplicity is a reflection of their previous training and research style. The revolutionary nature of their findings depended on the originality of their assumptions. However, given their background as physicists, trained in x-ray crystallography under Bragg, and their interest in biology, their directions of attack were almost predetermined. Thus, in these illustrative and illustrious cases, physicists and physics created an exciting field of research in biophysics, which has now been merged with biochemistry and molecular biology.

Determination of the structure of large biological molecules will continue to be an active field of research. In addition to x-ray crystallography, other physical techniques such as nuclear magnetic resonance, electron spin resonance, Mössbauer studies, and optical studies are being used more and more. These techniques, when used for structural studies, often complement x-ray data by giving information on a finer scale. This research is directed toward structural determinations of larger molecular aggregates, that is to say, membranes and membrane-mediated enzyme systems; ribosomes, which are the site of protein synthesis, composed of nucleic acids and protein, with a molecular weight of $\sim 10^6$; mitochondria, the membrane-bound volume in which the chemical energy of nutrients is converted to more usable forms by electron transfer reactions; and the photosynthetic unit in which photons are converted to chemical energy. In all of these systems, scientists are trying to understand biochemical functions in terms of the structure of the molecules and the physical interactions among them. Beyond the structures of isolated biological molecules lie the complicated questions of intermolecular interactions, which should challenge physical methods for many years. The collision techniques, which are only now beginning to be applied to the elucidation of elementary chemical reaction kinetics of simple inorganic molecular complexes, were developed in atomic and nuclear physics. Application of these approaches to molecular systems of biological interest is an exceedingly difficult but highly promising field.

When structure is examined at finer levels than the molecular, it is quite clear that quantum-mechanical understanding of the electronic structure of certain parts of biological molecules will become increasingly important. The advances made through electronic understanding of the molecules of interest to chemists and condensed-matter physicists show the promise of this approach. Recently, as experimental molecular physicists have studied biological molecules with the goal of understanding their electronic properties, the amount of systematic data has approached the point needed for theoretical synthesis and advances. This synthesis could lead in the future to a larger role for theoretical studies. Previously, the theorist's contribution to molecular biophysics has been very small, because, unlike the best experimenters, he generally has not learned enough biology to be able to ask good questions.

An exception occurs in the case of the theoretical models that are playing an increasingly important role in biology. This trend reflects the physicist's typically different viewpoint on biological problems. One of these differences is the physicist's desire for a simple, comprehensive model, capable of providing a first-order explanation of a wide variety of observations. It is often baffling to a physicist when biologists insist on the complexity of nature and the uniqueness of each result. It is, of course, equally unappealing to a biologist, struggling with complexities of DNA replication, to be informed by a physicist that the Ising model, or enough molecular quantum mechanics, would solve his problem. However, in the middle ground between these two extremes of oversimplification lies the productive application of physical models, based as always on experimental observations. For example, the concept of a genetic code proposed by physicists and theoretical chemists such as Crick, Orgel, Gamow, and Griffith appealed to a mind trained in physics. Various theoretical models were examined. One question that was proposed and answered was how much information had to be stored. The answer was that there were 20 amino acids that had to be coded by the DNA. Because DNA has only four possible bases as coding units, a minimum of three bases is required. Was the code overlapping? This question was answered negatively by considering the known mutations that had been observed.

Another useful model was that of Monod, Wyman, and Changeux, on allosteric proteins, whose function with respect to one small molecule can be affected by other small molecules. They proposed a generalized molecular basis for feedback in biological molecules and thus stimulated many experiments and analyses to determine the crucial facts.

Activity

Perhaps the most active research area at the interface between physics and biology is that involving the study and determination of the molecular bases for biophysical processes. This work has engaged some of the best people in the subfield and uses a variety of physical techniques and probes, from x-ray crystallography through nuclear magnetic resonance and Mössbauer techniques (see Figure 4.93) to nanosecond fluorimetry.

An older research area, but one that retains excitement and interest, is neural physiology. In part, this interest reflects the hope that such research can lead eventually to the understanding of the mysterious processes of human thought and memory, one of the remaining frontiers of man's understanding. In part, it reflects the physicist's assumption that when information is transmitted and processed by essentially electrical mechanisms, the problem should be amenable to physical analysis.

A striking example of physical reasoning in elucidating a particular property of a biological cell is the analysis of the electrical state underlying excitability in the giant axon of the squid. Its virtually unique diameter (500–100 μm) enabled Hodgkin and Huxley, in 1949, to conduct a series of fundamental electrical measurements that, in turn, made it possible to establish for the first time an adequate quantitative description of the electrical state associated with the nerve impulse. Both the design and execution of the experiments required a thorough knowledge of electronic circuits, in which feedback plays a crucial role, and of the theory of ionic electric currents. Moreover, the interpretation of the data demanded an ingenious mathematical analysis. This achievement was in large measure a product of Hodgkin's and Huxley's training in physics.

One of the basic findings of their analysis was that the ionic currents in

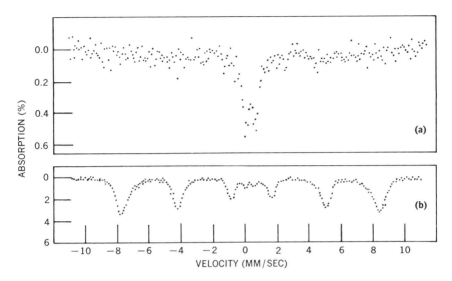

FIGURE 4.93 Human lung material from healthy lung (a) and from lung of hemosiderosis victim (b). Mössbauer spectrum of diseased lung indicates an abnormally large amount of iron (note the difference in absorption scales in the spectra), which appears to be in the form of a finely divided, low-molecular-weight compound. [Source: C. E. Johnson, "Mössbauer Spectroscopy and Biophysics," Physics Today *24*, 40 (Feb. 1971).]

the axonal membrane display strikingly nonlinear behavior in that the conductances are voltage-dependent and time-variant. The property of nonlinearity implies that, in a single neuron and chains of neurons (neural circuits), the elaborate calculations required to treat the excitable state can be carried out in general only with modern computers. And the statistical physics underlying the conductance changes in the cell membrane and at the junction between two cells (the synapse) presents a problem requiring highly sophisticated analysis.

The study of neural physiology is typical of the advanced research conducted on macromolecular aggregates in modern biophysics. It involves the design and development of new measurement techniques, computer simulation of neural behavior, and the study of signal-transmission characteristics of biological media. It continues to provide important surprises. Recent work on the brains of primitive animals, in which the brain contains at most only a few hundred cells, has shown that, even here, a remarkable symmetry of structure and function has developed.

Neural physiology stimulated some of the earliest physicists to move into biophysics; probably it will continue to attract them. Progress in neurobiology will demand advances in the biochemistry and ultrastructure of the neuron; but, in any event, the elucidation of physical mechanisms, for example, the analysis of excitability in the giant axon of the squid, will continue to play a crucial role.

A third major activity at the interface involves the interaction of radiation with high- and low-level biological systems. The types of radiation employed range from ultraviolet light to very-high-energy, heavy nuclear particles and mesons. The transfer of the techniques of nuclear physics—radioactive tracers, accelerator radiations, and nuclear instrumentation—has brought about a revolution in biophysics and in both clinical and research medicine.

Radioactive isotopes have contributed enormously to the general improvement in diagnosis, and a large number of radioisotopes are now in routine use. Isotopes commonly used include: ^{131}I, ^{125}I, ^{59}Fe, ^{113m}In, ^{99m}Tc, ^{51}Cr, ^{57}Co, ^{60}Co, ^{75}Se, ^{85}Sr, ^{197}Hg, ^{32}P, and ^{198}Au. These are used for visualization of the thyroid, brain, liver, lung, kidney, pancreas, spleen, heart, bone, and placenta and for a variety of physiological tests in which the rate of disappearance or rate of uptake of a particular labeled substance reflects the function of a given organ system (see Figures 4.94 and 4.95). In 1968, there were some four million administrations of labeled compounds to patients, and the rate of use has been growing rapidly. These isotopes are employed routinely in practically every hospital in the United States.

Increasing attention is being given to the use of very-short-lived isotopes, particularly ^{11}C, which should be especially useful because of the enormous potential for incorporation into a wide variety of biological compounds, with consequent extension of the range of diagnostic procedures. The short half-life, some 20 min, markedly limits the time for synthesis of the isotope into the desired compound and the time available for use. Thus the source of the isotope (an accelerator), facilities for rapid synthesis, and clinical facilities must be in close juxtaposition. A closely cooperating team of physicists, chemists, and clinicians is required for effective application. Use of this isotope, although still in its relatively early stages, is increasing rapidly and has great promise.

Another rapidly developing diagnostic procedure involves the determination of the entire amount of a given element, for example, calcium, in the body by means of activation analysis. The entire body is exposed to a beam of fast neutrons, which thermalize in tissue and are captured by the element in question. The patient is then placed in a whole-body counter and the total amount of the given element deduced from the total induced activity. The entire procedure can be accomplished with the delivery of only a small fraction of 1 rad to the patient.

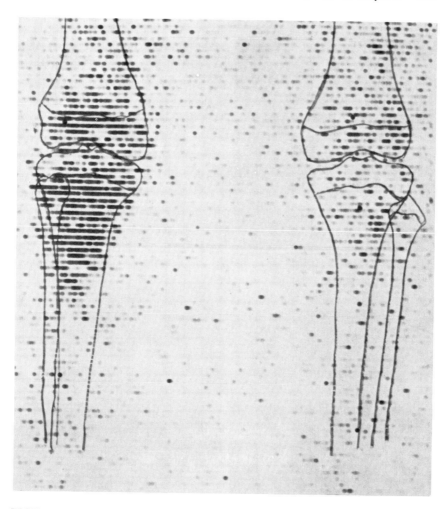

FIGURE 4.94 This is a bone scan made with short-lived strontium-87m (half-life 2.8 h). The patient was a 13-year-old girl with a bone sarcoma of the right tibia. The greater strontium uptake in the right leg indicates the presence of the lesion. [Source: J. H. Lawrence, B. Manowitz, B. S. Loeb, *Radioisotopes and Radiation* (Dover, New York, 1969).]

Isotopes are now used widely in radiotherapy in a variety of ways. High-intensity external sources, such as cobalt-60 and cesium-137, have in many instances replaced x-ray machines for routine radiotherapy. The depth–dose characteristics of radiations from these sources are more favorable than those of most x-ray beams, and the units have the advantage of

FIGURE 4.95 *Left:* The autofluoro-scope detector shown with its 2-in. lead shield removed. A bank of 293 sodium iodide crystals is in the lead-encased enclosure at bottom. This bank is separated from the 12 photomultiplier tubes by a 4-in. Lucite light pipe. The data are transferred electronically and recorded on Polaroid film.

Below: Four neck scintiphotos of different individuals made with the gamma-ray scintillation camera 24 h after administration of 25 to 50 μCi of iodine-131. *Upper left:* Normal, butterfly-shaped, thyroid gland. *Upper right:* Solitary toxic nodule in right lobe. This "hot" nodule takes up the iodine-131 to a greater extent than the normal thyroid tissue. *Lower left:* Degenerating cyst seen as a dark area in lower left of picture. This "cold" nodule is nonfunctioning, hence does not take up the radioisotope. *Lower right:* This patient had undergone a "total" thyroidectomy 2 years previously. The photo shows regrowth of functioning tissue, right, and also a metastasis, lower center.

[Source: J. H. Lawrence, B. Manowitz, and B. S. Loeb, *Radioisotopes and Radiation* (Dover, New York, 1969).]

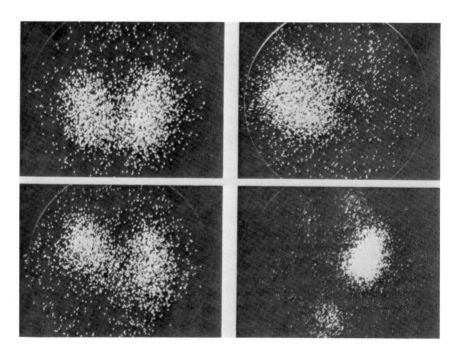

ease of operation and maintenance. As a specific example, the cure rate in cases of mammary cancer increased dramatically when cobalt-60 therapy replaced that with 100-kV x rays. The reason was that the higher energy ^{60}Co radiations were able to penetrate the sternum to a lymph node behind it, whereas the x rays could not. In addition, beta emitters such as strontium-90 are beginning to be used more frequently for the therapy of some superficial external lesions.

Physiological localization of radioactive isotopes is used in some forms of therapy, for example, iodine for treatment of hyperthymism and thyroid tumors and ^{32}P for treatment of some diseases of the bone marrow. These procedures represent optimal therapy only in a relatively few situations.

Accelerators have contributed greatly to the improvement of radio-therapy. Early accelerators allowed transition from the use of relatively low-energy x rays for radiotherapy to the use of supervoltage x rays, permitting the delivery of a relatively large dose to the tumor in depth, with a minimal dose to the intervening normal tissues. Electron accelerators such as the betatron have permitted an additional distinct improvement in the therapeutic ratio, or dose-to-tumor, dose-to-normal-tissue ratio.

Somewhat in the future is the therapeutic use of beams of negative pions, which currently are produced only in elementary-particle and very-high-energy nuclear-physics facilities. These have the enormous advantage of delivering not only their ionization energy but also their entire rest mass energy to their final destination in matter. Thus, while traveling to the therapy site, they do relatively little damage to surrounding tissues; their capture then releases some 200 MeV of energy at the treatment site.

Currently, there is much interest in the use of accelerators to produce beams of fast neutrons for radiotherapy. The rationale is that all tumors quickly develop small foci of poorly oxygenated or hypoxic cells. These hypoxic cells are markedly resistant to damage by x or gamma radiation but are much more susceptible to damage by neutrons or other densely ionizing radiations. Although a variety of reactions and neutron spectra might be used, the approximately 14-MeV neutrons from the D–T reaction are optimal in terms of penetrating characteristic and density of ionization. The procedure is experimental, and several years will be required to evaluate its efficacy. Man-made transuranium isotopic sources of neutrons, such as ^{252}Cf, have just recently become available and are also being used, with the same rationale.

Not only are radiations in the high-energy or nuclear realm used in such studies and applications but also ultraviolet and infrared radiation (see Figure 4.96). For example, Setlow and his collaborators at Oak Ridge very recently discovered that a particular species of skin cancer can be traced not only to a specific damage site in the human cell DNA induced by

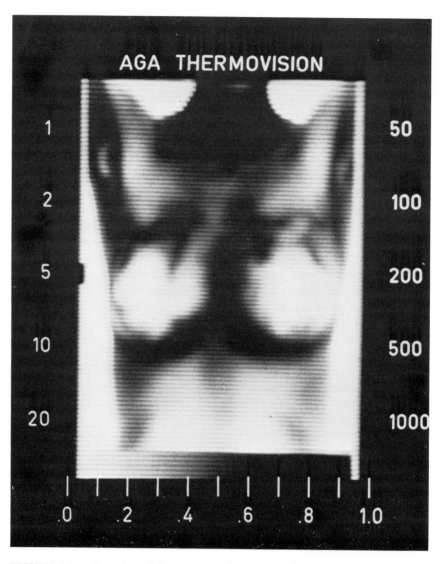

FIGURE 4.96   Detection of breast cancer by infrared thermography. [Photograph courtesy of the Lovelace Foundation for Medical Education and Research.]

ultraviolet radiation but also to the lack, in susceptible individuals, of a rather rare enzyme the function of which is to repair such damage. Already a test has been developed that can detect the total lack of this enzyme *in utero* and make possible therapeutic termination of the pregnancy; otherwise, with total lack of the enzyme, the normal life-span of a child would be brief. It is hoped that further research will result in a technique for the detection of, and compensation for, the partial lack of this enzyme in the population.

The use of infrared photography as a diagnostic tool is now rather well developed. Reflecting the increasing metabolism of cancer cells is a local temperature elevation that shows clearly in an infrared photograph; this technique is extremely simple, over 90 percent effective, and widely employed.

The long-term effects of very-low-level radiation on a population is, of course, a matter of continuing concern and controversy. The controversy arises largely because, at the levels now under consideration, even granting the validity of the mouse–man extrapolation, it has been estimated that an 8-billion-mouse colony would be required to yield statistically significant results. This situation shows the importance of understanding mechanisms not just doing statistical experiments. It is an example of a situation that Weinberg has defined as "trans-science," * a problem in which it appears superficially that in principle science should be able to give concrete answers but, when examined in greater detail, exceeds the scope of any economically feasible scientific study.

Although now of somewhat less importance than in past decades, the overall questions of thermodynamic energy balance and stability in biological systems continue to occupy the attention of a small group of physicists in biology. Although the broad outlines of the physics involved in the intricate energy-transfer mechanisms have been established, major questions remain unanswered.

In fluid physics and rheology there also are problems. The dynamics of human body fluids are still inadequately understood and can pose significant problems in vascular surgery and repair and in the more ambitious organ-replacement projects now in developmental phases. In rheology, a wide range of open questions in regard to bone growth, muscle attachment, and the like need answers.

Biomedical engineering has become a specialty in its own right, stimulated by the pressures for improved man–machine interface designs in supersonic aircraft, space vehicles, and precision industrial production lines, as well as by more prosaic needs such as improved kitchen appliances. The

* *Science, 174,* 546–547 (Nov. 5, 1971).

design of artificial organs, prosthetic devices, and the like is another part of this field. Progress in the development of long-lived portable power sources and parallel progress in ultraminiaturization of semiconductor electronic components should lead to a greatly increased capability to mitigate human infirmities. Much of this work represents applied physics at its best.

One of the most difficult tasks in the biomedical engineering areas of bacterial colony analysis, brain scintigram analysis, blood-cell identification, chromosome analysis, and heart image extraction from a cardioangiogram is the extraction of relevant objects from an irrelevant background. The principal reason is that the pictures in these fields are often complicated by unwanted background, and object images are poorly defined. In recent years, computer processing of radiographic images has emerged as a highly promising technique (see Figures 4.97 and 4.98).

Three aspects of biophysical instrumentation merit attention. The first involves instrumentation for clinical application in both diagnosis and treatment. Major progress is under way in clinical instrumentation; a modern hospital's intensive-care wing illustrates the vital role that physics research plays here. A second aspect of instrumentation involves biological and biochemical laboratories in which new techniques, devices, and approaches permit massive increases in both the speed and quality of measurement, thus making possible the extension of the most modern diagnostic aids to a much larger segment of the population. The techniques also facilitate ongoing research in biology, biochemistry, and biophysics. Third, advances in instrumentation open entirely new areas of biophysical research. Examples of devices that have led to major breakthroughs are the scanning electron microscope (see Figures 4.99 and 4.100), the ultracentrifuge, and nuclear magnetic resonance.

Institutions

Research at this interface takes place in universities and in a number of large federally supported laboratories or institutes. The principal supporting agencies are the National Institutes of Health (NIH) and the AEC. Some more specialized work is supported in-house by NASA and the DOD. A relatively small fraction of the total effort is in industrial organizations. The division between university and federal laboratory activity is roughly equal.

Academic research at this interface, in contrast to research at institutes, carries a heavy burden of departmental responsibilities. Two means have been used to broaden the scope of academic research. In many cases, temporary departments of biophysics were established and made into formal departments when it seemed desirable. This practice has worked well in a number of universities and less well in others. In the most successful cases,

FIGURE 4.97 Quantitative information to assess the functional state of the heart, especially the left ventricle, can be obtained from a cardiac cineangiogram—an x-ray motion picture of the heart. To collect such information, an essential task is to detect or outline the boundary of the left ventricular chamber on each frame of the cineangiogram—a laborious task, heretofore performed by humans. C. K. Chow and T. Kaneko (IBM, Yorktown Heights, New York) have developed a computer algorithm to detect the left ventricles automatically; thus making the automatic extraction of quantitative information feasible. *Above:* The scanned image of one frame of a cardiac cineangiogram; *below:* the boundary detected by the algorithm superimposed on the image. The detected boundary agrees reasonably well with human recognition.

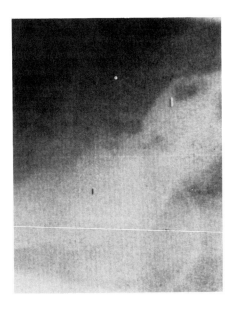

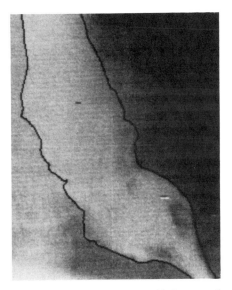

such biophysics departments became departments of modern biology and often have lost much of their initial chemical or physical orientation. However, the more physics-oriented aspects of biophysics still flourish in university departments of this sort, generally under a title such as "Biophysics and Molecular Biology." Advanced electron microscopy research, new

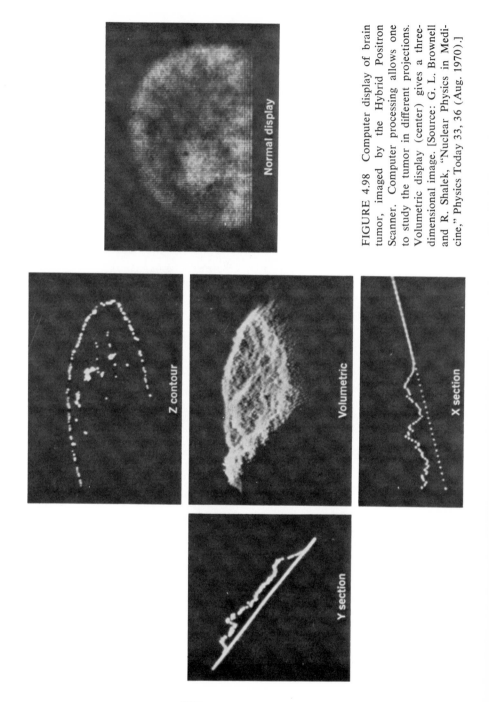

FIGURE 4.98 Computer display of brain tumor, imaged by the Hybrid Positron Scanner. Computer processing allows one to study the tumor in different projections. Volumetric display (center) gives a three-dimensional image. [Source: G. L. Brownell and R. Shalek, "Nuclear Physics in Medicine," Physics Today 33, 36 (Aug. 1970).]

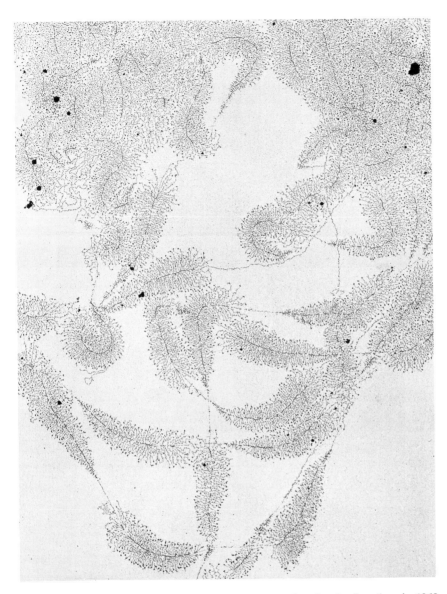

FIGURE 4.99 Photographs of genes in action were taken for the first time in 1969 at Oak Ridge National Laboratory in Tennessee. Single genes on the long central strands of DNA have thousands of shorter RNA strands peeling off of them. [Source: *Science Year. The World Book Science Annual.* Copyright © 1969, Field Enterprises Educational Corporation.]

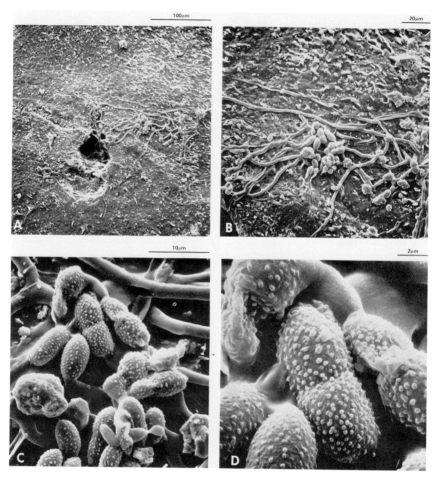

FIGURE 4.100   Surface of a Sturmer apple fungus and spores adjacent to a lenticel. Some hyphae appear to enter the lenticel, possibly indicating first stage of penetration into the apple. Part of a study concerned with the occurrence of lenticel spot. (a) Magnification ×160; (b) magnification ×400; (c) magnification ×1600; (d) magnification ×12,000. [Photographs courtesy of Stereoscan Micrograph—Cambridge Scientific Instruments Limited, England.]

techniques of radioactive tracers, magnetic resonance studies, and x-ray crystallography have prospered in these departments and profited from their interactions with the more biologically oriented activities. In practice, these departments do not interact to a large extent with physics deparments, although physics courses are increasingly important to their students. In this respect, it is generally felt that physics departments do less than chemistry departments to accommodate such students, among whom are an increasing number of premedical students who desire some physics courses. Revision of the physics curricula to accommodate the needs of these students would be one way in which the interactions of physics and the life sciences could be strengthened.

There are important opportuntities to further these interactions in the large national laboratories initially established for specific mission purposes. These laboratories have extensive facilities and trained staffs who could bring a unique competence and capability to bear on problems at the physics–biology interface. A pressure that is favoring larger research groups is the increasing cost of the equipment used in biophysical research. The cost of commercially available high-resolution nuclear magnetic resonance (NMR) equipment used to measure the proton NMR spectra of complicated organic molecules can be cited as an example. In 1957, this technique was first used to study the spectrum of a protein. During the past decade, stimulated in part by the possibility of extensive biological applications, this equipment has increased rapidly in sophistication and price. At first, the cost increase followed a semilog growth curve, but recently, because of the use of superconducting solenoids and Fourier transform computer techniques, the costs have been increasing more rapidly with time (see Figure 4.101). At the same time that the equipment is becoming too expensive for an individual scientist to afford, its usefulness in biochemical research has been increasing rapidly because of improvements in sensitivity, resolution, and experience. Clearly, large institutional research activities, similar to the national laboratories, will soon be more useful in biophysical research.

Manpower

The number of people identified with physics in biology, or biophysics, depends on how the interface is defined. The number of physics participants in the 1970 National Register of Scientific and Technical Personnel who indicated physics in biology as the subfield in which they were working was 465, or 1.3 percent of the 36,336 physicists who took part in the survey. Sixty percent (277) of the 465 held PhD's, representing 1.7 percent of all physics PhD's.

If the population of the interface is based on membership in the Biophysical Society, then it increases to approximately 2500. The Society was founded in 1957, and its growth has been extremely rapid.

The numbers of positions and graduates seem to be reasonably well balanced in this interface between physics and biology. There have been fewer positions in the past few years, as in all of physics, but the situation seems no worse in physics in biology than in physics in general.

The contributions of physics are essential to modern biological research

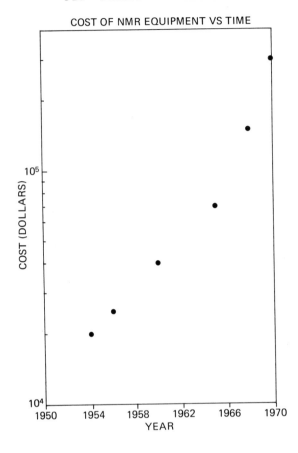

COST OF NMR EQUIPMENT VS TIME

FIGURE 4.101 Increases in the cost of nuclear magnetic resonance equipment, 1950–1970.

and are likely to increase rapidly. Research in biophysics is particularly rewarding in a personal sense, first, because of the exhilarating intellectual challenge of this wide open, diversified, and fast-moving field and, second, because of the humanistic aspects of biological research, with its long-range goal of improving human life. In the future, many more physicists probably will move into biology. In the Report of the Panel on Physics in Biology (see Volume II), a number of ways in which this movement can be facilitated and encouraged are suggested.

[T]he only certain means and instruments to discover
and anatomize nature's occult and central operations;
which are found out by laborious tryals, manual
operations, assiduous observations and the like.

JOHN WEBSTER (1610–1682)
Academiarum Examen, 1653 (page 106)

INSTRUMENTATION

Introduction

High-quality observations are the lifeblood of science; physics, in particular, because of its focus on fundamental causes, is critically dependent on the instrumentation that makes such observations possible. At the same time, physics provides a great many of the concepts and devices that underlie all modern instrumentation.

In surveys such as this one there is always a tendency to focus attention on the instrumentation applications of the most recent and spectacular scientific findings—for example, the laser—and on the instrumentation required to reach the frontiers of modern scientific research—for example, superconducting magnets in elementary-particle physics. Although these applications are very important, such a focus does a great disservice to the vital role of instrumentation throughout science, society, and the economy. Indeed, instrumentation is that channel through which the results of physics research are most directly brought to bear on the problems of greatest concern to the citizen.

In its survey the Committee attempted to avoid this too narrow view of instrumentation through the appointment of a panel of distinguished representatives of leading companies in the U.S. instrumentation industry. These companies and their products span the entire range from the most practical instrumentation used in heavy industry—for example, steelmaking monitors—to instruments of great delicacy and precision that are used in science and medicine—for example, atomic clocks and neurosurgical probes (see Figures 4.102–4.105). This section draws heavily on the report of this Panel on Instrumentation, which appears in Volume II. After presenting a brief historical background, we examine the role of instrumentation in both the scientific and economic life of the nation and draw certain comparisons with activities in selected foreign countries. This examination is followed by a brief discussion of the development process in the instrumentation industry, tracing, in selected cases, this development from basic physics through technical development to practical application. The section concludes with an examination of certain outstanding problems that the instrumentation industry faces at present.

FIGURE 4.102   Photograph of a modern airborne proton magnetometer used for making magnetic surveys, which are particularly useful to petroleum geologists in helping to identify formations in which oil might be found. [Photograph courtesy of Varian Associates.]

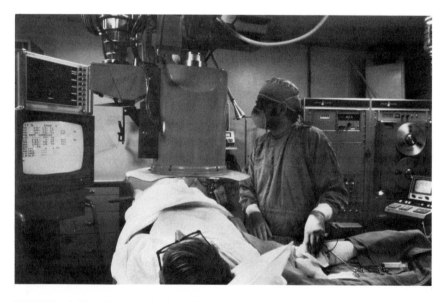

FIGURE 4.103   Computerized cardiac catheterization systems aid doctors in the prompt diagnosis and treatment of critically ill heart patients. [Photograph courtesy of Hewlett-Packard Company.]

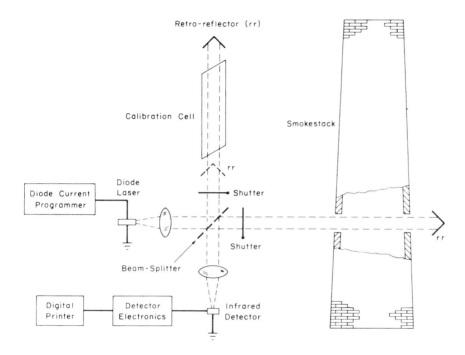

DIODE LASER SYSTEM FOR ACROSS-THE-STACK MONITORING

FIGURE 4.104   Diode laser system for across-the-stack monitoring of sulfur dioxide. This system is under construction for the Environmental Protection Agency and is based on the absorption of tunable laser radiation—yielding an average value for the pollutant concentration over the stack diameter. [Photography courtesy of Massachusetts Institute of Technology, Lincoln Laboratory.]

Applications of Instrumentation

Although there are a great many possible categorizations of instrumentation depending on use, precision, scope, underlying mechanisms, and the like, here we consider only three broad classifications. These are

1. Instruments for research, laboratory measurement, testing;
2. Instruments for industrial process management and control;
3. Instruments for analysis, which include types for both research and laboratory use and industrial plant on-line use.

Before examining these categories in detail, however, it may be useful to consider the applications of instrumentation more generally. A society

FIGURE 4.105   Lone Star Cement Control Center, Greencastle. [Photograph courtesy of Leeds & Northrup Company.]

without numbers, and, hence, a society without instruments, is inconceivable in today's world. There are very few types of human activity in which instruments are not necessary. Four activities involving major applications are the following.

Research   Instrumentation is a key and indispensable element in research, whether the research be attempts to advance man's knowledge of nature's laws, of the characteristics of nature's materials and energies or of life and biological processes; or to synthesize new materials or develop new techniques of energy conversion or new products and processes. Each major step forward in research creates a need for new and better instruments of greater resolution, capability, or flexibility, so that the next significant step forward in research can be undertaken and achieved.

The relationship between research and technology, as exemplified in the instrumentation and apparatus of research, has been essentially one of symbiosis. Advances in science have both spurred and made possible new technological advances; these, in turn, have made possible major advances in science. Examples of this symbiosis abound in this Report and in the panel reports in Volume II, and we return to this subject in a subsequent part of this section.

Industrial Processing   In the scientific and industrial sense, ours is a well-ordered world. Fortunately, physical and chemical properties of things or materials can be established by investigation and, under comparable conditions, can be repeated. The essence of production is to create materials and things—and in some cases to provide services—with fixed, predictable qualities, qualities that can be reproduced if the conditions and environment of production are properly established and maintained.

It is the function of instrumentation first to determine the magnitude of parameters that define the conditions of a process. Then, by automatic control means, incorporating in recent times computers of broad capability, appropriate variables are adjusted so that desired process conditions are achieved. Analysis instrumentation, either in the laboratory or on-line, depending on the prevailing state of the art, assists in regulating the process to yield products of desired quality while achieving economical process operation. The automation and process control that modern instrumentation has made possible have already had a profound impact on the quality of life throughout the world. The grinding monotony and drudgery that frequently characterized industrial working conditions for unskilled labor in the past are now largely gone, and, at the same time, the reliability and economic competitiveness of U.S. industrial products have increased significantly.

World Competition   Successful competition in the trade markets of the world is an essential element in maintaining a nation's standard of living and a correspondingly highly cultured life style. World competition is both technological and economic. Given steady trends toward equalization of basic technology, at least in some major fields, the competition in such fields resolves into economic comparisons, that is to say, largely comparative process and plant productivity.

The importance of productivity in world competition, in the face of other international economic trends, is illustrated in Figures 4.106–4.109. Figure 4.106 shows comparative wage increases over the past several years in the principal industrial producing countries of the Western world. Japan has had the highest rate of wage increases, with West Germany close behind. The United States, surprisingly, has had the lowest. Figure 4.107 provides an index to changes in consumer prices. Again, Japan is the highest, with

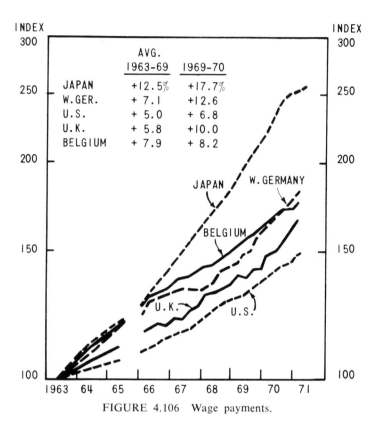

FIGURE 4.106   Wage payments.

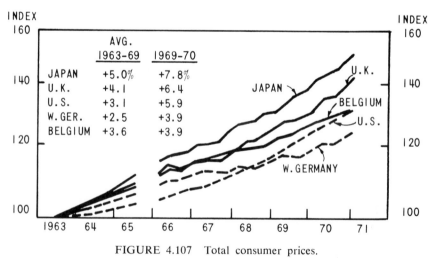

FIGURE 4.107   Total consumer prices.

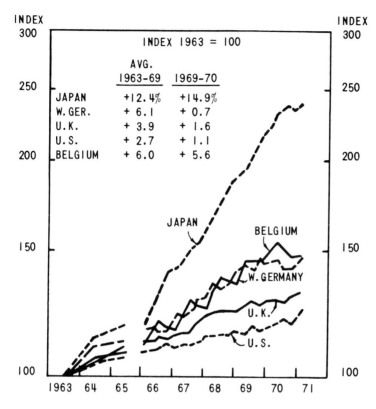

FIGURE 4.108   Changes in manufacturing output per man-hour.

the United States appearing between it and West Germany. Figure 4.108 then illustrates changes in manufacturing output per man-hour, which is to say, productivity. Japan shows the greatest improvement; the United States, the least. Figure 4.109 shows the net result of changes in wage rates and productivity, in terms of unit labor costs. It will be seen that Japan, despite the highest rate of wage increase, has had sufficient counterbalancing improvement in productivity so that its unit labor costs have been held substantially constant over the past few years. The United States, together with the United Kingdom, shows the highest increases in unit labor costs.

These curves demonstrate, quite emphatically, that U.S. productivity must be improved if this nation is to maintain a satisfactory position in world competition. Manifestly, the problem is not a simple one. A solution requires participation by, and significant contributions from, many sectors and groups. Advances in science and technology and related improvements

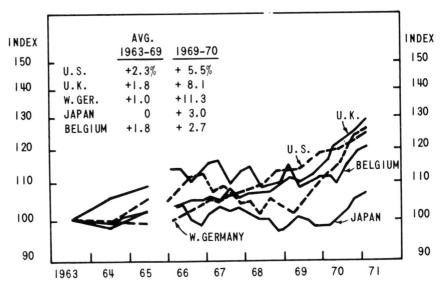

FIGURE 4.109   Changes in unit labor costs.

in process control by means of instrumentation can be important factors in achieving the requisite productivity improvements in U.S. industries.

Society   Throughout this Report and those of the panels we have emphasized the importance of the contributions that physics has made, and can make, to society. These range from the cultural and educational—both closely related to the extension of man's intellectual horizons in research— to the alleviation of human ills and suffering, through clinical and research medicine, to the satisfaction of material needs and desires of the nation's consumers, and to the assurance of safety and defense both locally and internationally.   In all these areas instrumentation plays a vital role in the interface between the scientific principles involved and the actual application or device.

   The extent of these contributions is great, but their existence and their benefits are difficult for the public to sense or see. One reason is that they frequently occur within the technical community and the capital equipment industry at stages where no consumer products or services are directly or visibly involved. Another is that the benefits are diffuse, composed generally of an extremely large number of individual improvements, efficiencies, and advances in an extremely wide variety of industries, institutions, and human activities. But they are there.

The U.S. Instrument Industry

Historical Background During the nineteenth century, instruments used in the United States were for the most part imported from abroad, primarily from Germany and England. At the turn of the century and in the decade that followed, domestic instrument companies were organized and developed. Also in 1901, the National Bureau of Standards was organized, and its pioneering work in the development of standards and measuring techniques profoundly influenced and stimulated the design, fabrication, and use of domestic precision and industrial instruments.

Thus, when World War I began, apparatus, standards, and techniques for measurement and control were available domestically in support of the nation's industrial and defense needs.

In the post-Depression years preceding World War II, a number of small companies, largely oriented to the emerging science of electronics, entered the instrument field. They, and other new groups, supplemented the efforts of the older instrument companies in providing major support to the nation's technical effort—in both research and production—during the Second World War. The technologies emerging from that war, including those based on the new nuclear science and on radar activity, found expanded use in world science and industry and stimulated considerable growth of U.S. instrument enterprises. Later, space programs, expanded defense needs, extensive government-sponsored research, expansion in industrial processing, and major scientific advances in solid-state physics and computer technology all contributed to new and larger needs for instrumentation, with corresponding growth in the U.S. instrument industry and—at least until the present time—worldwide predominance in this field. Until very recently, the U.S. scientist or engineer who visited laboratories in all parts of the world, except the Soviet Union or China, found that the instrumentation in use was familiar—in many cases identical to that in use in the United States and produced by the same manufacturer. This situation is changing rapidly to one of lesser dependence on U.S. instrumentation sources.

Present Instrumentation Industry Activity Four aspects of U.S. instrumentation activity merit particular attention.

1. *Size:* The current annual sales of the U.S. instrument industry are estimated at about $3 billion and are made by approximately 1800 establishments having over 165,000 employees. Company sizes vary widely: the largest has annual sales of about $350 million; numerous companies have annual sales in the range from $500,000 to $1 million. Small companies compete successfully with the larger ones on the basis of innova-

tive technology and aggressive applications activity in more narrowly selected product lines.

2. *Research and Development Activity:* Product lines are characterized by high technology content, and, although research and development activity elsewhere represents an important market for the industry, it ranks high among industries in the amount of sales reinvested in its own research and development. For individual instrument companies, the range of investments in new product development extends from 5 percent to 10 percent of sales. Leading companies are at the high end of this range. In 1969, the overall average for the instrument industry was 6 percent, with about 13,200 scientists and engineers employed in new product development activity at the end of that year.

3. *Contributions to the Balance of Trade:* World leadership in instrumentation technology coupled with sound business practices have enabled the U.S. instrument industry to achieve a high level of export sales. They amount to as much as 50 percent of the total annual sales in some products and average about 20 percent of the total annual sales for the industry as a whole. Exports of instrument products have exceeded imports in the ratio of 3.5:1, thereby contributing significantly to the nation's balance of trade.

4. *Professional Organizations:* The Scientific Apparatus Makers Association, with headquarters in Washington, D.C., now in its fifty-second year, is the principal national trade association for the industry. The Instrument Society of America, with headquarters in Pittsburgh, which was founded 27 years ago, is a professional society devoted specifically to instrumentation science and technology. It has 20,000 members and is represented on the National Research Council.

New Product Development in the Instrument Industry

Funding The U.S. instrument industry is a strong investor of its own funds in new product development. As previously noted, the industry invests, on the average, about 6 percent of its annual sales in in-house research and development, although some of the more successful companies spend as much as 10 percent or more of annual sales in such efforts. It should be emphasized that these rates of expenditure refer to corporate funds and do not include any special projects that may be funded within some companies by government agencies.

Time Scales in New Product Development There is a well-developed progression of steps involved in the conversion of a basic scientific principle into an acceptable reliable instrument or other product. Research, development, engineering design, tooling, testing, manufacture, and marketing are the requisite elements of the overall development process, from recognition of a need to implementation in the marketplace (see Figure 4.110).

The overall time for moving through these in-house steps will vary from perhaps two to five years, depending on the complexity of the product, the degree of innovation that it incorporates, the nature of the environments in which it will be applied, the capabilities of the purchaser's personnel, and the policies of the manufacturer in regard to field failures or inadequacies of his products. Generally speaking, new products will have been thoroughly tested and will be marketed only after the achievement of successful results in the environments of their proposed use.

An example of a relatively simple new product that required a much longer time for successful development than might have been anticipated initially is the expendable immersion thermocouple for molten steel (discussed in greater detail later in this section). A relatively simple product in appearance, based fundamentally on well-known concepts of classical physics, it nevertheless required several years of intensive innovative design work and testing before fully reliable units, adapted to the rough impersonal treatment and hostile environment of a steel mill and retailing for less than $1 each in spite of their noble-metal content, were produced.

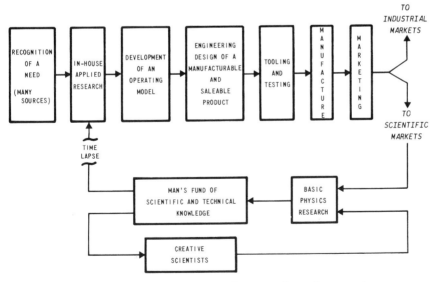

FIGURE 4.110 Steps and cycles in the development of new instruments.

The challenges to successful production and performance for such a device were great.

At the other end of the scale are highly complex analytical tools, such as nuclear magnetic resonance (which we discuss in greater detail later in this section), that, at least in their earlier period of introduction, are used in an atmosphere of sophistication by highly trained and sympathetic scientists or technicians who are conditioned to the uncertainties of research and are strongly motivated to have the equipment perform satisfactorily. Anticipating such an atmosphere of use may well encourage a manufacturer to reduce the development time required for initial versions of a product destined for these markets and anxiously awaited by potential users.

Risks in Product Development Frequently, the time span of successful product development proves to be much longer than anticipated, and the attendant development costs much higher than originally planned. Also, acceptance in the marketplace may be less than was expected. This lack of acceptance may result from unanticipated economic conditions, competitive breakthroughs and products, or overly optimistic estimates of what potential users really would be willing to buy. All these factors are risks in development, and, axiomatically, the greater the technical innovation, the greater the economic risk is likely to be. Nevertheless, equally axiomatically, the U.S. instrument industry is dedicated to taking these risks, recognizing that without them it cannot sustain growth and leadership in its selected markets.

As the industry performs its applied research in the hope of developing profitable new products, it has the comparable hope that in other appropriate areas there will be adequate support of basic research, for the results of such research will contribute to the industry, if not directly, then inevitably in time and in unanticipated ways. We have noted previously the great degree of symbiosis between the scientific and instrumentation communities. The research scientist is alternately discoverer, innovator, customer, and user, and in all these roles he advances the state of the art in the instrumentation industries. In all these roles, too, problems have developed, which we shall discuss.

Using the Results of Basic Physics Research

Possible Spinoff from Basic Research Creative scientists, basing their work on earlier discoveries, and frequently using instrumentation from commercial sources (as well as creating instruments of their own), engage in basic research, the results of which become significant additions to man's fund of scientific and technical knowledge. Such new knowledge is sought for its own sake as part of man's insatiable curiosity about his universe and his desire to penetrate and divine nature's mysteries and secrets. There is, in such basic research, no advance consideration of whether a practical use will be found for its results. But inevitably, and after varying lapses of time, such uses are found—often where least expected.

Because the instrument industry, like all other industries, is profit-oriented, its research and development programs have, with only rare exceptions, specific marketing objectives. Thus they are properly defined as applied rather than basic research. Basic research, in other words, is generally beyond the profit-oriented scope of instrument companies. Their research and development departments, however, do have an ongoing appetite for new knowledge. They are continually on the lookout for new scientific information available from other sources that might in time find a place in new products that are to be developed.

Examples In the Report of the Instrumentation Panel in Volume II, a large number of instrumentation systems deriving rather directly from basic physics research are listed, and case histories of selected ones are presented. We have selected two of these examples for purposes of illustration here. The first, the expendable thermocouple, represents an example of a simple device, based on classical physics, designed for use by unskilled personnel in an extremely hostile environment. The second, nuclear magnetic resonance, is an example of a sophisticated device based on quantum physics and designed for use by relatively skilled personnel in what might be characterized as a research environment. Despite the vast difference in these devices, however, they share the common attribute of making possible measurements that would be obtained otherwise only with great difficulty if at all and that have important economic and social consequences.

To manufacture steel economically, the steelmaker must know the temperature of the molten steel in the open-hearth, basic-oxygen, or electric-arc furnace so that the steel can be poured, or tapped, at the appropriate point. If the heat is tapped at too low a temperature, steel will solidify in the receiving ladle before it can be completely transferred to the ingots molds. This situation results in so-called skulls, which must be removed from the ladle, broken up, and remelted—obviously a wasteful procedure. If the steel is tapped at too high a temperature, it penetrates crevices in the mold, causing the ingot to adhere to its mold. This result is called a sticker. The mold must be broken to remove the sticker, again a costly procedure. Also, heating the steel to a temperature higher than is necessary lengthens the time of a heat and uses extra fuel, which increases the cost of the heat. In spite of intensive efforts over a period of five decades by steelmakers throughout the world, a fully satisfactory solution to the problem had not been developed by the mid-twentieth century.

Techniques tried during the 50 years of experimentation included thermocouples of various designs and radiation-responsive units. The high temperatures and hostile environment of the measurement militated against the attainment of successful results.

In 1958, a thermocouple unit sufficiently low in cost, despite the use of noble metals, to permit throwing it away after one use, and highly accurate in its single use, was introduced. It won rapid acceptance throughout the steel industry. It used a number of concepts derived from basic physics, although many of these concepts had been well known for decades. This device is an example of a lengthy time lapse before principles derived from basic physics were applied to practical instrumentation to fill a specific need. It emphasizes the challenges of in-house efforts to extend available knowledge effectively with innovative development and engineering activity to produce a reliable manufacturable product that can be sold at a profit.

Because the successful solution to the molten-steel temperature-measuring problem is a specially designed thermocouple used with a null-balance potentiometer recorder, it is of interest to sketch the historical background for these devices.

The thermocouple effect, that is, the development of a small dc voltage related to the temperature difference between the hot and cold junctions of two dissimilar metals, was discovered by Seebeck in Germany in 1821. At just about the same time, at Berlin University, Poggendorff devised the first known null-balance type of potentiometer, which he used to measure the output of electrochemical cells without drawing current from them. As far as is known, not then or for a considerable time thereafter, was there a common interest in or joint use of the thermocouple effect and the null-balance potentiometer. It took about 80 or 90 years for the two to be combined for industrial temperature measurement.

In about 1886, some 65 years after its discovery, practical application was made of Seebeck's thermocouple effect by Le Châtelier in France, who measured the output of the thermocouple with an indicating deflection millivoltmeter. Shortly thereafter, millivoltmeter-type recorders were adapted to temperature recording using thermocouple temperature detectors. But such recorders had limitations due to insensitivity and calibration drifts of millivoltmeter movements, the influence of length of leads on the temperature measurement, and difficulties related to reference junction compensation.

Meanwhile, along another path of temperature-measurement development, the first platinum resistance thermometer, depending on the temperature coefficient of platinum, was developed in Germany in about 1871. By 1890, a practical platinum resistance thermometer was developed in Great Britain.

In 1897, the Callendar Wheatstone bridge recorder, believed to be the first null-type recording instrument on record, was introduced. It used advanced concepts of electrical balancing but did not prove to be very practical in the hands of other than skilled operators.

Then, early in the twentieth century, individuals in industrial plants began to combine thermocouples with manually operated null-balance potentiometers to make precise measurements of temperature. This technique, which at balance drew no current from the thermocouple, had many clear advantages and stimulated efforts to develop an automatic null-balance potentiometer recorder to take the place of the manually operated instruments. By 1912, such a successful recorder was developed, and Seebeck's thermoelectric effect came into its own for industrial temperature measurement and control. Literally millions of thermocouples and tens of thousands of potentiometer recorders have since been applied to a wide variety of industrial processes. Nevertheless, despite the great success of this technique in other applications, it had not been applied successfully to the molten steel temperature-measuring problem by the mid-twentieth century.

Some of the problems associated with making a successful measurement of molten steel were the following:

1. The working temperature is 2900°F—high for an industrial measuring device.
2. An accuracy of 5 to 10°F is desired.
3. The molten steel is covered with a surface of slag.
4. Expensive radiation-responsive devices intended for multiple repetitive use require heavy protection, are cumbersome, and are subject to large errors.
5. Similarly, thermocouples intended for multiple repetitive use require heavy protection and limit the operator's ability to establish temperature trends before he has to tap the furnace. Also, repetitive exposure of the platinum–platinum, rhodium thermocouple to the high temperature of the bath causes significant calibration drifts.

What clearly was needed was a device that would overcome all these limitations and permit measurements to be made rapidly and accurately at a cost not exceeding $1 to $2 per measurement.

The unit introduced in 1958 that, together with a null-balance potentiometer recorder, successfully solved the molten-steel temperature-measuring problem is a thermocouple of very special design. Its features are these:

1. It uses a platinum–platinum, rhodium thermocouple and achieves the 5–10°F accuracy by being used only once and then discarded.

2. It achieves the economic targets by using only a very small length of the precious thermocouple wires.

3. Base metals with suitable thermoelectric effects are used within the thermocouple in place of longer extensions of the precious wires. The junction of these base and noble metals must be at less than 200°F even while the thermocouple is immersed in the 2900°F bath. This requirement posed interesting problems in regard to thermal conductivity to the thermocouple tip and thermal insulation for the noble metal–base metal junction. The combination needs to act like a complete noble-metal thermocouple for just the few seconds it takes to make the measurement.

4. The tiny thermocouple tip is protected from contamination and physical damage as it is plunged through the furnace atmosphere and the surface slag.

5. The thermocouple holder, or lance, is light enough to be handled easily by one man and yet heavy enough to pierce the slag and metal surfaces and withstand general rough steelmill use.

Extensive development and design work, tooling, and testing ultimately produced the present expendable devices that sell for considerably less than $1 each. In the 13 years since their introduction, over 100 million units have been used in steelmills throughout the world. This was a case, then, of the basic foundation being available from classical physics; the challenging applications problems awaited imaginative, innovative design work before they were resolved.

As a second example of the contributions of basic physics to instrumentation, nuclear magnetic resonance (NMR) provides a good illustration of sophisticated basic research that resulted in the development of instrumentation that, through its use in both further research and field measurement applications, has been of major assistance in areas of substantial social benefit. It has benefited the U.S. balance of trade by providing exports not only of NMR instrumentation but also of new products particularly in areas related to chemistry. It has found increasingly important applications in biology and medicine in uses as diverse as understanding biological structures at the molecular level and the routine search for drug addiction evidence in urinalyses. The invention, development of the technique, and its commercialization have been a cooperative endeavor among the scientists—in many cases physicists—the applications scientists, the business administrators, and the production workers.

The phenomenon of NMR was discovered by Bloch at Stanford and Purcell at Harvard during the winter of 1945–1946 and earned for them the joint award of the 1952 Nobel Prize for Physics. The interest of these physicists was the determination of magnetic moments and angular momenta of various nuclear isotopes. The measurement is performed by placing the atoms of interest in an intense magnetic field and exciting and detecting absorption of energy as a function of its frequency. By measuring the frequency at which energy absorption takes place and relating it to the strength of the magnetic field, it becomes possible to determine the ratio of magnetic moment to angular momentum of the particular nuclear species in question. Since this information provides a characteristic signature not only for the nucleus involved but also, as was soon discovered, for the particular atomic or molecular environment in which it finds itself, the basis for a highly useful analytical tool was at hand.

It was soon found, too, that the measurement could be used in reverse to determine the strength of a magnetic field by relating it to the absorption frequency of a nucleus for which the gyromatic ratio was already known. Thus, as a significant by-product, a very accurate magnetometer was born.

The greatest impact of the NMR technique was found in chemistry. It came about because of the discovery, again by a group of physicists, that the exact gyromagnetic ratio of a given atom depended to a small extent on the chemical molecule in which the atom was located. Should several like atoms exist in the same molecule, several slightly different resonance frequencies are found. This finding was designated the chemical shift because its value is determined by the chemical environment of the atom. Studies of proton NMR spectra have been used extensively since 1952–1953 to determine the structure of molecules and the composition of mixtures and to understand equilibrium reactions. As instrumentation was improved by engineers and scientists, the technique was used for a wide variety of elements, for example, fluorine, phosphorus, boron, and nitrogen, in addition to the original hydrogen, and is now one of the most valuable instruments for chemical research. In fact, after a chemist synthesizes a new product, he can determine within hours many of its structural characteristics. In the latest development with superconducting magnets, Fourier transform techniques, and signal averaging devices, the method is being used in the biochemical field to study the structure of very complex proteins for the identification of various hydrogenous and other components of the complex molecule. The present state of development indicates that work with ^{13}C will make it possible to identify the different carbon atom complexes and locations in complex organic molecules, again providing basic structural information for the research chemist and bio-chemist. The technique is now accepted as a standard method of chemical analysis and is used throughout the chemical industry providing a significant saving in time and cost.

Both NMR and its companion, electron spin resonance (ESR), discovered at about the same time by groups at Kazan and Oxford, are used to study biomolecules from simple amino acids and sugars through hormones, enzymes, and DNA. It is also used to study molecular conformation and structure, as well as metabolic reaction rates and mechanisms. A recent widely publicized application of NMR and ESR is for detection of a metabolic product of heroin that is found in the urine of a user, thus giving the first rapid and reliable screening test for drug addiction. This application marks only the beginning of widespread use of these techniques for similar pur-poses.

Since about 1955, NMR magnetometers have been used extensively to make airborne and ground-station measurements of the earth's magnetic field. These measurements are particularly useful to petroleum geologists in helping to identify formations in which oil might be found. Some of the early earth satellites carried proton magnetometers to determine the mag-netic-field strength at large distances from the earth and relate it to the various theories concerning the origin of the earth's magnetic field. Several companies produce laboratory magnetometers for determining the precise field strength of laboratory magnets.

Fundamental to the application of NMR has been the conversion of a complex and often temperamental laboratory instrument, which required tender loving care from the physicists and skilled technicians who used it, to a reliable, rugged, commercially available instrument that when turned on could be expected to work consistently without delicate adjustment or special care. Here again the skills of the applications physicist and engineer have been brought to bear to provide an enormously useful new device.

Estimates of the time needed for the cost savings to pay for an NMR instrument range from a few months to a few years. The usefulness to in-dustrial laboratories has been further demonstrated by the fact that prac-tically all major chemical companies have multiple NMR instruments, some of them being multiple copies of a given type and others having differing characteristics.

The NMR instrument market for chemistry has grown rather steadily since the first sales in 1953. The market size in 1971 is estimated at $25 million, with approximately 40 percent of this market within the United States. Prior to 1966, almost the entire market was supplied by one U.S. company. At the present time, major suppliers of this market are found in Germany, England, and Japan, as well as in the United States. The ratio of U.S. exports to imports of NMR equipment has dropped from approximately 10 in 1966 to approximately 1.5 at the present time.

Instrumentation Costs

Estimates of the escalation in the intrinsic costs of doing scientific research range from some 3.5 percent to 7.5 percent per year depending on the exact assumptions made. This figure is quite apart from any consideration of overall inflation or of salary escalation in the research enterprise; it relates only to the increased costs that reflect increased sophistication of the questions addressed and the measurements made. It has long been traditional wisdom in the physics community—and there are spectacular examples to support this belief—that much of this increase stemmed from the rapidly increasing cost of the necessary instrumentation. The Panel on Instrumentation examined this question and concluded that no such simple conclusion is warranted.

Well-Established Instruments Instruments that were already well established a few years ago, such as certain types of test instruments and recorders, successive versions of which simply reflect improvements in design details, have had inflationary price increases during the past ten years that range from 25 percent to 50 percent. Certainly for such instruments the impression of higher prices would be confirmed.

Innovative Instruments Instruments that a decade ago were relatively new in the marketplace and reflected major steps forward in complex instrumentation, such as NMR analyzers, have decreased steadily in price in recent years—for the same performance capability—moving significantly counter to inflationary trends. A basic assembly that sold for, say, $50,000 a decade ago might sell today for 60 percent of that amount. This pricing trend is in contradiction to the general impression of price increases. However, a research scientist might not be satisfied to purchase currently just such a basic instrument; rather, he might desire—or need—today's most advanced design to take advantage of its greater capabilities. Such an advanced unit might well sell at the same price or moderately higher than did the only unit available some ten years ago. In addition, it might be equipped with additional accessories, at a still higher price.

Computers Computers and their constituent elements are considerably less expensive today than they were a decade ago. Figure 4.111 illustrates the drop in price in core memories from about $0.07 per bit in 1966 to less

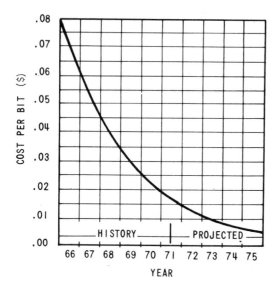

FIGURE 4.111   Core memory system cost.

than $0.02 per bit in 1971. Similarly the decrease in the price in mini-computers from about $27,000 each in 1963 to the $5000–$7000 range in 1971 is shown in Figure 4.112.

In each case, these decreases reflect advances in the technology of design and fabrication and substantial increases in production volume. Thus if a research scientist were to purchase a bare computer today he would be spending much less for it than he would have spent a few years ago. This, again, apparently contradicts the impression of increased prices for instrumentation. However, the current substantially lower price for a computer frequently encourages its purchase by a scientist who might not have considered such an acquisition when their cost was greater.

Advanced Systems Combinations   It is possible that advanced systems combinations have been most responsible for the impression of high current equipment costs. Automated system assemblies, incorporating computer direction and computer data processing, have been developed, with capabilities far beyond those available even a few years ago.

It is generally true that an individual researcher spends more for his instrumentation—for outfitting his individual laboratory—than was the case even a few years ago; but he may be getting a great deal more than he could have acquired at a lower cost a few years ago. In addition, he may be achieving secondary savings due to conservation of his own time and possibly the time of technician assistants through the automation of

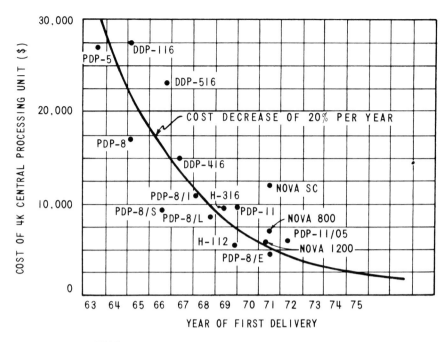

FIGURE 4.112   Decreasing costs of minicomputers.

computational and analysis steps. In terms of the cost per unit of new information, however inadequately this may be defined, there appears to be little question that a significant reduction has occurred in recent years. However, this reduction has been coupled with the need for more information to establish each new scientific finding as man probes ever deeper into nature. Despite the best efforts of the instrumentation industry to provide more capability for unit cost in its products, the cost of acquiring new information increases at the rates indicated earlier in this section.

Problems in U.S. Instrumentation Activities

As indicated above, U.S. instrumentation, at least since the mid-1940's, has enjoyed a position of international pre-eminence. Progress has been exceedingly rapid, spurred by demands from a burgeoning technological society and a vigorous scientific and engineering community. Both progress and pre-eminence are currently in danger.

The change in the growth rate of support for physics has had at least four important effects on instrumentation activity in the United States:

1. The scaling-down of research effort suggests that fewer new concepts and ideas fundamental to new instrumentation will become available in any given period; it also implies decreased pressure on the instrumentation industry for new instruments.

2. Although the instrumentation industry has a strong record of in-house research and development, the instrumentation groups in national laboratories and some of the larger universities have served traditionally as major sources of new ideas and devices that were then picked up, modified, and marketed by the appropriate industry. Under conditions of limited support these instrumentation groups, assembled patiently over many years, have been early casualties and this source of instrumentation innovation essentially terminated. Unless means are developed to reinstate such activities it will be difficult indeed to maintain the United States in a competitive position in the face of burgeoning activity in Europe and Japan.

3. With limited funding there is a natural tendency in any laboratory to live on instrumentation capital—to defer the purchase of a new instrument, however badly needed, in preference to discharging yet another research staff member. In the long term, this practice can have serious ill effects on the research activity; in the short term it can have a devastating effect, particularly on smaller, highly specialized instrumentation industries whose markets effectively evaporate. Many have been forced into bankruptcy in the past few years. Unless current trends are reversed, more such firms probably will face bankruptcy. These industries represent an important national resource and one whose loss cannot be viewed lightly.

4. In the past decade, there has been a discernable de-emphasis in the education and training of instrumentation scientists and engineers in U.S. universities and colleges. It is in just this activity of exploiting scientific phenomena in new engineering and technological areas that the United States has been enormously successful in the past; it is in this type of activity that this nation must continue to remain at the forefront if it is to preserve not only an internationally competitive economic posture but also the capability to respond to the challenge of improving the environment and the quality of life. There is a serious need for re-examination of U.S. education programs related to instrumentation development and application.

> Here, too, we may again repeat what we have said
> above concerning the extending of natural philosophy
> —so as to prevent any schism or dismembering of
> the sciences; without which we cannot hope to
> advance.
>
> FRANCIS BACON (1561–1626)
> *Novum Organum,* Book I, 107

THE UNITY OF PHYSICS

Introduction

This chapter has been concerned thus far with the individual subfields of physics, partly for convenience, partly to bring out the internal logic in each. To stop at this point, however, would be to neglect perhaps the most important aspect of physics—and one all too frequently forgotten: The study of natural phenomena proceeds simultaneously on many frontiers each of which supports and nourishes the other. There is a very broad spectrum of research activities in physics ranging from the most basic to the most applied—from almost exclusively intrinsic to equally exclusively extrinsic activities with all intervening gradations. This diversity, amid overall continuity, is the hallmark of a vigorous science and has long been recognized as such. Its application to physics is simply typical of its more general truth throughout all of science.* Preservation of this unity in science and in physics is of vital importance and is an essential consideration in any national planning for science.

With increasing specialization what tends to be forgotten is the remarkable extent to which even the most intrinsic techniques and concepts diffuse throughout physics and the remarkable degree of commonality that exists. The frequently discussed fragmentation of physics has been much exaggerated.

In the nineteenth century, the unifying concepts were those of Newtonian mechanics; in the twentieth century, the two unifying principles have been relativity and quantum mechanics. In fact, the latter has played by far the dominant role for the simple reason that, in the usual terrestrial physical phenomena, relativistic considerations appear as fundamental but nonetheless small corrections and perturbations; elementary-particle physics is clearly an exception. Increasingly throughout physics, and more recently in both chemistry and biology, quantum mechanics has provided the overall unifying concept.

To illustrate this unity we draw on data from the physics of condensed matter, atomic physics, nuclear physics, and elementary-particle physics.

---

* See Chapter 3.

The degree to which the latter three subfields show parallel spectroscopies was first emphasized by Weisskopf.* Here his discussion is expanded to include the physics of condensed matter as well.

The Four Spectroscopies

Figure 4.113 is a highly schematic illustration of the systems considered in this section. In condensed matter the exciton is selected as the elementary excitation. The exciton is simply the composite entity resulting

* See, for example, V. Weisskopf, "The Three Spectroscopies," Scientific American, *218*, 15 (May 1968).

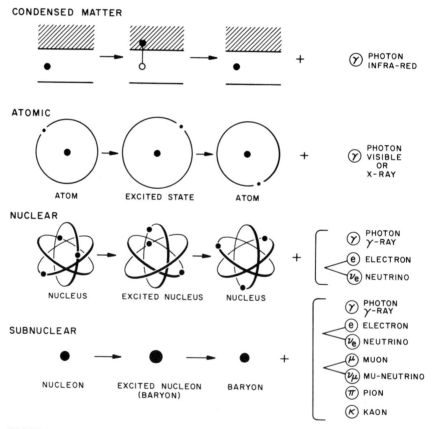

FIGURE 4.113   Four spectroscopies. [Source (for part of figure): V. A. Weisskopf, "The Three Spectroscopies," Scientific American, *218*, 16 (May 1968).]

from the interaction of an electron, in the conduction band of a semiconductor, bound to a hole in the band gap. When the electron and hole recombine, the energy involved is radiated as an infrared photon. In the condensed-matter and atomic examples, the energies involved are not sufficiently high to permit radiation of other than photons of appropriate wavelength. In the nuclear example, once the available energy exceeds 1.022 MeV (the rest mass of an electron–positron pair) it becomes possible to radiate such a pair instead of a photon; in certain situations in which angular momentum selection rules forbid photon emission (as from a O+ to a O+ state), pair emission is the only mode of de-excitation available to the nuclear system. Such decays clearly occur via the electromagnetic interaction.

In the nuclear system, the weak interaction—typically weaker by a factor of roughly 10^9, and one of the two new natural interactions first encountered in nuclear physics—can also act to de-excite nuclear states. Wherever it is in direct competition with the electromagnetic interaction, the electromagnetic decay modes are completely dominant. Under certain conditions, however, selection rules strongly hinder the electromagnetic interactions, and the weak-interaction decay modes, in which an electron and an antineutrino (or a positron and a neutrino) are emitted, become dominant.

In moving to the much higher energies characteristic of elementary-particle physics, clearly these electromagnetic and weak-interaction decay modes remain; however, as soon as the available energy becomes greater than the rest mass of the muon, the weak-interaction mode involving emission of the muon (heavy electron) and its corresponding mu-neutrino becomes possible. In going to larger available energies, as soon as the rest mass of the pion is exceeded its emission becomes possible; because pion emission takes place via the strong nuclear interaction (typically 100 times stronger than the electromagnetic interaction, and the second new natural force encountered in nuclear physics), it is a dominant decay mode unless, again, selection rules inhibit or preclude it. With still higher available energies, of course, it becomes possible, through the strong interaction, to emit the heavier kaons and other mesons as well. It may well be that with increasing energy an even heavier electron than the muon will be discovered under the mantle of the weak interaction. What is evident from this description is the systematic opening of new decay modes as ever more energy becomes available.

Next we illustrate the striking similarity in the actual spectroscopy of the different subfields. Figure 4.114 compares the atomic and nuclear spectra of sodium. Plotted here is simply the excitation energy of the quantum states of the atomic and nuclear systems—in essence the energies

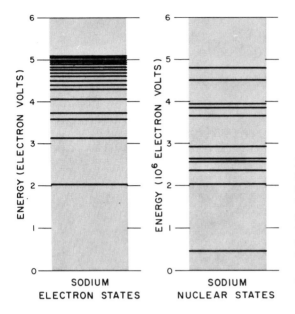

FIGURE 4.114 Atomic and nuclear spectra of sodium are similar in character. But the atomic spectrum (left) can be plotted on a scale whose units are electron volts, whereas the spectrum of nuclear states (right) requires a scale whose units are larger by a factor of 10^6. [Source: V. A. Weisskopf, "The Three Spectroscopies," Scientific American, *218*, 21 (May 1968).]

corresponding to the different solutions to the quantum-mechanical Schrödinger equation for each of these two systems.

What must be emphasized is that knowledge of these solutions comes from experimental observations, not from actual solution of such equations. The reason is obvious but perhaps bears repeating. Even assuming that physicists had complete knowledge of the forces involved—as, indeed, they do in regard to the atom, where the force is purely electromagnetic, but do not in regard to the nucleus, where as yet there is only approximate knowledge of the strong nuclear forces involved—the necessary mathematics for the exact formulation of the many-body problems involved does not exist. Of course, one can write approximate equations and, with the help of ever larger computers, obtain solutions of ever-increasing precision, granting the validity of the assumed forces. But in the case of the atom, because it is vastly simpler and less expensive, and in the case of the nucleus, because lacking better knowledge of the forces involved there is no option, physicists locate the quantum states, the solutions of implicit Schrödinger equations, experimentally in that branch of each subfield labeled spectroscopy. With the sophisticated techniques now available, it is possible not only to locate these solutions in energy (determination of the eigenvalues) but also to determine each of the quantum numbers labeling the individual solutions and, to an increasing extent, from a wide variety of measurements, to obtain increasingly detailed information on the structure of the state (determination of the eigenfunctions).

The spectra of Figure 4.114 illustrate, first, the remarkable similarity between the two excitation spectra, with an ordinate scale change of 10^6. They also illustrate a characteristic difference between spectra for the long-range atomic and the short-range nuclear forces: in the former there is a sharp break between the bound states and those in the continuum, whereas no such break occurs in the latter. Parts (a) through (d) of Figure 4.115 are corresponding presentations of the excitation spectra for exitons in cadmium selenide, a potassium atom, an aluminum nucleus (^{25}Al), and the baryon family of elementary particles (see also the Report of the Panel on Elementary-Particle Physics in Volume II for the corresponding spectra for the meson family). In each case, the spectrum has been decoupled horizontally to display more clearly some of the internal grouping of states according to certain of the dominant quantum numbers. The similarity, apart from the ordinate change from millielectron volts to electron volts to millions of electron volts and finally to billions of electron volts, is again striking and serves as an excellent illustration of the common approach used throughout wide areas of physics. Fundamental understanding comes as the physical models, which in essence are nothing more than mathematically tractable approximations to the unknown Schrödinger equations, are adjusted and modified—and occasionally scrapped—to reproduce the spectroscopic data. A model that accomplished only this task would be of little value; its value is based on the extent to which it permits new insights, suggests new studies, and successfully predicts the results of measurements yet unmade.

As might be expected from Figure 4.115, to the extent that all the emitted radiations in the different systems are the same, the experimental as well as the theoretical techniques have much in common throughout all of physics; increasing energy, of course, poses its own special detection problems.

Thus far we have emphasized the commonality of approaches and techniques that unify physics; we now turn to a different aspect of this unity, the way in which the results and insights of many branches of physics are required for fundamental understanding of a given physical situation or object. We have chosen to illustrate this characteristic with reference to one of the most exciting new objects found in nature in the recent past—the pulsar—because it has attracted the interest and activity of a very wide range of leading physicists and because it draws on so

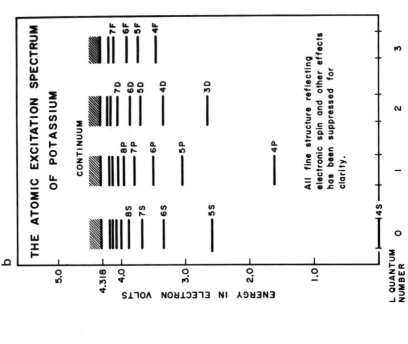

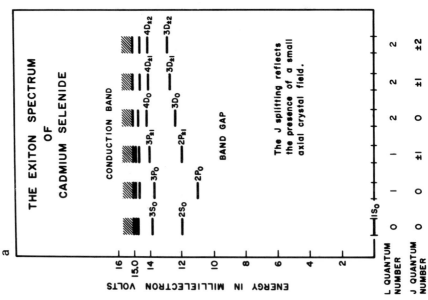

338

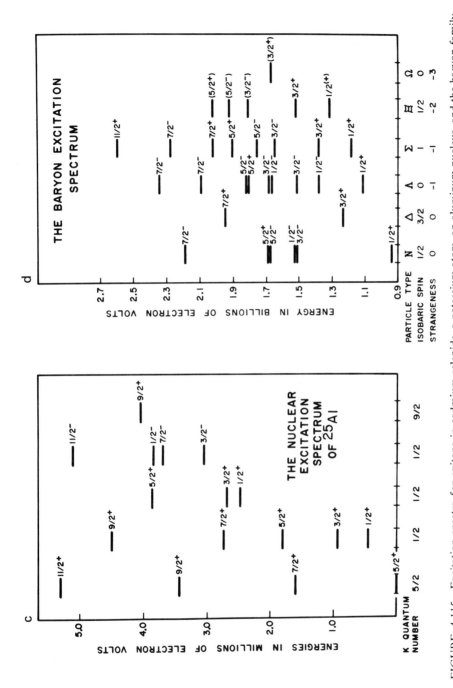

FIGURE 4.115 Excitation spectra for exitons in cadmium selenide, a potassium atom, an aluminum nucleus, and the baryon family of elementary particles.

339

many physical subfields for understanding of its structure and behavior. In considering this example we have drawn heavily on a recent article by Ruderman.*

The Pulsar

Perhaps the most exotic and varied forms of matter in the universe are to be found within the cores of dying stars. As a star evolves, its center grows hotter and denser. When it has finished burning its nuclear fuel, which supplied almost all of its radiated light energy, the star rapidly approaches its final state. Theory predicts three possibilities:

1. A star whose mass is less than about 1.4 times that of the sun can die as a familiar white dwarf. These are extremely dense stars about the size of the earth. Their central densities may exceed millions of grams per cubic centimeter. In such stars the enormous pull of gravity, which tends to crush the star, is balanced by the pressure from very rapidly moving electrons whose velocities may be comparable to that of light. Such high velocities are a direct consequence of the high density that forces electrons to be much more closely packed together than they are in the atoms of normal matter. The Pauli Exclusion Principle of quantum mechanics forbids identical electrons from getting close to each other unless their relative velocity is correspondingly large. This motion is the same as that which gives rise to the pressure that makes common solids difficult to compress and prevents the collapse not only of white dwarfs but of almost all forms of terrestrial matter. The different stages in a stellar collapse to and beyond the white dwarf phase are illustrated schematically in Plate 4. XVII.

2. A dying star may appear to contract forever toward a radius of a few kilometers and a density exceeding 10^{17} g/cc. This strange fate is predicted by the General Theory of Relativity for all stars more than twice as massive as our sun. Such relics are called black holes. Although they can attract to themselves whatever matter or radiation approaches them, the gravitational pull on their contracting surfaces has become so huge that not even light can escape. Black holes are probably common in the galaxy and throughout the universe, but they are hidden because of the great difficulties in observing them. The evolution of a black hole is shown schematically in Plate 4. XVIII.

3. Finally, the collapsing system may stabilize as a neutron star—a mass of a million earths compressed into a sphere barely capable of

* M. A. Ruderman, "Solid Stars," Scientific American, *224,* 24 (Feb. 1971).

Space Swallows a Star

The outside view of a collapsing star

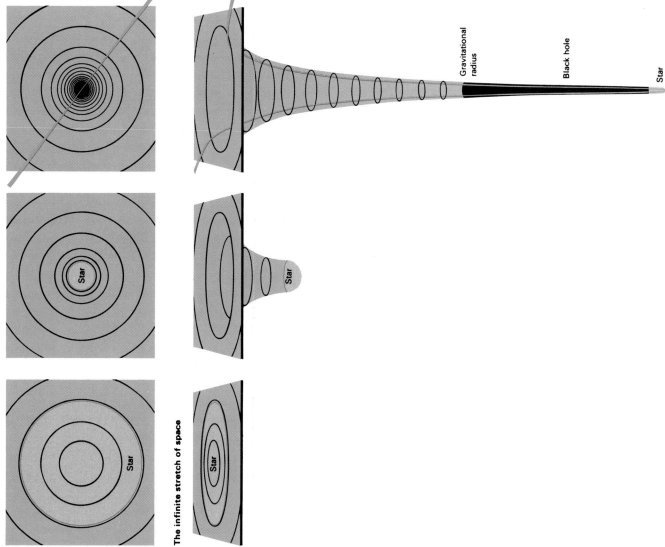

The infinite stretch of space

PLATE 4.XVIII Space swallows a star. Distances near a collapsing star are stretched by the star's rising density. To an outside observer, the star merely shrinks (top row). The originally flat surface of this simplified two-dimensional space is rapidly stretching, however (bottom row), increasing the distance to the star. Thus, the red ribbon in the right drawing is actually much longer than it seems from the outside. In three-dimensional space, the star would be shown within a cube, instead of a square, and as the star collapses, the distance through the center of the cube would stretch to infinity. [Source: *Science Year. The World Book Science Annual.* Copyright © 1968, Field Enterprises Educational Corporation.]

Matter's Struggle for Space in the Crush of Collapse

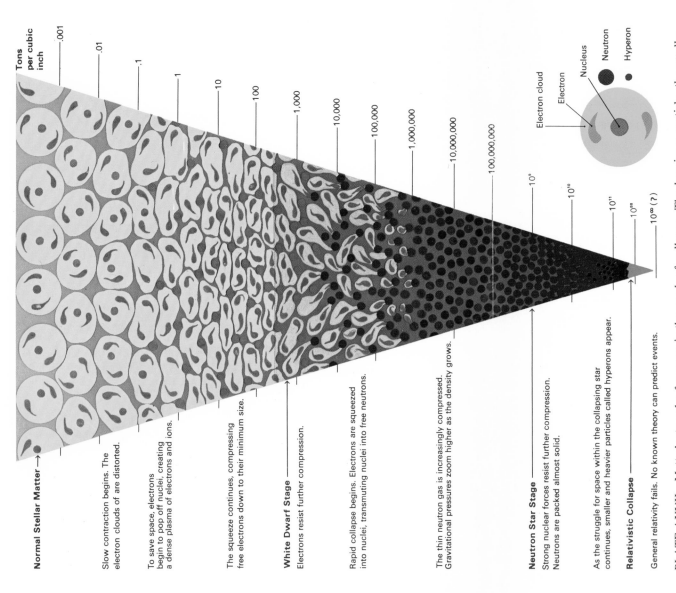

Tons per cubic inch

.001
.01
.1
1
10
100
1,000
10,000
100,000
1,000,000
10,000,000
100,000,000
10^9
10^{10}
10^{11}
10^{88}
10^{∞} (?)

Normal Stellar Matter

Slow contraction begins. The electron clouds of are distorted.

To save space, electrons begin to pop off nuclei, creating a dense plasma of electrons and ions.

The squeeze continues, compressing free electrons down to their minimum size.

White Dwarf Stage
Electrons resist further compression.

Rapid collapse begins. Electrons are squeezed into nuclei, transmuting nuclei into free neutrons.

The thin neutron gas is increasingly compressed. Gravitational pressures zoom higher as the density grows.

Neutron Star Stage
Strong nuclear forces resist further compression. Neutrons are packed almost solid.

As the struggle for space within the collapsing star continues, smaller and heavier particles called hyperons appear.

Relativistic Collapse

General relativity fails. No known theory can predict events.

Electron cloud
Electron
Nucleus
Neutron
Hyperon

PLATE 4.XVII Matter's struggles for space in the crush of collapse. The heavier a particle, the smaller the volume it is able to occupy. Atoms in a shrinking star are thus crushed, in turn, to electrons and nuclei, neutrons, and increasingly heavy hyperons. Whether each can counter gravity depends on the star's mass: less than 1.4 suns, free electrons stop the slow compression; less than 2 suns, neutrons will halt the collapse; more than 2 suns, no particle is able to resist gravity. [Source: *Science Year. The World Book Science Annual.* Copyright © 1968, Field Enterprises Educational Corporation.]

containing the area of a medium-sized city. The fantastic gravitational attraction at the surface of a neutron star (10^{11} times that at the earth's surface) is balanced by the same combination of nucleon motion and repulsive forces that keep atomic nuclei from collapsing despite the powerful attractive nuclear forces that hold them together. Since matter must be squeezed until different nuclei touch each other before such repulsive forces become effective in preventing stellar collapse, a neutron star must have a central density near to or greater than the matter within such nuclei. Therefore, the core density of a neutron star exceeds 10^{11} g/cc (or $\sim 10^8$ tons per cubic inch), an almost inconceivable density at which a speck of sand would be more massive than an ocean liner. If the mass of a neutron star is less than one sixth that of the sun, the central density is too low to support it stably against the pull of gravity, thus it would pop out to remain a white dwarf. If, on the other hand, it is considerably heavier than the sun, gravity will crush it toward a black hole no matter how repulsive the nuclear forces are at a short range. Here is the strange case in which the gravitational forces, intrinsically weaker than the nuclear forces by the enormous factor of 10^{40}, still dominate because of the great concentration of matter.

The rapidly growing body of evidence that suggests identification of pulsars as rapidly rotating neutron stars gives a clue to their abundance and genesis. The association of young pulsars with supernova remnants, as in the now familiar example of the Crab (see Figure 4.116), suggests that neutron stars are formed in violent supernova explosions that have already consumed about one thousandth of all the stars in our galaxy. Many or most such explosions result in neutron stars. When formed they are extraordinarily hot, well above 10^{11} deg. Various neutrino emission processes should quickly cool them so that within a few centuries their internal temperatures would drop to a few hundred million degrees. Remarkably, the behavior of this very hot stellar interior is in many respects exactly analogous to that of terrestrial liquid helium when it is cooled close to absolute zero; in terms of the phenomena expected within a neutron star interior—superconductivity and superfluidity—it is the coldest known place in the universe.

Constituents of a Neutron Star

A 10-km traverse from the surface to the center of a neutron star would take a traveler through all densities from a near vacuum to well above 10^{11} g/cc. The corresponding pressures would extend from zero to above 10^{28} atm. Present knowledge of how the pressure and constituents of

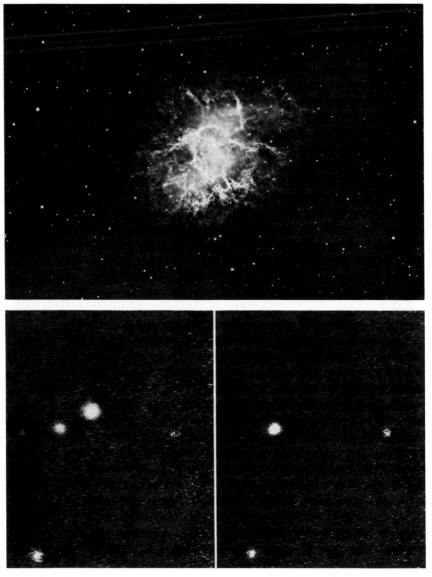

FIGURE 4.116   The Crab nebula. Within the Crab nebula (top), an exploded star is a pulsar that emits visible light—30 bursts a second. It was thought to be a normal star, but high-speed electronic imaging (below, left) shows that the source is off more than 97 percent of the time (below right). [Source: Photograph (top) courtesy of the Hale Observatories. Photographs (bottom) from: *Science Year. The World Book Science Annual.* Copyright © 1969, Field Enterprises Educational Corporation.]

matter vary with increasing density is represented in Figures 4.117 and 4.118. The different portions of the pressure versus density equation-of-state curve are contributed by physicists in varied specialties. From A (matter at negligible pressure) to B (about the pressure at the center of the moon), the equation of state can be measured in the laboratory and, of course, depends sensitively on the chemical nature of the material. From B to C, a combination of a mathematical model, evolved from one proposed about 40 years ago by E. Fermi and L. H. Thomas, together with the use of modern high-speed computers, yields an adequate theoretical description. Beyond the point C, matter is so compressed that the Pauli Exclusion Principle compels the electrons to have too much kinetic energy to remain bound to their nuclei as in the case of normal matter. These high-speed free electrons then contribute almost all of the pressure; the equation-of-state curve can be calculated accurately using modern techniques of many-body quantum theory. Near point D, the electron velocities approach that of light so that a correct description of such matter must utilize the Dirac relativistic quantum-mechanical equations. At a density of about 10^9 g/cc, the electron energy becomes so large that the tightly bound protons within the atomic nuclei absorb some of the more energetic electrons. Such protons are thus converted to neutrons (plus neutrinos, which easily escape) in a process that is the precise reverse of the normal process wherein a free neutron decays with a

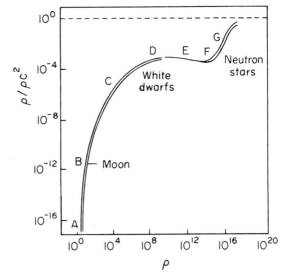

FIGURE 4.117 The dimensionless ratio of pressure (p) to density (ρ) times the speed of light squared (c^2) as a function of density, in grams per cubic centimeter. (The doubled portion of the curve corresponds to matter stiff enough to keep a star from collapsing under its own weight. Neutron stars have central cores whose pressure and density lie above the point F. White dwarfs, planets, etc., have corse below D.) [Courtesy of M. A. Ruderman.]

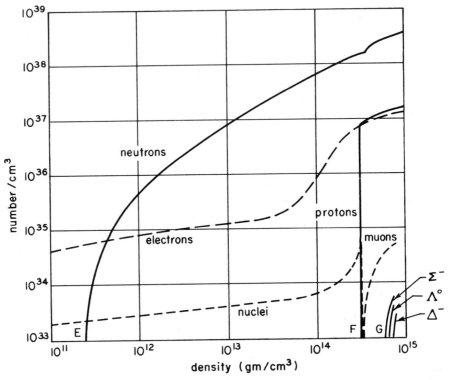

FIGURE 4.118 The constituents of matter as a function of density. The letters, E, F, G, refer to points on Figure 4.116 and are described in the text. Σ^-, Λ^0, Δ^- are the names of heavy unstable nucleons. The muon is a very heavy unstable electron. [Source: M. A. Ruderman, "Solid Stars," Scientific American, *224*, 24 (February 1971).]

lifetime of 12 min, to leave a proton and an electron antineutrino pair, and the nuclei in such superdense matter are much more neutron-rich than are stable nuclei in normal matter. As matter is squeezed to still higher densities, the electrons continue to be absorbed by protons, converting them to neutrons. The pressure from the rapidly moving electrons no longer rises strongly with increasing density because of their disappearance into the neutrons, so that in this regime matter is relatively more compressible than it is at lower densities. At the point E, corresponding to a density of 3×10^{11} g/cc, the nuclei are so neutron-rich that they "drip" neutrons into the interstitial volume between the nuclei. Beyond E, matter consists of free electrons and nuclei embedded in a sea of neutrons. These nuclei contain two or three times more neutrons than any isotopes found on earth. Continued compression causes further absorption of electrons by protons and their conversion into neutrons. At densities approaching 10^{11} g/cc, almost all of the matter density is composed of the free neutrons, which are about 25 times more abundant than the protons (all of which are bound into nuclei) or electrons. At 3×10^{11} g/cc (point F), the nuclei rather suddenly dissolve, and such ultradense matter consists mainly of a sea of neutrons interpenetrated by a much less dense sea of unbound protons and relativistic electrons. A further increase in density to point G results in the conversion of some of these particles to other kinds of elementary particles, which, on earth, are produced only in the largest particle accelerators. These particles, when isolated, are very unstable, with lifetimes between 10^{-6} and 10^{-22} sec. However, in the superdense environment within a neutron star, they can be stable because the Pauli Exclusion Principle prevents their decay products from being injected into an already existing sea of similar particles and the decays do not occur. The equation of state of this sort of matter is not yet known. For denser stars most of the total stellar mass may consist of these normally unstable strange particles. During the collapse, as illustrated also in Figure 4.117, the stellar matter passes from the atomic regime of condensed matter and atomic physics through the nuclear realm, where, incidentally, nuclear matter—long a favorite idealized concept of nuclear physicists—emerges in kilometer-scale natural samples, to the still mysterious realms of elementary particles —some already known and some yet to be discovered in the march to ever higher energies. In this one collapse, the star sweeps through much of modern physics.

The original name, neutron star, is generally used for these collapsed objects, even though other elementary particles may well be present, because a typical collapsed star would be expected to contain mainly neutrons, especially in its deep interior. The nuclear force interactions between pairs of neutrons are exactly analogous to those between pairs of electrons in a normal low-temperature solid that has become a super-conductor, except that they are much stronger, thus causing such a transition at enormously higher temperatures (Figure 4.119). Because the neutrons are uncharged, the neutron sea is expected to become only a superfluid, without spectacular electrical properties. It can conduct heat fantastically quickly, and it can flow as if without any viscosity. Normal liquid helium, when cooled to within a few degrees of absolute zero, has such properties. But the closest terrestrial analogue is a fluid composed of the helium isotope ^{3}He. However, it has not yet been possible to cool it to a low enough temperature (much less than a thousandth of a degree above

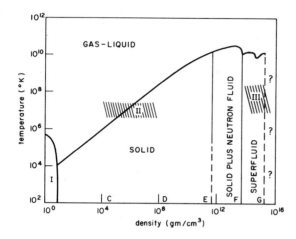

FIGURE 4.119 Phases of super-dense matter at various densities and temperatures. The nuclei are assumed to be the most stable ones for a given environment. (At densities above 3×10^{11} g/cc this would occur quickly even at zero temperature.) Region I is, roughly, that accessible in the laboratory. Region III corresponds to neutron star cores. Region II is that for the centers of white dwarfs. The point C is that for the center of our sun, which will ultimately move up and toward the right. [Source: M. A. Ruderman, "Solid Stars," Scientific American, *224*, 24 (February 1971).]

absolute zero) to mimic the behavior expected in the neutron fluid of a neutron star at temperatures of a few hundred million degrees. The free charged protons in the deep interior of neutron stars are expected to form a superconductor. Currents once set up maintain themselves forever; a magnetic field trapped within such matter will never die out. But even if the interior protons were not superconducting, the electrical conductivity contributed by the relativistic electrons in the crust and core is over a million times greater than that of copper. This characteristic alone is sufficient to hold an original magnetic field for over tens of millions of years. The entire physics of condensed matter contributes to the understanding of these phenomena.

At densities approaching 10^{15} g/cc, expected in the central core of heavier neutron stars, neither the composition nor the structure is yet fully understood. Both depend on a deeper knowledge of the properties of elementary particles and their interactions than is presently available. Here, if anywhere, is a region that might contain surprises.

Neutron stars are probably spinning very rapidly when formed, perhaps up to a thousand revolutions per second. Fast rotation rates are inferred from observations of present pulsar periods and their rate of increase, as well as from the conservation of angular momentum in a rapidly collapsing star. An interior neutron superfluid cannot rotate like a normal fluid. Rather it must set up an array of tiny vortices (whirlpools) whose rotational axes tend to be parallel to the rotational axis of the fluid as a whole. In a neutron star rotating about once a second—characteristic of pulsars—but with no external torques tending to slow its spin, the interior neutron superfluid would set up about 10^4 vortices per square centimeter in a triangular array. Averaged over many vortices, the motion of the fluid is essentially that of a rotating rigid body. But on a microscopic scale, the motion entails an enormous array of tiny whirlpools whose centers corotate with the star. Here many of the phenomena of fluid dynamics occur.

The laboratory analogue of the hot rotating neutron star is a cold hollow steel shell, with a thickness of about one tenth its radius, filled with low-temperature corotating liquid helium. At the center there may be a core with unknown and perhaps unexpected properties. The crust should be approximately $10°$ above absolute zero and the superfluid only a few hundredths of a degree. However, this rather remarkable model of a neutron star is complicated and enriched by an enormous magnetic field that threads the star and its immediate environment.

Structure

Nuclear physics and elementary-particle physics determine the composition of ultradense matter in neutron stars. But experiments and theories of condensed matter at temperatures close to absolute zero describe the organization and behavior of this matter (Figure 4.120). The outermost layer of a neutron star is a hot gaseous liquid whose density reaches 10^6 g/cc about a meter below the surface. Such matter is very much like that found throughout the interior of many conventional stars, especially white dwarfs. Below this region, and until a density of 3×10^{11} g/cc is reached, at which all nuclei have dissolved, is the solid crust in which the nuclei arrange themselves in a crystalline lattice much like those in terrestrial solids below their melting points. The computed melting temperature of superdense matter can exceed 10^{10} deg as long as nuclei remain to form a lattice. This enormous transition temperature is a consequence of the strong electrical (Coulomb) repulsion among nuclei that have lost their atomic electrons and are pushed very close together. Because the present temperature within a neutron star is far below the lattice melting temperature, the outer kilometer or so will form a solid crust. Relative to its melting temperature, the neutron star crust is much colder than that of the earth. It is up to 10^{18} times more rigid than a piece of steel and over 10^{20} times more incompressible. In the inner regions of the crust, where a neutron fluid fills the space between the nuclei, it is easy for the nuclei to change the number of neutrons they contain as well as their charge by electron emission or absorption. Therefore, any impurity

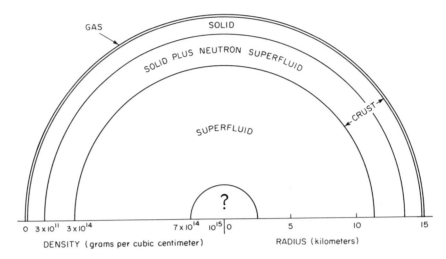

FIGURE 4.120   The structure of a neutron star. The thickness of the gaseous atmosphere is less than a thousandth of the stellar radius. The very lightest neutron stars, whose mass is near a sixth that of the sun, would probably have a crust that extends right to the center. Their total radii will also be much greater than that of the typical neutron star shown here. [Source: M. A. Ruderman, "Solid Stars," Scientific American, *224*, 24 (February 1971).]

nucleus whose charge or mass is inappropriate for the most stable nuclei at a given density will quickly change to one that is appropriate. In this way impurities disappear. A neutron star crust is thus the purest as well as the strongest matter known in the universe.

Dynamics

Even before pulsars were discovered it had been conjectured that, if a neutron star contained within itself the magnetic flux it had before its collapse, then it might possess a huge 10^{12}-G magnetic field. (The earth's field is less than a gauss.) The slow lengthening of pulsar periods suggests that neutron stars, at least those observed as pulsars, do have such enormous fields. Such magnetic fields can give sufficient radiation or coupling to the surrounding plasma to account quantitatively for the lost rotation energy. Ultimately this energy seems to appear in the production of very energetic cosmic rays; most of the cosmic rays that continuously bombard the earth may originate from such spinning neutron stars. The magnetic field profoundly affects the exterior, surface, and interior of a rotating neutron star. Outside the star it controls the coherent flow of charged particles that gives the fantastically powerful radio emission that makes pulsars observable. Understanding of the behavior of a relativistic plasma in such a huge field is still very limited. At the stellar surface, the 10^{12}-G field completely changes the nature of whatever atoms are present. The stellar rotation coupled to the magnetic field can generate electric fields approaching an astonishing 10^{12} V/cc between the poles and the equator. Such electric fields will pull charged particles from the surface into a surrounding plasma. Finally, such an enormous magnetic field will strongly couple all electrically conducting regions within and outside the star. The plasma, crust, and core electrons and protons all corotate and slow together. It is their period of rotation that is measured in pulsars. But an interior neutron superfluid would probably rotate slightly faster, because it is generally difficult to slow a rotating superfluid. A slight jump in the crust angular rotation velocity would cause it to rub on the neutron superfluid, but so weakly that it might take weeks or years to return to its original spin rate. There seems to be support in observed pulsar periods for such behavior.

The theory of a slowing neutron star is still very primitive. There are many crucial unanswered questions: Exactly how does the spinning crust couple to the neutron fluid? How does an unknown central core react? Do the parallel vortex lines in the neutron superfluid tend to bend and twist to form a pseudo-turbulent interior superfluid? How does the crust respond to the shape changes forced on it by the varying centrifugal forces as the neutron-star spin slows down? Does the crust creep and flow plastically, or does it crack and give starquakes? How exotic can the behavior of the crust be? Could there be analogues of terrestrial volcanoes and mountain building?

Very small sudden changes in pulsar period equivalent to those that could be caused by surface motions of 10^{-3} cm have been detected. Such marvelously accurate observations, together with more refined theories of the various neutron star regions, will ultimately lead to a more precise description of spinning neutron stars. But enough is already known to show that the neutron star has a unique and wonderful structure that causes remarkable behavior, like that of no other object in our universe, and that calls on almost every branch of physics for its understanding.

Conclusion

The physicist who explores the nature of his immediate world learns directly only about a very rare and special part of the universe that can support life. Most of the rest is far too hot or cold or dense or empty. Knowledge of the structure and phenomena within most of the universe is based on an immense extrapolation from immediate surroundings to regimes of density and temperatures far removed not only from those found on earth but even enormously beyond those achievable in the laboratory. Such an extrapolation is sustained by experiments and theory embracing most subfields of physics. An understanding of many fantastic objects observed beyond the earth depends, in part, on the physics of normal condensed matter at temperatures near and below the lowest ever achieved on earth. Knowledge of the astrophysical universe uses nuclear physics in almost all of the regimes accessible in the laboratory; it also exploits high-energy physics and the theory of elementary particles at energies even beyond those reached in the largest accelerators. Plasma and fluid physics, as well as the surface physics of tiny dust particles, and the general theory of relativity are necessary contributors to the history of a star, the former to its birth as a condensation from a primeval gas, the latter to its death as an ultradense stellar cinder.

We shall return to this question of the unity of physics, and of science, repeatedly throughout this Report. The examples chosen for inclusion here are particularly striking; they are by no means isolated cases.

APPENDIX 4.A.   PROGRAM ELEMENTS

In the following tables, subfields of physics have been divided into program elements (major research areas). Although the definition of this term is somewhat imprecise, the intent has been to identify separable components of the subfields—components sufficiently large to have some internal coherence and reasonable boundaries and for which it might be possible to estimate present funding levels and the PhD manpower involved. As discussed in Chapter 5, the purpose of this exercise was to divide the subfields into units of activity that the Committee could rate in terms of intrinsic, extrinsic, and structural criteria. The purpose of the ratings was to test the feasibility of arriving at a consensus regarding the desirable relative emphasis among subfields and among program elements within each subfield.

Data available to the Committee permitted identification of program elements, at least roughly, for all subfields except earth and planetary physics. In some cases, the program elements cover the major activities in the subfield; in others, they do not. It should be noted that much of the work in such subfields as optics and acoustics lies largely outside physics, and that the basic physics research in some of the program elements involves only a small part of the total dollars and manpower associated with the subfield.

Clearly, most of the program elements could be further subdivided, but, to keep the overall number for the Committee's consideration within manageable limits, the number per subfield was somewhat arbitrarily restricted to approximately ten. As expected, the subfield panels found it convenient to make the divisions into program elements along different lines. For example, in elementary-particle physics the division is made in terms of the small number of major facilities and associated programs, whereas in condensed matter the division accents specific areas of research such as superconductivity. The program elements for astrophysics and relativity identify emerging areas of research that will require greatly increased funding. In the projected program for this subfield, the costs of satellites and large facilities are included. Funding figures associated with program elements in the other subfields do not include construction costs of major facilities.

In developing these program elements, the Committee worked with the panel chairmen; however, in some cases the elements used here are not identical with those suggested by the panel chairmen. In elementary-particle physics and nuclear physics it was possible to assign funding levels and manpower rather precisely. Similar assignments for some of the program elements in the other subfields may be in error by a factor of 2.

Elementary-Particle Physics (1971)

Program Elements	Federal Support [a] (Operations and Equipment) ($Millions)	PhD Manpower [b]
1. Accelerator developments	3.6	— [d]
2. National Accelerator Laboratory	— [c]	—
3. Other major facilities (e.g., AGS improvement project)	1.5	— [d]
4. Stanford Linear Accelerator	27	99
5. Brookhaven AGS	27	99
6. Argonne ZGS	20	65
7. Berkeley Bevatron	26	100
8. Cornell Synchrotron	3	20
9. CEA Bypass Storage Ring	2.3	11
10. University groups	37	1245 [e]

[a] Construction costs not included.

[b] Includes approximately 300 scientists having the following specialties or combinations of them: computer employment in research, accelerator design and development, accelerator operation, device design and development, and emulsion experiments.

[c] No NAL figures are given for FY 1971 since the accelerator will not be in operation until FY 1972. When in full operation, operating plus equipment funds required are estimated at about $60 million.

[d] Manpower included in elements 4 to 9.

[e] Includes approximately 245 PhD's doing particle research but not supported directly by federal funds.

Description of Program Elements

1. These activities are an integral part of the ongoing work at each of the major accelerator laboratories. They have a creative content quite apart from the particle research itself, although neither can progress without the other. The technological requirements lead to innovation and development in such fields as radio frequency, engineering, superconducting magnets, ultrafast electronics, computer technology, radiation-detection instruments, pattern recognition, and particle orbit theory.

2. 200–500-GeV proton accelerator to be for some years the only controlled source of protons in the world for research * in the energy range above 80 GeV and the only one in the United States above 33 GeV. Also includes in-house research activities comprising a small fraction of the particle research to be carried out at the accelerator.

3. These are major additions to the capabilities of accelerators—other than NAL—that have been planned or under construction for some years and are now complete or nearing completion. Includes: the major modification of the AGS to increase its intensity and other capabilities, the SPEAR storage ring at SLAC and the 12-ft liquid hydrogen bubble chamber at the ZGS. Each facility offers unique opportunities to open new areas of research but will require incremental operating and equipment funds for the purpose. Construction costs are not included.

4. 22-GeV electron accelerator, which is the only controlled source of electrons in the world for research * in the energy range above 10 GeV. Also includes in-house research activities comprising a substantial fraction (about one half) of the research carried out with this acclerator.

5. The 33-GeV proton accelerator at BNL, which is the principal source in the United States for research * using protons in energy range 12–33 GeV. Includes in-house research comprising about 25 percent of the total research activity.

6. The 12.5-GeV accelerator at ANL, which is the principal source in the United States for research * using protons in the energy range 6–12 GeV. Includes in-house research effort comprising about 25 percent of the total research activity.

7. The 6-GeV proton accelerator at LBL, which is the principal source in the United States for research * using protons in the range of proton energies 1–6 GeV. Includes substantial in-house research activity.

8. A 10-GeV electron accelerator. This is a high-duty-cycle machine (in contrast to SLAC) for research * with electrons in the energy range 1–10 GeV. The in-house research activity is dominant, but there is potential for expansion to include more research by outside users.

9. A 6-GeV electron accelerator with a high duty cycle, which has recently been limited to activities associated with the development of a bypass to serve as a storage ring to study the collisions of 3.5-GeV electrons and positrons. It is the only such facility presently available in the United States and is currently under test.

10. University research groups are responsible for carrying out most of the experimental particle physics research at the major accelerator laboratories. Includes activities of professors, postdocs, graduate students, and associated technical services required to provide electronics, detection equipment, data handling and analysis systems, etc., to the extent that these aspects of the research can be mounted at the universities. Includes both experimental and theoretical physicists. User groups are partly funded under this item; they do not pay any accelerator use charges.

* Each accelerator is a source of the indicated primary particles and many beams of secondary particles (pions, K-mesons, neutrinos, muons, antiprotons, hyperons, etc.).

Nuclear Physics (Fiscal Year 1969)

Program Elements [a]	Federal Operations Support [b, c] ($ Millions)	PhD Manpower [a, d]
1. Nuclear excitation } 2. Nuclear dynamics }	33.7	695
3. Heavy-ion interactions	3.1	20
4. Higher-energy nuclear physics	7.9	95
5. Neutron physics	7.7	115
6. Nuclear decay studies	3.2	65
7. Weak and electromagnetic interactions	0.7	20
8. Nuclear facilities and instrumentation	3.8	50
9. Nuclear astrophysics	0.4	20
10. Nuclear theory	5.0	260

[a] These program elements do not include all current basic research activities in nuclear physics supported by the federal government. Approximately 140 PhD's are working in the areas of data compilations, nuclear chemistry, etc.
[b] Nonfederal support estimated at 25–30 percent of federal support, on the average, of those projects supported by the federal agencies.
[c] Construction costs of major facilities not included.
[d] Another 300 PhD's are working either in applied nuclear physics or are supported entirely by nonfederal funds.

Description of Program Elements

1. The study of the nuclear degrees of freedom with a broad spectrum of nuclear probes.
2. The study of the nature of nuclear reactions.
3. The study of the now largely unknown interactions of masses of nuclear matter.
4. The study of nuclei with short-wavelength electron, proton, and mesonic probes.
5. The study of nuclear phenomena with a neutral strongly interacting probe.
6. The study of nuclear states via the decay of radioactive nuclei.
7. The study of fundamental symmetries in the nuclear domain.
8. The tools of nuclear physics.
9. Identification of which nuclear reactions are of importance and measurement of the relevant nuclear data—primarily cross sections.
10. The theoretical aspects of all the above fields and their relations to the fundamental nuclear interactions.

Atomic, Molecular, and Electron Physics (1970)

Program Elements [a]	Federal and Industrial (Operations) Support [a] ($Millions)	Estimated PhD Manpower [b]
1. Gas discharge	1.6	35
2. Electron physics	4.0	80
3. Lasers and masers	5.0	100
4. Atomic and molecular spectroscopy	4.0	80
5. Atomic, ionic, and molecular beams	6.4	130

[a] A substantial amount of activity in these program elements is supported from sources outside the AME subfield, e.g., plasma physics, space and planetary physics, electrical engineering, and chemistry. If included, this may double most of the above numbers.
[b] Based on estimated level of activity and 1970 National Register data in which a total of 1065 PhD scientists identified with AME physics.

Definition of Program Elements

1. Gas discharge including low- and medium-density plasmas.
2. Electron physics including the low-energy electron diffraction techniques, electron optics, electron–atom collisions, high-vacuum techniques, surface properties.
3. Lasers and masers including time and length standards, higher-order electromagnetic interactions, photon statistics, nonlinear spectroscopy, coherent x rays.
4. Atomic and molecular spectroscopy including positronium and muonium spectra, tests of quantum electrodynamics, optical pumping, vacuum uv, far-infrared and radio spectroscopy.
5. Atomic, ionic, and molecular beams including colliding beams, beam-foil spectroscopy, highly excited molecules and atoms.

Condensed Matter (1970)

Program Elements [a]	Estimated Federal and Industrial (Operations) Support ($Millions)	Estimated PhD Manpower [b]
1. Crystallography, etc.	9.0	150
2. Surface physics	20.5	340
3. Semiconductors	33.0	550
4. Nonelectronic aspects	18.5	310
5. Luminescence, etc.	9.5	190
6. Electronic properties of solid or molten metal	15.0	260
7. Magnetic properties	25.0	430
8. Quantum optics	9.0	150
9. High magnetic fields	3.5	60

Condensed Matter (1970)—*Continued*

Program Elements [a]	Estimated Federal and Industrial (Operations) Support ($Millions)	Estimated PhD Manpower [b]
10. Superfluidity	4.5	75
11. Slow neutron physics	8.5	75

[a] These program elements do not include all current basic research activity in condensed-matter physics.
[b] Based on estimated level of activity and 1970 National Register data in which 4160 PhD scientists identified with condensed-matter physics.

Description of Program Elements

1. Structures of crystals, including studies of atomic arrangements by neutron, electron, and x-ray diffraction techniques.

2. Includes all the properties of surfaces and thin films, crystal growth from vapor or the melt, properties of solid–solid interfaces.

3. Includes all the electronic properties of nonmetallics with small bandgaps in their pure states and having appreciable conductivity in suitably doped states.

4. Includes all the properties of defects and dislocations in crystals that are usually described without invoking the quantum-mechanical behavior of atoms. Includes plasticity, rupture, internal friction diffusion, ionic conduction, phonons, and lattice vibrations.

5. Includes band-structure calculations, optical properties, optical effects and electronic levels of impurities, and other information bearing on the electronic levels of insulating crystals.

6. Includes all the electrical and thermal conduction phenomena due to electrons, optical properties of metals, band-structure calculations, plasma oscillations, and superconductivity.

7. Includes electron paramagnetic and nuclear paramagnetic resonance work, studies of static magnetic susceptibilities, and all phenomena connected with ferromagnetism.

8. Includes lasers and masers, nonlinear optical effects, and other effects that can only be studied by laser light.

9. Experiments that are done in condensed matter with fields in excess of 120 kG.

10. Work with superfluid liquid helium.

11. Work requiring use of moderated neutrons from a pile.

Optics (1970)

Program Elements [a]	Estimated Federal and Industrial Support [b] ($Millions)	Estimated PhD Manpower [a, c] (Physicists)
1. Metrology	0.8	15
2. Optical information processing	0.9	17

Optics (1970)—*Continued*

Program Elements [a]	Estimated Federal and Industrial Support [b] ($Millions)	Estimated PhD Manpower [a, c] (Physicists)
3. Optical band communication	0.1	2
4. Optical systems, lens design, etc.	2.2	40
5. Laser-related light source	7.5	137
6. Holography and information storage	4.0	73
7. Integrated optics	0.6	11
8. Nonlinear optics	3.0	55

[a] These program elements do not include all areas of basic research in optics. It is estimated that there are another 690 PhD physicists working in areas not included in these program elements.
[b] The average annual cost per PhD does not vary widely across the program elements and is estimated to be $55,000/PhD.
[c] These estimates do not represent the magnitude of the manpower effort in the various program elements. They represent a judgment on the number of personnel from the physics section of the National Register of Scientific and Technical Personnel and do not include the large effort made by engineers, which is uniformly and properly considered optics.

Description of Program Elements

1. Metrology is the science of measurement. With lasers, very precise measurements may be made of such things as the distance to the moon, the compression of the earth in earthquake zones, and the deformation of large structures. Useful new phenomena will certainly be discovered.

2. Optical information processing is used to reduce blur in photographs, to enhance contrast, smooth out grain, sharpen edges, etc. It is also possible to use optical techniques for automatic photointerpretation and character recognition.

3. Communication at optical band frequencies permits the transmission of tremendous amounts of information wherever a beam of light can be sent. Long-distance communication through glass fibers now seems possible with modulated laser beams.

4. Modern computers and system science have made it possible to design optical systems and instruments that are optimized. Very large improvements can be made, particularly when new laser sources and solid-state receivers are included in the design.

5. Lasers can be made to have extremely high energy or power or power density. Others have very precise and steady wavelength, and still others can be tuned to different wavelengths. Each new improvement makes new techniques possible and simplifies the solution of old problems.

6. Holography is a method of storing an image or other information in a photographic film by recording the interference pattern between the signal-carrying light and a coherent reference wave. It offers potential advantages over other compact storage methods for large amounts of information.

7. A beam of light can be trapped and guided in a thin film on a solid surface, rather like electricity in a wire. It can then be manipulated by acoustical, electrical, or other optical signals for computer logic, modulation, scanning, or signaling. The combination is called integrated optics.

8. Some materials, when illuminated very intensely, give off light of doubled fre-

quency. In other cases, two beams mixed in a crystal give light of several sum and difference frequencies. Knowledge can be gained about the material, and useful devices can be built.

Acoustics (1971)

Program Elements	Estimated Federal and Industrial Support [a, b] ($Millions)	Estimated PhD Manpower [c]
1. Noise, mechanical shock, and vibration	1.8	35
2. Underwater sound	4.9	90
3. Music and architecture	0.8	15
4. Ultrasonics and infrasonics	1.1	20
5. Electroacoustics and acoustic instrumentation	1.1	20
6. Hearing, speech, and biophysical acoustics	0.8	15

[a] Based on estimated level of activity of physicists doing basic research in acoustics that leads to publishable reports. Costs/PhD across the program elements is assumed to be $55,000/year.
[b] Federal and industrial support of applied research in acoustics is estimated at $50 million.
[c] Based on estimated level of activity and 1970 National Register data in which a total of 325 PhD scientists identified with acoustics.

Description of Program Elements

1. The field of noise and noise abatement is a huge one in modern technology. It covers the sounds from jet engines, sonic boom, airflows in ducts and cooling systems, unwanted sounds of all kinds in housing and working areas. Closely related are the vibrations and shocks produced by machines. This program element has a considerable overlap with the program element of turbulence in fluid dynamics and has a strong interest in the problems of fluctation theory. In both of these areas, physics has a role to play, but the relative importance of physics research to the entire field is small, and the share of physics research will probably remain similarly small. There is still need, however, for fundamental research on the way in which particular noises arise and on their transmission through various media.

2. The study of sound propagation in water, and more specifically, seawater, has been enormously stimulated by military needs. Most of the work supported in underwater sound has been in technology rather than physics. There is strong overlap between this program element and that of oceanography. In rating both this field and that of noise, this overlap should be kept clearly in mind. Underwater sound will continue to play a significant role in the development of the field of oceanography.

3. Music includes studies of the character of musical sounds and how they are produced, both naturally and synthetically. Architectural acoustical studies are aimed at elucidating the factors that govern the acoustical character of concert halls and other structures, determining how these factors are related and how this knowledge can be translated into the design and construction of enclosures of specified acoustical characteristics.

4. The study of ultrasonic propagation in gases and liquids has long been a major component of physical acoustics. To traditional fluids, one should add the study of sound propagation in quantum liquids and in plasma. The use of Brillouin scattering to extend the frequency range of study upward, the prosecution of studies in liquid helium and plasma, and the application of our knowledge to border areas in chemistry and oceanography make this part of acoustics an especially lively one today. Of major interest has also been the contribution of this research to our understanding of relaxational phenomena and chemical kinetics. The study of sound propagation in solids is usually classified elsewhere than in acoustics. Of more purely acoustical interest in studies of physics in solids are high-accuracy velocity change measurements. Spin waves and acoustic NMR and EPR have also been studied widely. The field of nonlinear acoustics has grown out of ultrasonic propagation studies in fluids and has high promise of applications in underwater sound and biophysical acoustics. Infrasound sources include volcanoes, aerodynamic turbulence, weather frontal systems and tidal waves, and studies related to the large-scale behavior of the atmosphere, with application to clear-air turbulence detection and storm and tsunami tracking systems in this growing field of physical acoustics.

5. Represents the range of use of electrical and electronic techniques for devices that are acoustical in character and include modern stereophonic systems, acoustic pulse generation and detection, much of signal processing, and the use of computers in acoustics. The degree of involvement of physics with electroacoustics varies from time to time and depends on the particular stage of development of the devices and applications. Today, the most promising areas, from a physical viewpoint, are those of the direct production of ultrasound from electromagnetic radiation of a metal, the emission of acoustic radiation from dislocation walls in crystals, the use of heat pulses as sources on acoustic waves in the $10^{11}-10^{12}$-Hz range, and acoustic thermometry. Many of these instrumentation studies are pioneering, and there is a substantial possibility of major advances in the production and use of sound.

6. While most of speech and hearing lie outside of physics, there is much that remains, such as models for speech production and analysis of the acoustic content of speech and the mechanism by which hearing takes place beyond the conversion of mechanical motions of the inner ear to nerve impulses, as well as the nonlinear behavior of the ear and its effect on hearing. Since all of hearing and speech can be classified as bioacoustics, it is convenient to particularize the rest of the field by the term *biophysical acoustics,* a field that goes beyond medical diagnosis and therapy. Biophysical acoustics overlaps with acoustic holography and includes the problem of communication of the deaf. Much of its basic thrust is in the development of ultrasonography and the use of sound waves and acoustic devices in medical treatment. As physics, the field is still small, but it is of growing significance, and the future potential is large.

Plasma and Fluids (1970)

Program Elements	Federal (Operations) Support [a] ($Millions)	Estimated PhD Manpower [b]
1. MHD power generation	1.0	20
2. Controlled fusion	30.0	410

Plasma and Fluids (1970)—*Continued*

Program Elements	Federal (Operations) Support [a] ($Millions)	Estimated PhD Manpower [b]
3. Fluid dynamics, plasmas, and lasers	24.0	475
4. Meteorology		
5. Computer modeling	8.5	210
6. Oceanography		
7. Lab and astrophysical plasma and fluids	0.5	10
8. Turbulence in fluid dynamics	3.0	65

[a] Approximately 55 percent of the scientists in plasmas and fluids are theorists. Annual support per PhD theorist is assumed to be $40,000.

[b] Based on estimated level of activity and 1970 National Register data in which 1110 PhD scientists identified with plasmas and fluids.

Description of Program Elements

1. It is possible using an intermediate state between plasmas and fluids, namely, a very-high-temperature conducting gas, to extract useful power by the flow of such a gas through a very strong magnetic field. The high-temperature gas may be the product of combustion, in which case, a very much higher temperature of combustion can and may be used as the initial starting state of a power-generating cycle. The feasibility of higher temperature in an MHD channel as compared to the limits imposed by boilers and turbine blades affords a possible significantly higher efficiency in power generation from the same fuel input, and, as a consequence, using MHD as a topping cycle affords the possibility of a significant improvement in the efficiency and simplicity of generating electrical power.

2. The goal of achieving useful power from controlled thermonuclear fusion of the heavy hydrogen isotopes requires the detailed and exhaustive understanding of the properties of high-temperature collisionless plasmas confined by various geometries of magnetic field. The thermal isolation afforded by various magnetic field configurations is limited by a complex hierarchy of instabilities whereby the high-temperature fusion plasma can escape and cool at the walls of the vessel. The understanding of these phenomena and work toward the solution of the applied goal represents the most advanced application and understanding of plasmas and of the physics of plasmas.

3. Fluid dynamics, plasmas, and lasers include the basic physical understanding of the properties of plasmas and fluids and the application of this knowledge. Because of its separate importance, turbulence has been excluded but lasers are mentioned to emphasize applications. An understanding of plasmas and fluids requires the very broadest knowledge of cooperative phenomena based on principles derived from the simplest individual particle interactions.

4. Meteorology is a specific branch of the physics of fluids because of the complexity of the water vapor, water, air, rotational centrifugal field, gravitational field of the earth–atmosphere system. Computer modeling, statistics, observation, and weather modification are the ingredients for understanding the earth's atmosphere.

5. Computer modeling of both fluids and plasmas has progressed to the state where the most complicated flow patterns, convection, partial turbulence, waves, instabilities,

and plasmas can now be modeled using finite-difference calculations on the more advanced computers. It is fair to state that the most advanced computer designs have, to a large extent, been motivated by the complexity of the modeling of fluid and plasma problems, particularly those associated with weapons design. In the future, we expect to see the problems of both controlled fusion, meteorology, and oceanography have an equal and dramatic bearing upon the evolution of computer complexity.

6. The fluid flow of the ocean is complicated by a set of constraints similar to that of the atmosphere, namely, rotation, gravitational field, and density stratification. In the case of the ocean, the thermohaline instabilities and density gradients lead to fluid-flow problems of great complexity. Oceanography in the context of fluid dynamics attempts to understand the fluid flow in the oceans due to the constraining forces as well as density gradients that lead to such exotic phenomena as the gulf stream, tides, and ocean waves. Understanding the interaction of the ocean and the atmosphere is a major objective of the physics of fluids of the earth.

7. Laboratory and astrophysical plasmas and fluids include the basic physical understanding of the properties of plasmas and fluids aside from turbulence, e.g., laminar flow, diffusion, transport coefficients, radiation properties, masers, lasers, and the application of this knowledge to the understanding of astrophysical phenomena.

8. Turbulence in fluids describes that quasi-random behavior that occurs when a highly coordinated flow breaks up into a series of partially correlated random fluctuations. In general, the fluid is characterized by the property that it may be infinitely extended quasi-statically with no restoring force. In addition, fluid turbulence exists without restoring forces; however, the one-body force that is included in fluid turbulence is gravity.

Astrophysics and Relativity

Program Elements [a]	1970		Proposed 10-Year Program	
	Estimated Federal Support [b] ($ Millions)	Estimated PhD Manpower [c]	Annual Federal Support ($Millions)	PhD Manpower
1. Gamma-ray detectors in astronomy	0.5	5	1.5	10
2. Digitized imaging devices for optical astronomy	0.5	5	1.5	30
3. Infrared astronomy generally	1.0	—	3.0	60
4. Very large radio array	—[d]	—	10.0[f]	75
5. Aperture synthesis for infrared astronomy	—	—	2.0	20
6. X- and γ-ray observatory	—[e]	—	40.0[f]	75
7. Gravitational radiation	0.3	5	1.0	10
8. Neutrino astronomy	0.3	5	1.0	10

Astrophysics and Relativity—*Continued*

Program Elements [a]	1970		Proposed 10-Year Program	
	Estimated Federal Support [b] ($Millions)	Estimated PhD Manpower [c]	Annual Federal Support ($Millions)	PhD Manpower
9. Theoretical relativistic astrophysics	1.3	50	2.5	100
10. General relativity tests	0.8	10	4.0	20

[a] These program elements at present include only a small fraction of the total research activity in astrophysics. They identify areas of research that are ripe for exploration.

[b] Total annual federal support for A&R is estimated at $60 million. The cost of space-based observations amounts to about three fourths of the total federal support. University support of the field is substantial.

[c] The number of PhD's working in A&R is estimated at 300.

[d] A very large array for which design studies are complete and funding is being sought.

[e] The High Energy Astronomical Observatory in space proposed by NASA.

[f] Construction costs to be amortized over a 10-year period.

Description of Program Elements

1. Gamma-ray detectors of greatly improved sensitivity, particularly in the 0.5–30-MeV region, are essential for understanding the history of nucleosynthesis in the universe. Also needed are better means of detecting gamma rays (> 10 GeV), which may be present as a result of a variety of energetic processes in exploding objects.

2. Equipping all large telescopes with digitized imaging devices would greatly aid work in cosmology by speeding up observations by a substantial factor .and by permitting electronic subtraction of atmospheric interference over a large dynamic range.

3. Infrared astronomy, still a young discipline, requires intensive development both in terms of conventional telescopes and the invention of new techniques to permit further exploration of such vast energy sources as radio galaxies and quasars.

4. There is now need for a very large radio array ($\sim$27 dishes) capable of achieving beam widths of the order of 1 sec of arc at centimeter wavelengths for studying the details of nearby bright sources with precision and for detecting faint sources out to the limits of the observable universe against the background imposed by many apparently brighter sources.

5. The technique for synthesizing a large aperture using small apertures, so successfully used in the radio range, is being tested in the infrared range using a system aimed at resolutions of 10^{-2} sec of arc or better in strong ir sources such as galactic nuclei. It is important to develop this technique to the ultimate extent possible, perhaps even to the limit imposed by the diameter of the earth (10^{-7} sec of arc).

6. Construction of a High Energy Astronomical Observatory in space for x and gamma rays would permit orders-of-magnitude improvement in sensitivity, position determination, spectral resolution, and variability measurements. Because x and gamma rays are emitted in great quantities by objects such as pulsars and quasars, it

is important to cosmology to determine whether the backgrounds of these radiations are intergalactic in origin or due to a large number of superimposed sources.

7. Recent experiments are yielding indications that gravitational radiation is emitted from astronomical sources. In view of the need to test the predictions of relativity and to identify the extreme conditions that must exist in any source capable of emitting such radiation, it is important to continue and refine such experiments.

8. The attempt to detect solar neutrinos is critically important because of its implication for the whole theory of stellar structure and evolution on which so much of astrophysics is based. It is necessary that attempts to detect solar neutrinos continue until decisive results are achieved.

9. Application of the equations of general relativity to astronomically observable objects is important to verify the correctness of the theory and clarify the basic processes that are occurring. As in all astrophysics, construction of theoretical models is the only way we have of interpreting the fragmentary information yielded by observations of relativistic objects. Therefore, in any balanced program, it is essential to increase our activity in theoretical model building in proportion to observational research.

10. Experimental tests of general relativity within the solar system have not achieved an accuracy adequate to distinguish Einstein's theory from competing theories of relativity. The advanced techniques and technology now available should enable clarification of this situation.

Physics in Chemistry (1968)

Program Elements	Estimated Federal and Nonfederal Support [a] ($ Millions)	Estimated PhD Scientists in the Physics–Chemistry Interface [b]	
		Physicists	Chemists
1. Molecular structure and spectroscopy	60	520	650
2. Kinetics and molecular interactions	73	850	600
3. Condensed phases	64	460	825
4. Surfaces	28	290	275
5. Other	25	300	200

[a] Funding for scientists in the physics–chemistry interface area comes from sources that traditionally support physics and sources that traditionally support chemistry. The federal physics-related funds have been largely included in the funding estimates for AME and CM physics. No attempt was made to quantify the funding of chemistry in the United States. The average annual cost per PhD does not vary widely from program element to program element and is estimated to be $50,000/PbD.
[b] Based on estimated level of activity, the 1968 National Scientific Registry, and information provided by the Data Panel of the Physics Survey Committee.

Description of Program Elements

1. Includes spectroscopy of any sort (when structural information is its aim), quantum-mechanical studies of molecular structure (whether they are the phenomenological studies common to microwave and magnetic resonance studies or *a priori* studies of electronic structure), and electronic structure of solids in the context of the physics–chemistry interface.

2. The aspects of chemical kinetics in general that are considered part of the physics–chemistry interface, rather than pure chemistry, tend to involve reactions in the gas phase at all energies but concern reactions in condensed phases primarily at high energies.

3. Includes some parts of solid-state physics and chemistry and large portions of amorphous phases and polymers. Includes structure and dynamical properties of polymers; mechanical and electrical properties of liquids, glasses, and liquid crystals; luminescence and photoconductive properties of amorphous phases and molecular crystals; and some efforts toward developing devices such as liquid and plastic scintillators and amorphous switching devices.

4. Includes heterogeneous catalysis, sorption and evaporation, high-vacuum techniques, and reactions on surfaces and gas–solid interactions such as channeling.

5. Miscellaneous unclassified areas of the chemistry–physics interface.

Physics in Biology (1971)

Program Elements	Costs and Manpower
1. Molecular bases for biophysical processes 2. Neural physiology 3. Radiation phenomena 4. Thermodynamics, energy balance, and stability	The Panel found it impossible to attach manpower or funding figures to the individual elements. The number of PhD physicists doing basic research in these areas is estimated at 280

Description of Program Elements

1. A very broad category involving use of almost the entire arsenal of physics probes from x-ray crystallography through NMR and Mössbauer techniques to nanosecond fluorimetry.

2. Typical of sophisticated areas of study of macromolecular aggregates. Involves major design of new measurement techniques, computer simulation of neural behavior, study of signal-transmission characteristics of biological media.

3. Includes effects of both low- and high-level radiation on biological systems—uv to high-energy heavy particles—regeneration and repair mechanisms, and long-term effects on populations.

4. Basic questions of energy utilization and control in biological systems.

APPENDIX 4.B.   MIGRATION TRENDS WITHIN PHYSICS 1968–1970

The following table and figures display migration data for PhD physicists obtained from the 1968 and 1970 National Register of Scientific and Technical Personnel data tapes. Unless otherwise indicated, the manpower numbers shown in these figures refer to the 1970 data.

These data cover a total of 12,609 PhD respondents to the National Register questionnaire, who provided relevant input information; they represent 79 percent of all PhD respondents to the 1970 Register survey. It is estimated that approximately 85 percent of all PhD physicists in the United States are covered by the Register.

One general conclusion follows directly from Table 4.B.1. Physics PhD manpower is very much more mobile than has been commonly believed. During the period 1968–1970, about one third of all these PhD's changed their subfields of major interest and activity. This is again a strong indicator of the unity of physics; the subfield interfaces are highly permeable.

Particularly striking in the table are the 60 percent and the 33 percent increases in astrophysics and relativity and in optics PhD manpower, respectively. In the first case, this increase is a measure of the frontier challenge of the subfield and of its rapid transformation into an experimental and laboratory-based as well as a theoretical area of physics. In the second, it reflects rapid growth in activity in quantum optics. The 20 percent increase in PhD manpower shown for earth and planetary physics mirrors increasing activity in geophysics and oceanography particularly, and, more generally, increasing interest in environmental questions.

TABLE 4.B.1. PhD Migration Data in the Physics Subfields 1968–1971 [a]

Physics Subfield	1968 PhD's	1970 PhD's	Percentage Change in Total	Number of PhD's Who Remained in 1968 Subfields	Percentage PhD's Who Remained in 1968 Subfields
Astrophysics and relativity	121	195	+60	82	42
Atomic, molecular, and electron	925	783	−15	440	56
Elementary-particle	1210	1064	−12	895	84
Nuclear	1674	1390	−17	1156	83
Fluid	367	441	+20	202	46
Plasma	458	396	−14	330	83
Condensed-matter	3759	3248	−14	2634	81
Earth and planetary	486	581	+20	323	56
Physics in biology	180	201	+12	112	56
Optics	637	848	+33	317	37
Acoustics	249	257	+3	171	67
Astronomy	658	484	−26	391	81
Astrophysics, relativity, and astronomy	779	679	−13	253	37

[a] Based on National Register responses from 12,609 U.S. PhD physicists.

Throughout the physics community, growing interest in biological problems and opportunities is indicated by the 12 percent increase in this interface.

The decreases, by some 12 percent to 17 percent, in the PhD manpower in elementary-particle physics, nuclear physics, condensed-matter physics, and atomic physics directly reflect the cutback in available support in these subfields.

Particularly noteworthy is the decrease, by 14 percent, in the plasma-physics manpower complement, which has occurred at a time when the momentum of the subfield and the potential for major scientific and technological success is high. In view of the potential societal benefits that would follow from these successes, the nation can ill afford a continuance of such trends.

This table also provides a measure of relative manpower stability in the different subfields during the 1968–1970 period—the period immediately following the sharp break in the growth of support for physics. It is somewhat surprising that the subfields show a marked grouping into

MIGRATION IN AND OUT OF PLASMA PHYSICS

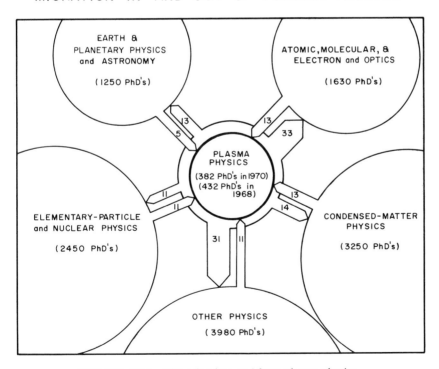

FIGURE 4.B.1   Migration into and from plasma physics.

those wherein some 80–85 percent of the PhD complement remained in its respective subfields and those in which approximately one half changed to other subfields. There is no obvious correlation in either group.

The specific interchanges between subfields are shown in Figures 4.B.1–4.B.11. It should be emphasized that the actual numbers in many cases are too small to have any statistical significance; however, the general trends indicated by these figures are of interest in establishing patterns of mobility among subfields.

MIGRATION IN AND OUT OF NUCLEAR PHYSICS

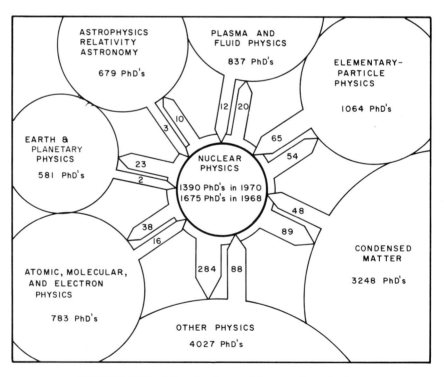

FIGURE 4.B.2   Migration into and from nuclear physics.

MIGRATION IN AND OUT OF ELEMENTARY-PARTICLE PHYSICS

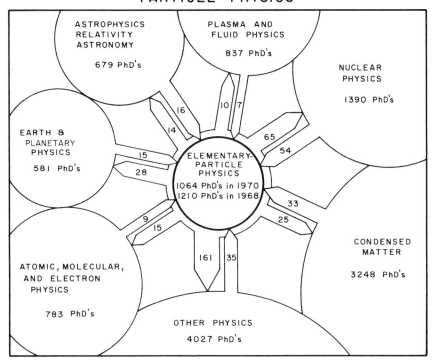

FIGURE 4.B.3   Migration into and from elementary-particle physics.

MIGRATION IN AND OUT OF ATOMIC, MOLECULAR, AND ELECTRON PHYSICS

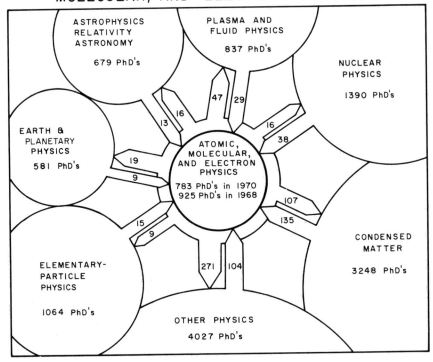

FIGURE 4.B.4   Migration into and from atomic, molecular, and electron physics.

MIGRATION IN AND OUT OF CONDENSED-MATTER PHYSICS

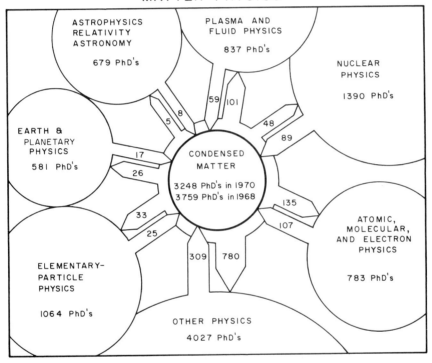

FIGURE 4.B.5   Migration into and from condensed-matter physics.

MIGRATION IN AND OUT OF ASTROPHYSICS, RELATIVITY, AND ASTRONOMY

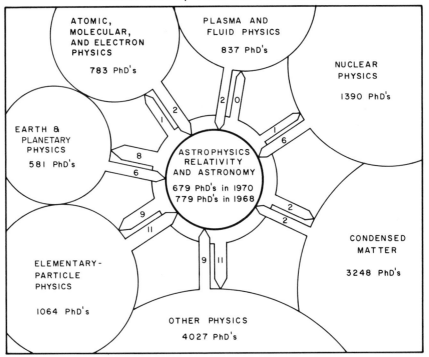

FIGURE 4.B.6   Migration into and from astrophysics and relativity and astronomy.

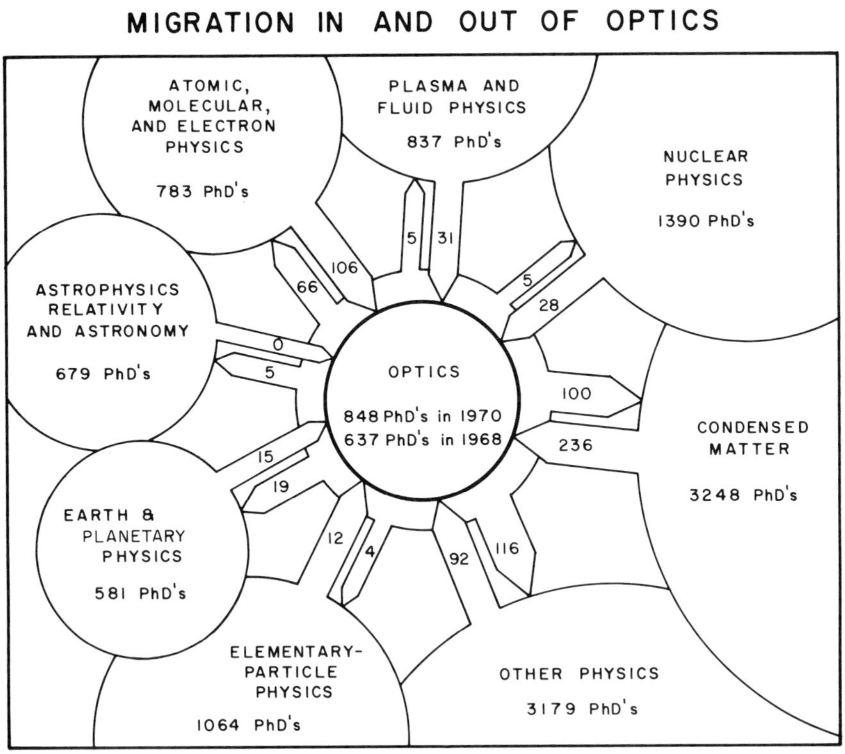

MIGRATION IN AND OUT OF OPTICS

FIGURE 4.B.7   Migration into and from optics.

MIGRATION IN AND OUT OF PHYSICS
IN BIOLOGY

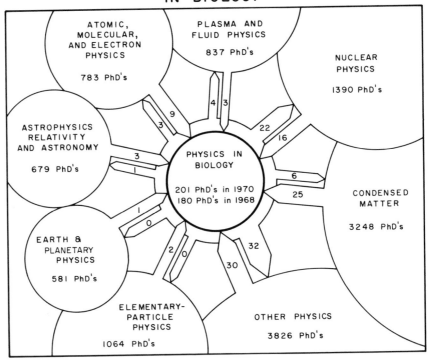

FIGURE 4.B.8   Migration into and from physics in biology.

MIGRATION IN AND OUT OF ACOUSTICS

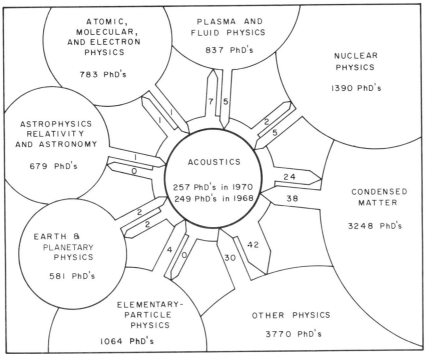

FIGURE 4.B.9   Migration into and from acoustics.

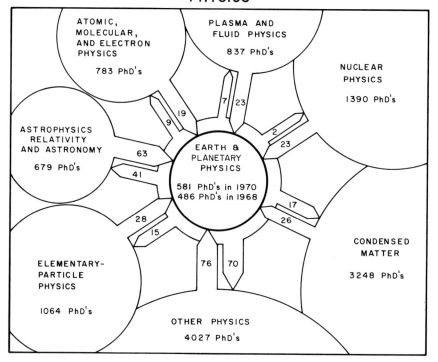

FIGURE 4.B.10   Migration into and from earth and planetary physics.

MIGRATION IN AND OUT OF FLUID PHYSICS

FIGURE 4.B.11   Migration into and from fluid physics.

APPENDIX 4.C.   SUPPLEMENTARY READING IN THE SUBFIELDS OF PHYSICS

Elementary-Particle Physics

Alvarez, L. W., "Recent Developments in Particle Physics," Science *165*, 1071 (1969).

Dyson, F. J., "Field Theory," Sci. Am. *188*, 57 (Apr. 1953).

Gell-Mann, M., and E. P. Rosenbaum, "Elementary Particles," Sci. Am. *197*, 72 (July 1957).

Glaser, D. A., "The Bubble Chamber," Sci. Am. *192*, 46 (Feb. 1955).

Hearings before the Subcommittee on Research, Development, and Radiation of the Joint Committee on Atomic Energy—"High Energy Physics Research"—March 2–5, 1965.

Joint Committee on Atomic Energy. "High Energy Physics Program: Report on National Policy and Background Information." (Washington, D.C., February 1965).

Morrison, P., "The Overthrow of Parity," Sci. Am. *196*, 45 (Apr. 1957).

Sachs, R. G., "Can the Direction of Flow of Time Be Determined?" Science *140*, 1284 (1963).

Segrè, E., and C. E. Wiegand, "The Antiproton," Sci. Am. *194*, 37 (June 1956).

Treiman, S. B., "The Weak Interactions," Sci. Am. *200*, 72 (Mar. 1959).

Weisskopf, V. F., "Physics in the Twentieth Century," Science *168*, 923 (1970).

Wigner, E. P., "Violations of Symmetry in Physics," Sci. Am. *213*, 28 (Dec. 1965).

Yuan, L. C. L., ed., *Elementary Particles and Society* (Academic Press, New York, 1971).

Yuan, L. C. L., ed., *Nature of Matter, Purposes of High Energy Physics,* Brookhaven National Laboratory (1964). Available from National Technical Information Service, U. S. Department of Commerce, Springfield, Virginia 22151.

Nuclear Physics

Baranger, M., and E. W. Vogt, eds., *Advances in Nuclear Physics* (Plenum Press, New York, 1968–   ). An annual series of review volumes providing authoritative treatments of contemporary nuclear research.

Bromley, D. A., "The Nucleus," Phys. Today *21*(5), 29–36 (1968). A 20-year survey review of nuclear physics.

Crewe, A. V., and J. J. Katz, *Nuclear Research U.S.A.—Knowledge for the Future* (Dover, New York, 1969). An elementary survey of U.S. nuclear research activity, profusely illustrated.

Cohen, B. L., *Concepts of Nuclear Physics* (McGraw-Hill, New York, 1971). An intermediate survey of nuclear structure and reaction research.

Cohen, B. L., *The Heart of the Atom* (Doubleday, New York, 1967). An elementary survey of nuclear-structure studies.

Mottelson, B. R., and A. Bohr, *Nuclear Structure* (Benjamin, New York, 1969). A three-volume advanced treatise on nuclear physics.

Seaborg, G. T., and W. R. Corliss, *Man and Atom: Building a New World Through Nuclear Technology* (Dutton, New York, 1971).

Atomic, Molecular, and Electron Physics

Bates, D. R., and I. Estermann, eds., *Advances in Atomic and Molecular Physics* (Academic Press, New York, 1965 to present), Vols. 1–5.

Levine, A. K., ed., *Advances in Lasers* (Academic Press, New York, 1966 to present), Vols. 1–4.

Marton, L., ed., *Advances in Electronics and Electron Physics* (Academic Press, New York, 1948 to present), Vols. 1–27 and Supplements 1–7.

* Ramsey, N. F., *Molecular Beams* (Oxford U.P., New York, 1956).

Schawlow, A. L., ed., *Lasers and Light.* Readings from the *Scientific American* (Freeman, San Francisco, 1969).

* Schawlow, A. L., and C. H. Townes, *Microwave Spectroscopy* (McGraw-Hill, New York, 1955).

Yariv, A., *Quantum Electronics* (Wiley, New York, 1967). An intermediate textbook with an account of masers, lasers, and nonlinear optics.

---

* These references are somewhat older, standard classics that provide a good introduction to modern atomic and molecular spectroscopy. Basic textbooks in quantum theory, statistical mechanics, optics, and electromagnetic radiation are both numerous and essential for a good grasp on this subfield.

Condensed Matter

Alexander, W. O., "The Competition of Materials," Sci. Am. *217*, 254 (Sept. 1967).

Charles, R. J., "The Nature of Glasses," Sci. Am. *217*, 126 (Sept. 1967).

Committee of the Solid State Sciences Panel, *Research in Solid-State Sciences: Opportunities and Relevance to National Needs,* Publ. 1600 (National Academy of Sciences, Washington, D.C., 1968).

Cottrel, A. H., "The Nature of Metals," Sci. Am. *217*, 90 (Sept. 1967).

Gilman, J. J., "The Nature of Ceramics," Sci. Am. *217*, 112 (Sept. 1967).

Ehrenreich, H., "The Electrical Properties of Materials," Sci. Am. *217*, 194 (Sept. 1967).

Ehrenreich, H., F. Seitz, and D. Turnbull, eds., *Solid State Physics* (Academic Press, New York), Vol. I, 1955–Vol. XXVI, 1971.

Huggins, R. A., R. H. Bube, and R. W. Roberts, eds., *Annual Review of Materials Science* (Annual Reviews, Palo Alto, Calif., 1971), Vol. 1.

Javan, A., "The Optical Properties of Materials," Sci. Am. *217*, 238 (Sept. 1967).

Keffer, R., "The Magnetic Properties of Materials," Sci. Am. *217*, 222 (Sept. 1967).

Kelly, A., "The Nature of Composite Materials," Sci. Am. *217*, 160 (Sept. 1967).

Kittel, C. *Introduction to Solid State Physics* (Wiley-Interscience, New York, 1971), 4th ed.

Mark, H. F., "The Nature of Polymeric Materials," Sci. Am. *217*, 148 (Sept. 1967).

Mott, N., "The Solid State," Sci. Am. *217*, 80 (Sept. 1967).

Reiss, H., "The Chemical Properties of Materials," Sci. Am. *217*, 210 (Sept. 1967).

Smith, C. S., "Materials," Sci. Am. *217*, 68 (Sept. 1967).

Ziman, J., "The Thermal Properties of Materials," Sci. Am. *217*, 180 (Sept. 1967).

Optics

Born, M., and E. Wolf., *Principles of Optics* (Macmillan, New York, 1964), 2nd revised ed. An exhaustive and highly mathematical treatment of all optical phenomena that may be treated by classical electromagnetic theory. It is a definitive reference work in optics.

Collier, R. J., C. R. Burckhardt, and L. L. Lin, *Optical Holography* (Academic Press, New York, 1971). This text begins with basic mathematics and optical concepts and then proceeds through nearly all aspects of the theory and art of holographic image formation with visible light.

DeVelis, J. B., and G. O. Reynolds, *Theory and Applications of Holography* (Addison-Wesley, Reading, Mass., 1967). Authoritative and well illustrated. Contains much advanced mathematics.

Goodman, J. W., *Introduction to Fourier Optics* (McGraw-Hill, New York, 1968). The work is directed toward an understanding of the applications of Fourier analysis and linear systems concepts to optics. It is particularly directed toward those familiar with network analysis but relatively weak in the principles of classical optics.

Jenkins, F. A., and H. E. White, *Fundamentals of Optics* (McGraw-Hill, New York, 1957), 3rd ed. A renowned college text by two distinguished teachers. It is especially recommended for its treatment of classical physical optics.

O'Neill, E. L., *Introduction to Statistical Optics* (Addison-Wesley, Reading, Mass., 1963). An introduction to classical statistical optics. It is suitable source material for seniors and first-year graduate students in physics or electrical engineering.

Schawlow, A. L., ed., *Lasers and Light*. Readings from the Scientific American (Freeman, San Francisco, 1969). This book contains 11 offprints of articles from the *Scientific American*, September 1968, and 21 offprints from earlier editions, making a splendid collection on light for college and high-school students and for the general reader.

Smith, W. J., *Modern Optical Engineering* (McGraw-Hill, New York, 1966). This book is intended for the individual whose background is in engineering or physics. It is designed to provide the information necessary for such an individual to pursue a career in optical systems design.

Wolf, E., ed., *Progress in Optics* (North-Holland, Amsterdam), Vol. I, 1961–Vol. IX, 1971. This series, which publishes an annual volume, contains review articles about current research in optics and related fields. Its purpose is to help optical scientists and engineers be well informed about advances in the field.

Wood, R. W., *Physical Optics* (Macmillan, New York, 1934), 3rd ed. Still unique is this entertainingly written and emphatically experimental description of the phenomena of physical optics, from which the mathematics of the electromagnetic theory was progressively removed as successive editions appeared.

Acoustics

Backus, J. M., *The Acoustic Foundations of Music* (Norton, New York, 1969). A look at the elements of music from the viewpoint of a physicist.

Békésy, G. V., *Experiments in Hearing* (McGraw-Hill, New York, 1960). An edited compilation of Békésy's published articles. Deals with physiological as well as with psychological aspects of hearing. Contains work that earned Békésy the 1961 Nobel Prize in Physiology and Medicine.

Beyer, R. T., and S. V. Letcher, *Physical Ultrasonics* (Academic Press, New York, 1969). A detailed discussion of prominent areas of ultrasonic research.

Flanagan, J. L., *Speech Analysis, Synthesis and Perception* (Academic Press, New York, 1965). A study of the acoustical properties of the voice and ear, techniques of speech analysis and synthesis, with mathematical accounts of models and systems.

Green, D. M., and J. A. Swets, *Signal Detection Theory and Psychophysics* (Wiley, New York, 1966). Fundamental text for much of modern psychophysics, especially psychoacoustics. The authors belong to a small group of theoreticians who adapted the general theory of signal detectability to psychophysics.

Harris, C. M., ed., *Handbook of Noise Control* (McGraw-Hill, New York, 1957). Articles on numerous aspects of noise, its effect on man and its control, written by more than a score of the leading workers in the field.

Harris, C. M., and C. E. Crede, *Shock and Vibration Handbook* (McGraw-Hill, New York, 1961). A three-volume set covering practical shock and vibration problems.

Kinsler, L. E., and A. R. Frey, *Fundamentals of Acoustics* (Wiley, New York, 1962), 2nd ed. The leading general text in acoustics.

Knudsen, V. O., and C. M. Harris, *Acoustical Designing in Architecture* (Wiley, New York, 1950). A classic in its field by two of the outstanding practitioners of auditorium design.

Kryter, K. D., *The Effects of Noise on Man* (Academic Press, New York, 1970). An extensive compilation of current knowledge concerning psychological and physiological effects of noise on people. Deals with fundamental and applied aspects of the problem.

Mason, W. P., ed., *Physical Acoustics* (Academic Press, New York, 1964–1970), Vols. 1–7. A comprehensive, continuing series on the physical side of acoustics. Now in seven volumes, many of which consist of two books.

Morse, P. M., and K. U. Ingard, *Theoretical Acoustics* (McGraw-Hill, New York, 1968). An exhaustive mathematical treatise, covering primarily vibrations, sound radiation, and scattering.

Truell, R., C. Elbaum, and B. B. Chick, *Ultrasonic Methods in Solid State Physics* (Academic Press, New York, 1969). A detailed presentation of the basic methods used in the investigation of the solid state by acoustical means.

Urick, R. J., *Principles of Underwater Sound for Engineers* (McGraw-Hill, New York, 1967).

Plasma and Fluid Physics

Batchelor, G. K., *The Theory of Homogeneous Turbulence* (The University Press, Cambridge, England, 1960).

Bishop, A. S., *Project Sherwood* (Addison-Wesley, Reading, Mass., 1958).

Courant, R., and K. O. Friederichs, *Supersonic Flow and Shock Waves* (Interscience, New York, 1948).

Glasstone, S., and R. H. Lovberg, *Controlled Thermonuclear Reactions* (Van Nostrand, Princeton, N. J., 1960).

Grad, H., "Frontiers of Physics Today: Plasmas," Physics Today 22, 34 (Dec. 1969).

Lamb, H., *Hydrodynamics* (Dover, New York, 1945).

Leslie, D. C., *Developments in the Theory of Turbulence* (Oxford U. P., Oxford, England, in press).

Lessing, L., "New Ways to More Power With Less Pollution," Fortune 82, 78–81 (Nov. 1970).

Oswatitsch, K., *Gas Dynamics* (Academic Press, New York, 1956).

Rosa, R. J., "Physical Principles of Magnetohydrodynamic Power Generation," Phys. Fluids 4, 182–194 (1961).

Spitzer, L., *Physics of Fully Ionized Gases* (Interscience, New York, 1962), 2nd ed.

Tennekes, H., and J. L. Lumley, *A First Course in Turbulence* (MIT Press, Cambridge, Mass., 1972).

Astrophysics and Relativity

Calder, N., *Violent Universe* (Viking, New York, 1969).

Gamow, G., *Creation of the Universe* (Viking, New York, 1952).

Frontiers in Astronomy, a collection of *Scientific American* articles (Freeman, San Francisco, 1970).

Menzel, D. H., F. L. Whipple, G. de Vaucouleurs, *Survey of the Universe* (Prentice-Hall, Englewood Cliffs, N. J., 1970).

Schatzman, E. L., *The Structure of the Universe* (World University Library, New York, 1968).

Earth and Planetary Physics

Commission on Marine Science, Engineering, and Resources, *Our Nation and the Sea: A Plan for National Action* (U.S. Govt. Printing Office, Washington, D.C., 1969).

Committee on Atmospheric Sciences, *The Atmospheric Sciences and Man's Needs: Priorities for the Future* (National Academy of Sciences, Washington, D.C., 1970).

Committee on Atmospheric Sciences, *Weather and Climate Modification: Problems and Prospects,* Publ. 1350 (National Academy of Sciences–National Research Council, Washington, D.C., 1966).

Committee on Earthquake Engineering Research, *Earthquake Engineering Research* (National Academy of Sciences, Washington, D.C., 1969).

Committee on Oceanography and Committee on Ocean Engineering, *An Oceanic Quest: The International Decade of Ocean Exploration* (National Academy of Sciences, Washington, D.C., 1969).

Committee on Seismology, *Seismology: Responsibilities and Requirements of a Growing Science* (National Academy of Sciences, Washington, D.C., 1969).

Committee on Solar-Terrestrial Research of the Geophysics Research Board, *Physics of the Earth in Space: The Role of Ground-Based Research* (National Academy of Sciences, Washington, D.C., 1969).

Cressman, G. P., "Public Forecasting—Present and Future," in *A Century of Weather Progress* (Am. Meteorol. Soc., Boston, Mass., 1970), pp. 71–77.

Division of Engineering, *Useful Applications of Earth-Oriented Satellites* (National Academy of Sciences-National Research Council, Washington, D.C., 1969).

National Aeronautics and Space Administration, *The Terrestrial Environment: Solid-Earth and Ocean Physics,* NASA CR-1579 (NASA, Washington, D.C., 1970).

Space Science Board, *Priorities for Space Research 1971–1980* (National Academy of Sciences, Washington, D.C., 1971).

Physics in Chemistry

Cooper, L. N., *An Introduction to the Meaning and Structure of Physics* (Harper and Row, New York, 1968).

Eyring, H., ed., *Annual Reviews of Physical Chemistry* (Annual Review, Inc., Palo Alto, California, 1950–).

Hinshelwood, C. N., *The Structure of Physical Chemistry* (Cambridge U. P., New York, 1961).

Prigogine, I., ed., *Advances in Chemical Physics* (Wiley, New York, 1958–).

Young, L. B., ed., *The Mystery of Matter.* The American Foundation for Continuing Education (Oxford U. P., New York, 1965).

Physics in Biology

Benzer, S., "Adventures in the rII Region," in *Phage and the Origins of Molecular Biology,* J. Cairns, G. S. Stent, and J. D. Watson, eds. (Cold Spring Harbor Laboratory of Molecular Biology, 1966).

Dyson, F. J., "The Future of Physics," Phys. Today *23,* 23 (Sept. 1970).

Goldman, L., and R. J. Rockwell, *Lasers in Medicine* (Gordon & Breach, New York, 1970).

Olby, R., "Francis Crick, DNA, and the Central Dogma," Daedalus *99,* 938–987 (1970).

Schrodinger, E., *What is Life?* (Cambridge U. P., Cambridge, Mass., 1944), reprinted 1967.

Wilkins, M. H. F., "Molecular Configuration of Nucleic Acids," Science *140,* 941 (1963).

5
PRIORITIES AND PROGRAM EMPHASES IN PHYSICS

The determination of priorities in science is a dynamic, complex, and subtle matter requiring a balance among many different considerations ranging from the quality of the people in a field to the estimated value of potential applications. It is sometimes asserted that the scientific community has no system for determining priorities within science, and that the Federal Government has no policy for allocating scientific resources. Neither of these statements is true.

The Physical Sciences—1970
A Report of the National Science Board

. . . Given adequate warning, academic sciences administrators are capable of judging priorities and shifting plans to meet overall limitations on Federal budgets.

Report of the Subcommittee on Science, Research and Development of the Committee on Science and Astronautics, U.S. House of Representatives February 25, 1970

381

INTRODUCTION

During the 1950's and early 1960's, U.S. science, reflecting generous public support, enjoyed a period of unprecedented growth in both quality and scope. In these years almost every competent scientist and almost every

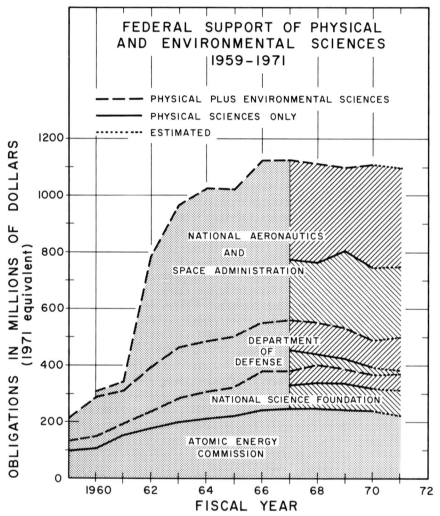

FIGURE 5.1 Federal support of physical sciences during fiscal years 1959 through 1971 by the AEC, DOD, NASA, and NSF. These four agencies provide about 90 percent of the federal support of the physical sciences. The NSF category, physical sciences, includes physics, astronomy, chemistry, and earth sciences. NASA funding includes charges for launch vehicles, launch operations, spacecraft development, and tracking and data acquisition.

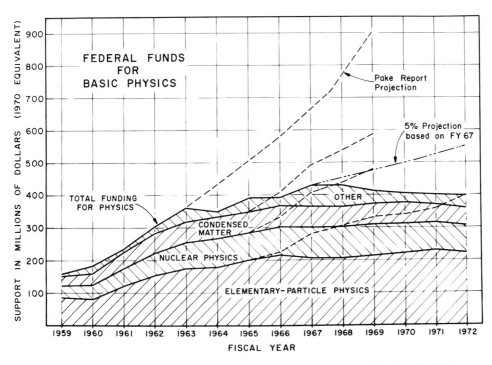

FIGURE 5.2 Federal funds for basic physics during fiscal years 1959 through 1972. Also shown for comparison are the Pake report projections for three major subfields and for the total support of basic physics and a 5 percent projection based on fiscal year 1967. Space physics and all of astrophysics and relativity have been excluded from these figures because of the definition problem involved in correctly allocating NASA funding to basic research. (These detailed values for subfield funding do not agree with those given in Table 5.11 or in the panel reports in Volume II, again because of questions of definition.) In this figure we have included construction funding for elementary-particle and nuclear physics, together with operating and equipment funding.

good idea could find support without undue delay. Annual growth rates of between 15 and 25 percent were not uncommon. The results were new knowledge, new technologies, and a large body of trained manpower commensurate with this national investment.

Such a growth rate could not continue indefinitely. Indeed, in the period since 1967 the support of many areas of U.S. science has become level or effectively declined (Figure 5.1); this is particularly true of physics (Figure 5.2). When science funding is increasing, questions of priority receive little overt attention because worthy new projects and new investigators can be supported with little detriment to work already under way. With ample funding for new initiatives and the capacity to exploit new opportunities, it is relatively easy to maintain the vitality of the scientific enterprise; it is much more difficult to do so under conditions in which not all good ideas can find timely support and many competent scientists cannot find professional opportunities that exploit even a part of their training. The nation is then in a position of being less able to gamble, and the cost of wrong choices becomes much higher. The question of priorities moves much more to the center of the stage and becomes much more critical to the successful performance of the scientific endeavor.

THE QUESTION OF PRIORITIES

Questions of priority, although often not made explicit, are an integral part of all human endeavor. Science is no exception. Scientists, science advisers, managers of science, and the scientific community must decide in one area or another what to do next and how to allocate energies and resources. What areas of science are most worth pursuing? Which are most deserving of encouragement and support? These are difficult questions to which there are neither obvious answers nor, indeed, obvious methodologies for obtaining answers. Yet, they are the real questions now faced by all who are concerned with science and who are forced to decide what to do next. Decisions imply priorities, judgments that it is better to follow one course than another—even if both alternatives have real merit. How then are priorities in science to be determined? What is the nature of such determinations?

It is necessary to make clear at the outset that determinations of scientific priorities are implied predictions. They are attempts to foresee the scientific and practical consequences of specific courses of action under conditions of uncertainty. Decisions on priorities must take into account not only the most probable outcome but also the consequences of alternative possible outcomes. They must also keep future options open in case matters do not turn out as predicted. Because of this high degree of uncertainty, the best decisions are usually made by those who have to live with their consequences. Each scientist must decide, day by day, what he is to do in the future—what problems he will take up, what approaches he will use, how he will deploy the resources and talents at his disposal—for his rewards, tangible and intangible, and his life work are at stake. In doing this he must take into account the decisions and findings of many other scientists, so that, in fact, each individual decision becomes a part of a collective judgment of much larger scope, involving an implicit consensus of a large community.

It has been suggested that the scientific community should be able to devise a rational system for determining priorities within scientific fields and among scientific disciplines. After extensive discussion, the Committee concluded that, although the matter can be stated rationally in principle, the information that could provide a completely rational system of decision making does not exist. In this respect the difference between science and many other activities is not so great as might be believed. Thus the nation does not have a rigorous basis on which to establish how much defense the country needs or how much education or how much health care or how much environmental protection. It is easy to state that one should continue to increase resources devoted to a given objective until the marginal return from such resources is less than that from alternative investments, but the calculation of such marginal returns in the future is guesswork, even when the uncertainties of prediction are much less than they are for scientific investigation into the unknown. In the absence of an analytical system, decisions are reached essentially by means of the complex of social processes within the scientific community and of social, political, and economic processes at the national level. The 1970 National Science Board report to the Congress reflects this situation:

The fact that much of science does not use a highly visible, centralized, priority-setting mechanism does not mean that other mechanisms do not exist. Actually, science uses a multiplicity of such mechanisms. One priority-setting mechanism operates when a scientist determines the problem on which he works and how he attacks it within the resources available. This determination is made taking into account other similar and related work throughout the world. Another mechanism operates as proposals of competing groups of scientists are evaluated and funded on the basis of systematic refereeing and advice of peer groups. Still another mechanism operates as aggregate budgets for various fields of science are influenced by the number and quality of research proposals received in that field. Like any market mechanism this system is not perfect and requires regulation and inputs from outside the system itself. Such inputs come from the mission-oriented agencies which balance their needs for new knowledge against their operating needs and from a whole host of outside judgments implicit in the budgetary and appropriation process. Trouble occurs either when these external judgments are completely substituted for the priority setting of the scientific community or when the priority setting of the scientific community becomes too autonomous.

In the affairs of science two forces are acting: those external to the science, which represent the aims of society, and those internal to it, which represent its natural development. Unless these forces are maintained in balance there is danger of collapse. If the external forces become too strong, the internal fabric—the unity of science, which is emphasized throughout this report—may be ruptured. Sir Brian Flowers has cited an example from the United Kingdom:

During the war the possibility of nuclear power was realised and immediately afterwards great efforts were made to produce economic nuclear power stations. Not un-naturally it was thought that the greatest lack of necessary expertise was in nuclear physics. Universities all over the country were urged to produce more nuclear physi-cists and they stoutly did so. Consequently other branches of physics were starved, and especially the physics of the solid state and of materials more generally. The fabric became distorted, but the aim was not attained. We know now that the reason for this lay partly in the fact that the economy of nuclear power depends as much upon the materials that are used as upon the nuclear physics. Moreover, we know that even the nuclear physics required by the power programme needed solid state devices which have only been developed in recent years. Now we do have economic nuclear power, but to attain it we had to compensate for the distortions of the fabric.

On the other hand, if the internal forces become too strong and science turns away from the society in which it is embedded, it runs the strong risk of becoming irrelevant.

Consideration of the external inputs from the mission-oriented agencies is essential in the evolution of priorities. There is an important internal input that also should be emphasized. Academic scientists are especially sensitive to the interests and concerns of students who join the scientific enterprise with new ideas and values not completely determined by the perspectives acquired by the senior scientists in the course of their working lives. The continuing entry of able and energetic students into the scientific enterprise tends to stimulate a continual re-evaluation of priorities among academic scientists and within the scientific community as a whole. The process of selection of faculty members for universities is itself another decentralized priority-setting mechanism.

Science is supported by the federal government and other institutions for a great many reasons. Physics directly and indirectly plays a role in such major national programs as defense, education, and industrial develop-ment. Decisions as to which subfields of physics are to be supported have direct impact on the lives of many people and can have major impact on the future of major national research facilities and national economic health. In any ultimate priority assessment these factors must also be taken into account.

Thus a discussion of the priorities has a value insofar as it illuminates the nature of the political debate that must ultimately determine the allo-cation of resources. This Committee can assign no priority system that in any way can, or should, completely circumvent that political process. Our view is that of a group of physicists appraising the needs of physics, admittedly from our particular viewpoint. We have tried, however, to look at issues and physicists from other points of view—for example, the needs of the nation, the needs of mankind—how successfully, we do not know. But information of this kind should be considered as one of the important elements that enters into the social, political, and market processes for deciding the allocation of resources.

APPROACHES TO THE ESTABLISHMENT OF PRIORITIES

As in so many other human affairs, a multiplicity of criteria must be brought to bear on scientific decisions. Thus, perhaps the best way to approach the question of priorities in science is to try to identify and develop the criteria by which they are made. There is now an extensive literature in this area; as a Committee we have studied this earlier work and have devoted much effort to the evolution of a set of criteria that we have found particularly useful in appraising the needs and potentials of the different subfields of physics. Before presenting these criteria, and our use of them, we have concluded that it could serve a useful purpose to summarize very briefly a number of the approaches to priority recommendations that we have examined, together with what we, as a Committee, consider to be some of their more important positive and negative aspects.

Base Priorities for Support on the Excellence of the People Involved and Their Record of Past Accomplishment

Positive Aspects   In any objective approach to scientific priorities, the support of excellence must play an overriding role. This is the end result of a properly functioning intellectual marketplace and provides the most direct route to the attainment of maximum internal progress in the science. In times of financial stringency, it helps the best and most productive scientists and scientific groups to survive and grow and thus to pass on the best traditions of science. In large measure this has been the approach underlying the support of U.S. science in postwar years, although the levels of available support, until very recently, did not require its application in too stringent a sense. It is largely responsible for the present U.S. leadership in world science.

Negative Aspects   This approach has no visible system of public accountability until after the fact and is open to misinterpretation as a somewhat cavalier dismissal of the very real demands of society for scientific relevance. Although excellence can and should include criteria of relevance, our traditions have not focused on other than intellectual brilliance. This approach can lead to a snowball effect in any given area, in that excellence attracts support and support attracts excellence, with consequent danger of overlooking and neglecting other fields. It can distort scientific manpower distributions by depriving areas of great relevance to society of talented scientists; it can even contribute to scientific unemployment by depriving these socially relevant areas of the intellectual leadership and enterpreneurial talent that could lead to much greater demand for adequately trained people.

Base Priorities on Marketplace Evaluation in Which Proposal Pressure and Peer-Group Review Play Dominant Roles

Positive Aspects   This approach relies on a combination of personal and project merit, with emphasis on the latter, and consideration of the former only to the extent that it has a bearing on the likelihood of success in achievement of the goals of the proposed project. It permits a fine tuning between the merit of people and of projects according to the circumstances, without being too explicit as to the relative weighting of the two for any given proposal. This approach facilitates progress along a broad scientific front and allows interdisciplinary effort. It allows changes in program emphases to occur as consequences of free discussions at the grass-roots level. The peer-group review process has worked well and provides a clear check at many stages.

Negative Aspects   Again, there is no visible mechanism for public accountability, since evaluation is conducted entirely by scientists in the same field, with a possible vested interest in the buildup of their field. It is difficult to compare proposals of markedly different size or scope in any objective fashion, and the approach can lead to both excessive fragmentation and dispersal of effort as well as to the snowball effect, with funding going to the most glamorous—or least self-critical—fields. This approach does not embody an explicit mechanism for taking into account relevance to societal needs, although this can to some extent be brought into peer evaluations, if the panels include scientists with a knowledge of these needs and the scientific questions deriving from them.

Base Priorities on Criteria of Intrinsic and Extrinsic Merit Similar to Those Evolved by Weinberg *

Positive Aspects   This approach provides a direct and obvious focus on the unity of the entire scientific enterprise and responds directly to perceived societal needs. It lends itself to a possible semiquantitative rationale for support of an area of science based on an appropriate combination of the intrinsic and extrinsic ratings, with slowly time-varying coefficients such that the support can change systematically to respond to new challenges or opportunities, without introducing large discontinuities and consequent dislocation. This approach, too, must play an important role in the evolution of scientific priorities.

[A framework within which to consider the usefulness of the various

* *Minerva, 1*(2), 159 (1963).

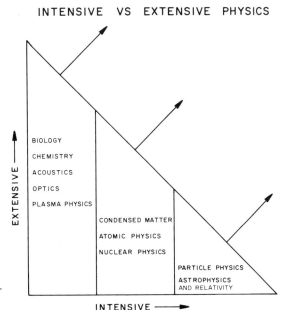

FIGURE 5.3 Intensive and extensive physics.

disciplines of physics has been developed by Weisskopf.* (See also Chapter 3.) He points out that the evolution of physics has not progressed along a line but across a plane as illustrated in the schematic presentation in Figure 5.3. The term "intensive" is used to describe research aimed at answering a small number of very fundamental and general questions and at the discovery of new laws and principles that have potential applicability to a broad range of phenomena and systems. The term "extensive" is used to describe efforts concerned much more with enlarging and deepening understanding of generally accepted physical principles. The further to the left a subfield is in intensive development, the more it is extended in the extensive direction, meaning that it is more and more involved with other questions, other scientific or technical activities. This view of physics, however, does not say anything about priorities, apart from the general implication that a higher quality cutoff point should be used for intensive than for extensive research. It is open to oversimplification and misinterpretation both within and outside of science. The intensive-extensive relationship is all too frequently translated as first–second or first–third class; and the consequences can be extremely divisive.]

* *Physics Today, 20*(5), 23 (1967).

Negative Aspects It is difficult to establish the proper relative importance of the intrinsic and extrinsic criteria; failure to do so leads to the imbalance discussed earlier in this section. Extrinsic criteria are difficult to apply. On the one hand, they are subject to wishful thinking or overoptimism; on the other, the most significant social applications of basic science are seldom foreseeable. There is little evidence, as yet, that the talents for projection of the scientific—or any other—community are such that application of extrinsic criteria have led to more or better or faster application of knowledge. The pursuit of chimerical applications can lead to exceedingly inefficient use of resources.

Base Priorities on the Identification of Disaster Areas and Determine What Rearrangement of Support Would Suffice at Each Total Funding Level To Optimize the Overall Health of the National Activity in the Science

Positive Aspects In this context a disaster area is one in which major and irreversible changes are anticipated in the absence of prompt incremental support. This approach has the advantage that it addresses directly some of the major identified problem areas in the nation's scientific activities. It emphasizes the unity of science by focusing on its overall health in an obvious fashion and could prevent what may well develop as a major waste of new facilities that were authorized during the last decade, under much more optimistic funding projections, and that now are, or may be, seriously underutilized because of inadequate operational support.

Negative Aspects The disaster area approach, which is essentially a method of incremental adjustment, is difficult to apply if the total science budget is regarded as fixed, so that every budgetary adjustment is a zero-sum game played among subfields requiring widely different total resources for viability. In this context, subfields that are heavily dependent on large, costly facilities would tend to be favored unduly. For example, a 10 percent increase for elementary-particle physics in one year would be highly disruptive if it came about entirely at the expense of small individualistic subfields such as atomic, molecular, and electron physics. Moreover, it would be difficult to apply such an approach to whole subfields as entities without a much finer-grained analysis on a project-by-project, or institution-by-institution, basis. It is difficult to balance many small negative but delayed effects spread across many subfields against a one-shot positive effect on a single large project.

*Base Priorities on the Outcome of Public Hearings before
a Scientific Congress in Which Competing
Proposals Could Be Presented and Judged on the Basis of
Standard Judicial and Adversary Procedures*

Positive Aspects This approach would remove any public concern over the possible manipulation of national scientific programs for the benefit of a self-perpetuating in-group. Such an approach, in principle, would give young or unknown physicists a better opportunity than is now available for personal presentation of their ideas and proposals to an impartial tribunal. If successful, it could also provide an important channel by means of which the activities of science could be made more accessible to public audiences.

Negative Aspects Inasmuch as scientific progress is basically a social process for arriving at a consensus on public knowledge, an adversary procedure could be counterproductive in substituting political strategy for scientific judgment and rhetoric for scientific merit. There are no obvious mechanisms by which the tribunal would be selected or authenticated, and the procedures could be extremely divisive in the community. Often very slight differences in quality between winners and losers would be greatly magnified, to the serious detriment of the latter. Serious difficulties could also arise through the premature enunciation of embryonic plans and ideas that public hearings could elicit or provoke. The approach would be applicable only to relatively large projects; smaller projects would have to be considered outside the tribunal to avoid unacceptable delays. The research of Zuckerman and Merton on refereeing, quoted briefly in Chapter 13, strongly suggests that at least this process of peer evaluation in physics is objective and fair and does not penalize young or relatively less well-known people. Whether the same conclusion would be reached in cases in which large amounts of money were involved is, of course, moot.

THE DEVELOPMENT OF A JURY RATING

After much discussion of the various approaches summarized in the preceding section, the Committee decided on an approach that combines many of their features: a jury rating of Committee members as to the appropriate emphasis that should be applied to a given activity within the next five years, taking account of both the internal, intellectual needs of physics and their assessment of the impact of these scientific developments on other sciences, technology, and societal problems generally. It must be emphasized that any such rating system has a value that is relatively short-lived, since science changes so rapidly. Moreover, any group as small as the Physics Survey Committee will necessarily represent certain prejudices or special interests that would be different for a differently selected but equally competent group of comparable size. The numbers are too small for nonobjective biases to cancel one another.

The development of this jury rating involved two aspects. First, the Committee as a whole devoted extensive discussion to the evolution of a list of criteria. Second, we worked with the chairman of each of the subfield panels to divide each subfield into a set of program elements that span the major areas of the subfield and could be evaluated in terms of the criteria. These program elements have been discussed in Chapter 4 and are presented in detail in Appendix 4.A of that chapter.

The purpose of the ratings was to test the feasibility of arriving at a consensus regarding the desirable relative emphasis among subfields of physics and among program elements within each subfield. Such judgments might then guide decisions as to increased or decreased support for each program element and subfield within whatever total might become available for physics as a whole. It is the Committee's view that the outcome of this exercise is properly described in terms of program emphases rather than priorities.*

In the light of recent experience, the disruptive consequences of rapid changes in the amount and distribution of physics funding demand attention. For this reason the Committee's Report has tended to emphasize the importance of continuity in support and of sufficiently long-range funding commitments to permit orderly planning at the working level. This stress on continuity comes not from any desire to maintain the *status quo* but to avoid ineffective use of the talents, investments, and specialized facilities already committed to the field by earlier decisions. Over a sufficient period of time, the physics community can adjust to relatively large changes in support, but this requires planning toward new levels, with lead times comparable to some of the characteristic time cycles of the field—for example, the time required for training a PhD or the interval between authorization and commissioning of a new facility (see Chapter 6).

Thus our goal has been to identify those program elements in physics that, on the basis of our criteria, should experience large *relative* growth rates. The questions of *overall* growth rates and support levels for physics

---

* In reaching the ratings on each individual program element, the Committee drew on all the information available to it, especially the detailed technical reports of the subfield panels (Volume II) and summary presentations to the Committee by an advocate of each subfield.

and development of contingency alternatives designed to respond most effectively to different levels of such total support are considered later in this chapter.

THE DEVELOPMENT OF CRITERIA

In developing our criteria, three general categories appeared useful: *intrinsic merit, extrinsic merit,* and *structure.*

Intrinsic merit we define as criteria internal to science. *Extrinsic merit* relates to impact on technology and on the resolution of human problems. *Structure* is concerned with impact on the national capability to do physics. These three categories are not truly independent of each other. If science does not progress at a sufficient rate in terms of its own internal logic, it will contribute less to society, or its contribution will come either at much higher cost or with unexpected negative side effects, reflecting the undertaking of technological projects with inadequate understanding. If science fails to contribute to technology, it will lose an important source of intellectual stimulation and may fail to attract some dedicated and socially motivated people. The viability of the institutions of science depends on the intellectual thrust of their accomplishments and *vice versa.*

Intrinsic Merit

Here the measure of merit is primarily scientific opportunity. What is the probability that work in a field will have a major impact on man's understanding of his world or the universe? More precisely, which of several possible scientific strategies will be most likely to result in the greatest increment of insight or understanding for a given expenditure of resources (effort, money, and talent)?

Because of new concepts, new questions, new experimental or theoretical approaches, or new instrumentation and observational techniques, some fields promise more immediate rewards from exploitation than others; this is clearly one of the most important elements of intrinsic merit. A related question, reflecting the unity of science, is the degree to which a particular program has the potential for illuminating the broader area of science of which it forms a part.

Will a new solution for the relativistic field equations provide the basis for new understanding of the origin and nature of mysterious astronomical objects? Will a better theory of superconductivity in metals throw light on the theory of nuclear structure? Will an improved understanding of nuclear structure permit a better understanding of stellar evolution and the origin of the elements? More rare, but of vital importance, are those investigations that might open whole new areas of investigation, as would be the case if experimental detection of gravitational radiation is confirmed or as happened following detection of isotropic background electromagnetic radiation in space. Thus, under the rubric of intrinsic merit, each scientific activity requires consideration not only of its own frontiers but also of the opportunities it offers for strengthening the whole fabric of scientific understanding.

The intrinsic merit of a field is also reflected in the quality of the scientists it attracts. In fact, the history of science shows that outstanding individuals make important contributions in any area that arouses their enthusiasm. Such individuals often make major contributions to several quite different areas during the course of their careers. The movement of a few outstanding individuals into a new area may be one of the surest indications that this field has become ripe for exploitation. This criterion recognizes that there are some individuals with such powerful scientific vision that their individual choices can be better trusted than those of any jury of experts. The mere fact that they are prepared to commit their own careers and reputations to a field is a compelling index of its intrinsic merit.

Extrinsic Merit

The contributions that one scientific field can make to others in terms of new fundamental insights has been noted.* There are extremely important but more localized interactions that are worth mentioning. The application of radioactive-tracer and stable-isotope techniques to the study of the circulation of atmospheric pollutants is one example. Another example is the transfer of concepts in the dynamics of nuclear reactions to the study of elementary reaction processes involving individual energy states of atoms and molecules in chemistry. A better theoretical model for turbulent fluid flow would illuminate large areas of the environmental sciences and of technology that depend on the flow of fluids. These examples speak of scientific opportunity, but the potential benefits and new insights are primarily external to the particular area of science under discussion. In a real sense, these are the beneficial externalities of fundamental scientific investigation, which are more often than not unpredictable

---

* Volume II provides many illustrations of the way in which increases in fundamental understanding of nature have influenced the capability for attacking practical questions of concern to society; some of these have been discussed in earlier chapters of this Report.

or can at best be anticipated through inspired hunch rather than logical extrapolation.

In addition to its impact on adjacent or even distant fields of science, an important criterion of extrinsic merit is its potential contribution toward opening new technological opportunities. Is a proposed program likely to have an important influence on engineering development and design, manufacturing processes, protection or enhancement of the environment, medicine, or some other area relevant to human welfare?

The assessment of the technological opportunities arising from science has two aspects. First, a given scientific activity acts as a source of concepts and experimental techniques. For example, semiconductor technology, and ultimately the sophisticated techniques of integrated circuits and microelectronics, has emerged from deeper understanding of many different aspects of condensed-matter physics and chemistry. The laser, with its host of new technological possibilities ranging from precise measurement to fabrication techniques, has been the direct outgrowth of work in both atomic, molecular, and electron physics and condensed-matter physics.

A second aspect, the more immediate symbiotic relationship between physics and technology, appears when a field of investigation draws heavily on adjacent areas of science and technology for concepts and techniques, often stretching the existing state of the art and providing an incentive for technological development that would not be supported initially for its own sake, because the potential applications are too distant, speculative, or actually unforeseen. For example, both nuclear and elementary-particle physics have drawn heavily on computer techniques, high-power radio-frequency technology, cryogenic engineering, techniques for the production and measurement of high vacua, and the development of high-intensity ion sources and ion optics. Through making lasers available, and in the course of using laser beams for studies of the properties of matter, atomic, molecular, and electron physics has contributed importantly to new developments in optical technology, including more efficient and accurate methods of lens design, the evolution of a practical system of holography, nonlinear optical techniques, and techniques for modulation, frequency mixing, and other nonlinear manipulations of signals at optical frequencies.

Another important criterion of extrinsic merit relates to the potential for rather immediate applications in other areas of science, engineering, or technology. The Doppler scattering of a laser beam from a high-energy electron beam to produce polarized gamma radiation is an illustration of application to several areas of science. Other current examples are the use of laser backscattering for studying atmospheric turbidity and the use of colliding atomic beams for the study of elementary reactions involved in upper atmospheric processes resulting from aircraft exhaust emissions at high altitudes.

The potential contributions of physics to national security cannot be ignored. These occur not only directly in the area of defense capabilities but even more significantly in the area of disarmament and both unilateral and multilateral inspection techniques for the monitoring of arms-control agreements. Successful and easily explained and visible achievements in fundamental science also enhance the potential prestige of the United States and give it a favorable image in the eyes of decision makers and opinion leaders in other countries. In addition, science, especially certain areas of physics, provides mechanisms for international cooperation that can lay the groundwork for cooperation or negotiation in politically more delicate areas.

Finally, there is the degree to which a given scientific activity may contribute toward public understanding of science and the extent to which it may lead itself to broad educational functions. For example, certain areas such as astronomy or space physics have a natural appeal to the public imagination that can serve as a means for communicating deeper understanding of physical principles generally. Similarly, areas such as bioacoustics, which deal with phenomena familiar to everyone, contribute to formation of bridges of understanding between science and a naturally receptive public.

Structure

The criteria discussed so far implicitly assume that the proper percentage distribution of support across the subfields can be determined independently of the total amount available, or that the same relative distribution will optimize the research output for any total amount. This is obviously not true, at least in fields heavily dependent on large, costly facilities. Thus if intrinsic and extrinsic criteria alone are applied, the result is a distribution that assumes that the level of activity in the various research components comprising a subfield can be scaled up or down without affecting the viability of the subfield as a whole. Furthermore, there is here an implicit assumption that not only the subdivisions but the subfields themselves are at most weakly coupled to each other.

On the contrary, the pursuit of scientific goals has resulted in a complex and interdependent social system. Disturbances in one part of the system are often communicated throughout all scientific and technological activity, even in the absence of obviously direct connections. Thus support decisions must be viewed in terms of their impact on the institutional and communication systems of science and consequent changes in future national capability and capacity to respond quickly to new opportunities and needs. Some of the kinds of issues of concern are the following:

1. In the last few years several unique facilities requiring a large national investment of both talent and money—for example, the Stanford Linear Accelerator (SLAC) and the MIT National Magnet Laboratory (NML)—have been operated on greatly reduced schedules to save electrical power and other costs. Such decisions to reduce the rate of exploitation of past investments in a facility should be made deliberately for carefully evaluated reasons, not as an incidental by-product of sudden pressures for economy.

2. The physics community responded to the initial reduction in support of physics, beginning in 1967, as though it were temporary. Short-term emergency measures were invoked to preserve capabilities in anticipation of renewed support in the future. Among these measures were deferral of exciting but highly speculative experiments, postponement of the upgrading of instrumentation, and reduction in the number of young scientists admitted to participation in frontier research activities. Rather than the widespread closing of laboratories and disbanding of research groups and graduate programs, which would have resulted in the concentration of support in a smaller number of remaining installations, the physics community responded by a more uniform belt tightening, a response that maximized future options. Over an extended period, however, the effort to keep future options open may result only in a uniform decline toward mediocrity, if the expected restoration of support fails to materialize, with no single group remaining competitive with its international counterparts. This would be a situation in which the total U.S. support of physics continued to appear large by international standards, but no one group or activity would receive a sufficient fraction of the total to maintain a leading, or even competitive, position.

3. In certain fields, major national facilities—the Los Alamos Meson Physics Facility and the National Accelerator Laboratory, to give examples in nuclear and elementary-particle physics, respectively—were approved in the mid-1960's, with the expectation that the operating funds required to exploit their research capabilities at the frontiers of these subfields would be primarily incremental to the ongoing programs. During this coming year, these facilities will begin large-scale research operation, and, while the Atomic Energy Commission's (AEC) funding projections reflect the need for incremental support, such funding has not become available. Two possibilities confront the nation: Either these newest facilities remain largely underused or the necessary funding is removed from other programs in the respective subfields. Inasmuch as costs per

PhD research scientist at these new facilities are substantially higher than at older and smaller facilities, the latter course could cripple large segments of these subfields. Although the availability of new frontier facilities will naturally result in the transfer of some scientific activities from older facilities, the cost savings do not compensate for the higher operating costs of new installations. Furthermore, the older facilities are often essential for obtaining data complementary to those obtainable with the new ones. Either of these possible solutions would be extremely wasteful of both talent and facilities. These again are structural considerations that must be dealt with in the evolution of any coherent national scientific program.

In view of these considerations, an important structural question can be recognized: To what extent is incremental support, beyond the current level, required to balance the need to capitalize on the national investments in large facilities and the equally important need to maintain the viability of the other components of the subfields in question. This is a critical time, in both respects, in the history of U.S. physics.

The needs of each subfield for major new facilities, both to maintain momentum of progress and to avoid loss of a world competitive position, merit consideration. In astrophysics, for example, there is a need for new arrays and dishes for radio and millimeter-wave astronomy. Infrared astronomy is just beginning to develop. In nuclear physics, heavy-ion accelerators offer access to an entirely new range of nuclear phenomena. This subfield was pioneered in the United States but is now being more actively pursued abroad. In recognizing the need for such new facilities, it is vitally important that full consideration be given at the time of authorization to the provision of adequate future operational support to enable full utilization of the new facility and full scientific return on the major investments, in both money and manpower, that these facilities reflect.

Another structural criterion has to do with the effective utilization of instrumentation and the effective investment, per using scientist, in such instrumentation. Much more expensive instrumentation is required to reach the frontiers of research in some subfields than others, and the costs per scientist may be correspondingly large. If various lines of scientific effort are judged to be equally valuable on the basis of other criteria, those that require a smaller investment per scientist would, on a structural basis, tend to receive preference for funding. Such costs are difficult to quantify, however, and final judgment would probably remain qualitative. If cost per scientific paper is to be used as a measure, as sometimes is proposed, are all papers to be treated as of equal value? How is the cost of instrumentation to be allocated among students, faculty, and professional researchers? Is each use to be given the same weight? Somewhat arbitrarily, perhaps, the average cost per PhD scientist man-year in each subfield has been chosen here as reasonably indicative.

Table 5.1 presents estimates of these costs—operating and equipment—per experimental PhD physicist man-year in the different subfields of physics. Experimental research costs have been selected because the unit costs in theoretical work do not vary significantly among subfields; an average unit cost for a theoretical physicist of $35,000 per year appears reasonable.

These costs must be viewed in the context of the importance of maintaining the essential unity of physics. As a parallel, within the Department of Defense (DOD) the unit costs involved per air crew member of one of the more advanced aircraft vastly exceed those per member of an infantry corps; it would indeed be the height of folly to suggest that the former be eliminated in favor of more of the latter. Such questions, nonetheless, cannot be ignored entirely in the overall consideration of establishing priorities.

Two additional structural questions concern the manpower now available and now in training in each subfield. Any projected program must be predicated on the availability of the necessary skilled manpower in that area. More important, however, is the question of balance between the rate of production of manpower in a subfield and the opportunities that are projected for the available manpower. This question is addressed

TABLE 5.1 Approximate Costs per PhD Man-Year in Experimental Physics

Subfield	Operations and Equipment Costs [a] (Thousands)
Acoustics	55
Atomic, molecular, and electron	50
Condensed matter	70
Elementary particles	175
Plasmas and fluids (excluding controlled fusion)	60
Controlled fusion	150
Nuclear physics	80
Astrophysics and relativity	55 [b]

[a] Costs do not include amortization of major facilities.
[b] When space-based research is included, the cost per year per PhD is approximately $200,000. Thus research in this subfield, in common with astronomy generally and with high-energy physics, is relatively expensive.

in detail in Chapter 12 and is one of vital importance to the future of the U.S. scientific community.

Perhaps less obvious, but also important, is the extent to which a given program element or subcomponent of a subfield is essential to the maintenance of the health of the scientific area of which it is a part. To cite a specific example, although atomic optical spectroscopy of itself might be given a relatively low competitive rating at the present time, its vital importance to much of atomic physics, plasma physics, and astrophysics requires that it be maintained in a healthy state. Similarly, the development of lithium-drifted germanium crystals might receive a relatively low rating in condensed-matter physics, but it is essential to large areas of contemporary nuclear physics as the basis for radiation detectors of unparalleled resolution.

CRITERIA FOR PROGRAM EMPHASES

Three sets of criteria emerged from our discussions and were refined through application to program elements in the various subfields.

Questions used to determine *intrinsic merit* were the following:

1. To what extent is the field ripe for exploration?

2. To what extent does the field address itself to truly significant scientific questions that, if answered, offer substantial promise of opening new areas of science and new scientific questions for investigation?

3. (a) To what extent does the field have the potential of discovering new fundamental laws of nature or of major extension of the range of validity of known laws?

    (b) To what extent does the field have the potential of discovering or developing broad generalizations of a fundamental nature that can provide a solid foundation for attack on broad areas of science?

4. To what extent does the field attract the most able members of the physics community at both professional and student levels?

To assess *extrinsic merit,* we asked:

5. To what extent does the field contribute to progress in other scientific disciplines through transfer of its concepts or instrumentation?

6. To what extent does the field, by drawing on adjacent areas of science for concepts, technologies, and approaches, provide a stimulus for their enrichment?

7. To what extent does the field contribute to the development of technology?

8. To what extent does the field contribute to engineering, medicine, or applied science and to the training of professionals in these fields?

9. To what extent does the field contribute directly to the solution of major societal problems and to the realization of societal goals?

10. To what extent does the field have immediate applications?

11. To what extent does the field contribute to national defense?

12. To what extent does activity in the field contribute to national prestige and to international cooperation?

13. To what extent does activity in the field have a direct impact on broad public education objectives?

Questions to establish *structural* criteria were:

14. (a) To what extent is major new instrumentation required for progress in the field?
(b) To what extent is support of the field, beyond the current level, urgently required to maintain viability or to obtain a proper scientific return on major capital investments?

15. To what extent have the resources in the field been utilized effectively?

16. To what extent is the skilled and dedicated manpower necessary for the proposed programs available in the field?

17. To what extent is there a balance between the present and envisaged demand for persons trained in the field and the current rate of production of such manpower?

18. To what extent is maintenance of the field essential to the continued health of the scientific discipline of which it is a part?

APPLICATION OF THE CRITERIA

One of the difficulties in applying these criteria is that of establishing the relative weighting of the intrinsic and extrinsic criteria. These were again subjected to a jury procedure in the Committee in a variety of ways involving both weighting of the individual criteria and of the three areas—

intrinsic, extrinsic, and structural—more generally. A consensus that was surprising to the participants was found. Although considerable spread in the weighting of individual criteria occurred, the intrinsic and extrinsic dimensions were assigned roughly equal weight, and structure roughly one third the value of the other two. Recognizing the inherent lack of precision in any such process, the Committee gave unit weight to each question within each of the three categories. When a question had two parts, as in 3 and 14, each part was assigned this same unit weight; indeed, the reason for listing these questions together rather than as separately numbered entries was to emphasize the close connection between them rather than to reduce the weighting of each individually. Inasmuch as the number of extrinsic criteria is roughly double the number of intrinsic ones, a relative weight of one half was assigned to each extrinsic criterion in our jury rating exercise; likewise, each structural criterion was assigned a relative weight of one third.

To illuminate the criteria through an attempt to apply them to actual cases (and it should be noted that this process resulted in extensive modification of the Committee's initial criteria), and to discover the degree of consensus that existed within the Committee, matrices bearing the program elements of each subfield (see Chapter 4) as rows and the intrinsic and extrinsic criteria as columns were prepared.* The structural criteria were not included in this exercise, because, to a much greater extent than those of an intrinsic and extrinsic character, the structural questions are of a detailed nature, applicable to each specific project, and they change rapidly with time. In arriving at ultimate decisions on program emphases, these structural criteria must be given due weight; but to apply them effectively a detailed study of the individual program elements, and of the individual research projects in them, is required. Certain exceptional cases in which the structural criteria have direct bearing on our recommendations are discussed below.

With the reports of subfield panels in hand, and following a brief presentation of the program elements by an advocate drawn from the Committee membership, the matrix elements were rated on a 0–10 scale by each

---

* This procedure was applied only to acoustics; astrophysics and relativity; atomic, molecular, and electron physics; condensed-matter physics; elementary-particle physics; nuclear physics; optics; and plasmas and fluid physics. The interfaces such as chemical physics, biophysics, and earth and planetary physics present special problems and were not considered amenable to this approach. In large measure, this conclusion reflects the fact that the physics components of these interfaces, although very important, are not dominant. Lacking a more comprehensive survey to place the physics elements of these interfaces in better perspective, we have not considered ourselves competent to carry out a similar jury rating.

Committee member. These ratings were subsequently combined to obtain the averaged matrices for each subfield.

The results of these ratings were examined from a number of viewpoints. Obviously, a first overall check on this approach is the extent to which the results were in accord with the overall intuition of the Committee members. To examine these questions a variety of histograms were drawn from the averaged rating data (Figures 5.4 through 5.12).

Each of these histograms presents the averaged Committee rating according to the various intrinsic and extrinsic criteria. To correspond to the relative weighting to be assigned to the two sets of criteria, the extrinsic criteria are plotted with one half the column width used for the intensive criteria.

Further, inasmuch as questions 4, 11, and 13 are of a somewhat different character from the remaining ones, they are presented in a separate histogram to the right of each figure.

Figure 5.4 is the average rating histogram plotted in this fashion for the above-mentioned eight core subfields of physics. Several interesting checks, which give a characteristic signature for each subfield, emerge immediately from inspection of this figure. Not unexpectedly, acoustics and optics have a signature that strongly emphasizes the extrinsic criteria, whereas astrophysics and relativity and elementary-particle physics emphasize the intrinsic criteria. The remaining subfields fall between these extremes in ordering; this result was entirely consistent with the intuition of the Committee members. Again, in terms of the ancillary histogram representing questions 4, 11, and 13, astrophysics and relativity and elementary-particle physics were given the highest rating consistent with the corresponding subfield signatures.

Figures 5.5 through 5.12 are the corresponding histogram presentations for each of the core subfields in terms of the averaged Committee rating of each of the program elements. Again the spread in characteristic signatures among the program elements in each of the subfields (except astrophysics and relativity) is relatively large, reflecting a healthy balance within each.

Most important was that, despite widely different backgrounds and interests, the spread in the ratings of the Committee members on individual matrix elements was small. The relatively large number of criteria that the Committee chose to use fostered a more objective evaluation of each individual criterion, which was reflected in the consensus obtained.

To obtain a characteristic score for each of the program elements within an intrinsic–extrinsic framework (but without inclusion of structural criteria for reasons given above), the averaged Committee ratings (with the rela-

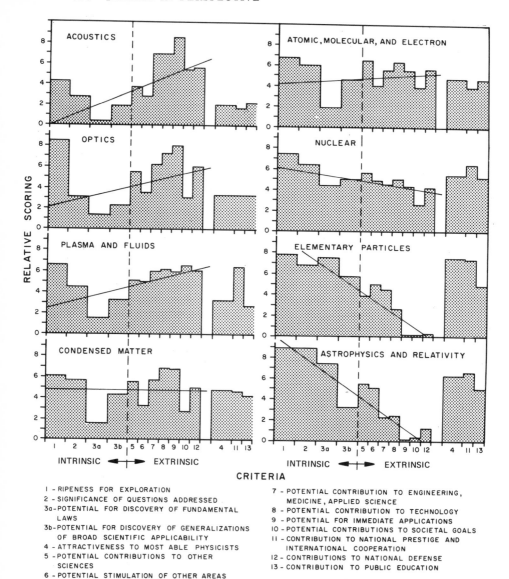

I - RIPENESS FOR EXPLORATION
2 - SIGNIFICANCE OF QUESTIONS ADDRESSED
3a-POTENTIAL FOR DISCOVERY OF FUNDAMENTAL
    LAWS
3b-POTENTIAL FOR DISCOVERY OF GENERALIZATIONS
    OF BROAD SCIENTIFIC APPLICABILITY
4 - ATTRACTIVENESS TO MOST ABLE PHYSICISTS
5 -POTENTIAL CONTRIBUTIONS TO OTHER
    SCIENCES
6 - POTENTIAL STIMULATION OF OTHER AREAS
    OF SCIENCE

7 - POTENTIAL CONTRIBUTION TO ENGINEERING,
    MEDICINE, APPLIED SCIENCE
8 - POTENTIAL CONTRIBUTION TO TECHNOLOGY
9 - POTENTIAL FOR IMMEDIATE APPLICATIONS
10 - POTENTIAL CONTRIBUTIONS TO SOCIETAL GOALS
11 - CONTRIBUTION TO NATIONAL PRESTIGE AND
     INTERNATIONAL COOPERATION
12 - CONTRIBUTIONS TO NATIONAL DEFENSE
13 - CONTRIBUTION TO PUBLIC EDUCATION

FIGURE 5.4 Histograms of the Survey Committee average jury ratings of the core physics subfields in terms of the intrinsic and extrinsic criteria developed in this Report. The straight lines superposed on the histograms are drawn simply to provide a characteristic signature for each subfield. It is interesting to note that these signatures divide naturally into three classes, with emphasis shifting from intrinsic to extrinsic areas as the subfield matures.

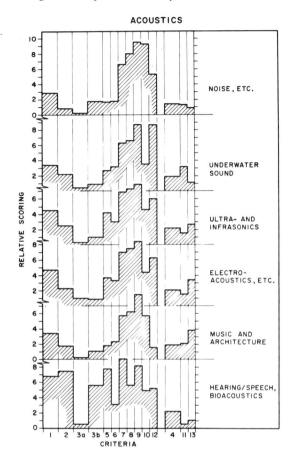

FIGURE 5.5 Histograms of the Survey Committee average jury ratings of the program elements in acoustics. (The detailed legend of Figure 5.4 applies to this figure as well.)

tive weights of 1.0 and 0.5 applied to the intrinsic and extrinsic criteria, respectively) were summed. Appendix 5.A to this chapter is the ordered listing so obtained. We would emphasize that the overall scoring was spread rather uniformly over the entire range covered by this listing; therefore, *within any restricted area of the listing, the relative ordering should not be considered significant.*

Inasmuch as the first four items on this overall score listing alternate between those having marked extrinsic and intrinsic ratings, respectively, we were again encouraged to believe that application of our criteria gave due weight to both extrinsic and intrinsic considerations. Figure 5.13 shows the distribution of these overall weighted scores within the different subfields. It is gratifying to note that the average overall scores for the different subfields all fall within a relatively narrow range from

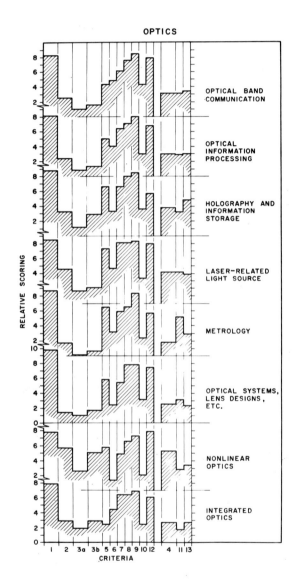

FIGURE 5.6 Histograms of the Survey Committee average jury ratings of the program elements in optics. (The detailed legend on Figure 5.4 applies to this figure as well.)

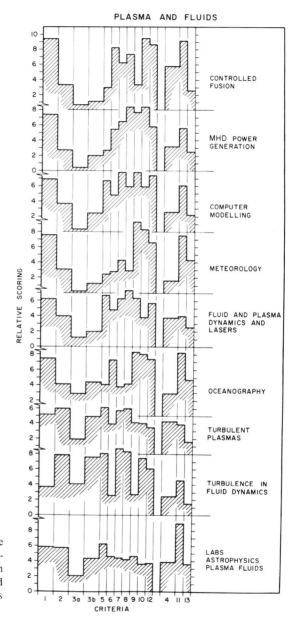

FIGURE 5.7 Histograms of the Survey Committee average jury ratings of the program elements in plasma and fluids. (The detailed legend of Figure 5.4 applies to this figure as well.)

CONDENSED MATTER

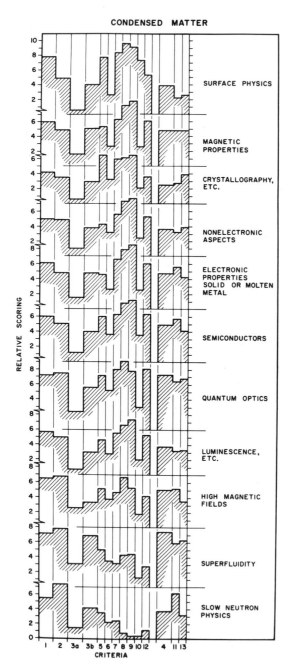

FIGURE 5.8 Histograms of the Survey Committee average jury ratings of the program elements in condensed matter. (The detailed legend of Figure 5.4 applies to this figure as well.)

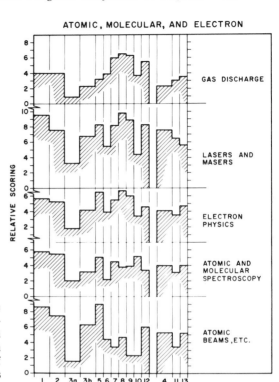

FIGURE 5.9 Histograms of the Survey Committee average jury ratings of the program elements in atomic, molecular, and electron physics. (The detailed legend of Figure 5.4 applies to this figure as well.)

3.3 to 5.0. This finding can be interpreted as an indication of a reasonable balance within physics.

Recognizing that different support agencies and different readers of this Report will wish to apply their own relative weightings to the intrinsic and extrinsic criteria, ordered listings of the scoring for each program element in terms of these criteria considered individually are included as Appendixes 5.B and 5.C. Again *the same qualification concerning the significance of relative ordering in any restricted section of the listing applies.*

In Figure 5.14 the extrinsic versus the intrinsic scores are plotted for each of the 69 program elements of the core subfields that we considered. Again the breadth of the distribution can be interpreted as a measure of the relative health of the total U.S. physics enterprise. It is interesting to note, however, that the distribution on such a plot is not uniform but rather has two regions of relatively higher density. If, for purposes of discussion, a quantity EIR is defined as the ratio of the extrinsic to the

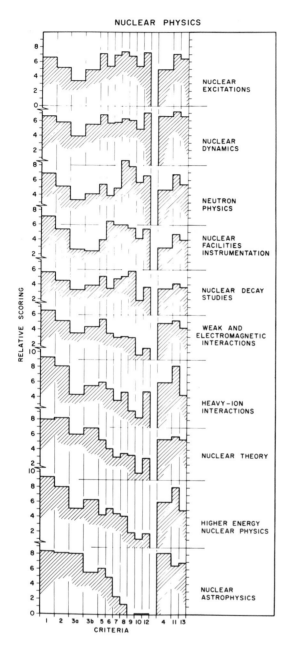

FIGURE 5.10 Histograms of the Survey Committee average jury ratings of the program elements in nuclear physics. (The detailed legend of Figure 5.4 applies to this figure as well.)

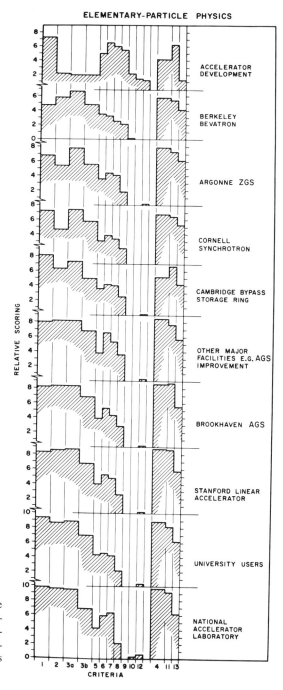

FIGURE 5.11 Histograms of the Survey Committee average jury ratings of the program elements in elementary-particle physics. (The detailed legend of Figure 5.4 applies to this figure as well.)

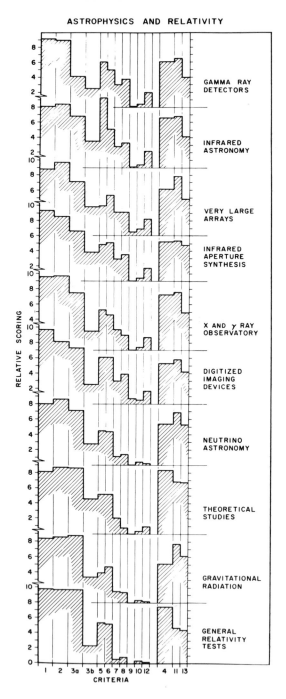

FIGURE 5.12 Histograms of the Survey Committee average jury ratings of the program elements in astrophysics and relativity. (The detailed legend of Figure 5.4 applies to this figure as well.)

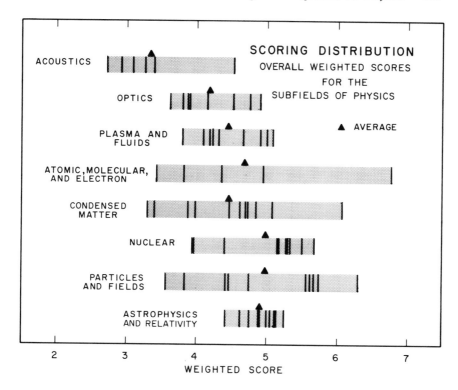

FIGURE 5.13 Distribution of overall weighted scores resulting from the Survey Committee jury ratings of the core physics subfields. Each vertical line represents a given program element; double- and triple-width lines indicate that two or three program elements received the same effective overall score.

intrinsic scale for a given program element, these regions center on EIR values of 1.2 and 0.5, respectively. This finding can be interpreted as evidence of a tendency, noted elsewhere in this Report, for the development of a separation between basic and applied aspects of physics that is not in the best interests of either physics or society. Within the uncertainties inherent in any such plot, the Committee believes that a more uniform distribution of program elements between these two regions would be desirable.

Appendix 5.D presents an ordered listing of the program elements in terms of their EIR rating as defined above. This list adds yet another dimension to our examination of the activity in the core subfields of physics; it is included as a possible input to the program-development process in different agencies and as an item of interest to members of the scientific

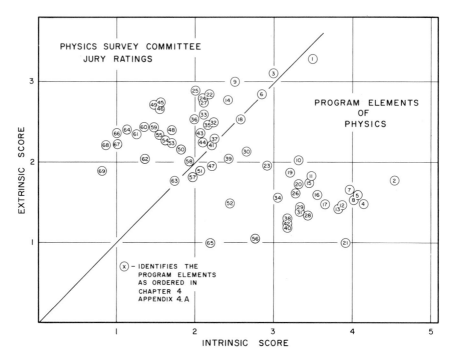

FIGURE 5.14   An extrinsic versus intrinsic map locating the jury ratings of the Survey Committee for each of the program elements in the eight core subfields of physics. The numbered circles are keyed to the overall score-ordered listing in Appendix 5.A.

community. Again *relative ordering in any restricted area of the listing is not significant.*

Figure 5.15 corresponds to Figure 5.13 and presents the distribution of EIR rating within the different core subfields.

We must emphasize once again that these ratings represent the result of a single exercise by the members of one committee and unquestionably reflect to some extent their individual biases and special interests. Nonetheless, the degree of unanimity that was achieved was surprising.

It also cannot be emphasized too strongly that priorities change with time—often very quickly—as suggested in the 1970 National Science Board report to the Congress:

Dynamic, complex, and subtle systems for setting priorities are common in everyday life. A fire in the home or a sick child may instantly change a man's priorities. Such effects also exist in our political sector.

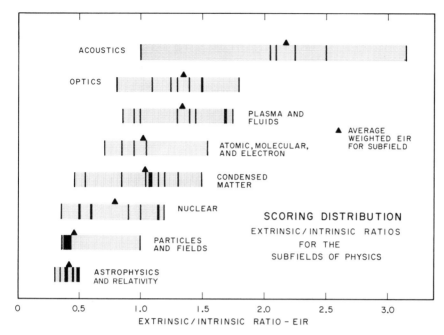

FIGURE 5.15   Distribution of Survey Committee EIR ratings for the core physics subfields. Each vertical line represents a given program element; double, triple, or greater width lines indicate that two, three, or more program elements have received the same EIR.

We believe that the approach we adopted and the criteria we evolved may have rather general utility in providing a somewhat more coherent and objective evaluation of the subjective intuition and folklore of the scientific community that have played an important role in the evaluation of program emphases. The specific listings, and indeed the selection of the program elements in the different subfields, are all clearly open to discussion and argument; they should be regarded as representing only an illustration of this approach to program element evaluation. To the extent that they are based on informed judgment by a group of relatively experienced physicists—drawing heavily on the detailed technical subfield reports—they may be of use in establishing program emphases within the national scientific enterprise.

The Committee's inability to formulate priority allocations based on some fundamental policy or underlying rational scheme does not result from a desire to evade this most critical question. However, after discussing the matter at length and reading the relevant literature, we conclude that an objective, rational, and systematic basis for the allocation of resources to science or within science does not exist at the present time, just as it does not exist in most areas of resource allocation for the public sector. The best that can be produced is an informed judgment based on experience and knowledge of possible developments in the various subfields. The conclusions presented are to be considered as those of a jury, informed but not necessarily impartial. (Utilization of these criteria will be discussed further following the next section of this chapter.)

NATIONAL SUPPORT LEVELS FOR PHYSICS

Thus far we have considered the distribution of a total level of funding, allocated to the national physics enterprise, among the subfields of physics, without explicit consideration of what that total level might be. It is abundantly clear, however, that the distribution is inevitably a strong function of this total level.

Support Levels for Science and the State of the Economy

There has been much discussion in recent years of possible mechanisms for establishing levels for long-term federal support of science; a number of them have been summarized by York.* More detailed discussions of the many problems involved may be found in the series of essays in *Basic Research and National Goals.*† Although now some seven years old and written near the end of a period of unprecedented growth, these essays retain a remarkable validity under present conditions.

What is clear, and a prime topic of this Report, is that the preparation of a longer-range plan for the support of science than has been available until now should be considered of great importance at the highest levels of government if the nation is to damp the transients in both manpower and funding induced by discontinuous changes in federal support such as those dating back to 1967.

Among the most frequently discussed mechanisms for the establishment of long-term support levels for science have been these:

---

* C. York, *Science, 172,* 643 (1971).
† Committee on Science and Public Policy, National Academy of Sciences, *Basic Research and National Goals,* A Report to the Committee on Science and Astronautics, U.S. House of Representatives (U.S. Government Printing Office, Washington, D.C., 1965).

1. Tie the support of science to the gross national product (GNP). (See discussion of the GNP in Chapter 7.)

2. On the basis that the scientific community was in a state of robust health in 1967, tie projected support levels to the GNP beginning at that time; this procedure would lead to an obvious present deficit, which would require rectification by step-funding increments in the near future.

3. On the basis that a healthy U.S. scientific enterprise is of particular importance to the well-being of high-technology industries, and through these, in turn, to the economic health of the nation, tie projected support levels to the productivity of the high-technology sector of the national economy.*

The first two of these mechanisms assume that the support level of an activity such as basic research, which directly or indirectly derives a major fraction of its support from public taxation, should be coupled to the GNP or some such indicator of the state of the national economy. For an activity such as physics, which is a major contributor to many aspects of U.S. society and is related to the GNP through its linkages with technology, for example, such a simple coupling may be wrong, not only in magnitude but even in phase. Two simple examples are pertinent. If the GNP were to drop steadily over a few years, the diagnosis could well be that this reflected the failure, on the part of the nation, to maintain a suitable level of development of new technology—a level that is increasingly linked to the health of the research enterprise. Such a diagnosis would suggest that a *decreasing* GNP should be reflected in *increasing* support of basic research. On the other hand, if the GNP were to rise rapidly for a few years, a tight, in-phase coupling of physics support to it could well result in the unstable dynamic situation in which U.S. physics now finds itself following the rapid growth in the early 1960's. Thus any such direct coupling of physics support—or indeed that of any science— to the GNP is overly simple; when the GNP is not increasing at a reasonably steady rate, such a coupling could have strong negative consequences.

Similar arguments apply to the possible coupling of support levels to the high-technology component of the economy. In this case, an additional problem is the definition of what constitutes high technology; in some discussions the argument has come full circle and high-technology areas are implicitly defined as those that enjoy a positive balance-of-trade posture!

Although a number of areas in physics can be correlated with industrial

---

* *Technology and International Trade: Proceedings of the Symposium Sponsored by the National Academy of Engineering at the Sixth Autumn Meeting, October 14 and 15, 1970* (National Academy of Engineering, Washington, D.C., 1970).

activity, and such concepts as return on investment can be used, examination of the above suggestions in turn reveals no objective mechanisms for determining an appropriate level of resource allocation even in these cases. Even more serious difficulties are encountered in assessing the proper level of research to be undertaken in areas at the forefront of scientific exploration (for example, elementary-particle physics or astrophysics and relativity, which have no present or even foreseeable direct impact on technology and industry). There it becomes necessary to rely on the difficult process of estimating what resources are required to maintain these subfields at an appropriate level of vigor. These subfields interact in a complex manner with other subfields of physics that are much closer to the technological and industrial enterprise, but it is impossible to do more than make informed estimates as to what they require to maintain a healthy and vigorous state.

Other Considerations

The problems are compounded by the fact that, in practice, funds made available to the subfields of physics are not interchangeable. Lack of recognition of this fact has already led to tensions within the physics community and even within subfields of physics. There have been recent publications that both implicitly and explicitly questioned whether a substantial component of the present funding for space and elementary-particle physics should not much better be redirected toward subfields such as condensed-matter or atomic physics in which the connection to societal needs is much more obvious. In the latter case, elementary-particle physics, it has been argued by some that by deferring the start of the National Accelerator for some months it might have been possible to avoid the closing down of the Princeton–Pennsylvania Accelerator and the operating restrictions that have been necessary at the Stanford Linear Accelerator. Similarly, in nuclear physics, some have argued that by deferring start-up of the Los Alamos Meson Physics Facility it would have been possible to avoid the recent termination of a number of federal grants and contracts supporting large existing facilities in this subfield. We believe that such arguments are unrealistic, and that they do not properly reflect the extent to which the Congress responds to individual opportunities in science rather than to science as such. Moreover, in the two specific subfields cited in these examples, the panels have reaffirmed the importance, from the viewpoint of the internal logic of their disciplines, of moving forward vigorously on the new frontiers that these major new facilities make accessible.

A further important point is illustrated in nuclear physics in which a significant fraction of the basic research is supported by the Reactor

Division and the Division of Military Applications of the AEC, in fulfill-
ment of their long- and short-range mission objectives. In any recommendation concerning the redistribution of funds aimed at optimizing the
overall health and balance of nuclear physics, redistribution of this rather
specific mission support might also appear desirable. This approach,
however, may frequently be unrealistic because of the limitations on
overall program flexibility; a significant portion of the work stems from
mission requirements, and some of it must be done in-house. The problem
of competing objectives is in no sense unique to nuclear physics.

However, this point does raise a central issue: It is extremely important
for science, physics, and the nation that a multiplicity of support channels
be maintained. Thus, while much research, particularly that which is
more extrinsic in nature, will find support from one mission agency or
another, the National Science Foundation (NSF), the mission of which
is the support of basic science, is able to support the more highly intrinsic
research so necessary for a balanced program and the health of science.

There are other problems inherent in any attempts to make recommendations for the long-term support level for physics—or indeed for
any science. To what aggregate of activity does the recommended total
refer? If it is all academic science, how is the funding of national centers
to be included? Is funding for user groups at these national centers
included in the total, or is only that fraction of the funding used by academic groups included? How is industrial support to be taken into account
—as in condensed-matter physics, for example, in which industrial and
federal support are roughly equal in the United States, or in optics and
acoustics, in which the major support is from industrial sources? If part
of the funding of the National Cancer Institute were to be used to support
a pion irradiation facility for a university-affiliated hospital, would this
count as academic physics funding? How is funding to universities under
the NSF–RANN (Research Applied to National Needs) program to be
counted, when some of this funding supports basic as well as applied
goals?

Because information on the level of industrial funding of the different
subfields is frequently unavailable, reflecting internal policy decisions
generally related to the handling of proprietary information, the Survey
Committee has directed its attention to federal support of basic physics
research, which is the major source of funds.

History of Physics Funding

Figure 5.2 presents the level of federal support of physics during the
fiscal years 1959 through 1972, with a partial breakdown into the major
subfields. Also shown for comparative purposes are the recommendations

TABLE 5.2   Operating Costs for U.S. Basic Physics Subfields in Fiscal Year 1970 [a]

Physics Subfield	Federal Funding ($Million)	Percentage of Total Federal Funding (%)	Estimated Industrial Funding ($Million)	Total Federal and Industrial Funding ($Million)	Percentage of Total Federal and Industrial Funding (%)
Acoustics	14	3	1	15	3
Astrophysics and Relativity [b]	60	13	0	60	11
Atomic, Molecular, and Electron [c]	13	3	7	20	4
Condensed-Matter	56	12	80	136	24
Nuclear [d]	73	16	2	75	13
Elementary-Particle	150	33	0	150	27
Plasma and Fluids	77	17	10	87	16
Optics	12	3	7	19	3

[a] The interfaces, physics in chemistry, earth and planetary physics, and physics in biology, are not included in this table because the panels have been unable to develop equivalently well-documented funding estimates for these fields. In physics in biology, only rough limits can be established. The membership of the Biophysical Society now stands at 2500, and at $50,000 each, this would correspond to an annual total (federal and industrial) funding of $125 million. Most of this work is applied, however. In the Physics Section of the National Register some 200 persons are identified as biophysicists; on the same basis this would put a lower limit of $10 million on the funding in this area. The Panel on Physics and Chemistry places the total support of chemical physics between $150 million and $200 million. Federal funding of the physics component of earth and planetary physics may be of the order of $200 million.

[b] Only that part of the subfield which is heavily physics-related is included.

[c] A substantial amount of activity is supported from sources outside the AME subfield, e.g., plasma physics and chemistry. If included, this may double the above numbers.

[d] The federal entry includes $12 million in funding for nuclear physics supported under chemistry.

of the Pake survey report,* published in 1966, and the funding level that would have been reached following fiscal year 1967, had an annual growth rate of 5 percent been possible.

Table 5.2 is a more detailed listing of the distribution among subfields of *federal* funds for physics in fiscal year 1970. In three interface areas—earth and planetary physics, physics in chemistry, and physics in biology—the best available data pertinent to the components related specifically to *physics* are included in a footnote to the table. At best, these data are only approximate. In earth and planetary physics, small changes of

* Physics Survey Committee (G. E. Pake, chairman), *Physics: Survey and Outlook* (National Academy of Sciences–National Research Council, Washington, D.C., 1966).

definition within the NASA program—for example, whether the cost of space vehicles is included in the direct research funding—can give the appearance of very large changes in the total support of research at this interface.

Distribution of Funds among the Subfields

Again as an exercise, the Committee members voted independently on the percentage distribution of support that they would wish to see in fiscal year 1977. This exercise was intended as a search for consensus and, to the extent that such consensus existed, as guidance in the formulation of more specific funding recommendations for the fiscal years 1973–1977. A remarkable degree of consensus was found. When faced with the requirement of recommending a specific distribution of available funds to maintain the overall health of U.S. physics, the Survey Committee votes indicated a distribution surprisingly similar to that in fiscal year 1970 and consistent with the rather narrow spread in the average score assigned by the Committee to the different subfields. This result does not mean that the Committee advocates the *status quo* but rather suggests that the complex political and marketplace processes that have shaped the existing balance in U.S. physics have worked rather well. The Committee did recommend small increases in support for astrophysics and relativity; plasma and fluids; and atomic, molecular, and electron physics at the expense of corresponding decreases in the three subfields with the largest fractions of federal support in fiscal year 1970. This recommendation reflects quite different judgments about the three subfields for which increases were suggested.

For astrophysics and relativity, the increase reflects a Committee judgment concerning the potential for major new and fundamental understanding of the physical universe. In plasma and fluid physics it reflects a corresponding judgment, particularly in the area of controlled fusion, that the subfield is ripe for exploration and that major societal advantages are possible. In atomic, molecular, and electron physics it reflects a Committee judgment that, having experienced a decline prior to the discovery and development of lasers and masers, this subfield still has not returned to a level of support commensurate with its potential for both new discoveries and, particularly, practical applications.

Development of Contingency Alternatives

It was recognized from the outset of the Survey that, in view of competing claims on the discretionary component of federal resources in any given year, it may not be possible to allocate to any given field the support that would permit it to make optimum progress. Therefore, a range of contingency alternatives in each subfield has been developed, representing an assessment from the physics community of means of obtaining the most effective utilization of whatever support becomes available—most effective from the viewpoint of the overall health of physics and of the contribution that physics can make to U.S. society.

Accordingly, the initial charge (Appendix B to this Report) to subfield panels requested that they develop programs as detailed as possible for their subfields under various assumed funding projections: (a) a so-called exploitation budget that attempts to exploit all perceived opportunities, both intrinsic and extrinsic; (b) a level budget—level in dollars of constant buying power; and (c) a declining budget—declining at an arbitrarily established rate of between 6 and 7.5 percent per year. To obtain the hoped-for interpolation possibilities, it was necessary to evolve an intermediate growth-rate budget between the exploitation and the level budgets. In each case the panels were asked to emphasize the costs to science and the nation of reductions to levels below the exploitation budget. This exercise required very detailed examination of the internal structure of each subfield and of its opportunities and needs. It also required sharp scrutiny of the internal priorities of the subfield.

Development of such budgetary projections was more easily accomplished in some subfields than others. In subfields such as elementary-particle physics and astrophysics—and to an increasing extent in nuclear physics—research activity is largely centered on major facilities. It is characteristic of such facilities that a large fraction of their total operational costs are invariant to the extent that the facility is maintained in an operational state; support and developmental staffs, power for magnets, and radio-frequency and other systems must be provided unless the facility is closed down. This situation is reflected in a very large leverage factor; what appear to be very small percentage changes in the overall operational budgets of such facilities can be reflected as major fractions of the discretionary component of these budgets—the fraction that is applied directly to the pursuit of research and not simply to keeping the doors open.*

In such heavily facility-based subfields, reductions below the exploitation budget typically involve the closing down of entire facilities or at least major changes in the style and scope of operation that is possible. Such

---

* An excellent discussion of some of the problems involved here is given in the essay by Brooks in the above-mentioned *Basic Research and National Goals;* the problems that he addressed in 1964 are simply much more acute in 1971. Additional discussion of this same topic—but from a quite different viewpoint—appears in the essay by Kaysen in the same publication.

results lead to a corresponding reduction of the manpower that the sub-field can accommodate, quite apart from possible opportunities for new personnel now being trained. The dislocation and career disruptions involved here for excellent scientists and support personnel (detailed in Chapters 6 and 12) are a waste of resources that in our opinion the nation can ill afford.

In subfields such as condensed-matter and atomic, molecular, and electron physics the effects of budgetary reductions are less obvious and the manpower problems less acute. Because the research is much less facility-intensive, reduced funding means that the objectives of each scientist or scientific group are lowered—less work is done, fewer challenges are met, and the subfield slows down. While this process can proceed for a time without overt symptoms of serious trouble, trouble is there; the morale and enthusiasm dwindle, and the subfield is less able to respond to challenges or opportunities. Quite apart from these differences, however, all subfields (see Volume II) have concluded that a budgetary level declining at 7.5 percent per annum would, within five years, bring them below that critical point at which productivity, however measured, falls dramatically. An analogy or two are illustrative.

Following World War II, Montgomery Ward, in its continuing competition with Sears Roebuck in the retail business sector, made the decision to develop a strong cash position through strict reductions of both operating and capital expenditures. This step included reduction of expenditures on marketing and advertising and the elimination of plans for the construction of new stores. The result was not what Montgomery Ward anticipated, for a major fraction of the company business was lost irretrievably to Sears Roebuck, and Montgomery Ward found itself essentially out of the mail-order business.

Another useful analogy is one used in the past by the Soviet physics community in presenting their case for Soviet governmental support to enable them to compete with U.S. activity in elementary-particle physics. They described two ships exploring an unknown river, both traveling at almost the same speed but with one slightly ahead. The ship that was ahead saw all the new scenery first and made all the new observations.

It is most important to keep this analogy firmly in mind in planning for science. The point is that a slight extra margin of capability or investment may make all the difference between making new discoveries and merely confirming or refining the discoveries made by others. The productivity of the scientific enterprise is highly nonlinear, just by virtue of the fact that it is competitive. It need hardly be emphasized that this nonlinearity not only applies to international competition but also determines, to a marked extent, the productivity of any scientific group inside or outside of any country. In terms of the values and norms of the scientific social system, the marginal utility of a confirmatory discovery is very much less than that of an original discovery; unless there is a level of support for a field sufficient to enable it to keep pace with world-wide standards of performance, its chances of making original discoveries are small indeed.

Psychologically, it is very difficult to maintain a position of second best if that position is reached through falling back from the first position rather than attaining it during continuing progress toward the leading position, as is currently characteristic of Europe, Japan, and the Soviet Union in relation to the United States. We believe that the interests of both the nation and physics will continue to be best served by the maintenance of physics as a vigorous enterprise at or near the frontiers of activity in each of its various branches.

In physics as in any science, a nation's standing in the international community at any point in time is largely determined by its contributions to the development of new areas of research. Clearly, the United States should contribute to the development of such new areas of research, or growth points in physics, and exploit them, but it cannot do so when budget constraints force drastic contraction of the base from which these growth points emerge.

We return then to the specific question of the appropriate support level for a scientific field such as physics. In summary, as indicated previously, there does not appear to exist any objective mechanism that can be used to define this level properly in terms of considerations wholly external to the field. Therefore, we have examined carefully the levels that are indicated by internal considerations, applying in regard to each all the criteria—intrinsic, extrinsic, and structural—discussed earlier in this chapter.

An Exploitation Funding Level

Analysis leading to this funding level has been carried out in detail by the Panels on Elementary-Particle Physics and Nuclear Physics. These subfields, in which activity centers on relatively large accelerators, are ones for which the consequences of a given level of support are easier to specify than for subfields such as condensed-matter physics, in which research requires less dependence on major, shared facilities. Many aspects of astrophysics and relativity have the same facility-oriented character as do elementary-particle and nuclear physics, but in the former it is difficult to separate the physics and astronomy parts of the funding, to say nothing of the difficulties in properly allocating NASA support in this subfield. Accordingly, the discussion focuses initially on elementary-particle and nuclear physics, for which statistical data are complete and budget projections most detailed.

Both the Panels on Elementary-Particle Physics and Nuclear Physics have arrived at exploitation budgets growing at the rate of about 11 percent per year for the period through fiscal year 1977; the detailed analyses leading to this common level have been distinctly different in the two subfields (see the panel reports in Volume II). In both cases, however, a significant shift of activity to major national facilities is envisaged—to the National Accelerator in elementary-particle physics and to the Los Alamos Meson Physics Facility and a hoped-for national heavy-ion science facility in nuclear physics. In both cases, too, this shift would involve effective or actual termination of smaller facilities on university campuses and a corresponding change in the style of the typical research program.

During the five-year period for which these panels prepared detailed forecasts and descriptions, the exploitation funding budget would bring the level of activity in these subfields to about that which could have been sustained if support had increased at a rate of 5 percent per year since 1967. The annual increase of 11 percent should be regarded as a rebound or catch-up phenomenon to allow these subfields to regain a state of health and vitality sufficient to optimize their contributions to society and to provide a solid base for future activity and productivity. As discussed below, and as is immediately obvious, such a rate of growth could not continue indefinitely.

Exploitation of the opportunities now available in both cases requires vigorous exploration of the new frontiers and aggressive capitalization on the possibilities now in the early stages of development. While the programs underlying the exploitation budgets in these subfields were evolved without direct reference to the availability of trained manpower, it is clearly essential that this aspect receive full consideration in determining the extent to which the proposed exploitation budgets are realistic. A detailed manpower study, based on actual counts of students now in the educational pipeline and to be graduated during the period ending in 1977, and the assumption that the fraction of those physicists trained in these fields who remain in them would remain the same as in 1967 (see Chapter 12) if adequate funding levels were restored, leads to an annual increase of 8 percent in the number of scientific man-years in each field.

The residual 3 percent difference between the 8 percent manpower growth rate and the projected 11 percent funding growth rate may be viewed as reflecting the effective escalation factor of the real cost of doing research. The available statistical evidence bearing on this question is not complete; however, a recent NSF study indicates that price inflation accounted for about a 50 percent increase in the *direct* costs of academic research and development over the period 1961–1971.* Most of this increase occurred during the last five years when the average inflation rate was approximately 5–6 percent a year—a faster rate than the GNP deflator until 1969, about the same or slower since. The main factor in the change has been the decline in PhD salary increases since the emergence of a much more competitive employment market. Since PhD's may well remain in relative surplus, or at least not in tight supply for the next several years, research cost inflation is likely to be less than that in the cost of living, because a major fraction of the total cost appears in the form of salaries, and these almost certainly will decline relative to the general wage level. In some instances, particularly in big science, this factor may be offset by rapid escalation of power costs due to environmental considerations. This factor will apply to accelerators and other installations having very high proportions of power cost.

Reflecting the greater flexibility that characterized the remaining core subfields of physics, in which, in the absence of major facilities, relatively large fluctuations in support are reflected in equivalent expansion or contraction of the scope of individual or research group activities, the remaining panels have not found it possible to evolve correspondingly detailed budgetary projections. Inasmuch as elementary-particle and nuclear physics together represent some 49 percent of the total federal funding for subfields of physics indicated in Table 5.2, it appears reasonable to adopt the 11 percent per annum exploitation budget level recommended by these panels as appropriate to *all* physics.

As a first check on the more general appropriateness of this growth rate, the projected overall physics manpower situation was examined, drawing on the discussions in Chapter 12 and on the study of the Grodzins Committee.†

In considering an exploitation funding level, the situation characteristic of 1967, when some 90 percent of the new PhD physicists could be considered as a new increase of those active in the field, serves as an appropriate base. That the number of new PhD's in physics in the United States will stabilize at roughly 1500 per year during the period through 1977 is also a reasonable assumption in view of the number of students working toward PhD degrees.

Grodzins has found that of the 20,000 PhD physicists in the United

* *Science Resources Studies Highlights,* NSF 71–32 (National Science Foundation, Washington, D.C., November 1, 1971).
† *The Manpower Crisis in Physics,* Special Report of the Economic Concerns Committee (American Physical Society, New York, N.Y., April 1971).

States, 10,000 are in universities, 5000 in industry, and 5000 in government and other laboratories and federally funded research and development centers. For the purpose of developing rough estimates we considered those in universities and governmental laboratories to be engaged in basic research and those in industry to be not thus engaged; we recognized that this is clearly *not true,* but we assumed that the basic researchers in industry are effectively balanced by the applied researchers in governmental laboratories. In converting to effective scientific man-years (SMY), a factor of 0.5 was used for university workers and 1.0 for those in government laboratories, resulting in a total effective 10,000 SMY (PhD level) now engaged in basic physics.

Assuming that 90 percent of the 1500 new PhD graduates per year are retained in physics, we have a total of 1350 new physicists annually. Converting to scientific man-years by the same factor (10/15) used above results in the 900-SMY effective increase per year—a 9 percent manpower increase overall as compared with the 8 percent quoted above for elementary-particle and nuclear physics. With such a 9 percent manpower increase, the projected exploitation budget growth rate of 11 percent per annum then corresponds to a minimal 2 percent allowance for escalation in the actual costs of doing research.

The net outcome of this exercise, however, is that the 11 percent annual growth rate projected for *all* physics, on the basis of the detailed elementary-particle and nuclear physics panel studies, is consistent with the trained manpower now identified as potentially available for the period through fiscal year 1977.

It is pertinent to ask whether such a growth rate would not again induce the type of oscillation in support and manpower that characterized the mid-1960's and underlies some of the current problems now facing physics in the United States. Clearly, too, it is unrealistic to anticipate that such an annual growth rate could, or indeed should, be maintained over any extended period unless major steady growth of the national economy were to occur over an equivalent period.

At the same time, it is extremely important to emphasize that, as shown in Figure 5.16, there has been a dramatic turnover in the operational support of U.S. physics since 1967, with a myriad of consequences that are considered in Chapter 6 (and elsewhere) in this Report. Also shown in Figure 5.16 is the fact that, even by 1977, a growth rate of 11 percent per annum will not bring physics to the level that it would have reached had a modest 5 percent annual increase been possible over the period from fiscal year 1967 (even with the inclusion of certain fiscal year 1973 step increases discussed below). The 11 percent growth rate during this period would pay handsome dividends to society in both the

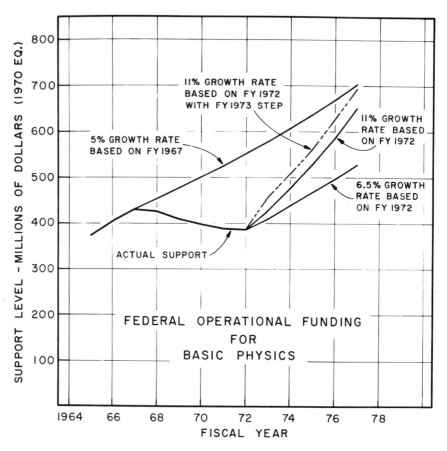

FIGURE 5.16   Total federal support for basic physics; actual expenditures for fiscal years 1965 through 1972 with projections through fiscal year 1977. Also shown for comparison is a 5 percent projection based on fiscal year 1967.

long and short range. Detailed projections beyond 1977 have not been attempted, for physics and its opportunities change so rapidly that such an attempt would be largely pointless.

Long before 1977 there will be changes of which the Committee has no present indication, and it will be necessary to maintain a continuing watch and re-evaluation of the appropriate projected growth rates. With a healthy economy it appears that a new effective growth rate of 5 percent per annum over an extended period would be a reasonable one for science and for physics. The present 11 percent exploitation budget growth rate during the coming five years is an attempt to regain some of the ground lost by U.S. physics in the past five years. Thereafter a smooth transition to something more like the 5 percent figure, or whatever other figure may be more appropriate to the needs and health of the national economy, could be effected.

Because of the difficulties in long-range projection, the Committee has not attempted to depict in detail the exact transition between the rebound growth rate and a more nearly steady-state condition. Clearly this is one of the most important problems that must be addressed by both the physics community and the science-policy components of the federal government within the next four years in the light of developments during the intervening period.

A Funding Level Floor

Can a minimum funding level be established below which substantial severe damage would be done to the U.S. physics enterprise?

Again, the heavily facility-oriented subfields of elementary-particle and nuclear physics, in which the effects of budgetary reductions have been particularly severe and identifiable, are instructive. Under continuing level operating budgets (in dollars of constant value) the consequences would be severe indeed. By 1977, such operating budgets would reduce the effective SMY numbers in the two subfields by about 22 percent and 30 percent, respectively, and would remove both subfields from many important areas of research in any internationally competitive sense. Operating budgets declining at the rate of 6 percent per annum (documentation is presented in the respective panel reports of Volume II) would force termination of large segments of the U.S. enterprise in both subfields by 1977. U.S. aspirations, as suggested elsewhere in this Report, would be reduced to a qualitatively different level. Instead of working at the forefront of these subfields and participating in many new discoveries, the physics community would be reduced to holding together the best possible response capability so that it hopefully would be able to understand the new discoveries made elsewhere and use them for the benefit of U.S. society.

The impact on the other subfields would be almost equally great despite their greater flexibility. One of the important components of this reduction would be the demoralization of the overall field and the dismantling and dispersal of important and able research groups that have been developed and even now are being held together in many cases only by the most stringent of emergency measures. Once dispersed, these groups could not be reconstituted in less than a substantial period of years; frequently they could *never* be reconstituted, as the members would respond to other individual opportunities.

It should be emphasized, too, that in the past five years, following the growth period of the 1960's, there has been a continued period of belt tightening, of readjustment, and, more important, of development of emergency measures based on the hope that difficulties were temporary. These cannot continue any longer without permanent damage.

It is important to note here that in assessing the support of research in physics the tendency has been to talk about *total* support, including indirect costs and fringe benefits. When support was growing, direct costs tended to remain approximately proportional to total costs, and indirect cost rates remained reasonably constant. Since the period of declining budgets began, direct costs have been falling considerably faster than total costs. There is also considerable inertia built into indirect costs, because many of them represent long-term commitments based on an enterprise of a particular size. In general, when the federal agencies have a fixed amount of money and indirect cost rates go up, the investigator has no choice but to reduce direct charges to accommodate the increased university overhead rate. The consequence of this practice is that the direct cost base on which overhead charges for the following year are calculated has been overestimated each year of declining budgets, with the result that the indirect cost escalates in the following year. Should present trends continue, the effect on the direct cost base would be disastrous.

For all these reasons the Committee believes that it would be abdicating its responsibility to the nation, quite apart from any responsibility to physics, were it not to make the strongest possible case that the nation cannot afford to allow the effective support of physics to continue to decrease. Thus the level funding situation is regarded as a minimal floor below which U.S. activities in physics would no longer be in any sense competitive and which would be totally inadequate to the role they have long played as a source of technology, a fundamental base for and stimulator of other sciences, and a vital component of education.

Situations in Which Structural Criteria Have Major Impact

Certain particularly troublesome problems are at hand. Because of continuing funding limitations and commitments made in a more expansionary climate in the mid-1960's, at least two subfields of physics will be in extremely difficult situations in fiscal year 1973 unless special and immediate incremental funding can be made available. These situations occur in the most facility-intensive subfields and involve the largest and most costly facilities in each of these—the Los Alamos Meson Physics Facility (LAMPF) in nuclear physics and the National Accelerator Laboratory (NAL) in elementary-particle physics.

The Los Alamos Meson Physics Facility The LAMPF is an 800-MeV linear proton accelerator with an average proton beam intensity of 1.0 mA and major beam power on target of 1.0 MW. As such, it will be the world's highest-intensity source of mesons and neutrinos in its energy range. Its high-intensity proton beams will be uniquely powerful probes for previously suspected but inaccessible nuclear phenomena. Its unique research capabilities will range from entirely new frontiers such as basic nucleon–nucleon interactions, the correlations of nucleons in nuclear matter, and pionic and muonic atomic studies to extremely applied areas such as clinical treatment of cancer with pion beams, very-high-energy neutron studies of great relevance to national defense, much of contemporary nuclear physics, and lower-energy elementary-particle physics.

Construction of LAMPF, at a cost of $56 million, was authorized in fiscal year 1968, following the positive recommendation of an Office of Science and Technology panel chaired by H. A. Bethe, and construction was initiated in the third quarter of fiscal year 1968. An essential part of the Bethe panel's recommendation leading to this endeavor is the following:

We feel however that this [the Meson Facility] should not reduce the support of nuclear structure laboratories now in existence or under construction and we, therefore, recommend an increase in the total support for nuclear structure physics.

As yet such incremental operating funding has not become available, and nuclear physics is faced with an extremely difficult situation in fiscal year 1973 when LAMPF is expected to become operational. A full experimental schedule is anticipated by the middle of fiscal year 1974. Figure 5.17 shows the actual current costs and projections for operating plus capital equipment costs for LAMPF and its users for the period fiscal year 1969 through 1977 and the associated construction costs during this same period. A large part of the $7.0 million increment in operating costs between fiscal years 1972 and 1973 reflects electrical power and maintenance costs; these naturally increase sharply when the facility is first turned on and can be expected to rise only slowly in subsequent years.

The first half year of operation (last half of fiscal year 1973) will involve only an advance guard of the number of users ultimately expected. In fact, of the $7 million increase between 1972 and 1973, only $1.5 million is allocated to increased user operations. Since the users will be transferring from much less expensive installations, the saving from closing the latter, when this occurs, will be only a fraction of the $1.5 million and very small compared with the $7.0 million increment required.

The present total budget of the community of which the LAMPF operation will be a part is about $50.3 million (AEC low-energy, medium-energy,

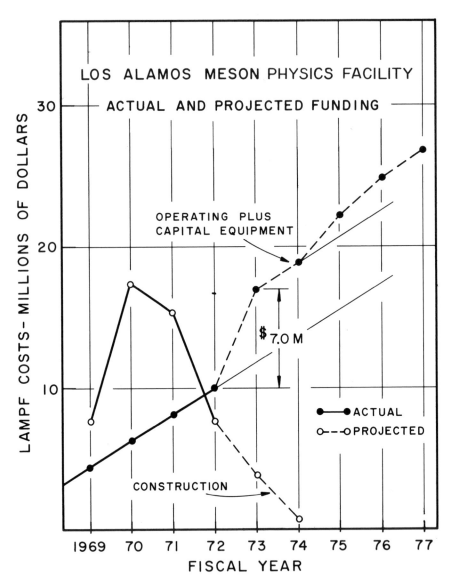

FIGURE 5.17 Operating and construction costs—actual and projected—for the Los Alamos Meson Physics Facility.

NSF nuclear physics, operations plus capital equipment). To attempt to extract $7 million, or 14 percent, from the budget in one year, following the stringent reductions that have been implemented in this subfield in recent years, would be highly detrimental to the whole nuclear-physics enterprise and would raise the question of whether it would be better to abandon LAMPF instead. Because of the complementarity between small and large facilities, it is scientifically unsound strategy to phase out all small facilities to finance new frontier equipment. The smaller facilities are needed to prepare and test experiments before they are committed to the large facilities and also to exploit in more quantitative and painstaking detail some of the discoveries first made with frontier facilities. In other words, new and more powerful facilities are at most only a partial substitute for older facilities. The situation is not like a transistor in place of a vacuum tube, where the new functionally replaces the old. Rather, the new frontier equipment adds a new capability. The early discovery of pulsars by radio telescopes was followed up in large measure by old-fashioned optical telescopes. The first clues to the failure of parity were discovered on a frontier machine at Brookhaven, but the matter was clinched by a quite classical investigation in beta decay.

On a smaller scale, other new facilities authorized in the mid-1960's, and currently in final stages of construction or ready to operate, face the same dilemma. Including the major new facilities—the Super-HILAC at the Lawrence Berkeley Laboratory, the double MP Tandem at Brookhaven, the ORELA at Oak Ridge, and the MIT Electron Linac, in addition to LAMPF—the increment of required operating funds between 1972 and 1973 is $8.7 million, and in 1973–1974, $2.5 million. Again, a large portion of these increments occurs as operating costs for the new machines, not as direct research costs, and only a small fraction will be saved through shutdown of other facilities, which otherwise would increase the productivity of the new facilities.

National Accelerator Laboratory   The NAL is a nominal 500-GeV proton synchrotron and as such is by almost an order of magnitude the highest-energy proton accelerator yet built anywhere. It opens the entire higher-energy frontier of man's search into the ultimate structure of matter. If history is any guide, major progress as well as startling new developments can confidently be expected when this facility becomes operational.

It was authorized on July 11, 1967, as a 200-GeV accelerator facility, at a total construction cost of $250 million, following the recommendation of the GAC–PSAC Panel on High Energy Accelerator Physics, chaired by Norman Ramsey, and construction was initiated on December 1, 1968. That problems would arise in the support of so large a new facility had been recognized from the outset. The Ramsey Panel reported in 1963 as follows*:

The methods of high energy physics research are undergoing profound changes at this time, and decisions will have to be weighed most carefully to set a wise course. In general terms, the difficulty is that the results of the past point toward more complex installations, with resultant problems in management and great increase in cost. Because of their high costs in money and manpower there will be fewer leading centers than there are now. For this reason, with the increasing number of very able research workers entering the field, it appears that more and more high energy physicists will participate in the work of fewer centers. At the same time, the direct participation of our foremost universities, which have played a key role in high energy physics in the past, become progressively harder to maintain. These conclusions are unavoidable, but the negative effects can and should be minimized by the following steps:
   1. Preserving the vitality of existing centers by updating their facilities wherever possible and desirable.
   2. Designing and constructing more extensive facilities ancillary to the principal accelerators to permit more noninterfering uses.
   3. Preserving university ties as closely as practicable by locating such centers where possible near a large university and by strong support of "university users" groups.
   4. Providing a balanced program, with different centers equipped to explore those frontiers of research opened up by extension of different accelerator parameters (energy, intensity, kinds of particle, etc.).
   The intent is to present a balanced program for the next decade and beyond which contemplates support of presently operating accelerators, improvement programs for presently operating machines, new design and construction starts, and support of new accelerators. The program also foresees the eventual termination of some current projects.

As in nuclear physics, however, this subfield is faced now with extraordinary difficulties because of the funding requirements of NAL, which has been scheduled to begin research operation early in fiscal year 1973. Figure 5.18 shows the actual and projected funding for NAL for fiscal years 1969 through 1977.

To provide the $19 million of operating and capital equipment funding required for fiscal year 1972, within the total available for elementary-particle physics from the AEC, it has been necessary to continue the reduction in funding for the AEC's remaining major facilities, a steady reduction that has gone on during the past five years. These other major facilities have experienced a decrease of some $14 million in real dollars since

---

* See Appendix 1 of High Energy Physics Program: Report on National Policy and Background Information, Report of the Joint Committee on Atomic Energy to the Eighty-Ninth Congress, 1965.

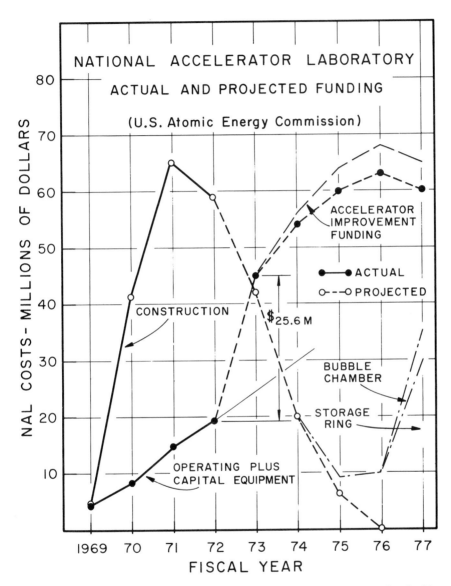

FIGURE 5.18 Operating and construction costs—actual and projected—for the National Accelerator Laboratory.

fiscal year 1970. As shown in Figure 5.18 an increment of some $20 million in operating and capital equipment funding will be required in fiscal year 1973 for NAL when its research program becomes fully operational. If the additional funds are provided entirely at the expense of the operating budgets of the remaining facilities, the effect on them will be drastic indeed. It is difficult to see how the ZGS and the Bevatron could survive; such a loss of two of the nation's four proton accelerator facilities would severely damage the national effort in elementary-particle physics. These facilities are needed for continuation of fundamental research with beams at energies that cannot be provided economically otherwise. Furthermore, they play critical roles in the development of the subfield in terms of training and innovation. Damaging though it would be, however, the shutdown of the ZGS and the Bevatron would be less detrimental to the high-energy physics program than the continued starvation of NAL, Brookhaven, and the Stanford Linear Accelerator.

In these two instances alone (NAL and LAMPF), unless $7 million and $20 million can be made available to the two newest, largest, and internationally unique facilities in their respective subfields, they will be unable to operate at anything resembling full capacity, and the return on the major investments that these facilities represent, in terms of both money and scientific and engineering talent, will be very much reduced. These two subfields have already made, and are prepared to make, very difficult decisions to cut deeply into their ongoing programs at other facilities to permit as broad an exploitation of the new facilities as possible under existing funding limitations. However, because the increment required in fiscal year 1973 for these facilities represents such large fractions of the total funding for the subfield activities, unless substantial incremental funding is made available, over and above that required by the presently much reduced programs, both subfields will suffer irretrievable losses in manpower, facilities, and capability. Even though more adequate funding may become available later, it would take years, and a much greater expenditure, to regain a level approximating the present relative position of the United States in the international community.

High-Leverage Situations

Small changes in funding—either increases or decreases—can sometimes be reflected in disproportionately large changes in scientific productivity. (This concept of leverage is discussed in more detail in Chapter 6.) In the case of major facilities, such a large fraction of the total funding is required to keep them in operation that even small fractional changes in support are reflected in very large changes in the research component, to which scientific productivity is much more directly coupled. In subfields in which new breakthroughs, either in concepts or in instrumentation, have occurred, new frontiers are opened and investment in research at those frontiers can be expected to yield a high scientific return. In other subfields, again because of breakthroughs in instrumentation or ideas, or because the internal development of the subfield has reached a state at which further investment can be expected to yield returns of major importance to society, the leverage is high.

The relative weighting or importance assigned to each of these types of leverage will vary among subfields and from one support agency to another. This is healthy and proper. In examining program elements as candidates for high-leverage consideration, structural criteria play an important role. It is here, for example, that continuity considerations enter explicitly.

As illustrations of various types of high-leverage situation and the use of our Committee-ordered listings in combination with the subfield panel reports, the Committee has selected 15 program elements from the subfields whose growth potentials warrant high priority for their support in the next five years. These are arbitrarily presented in an order that reflects the PhD manpower employed in the various subfields—for example, condensed-matter physics employs the greatest number of PhD physicists, astrophysics and relativity the least.

It should be emphasized that the increased support recommended for these program elements should not be at the expense of other activities in the subfields, although clearly some readjustment is not only necessary but healthy as the various program elements attain different levels of scientific maturity. At the same time, it should be recognized that should only the selected program elements be supported, the overall physics research program would be totally unbalanced.

Macroscopic Quantum Phenomena This title includes superfluidity and superconductivity. With the development of a comprehensive theory of macroscopic quantum phenomena in solids and liquids, this area has attracted renewed activity because of its intrinsic interest and its insight into the behavior of a many-body system—one of the central open problems in basic physics, with broad applications in many other subfields. It also has important potential for utilization of these phenomena in such areas as low-loss power transmission in superconducting transmission lines, rapid transportation by means of magnetic levitation, and very compact, high-efficiency motors. Measurements on large-scale superconducting systems, until now primarily studies in connection with possible new accelerator designs, have revealed unexpected questions and problems. This is a situation in which increased activity can bring high returns, both applied and fundamental.

Quantum Optics   This area is closely related to that of lasers and masers and shares similar advantages and potentials. Those aspects that are peculiar to condensed matter, however, hold very high promise of important new applications in miniaturized devices, extraordinarily wide-band communications, and high-speed computers, to cite only a few obvious examples. This potential again is entirely apart from the fundamental new insights already gained—and to be gained—from investigations of the basic structure of both solids and liquids.

Scattering Studies on Solids and Liquids   Here are grouped several of the listed program elements in condensed-matter physics—studies involving scattering of neutrons, photons, and phonons in liquids and solids. New techniques and more intense sources have opened entirely new ranges of phenomena, and recent progress toward understanding microscopic short- and long-range order in condensed matter has been rapid.

The urgency of this situation is dramatized by the imminent start-up of the Franco-German center for neutron research at Grenoble. This $75 million facility devoted primarily to slow-neutron interaction with condensed matter suggests that in Europe this subfield is regarded as of prime importance and promise. Although this research originated in the United States, unless drastic action is taken, supremacy will pass to Europe within the next five to ten years.

Heavy-Ion Interactions   Internationally this area of nuclear physics is attracting the greatest interest, effort, and support. Involving the interactions of large pieces of nuclear matter, this research makes accessible, for the first time, entirely new modes of nuclear motion and dynamics and permits study of more familiar phenomena in entirely new regions of angular momentum and other parameters. It also makes accessible new nuclear species—through moving away from the nuclear valley of stability to isotopes as yet unknown and proceeding upward along this valley to possible new supertransuranic elements. Quite apart from the very great intrinsic interest in these new areas, there are potential applications for these new species in medicine, power generation, and national defense. Initiated in the United States, this work is being pursued vigorously by the Soviet Union, Germany, France, and other western European countries, and indeed in all the major nuclear centers around the world, with a wide variety of major accelerator facilities newly under construction. Unless the U.S. community can move rapidly toward the establishment of a competitive national facility, we will cease to have a significant role in this important field.

Higher-Energy Nuclear Physics   In a very real sense, the detailed microscopic study of the nucleus, up to the present time, has focused on the behavior of the outer nucleons. Although extrapolation of these surface findings deep into the nuclear interior has provided useful insight into many nuclear phenomena, until recently the nuclear interior has not been accessible to careful experimental scrutiny. Similarly, measurements in the past at typically available energies have not been able to provide unambiguous information on the very-short-range behavior of the fundamental nucleon–nucleon interaction or on the importance of three-body or more complex possible interactions. With the new facilities, typified by LAMPF and such facilities as the Brookhaven AGS, these phenomena can be subjected to critical study. They are of fundamental importance to further understanding of nuclear phenomena. So also are the experiments in which new secondary meson and hyperon probes are used to study nuclear systems.

The National Accelerator Laboratory and Its Program   As the world's most powerful proton accelerator, this facility truly represents a frontier salient in man's understanding of the ultimate structure of matter. It holds high promise of discovering fundamentally new aspects of nature that can have ramifications throughout science. It represents a cutting edge of science and has attracted the talents of some of the world's most distinguished physicists. Although the coupling to more extrinsic sciences and technology is still relatively remote, it would be shortsighted indeed to conclude that such coupling cannot come about in the future.

The Stanford Linear Accelerator and Its Program   This facility is the world's most powerful electromagnetic probe for study of the structure of matter and is complementary to the NAL, with its strongly interacting proton beams. During the past year, it has been forced to operate substantially below full research capability and, indeed, was closed for two months to keep costs within the limits of available funds. Reflecting the high-leverage factors inherent in any such facility, even small fractional increases in funding will have disproportionately large returns in research productivity and effective utilization of highly trained manpower. The United States has already made a major investment in this accelerator, and its extensive ancillary instrumentation should be fully exploited, for it provides one of the most promising approaches to totally unknown realms of natural phenomena.

Controlled Fusion In view of impressive recent progress, this work warrants greatly increased support. It holds high promise for the development of a new power source with reduced undesirable side effects. With an expected minimal impact on the environment and an inexhaustible fuel supply, availability of fusion power would have enormous beneficial consequences for man. We believe that a major commitment to the achievement of economic fusion power at the earliest date consistent with the orderly progress of the research and development activity in the subfield is justified (see also Chapter 2). Again, this is a highly competitive subfield, with the United States and the Soviet Union as major contenders. Without increased support, it will be difficult, if not impossible, for the United States to maintain a competitive position or even to take advantage of developments elsewhere.

Turbulence This is an area of extreme complexity and difficulty but one of correspondingly great importance in all fields involving fluid flow. The subject has an impressive range from global circulation problems in meteorology and oceanography through phenomena involved in supersonsic flight and shock-tube phenomena to the flow of blood in human circulatory systems. The present level of activity in this area is relatively low. Because of its broad range of potential applications, both within and outside of science, increased activity could yield impressive returns.

Nonlinear Optics This is an area of high leverage because it can provide an effective interface between atomic and molecular physics, condensed-matter physics, and major areas of technology. In addition, it has its own intrinsic potential for progress and new developments. It is important to emphasize, too, the extent to which this work in optics and that in the other areas of quantum optics and laser phenomena are symbiotic, with major progress in one frequently opening opportunities for equivalent progress in the others. Applications from work in this area have only begun, and increased activity holds high promise of both extrinsic and intrinsic rewards.

Lasers and Masers Quite apart from the fundamental new physics intrinsic in these devices and their underlying theoretical understanding, applications and implications of studies of lasers and masers have been remarkably pervasive throughout much of science and technology—and in fields as disparate as medicine and the fine arts. The exploitation of these new devices has only begun.

Atomic and Molecular Beam Studies Work in this area has undergone a renaissance since the so-called atomic or chemical accelerators have become available, providing beams of atomic and molecular species at the electron-volt energies of interest to atomic physics, chemistry, and biology, and since the transfer from nuclear and particle physics of a large body of techniques, both experimental and theoretical, relating to the study of elementary quantum collisions and interactions has occurred. Studies in this area could provide fundamental information on molecular structure and the basic mechanisms by which atomic and molecular species interact. From a practical viewpoint, this capability would place the understanding of chemical reaction mechanisms on an entirely new and more fundamental basis, with great potential return. It also can provide vital insights into mechanisms of major interest to molecular biology.

Biophysical Acoustics As is true of the preceding program element, recent progress in biophysical acoustics has yielded new fundamental understanding of the physics involved in speech and hearing functions. Here again, increased activity holds high promise of alleviating a wide variety of incapacitating human ills within a relatively short time. Research has drawn on a wide variety of techniques from other disciplines of physics ranging from the Mössbauer effect from nuclear physics, used in measuring microscopic motions of the components of the inner ear, to the use of miniaturized, implanted solid-state transducers and precision optical interferometric devices. It provides an excellent example of the application of such techniques to biophysical problems. As a part of physics, the subfield is still small, but it is of growing significance, and the future potential is large.

Very Large Radio Array We concur in the conclusion of the Astronomy Survey Committee * that the provision of very large radio telescope arrays holds high promise of major new discoveries concerning the structure of the universe. Most of the recent astonishing discoveries in astrophysics—quasars, pulsars, cosmic background radiation, interstellar masers—were made by radio telescopes. To exploit these discoveries, a major new instrument capable of producing sharp images of the radio sky is needed. A large radio array, which produces an image as sharp as that of the 200-in. optical telescope (1 sec of arc) by means of a principle known as aperture synthesis, can be built for $62 million. This instrument would be far superior to any available abroad for many years to come. Three-antenna prototypes of the system (which would consist of 27 individual antennas)

---

* Astronomy Survey Committee (J. L. Greenstein, chairman), *Astronomy and Astrophysics for the 1970's* (National Academy of Sciences, Washington, D.C., 1972).

have been built and have demonstrated the system's potential to distinguish powerful quasars at the limits of the observable universe and to make sharp pictures of nearby exploding stars and galaxies. It is essential that a large array be built if U.S. radio astronomers, now doing excellent competitive work with present instruments, are to continue the intensive study of such objects and to participate in the exciting discoveries certain to be made, if not in the United States, then with instruments now under construction overseas.

X-Ray and Gamma-Ray Astronomy   We again concur with the recommendation of the Astronomy Survey Committee * that the High-Energy Astronomical Observatory (HEAO) also should be an important part of the national effort in astrophysics and astronomy. Because of the opacity of the earth's atmosphere to both x and gamma radiation, these windows on the universe have only recently been opened through rocket and satellite astronomy. Any reasonable extrapolation from the preliminary soundings that have been possible so far suggests a large return in fundamental insight into the structure and history of the universe. In the total national physics program, the estimated cost of this facility—$400 million—is extremely high. To be considered in proper perspective it must be viewed in the context of the total expenditures of the U.S. space program. In anticipated scientific return—both short- and long-range—it merits high priority in that program.

Further Considerations   The selection of the foregoing program elements for special emphasis is intended to illustrate the use of the jury-rating procedure. In particular, it shows that to base such recommendations on the order established through evaluation along any one dimension, however carefully that ordering may have been developed, would be unacceptable. Further, it demonstrates that a wide range of input considerations, of some of which the Committee may be completely unaware, must be included in arriving at a final balanced program.

Program emphases at one overall level of funding may be quite different from those at another. The instinct for survival takes precedence over any objectively evolved, balanced program when the necessity for such consideration is forced on any field. We would not recommend that it be otherwise. It is vitally important that, even in the least favorable situations, this nation retain a rebound capability so that, with an improved economy, the scientific enterprise remains viable and capable of renewed expansion without unavoidable delay. Science must be able to respond to the available opportunities at whatever level of effort the overall play

---

* Astronomy Survey Committee (J. L. Greenstein, Chairman), *Astronomy and Astrophysics for the 1970's* (National Academy of Sciences, Washington, D.C., 1972).

of political processes may make possible. In view of the increasingly close relationship between the health of the nation's scientific and technological enterprise and the national economy, any direct coupling of the support of this enterprise to the economy, particularly in times of a depressed economy, could have elements of disaster. Illustrative of this point is the experience of the General Electric Company, which made the corporate decision during the economic depression in the early 1930's to avoid the all too common drastic cutback in research and developmental activities. As a result, when the economy rebounded, General Electric had a backlog of new ideas and devices and an internal strength in experienced and dedicated manpower that played a very significant role in the attainment of the competitive position that the company now enjoys.

CONCLUSION

For the various reasons outlined in this chapter, the Committee has not attempted to present any specific detailed national program for physics in the range between the exploitation (11 percent growth) and flat (0 percent growth) levels of investment, in dollars of constant purchasing power, for the next five years. Within each subfield, to the extent that the subfield panels have found it possible, detailed programs for each funding level have been prepared (Volume II). The procedure was more feasible in some subfields than others.

That we do not recommend a detailed national physics program appropriate in our view to different possible levels of funding does not reflect unwillingness to face the difficulties inherent in any such attempt. Rather, it reflects the conclusion that it is impossible for any group such as the Physics Survey Committee to develop either adequately complete information or the insight to make such a detailed attempt meaningful. It is unrealistic to regard the total support of U.S. physics as an effective reservoir from which funds for individual program elements may be distributed without cognizance of all the internal and external pressures and constraints within both the different funding agencies and the physics community. These, moreover, change rapidly with the magnitude of the overall support available. Any system of funding for science must allow for a considerable degree of initiative and new directions originating at the working level. An *a priori* allocation system that divides a fixed amount of total funding among predefined fields of science is likely to stifle initiative and innovation.

The subfield panel reports provide a detailed discussion of opportunities and needs viewed primarily in terms of intrinsic and structural criteria, which are internal to the subfields. As a Committee we have attempted to evolve a procedure for viewing the program elements of these subfields within a broader context including more explicit extrinsic criteria. We have applied these criteria to the individual program elements and, to illustrate their use, have carried out a detailed jury rating in terms of the intrinsic and extrinsic criteria. This exercise has resulted in a series of ordered listings of the program elements, weighting the different classes of criteria in different fashion. These might be of interest as evidence of the consensus of the Survey Committee, but we would emphasize again that relative ordering in any restricted section of these listings is *not* to be considered as significant.

It is hoped that this illustrated approach to a more general evaluation of the program elements in physics, together with the more detailed documentation in the subfield panel reports, may provide a major input to the development of program emphases at whatever level of support can be made available to the field.

We have discussed two particularly urgent situations (LAMPF and NAL) that must be addressed in fiscal year 1973 and have selected some 15 program elements that hold promise of unusual productivity, with increased support, in the period through fiscal year 1977.

Our exploitation budget—a budget increasing at the rate of 11 percent per year—would permit exploitation and exploration of the outstanding opportunities and challenges foreseen through fiscal year 1977. It would enable the physics community to regain much of the strength and vigor that it could have retained had it been possible to maintain a steady growth of even 5 percent per annum since fiscal year 1967. It would permit effective national utilization of the available trained manpower and those already in the educational pipeline who will join the physics community during this same period ending in fiscal year 1977. Although it may not be possible, in view of competing demands on national resources, to permit this growth rate, it must be emphasized that cutting back from it is done at a cost, both long and short range, that can have very serious consequences in relation to new national goals, the health of the national economy, a competitive position in the international marketplace, and national defense.

APPENDIX 5.A ORDERED LISTING OF PROGRAM ELEMENTS— OVERALL SCORING

PROGRAM ELEMENT	PHYSICS SUBFIELD
1. Lasers and masers	Atomic, molecular, and electron
2. National Accelerator Laboratory	Elementary-particle
3. Quantum optics	Condensed-matter

APPENDIX 5.A—*Continued*

PROGRAM ELEMENT	PHYSICS SUBFIELD
4. University groups—EPP	Elementary-particle
5. Stanford Linear Accelerator	Elementary-particle
6. Nuclear dynamics	Nuclear
7. Major facilities—EPP, AGS improvement, etc.	Elementary-particle
8. Brookhaven AGS	Elementary-particle
9. Nuclear excitations	Nuclear
10. Heavy-ion interactions	Nuclear
11. Higher-energy nuclear physics	Nuclear
12. Nuclear astrophysics	Nuclear
13. Theoretical relativistic astrophysics	Astrophysics and relativity
14. Neutron physics	Nuclear
15. Nuclear theory	Nuclear
16. Very large radio array *	Astrophysics and relativity
17. X-ray and gamma-ray observatory *	Astrophysics and relativity
18. Turbulence in fluid dynamics	Plasma and fluids
19. Superfluidity	Condensed-matter
20. Infrared astronomy *	Astrophysics and relativity
21. General relativity tests	Astrophysics and relativity
22. Oceanography *	Plasma and fluids
23. Atomic and molecular beams	Atomic, molecular, and electron
24. Laser-related light sources	Optics
25. Controlled fusion	Plasma and fluids
26. Digitized imaging devices for optical astronomy	Astrophysics and relativity
27. Surface physics	Condensed-matter
28. Gravitational radiation	Astrophysics and relativity
29. Aperture synthesis for infrared astronomy	Astrophysics and relativity
30. Nonlinear optics	Optics
31. Argonne ZGS	Elementary-particle
32. Magnetic properties of solids	Condensed-matter
33. Semiconductors	Condensed-matter
34. Gamma-ray detectors in astronomy	Astrophysics and relativity
35. Electronic properties of solids and liquids	Condensed-matter
36. Holography and information storage	Optics
37. Hearing, speech, and biophysical acoustics	Acoustics
38. CEA bypass storage ring	Elementary-particle
39. High magnetic fields	Condensed-matter
40. Cornell synchrotron	Elementary-particle
41. Laboratory astrophysics, plasma and fluids	Plasma and fluids

* These program elements involve large funding and activity in adjacent sciences; in this Survey attention has been focused on the physics component of each.

APPENDIX 5.A—*Continued*

PROGRAM ELEMENT	PHYSICS SUBFIELD
42. Neutrino astronomy	Astrophysics and relativity
43. Nuclear facilities and instrumentation	Nuclear
44. Electron physics	Atomic, molecular, and electron
45. Computer modeling	Plasma and fluids
46. MHD power generation	Plasma and fluids
47. Turbulent plasmas	Plasma and fluids
48. Optical band communication	Optics
49. Fluid and plasma dynamics and lasers	Plasma and fluids
50. Nonelectronic aspects of solids and liquids	Condensed-matter
51. Nuclear decay studies	Nuclear
52. Weak and electromagnetic interactions	Nuclear
53. Optical systems and lens designs	Optics
54. Luminescence, etc.	Condensed-matter
55. Optical information processing	Optics
56. Berkeley Bevatron	Elementary-particle
57. Atomic and molecular spectroscopy	Atomic, molecular, and electron
58. Integrated optics	Optics
59. Meteorology *	Plasma and fluids
60. Metrology *	Optics
61. Accelerator development	Elementary-particle
62. Gas discharges	Atomic, molecular, and electron
63. Crystallography, etc. *	Condensed-matter
64. Electroacoustics and acoustics instrumentation	Acoustics
65. Slow neutron physics	Condensed-matter
66. Ultrasonics and infrasonics	Acoustics
67. Underwater sound	Acoustics
68. Noise, mechanical shock, and vibration	Acoustics
69. Music and architectural acoustics *	Acoustics

* These program elements involve large funding and activity in adjacent sciences; in this Survey attention has been focused on the physics component of each.

APPENDIX 5.B ORDERED LISTING OF PROGRAM ELEMENTS—EXTRINSIC SCORING

PROGRAM ELEMENT	PHYSICS SUBFIELD
1. Lasers and masers	Atomic, molecular, and electron
2. Quantum optics	Condensed-matter
3. Nuclear excitations	Nuclear
4. Controlled fusion	Plasma and fluids
5. Nuclear dynamics	Nuclear

APPENDIX 5.B—*Continued*

PROGRAM ELEMENT	PHYSICS SUBFIELD
6. Oceanography *	Plasma and fluids
7. Laser-related light sources	Optics
8. Neutron physics	Nuclear
9. Surface physics	Condensed-matter
10. Computer modeling	Plasma and fluids
11. MHD power generation	Plasma and fluids
12. Semiconductors	Condensed-matter
13. Holography and information storage	Optics
14. Turbulence in fluid dynamics	Plasma and fluids
15. Optical band communication	Optics
16. Magnetic properties of solids	Condensed-matter
17. Meteorology *	Plasma and fluids
18. Electronic properties of solids and liquids	Condensed-matter
19. Fluid and plasma dynamics and lasers	Plasma and fluids
20. Hearing, speech, and biophysical acoustics	Acoustics
21. Metrology *	Optics
22. Nuclear facilities and instrumentation	Nuclear
23. Optical information processing	Optics
24. Electroacoustics and acoustics instrumentation	Acoustics
25. Optical systems and lens design	Optics
26. Electron physics	Condensed-matter
27. Noise, mechanical shock, and vibration	Acoustics
28. Ultrasonics and infrasonics	Acoustics
29. Laboratory astrophysics, plasma and fluids	Plasma and fluids
30. Underwater sound	Acoustics
31. Nonelectronic aspects of solids and liquids	Condensed-matter
32. Gas discharges	Atomic, molecular, and electron
33. Nonlinear optics	Optics
34. Heavy-ion interactions	Nuclear
35. Luminescence, etc.	Condensed-matter
36. Atomic and molecular beams	Atomic, molecular, and electron
37. Crystallography, etc. *	Condensed-matter
38. High magnetic fields	Condensed-matter
39. Integrated optics	Optics
40. Turbulent plasmas	Plasma and fluids
41. Nuclear decay studies	Nuclear
42. Music and architectural acoustics *	Acoustics
43. Superfluidity	Condensed-matter

* These program elements involve large funding and activity in adjacent sciences; in this Survey attention has been focused on the physics component of each.

APPENDIX 5.B—*Continued*

PROGRAM ELEMENT	PHYSICS SUBFIELD
44. Higher-energy nuclear physics	Nuclear
45. Accelerator development	Elementary-particle
46. Atomic and molecular spectroscopy	Atomic, molecular, and electron
47. National Accelerator Laboratory	Elementary-particle
48. Nuclear theory	Nuclear
49. Infrared astronomy *	Astrophysics and relativity
50. Major facilities—EPP	Elementary-particle
51. Digital imaging devices for optical astronomy	Astrophysics and relativity
52. Stanford Linear Accelerator	Elementary-particle
53. Very large radio array *	Astrophysics and relativity
54. Gamma-ray detectors in astronomy	Astrophysics and relativity
55. Brookhaven AGS	Elementary-particle
56. Weak and electromagnetic interaction	Elementary-particle
57. X-ray and gamma-ray astronomy *	Astrophysics and relativity
58. University groups—EPP	Elementary-particle
59. Nuclear astrophysics	Nuclear
60. Theoretical relativistic astrophysics	Astrophysics and relativity
61. CEA bypass storage ring	Elementary-particle
62. Argonne ZGA	Elementary-particle
63. Aperture synthesis for infrared astronomy	Astrophysics and relativity
64. Cornell synchrotron	Elementary-particle
65. Gravitational radiation	Astrophysics and relativity
66. Neutrino astronomy	Astrophysics and relativity
67. General relativity tests	Astrophysics and relativity
68. Berkeley Bevatron.	Elementary-particle
69. Slow neutron physics	Condensed-matter

APPENDIX 5.C ORDERED LISTING OF PROGRAM ELEMENTS— INTRINSIC SCORING

PROGRAM ELEMENT	PHYSICS SUBFIELD
1. National Accelerator Laboratory	Elementary-particle
2. University groups—EPP	Elementary-particle
3. Stanford Linear Accelerator	Elementary-particle
4. Brookhaven AGS	Elementary-particle
5. Major facilities—EPP	Elementary-particle
6. General relativity tests	Astrophysics and relativity
7. Nuclear astrophysics	Nuclear
8. Theoretical relativistic astrophysics	Astrophysics and relativity

* These program elements involve large funding and activity in adjacent sciences; in this Survey attention has been focused on the physics component of each.

APPENDIX 5.C—Continued

PROGRAM ELEMENT	PHYSICS SUBFIELD
9. X-ray and gamma-ray astronomy *	Astrophysics and relativity
10. Lasers and masers	Atomic, molecular, and electron
11. Very large radio array *	Astrophysics and relativity
12. Higher-energy nuclear physics	Nuclear
13. Nuclear theory	Nuclear
14. Gravitational radiation	Astrophysics and relativity
15. Argonne zgs	Elementary-particle
16. Infrared astronomy *	Astrophysics and relativity
17. Aperture synthesis for infrared astronomy	Astrophysics and relativity
18. Heavy-ion interactions	Nuclear
19. Digitized imaging devices for optical astronomy	Astrophysics and relativity
20. Superfluidity	Condensed-matter
21. Neutrino astronomy	Astrophysics and relativity
22. cea bypass storage ring	Elementary-particle
23. Cornell synchrotron	Elementary-particle
24. Gamma-ray detectors in astronomy	Astrophysics and relativity
25. Quantum optics	Condensed-matter
26. Atomic and molecular beams	Atomic, molecular, and electron
27. Berkeley Bevatron	Elementary-particle
28. Nuclear dynamics	Nuclear
29. Turbulence in fluid dynamics	Plasma and fluids
30. Nonlinear optics	Optics
31. Nuclear excitations	Nuclear
32. Weak and electromagnetic interactions	Nuclear
33. High magnetic fields	Condensed-matter
34. Neutron physics	Nuclear
35. Hearing, speech, and biophysical acoustics	Acoustics
36. Slow neutron physics	Condensed-matter
37. Magnetic properties of solids	Condensed-matter
38. Turbulent plasmas	Plasma and fluids
39. Laboratory astrophysics, plasma and fluids	Astrophysics and relativity
40. Electronic properties of solids and liquids	Condensed-matter
41. Oceanography *	Plasma and fluids
42. Surface physics	Condensed-matter
43. Laser-related light sources	Optics
44. Semiconductors	Condensed matter
45. Electron physics	Atomic, molecular, and electron

* These program elements involve large funding and activity in adjacent sciences; in this Survey attention has been focused on the physics component of each.

APPENDIX 5.C—Continued

PROGRAM ELEMENT	PHYSICS SUBFIELD
46. Nuclear decay studies	Nuclear
47. Nuclear facilities and instrumentation	Nuclear
48. Atomic and molecular spectroscopy	Atomic, molecular, and electron
49. Controlled fusion	Plasma and fluids
50. Holography and information storage	Optics
51. Nonelectronic aspects of solids and liquids	Condensed-matter
52. Integrated optics	Optics
53. Luminescence, etc.	Condensed-matter
54. Accelerator development	Elementary-particle
55. Fluid and plasma dynamics and lasers	Plasma and fluids
56. Optical band communication	Optics
57. Optical systems and lens design	Optics
58. Computer modeling	Plasma and fluids
59. MHD power generation	Plasma and fluids
60. Optical information processing	Optics
61. Meteorology *	Plasma and fluids
62. Crystallography, etc.*	Condensed-matter
63. Gas discharges	Atomic, molecular, and electron
64. Metrology *	Optics
65. Electroacoustics and acoustics instrumentation	Acoustics
66. Ultrasonics and infrasonics	Acoustics
67. Underwater sound	Acoustics
68. Music and architectural acoustics	Acoustics
69. Noise, mechanical shock, and vibration	Acoustics

APPENDIX 5.D ORDERED LISTING OF PROGRAM ELEMENTS—EXTRINSIC/INTRINSIC RATIO

PROGRAM ELEMENT	PHYSICS SUBFIELD
1. Noise, mechanical shock, and vibration	Acoustics
2. Underwater sound	Acoustics
3. Music and architectural acoustics	Acoustics
4. Ultrasonics and infrasonics	Acoustics
5. Electroacoustics and acoustics instrumentation	Acoustics

* These program elements involve large funding and activity in adjacent sciences; in this Survey attention has been focused on the physics component of each.

Appendix 5.D—*Continued*

PROGRAM ELEMENT	PHYSICS SUBFIELD
6. Metrology *	Optics
7. Meteorology *	Plasma and fluids
8. Computer modeling	Plasma and fluids
9. MHD power generation	Plasma and fluids
10. Gas discharges	Atomic, molecular, and electron
11. Crystallography, etc.*	Condensed-matter
12. Optical band communication	Optics
13. Optical information processing	Optics
14. Controlled fusion	Plasma and fluids
15. Fluid and plasma dynamics and lasers	Plasma and fluids
16. Optical systems and lens design	Optics
17. Surface physics	Condensed-matter
18. Laser-related light sources	Optics
19. Oceanography *	Plasma and fluids
20. Holography and information storage	Optics
21. Neutron physics	Nuclear
22. Nuclear facilities and instrumentation	Nuclear
23. Nonelectronic aspects of solids and liquids	Condensed-matter
24. Nuclear excitations	Nuclear
25. Semiconductors	Condensed-matter
26. Electronic properties of solids and liquids	Condensed-matter
27. Magnetic properties of solids	Condensed-matter
28. Integrated optics	Optics
29. Luminescence, etc.	Condensed-matter
30. Electron physics	Atomic, molecular, and electron
31. Quantum optics	Condensed-matter
32. Laboratory astrophysics, plasma and fluids	Astrophysics and relativity
33. Hearing, speech, and biophysical acoustics	Acoustics
34. Accelerator development	Elementary-particle
35. Nuclear dynamics	Nuclear
36. Lasers and masers	Atomic, molecular, and electron
37. Turbulence in fluid dynamics	Plasma and fluids
38. Nuclear decay studies	Nuclear
39. Turbulent plasmas	Plasma and fluids
40. High magnetic fields	Condensed-matter
41. Atomic and molecular spectroscopy	Atomic, molecular, and electron
42. Nonlinear optics	Optics
43. Atomic and molecular beams	Atomic, molecular, and electron
44. Heavy-ion interactions	Nuclear

* These program elements involve large funding and activity in adjacent sciences; in this Survey attention has been focused on the physics component of each.

APPENDIX 5.D—*Continued*

PROGRAM ELEMENT	PHYSICS SUBFIELD
45. Weak and electromagnetic interactions	Nuclear
46. Superfluidity	Condensed-matter
47. Higher-energy nuclear physics	Nuclear
48. Nuclear theory	Nuclear
49. Infrared astronomy *	Astrophysics and relativity
50. Digitized imaging devices for optical astronomy	Astrophysics and relativity
51. Gamma-ray detectors in astronomy	Astrophysics and relativity
52. Very large radio array *	Astrophysics and relativity
53. Aperture synthesis for infrared astronomy	Astrophysics and relativity
54. Slow neutron physics	Condensed-matter
55. National Accelerator Laboratory	Elementary-particle
56. Stanford Linear Accelerator	Elementary-particle
57. Brookhaven AGS	Elementary-particle
58. Major facilities—EPP	Elementary-particle
59. X-ray and gamma-ray observatory *	Astrophysics and relativity
60. Gravitational radiation	Astrophysics and relativity
61. Argonne ZGS	Elementary-particle
62. Neutrino astronomy	Astrophysics and relativity
63. CEA bypass storage ring	Elementary-particle
64. Cornell synchrotron	Elementary-particle
65. Berkeley Bevatron	Elementary-particle
66. University groups—EPP	Elementary-particle
67. Nuclear astrophysics	Nuclear
68. Theoretical relativistic astrophysics	Astrophysics and relativity
69. General relativity tests	Astrophysics and relativity

* These program elements involve large funding and activity in adjacent sciences; in this Survey attention has been focused on the physics component of each.

6

THE
CONSEQUENCES
OF
DETERIORATING
SUPPORT

In Chapters 4 and 5 the decline in federal support for research in physics and its subfields during recent years has been discussed in some detail. Decreasing support has affected the various institutions of physics in different ways, and their responses to reductions in funds also have differed. Many subtleties that can affect the health of the overall physics enterprise—some of them profoundly—are involved. Ultimately, reducing expenditures affects the livelihood of highly trained people. Some of the effects are described in this chapter, and detailed data appear in Chapter 12. To a considerable degree, of course, these effects apply to any field of scientific research and are not unique to physics. Although our discussion is rooted in physics, the problems we address have much broader relevance.

Some Effects of Reduced Research Expenditures

Various measures have been adopted to achieve the necessary reductions in agency expenditures. A number of research projects have been terminated. Activities in others have been slowed as a consequence of reduction in service personnel and in monies for new equipment and the improvement of old equipment. According to the National Science Founda-

453

tion (NSF),* 51.4 percent of the university physics departments polled in a recent survey reported projects that were permanently halted, while another 13.9 percent reported temporary halts in projects. In all, three fourths of all reporting departments indicated reduced research activity. These reductions are most visible in the subfields that depend heavily on large facilities, such as accelerators, that require a certain level of minimum support if operation is not to be greatly curtailed or halted. Examples in elementary-particle physics include the reduction in the operation of the Cambridge Electron Accelerator and the Stanford Linear Accelerator. In the "medium-energy" nuclear-physics program, a number of accelerators have been closed or have experienced sharply reduced budgets. Although the effects in "little science" subfields are generally less dramatic than those associated with the closing or severe cutback in operation of a major facility, they are no less real.

The decreasing size and scope of research programs have meant that fewer graduates and postdoctorals could be accommodated. The first-year graduate classes are significantly smaller. In the 11 leading private institutions listed by the American Council on Education, the entering 1970 class is only 59 percent as large as that of 1967. In state-supported institutions, the corresponding figure for the same period is about 83 percent. In an NSF-conducted survey,* 34.7 percent of the departments that responded reported a reduction in the size of their graduate programs. Evidence indicates that the strongest academic science centers, particularly those in private institutions, have instituted the greatest reductions.

As one might expect, reductions in the number and size of research projects have been accompanied also by reductions in the number of support personnel. This situation is all too frequently neglected in discussions of manpower problems. The number of science and engineering technicians in the United States dropped by 5.7 percent in 1968–1969 and by 4.4 percent in 1969–1970; these decreases can be compared with the corresponding reductions for all sciences of 1.5 percent and 0.5 percent. Seemingly, support personnel are affected to a greater extent, and perhaps sooner, by reduced budgets than are other types of research personnel.

In regard to physicists, the national laboratories and industry have begun to discharge sizable groups, while the number of university tenured positions has been sharply reduced. This decrease in job opportunities has been reflected in the growing number of unemployed (and underemployed) physicists. Preliminary figures, as of June 1971,† indicate an unemploy-

---

* National Science Foundation, *Impact of Changes in Federal Science Funding Patterns on Academic Institutions, 1968–70*, NSF 70–48 (U.S. Government Printing Office, Washington, D.C., 1970).
† National Science Foundation, *Unemployment Rates for Scientists, Spring 1971*, NSF 71–20 (U.S. Government Printing Office, Washington, D.C., 1971).

ment rate for physicists, including both PhD's and non-PhD's, of 3.9 percent, compared with 1 percent in 1968 and 2.2 percent in 1970. The fraction unemployed is considerably greater among those not having a PhD degree than among PhD's. Unemployment also is proportionately greater among those who work in industry than among those who work in educational institutions. (See Chapter 12 for a detailed discussion of these data.)

An examination of the age distribution of the unemployed shows that the group under 34 years of age is the most severely affected. Younger scientists, especially physicists, appear to be bearing the brunt of the contraction produced by the tighter budgets and reduced funding. Among all the sciences, "[I]t appears that physics, the discipline most dependent upon Federal support, suffered most unemployment, chemistry was next." *
A numerically smaller problem but one with profound social and humanitarian overtones is that of the midcareer physicist, particularly in industry, who suddenly finds himself unemployed. We discuss this problem in greater detail in a subsequent section of this chapter.

NEW SOURCES OF SUPPORT

To compensate for reduced support from the traditional federal sources, funds have been sought elsewhere. Among universities, the chief sources of nonfederal "compensating funds" have been student tuition and fees, endowment income, and foundation and individual gifts. Public institutions have also turned to state and local governments. Some universities have drawn on endowment capital. In the NSF survey, about 37 percent of the physics departments † that responded had compensated in these ways for the entire reduction in federal funds. But these measures can be only temporary expedients. According to a study sponsored by the Carnegie Commission on Higher Education,‡ all the major universities have serious financial difficulties; seven of the eight Ivy League institutions operated at a deficit during 1969–1970. The results for "research universities" are shown in Table 6.1.

New endowments are decreasing; increases in tuition fees may have reached the saturation point; and some state governments, reflecting the mood of many voters, are retrenching. Thus there appears to be little

---

* B. C. Henderson, "Unemployment Among Scientists," *Geotimes, 16*(9), 16 (1971).
† National Science Foundation, *Impact of Changes in Federal Science Funding Patterns on Academic Institutions, 1968–70,* NSF 70–48 (U.S. Government Printing Office, Washington, D.C., 1970).
‡ E. F. Cheit, *The New Depression in Higher Education, A Study of Financial Conditions at 41 Colleges and Universities,* A General Report for the Carnegie Commission on Higher Education and the Ford Foundation (McGraw-Hill, New York, 1971).

TABLE 6.1   How Research Institutions Are Faring Financially

In Financial Difficulty	Headed for Trouble	Not in Trouble
I. National Research Universities		
Stanford	Harvard	U. of Texas, Austin
Columbia	Chicago	
U. of California, Berkeley	Michigan	
	Minnesota	
II. Leading Regional Research Universities		
New York U.	Ohio U.	U. of North Carolina, Chapel Hill
St. Louis U.	Syracuse	
Tulane	U. of Missouri, Columbia	
	U. of Oregon	

likelihood that the universities, particularly the major research institutions, will be able to compensate for further effective decreases in federal funding by developing new nonfederal sources of revenue.

Not only universities but all institutions of physics, including industrial organizations and the national laboratories, are attempting to broaden their base of support. Many believe that, in the future, significant support must come from those federal agencies with missions directly related to the solution of major societal problems—health care, medicine, transportation, and, more generally, communication, preservation of the environment, new energy sources, disarmament, and education (see also Chapter 10).

Many of these agencies are still in the process of developing appropriate research programs and are unsure of the extent to which physics can or should play a part in them. They do not appear to appreciate fully, as yet, the long-range importance of research in physics, although in general they accept the proposition that the solution to many of their problems lies in the applications of science and technology. Nor do they appear to be aware of the possible benefits to problem identification, analysis, and solution that could accrue from the appropriate use of well-trained physicists in agency staff positions (see Chapters 7 and 10). Changes in this situation might be accelerated by various means, such as legislation, directed funding, and proposal pressure, but physicists themselves must take much of the initiative.

Industry on the whole does little basic research. Although in principle it could benefit from a substantially greater basic research effort, the link to a commercial end product is too tenuous and the time interval from initial discovery to feasible application often too great for most organizations to justify large-scale programs of basic research. Such programs must be supported from company profits and, when these are low, basic research, being less directly related to a profit-making end product, tends to be reduced. As might be expected, the depressed state of the national economy over the past two or three years has brought about extensive cutbacks of this sort.

IMPACT OF DECREASING SUPPORT ON SCIENTIFIC PRODUCTIVITY

Regardless of the expedient adopted, decreased expenditures act in the long run to reduce scientific productivity and, more significantly, can reduce the quality of the scientific product. There is no immutable law involved. Rather, reduction in quality occurs all too easily unless special care is taken to preserve the general excellence of a field as effort contracts. With diminishing funds and with the trend toward reducing the number of facilities, scientists, and support personnel, scientific programs tend to become conservative. Experiments with a high probability of success are more likely to be performed than those that are bold and daring but of less certain outcome. Such risk-taking may bring about major advances or result in failures. Both are a necessary part of the growth of any scientific field. Uncertainties also militate against the systematic studies that must be pursued for a comparatively long period of time before they can be expected to yield complete and applicable results.

Confronted by a funding reduction, the first response of a research group or laboratory may well involve temporary expedients, which will, for a short time, maintain the momentum of the program and keep the full research capability intact in the expectation that the funding situation will return to "normal" in a reasonably short time. Under these circumstances, changes or improvements in experimental programs are held to a minimum, and there is an effort to realize as much data as possible from present facilities and projects. The more routine experiments tend to be performed. Earlier results that were passed over as less interesting or important are now analyzed and prepared for publication. In fact, for a short time the production and analysis of data may exceed the normal rate. Thus at the inception of a period of reduced funding the number of papers generated by a laboratory may increase. Superficial observations may lead to the erroneous conclusion that scientific productivity has, if anything, increased! Indeed, the experiments and data analyzed are worthwhile, but the whole procedure is one of confining the research to a holding pattern. More limited in scope as it tends to be, such a program eventually dies as the number of interesting and unexplored phenomena accessible to it for experimental study becomes exhausted.

If after a short time an adequate funding level is not restored, a more enduring strategy must be adopted for preserving the excellence of the field. The laboratory may be able to make up for its loss in funds by further reductions in expenditures. One way is to narrow its scope. In the larger institutions, particularly, this measure may involve the consolidation of research groups. If the funding cuts are sufficiently large, operation of the facility or the performance of sufficiently challenging research may no longer be possible; thus the program is terminated.

When expenditures must be reduced, the wise administrator makes every effort to conserve those elements of a program that are essential for scientific productivity. Reductions in the less senior personnel are made, but those needed for the completion of research programs in progress and for the creation and mounting of new ones are retained. The younger, less experienced, nonprofessional staff members are discharged first. Next there are decreases in the number of personnel who help to develop and assemble the needed electronics, to carry out machine-shop fabrication, and to assist in setting up experiments.

It should be emphasized that this support staff, particularly in universities, typically is assembled carefully over an extended period and includes persons who have broad ranges of complementary skills. Most could command much larger salaries in other environments, but they enjoy the atmosphere and freedom from routine characteristic of the university. Most have developed a sensitivity to the needs of exploratory research, with its vagaries and rapid changes. As a result, they can participate as creative partners in the ongoing research program. Once dissolved, such groups cannot be reassembled except through a slow and painstaking process of selection continued over a period of years. This is a vitally important but often neglected aspect of the "faucet effect" in the support of research.

As budget constraints increase, the number of graduate students, young PhD's, non-PhD scientists, and older scientists whose scientific output may have been decreasing are also reduced. Each institution will decide on some mix depending on its scope and character and the philosophy of the laboratory management. One fairly common procedure is to decrease the number of graduate students and young PhD's who are employed. However, in some universities the graduate student has come to play such an essential role in undergraduate education that this tactic is regarded as impractical. In national and industrial laboratories it is easier to discharge older scientists whose interests and effectiveness no longer conform with the mission of these institutions. (Trends in the number of support personnel and the general contraction of the scientific corps are supported by the statistics found in Chapter 12.)

Reductions in support also make it less likely that significant sums will be spent on the improvement of old equipment or the purchase of new. To the extent feasible, the laboratory makes do with what it has, improvising when necessary. As a consequence, some experiments will be delayed, some will be eliminated, and in those performed the data may be less useful. The quality of the research depends heavily on the quality of the required instrumentation.

The number of operating hours of a facility may be reduced, with the result that the number of experiments that can be performed is also limited. This is a particularly pernicious procedure, since the full scientific return on the large capital investment represented by the facility is not realized.

Travel to scientific meetings as well as visits to other institutions also will be reduced. Each scientific group may tend to become a bit isolated, and the flow of information to and from other groups is likely to decrease. (See also Chapters 8 and 13.)

Large reductions in personnel and equipment of a laboratory are likely to lead to the consolidation of some research groups. A narrowing of the scope of the research program is almost inevitable. Initially, the less productive programs are discontinued, but ultimately the laboratory management is likely to have to choose among several experimental programs regarded as equally productive and significant.

One possible result of consolidation is consensus physics, that is, concentration on programs that require the agreement of many investigators or of several groups rather than being the conception of just a few. In such circumstances, the majority opinion tends to reflect the mainstream and the chance that an offbeat idea will be acted on is substantially smaller.

In a variety of ways, funding cuts conspire generally to increase the time needed for the completiton of a research program. It may be argued that such a slowdown causes no great harm. But, in fact, many programs will not be initiated if the prospect for their completion lies too far in the future. Under the best circumstances, embarking on a program involving a lifetime of effort for its completion and several generations of graduate students will require singular dedication on the part of the research staff and unusual understanding on the part of the funding agency. Initiation of such a program will be even less likely if the significance of the results cannot be guaranteed in advance. Thus because of a slowdown of effort resulting from lack of funds, a whole class of potentially valuable experiments may never be attempted.

When the funding cuts become sufficiently great, the termination of projects or closing of facilities must be faced. The point at which this threshold is reached depends on whether one is dealing with big science or little science.

Big science is characterized by a large capital investment in instruments or facilities. Some physics subfields depend more heavily on these than do others. For example, facilities in elementary-particle physics include large accelerators, detectors, and computers that are employed to process the data as well as to search for a meaningful pattern among the results. In solid-state research the magnet laboratory and the high-flux reactor are major large facilities. Big-science facilities in astrophysics include radio telescopes and the planned orbiting telescope. In plasma physics there are large devices for the study of controlled fusion. As a final example, in nuclear physics there are the large linear accelerators as well as the multistage electrostatic accelerators. (See Chapter 4 for data on activities and costs in various subfields.) Each of these facilities requires not only a large expenditure for construction but also a large ongoing expenditure for operation. As a consequence, a plot of scientific productivity versus funding yields the S-shaped curve shown in Figure 6.1. In or near the region of the steep rise in the curve, small changes in support produce

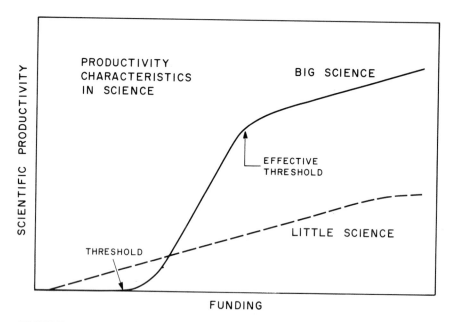

FIGURE 6.1   Scientific productivity as a function of funding in big science and little science.

large changes in scientific productivity. This is referred to as the leverage effect. (Chapter 5 discusses and gives examples of such situations in various physics subfields.) As an illustration, reducing the number of shifts during which experiments can be done from ten to eight per week brings about a reduction in scientific productivity of approximately 20 percent. However, the money saved will be a much smaller fraction of the operating costs. Obviously, the tactic of reducing the number of shifts, which several laboratories have been forced to adopt, is in the long run exceedingly wasteful. Among other things, it results in a large increase in the cost per experiment. One also recognizes from the curve that there is an effective funding threshold—the cost of turning a machine on. If funds for this purpose are not available, or if the excess above this funding threshold is so small that only very few experiments can be done, and at a very large cost, the operation of the machine must cease. The arrows in Figure 6.1 indicate the effective funding threshold above which the installation can be operated fruitfully and efficiently and below which its operation is inefficient and ineffective.

New big-science facilities can disrupt violently the funding pattern of a given subfield. Indeed, the problems in both elementary-particle physics and nuclear physics (discussed in Chapter 5) stem in part from such disturbances. In both these subfields the construction of several very large and costly facilities, separately funded, has just been completed, but the additional funds for their operation have not been forthcoming. Either incremental funding must be obtained or the elimination of a number of other projects in these subfields must occur. In recent years, because of diminishing support, the second mechanism has been operative. A number of accelerators have been, and others will be, shut down, while the operation of still others has been reduced, with the possibility of their early termination in order to accommodate the operating costs of the Los Alamos Meson Physics Facility and the National Accelerator Laboratory. One result has been a trend toward an increasing centralization of facilities into fewer installations, many of which are accessible to university physics departments only if they are willing to do their experiments on a "user-group" basis (see discussion in Chapter 9). Also, because the newer facilities are generally more expensive, expenditures per scientific man-year are greater; hence, level funding, and *a fortiori* diminishing funding, leads to a reduction in the number of scientific man-years to be supported and thus of the number of positions.

The choice between operating a new facility that is clearly at the frontier of research activity or operating older facilities whose capabilities have only been partially realized can be particularly difficult for fields that are evolving from a little-science status into one with some mix of big and little science. Unfortunately, the number of little-science projects that must be terminated to release sufficient funds for the operation of a big-science facility is indeed large. Either choice, postponing the use of the new facility or terminating partially exploited older facilities, is wasteful. Long-range planning is clearly needed.

To summarize, in the subfields with big-science components, relatively small decreases in funding, however they are brought about, may lead to the premature termination of facilities and projects and to centralization of research facilities, with consequent changes in the style in which research is conducted as well as in the character of the research institutions—the universities and the national and industrial laboratories.

Decreases in funding for little science result in corresponding reductions in scientific productivity when the level of support places the subfield in the linear region of Figure 6.1. A reduction in funds simply causes a proportional reduction in activity, the main consequence thus being the longer time interval needed to complete a study. Few terminations are necessary and the major effect is a decrease in the number of long-range programs undertaken.

IMPACT OF DECREASING SUPPORT ON PHYSICISTS

The manpower and employment picture is described in detail in Chapter 12. As with any bright young person who has devoted many years of hard work to training himself for an interesting and moderately well-paying profession, the reaction to unemployment or underemployment may be disillusionment and bitterness. How widespread and how important this reaction is for the future of physics is difficult to estimate. But that it exists in some degree is clear from both published letters and discussion in *Physics Today* and *Science*.

The effect on the climate in which physics research is performed can be significant. An attitude of increasing competitiveness is likely to result, which, though often beneficial, can also have an adverse effect. To stay in physics, to be promoted, or to maintain one's position in a laboratory when the reduced financial resources of the institution require a staff reduction will increasingly demand demonstrated superiority with respect to one's colleagues. This requirement can put a premium on visibility and verbal ability. It may also encourage conformity, exploitation, and sycophancy. Of course, whether this situation develops in a given institution, and to what extent, will depend critically on the personalities and philosophy of management of the supervisory personnel. But clearly such concerns can detract from the dedication to research and to teaching in academic environments. They can inhibit the spirit of cooperation and the free discussion of research ideas and problems.

Ultimately the program officer of the funding agency decides the level of support of a given group. When funds are level or decreasing and greater selectivity is necessary, he may feel a greater vulnerability because a failure among the research projects he supports will reflect more on his judgment.

The reaction of growing demands for support under conditions of decreasing funds varies widely among program officers. Some tend to become more involved in the programs they support and to make managerial decisions rather than leaving these matters to the principal investigator. In such cases the program officer may become too committed to the program, with the obvious dangers that this situation would imply.

The experienced research physicists, and the research projects that are more certain to yield significant results, are more likely to be favored. The young physicist may well suffer under such a regime, since his scientific credentials may not yet have been fully established. Probably he will find it much easier to obtain support, machine time, and the like if he is associated with an older and better-known figure.

Much of modern physics is a team effort. Teams are required to run the large machines, and teams are involved in the conduct and interpretation of experiments. The importance of these cooperative efforts cannot be overemphasized. The operation of a large machine, the maintenance of the quality of its performance, the periodic installation of improvements, and the servicing of the experimental groups require many diverse talents acting in concert. Research groups will typically involve a mix of senior physicists, younger PhD's with more free time but much less experience, graduate students, and technicians, who play a most essential role in preparing and mounting an experiment.

Groups of this kind, whether research- or facility-oriented, are remarkably fragile and not easily built up. A happy concatenation of circumstances is required, for not only must there be perceptive entrepreneurs who take the responsibility for the creation of the team but also its members must be talented and mutually congenial. Further, their talents must be mutually reinforcing. Even under the best of circumstances several years are required for the creation of such a team. Its esprit de corps and effectiveness can be maintained for decades. Schools of physics of great productivity and influence are generated in this way. The importance of such coherent teams in national emergencies or needs is obvious. They were responsible, initially at least, for the successes of American scientists in World War II.

With the exception of those physicists who have secure positions and for whom the major effect of reduced support is decreased research opportunities, there are essentially three groups of physicists who are most directly affected by decreases in funds. Consider first the university situation. In a new graduating class those who are not rated highly probably will not be able to find positions within the physics community. They are the first casualties. The problem they pose must not be dismissed simply because they are not among the best; they are still talented.

The highest-ranking graduates will obtain postdoctoral fellowships or positions in the lower echelons of the faculty or of national or industrial laboratories. Every effort will be made to keep those regarded as brightest. For them the problem of obtaining a more secure position is postponed.

The crucial problem arises when the individual has reached the point in his career at which he should have tenure or permanent status. As discussed in Chapter 12 such positions are in increasingly short supply and difficult to obtain, for a variety of reasons mostly related to the tight financial situation.

As noted earlier, a third group that suffers when funding decreases force retrenchment is composed of fully qualified physicists in their late 40's. Although productive and effective they have been in their organization for a length of time sufficient that administrators may conclude that they are not likely to make spectacular discoveries or rise to top-level management positions. Because of their seniority, management finds it economically attractive to replace them with young scientists who command lower salaries and offer the possibility of greater productivity and effectiveness than the older men they replace. The economic and psychological plight of these older men who typically have outstanding mortgages, established life styles, and children in college is indeed sad, and the nation can ill afford the waste of such talent and experience.

The foregoing qualitative remarks are substantiated by recent quantitative studies made by Grodzins for the American Physical Society's Economic Concerns Committee (EEC) and discussed in Chapter 12. These studies and the situations they depict will not go unnoticed by the young people who are now, or soon will be, choosing a career, as well as by their advisers. One of the impacts of level or diminishing support will undoubtedly be a decrease in the number of graduate students and in the number of PhD's granted. To a point this reduction is justified at present. However, the danger of a possible overreaction must be seriously considered. Even more significant is the question of the quality of those who in the future choose physics. Will a considerable fraction of the more promising young people continue to choose physics as a career? Will its intellectual challenge continue to attract the more brilliant student as it has in the past? There are many circumstances that will influence his or her decision. One circumstance will certainly be reasonable assurance that opportunities for creative and useful careers are open to those who have the necessary combination of talents.

Impact of Decreasing Support on Research Institutions

National laboratories seem more vulnerable to reductions in support than do universities or industries, but this difference may be more apparent than real. These laboratories are usually considered integral parts of a particular funding agency, which naturally will make special efforts to keep them viable. Most of them exist to help the agencies to which they are appended fulfill their missions. With changing national priorities and missions, their importance changes. Unless laboratories can respond quickly to such changes they may find themselves with greatly reduced or no support. Agency missions may include the operation of a number of big-science facilities for the benefit not only of the local researcher but also for users from other parts of the United States. Missions also may include the conduct of research in particular areas of technical or academic interest. Even when only a general reduction in funds occurs, a laboratory's only response is to restrict its activity. As a rule it does not have another source of funds or another activity it can immediately turn to. Its only recourse in the long run is to reduce its staff.

Under the adverse conditions of recent years some laboratories have tried to diversify, to find new missions and new support, sometimes from several federal agencies. Large laboratories are more likely to succeed in this attempt, since they have a broad range of skills and know-how that might be applied to many differing problems. Such a course may be the evolutionary path these laboratories will have to follow in order to survive.

The financial crisis that faces universities has already been described. The declining support of science in general multiplies the difficulties. Contributions to new endowments are decreasing, some state legislatures have already begun cutting appropriations, and tuition has been increased to the point of diminishing returns. Thus ways must be found to cut running expenses. A first move in this direction is to reduce staff, primarily by reducing the nontenured faculty and by not replacing members of the tenured faculty who depart to fill positions elsewhere or to retire. Such policies have been only partially adopted so far. The young physicist, with a recent PhD, and the untenured experienced PhD will bear the brunt of these tactics. One consequence is an aging faculty and research staff.

Other retrenchments may take the form of increased teaching loads and reduced salaries. The evidence is that these measures have not yet been employed extensively.

The reduction in graduate support has been made even more severe in recent budgets. Predoctoral fellowships have been eliminated in several federal programs and abruptly reduced in others. To this must be added the reduction of graduate student support, which has occurred because of the reduction in research support. Universities have responded by using their available funds, which are not too plentiful, to support those students who were well on the way to finishing their graduate education. One third of the departments participating in a recent NSF survey needed to increase their use of nonfederal funds for the support of graduate students. The sharp reduction in the number of entering students has been even greater than was anticipated at the time the cutbacks in graduate support were instituted. These effects are most immediately visible in the universities with the most advanced programs in physics; eventually they will be felt in all institutions.

Some universities are in greater difficulties because they have used contract funds to finance research salaries during the academic year. The ratio of research time to teaching load was correspondingly increased for some faculty members. Consequently, additional faculty was needed and hired. When funding is reduced, it is not possible to reverse this process if the additional faculty has acquired tenure in the meantime. The need for long-range planning is demonstrated once more.

Industrial research is for the most part little science. Industry can thus retrench by reducing its research staff in proportion to the reduction in support. Support, at least in terms of direct funding, is less dependent on federal funds than the university or national laboratory, but it is very sensitive to the general economic health of the nation and is closely tied to profits. Even when the profits originate in government contracts, there is a degree of decoupling of laboratory support from federal funding that permits an additional flexibility.

Industrial laboratories tend to be insulated from fluctuations in government support to the extent that (a) they are general-purpose company laboratories financed essentially from past company profits, and (b) the management regards them as necessary for the invention and development of commercially profitable products or services. When their work is directly motivated by a federal mission and lacks commercial exploitability, they are vulnerable to a change in the mission's priority and the accompanying reduction in federal support. Like other research institutions, an industrial company must weigh the short-range financial advantage gained by decreasing the staff against the long-range loss in capability that results when effective research teams are dissolved. Once these groups are allowed to disintegrate, they are likely to prove exceedingly difficult to build up again when financial conditions improve. In their worst form these and other factors tend to make positions in the national and university laboratories look more desirable even if their salaries are considerably lower than those of industry.

Recent reports * indicate that between October and December 1971, four of the largest U.S. corporations made major cuts in staff or changes in the direction of their basic research laboratories. The companies and their laboratories are U.S. Steel Corporation's Edgar C. Bain Laboratory for Fundamental Research; RCA's David Sarnoff Research Center; Shell Oil Company's Emeryville Research Center; and Ford Motor Company's Scientific Research Staff at the Research and Engineering Center. Approximately 30 percent of the scientists employed by each of the first three have been dismissed. While there is no plan involving such a staff reduction at Ford, reorientation of work is expected. A primary motive appears to be short-term economies brought about as a result of a decline in profits.

CHARACTERISTIC TIMES AND THE IMPACT OF FUNDING CHANGES

The interrelationships among institutions, facilities, people, programs, and support mechanisms on which the physics enterprise depends are many and complex. The extent to which any of these elements or relationships can be changed without producing unduly wasteful perturbations in others depends on time. As an example, a certain minimum amount of time (characteristic time) is required for the construction of a modern accelerator; another characteristic time is required for examining the new range of experimental variables it makes accessible; and finally there is a third characteristic time after which the accelerator is considered obsolete. In general these characteristic times are measured in terms of several years. When, because of a substantial change in the budget, a particular element must be modified in a major way and in a time that is small compared with its characteristic time, it is likely that substantial waste will result. There will also be secondary effects, the long-term influence of which is difficult to predict. It is the great disparity between the *annual* budget changes and the characteristic times that are of the order of several years that produces the kinds of problem described in the preceding sections of this chapter. Such problems could be avoided if changes in levels of support occurred gradually over a time comparable with the characteristic times or if adequate advance warning were available. As an aid to future planning, we will discuss briefly the characteristic times of a number of elements.

* D. Shapley, "Industrial Laboratories: Whither Basic Research," *Science, 174,* 1214 (Dec. 17, 1971).

Manpower

Here the question is how much time is required to adjust the number of trained physicists to a new funding level. If this time is long compared with the time during which a funding change is imposed, a serious adjustment problem results. In essence, the time required to adjust the number of trained physicists depends on the time required to train a PhD, as well as the time during which a trained man is productive. The first of these time periods gives us some estimate of the number in the "pipeline"; the second tells us about how long physicists already employed will remain active.

On the average, a graduate student requires about 5.5 years after the baccalaureate to obtain his PhD. (The median registered time in 1966 was 5.6 years.) In addition, one must add a number of years of specialized undergraduate study, so that the total is at least 9 years. The time during which a man is productive varies a great deal with the individual. But the time during which he is employed as a physicist is generally of the order of 30 years. (Questions of training and manpower are discussed in detail in Chapters 11 and 12.)

The two characteristic times of approximately 9 and 30 years make it difficult to adjust quickly to a budget change in either direction without inducing serious perturbations. To increase immediately the number of PhD physicists would require drawing on other fields of science and on PhD's from other countries. To achieve a decrease, the only prospect is to discharge physicists or convince as many as possible of the advantages of early retirement. But this step is not sufficient since there are already several thousand graduate students in the pipeline, many having made the decision to become a physicist as much as a decade ago. As described earlier, the impact of recent funding changes already is being felt in the reduced number of entering graduate students. But it has not dropped to zero. As a consequence, unemployment will rise. (Various estimates and projections of the extent to which supply will exceed demand appear in Chapter 12.)

Big-Science Facilities

The construction of the facilities of large science has in recent times taken from five to ten years. Therefore, authorization of such projects implies relatively long-range commitments for appropriations. It also implies a much longer commitment for provision of support for operation after construction has been completed. Otherwise, the new resource is not likely to receive full exploitation.

The scientific lifetime of a new device is often difficult to estimate, for the facility may become useful for studies other than those for which it was originally built. Cyclotrons, which originally were built to study nuclear reactions, later have been primarily used for isotope production. Small electrostatic accelerators might be used to study various aspects of atomic spectra, as in beam-foil spectroscopy, or aspects of condensed-matter physics, as in ion-implantation studies. The elementary-particle physics accelerators are ultimately useful for high-energy studies of nuclei. In overall scientific planning it is clearly more efficient to expect and encourage such shifts in direction of effort.

Large devices used for the purpose for which they were originally constructed have two such lifetimes. One is for the rapid exploration of the new ranges of experimental variables they make available, a process called skimming the cream, which requires about five years. The other is for the more painstaking and quantitative studies these earlier results suggest. Disregarding a number of possible perturbations, a reasonable figure for the total lifetime of large devices, unless major upgrading or conversion is feasible, is of the order of ten years.

Research Programs

The characteristic times associated with research programs are the time required to plan and perform a single experiment and the time for carrying out a program of experiments designed to expose a given phenomenon in all of its aspects. Although the styles of big science and little science differ in many important ways, the latter time is of the same order for both. The fundamental reasons are human ones: It is difficult to maintain a program of experiments that does not provide the reward of some successes within a reasonable period of time—approximately two to three years or less. Within this interval, a graduate student can get his degree or a young PhD can obtain some hard evidence to help him as he starts up the professional ladder. The older and more secure researcher might find a longer period too much of a gamble, and so might the supporting agencies.

Although there are differences, research programs conducted at large facilities tend to be run in the same way as smaller-scale efforts, for a major requirement following from the large investment involved is to maximize their productivity. In the case of elementary-particle physics, the planning and execution of an experiment may take approximately one to two years; it may involve a team of five to ten PhD physicists and an expenditure of as much as $1 million. These experiments must be thoroughly planned, they must be approved by the appropriate administrative arm of the facility, and they must be ready to run when the scheduled time becomes available. There is little room for error! The analysis of the data may require a year or more and will lead to a number of publications.

In little science the time required for a single experiment is much less. Yet the time required for a series of experiments to make an impact on a field similar to that of the single run at the big-science facility is not too different from the three-year span needed for the latter. Usually, little science requires a systematic study of many cases and circumstances before an effect can be clearly established. There are exceptions, of course. But the genuine flash of intuition that permits the design of a single critical experiment unfortunately occurs rather rarely.

Theoretical physics should be included in the little-science category, although some aspects require the use of large computers and thus can be more expensive. Again, although a single publication may at times require only a relatively short period of work, establishing a useful concept, if in fact there is one, may require several years.

To summarize, the characteristic times that must be considered in the planning and support of physics are (a) the time required to produce a PhD, which, beginning with the time he decides to enter physics, is about eight to nine years; (b) the productive life of a professional physicist, which is about 30 years; (c) the time required to construct a large facility, which is approximately five years, and to exploit its capabilities, which is about ten years; and (d) the time required to carry out a research program directed toward elucidating a sufficiently coherent group of phenomena, which is about three years. These numbers are only rough approximations and should not be taken too literally. However, clearly, all the various characteristic times are much larger than the one-year funding cycle.

LONG-RANGE PLANNING AND EFFICIENT RESOURCE UTILIZATION

We have described briefly a number of the effects of declining support on physics and physicists. Some of the more immediate ones are already evident; others will become perceptible only in the future. Much could be done to minimize or eliminate all these effects if a greater degree of stability in funding could be achieved or if sufficient forewarning of necessary funding changes were available. The importance of stability to other sectors of the economy has been recognized; the Economic Act of 1948, which was directed toward industry and agriculture, is an example. It is time that ways be found to apply longer-range planning to science. As a step in this direction, federal agencies supporting physics should seek an increase in their appropriations and any necessary authority such that, say, one third of their support could be assured for a three- to five-year period. Forward funding such as this does not imply increased support levels or treasury withdrawals in any one year. It would, of course, have to be administered under appropriate guidelines.

Essentially, money is but a medium of exchange that purchases in the ultimate two things: natural resources (renewable and nonrenewable) and human resources. Viewed from the standpoint of a complete, autonomous system, the use of these resources is all that matters. The United States is almost, though not quite, such a system. To that (or even a higher) approximation, the cost to the country of a given activity should be measured only in terms of its use of resources. To do research requires little use of natural resources—still less of nonrenewable or recycleable resources. Furthermore, not to use available human resources, for example, to have unemployed scientists, technicians, and the like, is to waste resources, not to conserve them. To underemploy them has a similar but lesser effect, for others could replace them in what they are doing. The human resources required outside the scientific laboratory are also in oversupply. Hence, to the *nation as a whole,* research costs exceedingly little in the fundamental sense. It generally uses resources that would otherwise be wasted.

Thus, the only reason *not* to support worthwhile research within manpower limits is to do something more urgent—as was the case, for example, during World War II.

It is an obsession with the measurement of costs in terms of money that causes the trouble. Individuals or institutions must measure in these terms because to a high degree of approximation they are *not* closed systems, and money is the medium of exchange across their boundaries. But fundamentally, the United States, taken as a whole, does not have to measure costs in this way.

7

PHYSICS

IN

U.S. SOCIETY

But as soon as I had acquired some general
notions concerning Physics, and a beginning to
make use of them in various special difficulties,
I observed to what point they might lead us and
how much they differ from the principles of
which we had made use up to the present time.
I believed that I could not keep them concealed
without greatly sinning against the law which
obliges us to procure, as much as in us lies, the
general good for all mankind.

RENÉ DESCARTES (1596–1650)
Discourse, Part VI

One lesson of man's history is unmistakable: the
crucial element in the rise of our material well-being
has been the progressive utilization of our ever-
growing store of knowledge of the world in which
we live.

Economic Report of the President,
Washington, D.C., 1964

472

INTRODUCTION

The twentieth century is often said to be the Age of Science. Whether an expression of admiration or apprehension, this phrase implies that science, in conjunction with technology, has played a major role in shaping contemporary society. Not only have man's material surroundings been transformed by the products of science and technology but also what he thinks and how he analyzes many problems. Physics, one of the most fundamental of all sciences, has played a critical part in the transformation. The contributions of physics have come about in four primary ways: discovery of physical phenomena and general laws underlying these phenomena; application of the information and insight that emerge in the orderly development of physics; devices and inventions that are offshoots of the apparatus and machinery designed for scientific investigations; and ingenuity of physicists in applying their skills and methods to problems of this technologically advanced society. In this chapter we review the relationships between physics and contemporary U.S. society, emphasizing those aspects not covered in Chapters 3 (The Nature of Physics) and 11 (Physics in Education and Education in Physics).

The primary objectives of physics are the study, analysis, and understanding of the physical universe; however, the cumulative results of this effort have profound social significance. In the past, physicists were often thought of as men who were preoccupied with the attempt to understand certain physical phenomena and who worked in laboratories and institutions that were divorced from the mainstream of American life. This "ivory tower" view of the physics community continues to acquire added acceptance because there is still little public awareness and understanding of the nature of physics research and the special facilities required for its conduct. But this view of physics ignores the impact that the results of research in this field have had on society. Of course, what physicists produce must be transformed through industrial organizations and educational institutions into devices and techniques that are directly relevant to some practical end. In this country, an elaborate system has developed to convert research results into goods and services; and the standard of living, which we in the United States take largely for granted, is, in significant measure, a reflection of the success that has been achieved in effecting this transformation.

In recent years, the cost that large-scale exploitation of technology exacts in terms of what the economist calls "externalities" has become strikingly evident through, for example, smog, polluted streams, and congested traffic. Not surprisingly, therefore, a tendency has developed to question science and technology in rather simplistic ways as having failed generally to meet current needs or, even worse, as having seduced U.S.

society into trading its natural birthright of a clean and open environment for a mess of technological pottage. That most pollution is due not to advanced technology but to old, and largely empirical, technology is largely ignored. Examples are paper production, smelting of metal ores, fossil-fuel power production, and automobiles. These technologies evolved during an earlier era when there was little concern about the environment. Pollution became a serious problem as they were applied on an ever larger scale, without sufficient modification based on scientific understanding. In general, advanced technologies are much less polluting. If piston aircraft were used today to produce the number of passenger miles now generated by jets, the resulting noise and pollution problems would be much more intolerable than those now imputed to jets. Thus it is a complete perversion of reality to attribute pollution to advanced technology; it is almost wholly due to technology that has *not* advanced or, at least, not advanced fast enough to adapt to increasing scale. Except for thermal pollution, nuclear-power plants are less polluting than fossil-fuel plants, and the only reason that the thermal pollution is worse is that it employs an older technology of steam conversion, that is, the power conversion process is technologically retrogressive. This type of pollution is not inherent in nuclear power but merely the result of an unwise decision made 20 years ago, which many people objected to at the time. In the case of the automobile, the internal combustion engine was perfected empirically without benefit of fundamental research on it or on possible alternatives; consequently, automobile propulsion could not be adapted quickly to changing scale.

The following quotation concerning science in England has a familiar sound but was published by Charles Babbage * in 1830, also a period of rapid technological change:

Many attacks have lately been made on the conduct of various scientific bodies, and of their offices, and severe criticism has been lavished on some of their productions. Newspapers, magazines, reviews and pamphlets have all been put in requisition for the purpose.

And, a century later, the following quotation, pertaining especially to U.S. physics, appeared, not in the 1970's but in the mid-1930's:

Physics has enjoyed a place in the sun which it cannot expect to hold permanently. . . . Physicists would be more than human if they were not somewhat spoiled by the popularity they have enjoyed. . . . Physics in [the United States] has simply growed like Topsy and, unless some thought is given to these matters, we may have an autopsy on our hands.†

* C. Babbage, *Reflections on the Decline of Science in England and on Some of Its Causes* (B. Fellowes, London, England, 1830).
† C. Weiner, "Physics in the Great Depression," *Physics Today, 23*(10), 31 (Oct. 1970).

More recent remarks of like tenor prompted this Committee's effort to place physics in context in U.S. life and society.

In the past 300 years society has shifted from man and other animals as sources of power to the steam engine, fossil-fuel prime movers, and, now, nuclear power. And society is undergoing yet another revolution, the reduction of tedious repetitive work through a variety of control and sensing mechanisms. During these centuries, new materials—metals, chemicals, and synthetic products—have come into use and new instruments and measuring techniques have evolved. These developments have drastically changed the way of life in the United States and much of the world. In addition, man's curiosity and need to explore led him to climb the highest mountains, circumnavigate the earth on the seas and in the air, and land on the moon. Crucial to each of these endeavors were certain instruments and the state of knowledge of fundamental physical laws. Columbus used a magnetic compass; the astronauts employed radar, laser beams, and computers. All these aids that helped to ensure the success of their explorations had their origin in the laboratories of physicists. An interchange of the observations of the explorers and the findings of the physicists was essential to continued progress.

There is no sharp chronological boundary between the era of society symbolized by power—for example, the nuclear generation of electricity —and that characterized by computers and communication and control mechanisms. That science changes the society that supports it is well known. The process is as old as science itself. However, in the United States, the application of science to industrial procedures and products is particularly highly developed; during the past three decades this country has exploited science more rapidly and successfully than has any other nation. The results of physics research have led to the development of new industries in electronics, computers, communications, and energy production. Many of these industries, although facing strong competition from abroad, still maintain a leading position in international trade because of the ability to couple research effectively with industrial productivity.

Although industry and government have steadily improved their capability to develop and apply science since the beginning of the century, World War II marked a major turning point in the relationship of science to society. The mobilization of U.S. scientists as part of the war effort resulted in the development of sophisticated radar, the nuclear weapon, and many other improvements in technology and provided an important lesson for the country. The U.S. public discovered that scientists, when appropriately organized and financed, could make major contributions to pressing national needs. Consequently, an elaborate network of governmental and industrial research and development organizations has come into being to exploit science for national purposes. At the same time, support of physics in the universities has grown steadily to provide the manpower and new knowledge that are especially important for science-based industries.

With growing support and manpower, physics research has developed new analytical procedures, for example, the mathematical formulation of theories and techniques for experimental testing, that are widely applied in business and government. Again, World War II marked a turning point in the evolution of these methods. During the war, physicists played a significant role in the development of operations research, in which the techniques of scientific experimentation and analysis were applied to problems of military strategy and tactics. The objective was to remove some of the guesswork from the decision process. The new analytical techniques led also to new approaches to engineering problems and the development of systems engineering. Further, they became an important part of the training of business executives, since they offered improved way of analyzing and comparing options.

Such contributions of physics and physicists to U.S. society are the main concern of this chapter. Clearly, the translation of basic discoveries in physics to industrial application and public use often requires the efforts of many people working in a logical sequence of activity beginning with applied research and continuing through engineering design, manufacturing, marketing, and the sales part of putting science and technology into everyday use. Exploiting the phenomena of fission required a national commitment of manpower and other resources. The physicist would claim only that he is an essential member of the team. The following section describes briefly some characteristics of physicists and the nature of their work in their own and other scientific disciplines as well as in government and industry. The next section gives examples of the impact of the tools and findings of physics. Two sections then examine basic and applied research in physics in relation to society. Next, some economic implications of physics research are explored. The chapter concludes with a brief assessment of what physics can contribute to the achievement of national goals.

PHYSICISTS: THEIR CHARACTERISTICS AND WORK

Characteristics of a Physicist

Physics gives scope for widely differing intellectual styles within its major experimental and theoretical subdivisions. Among experimental physicists are some meticulous measurers for whom the next significant figure in a physical constant is a challenge like an unclimbed peak. Their work is essential in creating a sound and ever-growing base of scientific knowledge, and it is not devoid of surprises. Many are ingenious inventors of new instruments. For a highly inventive mind the physics laboratory is a congenial setting; nothing is too "far out" to try, and even small inventions can bring about immediate results. Some physicists enjoy the strategic planning of a large group enterprise, even the administrative work. Others like to concentrate on a particular problem. Other styles could be identified by such tags as: long-shot takers, dogged pursuers, critical collators, experimental opportunists, or systematic explorers. Their common goal is to contribute to the body of universally accepted knowledge.

Among theoretical physicists, distinctions are perhaps less obvious and a tendency to follow prevailing fashion more noticeable. But it is still possible to recognize substantial differences in attitude and interest. Some physicists are attracted to problems in which no rigorous theory yet exists and a phenomenological approach is the only way to interpret observations meaningfully. Others are occupied with the theoretical foundations. Some theorists enjoy uninhibited speculation; others proceed more methodically and critically. A few are remarkably powerful analysts. Some are lucid expositors and will make their major contributions to the development of physics as critics and teachers. And some of the most brilliant minds fit either in none of these classes or in most of them.

In spite of their diversity of interests, physicists must have certain traits in common—curiosity about how things work, confidence in reason, and a belief in an underlying order in the physical universe that man can and ought to discover. This last trait is a philosophical position, of course, and one not without antagonists in the history of human thought, who are being heard again today. However, these things could be said about any scientist. What marks the physicist as a physicist could be merely that he finds the very simplest phenomena in nature the most interesting—using, of course, his own definition of simple! The scattering of a photon by an atom is fairly simple; a sunset is more complicated.

Like any science, physics sternly enforces respect for facts. Mistakes are pitilessly revealed. Ingenious theories can be contradicted by new observations and must then be discarded. Experiments can fail, too. The course of actual discovery is seldom a smoothly converging sequence of hypothesis–test–hypothesis–test. Instead, it abounds in puzzlement, even confusion. There may be uncertainty about which of some apparently contradictory observations should be taken seriously, uncertainty that perhaps will be illuminated by an unexpected experimental result or a bright idea. Nevertheless, physics, perhaps to a unique degree, has the particular quality of logical and elegant simplicity and great generality, especially when new discoveries are viewed retrospectively.

Physics is difficult, the competition is rigorous, and the joy of discovery is elusive. Notwithstanding the glamour and prestige that attached to physics for some years, most of the people who remain committed to careers in it are in some way deeply dedicated to this science. But what kind of people are they in other respects, and do they contribute to society in other ways?

First, they are generally very bright people—among the most intelligent, by all available objective criteria, that this society produces. This assertion does not answer the above question, nor does it imply that they necessarily have wisdom to match this intelligence, but it is a reminder that the way in which this human resource is used should be a matter of general concern. They are also relatively young, half of all the active physicists being less than 37 years of age. It is they who will principally determine the contribution of physics to society in the next decade.

Physicists have tended to be fairly versatile as a result of both the general applicability of basic physics and an uninhibited resourcefulness developed through experience in the laboratory. The evaluation of evidence, including statistical evidence, intelligent use of models, distrust of conventional assumptions, and readiness to try a novel approach— all these come naturally to a working physicist. Outstanding in the career pattern of any successful physicist is a characteristic style and approach to problems, an ability to focus on basic issues, and an economy of thought. They are important qualities for dealing with the problems of society.

Other sciences, of course, also foster such qualities. Perhaps physicists are only more self-confident. In any case, from World War II until today, among the scientists taking an active part in issues of public concern, physicists have been conspicuous leaders. There were powerful historical reasons for their early involvement. But they continued to work for some 20 years in both national and international councils for the rational application of science and technology.

Now the United States faces a resurgence of concern. The crucial issue of arms control is still present, together with other agonizing problems of society. A new generation of physicists, basically as idealistic as their predecessors, and certainly as bright, wants to do something about such problems. Many of them, probably most, cannot share the attitudes and relations to the policy-making establishment that the older physicists developed. But they represent a truly extraordinary public resource in mind and spirit.

New PhD physicists represent an extremely small fraction of the total

U.S. technical manpower. Because of the fundamental character of their training, these physicists could have great impact on many types of scientific and technical endeavor, a number of which have greater social and economic implications than does physics *per se,* as traditionally defined. Unfortunately, recent trends in physics education and research, and the customary patterns of recruitment of technical people in other fields, conspire to interfere with the diffusion of physicists into nontraditional types of employment. Yet, only a small percentage expansion of opportunities for physicists in nontraditional job markets could have an enormous impact on both the employment opportunities for physicists and the vitality and capacity for innovation of the fields they enter.

In the last 30 years, many physicists have contributed in a major way to technological developments and problem-solving outside their main fields of research interest by working on a part-time basis. This contribution has occurred largely in relation to space and defense projects and in the nuclear programs. It is important to find analogous ways of enabling physicists, especially the younger generation of physicists, to contribute to the analysis and solution of societal problems without necessarily giving up their careers as basic physicists. Although sometimes naive and arrogant in approach, they can bring fresh points of view to the analysis of problems outside their nominal fields of competence.

This problem requires a systematic and continuing attack on the part of universities, industry, and the government. Possible mechanisms to encourage broader application of the talents and capabilities of physicists include (a) fellowships to enable PhD physicists to enter new fields and (b) meetings, conferences, and continuing education programs designed to expose physicists to other types of research. These mechanisms also provide an opportunity for scientists in other disciplines to familiarize themselves with the perspective and research style of the physicist.

The Physicist in His Discipline

Physicists, whether employed in industry, government, or academia, have a common educational experience from undergraduate major through postdoctoral apprenticeship. Although a physicist may not always work in what may be considered the most challenging branch of his discipline, his training makes him a valuable member of the physics, and of the total, community.

The activities of physicists during World War II provide perhaps the best-known example of the contributions they can make to U.S. society. This story, documented in many reports and books, demonstrates the way in which a nation's intellectual resources can be made available for the common good during a period of national need, a need that is recognized and accepted by the country as a whole. For example, physicists left their own research and flocked to the Radiation Laboratory in Cambridge, Massachusetts, where, for military purposes, they perfected microwave radar. The development of nuclear weapons was the work of physicists, chemists, and engineers in Chicago, Los Alamos, and Oak Ridge. Groups also were established in Massachusetts, New York, Connecticut, California, and elsewhere to work with other scientists and engineers on sonar and rocket-type armaments with solid propellant fuel.

In the various wartime research efforts there was a close working relationship between physicists and engineers in industry and academic physicists who had temporarily left their research. The military departments that used the new weapons and methods also were closely coupled to this large cooperative enterprise. Because physicists played a leading role in design and production groups, user groups, industry, and the military, a common understanding and vocabulary began to emerge.

When the war ended, most physicists returned to their universities, some remained in government, and others re-established and expanded the laboratories created during the war and channeled their efforts in new directions. Industry also expanded its physics activities, both basic and applied, and a new generation of powerful industrial laboratories was developed.

The Physicist and Other Sciences

An important aspect of the overall contribution of physicists is their impact on other sciences, which, in turn, affect society and social well-being. Those trained in physics are frequently attracted to challenging problems in other disciplines. The names of a few of these disciplines are indicative of their basic relationship to physics: astrophysics, biophysics, geophysics, physical chemistry, physical oceanography, and physical meteorology. The physicist working in such disciplines applies his training to the generation of new ideas and new interpretations of observations and to the improvement of instrumentation and the processes of analysis and prediction. A brief description of the nature of research in these interdisciplinary fields and the influence and involvement of physicists appears in Chapter 4, and a more detailed description in the Panel Reports in Volume II. In addition, Chapter 4 and the Panel Reports describe the interaction of each of the core subfields of physics with other sciences and with technology. We have selected one area of special concern to society, medicine and health care, to illustrate the contributions that physicists can make to other types of endeavor; even here we further restrict consideration to nuclear medicine. The results achieved have often been particularly dramatic. Other examples appear throughout this chapter, particularly in the final section on Physics and National Goals.

Physicists first observed x rays and their power to penetrate optically opaque matter at about the turn of the century. To the physicist, these rays offered a tool to probe atoms, molecules, and solids; to the medical practitioner, they offered a new diagnostic and therapeutic procedure. But the widespread application of x rays in medicine and in other fields, such as microbiology, depended on the work of physicists. Their research was prerequisite to the identification and characterization of the electron, the attainment of increasingly high vacuum, the design of high-voltage machinery to accelerate the electron, and the identification of the x rays characterizing various materials. A wide variety of industrial firms took up the manufacture of x-ray machines and tubes. Not only are physicists active in all of these companies, but also every large U.S. hospital has physicists in its radiation department working with physicians in both diagnosis and therapy.

The existence of such departments and the presence of physicists in them have accelerated medical applications of other radiation sources that the physicist has identified and learned to control. Gamma radiation neutrons, and other particles that physicists have used in probing the structure of nuclear matter interact characteristically and specifically with biological tissue, permitting the design of specific clinical treatments. It is now possible, for example, to irradiate sites deep in the body with a minimum of potential damage to normal and overlying tissue. (See Chapter 4 for a fuller discussion of these techniques.) Radioactive isotopes are a common tool in diagnosis and therapy. As new isotopes become available and knowledge of human physiological reactions grows, these diagnostic and therapeutic techniques can be more fully developed and refined.

Most of these isotopes are produced artificially; the widely used element, technetium, for example, appears naturally only in certain stars, but its isotopes are extensively used in both diagnosis and treatment of human cancers. Precise measurements in physics and chemistry laboratories of the energy and amount of radiation produced by hundreds of known artificially produced isotopes make possible selection of the most effective one for each particular application.

The isotopes are produced in nuclear reactions induced in accelerators or nuclear reactors. A substantial fraction of the entire isotope program at the Oak Ridge National Laboratory, a major national facility, is devoted to the manufacture of molybdenum-99. It is used in hospitals as a source of technetium-99, the isotope that is frequently most effective in the diagnosis and treatment of surgically inaccessible brain tumors. Over 100,000 patients received such treatment in 1971, and this number is growing rapidly, with improved techniques for production and administration of the isotopes to patients.

A few additional statistics afford a perspective on the benefits to medicine of just this small segment of physics. In 1968, some four million people in the United States received radiation therapy for cancer; this number, too, is growing rapidly. During the past ten years, improvements in radiotherapy have been astonishing. Hodgkins disease, at one time believed to be incurable, is now cured in about 70 percent of all cases through the use of high-voltage x-ray and cobalt-60 gamma-ray therapy. The Anger camera, which pinpoints lesions deep in the brain and in other inaccessible locations in the body for surgical removal, is widely used. Currently, some ten million scans are performed each year, and the number is growing by some 15 percent per year. New x-ray treatment units, based on an advanced linear accelerator design developed for the Los Alamos Meson Physics Facility, provided the high-voltage x radiation required to reach deep lesions; over 500,000 treatments were recorded in 1970, and the rate of cure for breast cancer has been dramatically increased as the higher-energy radiation penetrates to deep-seated lymph nodes, which escaped radiation by lower-energy x radiation. Some 300,000 patients received direct charged-particle accelerator therapy for cancer in 1970. Here again the cure rates have risen dramatically because of the possibility of programming, with greatly improved precision, the location and character of the delivered radiation.

The application of measurement techniques developed in physics also benefits medical science in other ways. For example, a recent device to measure the number and kinds of cells in blood samples is based on the light scattered by individual cells as they pass under a sharply focused laser beam. Particular kinds of cells are selected by using a beam of ultrasonic radiation to break the stream of blood flowing through the instrument into microdroplets and then, after they have been electrostatically charged, deflecting the droplets containing the special, or aberrant, cells of interest into collectors so that adequate quantities of the different cell populations can be accumulated for study or further culturing.

Medical applications of acoustics, the physics of sound, are many and are increasing rapidly. A cardiogram taken with ultrasound can disclose the motion of individual heart valves. A scalpel has been developed that uses a beam of high-frequency sound to form the very fine cutting edge required in brain surgery. And acoustic holography offers hope of eliminating the major dangers inherent in x-ray examination of the human body—that is to say, exposure to potentially harmful radiation—and of obtaining a depth of image not possible with x rays. This development is of particular importance in obstetrical practice in which acoustic techniques permit fetal study without any possible danger from x or other high-energy radiation.

Increasingly, instruments that once were used only in a physics laboratory are proving their effectiveness not just for diagnosis and therapy but for on-line monitoring of patient treatment in hospitals. These instruments are typically connected to a central computer that permits rapid interpretation and correlation of the various data it records, thus giving the attending physician a rapid summary of the various processes occurring in the patient. In medicine as in physics, such systems present a vast and overwhelming array of analog information in such a format that the essential decisions in treatment, as in experiments, can be made quickly and effectively.

Chapter 4 presents a wide variety of other applications of similar character; basic to all of them is the fact that imaginative use of physics instrumentation has provided medical scientists with probes of unprecedented power and specificity for both direct treatment of human ills and better understanding of the underlying biological and physiological processes.

The Physicist in Government Programs

Since the mid-1940's the federal government has been the main source of support for physics research in the universities (see Chapters 9, 10, and 12). This support was motivated by both recognition of future technological needs and the often justified belief that all research ultimately has an impact on technology. This contact with academic science provided a flow of information on technical problems between physicists and government departments that has continued and grown. Physicists serve on many advisory committees and participate in many of the *ad hoc* analyses on which national policies are based; in the years since the war they, perhaps more than any other scientists, have played an important role in the development of these policies.

In recent years, as concern with problems of the environment has increased, some of the most distinguished physicists have played important roles in working with geophysicists, meteorologists, and others to acquire an understanding of the problems articulated in the growing public debate and to learn what physics and physicists can contribute to their solution. Physicists have actively participated in a number of recent NAS-NAE Environmental Studies Board studies, including those on the proposed

Everglades jetport and on the proposed new jetport for the New York metropolitan area. Although increasing numbers of physicists are prepared to leave the relative isolation of their personal research and work in such newly defined social problem areas, the speed at which such transfers can take place depends not only on the availability of financial support but, more importantly, on the existence of institutions that facilitate such work and assure its translation into practice.

The scientific community has a responsibility to explain the implications of all its work, drawing insofar as possible on its special professional expertise. Scientists should participate in the social and political processes that determine how their work will ultimately be used, although they have no right, or special wisdom, to dictate to society the social uses of this work. Physicists, however, do have a strong tradition of concern about the social consequences of their discoveries, as exemplified by their central role in the development of plans for the control of nuclear energy after World War II. In more recent times physicists have been much involved in the movement for scientific study of arms control, and many individual physicists played prominent parts in both the informal and formal negotiations leading to the nuclear test ban treaty. The calculation of nuclear weapons effects, methods for the detection of underground or atmospheric nuclear tests, and methods for monitoring the production of materials for weapons all depend on instruments familiar to physicists and detailed application of modern physical theories. Physicists also have contributed substantially to educating the public and delineating the scientific issues that are relevant to public-policy debates on disarmament.

To develop arms-control policy or to serve on the staff of any such policy group, an understanding of the total problem must include its political context. A thorough knowledge of physics is a necessary but far from a sufficient qualification. Moreover, a rare talent of wisdom and self-confidence combined with intellectual humility is sorely needed; this trait is not always characteristic of physicists.

In an increasingly technological civilization, the number of major national decisions that have critical scientific and technological components (witness the recent SST debates) grows rapidly. Arms control is only one of many such major issues to which physicists can make significant contributions. But in doing so it is essential that they be prepared to devote the necessary time to appreciate not only the political and social subtleties of the situation but also the essential inputs required from all the sciences other than physics. All too frequently, physicists have tended toward overly simplistic views of national problems, and their effectiveness as partners in the solution of these problems has been correspondingly reduced.

The Physicist in Industry

Some 30,000 to 50,000 scientists are employed in about 150 industrial research laboratories; this total does not include a limited number of major development and engineering laboratories (such as those of IBM, Bell Telephone Laboratories, General Electric, and the like), although the boundaries between these types of activities in industry are difficult to define. We estimate that there are approximately 3000 physicists in these 150 research laboratories, or about 6 to 10 percent of the total scientific population in industrial research (see also Chapter 9).

The training of industrial physicists is usually the same as that of their colleagues who follow careers in academic, governmental, or nonprofit institutions. But the industrial environment differs from other professional environments; this difference stems from the main function of industry—that of providing products and services to society. These products and services focus the creative talents of the industrial physicist on ways to improve the performance of a desired or specified function; possible new functions and services that would benefit society; and means of reducing the cost of a product, function, or service to make it more readily available and useful to society. Each industrial laboratory has its own characteristics, which are determined by the nature of the industry it serves, the orientation and goals of the specific enterprise, and most important, the people who work in the laboratory and the direction or lack of direction that the individual research experiences.

A major responsibility of the industrial physicist is that of being aware of developments in other laboratories and research areas that could have potential impact on the products and services of his own company. There are many methods of monitoring the outside world, including reading and extracting pertinent information from published research reports. But this approach, as we discuss in Chapter 13, is not sufficient. To be alert to the subtleties of research results, one must work in at least a related field.

The industrial physicist must interact with two worlds, that of his professional peers in universities and other firms and industries and that of his particular industrial organization with its special goals, products, and problems. He looks to both worlds for reward and recognition, which can present difficult problems, for the two sets of goals are not always compatible. His professional peers recognize open publication and contributions to the conceptual development of physics as a science. The goals of his firm often require that proprietary restrictions be placed on information. Further, industrial rewards depend largely on practical contributions to the improvement of products or to the development of new products that enhance an industry's growth and profits. To combine both goals successfully takes a rare talent, but that it can be done is shown by the large number of successful industrial physicists who have also achieved recognition for their contributions to physics as a discipline; among them are such names as Coolidge, Hull, Germer, Brattain, Maiman, Rose, Gunn, Giaever, Matthias, and many others.

The role of the physicist in industry is not confined to the research and development laboratory. Some move from the laboratory to the corporate staff and the upper levels of management. Some who have helped to develop new materials and instruments form their own companies or join recently established firms to exploit their ideas.* Because many physicists prefer to maintain contact with, or responsibilities on, university faculties and to participate in industrial ventures on a part-time basis, such new industries frequently are located in close proximity to leading universities. Examples are optical companies producing lasers, fabricators of transistors and large-scale integrated circuits, manufacturers of energy equipment, and producers of various types of specialized instruments. Route 128, with its close intellectual ties to the academic communities of the Boston area, is an outstanding example of this symbiosis.

To start a new company, or expand company resources to explore new technological concepts, requires capital. To provide support for such an enterprise, a bank or lending institution must have some evaluation of the technical validity of the endeavor and its potential. Increasingly, a role that the physicist fulfills is that of full- or part-time consultant to provide information and judgments on the character of the risk. Increasingly, too, physicists are finding that their education and training in fundamental science provides an effective basis for legal training and for careers in major brokerage institutions.

The Physicist in Education

Surprisingly, the contribution of physicists to their society through their participation in general education is often ignored or misunderstood. We consider these questions at greater length in Chapter 11, but it seems appropriate to raise some of these issues here as well.

Increasingly, U.S. citizens experience a sense of frustration, and often unease, because their education has not equipped them either to understand or appreciate the major scientific and technological components of

---

* For an extensive discussion of mobility and innovation see D. Shimshoni, "The Mobile Scientist in the American Investment Industry," *Minerva*, 7(1) (January 1970).

national concerns. Because of its emphasis on fundamentals and economy of thought—with which much of the entire range of natural phenomena may be understood, at least in broad-brush outline, on the basis of a very limited number of general principles—physics should form part of the intellectual arsenal of any educated citizen. This suggestion in no way implies any necessary degree of professional expertise, but it does imply a sufficient acquaintance with the principles and approaches of physics to provide a framework within which the citizen can begin to appreciate—and enjoy—some of the subtleties of natural phenomena and can come to grips with at least the essential technological elements of national problems.

Physicists have not been adequately concerned with their responsibilities either in public education or in helping the general public to develop an awareness of the substance and goals of their science. It is essential that this situation be changed. (These questions are addressed more fully in Chapters 2, 11, and 14.)

SOME TOOLS AND FINDINGS OF PHYSICS: THEIR IMPACT ON SOCIETY

Although the connection between specific devices or findings of physics research and the various products and services that flow from them is not always apparent, there are some cases in which it is striking and obvious. The relationship of the huge modern electrical power generator to the simple coils and magnets of Faraday's laboratory is easy to see. And the connection is clear between the television set and the primitive electron tubes that Thomson used in his laboratory to elucidate the nature of the electron. But transforming the handmade "crotchety" curiosities of the research laboratory into broadly useful and reliable items and services requires the ingenuity and skill of many engineers and physicists. The costs of this transformation process usually are far greater than those of the original research effort and many more people typically are involved.

Familiar examples of devices commonly associated with physics, such as transistors, x-ray machines, and lasers, tend to obscure hundreds of less obvious examples. As an illustration, the thermos bottle, a common object in the U.S. home, was invented by Sir James Dewar early in this century in connection with his work on the liquefaction of gases. Indeed, the modern thermos bottle, apart from its metallic protective can, looks very much like the one that Dewar constructed for his own research. Another example of the ubiquity of the devices of physics is the analyzing instrument now used in urinalysis for the detection of heroin addiction;

it is a nuclear magnetic resonance device developed in the purest of physics research programs 25 years ago, for which its inventors received the Nobel Prize in physics. It has now become a standard tool throughout chemical and biological research. Liquid oxygen, now manufactured in enormous quantities for widespread industrial use, was first produced in a physics laboratory in minute quantities nearly 100 years ago; it was then nothing more than a strange blue magnetic liquid—a laboratory curiosity. In the early nineteenth century, even the thermometer was regarded as only a scientific instrument.

The discovery of the electron, the production of high vacua, and the study of gas discharges are directly related to such commonplace processes and devices as illumination in homes and streets, communications, computers, and photography. New industries and social institutions have come into being, and curricula in schools and colleges are continually modified because of such discoveries in fundamental physics. And with these developments society's need for people educated in physics has increased.

The vacuum tube is the fundamental device that made possible radio and television broadcasting and reception and permitted telephone conversations between continents. It also was the basis for the development of radar, which is now the tool for the control of air traffic and for the tracking of storm clouds.

Vacuum tubes also became the building blocks of the first electronic digital computers, and, as a by-product, instruments such as the electron microscope came into being. As we discuss in Chapter 4, this has had the most profound effect on our knowledge of the microscopic world. With such instruments as linear electron accelerators operating at 1–20 GeV, it is possible to observe and deduce individually the structure of the nucleus of the atom and of its component parts. At lower energies, the electron microscope has become a common tool in chemical, biological, physical, and medical laboratories. At present, more powerful and versatile electron microscopes are being designed; and not only does the ancient dream of seeing individual atoms and molecules appear close to realization but also, with the new scanning electron microscopes, a degree of three-dimensional realism has been attained that provides all scientists with unprecedented information and new understanding of their microscopic worlds of study. Physicists working in research and engineering laboratories continue to make major contributions.

Electrons moving through a partial vacuum produce light. Various sources of light for illumination, therapy, disinfection, and display derive from discoveries in the physics of gas discharge, which is an active field in many industrial research laboratories. The study of fluorescent material is closely related to solid-state physics. Historically, the identification of the electron and the development of quantum mechanics formed the basis for contemporary work in solid-state physics. The initial theoretical work on the solid state, which occurred in the late 1920's and early 1930's, concentrated on metals—copper, gold, iron, tungsten. These studies led to greatly improved understanding of the electrical resistance, specific heats, and many magnetic properties so important now in magnetic tape used in recording, computer systems, and other industrial applications. They also led to vastly improved understanding of the strength of materials, the causes of structural failures, and the possibility of entirely new designs using new materials and old materials in new ways.

Continued experimentation and theoretical speculations brought other basic concepts into being; these included imperfections, band structures, and junction phenomena. Much more experimentation and theoretical work are necessary to achieve a thorough understanding of these concepts. However, the present state of knowledge has been sufficient for the development of the transistor and the laser, to give only two examples, and for the widespread use of ferromagnetic and ferroelectric materials in such devices as copying machines and photographic systems. Here again Chapter 4 provides many additional illustrations.

Superconductivity was observed early in the twentieth century. At very low temperature—20 deg, or less, above absolute zero—many materials lose their electrical resistance entirely; electrons can flow without dissipating energy. Fifty-six years after these early observations were made, applications are now emerging. These applications relate, for example, to new instruments, such as stronger magnets, requiring very little power and more efficient ways of transferring energies to charged particles in linear accelerators. Increasing skill in the fabrication of such instruments could lead in the future to transmission of electric power by means of superconducting cables, vastly increasing the efficiency and lowering the cost of electrical transmission. This development could have profound social consequences in making feasible the siting of large electrical power generation plants remote from the urban areas they serve. This example is typical; observations in the laboratory, resulting from continuing experimental and theoretical probing, often lead to the discovery of new properties of matter and their application. Applications in laboratory instrumentation usually are found first, followed by applications in industry and broad social utilization.

The digital electronic computer could not have developed to a capability approaching its present one without the many inputs arising from

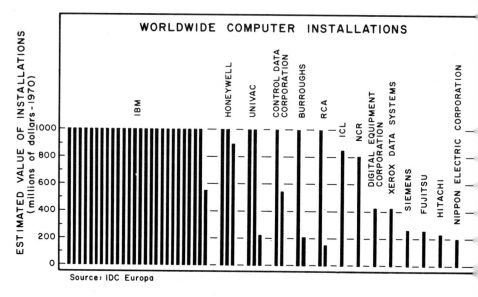

FIGURE 7.1 Computer installations worldwide, by company. [Source: *The Economist* (October 2, 1971).]

basic research in solid-state physics, but many other intellectual efforts were required to incorporate the physics research results into the conceptual design of an operable computer. This industry is still growing throughout the world—Figure 7.1 gives some indication of its volume—and its social impact has yet to be fully felt and understood. In 15 years, the basic building blocks of the computer industry have progressed from vacuum tubes for circuits and cathode rays for memory devices to solid-state memory and circuit elements and from discrete solid-state devices connected to printed circuit boards to completely integrated circuits, with all components and interconnections fabricated within the same tiny wafer of silicon or germanium crystal. In the transition, memory devices changed from cathode rays to ferrites to semiconductors. Similarly, the input and output of computers no longer are limited to such media as cards and printers but include cathode-ray-tube displays, three-dimensional and holographic display systems, and the like. As each of these technological transitions developed, the computer became more powerful, less costly per computation, and smaller; it also required progressively less

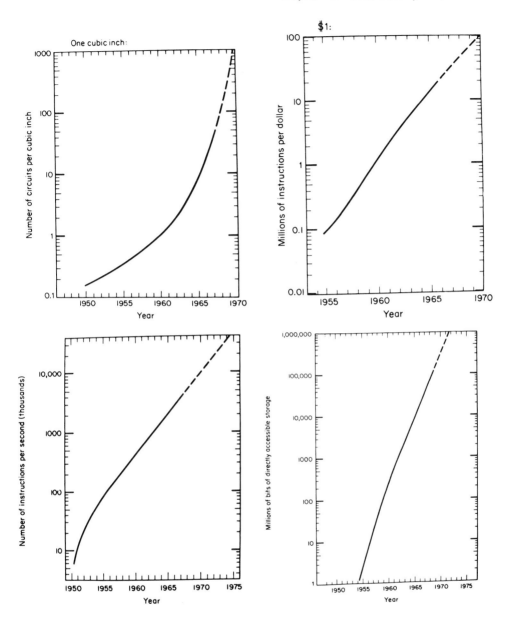

FIGURE 7.2   Improvement in computer efficiency and cost. [Source: J. Martin and
A. R. D. Merman, *The Computerized Society* (Prentice-Hall, Englewood Cliffs, N.J.,
1970).]

power. This trend continues, almost exponentially, as Figure 7.2 shows. The rate of advance depends on research in industry, universities, and government laboratories, and especially on a strong capability in condensed-matter physics. This capability also will have increasing, and perhaps revolutionary, impact on communications.

A final example of an instrument developed by physicists that is finding ever-wider application, with ever-growing impact on society, is the laser. Until 1960, all known light sources were broadband, that is, they contained a mixture of wavelengths. The first means of producing a sharply tuned source emitting only one wavelength was developed by C. H. Townes and A. L. Schawlow in 1958, when they applied the principle of stimulated emission to the problem (*light amplification by stimulated emission of radiation*, or laser).

In the last decade, research has increased the power of laser pulses over 100,000 times, to a million million watts—on an instantaneous basis equivalent to the entire output of all electrical power plants on earth! (This power, however, is sustained for only one million millionth of a second, so only 1 J of energy is involved in total.) The energy emerges as a small button of radiation only 0.001 in. thick, but in that disk the electromagnetic field strength is so high (over a million times larger than was ever attained previously) that a host of new physical phenomena are observed as the pulse moves into ordinary matter. Among other effects, it should be possible with such a laser to heat matter to temperatures approaching 100 million deg, at which thermonuclear reactions can be initiated; this capability possibly has very important applications to the development of controlled fusion power.

Currently, lasers can be used for welding, drilling, and cutting in precision work; for extremely accurate measurement of distances on the surface of the earth and between the earth and moon (recent work achieved a precision of less than 6 in. between the earth-based laser and the reflector systems left on the lunar surface by the Apollo astronauts); and for various military purposes, both passive (as in range finding) and active (as in antimissile defense). Potential applications in dentistry and medicine also are being explored; already lasers are widely used for repair of retinal detachments and in exploratory fashion to replace the traditional dental drill. In combination with many other optical techniques, lasers hold great promise for improved communications, because the information capacity of a system is inversely proportional to the wavelength. Thus, a factor 10,000 in capacity is realized in going from a relay link based on microwaves to one based on visible laser radiation. The rate of increase in demand for communications is such that this high-information bandwidth will almost certainly be required before the end of the century.

BASIC RESEARCH IN PHYSICS: ITS IMPLICATIONS FOR SOCIETY

We live in a world shaped by the application of discoveries in physical science. However, it remains difficult, if not indeed impossible, to foresee the ultimate applications and social significance of any particular piece of research at the time it is begun. The revolutionary social consequences of Einstein's work on energy and matter were not anticipated at the time it was done, nor were the implications and applications of Maxwell's work on the electromagnetic field immediately evident. In the mid-1930's it was far from obvious, even to Rutherford, the dominant figure in the discovery of the atomic nucleus, that understanding of nuclear phenomena could lead to energy sources transcending in magnitude anything then even imagined. It was only when the fission process in heavy nuclei was discovered (1938) that the possible accessibility of nuclear energy became apparent. The subsequent development of nuclear energy has had a profound effect on society and currently is the basis of an industrial enterprise in the United States involving over $2 billion per year.

This inability to foresee the ultimate results and consequences of physics research has important implications with regard to the allocation of resources and the setting of priorities and program emphases in physics. Much of physics research involves pure exploration, and the return to society from investment in any kind of exploration is always difficult to quantify. It is not possible to guarantee that any particular piece of research will lead to new inventions, just as oil companies cannot guarantee that any particular drilling project will result in a new source of oil.

The invention of the transistor in 1947 by three Bell Telephone Laboratory physicists,* as we have mentioned previously, well illustrates the impact that basic research can have on society. An enormous variety of solid-state devices has evolved since. The manufacture of integrated circuits, for example, now constitutes a U.S. industry that grosses about $500 million per year. This development came as a natural outgrowth of research in basic physics.

The principle of the transistor was discovered inadvertently. Researchers at Bell Telephone Laboratories, Purdue University, and several other laboratories were attempting to understand how to modify conductivity and how, by means of an electric field, so-called surface states of a material affect this phenomenon. On one occasion, the Bell workers made a technical error, such that the device for applying the surface field did not function as intended, and an unexpected phenomenon resulted. The effect achieved was just what was needed for signal amplification; thus, the transistor was born.

---

* W. Shockley, J. Bardeen, and W. H. Brattain, in *Nobel Lectures in Physics 1942– 1962* (Elsevier, Amsterdam, 1964).

Indeed, the importance of these totally unplanned events in the history of science cannot be overestimated. What characterizes the scientist, however, is his ability to recognize the unexpected and unknown phenomenon as potentially interesting so that it is examined critically rather than ignored. In just this way Fleming discovered penicillin; the list of such accidents in research is long. This sensitivity comes with long and intensive experience in a field so that the background into which a new phenomenon will fit is already understood.

Although the discovery of the transistor was inadvertent, two years of patient research that was to explain the basic mechanism involved had preceded the discovery. And even this research was based on a much longer series of theoretical advances in, and experimental tests of, electronic conduction in solids that extended back to 1926.

When the connection between physics research projects and a perceived social need is obvious, as it is in some cases, it is reasonable to expect that such projects will receive a significant share of the resources available for the support of physics research. In fact, as discussed in Chapter 10, a part of the present system of allocating funds to research tends to favor projects with probable applications. Many institutions that support physics research—especially mission-oriented government agencies and industry—appropriately and necessarily are biased toward research with a high probability of short-term payoff.

However, a balanced long-range national program of physics research cannot be achieved by providing funds for only those research programs that appear to have immediate applications. Some portion of the available resources must be devoted to exploration, the goal of which is not only the discovery of new phenomena but also better understanding in general. In many cases, the new phenomena have been predicted on the basis of existing theory or as the result of a deeper insight into theory. The classic case is the Josephson junction, involving the tunneling of paired electrons through a thin insulator separating two superconducting metals. Once Bardeen, Cooper, and Schrieffer had provided a fundamental understanding of the phenomenon of superconductivity, it was possible for Josephson to predict, with remarkable precision, the behavior of these junctions, which now bear his name and which have such widespread applicability both as tools in fundamental research and as components in computers and other electronic devices. In other instances, only part of a phenomenon was predicted theoretically but, when studied experimentally, proved to be much richer and more interesting than originally foreseen; this was the case in work on the tunnel diode, for example.

Although it is essential that physics research that improves existing technology (for example, electron-tube sources of microwave radiation) be maintained, the exploration of new phenomena that might yield radically new approaches and methods (for example, a new way to produce microwave radiation) must not be shortchanged. Had exploratory research been neglected in the past, the transistor, laser, and nuclear power sources, which have revolutionized the technologies to which they have so far been applied, would not have come into being. Even if one were to adopt the extreme view that the only purpose of physics is to produce material goods for society, basic, exploratory research would still have to provide the foundation on which physics would build.

This point was emphasized in symposium presentation * by H.B.G. Casimir of N. V. Philips Gloeilandenfabrieken:

I have heard statements that the role of academic research in innovation is slight. It is about the most blatant piece of nonsense it has been my fortune to stumble upon.

Certainly, one might speculate idly whether transistors might have been discovered by people who had not been trained in and had not contributed to wave mechanics or the theory of electrons in solids. It so happened that inventors of transistors were versed in and contributed to the quantum theory of solids.

One might ask whether basic circuits in computers might have been found by people who wanted to build computers. As it happens, they were discovered in the thirties by physicists dealing with the counting of nuclear particles because they were interested in nuclear physics.

One might ask whether there would be nuclear power because people wanted new power sources or whether the urge to have new power would have led to the discovery of the nucleus. Perhaps—only it didn't happen that way, and there were the Curies and Rutherford and Fermi and a few others.

One might ask whether an electronic industry [would have] existed without the previous discovery of electrons by people like Thomson and H. A. Lorentz. Again, it didn't happen that way.

One might ask even whether induction coils in motor cars might have been made by enterprises which wanted to make motor transport and whether than they would have stumbled on the laws of induction. But the laws of induction had been found by Faraday many decades before that.

Or whether, in an urge to provide better communication, one might have found electromagnetic waves. They weren't found that way. They were found by Hertz who emphasized the beauty of physics and who based his work on the theoretical considerations of Maxwell. I think there is hardly any example of twentieth century innovation which is not indebted in this way to basic scientific thought.

The exploitation of basic physics research for the production of goods and services for society has proceeded at a phenomenal pace, especially since World War II. Current emphases in physics research suggest that this process will continue at an accelerated rate (Table 7.1). In nearly every subfield of physics there is promise of new technological developments. A few examples selected at random are the following:

---

* Symposium on Technology and World Trade, National Bureau of Standards, Washington, D.C., November 16, 1966.

TABLE 7.1   Decreases in the Interval between Discovery and Widespread Application [a]

Device or Process	Years from Discovery to Exploitation	
	Baker Data	Hafstad Data
Electric motor	65	—
Vacuum tube	33	—
Radio	35	35
X-ray tubes	18	—
Nuclear reactor	10	—
Radar	5	15
Atomic bomb	7	9
Transistor	3	5
Solar battery	2	—
Stereo rubbers	3	—
Photography	—	112
Telephone	—	56
Television	—	12

[a] W. O. Baker and L. R. Hafstad have provided supporting data for the assumption that the interval from discovery to widespread application is shrinking. Baker's data show a decrease from 65 years for the electric motor to 2 years for the solar battery. Hafstad's data show a decrease from 112 years for photography to 5 years for transistors. The apparent discrepancies in the table simply reflect the use of somewhat different criteria in the two studies.

1. New sources of energy based on nuclear fusion;

2. Direct generation of electrical power by magnetohydrodynamic means;

3. Improvement of communications through new solid-state sources of microwaves and the development of small portable radars for personal traffic guidance and control;

4. New metal fabrication methods based on the laser;

5. Increased application of nuclear radiation and instrumentation and of radioactive tracers to health care and to the solution of agricultural problems;

6. Improved instrumentation for monitoring environmental quality;

7. Increased possibility of weather control;

8. Earthquake warning and prediction through laser-based precision measurement of crustal motions;

9. New large-scale excavation techniques based on new very-high-temperature physics and technology;

10. Low-loss transmission of electrical power and new transportation mechanisms, both based on superconducting systems.

THE NATURE AND ROLE OF APPLIED PHYSICS RESEARCH

It is almost impossible to draw a meaningful dividing line between basic and applied physics research, although there are activities at either end of the research continuum that can be classified as either basic or applied physics. Further, the line between basic and applied research changes constantly as physics finds new applications and, at the same time, new fundamental questions are raised by technological developments. There is some applied physics in almost every subfield of physics, with some including much more than others. The largest effort in applied physics is closely related to the physics of condensed matter and includes metal physics, high-polymer physics, much of thermal physics, and solid-state device physics. A sizable fraction of atomic, molecular, and electron physics and of plasma physics and the physics of fluids also can be considered as applied physics. And virtually all of optics and acoustics can be so considered. In fact, acoustics and optics have been regarded largely as applied physics for many years, since the basic principles have long been well understood and most work has been directed toward the application of these principles in a variety of specialized situations or configurations of potential practical value.

The relationship between basic and applied physics can be illustrated by the history of nuclear physics, applied nuclear science, and nuclear engineering. Until the discovery of fission at the end of the 1930's, nuclear physics was primarily of intellectual interest. Its development was guided by the effort to discover empirical regularities in nuclear structure and nuclear reactions, leading to general laws, and by the overriding effort to uncover the nature and quantitative description of nuclear forces. But, immediately following the discovery of fission and the invention of the concept of the neutron chain reaction, much physics already in existence, which had been of primarily intellectual interest, suddenly became applicable. At the same time, many experiments and theories that had been carried as far as seemed rewarding on intellectual grounds suddenly needed to be extended greatly in detail, accuracy, and scope to meet the needs of a practical objective. This activity was applied nuclear physics, though that classification was not given it at that time. Subsequently, growth in the use of radioisotopes and the development of nuclear power reactors created a demand for people familiar with the techniques and concepts of nuclear physics to adapt them to the needs of newly emerging technologies. New lines of investigation were opened in neutron physics, radiation biology, and nuclear chemistry and on the effects of pile radiations on crystalline and other materials used in reactor construction. New and more reliable particle-counting techniques had to be developed. The more systematic study and detailed measurement of the properties of radionuclides had to be undertaken. Little of this work would have been regarded in advance as having sufficient intellectual importance to the conceptual structure of nuclear physics to have justified a large investment of people and money, although this work did, in retrospect, have considerable impact on the conceptual development of nuclear physics. For example, it led to the concept of nuclear shell structure. The theoretical and conceptual impact was, however, partly a by-product of a primarily applied effort. It is impossible to designate an exact point at which nuclear and neutron physics became applied nuclear physics or nuclear chemistry or a point at which applied nuclear science became the new discipline of nuclear engineering.

The motivation for much of the present research in physics is mixed. A deeper and more detailed understanding of nature is sought both for its own sake and because, as we have noted above, the phenomena revealed, or better understood, can lead to a new technology, the modification and improvement of old technology, or the development of ancillary technology needed to make a recent invention practicable. As an example, the work on the physics of crystal growth and crystal imperfections was directed partly toward obtaining better controlled materials for transistors and other semiconductor devices. Yet, even such practical work in applied physics is characterized by a mixed strategy involving both a practical and an intellectual aspect. The choice of a problem is motivated by the need for an answer for a particular purpose and the ripeness of the problem for solution in the existing state of science. Both opportunity and need coincide. It is, perhaps, this search for coincidence between scientific opportunity and technological need that distinguishes applied physics from engineering technology. In engineering, the practical need can be so urgent that it may be necessary to be content with an empirical answer, even in the absence of full understanding of all the parameters of the problem. In applied physics, one is still seeking a practical solution, but usually a generic one that can be applied beyond the specific case in question because a genuine understanding of the parameters of the problem and their interrelationships has been achieved. If the practical problem is sufficiently urgent, it may be necessary to attack it simultaneously with an empirical and an applied physics approach. There is no guarantee that the applied physics approach will succeed first, but when and if it does succeed, it is likely to be much more valuable in terms of long-range impact and applications and more economical of total effort.

And frequently research projects that have very specific applied and practical goals result in the discovery of totally new fundamental phenomena or behavior. Two examples of major fundamental discoveries that have resulted from work in applied physics are the original discovery of radiations from space in the radio spectrum and the more recent discovery of the cosmic background microwave radiation. Both of these discoveries were made in the course of studies of sources of radio interference for radio and microwave communication. Both were made by scientists at the Bell Telephone Laboratories, who had to deal with and try to solve the practical communications problems raised by such interference. From the applied viewpoint, the characterization of the noise in frequency, intensity, and directionality and its correlation with environmental parameters was sufficient. But each of these discoveries actually opened entirely new areas of fundamental science, which have subsequently been pursued beyond the need for practical application because of their intrinsic intellectual importance. However, there are also examples of situations in which a less talented scientist might well have simply ignored or dismissed the new phenomena as irrelevant to the main practical thrust of his research. It is this sense of the importance of the unexpected that often leads to the most far-reaching discoveries.

Applied physics takes place for the most part in industrial and government laboratories. In general, the universities work on the less applied, less problem-oriented aspects of physics. However, an important recent development is the growing emphasis on applied physics in engineering training. A much more thorough background in fundamental physics is now required of most engineers, especially electrical and electronics engineers. These two engineering disciplines rest heavily on condensed-matter science and other modern physics; thus the distinction between applied physics and electrical and electronics engineering, especially at the doctoral level, is becoming tenuous. Evidence of the close coupling between physics and these engineering disciplines is that 40 percent of the citations in articles in the *Proceedings of the IEEE* are to articles published in the journals of the American Institute of Physics.

The electrical, aerospace, and instrument industries depend especially heavily on physics. There is also a considerable amount of applied physics performed in the oil industry in connection with the development of geophysical exploration techniques. In government laboratories, most applied physics takes place in the Department of Defense laboratories, the centers of the National Aeronautics and Space Administration, and the National Bureau of Standards. An increasing amount of applied physics is performed in connection with environmental sciences, such as meteorology, oceanography, and geophysics, but applied physics constitutes at the present time only a small fraction of the overall work in these fields, except in the case of planetary physics. There is scope and need for the capabilities of applied physicists in these fields that impinge directly on problems of national concern.

ECONOMIC IMPLICATIONS OF PHYSICS RESEARCH

The United States currently invests more than $500 million a year in physics research. It does so in the expectation of a major return in benefits to society. To predict and quantify the return on this investment with any degree of certainty is impossible at the present time simply because the connections between basic research and its applications are too complex and subtle for present modes of analysis.

Though physics research often is associated with the initiation of new processes for industry, it seldom is directly connected with an end product. Physics research leads to the development of new devices, new materials, and new processes as well as improved understanding of existing processes, which permits better control and more efficient production. To see the impact of physics research in large industries, such as the electronics, power, or computer industry, is easy. To assess its contribution quantitatively is another matter.

The factors that determine economic growth are many and complex. Economists generally agree that research and development play a significant role, but they differ in their estimates of the actual magnitude of this role. The one clear conclusion that emerges is that the contribution of research and development to economic growth is positive, significant, and high. The rates of return on research and development are almost certainly greater than the rates of return on capital investment.*

Nelson † has pointed out that the standard measures used by economists as indices of increased economic performance do not adequately take into account the contributions to improvement in human life made by many kinds of scientific and technological advance.

Economists used the gross national product (GNP) (deflated to take price changes into account) as the standard measure of improvement in the ability of the economy to meet society's wants and needs. The rate of growth of the GNP (or of the related concept, national income) is often designated as the rate of economic growth, and the growth of GNP per capita (or per capita income) is interpreted as the rate of improvement in the standard of living. Although these indicators probably are the best

---

* *A Review of the Relationship Between Research and Development and Economic Growth/Productivity* (National Science Foundation, Washington, D.C., Feb. 1971).
† R. R. Nelson, *Science, The Economy, and Public Policy,* RAND Rept. P-2903 (The RAND Corporation, Santa Monica, Calif., 1964).

currently available, they fail to show many of the more intangible benefits to society resulting from scientific and technological change as well as some of the adverse effects.

The GNP in any year is (generally speaking) the weighted sum of the goods and services purchased by consumers, provided by the government, or purchased by industry to increase its productive capacity. The output of goods that are sold on the market is valued at market prices. Those benefits, such as defense and public education, that are not provided to the public through market channels are valued at cost. There are a number of conceptual and practical difficulties with the GNP measure; for example, defining "final goods" and deciding what should and should not be included, taking adequate account of changing quality, and devising an appropriate weighting system so that the rate of growth of GNP is meaningful even when the different goods and services that comprise it are growing at different rates. All these problems would exist in the absence of the introduction of new products and services; however, introducing new products and services into the economic system complicates the problems of analysis. Comparison of the effects on the GNP of two major technological developments provides an example.

Consider first the contribution to economic capability made by the invention and development of processes for catalytic cracking of hydrocarbons. From the point of view of both final consumers and the GNP, catalytic cracking does not lead to the production of anything new. What it does is to increase the amount of gasoline and other high-value products that can be produced by a given quantity of human and material resources. In other words, catalytic cracking increases productivity. If, to simplify the discussion, it is assumed that there is no change in total resources used in the production of gasoline, the value of the increased output made possible by catalytic cracking will be counted as an increase in GNP. And, in this case, this measure seems a reasonable first approximation to the contribution to the ability of the economy to produce wanted goods and services that resulted from the invention and development of catalytic cracking.

But consider also the invention and development of the airplane. Unlike the invention and development of catalytic cracking, the airplane provided an entirely new service. Without airplanes it would be impossible, for example, to go from New York to Berlin in less than four days, much less eight hours—not just more costly (in resources and manpower), or less convenient, but impossible. How much does U.S. society value this new capability? Certainly by much more than the dollar amount of air travel fares. Yet, this latter is what the GNP, as presently calculated, counts.

Sometimes a major advance has a seemingly adverse effect on the GNP. For example, the longer life of automobile tires that resulted from the use of synthetic rubber did not add to the GNP. In fact, the GNP was decreased, since fewer tires had to be purchased at the same price to get many more passenger miles. Yet tires count as final goods.

Technological advance clearly has played a major role in expanding the range of the possible and increasing the potential of the economy to meet the material needs of society. Often, as with catalytic cracking, technological change increases the quantity of existing goods and services that the economy is capable of producing. Frequently, too, as in the case of the airplane, technological advance permits needs to be met more effectively than before or satisfies needs not previously recognized. This broadening of the options is scarcely considered in GNP calculations. However, the GNP and the balance of payments (which suffers from the same difficulties of analysis) provide convenient gross indicators of the health of the national economy.

The unprecedented standard of living that the U.S. citizen enjoys is strikingly depicted in Table 7.2 in which a wide variety of countries are

TABLE 7.2   Gross National Product per Capita [a]

GNP/Country (per Annum)	Country
Over $3000	United States
Over $2000	Canada, Sweden, Switzerland
$1000–1999	Australia, Austria, Belgium, Czechoslovakia, Denmark, Finland, France, Germany, Iceland, Israel, Luxembourg, Netherlands, New Zealand, Norway, United Kingdom, U.S.S.R. (and, by inference, the Baltic countries absorbed by the U.S.S.R. after World War II)
$500–999	Argentina, Bermuda, British West Indies, Bulgaria, Chile, Cuba, Cyprus, Greece, Hungary, Ireland, Italy, Japan, Malta, Mexico, Okinawa, Panama, Poland, Romania, Spain, Republic of South Africa, Uruguay, Venezuela
$200–399	Central and South America (other than those countries included in other categories in this table), Guam, Hong Kong, Iran, Iraq, Jordan, Lebanon, Libya, Malaysia, Singapore, Portugal, Samoa, Turkey, Yugoslavia
$100–199	Algeria, Bolivia, Brazil, Cambodia, Ceylon, Egypt, Gaza Strip, Korea, Liberia, Morocco, North Borneo, Paraguay, The Philippines, Saudi Arabia, South Vietnam, Sudan, Syria, Taiwan, Thailand, Tunisia
Below $100	Afghanistan, Borneo, Burma, India, Indonesia, Laos, Nepal, Pakistan, Haiti, Africa south of the Sahara (except the Republic of South Africa)

[a] This table is based on data from *Mobility of Ph.D.'s—Career Patterns Report Number Three* (National Academy of Sciences, Washington, D.C., 1971).

grouped in terms of GNP per capita, obtained as the ratio of reported GNP to total population.

International trade provides another economic indicator. Table 7.3 and Figure 7.3, taken from a report by M. Boretsky,* show trends in U.S. trade with other countries, in 1951–1969, by broad commodity groups. Particularly striking is the increasing extent to which the retention of an overall positive balance-of-payments posture (exports–imports) has depended on the favorable balance in the so-called technology-intensive products. As this favorable balance in technology-intensive products increased from $5.8 billion to $7.8 billion between 1962 and 1969, the overall balance declined from $3.2 billion to $1.7 billion. Early in 1971, indications were that for the first time in the postwar period this overall balance could become negative. About 40 percent of the favorable trade balance in high technology arises from electronics, computers, and aircraft—all industries that are heavily based in, and dependent on, physics. Even in high-technology industries, however, the recent trends are not reassuring. The balance still is favorable, but imports are growing more rapidly than exports. After a period of substantial gain in the 1950's and 1960's, the U.S. trade balance in these industries has leveled off. The rate of development of technology-intensive industries abroad, especially in Western Europe and Japan, suggests that our balance will deteriorate. The net flow of trade with Japan in such products has shown a deficit since 1965. The average annual growth of exports and imports in excess of the average growth of the GNP in 1968–1969 is shown in Figure 7.4, also taken from the Boretsky presentation.

The U.S. lead in high-technology trade does not occur for all products. For example, in electronics the United States has lost its lead in radios, television (except color), tape recorders, sound-reproduction systems, and the like. This situation developed because these are mature branches of high technology, which can be relatively easily adapted by other nations with lower labor costs. The U.S. trade superiority lies principally in products that are actually labor intensive, but in which the necessary skills are a unique possession of the United States because they are at the forefront of advancing technology. As a case in point, the United States currently supplies well over half of the world's consumption of integrated-circuit components while supplying very much less than 50 percent of the consumer goods based on these components. As the technology matures, the special skills are learned by others or are essentially embodied in standardized capital goods.

It is obvious that research and development play an especially important

* M. Boretsky, "Concerns about the Present American Position in International Trade," in *Technology and International Trade* (National Academy of Engineering, Washington, D.C., 1971).

TABLE 7.3 Trends in U.S. Trade with All Countries in the World by Defined Commodity Group [a,b]

	Transactions ($Millions), Average for						Average Annual Growth, Percent	
	1951–1955	1957	1962	1964	1968	1969	1951–1955 to 1962	1962 to 1969
1. Agricultural Products:								
U.S. Exports (Gross)	3,247	4,643	5,034	6,348	6,227	5,936	5.0	2.0
U.S. Imports	4,450	3,872	3,869	4,143	5,054	4,954	−1.6	4.0
Gross Balance	−1,203	771	1,165	2,205	1,173	982	...	...
Noncommercial Exports	n.a.[b]	n.a.	1,446	1,619	1,178	1,018	...	−5.0
Commercial Balance	n.a.	n.a.	−281	586	−5	−36	...	...
2. Minerals, Unprocessed Fuels, and Other Raw Materials:								
U.S. Exports (Gross)	1,611	3,252	2,742	3,420	4,154	4,741	5.3	6.7
U.S. Imports	3,660	4,978	4,946	5,500	7,548	8,077	3.4	7.2
Gross Balance	−2,049	−1,726	−2,204	−2,080	−3,394	−3,336	...	...
Noncommercial Exports	...	...	...	...	...	...	...	...
Commercial Balance	−2,049	−1,726	−2,204	−2,080	−3,394	−3,336	...	...
3. Nontechnology-Intensive Manufactured Products:								
U.S. Exports (Gross)	3,711	4,045	3,452	4,419	5,419	6,210	−0.8	8.8
U.S. Imports	1,884	2,900	5,107	6,038	11,220	11,689	11.5	12.6
Gross Balance	1,827	1,145	−1,655	−1,619	−5,801	−5,479	...	...

Noncommercial Exports	n.a.	n.a.	246	211	90	92	⋯	−17.0
Commercial Balance	n.a.	n.a.	−1,901	−1,408	−5,891	−5,571	⋯	⋯
4. Technology-Intensive Manufactured Products:								
U.S. Exports (Gross)	6,630	8,752	10,216	12,110	18,399	20,575	5.0	10.0
U.S. Imports	897	1,570	2,542	3,068	9,404	11,323	12.3	24.0
Gross Balance	5,733	7,182	7,674	9,042	8,995	9,252	⋯	⋯
Noncommercial Exports	n.a.	n.a.	1,816	1,922	1,420	1,440	⋯	−4.0
Commercial Balance	n.a.	n.a.	5,858	7,120	7,575	7,812	⋯	⋯
5. All Commodities:								
U.S. Exports, Including Re-exports [c] (Gross)	15,336	20,871	21,713	26,650	34,636	38,006	3.9	8.4
U.S. Imports	10,961	13,418	16,464	18,749	33,252	36,043	4.7	12.0
Gross Balance	4,375	7,453	5,249	7,901	1,384	1,968	⋯	⋯
Noncommercial Exports	n.a.	n.a.	3,508	3,752	2,688	2,550	⋯	−8.0
Commercial Balance	n.a.	n.a.	1,741	4,149	−1,304	−582	⋯	⋯

[a] This table is taken from M. Boretsky, "Concerns About the Present American Position in International Trade," in *Technology and International Trade* (National Academy of Engineering, Washington, D.C., 1971).

[b] ⋯ = nil; n.a. = not available.

[c] The value of all commodities consists of the value of the four specified commodity groups, plus the small value (1.2 to 1.5 percent of the total) of re-exports not reported by commodity group. The value of these re-exports to all countries in the world was as follows ($Millions): 1951–1955 = 137; 1957 = 180; 1962 = 269; 1964 = 353; 1968 = 437; 1969 = 544.

Sources:

All Transactions: 1951–1955 to 1962—Computed from the data on U.S. merchandise exports and imports by class of commodity, rearranged to fit the classification of commodities used in this table, as reported by the Bureau of the Census in *Historical Statistics of the United States, Colonial Times to 1957 and Continuation to 1962* (Series U51 through U72); and *Statistical Abstract of the United States*, editions for 1960 (Tables 1182 and 1183), 1963 (Tables 1205 and 1206) and 1965 (Tables 1238 and 1239). *1962–1968*—Bureau of the Census, excerpted and checked for intertemporal consistency by International Trade Analysis Division, U.S. Department of Commerce.

Noncommercial Transactions (military grant/aid shipments, exports financed by U.S. foreign aid programs and sales of agricultural products for nonconvertible currencies under Public Law 480), estimated from data of Department of Defense, Agency for International Development, and the Department of Agriculture.

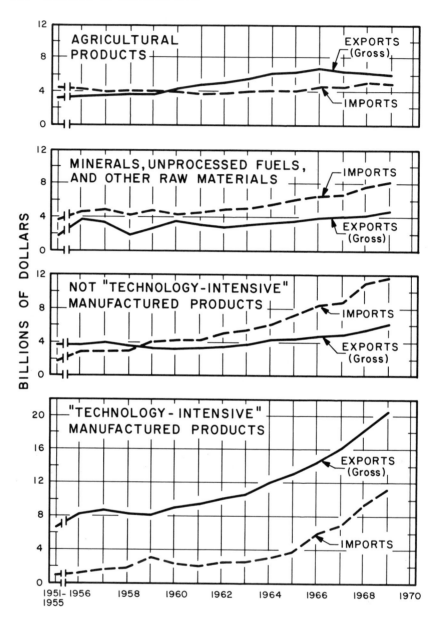

FIGURE 7.3 Trends in U.S. trade in four defined commodity groups with all countries in the world, 1951–1955 to 1969. [Source: M. Boretsky, "Concerns About the Present American Position in International Trade," in *Technology and International Trade* (National Academy of Engineering, Washington, D.C., 1971).]

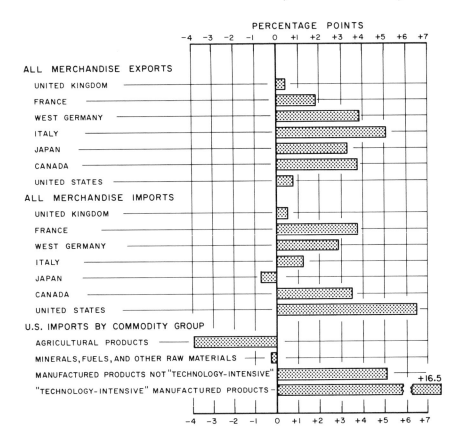

FIGURE 7.4 Average annual growth of exports and imports in excess of average growth of GNP in 1963–1969 (current prices). [Source: M. Boretsky, "Concerns About the Present American Position in International Trade," in *Technology and International Trade* (National Academy of Engineering, Washington, D.C., 1971).]

role in the sector of industry that produces technology-intensive products, such as electronic devices, computers, scientific instruments, aircraft, and the like. Figures 7.5 and 7.6 provide a striking illustration of this in the case of computers, in which the strong and broadly based research and development activity of the IBM Corporation has helped it to take a commanding lead in world computer activity. In these industries, innovation is essential for success. Although these industries are vital for the health of the domestic economy, they occupy a particularly sensitive position in relation to exports; consequently, they have a major impact on the

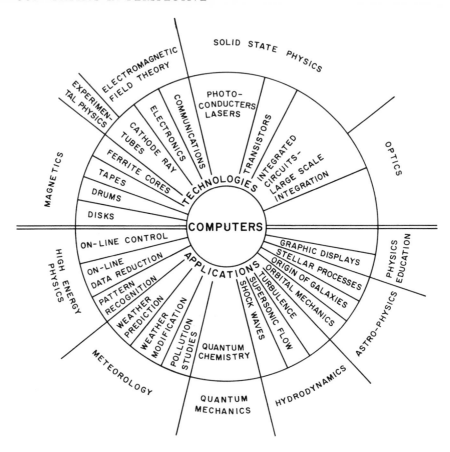

PHYSICS AND COMPUTERS

FIGURE 7.5   Physics and computers: contributions to the development of computers and applications of computers in physics.

balance of U.S. trade with other countries.   The technology-intensive industries provide the largest component—more than half—of all U.S. commodity exports; during the past two decades this component was the only one in which exports always exceeded imports.   Maintaining this favorable balance is a matter of increasing concern, especially in view of the recurring rate deficits that have characterized this country's international trade in raw materials and manufactured products that are not technology-intensive.

Research and development are not the only factors that determine the

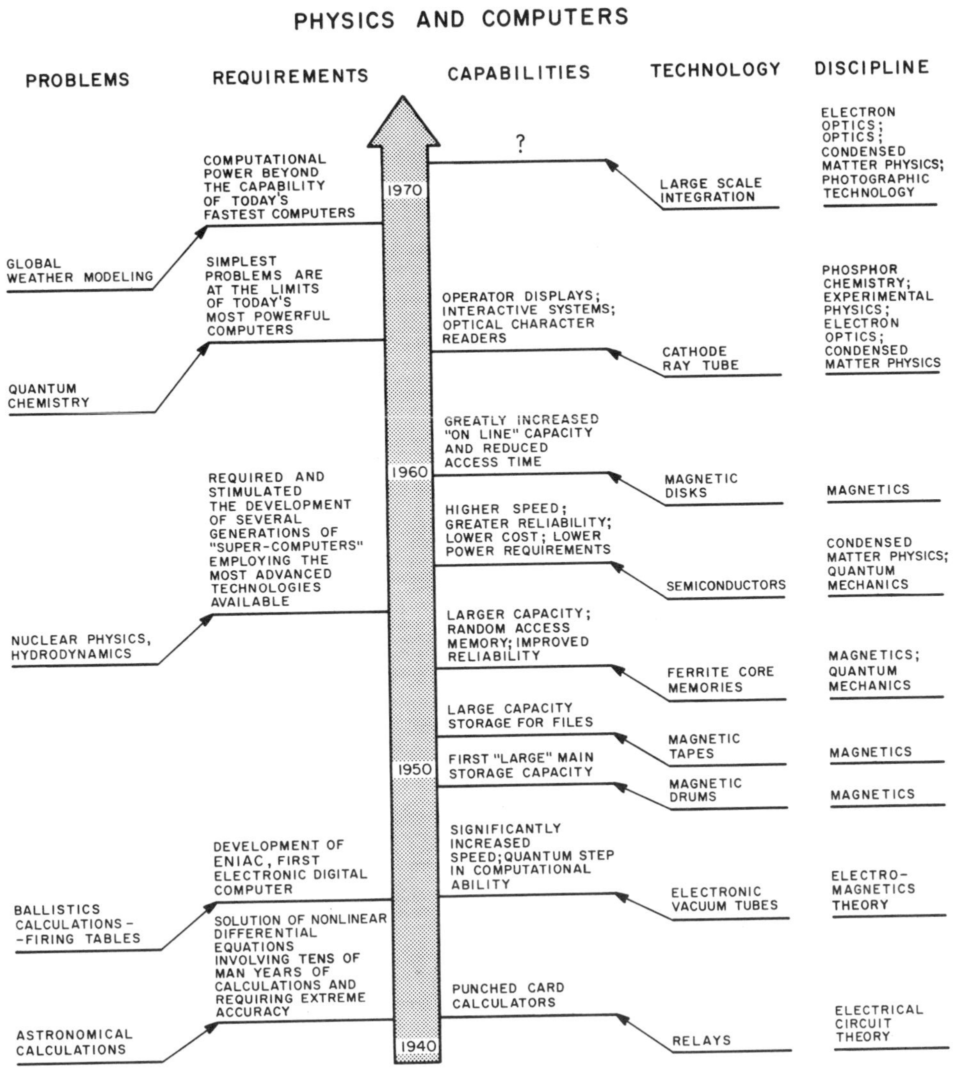

PHYSICS AND COMPUTERS

FIGURE 7.6 Physics and computers.

U.S. ability to trade in high-technology products. Productivity and market-ing could be more important. Clearly, however, a strong research base is essential for the success of such industries. The deterioration of many U.S. research and development institutions and of industrial involvement in research and development in recent years will diminish this country's ability to compete in this critically important sector of world trade. The consequences for the U.S. domestic economy could be serious.

Recognition of the pressing need to stimulate research and development in U.S. industry underlies the broad range of studies comprising the New Technological Opportunities Program, initiated at the request of the President of the United States; this need was also the keynote of recent hearings * held by the Subcommittee on Science, Research, and Development of the Committee on Science and Astronautics, U.S. House of Representatives, chaired by John W. Davis. The problem is receiving intensive study in the Department of Commerce and is of major concern to the Office of Management and Budget, an official of which recently stated †:

There is no doubt that R&D has a positive effect on economic growth—the problem is to figure out the financial mechanisms for increasing R&D.

More recently, President Nixon in his State of the Union message on January 20, 1972, noted as follows:

As we work to build a more productive, more competitive, more prosperous America, we will do well to remember the keys to our progress in the past. There have been many including the competitive nature of our free enterprise system; the energy of our working men and women; and the abundant gifts of nature. One other quality which has always been a key to progress is our special bent for technology, our singular ability to harness the discoveries of science in the service of man.

At least from the time of Benjamin Franklin, American ingenuities enjoyed a wide international reputation. We have been known as a people who could "build a better mousetrap"—and this capacity has been one important reason for both our domestic prosperity and our international strength.

In recent years America has focused a large share of its technological energy on projects for defense and for space. These projects have had great value. Defense technology has helped us preserve our freedom and protect the peace. Space technology has enabled us to share unparalleled adventures and to lift our sights beyond earth's bound. The daily life of the average man has also been improved by much of our defense and space research—for example, by work on radar, jet engines, nuclear reactors, communications and weather satellites and computers. Defense and space projects have also enabled us to build and maintain our general technological capacity, which—as a result—can now be more readily applied to civilian purposes.

Physicists have a special opportunity and challenge in this accelerated attention to technologies applied to civilian purposes and in improving the U.S. competitive position in international technology.

During World War II, physicists demonstrated, in striking fashion,

* Subcommittee on Science, Research, and Development of the Committee on Science and Astronautics, U.S. House of Representatives, *Science, Technology, and the Economy* (U.S. Government Printing Office, Washington, D.C. 1971).
† *Science, 173*, 794 (1971).

their abilities in converting fundamental science into technology. During this period they were evolving entirely new approaches to old problems, identifying entirely new problems, and doing both on a crash basis. More recently, however—in part reflecting the fact that engineers now often receive much the same education that physicists received a decade or more ago—the traditional wisdom in much of the industrial sector of the economy is that engineers are much more flexible, more adaptive in their approach than are contemporary physicists. Thus, they are assumed to be more able to carry the responsibility for converting fundamental science into industrial technology than are the physicists.

In a very real sense, our international competitiveness in high-technology industry has reflected two facts—the fact that, in the tradition of the Yankee peddlers, U.S. industries were characterized by superior salesmanship and the fact that U.S. industries were, on the average, more effective in making the transition from fundamental science to technology. Since World War II, these factors have served to preserve the healthy balance of payment posture in high-technology industry shown in Figure 7.3. It should also be emphasized that this has been achieved in competition with foreign industrial operations characterized generally by much lower labor costs.

In recent years there has been ample evidence that the United States is losing its competitive salesmanship advantage to the Japanese and West Germans, if not to others as well. To remain competitive, U.S. industries must concentrate on even greater effectiveness in the transfer of basic science to technology, on substantial increase in the productivity of the individual worker, and on reducing the time interval between initial discovery and its industrial application.

This will require totally new approaches, as well as evolutionary improvement of old ones. It will require much the same style of approach that physicists demonstrated in the face of national emergencies in the 1940's. The Committee believes that physicists can and will rise to these challenges of the 1970's if given the opportunity to do so. Making this possible will require not only general recognition of the problem but also effective joint action by all three—the academic, industrial, and federal—communities.

PHYSICS AND NATIONAL GOALS

When defense and space were among the top-priority areas in the nation's overall program of action, a significant fraction of the overall U.S. physics research activity was oriented toward meeting the goals established in these areas. Now national priorities have shifted. The nation faces a range of critical problems qualitatively different from those of the 1950's and early 1960's. Among the major concerns of the present decade are improving the quality of the environment, providing adequate medical care for all citizens, extending educational opportunities to all qualified persons, evolving an adequate national transportation system, protecting the privacy of individual citizens, and providing energy to meet the burgeoning demands of U.S. society. Does physics have a vital role to play in achieving these new national goals?

To be effective in these areas, physics research must be part of a coherent national effort. Unlike the problem of national survival in time of war, when a consensus is easily achieved, these new problems do not as yet enjoy a firm and agreed-on basis for action. A major task for political leaders is to shape a consensus so that a coordinated, sharply focused program of action becomes possible. In this particular task, physics obviously is of minor importance.

However, physics clearly does have a role to play in the achievement of many of these goals. New ideas, materials, and instruments resulting from physics research will contribute to the development of the technology necessary to cope with such problems as identifying and measuring sources of pollution. Physicists can contribute to many phases of transportation and housing design. And physicists, with their colleagues in other sciences in the universities, will work to design a variety of new educational programs. Clearly, most of the problems of current concern relating to urban America and the environment are not amenable to purely technological solutions, because they do not have purely technological origins. But science and technology, carefully directed, can help in the solution of many national problems and the realization of national goals. Three types of activity to which physics already is making a major contribution provide examples of its role in contemporary society.

Weather Prediction

The economic and social importance of predicting storms as well as ordinary weather conditions has long been recognized. The U.S. Weather Service maintains a large network of stations designed to collect and correlate local meteorological data in an effort to make accurate predictions several days in advance. At present, the task of data gathering and numerical prediction is being organized into an international program, the World Weather Watch (data collection) and the Global Atmospheric Research Program (prediction). The goal is to achieve valid weather predictions over the entire surface of the earth for two weeks in advance. Hopes are based on the new satellite technology (see Figure 7.7) and the advent of

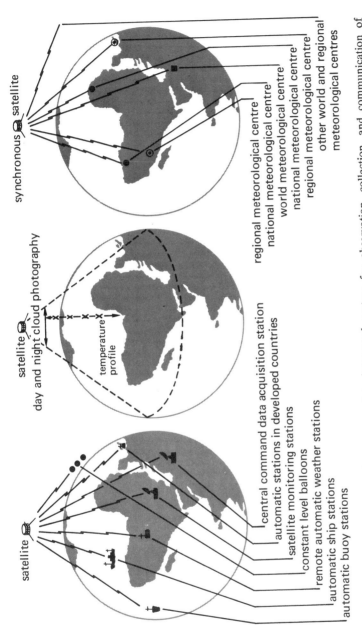

FIGURE 7.7 Satellites can form a complete integrated system for observation, collection, and communication of meteorological data. A satellite can make direct observation of, for example, cloud cover or temperature profiles in the atmosphere (center) or can locate and interrogate other data-observing platforms on the ground or in the atmosphere (left). As satellites are capable of carrying both systems, this offers a ready-made communication system for centralization of weather data at meteorological centers around the world (right).

even larger digital computers, together with the rapid advance in knowledge of fundamentals that has characterized meteorology since World War II. Although this program will continue into the 1970's, preliminary experiments have already succeeded in reproducing in the computer several well-known complex patterns that recur in the world's weather, including the maintenance of east–west prevailing winds by cyclonic waves. As an example of improved accuracy, in the last 20 years the occurrence of errors of 10° or greater in the Salt Lake City forecast has dropped from 60 percent to 24 percent.

The value of extended and accurate weather forecasts has not been accurately assessed. There is, however, a consensus that impressive savings could accrue to many segments of our national life. Great economic benefits would flow to the following industrial or public sectors: agriculture (for example, in planting and harvesting crops); construction (for example, optimum scheduling of the work force, materials, and machinery); water management and conservation (for example, flood and drought protection and irrigation); and public utilities (for example, more efficient facility repair, maintenance, and switchover of capability). Substantial savings would accrue to society through more efficient management of air, highway, and water traffic routing and scheduling, decreased spoilage of perishable commodities during shipping and storage, and improved planning of recreational activities. As an example, if the efficiency of utilization of men and equipment in construction projects could be improved by 5 percent through better scheduling based on reliable weather forecasts, more than a billion dollars would be saved.

The achievement of reliable numerical forecasts for a period of two weeks is an attainable goal. Physicists began to make a contribution toward this goal in 1946, when J. von Neumann realized that the advent of high-speed computers made it possible, at least theoretically, to predict weather with far greater precision. To carry out such a program the following steps are necessary:

1. The equations governing the flow of air and its accompanying water vapor around the earth must be derived from basic physical principles.

2. A close network of measurements of temperature and pressure must be made throughout the entire earth in the period immediately preceding that for which the prediction is to be made.

3. The equations must be solved, using the measurements as input data, to yield predicted temperatures, pressures, and wind speeds, for some future period of time.

Work has progressed on these steps since 1946. The problem is one of awesome complexity. Although the basic equations—those of fluid mechanics and thermodynamics—have been known for more than a century, simplifying them for application to this particular task has been exacting. The problem is to include all relevant phenomena, such as the mean stresses set up when a wind blowing across a mountain range causes turbulence, without being overwhelmed with an enormous quantity of useless or irrelevant information, such as the exact size and location of every turbulent eddy generated in this process.

Current efforts at improving numerical forecasts are of three general kinds, all of which are essential to progress. The need for a vast network of measurements is being met by an ingenious development using sensors in earth-orbiting satellites. From satellites, temperature information as a function of altitude, water-vapor distribution, cloud height, horizontal winds, and many other parameters can in principle be measured. The development and testing of these methods is of major importance to solving the prediction problem.

The second way in which improvements must be sought is in the dynamical equations of the weather models. The subgrid motions, and other phenomena that cannot be included explicitly, should be treated as faithfully as possible, consistent with economy of computation. Given a computer capacity and speed, there are many ways to select the meteorological variables that appear explicitly in the equations, and continuing research is needed to find the optimum choices. This problem involves many aspects of the physics of the atmosphere, oceans, and earth. (See Figure 7.8.)

The need for rapid computation, of course, is met by an electronic computer, although the complexity of the problem is such as to exceed the capacity of even the very largest and most sophisticated computers yet built. For example, a present-day computer may represent an instantaneous state of the atmosphere by perhaps 100,000 numbers; effectively it solves 100,000 simultaneous ordinary differential equations. The numbers might be the values of five meteorological variables at each of five elevations at each point of a grid of 4000 points distributed over the earth's surface. Therefore, each grid point, or weather station, must account for conditions over the order of 125,000 sq km. Systems of thunderstorm size generally will be lost between stations, while somewhat larger systems will be recognized but poorly described. Thus considerable compromise is necessary. The computing power needed to solve "conveniently" problems now under study exceeds, by a factor of 50, the power of the fastest of the present computers.

Obviously, continued effort is needed to increase the memory capacity and speed of computers. As noted in Chapter 4, recent developments in basic physics promise orders-of-magnitude improvement of both these

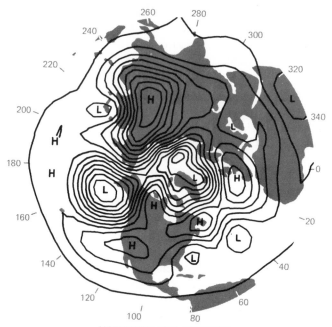

JANUARY 15, 1958: OBSERVED

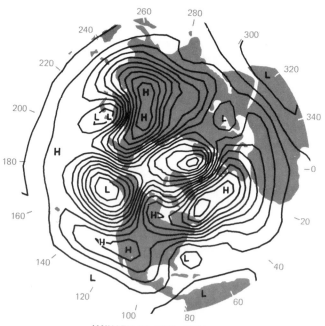

JANUARY 16, 1958: FORECAST

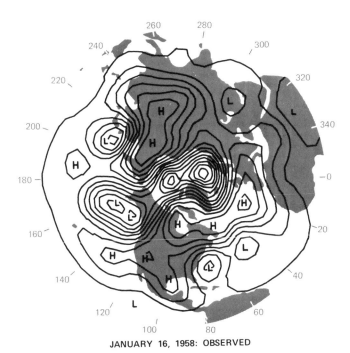

JANUARY 16, 1958: OBSERVED

FIGURE 7.8 A 24-h forecast is shown above on a polar projection of the NCAR atmospheric model. First two maps (*opposite page*) show sea-level pressure on consecutive days, the second map being computed by the two-layer model. Final map (*above*) shows the observed pressures on the second day. Model correctly predicted the generally eastward advance of low-pressure regions.

characteristics. It is also necessary to improve the techniques of getting data from the measuring stations to the computation center. The needed improvements in speed are a matter not only of reducing the time for individual arithmetic operations but also of designing the machine to take optimum advantage of the sequencing of operations called for in the solution.

Of enormous potential importance is the possibility of weather and climate control. Tornadoes kill several hundred people a year and do several hundred million dollars' damage to property. Any practical method for reducing the frequency or severity of these intense storms would be of great value. However, at present, there is no scientifically sound technique for modifying tornadoes. Hurricanes are now far better understood, and there is some evidence that they can be modified. In this regard, a recent study done by the Stanford Research Institute points out that a modification of hurricanes by only 15 percent could decrease property damage by at least 5 percent and save lives in the process. There are many more possibilities. An NAS–NRC report * outlines the many possibilities for small- and large-scale modification and reports some apparent successes. It also finds that any real advance depends on a fundamental understanding of weather phenomena. Although this understanding does not yet exist, it is the target of many research programs, for good or ill. In this decade, man may have substantial control over some aspect of the weather.

Electrical Power

The United States will face an extremely serious problem toward the end of this century in meeting the national requirements for electrical power without at the same time introducing severe environmental change. Between now and the year 2000, the nation will consume more energy than it has in its entire history. By the turn of the century, the U.S. demand for power will have more than doubled, while the world demand will have nearly tripled. (See Figure 7.9.) In calendar year 1970, the U.S. power consumption amounted to 69×10^{15} British thermal units (Btu); of this total 95.9 percent was derived from fossil fuels, 3.8 percent from hydroelectric installations, and 0.3 percent from uranium fission. This consumption of over 300 million Btu per capita can perhaps be better visualized as corresponding to 13 tons of coal or 2700 gallons of gasoline. (See Figure 7.10.)

Science and technology will be sorely pressed to meet the increased demands for power—and in ways acceptable to society. Fresh creative thought must be brought to bear on the possibility of order-of-magnitude improvement in technologies for generation of power and protection from damage resulting from the generation of power. The challenge must be met, for national welfare, the health of the economy, and national defense are vitally linked to energy. This section treats briefly some of the methods that involve physics and physicists in an obvious way—fission, breeder, and thermonuclear reactors, magnetohydrodynamic generators, and solar-energy plants.

The nature of the energy problem becomes clearer when demand is considered against the supply of economically recoverable fuel supplies. Since the earth's deposits of fossil fuels are finite in amount, energy from these sources can be obtained only for a relatively short time. Nuclear energy is,

---

* *Weather and Climate Modification: Problems and Prospects*, NAS–NRC Publ. 1359 (National Academy of Sciences–National Research Council, Washington, D.C., 1966).

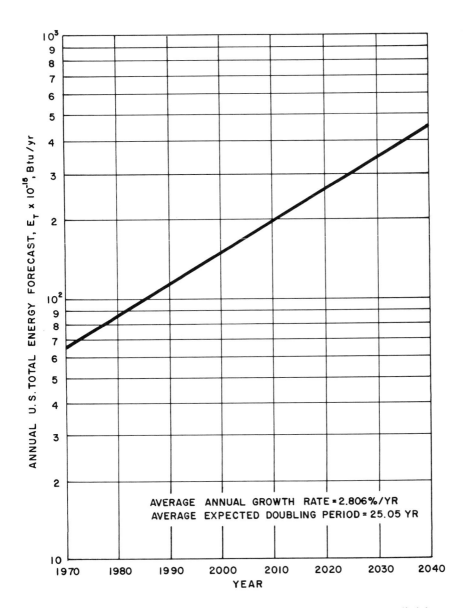

FIGURE 7.9   Annual U.S. total energy (E_T) forecast, 1970–2040 as compiled from 56 separate forecast estimates. The average annual growth rate shown is about 2.8 percent, resulting in a sevenfold increase in demand in 2040—a 25-year doubling time. [Source: *The U.S. Energy Problem (Volume 1): Summary,* ITC Report C 645 to the National Science Foundation (November 1971).]

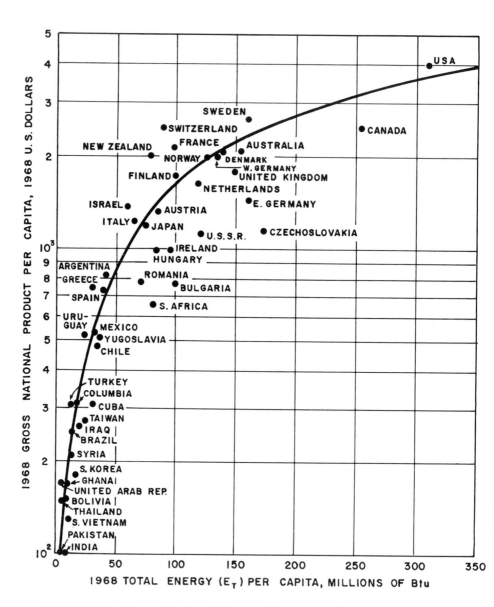

FIGURE 7.10 Gross national product per capita versus total energy (E_T) per capita consumption, 1968. [Source: *The U.S. Energy Problem (Volume 1): Summary,* ITC Report C 645 to the National Science Foundation (November 1971).]

however, a source capable of meeting the world's future energy needs at either present or projected rates of consumption. On the far horizon is the possibility of developing a continuous source of energy from solar sources (see Table 7.4).

TABLE 7.4 Estimates of Economically Recoverable Depletable Fuel Supplies and Continuous Energy Supplies [a, b]

Depletable Supply (10^{12} Watt-Years)	World	United States
Coal	670–1000	160–230
Petroleum	100–200	20–35
Gas	70–170	20–35
SUBTOTAL	840–1370	200–300
Nuclear (ordinary reactor)	3000	300
Nuclear (breeder reactor)	300,000	30,000
Cumulative demand, 1960–2000 (10^{12} watt-years)	350–700	100–140

	World		United States	
Continuous Supply (10^{12} Watt-Years)	Maximum	Possible by 2000	Maximum	Possible by 2000
Solar Radiation	28,000	—	1600	—
Fuel wood	3	1.3	0.1	0.05
Farm waste	2	0.6	0.2	0.00
Photosynthesis fuel	8	0.01	0.5	0.001
Hydropower	3	1	0.3	0.1
Wind power	0.1	0.01	0.01	0.001
Direct conversion	Unknown	0.01	Unknown	0.001
Space heating	0.6	0.006	0.01	0.001
Nonsolar				
Tidal	1	0.06	0.1	0.06
Geothermal	0.06	0.006	0.01	0.006
TOTAL SOLAR AND NONSOLAR	18–	3	1.2	0.2
Annual demand, year 2000 (10^{12}W)	15		5–6	

[a] Fossil-fuel supplies are equivalent to only about twice the cumulative demands for energy over the period 1960–2000. These fuels should not be exhausted for the generation of electricity and other purposes for which substitutes can qualify. Even nuclear fuels are limited if only conventional reactors are used. Breeder reactors now under development will lift this restriction. If a method can be developed for converting sunlight directly to electrical power at an efficiency of 12 percent, the sunlight that falls on a few percent of the land area of the United States would satisfy most of the energy needs of the nation in the year 2000.

[b] Source: Data in this table are based on C. Starr, "Energy and Power," *Sci. Am.*, 224(3), 36 (Sept. 1971).

A shift to nuclear power plants is already under way to avoid the chemical pollution and rapid fuel depletion inherent in the use of fossil fuels. It has been estimated that by 1990 nuclear energy will be furnishing about 40 percent of the total electrical energy required in the United States. There are special problems associated with these plants; however, there is a solid basis for optimism that the benefits of nuclear power can be realized without unduly affecting the environment. Management of the large amounts of radioactive wastes that will be produced is a major consideration; these wastes must be contained and isolated from man and his environment. Thermal pollution must be kept within acceptable limits. Public concern over safety requires that nuclear plants be located at relatively large distances from cities, thus creating an acute need for the efficient transmission of electric power from plants to the urban centers that they serve. Moreover, current nuclear-fission plants generally operate at lower thermal efficiencies than corresponding fossil-fuel plants because of current limitations imposed by available structural material. To be economical, they also have to be larger in terms of the total electrical power output of each individual plant. Fundamental thermodynamic principles require that any generating plant must be much less than 100 percent efficient in converting the available heat energy into electrical energy, but the efficiency rises almost linearly with the temperature at which it is possible to carry out this conversion. In all cases, however, the thermal energy that cannot be converted, because of these fundamental limitations, must be released to the environment as waste heat. Because a given nuclear plant characteristically provides more electrical energy than a fossil-fueled one, and because present nuclear plants are inherently somewhat less efficient because they must operate at somewhat lower temperatures, the nuclear plant is doubly at a disadvantage in terms of the amount of thermal energy that it must discharge into its environment. This thermal pollution, however, is a relatively minor problem. It is serious only because it was not anticipated, and plants were faced with the problem after they were already constructed. The thermal pollution problem arises primarily because nuclear plants have to be so big to be economic, not because of their lower efficiency. This situation has to do with the balance between fuel-cycle costs and capital costs. As the capital-to-operating ratio rises, the optimum size plant gets larger.

Although the light-water reactors, already in use, will be the workhorses of the nuclear power industry for some years, the hope for the future rests on the development of new kinds of reactors—breeder reactors that can simultaneously generate useful power and breed fissile material from very abundant but previously unusable uranium and thorium resources. A major development effort is being directed toward commercial utilization of these reactors. There are many problems; physics and physicists can contribute in major ways to their solution. It should be noted that the conventional fission reaction extracts only 1–2 percent of the energy potentially available from its uranium fuel and is dependent on the availability of large amounts of low-price ores to keep its power costs economically competitive with those of fossil-fuel plants. By contrast, breeder-reactor power costs are expected to be relatively insensitive to the cost of uranium and thorium, thus opening the economic use of low-grade ores.

The benefits to society are manifold. Nuclear reactors promise massive energy generation, with additional benefits such as reduced air pollution, virtual elimination of the need to transport large amounts of fuel, and the conservation of fossil fuels for future petrochemical purposes. At this time we have only a slight indication of possible future developments; possibilities include large-scale desalinization installations, electrification of the metal and chemical industries, and more effective means of utilizing the waste products of man's activities.

The question of efficiency lies at the heart of any serious consideration of the roles of physics in meeting the energy crisis. Although this fundamental limitation always applies, in fact, much of present technology is far from the achievable limits. At the same time, it is useful to recall that impressive progress toward more efficient systems has been made.

On the average, between 1900 and 1970, the efficiency with which fuels were consumed in the United States, for all purposes, increased by a factor of 4. Indeed, if it were not for this increase, the nation would be consuming energy at the rate projected for 2025 or thereabouts and an energy crisis would have appeared some time ago.

When wood or coal are burned in an open fireplace, the conversion to useful thermal energy is at most 20 percent efficient; a well-designed home furnace has a corresponding efficiency of 75 percent for space heating. The average U.S. efficiency for space-heating systems is currently between 50 and 55 percent.

Progress in the efficiency of conversion of fuel energy into electrical energy has been more dramatic. In 1900, less than 5 percent of the fuel energy appeared as electricity. Currently, the best fossil-fuel plants have an efficiency of about 40 percent, while present nuclear-fission plants have a 30 percent efficiency reflecting their lower operating temperatures (Figure 7.11).

Even now, however, substantial progress in material science is permitting the design of new nuclear plants, with efficiency comparable with the most modern ones using fossil fuels, and imaginative use of the waste thermal energy is being discussed widely. Application of the physics of the superconducting state may also alleviate one or both of the above problems.

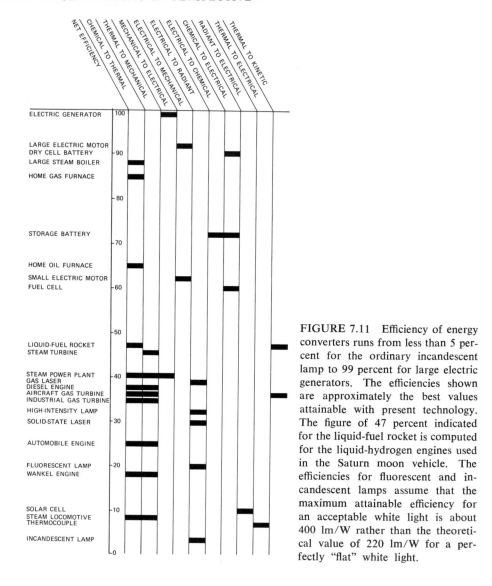

FIGURE 7.11   Efficiency of energy converters runs from less than 5 percent for the ordinary incandescent lamp to 99 percent for large electric generators. The efficiencies shown are approximately the best values attainable with present technology. The figure of 47 percent indicated for the liquid-fuel rocket is computed for the liquid-hydrogen engines used in the Saturn moon vehicle. The efficiencies for fluorescent and incandescent lamps assume that the maximum attainable efficiency for an acceptable white light is about 400 lm/W rather than the theoretical value of 220 lm/W for a perfectly "flat" white light.

Superconductors, as noted above, are metals that lose all resistance to electric current when cooled to very low temperatures (typically 5–20 deg above absolute zero). Power lines constructed of these materials should give negligible loss over distances of thousands of miles. As this power loss is a main obstacle to locating power plants in remote areas at the present time, the perfection of superconducting power lines would open vast new opportunities for the remote location of nuclear plants. A location as far north as the Arctic Circle might even be feasible. It should also be noted that such low-loss transmission lines would make it much more attractive economically to use electrical energy from remote hydroelectric installations such as the Churchill Falls site in Labrador.

Thus far in this discussion, the focus has been on conventional and present technology. Beyond this the role of physics becomes more challenging and the potential benefits even greater. A number of "novel" energy converters are being designed to exploit a variety of energy sources, as shown in Figure 7.12.

On a longer time scale, it may well be possible to change from fission to fusion nuclear power. Fusion power, based on a virtually unlimited supply of deuterium in the hydrosphere, or on the much more limited supply of lithium, as required in the deuterium–tritium fusion system, would yield less residual radioactivity than does the fission reactor. Direct conversion of nuclear energy to electricity is, in principle, possible, thus eliminating the thermodynamically imposed lower conversion efficiencies and permitting negligible thermal pollution. Producing a controlled fusion reaction is a major task for plasma physics (see the Report of the Panel on Plasma Physics and the Physics of Fluids in Volume II). The need for extremely large and stable containment magnetic fields is obvious. Nuclear fusion reactions take place at temperatures up to 100,000,000 deg, so that the fuel gases must be shielded from any material environment by a magnetic bottle. The ordinary method of producing a magnetic field in the necessary shape—passing electric current through copper coils—is too inefficient for this application, because the power dissipated in the coils would be comparable with that produced by the fusion reaction in the bottle. A possible solution is to make the coils of superconducting metals, which dissipate essentially no power. Even after plasma containment is achieved, many problems of a somewhat more engineering nature follow, for example, fast neutron damage and containment of large amounts of tritium at high temperature. The solution to most of these problems will require extensive input and knowledge from many branches of physics.

As an alternative to magnetic confinement, growing attention is being focused on laser ignition of small pellets of solid deuterium–tritium mixtures. Feasibility of a self-sustaining fusion system depends upon exceeding a so-called Lawson limit $n\tau = 10^{11}$, where n is the plasma density of the interacting nuclear species and τ is the containment interaction time. Since n is raised by many orders of magnitude, τ can be correspondingly decreased in going to these systems based on ignition of solid fuel pellets.

Energy outputs from currently available laser systems are almost adequate to achieve ignition in practical systems; containment here is achieved

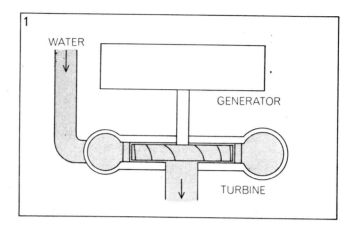

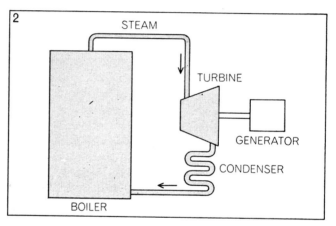

FIGURE 7.12 Electric-power generating machinery now in use extracts energy from falling water, fossil fuels, or nuclear fuels. The hydroturbine generator (1) converts potential and kinetic energy into electric power. In a fossil-fuel steam power plant (2) a boiler produces steam, the steam turns a turbine, the

effectively by the simple inertia of the fuel itself following ignition. Construction of efficient power lines for use with any or all of the systems and electromagnets requires large amounts of material that must be cooled to very low temperature to become superconducting. Finding materials that have the highest possible transition temperature and are capable of sustaining superconductivity even in the presence of strong fields and currents will be critical, for the expense rises rapidly as the required temperature decreases. Major effort in all these present and projected reactor systems

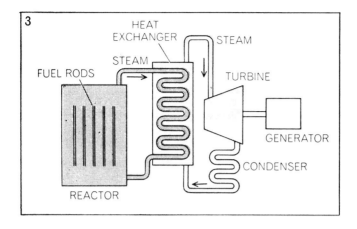

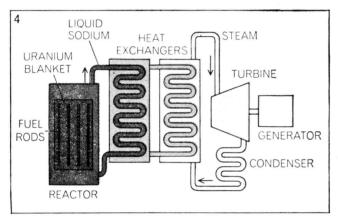

turbine turns an electric generator. In a nuclear power plant (3) the fission of uranium-235 releases the energy to make steam, which then goes through the same cycle as in a fossil-fuel power plant. Under development are nuclear breeder reactors (4) in which surplus neutrons are captured by a blanket of nonfissile atoms of uranium-238 or thorium-232, which are transformed into fissile plutonium-239 or uranium-233. The heat of the reaction is removed by liquid sodium.

must be directed toward understanding and solution of radiation-damage problems affecting all structural components. Here again physics must play a central role.

The future will undoubtedly see a mix of methods for electrical power generation—the magnetohydrodynamic (MHD) generator as a topper for

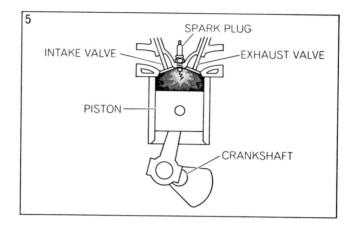

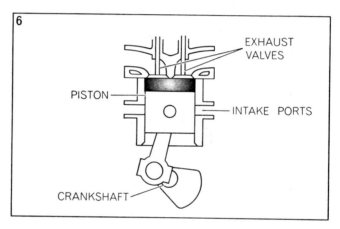

FIGURE 7.12 (Continued) Propulsion machinery converts the energy in liquid fuels into forms of mechanical or kinetic energy useful for work and transportation. In the piston engine (5) a compressed charge of fuel and air is exploded by a spark; the expanding gases push against the piston, which is connected to a crankshaft. In a diesel engine (6) the compression alone is sufficient to ignite the charge of fuel and air. In an aircraft gas tur-

the steam cycle appears promising and should be thoroughly explored and researched in all fuels. Much of the excellent work of a pioneering nature in the field has been carried out in this country but has been allowed to lose its momentum due mainly to lack of leadership. If the technology can be developed, it should be possible to design fossil-fuel plants with an efficiency of about 50 percent. Problems associated with gas conductivity, seed

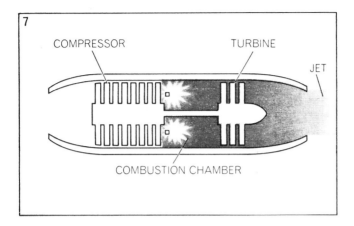

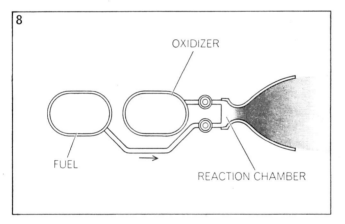

bine (7) the continuous expansion of hot gas from the combustion chamber passes through a turbine that turns a multistage air compressor. Hot gases leaving the turbine provide the kinetic energy for propulsion. A liquid-fuel rocket (8) carries an oxidizer in addition to fuel so that it is independent of an air supply. Rocket exhaust carries kinetic energy.

recovery, and materials, however, must be solved first. Since MHD requires very high temperatures (4000 to 5000°F), it is not suitable for use with nuclear reactors. Superconducting electric generators for power-station use have an exceptionally high potential. Low-temperature operation of a power-plant alternator offers the advantages of marked decrease in size, increase in power output, and decrease in unit cost.

There have been many suggestions in the past as to alternative sources of

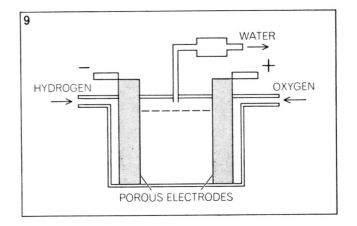

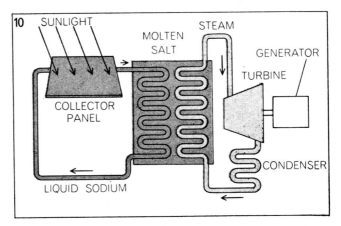

FIGURE 7.12 (Continued)   Novel energy converters are being designed to exploit a variety of energy sources. The fuel cell (9) converts the energy in hydrogen or liquid fuels directly into electricity. The "combustion" of the fuel takes place inside porous electrodes. In a recently proposed solar power plant (10) sunlight falls on specially coated collectors and raises the temperature of a liquid metal to 1000°F. A heat exchanger transfers the heat so collected to steam, which then turns a turbogenerator as in a conventional power plant. A salt reservoir holds enough heat to keep generating steam during the night and when the sun is hidden

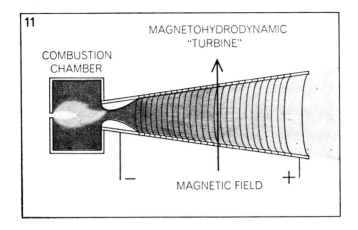

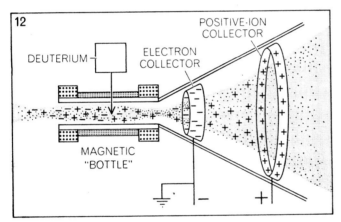

by clouds. In a magnetohydrodynamic "turbine" (11) the energy contained in a hot electrically conducting gas is converted directly into electric power. A small amount of "seed" material, such as potassium carbonate, must be injected into the flame to make the hot gas a good conductor. Electricity is generated when the electrically charged particles of gas cut through the field of an external magnet. A long-range goal is a thermonuclear reactor (12) in which the nuclei of light elements fuse into heavier elements with the release of energy. High-velocity charged particles produced by a thermonuclear reaction might be trapped in such a way as to generate electricity directly. [Source: C. M. Summers, "The Conversion of Energy," *Sci. Am.*, 224(3), 149 (Sept. 1971).]

energy. In the 1930's, the Passamaquoddy tidal project was much discussed and is now being revived. A projected output of 300 MW was to be obtained by exploiting an average tidal rise of 18 ft in the Bay of Fundy between Maine and Nova Scotia. With the development of more efficient low-head hydraulic turbines such projects may well play important local roles. But they cannot assume any national significance because of their intrinsic output limitations. If all favorable bays and inlets on the U.S. coastline were to be so harnessed, it has been estimated that at most a capacity of 100 GW (10^9 W) might be realized. This is less than 10 percent of the new capacity that will be required in the United States by 1990.

Some effort has been devoted to exploring wind power. In the U.S. Midwest, the average wind energy is some 18 W per square foot of area perpendicular to the wind direction. A propeller-driven system can convert this energy into useful electricity with some 70 percent efficiency. But the crux of the matter here is storage. Wind power is, if anything, erratic, and modern technology has yet to evolve an adequate storage medium or mechanism. There are major research opportunities in this area. Among the systems under active consideration are those wherein the wind-produced electricity is used to dissociate water and the subsequent hydrogen is burned in a very high temperature turbine–generator system. Overall efficiencies of 55 percent have been obtained in pilot-plant situations.

This question of much-improved electrical-energy storage merits further comment. Pioneering work on lithium and lithium silicon batteries in several U.S. laboratories shows promise of tenfold improvement in energy per unit weight or volume of battery. This alone can have dramatic consequences in making much more feasible self-contained electrical propulsion units for both private and mass transport. Both physicists and chemists are actively involved in current research to this end.

Again, in the U.S. Southwest some 18 W of solar energy falls on each square foot of land area, averaging the available sunlight over all weather conditions and over 24 h per day. Major attention is currently being devoted to development of more efficient systems for collection and utilization of this energy. It has been estimated that the 1000 GW of electrical capacity required in the United States by 1990 could be obtained by collecting the solar energy falling on only 14 percent of the desert areas of the U.S. Southwest. Projected designs now entering the demonstration phase envisage energy storage at 1000°F, with subsequent conversion to electrical energy at a 30 percent efficiency. In Chapter 4 some of the current problems in the thin-film physics required to optimize the collection efficiency of these solar-energy systems are discussed. Here again, much research remains to be accomplished. Even more exotic schemes involving large orbiting panels of solar cells, continuously exposed to sunlight and transmitting their collected energy to earth via microwave links, are under study. A 10-GW system, adequate to supply current electrical demands of New York City would require a collector 5 miles square with presently available conversion efficiencies in this system. Currently, the weak link here is the solar cell in which only between 15 and 20 percent of the incident radiant energy is converted to electricity. The remaining conversions to and from microwaves currently exceed 70 percent, although here again improvement is possible.

These are but a few examples of entirely new energy sources in which physics research can be expected to make very significant contributions. In the period between now and 2000, U.S. utilities will spend some $350 billion in development of such energy sources. Even a tiny fraction of this devoted to research in the fundamental and applied aspects of the mechanisms and materials involved can be expected to yield handsome dividends. It would be shortsighted, indeed, if such programs are not vigorously pursued by utility organizations. At present very little is being accomplished.

The search for greater efficiency is not limited to the generation phase of the energy cycle alone. At present, perhaps, the least efficient electrical utilization is in the production of light. Some 6 percent of the entire U.S. energy demand is so used, corresponding to about 24 percent of the entire

At present a 100-W incandescent lamp converts only some 5 W to visible light, with the rest being dissipated as heat; in 1900, however, only 1 W was converted to light in such a lamp. Current fluorescent lamps comprise some 70 percent of the nation's lighting, and these typically have a conversion efficiency of 20 percent. The average electricity to light conversion for the nation is about 13 percent, and, when combined with typical generation efficiencies, this yields 40 percent as the overall fuel-to-light conversion efficiency. This is depressingly low in a time of energy crisis, and entirely new conversion processes are needed. Some solid-state devices already show promise in this area, but again much additional research, both fundamental and applied, is needed. Again, too, the potential economic return from even a small improvement in efficiency is so great that research in areas such as this should represent an enormously attractive investment for industrial funds.

Environmental Quality

It is important to emphasize that in the majority of environmental problems, science, technology, and invention are less important than economics, sociology, politics, and the law. Nevertheless, there frequently are important, if not vital, technical elements to these situations, and opportunities for physicists to contribute are indeed large in number. For example, as noted above, despite the progress that has been made in converting primary into electrical energy, the maximum efficiency attained in any thermal process to that end is only about 40 percent. Thus energy is generated at a prodigal waste of primary input—coal, gas, oil, and nuclear. The amount of energy rejected to the environment is one and one half times greater than the amount available for useful purposes!

There are several distinct roles for physicists. The most obvious is in the development of instrumentation and techniques to be used for monitoring, studying, and restoring the environment and in research based on fundamental physical principles. Second, much research based on fundamental principles is needed in connection with air and water pollution, solid-waste disposal, recycling of raw materials, and utilization of infrared technology in satellite photography of earth-resource and crop management, to name just a few areas. There is opportunity for entirely new examination of accepted or traditional technologies and approaches. What is needed in any instance is a totally new concept, and the probability of this arising is greatly enhanced if problem situations can be viewed without preconceived boundaries or limitations.

In any program for control and abatement of atmospheric pollution, it is essential to provide for a continuous, remote method for the sensing of pollutants in ambient air. Present methods for making these measurements are too slow or too inconvenient in some instances, too costly and inaccurate in others. New systems are needed. A technique with exceedingly high potential for future monitoring systems is based on the spectroscopic analysis of laser light scattered from the atmosphere. Although the application of this technique in the field has only begun, it appears certain that it can be developed to a degree that will make it superior to any other procedure presently in use. Calculations indicate that it will be feasible to detect pollutant concentrations in the atmosphere of one part per million at distances of several kilometers from the laser source. In principle, it is possible to make a three-dimensional map of the concentrations of various molecular constituents. This would be an obvious advantage over other methods that provide information only on the average concentration over a long path.

Satellite observational techniques using ultraviolet, visible, and infrared spectroscopy may make it possible to monitor continuously the global distribution of natural and man-made variations of specific chemicals in the earth's atmosphere and to have almost instantaneous knowledge of changes. These techniques provide an immediate way in which a physicist can apply his specialized knowledge to problems of pollution and ecology.

Because of the proliferation of elements such as lead, mercury, beryllium, and cadmium in the environment, and their potential danger to the biosphere, there is need for extensive monitoring of such elemental contaminants in most areas of the environment. Methods developed for this purpose must be simple to carry out, reliable, and inexpensive. A group at the University of California at Davis has been highly successful in the application of the method of charged-particle-excited x-ray fluorescence to the analysis of trace elements in aerosols and in developing accurate, low-cost instrumental analysis using the characteristic x-ray fluorescence from elements excited by charged particles. Measurement accuracies have been extended by a factor of 10, and costs of analysis have been reduced by a factor of 5 compared with other methods. At the present time, their instrumentation is capable of measuring trace quantities (parts per billion) of most atomic elements above sodium in the atomic table. The high sensitivity and inherently quantitative nature of this technique promise to revolutionize large-scale environmental monitoring. It appears that any accelerator of charged particles with an energy greater than a few MeV can be used for such analysis; consequently, dozens of academic and industrial groups could participate in this important work.

The diffusion of fluids and particulate contaminants discharged into the air and into natural waters is a lively branch of fluid dynamics application. Diffusion and particle fallout rates must be estimated well, because they affect the concentration of contaminants. In 1970, practical meterologists and engineers were still trying to estimate pollutant diffusion by indiscriminate formulas, which are only slightly improved versions of those used 40 years ago. Improved understanding of turbulent flow will pay large dividends.

A major strategy for minimizing pollution will be to extract more of the noxious matter from gas or liquid effluent before it enters the environment. The technological development of improved chemical conversion and centrifugal or electrostatic precipitation devices will require a more detailed understanding of the associated flow phenomena. Improvement of electrostatic precipitators may rely on electrohydrodynamics as well.

The problem of pollution due to oil released at sea has developed because of the ever-increasing scale of offshore drilling and oil transport. At present, a single ship or oil well has an awesome capacity to pollute vast areas of seashore, a capacity documented by such incidents as the *Torrey Canyon* or the oil well in the Santa Barbara Channel.

When these accidents first became newsworthy, and hence drew public attention, the only method of dealing with the problem was to clean the beach by removing the oil as it came ashore. Several methods were tried, such as absorbing the oil on straw or washing the beach with detergent. However, experience soon showed that these methods, conceived for cleaning up small amounts of oil, did little to protect the environment from massive oil spills.

Because additives, such as detergents, are generally toxic in the marine environment, attention has been directed to simple mechanical means of containing and removing the oil. However, it was soon apparent that the various booms and barriers proposed to contain the oil, as well as devices to collect it, simply did not work in practice.

There are a number of reasons for this failure of existing technology, but the most basic was the lack of physical understanding of how one fluid, oil, spreads and is collected while floating over the other fluid, water. More fluid-mechanical research is required now to develop a workable technology for present oil-pollution problems, and more will be required in the future in order to assess properly the potential damage to the environment as the scope of offshore oil activity expands.

The techniques of magnetic filtration promise widespread application in the decontamination of waste water, sewage, drinking water, and waters used for recreational purposes. For example, the turbidity of some waters is such as to exceed federal safety standards regarding their use for swimming instruction. The particles causing this condition are not removable by conventional filtration, and a significant fraction even passes through millipore filters. However, magnetic filtration tests have shown that removal of virtually all particulates is possible at realistic flow rates and at fields of the order of 5 kG. High-intensity, high-gradient magnetic separation techniques appear highly promising for such diverse problems as decontamination of fuel oil and coal and the beneficiation of nonmagnetic taconite. In the latter case, this country has been living for 150 years on the magnetic taconite of the Mesabi Range in Minnesota. The supplies of magnetic taconite are running out, and the material now exposed throughout most of the Mesabi Range is oxidized into a mixture of nonmagnetic states. Laboratory-scale tests at the National Magnet Laboratory, using high-intensity magnetic separation, achieved 90 percent recovery at flow rates of 250 gal ft^{-2} min^{-1} and a process cost of 20–25 cents per ton. By contrast, the best system thus far proposed is chemical and has a cost of $1 per ton, at poor total yield.

Tracer techniques are expected to play a major role in studies of groundwater flow, mixing problems in lakes and rivers, aerosol behavior, tracing of pollutants, and studies of air masses. The tracer techniques may not involve the direct use of radioactive tracers but rather the use of stable element tracers followed by an analytical technique such as activation analysis to avoid introducing radioactivity into the environment.

Clearly needed is further development of, for example, small mass spectrographs, laser light-scattering devices for measuring particulate distributions, magnetic filtration and cryogenic trapping techniques, sensors, and techniques of data processing. Theoretical and experimental research in meteorology, thermal pollution, energy sources, and electrical power transmission is greatly needed, again to single out typical topics.

Basic questions of chemical kinetics in reactions that cause smog need further clarification. To be able to predict the effect of releasing various gases in the atmosphere, for example from aircraft operating in the stratosphere, it must be known what end products and processes will result and the nature of the process that determines the residence times of such gases in the atmosphere. It is possible that release of water and oxides of nitrogen from such operations may catalyze the destruction of atmospheric ozone and thereby increase the amount of harmful ultraviolet radiation reaching the surface of the earth. A knowledge of the distribution and amount of energy in the solar spectrum reaching the surface of the earth is important for understanding the possible long-range ecological dislocations that might take place. To obtain this information, and other information pertinent to better understanding of pollution problems, requires the determination of electronic ionization and recombination cross sections, photoionization cross sections, and many two- and three-body reaction rates for neutral–neutral and ion–molecule reactions. The techniques for accurate cross-beam experiments near thermal energies will be highly important in making the necessary laboratory measurement. The situation in the *D*-region of the ionosphere, in which ions involving clusters with water molecules recently have been found to be important, could well be indicative of the complex problems apt to be encountered in attempting to understand pollution.

The techniques of scientific analysis and mathematical modeling are proving useful in the development of policy and legislation related to the control of pollution. Application of the techniques of scientific analysis can show whether proposed regulations and legislation will achieve the goals for which they are designed. An example is the regulation of air pollution. The states are required to develop plans to implement the Federal Air Quality Standards, consistent with the Clean Air Act of 1967 and the Air Quality Act of 1970. The standards merely specify how much of certain pollutants can be permitted in ambient air; it is up to the states to determine how to control emissions to meet the standards.

During the past four years, several laboratories (the Argone National Laboratory is one) have developed mathematical models for computer use that simulate the production, transport, and dispersion of air pollutants. Scientists at Argonne have worked closely with the Illinois Pollution Control Board during 1970 and 1971 to develop a large and increasingly refined set of trial regulations. These regulations are programmed for a computer; then, with possible realistic meteorological data and pollutant emission levels relating to the regulations being tested, the computer determines what happens to the pollutants, what the overall air quality would be, and the extent to which the regulation could be met throughout the region being studied. The results of this interaction are not yet complete; however, two sets of regulations have been tested, each representing a wide variety of different strategies of control. After each set of computations was analyzed, the Pollution Control Board prepared model regulations and held public hearings on them. Preparation of a third and possibly final set of computations is in progress. Presumably it will be followed by final hearings and the adoption of regulations.

Such modeling is vitally important in the development of achievable regulatory goals in such complex problems. Obviously no valid point is served in promulgating regulations that are impossible to fulfill within realistic lower limits on input levels of pollutants; but, by the same token, no unnecessary leeway should be left in the regulations if they are to achieve the desired control level. This type of modeling comes naturally to the trained physicist, and it seems clear that this is one of the areas in which physicists can make important contributions to the overall problems of improving environmental quality.

Other problem areas in which this type of analysis can be applied effectively include water-pollution control, assessment of the effects of dredging, airport site location, and control of transportation flow.

It is important to recognize that scientific modeling of proposed legislation does not indicate what laws should be passed; rather, it shows what laws or regulations are *not* feasible and should *not* be enacted. It is an effective means of testing alternative approaches to a well-defined goal and of showing which procedures will not achieve that goal. The task of choosing among the approaches not eliminated by the analysis remains. This choice is determined by the application of social criteria and by the usual political processes.

A related but quite different example of the application of physics to pollution control occurs in the application of fundamental physics analysis to a complete technological cycle. An example recently studied by Berry concerns the question of the desirability of recycling the metallic components of discarded automobiles as part of a national campaign to eliminate a rapidly increasing environmental pollutant. Obviously, economic considerations are of great importance here, and these in turn rest upon the technologies that enter at each stage of the cycle of production and use of the automobile, from the mining and smelting of the original ores, through the fabrication and use, to the disposal or recycling of the discarded unit. Within the framework of a very crude thermodynamic model it is possible to examine the relative absolute efficiencies of each stage of this cycle. In doing so, Berry finds that the major thermodynamic inefficiency is in the area of the recovery of the metals from their ores—an area in which U.S. technology, and indeed world technology, has lagged. If even a small part of the large thermodynamic inefficiency present in this stage of the cycle were to be removed by new technology, the economics of the cycle could change substantially, thus making possible imaginative recycling possibilities.

This is simply an example of a much-neglected area—the use of fundamental physics analyses to isolate weak points in major technological and other systems and cycles, to highlight areas where new technologies or approaches could have very great return or where the actual processes involved are operating at efficiencies anomalously below the limitations, which might be improved at any fundamental level. This technique may well become a new and powerful tool in systems and operations research analyses of highly complex social and economic enterprises.

A third role for physicists is perhaps the most important, although it is difficult to describe or quantify. The most effective attacks on environmental problems will almost surely be carried out by problem-oriented centers whose personnel will be made up of scientists, economists, lawyers, doctors, ecologists, sociologists, and the like. A physicist working with such a team may never write Maxwell's equations or operate a counter of any sort. What he can contribute is a style of research and a confidence in man's ability to solve problems. There is a definite need for physics and for physicists and, it is hoped, in the future many physicists will turn their attention to this important problem of our times.

8

INTERNATIONAL ASPECTS OF PHYSICS

Science, especially in its basic aspects, is one of the few truly international activities. It knows no geographic limitations, for the pursuit of knowledge about nature is characteristic of all nations and men. Thus, it was not surprising that at the first Atoms for Peace Conference in 1955, when Soviet and U.S. physicists compared in detail their results on neutron-induced fission—work that had been entirely secret until then—they found full correspondence between their studies. Indeed, as the exchange of information developed, it was apparent that the two groups had used similar ideas and methods and even similar mathematical symbols. The Second International Conference on Peaceful Uses of Atomic Energy (1958) emphasized the more applied aspects of physics in fission and fusion. Here again, the universality of the fundamental laws of plasma physics showed no more respect for national boundaries than had the earlier revelation of the cross sections of nuclear physics.

Though data on the impact of scientific exchanges and communication on international understanding are few, there are many indications that international visits, meetings, and communications among scientists play a sizable part—a larger part than the relatively small numbers of scientists involved would suggest—in determining the intellectual climate of opinion in one country in regard to another. An indication of the ubiquity of scientists in international exchanges was found in a random sampling of entries in the British *Who's Who* performed by a member of the Statistical Data Panel of the Physics Survey Committee. The sample included persons whose biographies indicated that they had spent an interval of at least a year in the United States. The results showed an equivalent distribution among three categories—diplomats, entertainers, and scientists.

In this chapter we shall discuss briefly some of the aspects and implications of international scientific interaction, illustrating this discussion with some examples from physics. The discussion is divided into three parts: scientific activity in other countries, international exchanges and interaction in physics, and international organizations concerned with cooperation. What follows, however, particularly in the first two sections, must be considered as suggestive of magnitudes and directions. More data are needed, and the topics merit detailed study by scholars in the economic, social, and educational fields of endeavor even more than by those in the sciences. Yet we believe that such data and experience as were available to us warrant the general statements offered here.

SCIENTIFIC ACTIVITY IN OTHER COUNTRIES

General Comparisons

It is not possible to assess the state of physics in other countries in the same detail as domestic physics because sources of information comparable with those provided for this country by the National Science Foundation (NSF) and the American Institute of Physics (AIP) are not available. Consequently, the comparisons made in this section relate largely to the overall research and development enterprise, and even here many factors prevent the making of meaningful quantitative assessments.

In many instances, dollars expended for research and development are available, but these figures do not represent a true measure of the amount of research performed or of the input of resources to research. One dollar buys much more research in some countries than in others. Persons working on the collection and interpretation of data for the Organization for Economic Cooperation and Development (OECD) indicate that funding figures for different countries are only meaningful to within ± 30 percent, which is more than sufficient to obscure small differences and make comparisons difficult. Estimates of research purchasing power in various countries made by the OECD in 1962 appear in Table 8.1.

What these dollar figures are attempting to show is input, although output would probably be a more significant statistic. Unfortunately,

TABLE 8.1   Ratio of Research and Development Costs in the United States to Those in Various European Countries (1962) [a]

Country	Research and Development Cost Ratio
United States	1
United Kingdom	1.8
The Netherlands	1.9
Belgium	1.7
West Germany	1.7
France	1.5

[a] Source: *A Study of Resources Devoted to R and D in OECD Member Countries in 1963/4* (OECD, Paris, 1968), Vol. 2, p. 29.

there is no valid and convincing way to measure output, even for the United States. One of the best approaches to measuring the output of science is to obtain data on publications. An important point in regard to any measure is that equal inputs to research do not necessarily produce equally valuable outputs.

Another datum frequently used as a measure of research and development effort is manpower. The significance of manpower data is probably not as doubtful as that of monetary data; however, degree structure and the definition of what constitutes a qualified research scientist vary among countries and detract somewhat from comparisons.

Two rough measures of the size of science in various countries—the percentage of gross national product (GNP) devoted to research and development and the fraction of the population who are research and development scientists and engineers—are plotted against one another in Figure 8.1. The figure shows that the relative role of science in a nation's activities varies considerably. There is a rough proportionality between the two quantities plotted that can be described by the statement that the research and development expenditure per scientist-engineer is 15 times the GNP per capita. Because the GNP per capita varies widely among countries, one dollar, or its equivalent, does indeed buy quite different amounts of research and development in different countries.

Growth in the size of science with time in various countries is depicted in Figure 8.2, in which research and development expenditures are plotted against years from the early 1950's to 1970. Virtually all the world experienced a steady growth of expenditures for research and development throughout the past decade or so. However, these figures are not corrected for inflation; for example, in the United States the correction

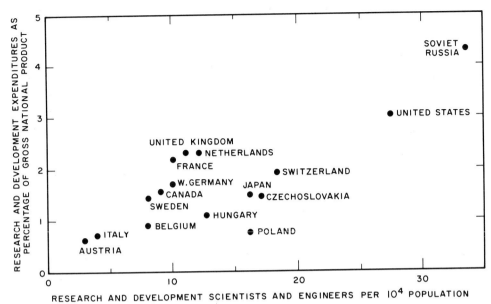

FIGURE 8.1 Measures of the size of the research and development enterprise in various countries. Research and development expenditures as a percentage of GNP are plotted against the number of research and development scientists and engineers per 10⁴ population.

for inflation, which is used in the other chapters in which comparison with other countries is not involved, actually produces a decrease in expenditures in real dollars during the past few years.

Data collected by the OECD indicate that among six highly developed countries, the most rapid growth rates in total (government plus private) expenditures on research and development between 1961 and 1967 occurred in Germany, France, Japan, and Canada, in which such expenditures more than doubled. Such expenditures in the United States rose 60 percent in this interval, and those in the United Kingdom, 40 percent. Thus, research and development in the countries that started lower grew twice as rapidly as in the United Kingdom and the United States. The estimated annual increase for 1967–1970 is in all cases less than the average for the preceding years, though only slightly less so for Japan. Expressed as a fraction of each country's GNP, the rate of increase in research and development expenditure in 1959–1970 ranged from 50 percent to 100 percent for Canada, Japan, and Germany; only a slight increase occurred in the United States, and none in the United Kingdom.

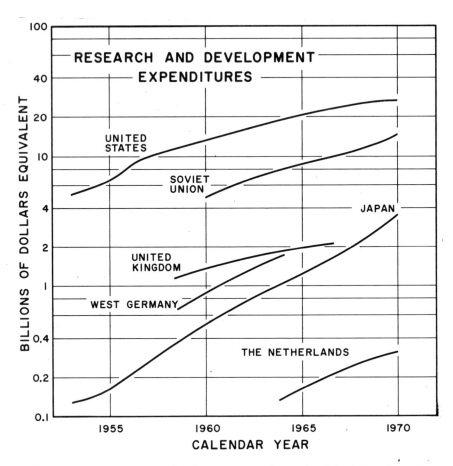

FIGURE 8.2   Research and development expenditures in U.S. dollars in various countries.

Many authors have commented on the steady and rapid growth of research that is illustrated in Figure 8.2. It has been evident for some time that world science has been growing at the enormous rate of a doubling every seven to ten years. But Figure 8.2 also shows that the rate of growth has slowed in most countries, and especially in the United States.

Figure 8.3 shows the rate of growth in research and development plotted against research and development expenditures as a percentage of the

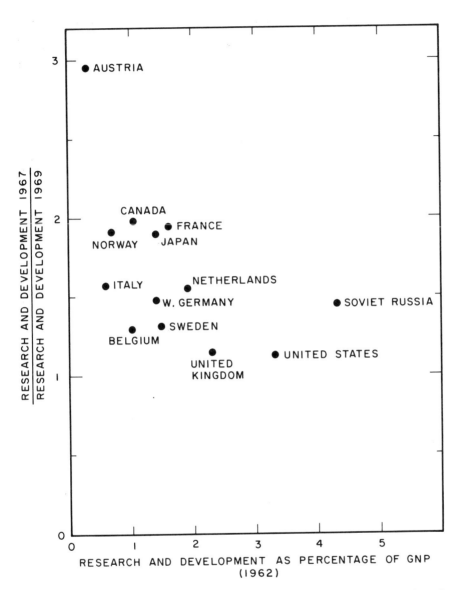

FIGURE 8.3 The growth of research and development funding between 1962 and 1970 plotted as a function of research and development as a percentage of GNP in 1962.

GNP. It indicates that growth tends to be more rapid when the enterprise is small. A notable exception to this trend, the Soviet Union, is discussed in greater detail in a subsequent part of this section.

Another measure of the size of the research and development effort of a country is its output of scientific publications. Table 8.2 presents the total number of scientists and engineers, the number of authors of scientific and technical papers, and the number of physics papers originating in various countries. A plot of the number of authors found in an

TABLE 8.2   Scientists and Engineers in Research and Development, Authors, and Physics Papers in Various Countries

Country	Scientists and Engineers (Thousands)[a]	Authors[b]	Papers in *Physics Abstracts* Sample[c]
United States	537	52,195	694
Soviet Union	760	10,505	408
United Kingdom	59.6	13,103	174
West Germany	61.6	8,398	127
France	49.2	6,862	113
Japan	158	5,702	96
Canada	19.4	3,997	66
India	—	2,882	63
Italy	19.7	2,733	55
Australia and New Zealand	—	2,991	30
Netherlands	15.7	1,412	24
Switzerland	11.0	1,767	22
Poland	50	1,305	20
Czechoslovakia	23.5	1,718	17
Sweden	7.4	1,650	15
Israel	—	1,125	18
Hungary	12.4	1,039	9
Belgium	7.9	924	9
Austria	2.4	646	5
Finland	1.5	447	5
Norway	3.0	432	0
Spain	10	277	4

[a] From OECD and UNESCO Science Policy Study No. 18 (1967), *The Role of Science and Technology in Economic Development.*

[b] Institute for Scientific Information count (see D. J. de S. Price, *Proc. Israel Acad. Sci. Humanities, 4,* 98 (1966).

[c] Statistical Data Panel sample of 2492 abstracts.

Institute for Scientific Information study * against the estimated number of research and development scientists and engineers appears in Figure 8.4. The ratio of authors to scientists and engineers varies greatly among nations. The average ratio is about one to ten, and the United States apparently is typical in this respect. Various explanations of the wide

* A study conducted by the Institute for Scientific Information of first authors of papers in *Current Contents*. See the *International Directory of Research and Development Scientists 1967* and the paper by D. J. de Solla Price in the *Proc. Israel Acad. Sci. Humanities, 4,* 98 (1966).

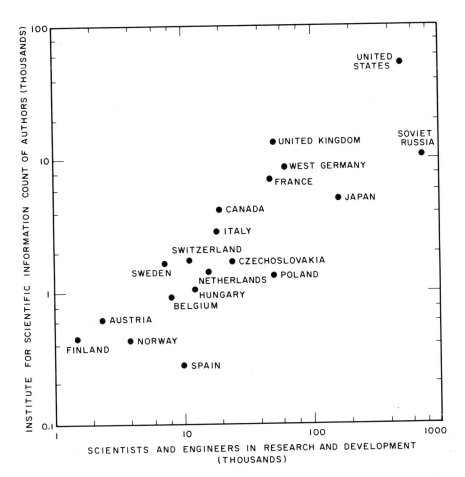

FIGURE 8.4 The number of first authors in the journal literature covered by the Institute for Scientific Information in 1967 plotted against the number of research and development scientists and engineers in 1964.

differences have been suggested. For example, it is possible that the productivity per worker varies from country to country. Clearly, it is likely that research productivity would increase with increasing level of support per worker. In addition, the results obviously would be greatly affected by the completeness with which the Institute for Scientific Information covers the journal literature of these countries. Preliminary results of an independent tabulation being undertaken by a member of the Statistical Data Panel suggest, for example, that the difference between the Soviet Union and the United States in the ratio of papers to authors is far less than that shown in Table 8.2.

Figure 8.5 plots support level, as measured by the number of scientists and engineers engaged in research and development divided by the GNP per capita, against the author to scientist-engineer ratio derived from Figure 8.4. There is a strong correlation, and it is tempting to interpret

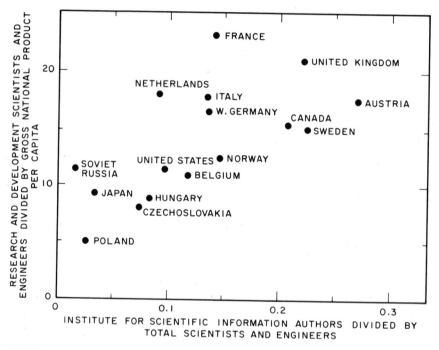

FIGURE 8.5 The ratio of research and development expenditure per research and development scientist and engineer to the GNP per capita plotted against the ratio of the number of science authors to number of research and development scientists and engineers.

this as an increase in productivity with increasing levels of support. However, other factors must be considered.

For example, publication is characteristic of scientific research, that is, work motivated to produce new understanding on which future research can build. Publication of technological work in the same format and detail as scientific work is not customary or useful. The outcome of development or engineering work usually is measured in terms of economic or social products, whereas the value of scientific research is more appropriately measured by its impact on scientific concepts and its acceptance by the scientific community after critical scrutiny and review. Thus, a high author-to-researcher ratio may indicate a larger emphasis on basic science as compared to development or mission-oriented research. In this context, it is interesting to note that the United States occupies a central position in the distribution shown in Figure 8.5.

Another possible source of the differences among nations that appear in Figure 8.5 is pressure to publish. In some countries the standards of measurement of scientific productivity (and prestige) may be more closely related to publication in the leading journals of another country than to publication in domestic journals. Competition for funds, which is generally a strong incentive for publication, also is likely to differ greatly among countries. In addition, the data on which Table 8.2 and Figures 8.4 and 8.5 are based could affect the results because of the inclusion of only first authors; methods of determining the order of names on a paper probably vary in different countries.

The average annual rate of increase in the number of scientists and engineers from 1963 through 1967 was about 7 percent in the United States, 14 percent in Canada, 11 percent in France, and 8 percent in Japan. (Comparisons with the Soviet Union and the United Kingdom appear in subsequent parts of this section.)

In almost all nations, the most important justification for large-scale support of research and development is the belief that these activities have the potential for contributing substantially to economic growth. Once research and development begin to represent a sizable fraction of the national economy, their rapid growth can be sustained only by rapid economic growth. Thus, science can flourish when it feeds back into nonscientific activities to create a favorable economic climate that will enable it to attain the growth to which it is geared. Maintaining this feedback from science to the economic system is largely a function of development and engineering, but science plays a part in that many scientists participate in the applied research and development aspects of technological activity, and science constitutes an important part of the training of those who perform engineering and development. Consequently, the framers of science policy in the various countries must look with care at the feedback that ensures that science will contribute to the economy that, in turn, creates the climate in which science is healthy and vigorous.

Japan is an example of a country in which a large fraction of research and development is motivated by economic factors, and Japan's economy is growing rapidly. It would be a mistake, of course, to assume that investment in research and development directed toward enhancing the health of the economy is the sole requirement for economic growth;

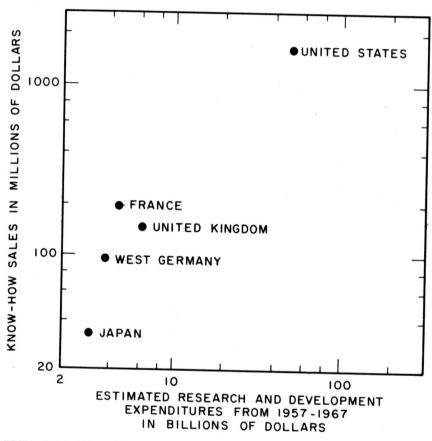

FIGURE 8.6 Sales of know-how as a function of estimated total research and development expenditures in 1957–1967 for various countries.

large investments of capital are necessary, and in Japan the purchase of results of research and development performed in other countries has also been a major contributing factor. Probably the results of research and development are nowhere more evident than in the trade in know-how. Figure 8.6 depicts know-how sales versus integrated research and development expenditures in several countries during the past decade; here the net research and development experience produces an extremely favorable balance.

The amount of the total (government and private) research and development expenditure devoted to industry is another related indicator. Most OECD countries devote more than 30 percent of the overall research and development expenditure to industrial research and development; the percentages range from 29 percent in the United States to 79 percent in Switzerland. The U.S. figure is low because this country is the only one without an explicit government contribution to industrial research and development. Much of the part played in other OECD countries by ministries of industry is assumed in the United States by spinoff from contract work for defense, nuclear, space, and other missions. In all the major OECD countries, the industry contribution to research and development has been growing as a percentage of both GNP and gross fixed capital investment.

The Soviet Union

Because of the great diversity among the nations and their economies, stages of development, and histories, it is difficult to make comparisons and draw sweeping conclusions about the role of research and development. A few specific comparisons of the United States with other leading nations may be useful. Probably the most appropriate comparison is with the Soviet Union, which is a competitor of the entire Western world in many areas related to science and technology, such as foreign trade, space exploration, and military activities. The anomalous position of the Soviet Union in many of the graphs cited in this chapter is obvious. For example, it is the only nation that spends a larger fraction of its GNP on research and development than does the United States. And support for research and development is continuing to grow at a steady rate, thus making the Soviet Union an important exception to the conclusion drawn from Figure 8.3. The continued growth of the Soviet research and development effort is indicated not only by the data depicted in Figure 8.1 but also by estimates of the number of persons engaged in research and development work. Figure 8.7 shows the number of research and development scientists and engineers in the United States and the number of graduates in research and development in the Soviet Union as functions of time.

During the latter part of the 1950's, the number of scientists and engineers engaged in research and development in the Soviet Union was approximately the same as in the United States. However, in the 1960's trained research and development manpower in the Soviet Union grew at a more rapid rate than in the United States, so that by 1969 Soviet scientific and engineering manpower exceeded that in the United States by some 50 to 75 percent.

The Soviet Union also emerges as an exceptional case in Figure 8.4, which shows numbers of scientific authors. The number of authors among practitioners of research and development in Russia is small compared with other nations. This finding could result in part from less

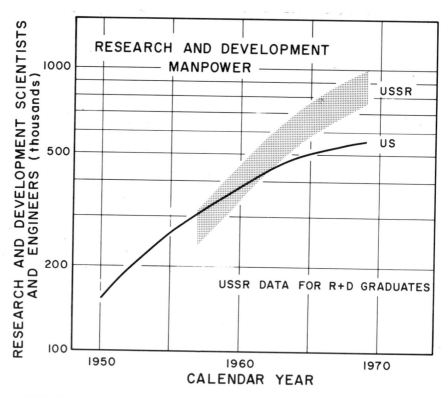

FIGURE 8.7 Growth of the number of research and development scientists and engineers in the United States and the Soviet Union.

adequate coverage of the scientific literature of Russia; however, it also could indicate a preoccupation of most Soviet research and development effort with technology, which leads less often to publishable work. That the Soviet researcher is well supported is evident in Figures 8.2 and 8.5.

A comparison of Table 8.3 with Table 8.2 shows that physics plays an unusually large part in the scientific enterprise of the Soviet Union. The ratio of *Physics Abstracts* recorded from the Soviet Union in the sample presented in Table 8.3 to the number of authors shown in Table 8.2 is almost double that of any other nation. (This result is only suggestive because of possible inadequacy in the coverage of the Russian literature and differences in the inclusion or ordering of authors' names on papers.)

It is also interesting to note that the Soviet Union has a significantly higher percentage of authors of physics papers who are women than does the United States. A recent small study of authors of articles in the *Physical Review* and in two comparable Soviet publications * showed that 2½ percent (± 1 percent) of U.S. authors were women as compared with 16 percent (± 2 percent) of the Soviet authors.

From these various comparisons we conclude that the Soviet Union has continuing faith in the value of science and engineering. Further, it appears that research and development are regarded as avenues to economic progress. The data on publication suggest that engineering and development receive somewhat greater emphasis in the Soviet program than in other countries. Most of the new developments that have emerged from the Soviet Union, such as the first earth-orbiting satellite and the first man in space or the advances in controlled nuclear fusion, are technical rather than scientific achievements.

The United Kingdom

The United Kingdom is next in order after the Soviet Union and the United States in proportion of the GNP devoted to research and development. Figure 8.3 shows that research and development have grown very slowly in the United Kingdom in the past decade. Apparently there is great emphasis on science in the United Kingdom, for the ratio of authors to scientists and engineers is exceeded by few other nations and is nearly double that of the United States. As in the United States, there is serious questioning of the value of science and science education, and currently there are difficulties in providing for the productive

---

* *Zhurnal Eksperimental'noi i Teoreticheskoi Fiziki Pisma v Redaktsiiu* and *Fizika Tverdogo Tela.*

TABLE 8.3 Physics Papers from Countries That Are the Leading Contributors to the Physics Literature in the Various Subfields [a]

Country	A&R	AME	EP	Nu	Pl	Fl	CM	E&P	P in B	Ac	Op	Other	Total	Theoretical (%)
United States	12	73	54	53	26	23	284	84	3	8	42	82	694	0.46
Soviet Union	1	30	22	33	24	22	174	47	0	1	40	14	408	0.42
United Kingdom	2	22	12	4	9	9	80	19	1	4	7	5	174	0.41
West Germany	0	12	11	20	2	3	56	6	0	1	13	3	127	0.39
France	3	17	4	9	5	3	46	9	0	3	8	6	113	0.35
Japan	0	9	6	7	4	6	48	7	0	1	6	2	96	0.40
Canada	2	8	1	8	3	3	22	9	0	1	6	3	66	0.55
India	1	9	5	3	0	6	30	4	0	0	3	2	63	0.44
Italy	2	1	15	7	3	1	19	1	0	1	3	2	55	0.51

[a] See Chapter 4 for a description of the various subfields.
Key: A&R, astrophysics and relativity; AME, atomic, molecular, and electron; EP, elementary particle; Nu, nuclear; Pl, plasmas; Fl, fluids; CM, condensed matter; E&P, earth and planetary; P in B, physics in biology; Ac, acoustics; Op, optics.

employment of all scientific and engineering manpower. In fact, these problems are probably somewhat more pressing in the United Kingdom than in the United States. The support of university research is not sufficient to meet the demands of the academic scientists. There are strong feelings that university education is too specialized and oriented toward basic science and as a result does not fit the graduate for employment in nonacademic institutions. There also are serious questions raised about the wisdom of the current level of production of scientists and engineers and the capacity of science to contribute to better technology and economic growth. Predictions of future manpower shortages led to a great expansion of advanced training in science in the United Kingdom in the early 1960's; the failure of the demands to match the expansion of education has led to employment difficulties similar to those found in the United States (see Chapter 12).

International Interaction and Exchange of Personnel

For the past three decades, U.S. physics has held a position of leadership in international activities in physics. This leadership has been demonstrated in many ways. Among the contributions of U.S. physicists are the development of lasers, the nuclear chain reaction, the concept of transistor action, the theory of superconductivity, magnetic cyclotron resonance, a majority of the most significant advances in elementary-particle physics, and theories of quasars and pulsars. The United States also has contributed to physics education through such internationally accepted works as the Physical Science Study Committee (PSSC) program in physics and the Feynman Lectures on Physics (see Chapter 11). The U.S. development of increasingly sophisticated tools for physics has also been significant. Perhaps best recognized are the particle accelerators of ever-increasing energy that the United States has pioneered in developing. Of almost comparable significance have been such associated tools as immense bubble chambers, spark chamber detectors, apparatus for nano-second time discrimination, superconducting magnets, electron microscopes of increasingly high resolution, and, of particular importance, the widespread development and application of digital computers for research purposes.

One measure of the role of the United States in world physics is the large number of foreign scholars who have been attracted to the United States as students and visitors. Another frequently quoted numerical measure is the award of Nobel Prizes; in physics the totals for the past 20 years listed by country in which the research was accomplished are

United States, 21; Soviet Union, 6; West Germany, 4; other countries, 7. U.S. physicists have received 55 percent of the awards although they produce only about 30 percent of the world's literature in physics.

A third measure is found in publications. The Statistical Data Panel of the Physics Survey Committee has determined the characteristics of physics publications of the world by a sampling of the 1969 issues of *Physics Abstracts*. (The details of sampling procedure and the nature of the sample are described in Chapter 13 and the special Panel Report on the Dissemination and Use of the Information of Physics.) Table 8.3 presents a summary of the publication picture for physics in various countries as obtained from this sample. The estimated number of publications in *Physics Abstracts* for each subfield of physics are given for the countries that are the ten largest producers of physics literature in the world. The United States is not only the largest contributor to the world's physics literature, it is also the largest source of the literature of each subfield. In fact, the distribution of the physics literature among the various subfields is relatively independent of the country involved. For example, about half of the papers from all countries deal with condensed matter. Perhaps the major difference among countries is in the relative emphasis on theoretical as opposed to experimental work. The percentage of the total papers in physics that are purely theoretical, that is, that contain no new experimental information, also appears in Table 8.3.

Yet another indication of national position in science is facilities. It is usually difficult to determine accurately the extent of facilities in a given country, but in one subfield, high-energy physics, the determination is especially easy. The extremely large size of some of these facilities makes them essentially unique; thus scientists from one nation are often anxious to use the facilities of another, making these installations exceptionally important points of international contact and interaction. As a result, the high-energy facilities of the world and their characteristics are widely known. Figure 8.8 presents a plot of the world's principal high-energy facilities, described by their beam current and energy, according to the country in which they are located.

In contrast to the dominant position of the United States in many other subfields of physics, fusion research in this country has been conducted on a level approximately equal to that of a number of other countries in manpower, originality of experiments, and results. A situation that appears to many to be the highest fruition of international cooperation has resulted. Periodically the results of Soviet research far exceed those of U.S. efforts. The Soviet development of Tokamak produced results that have revitalized the entire international fusion effort. Once secrecy was lifted in 1958, healthy scientific rivalries soon developed.

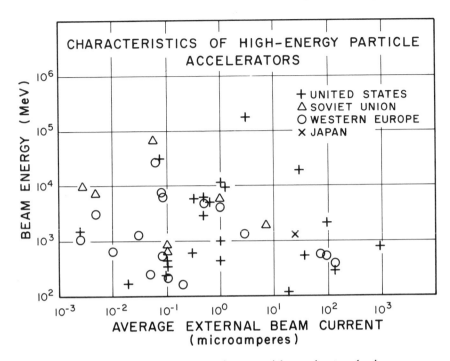

FIGURE 8.8 Locations of the world's large particle accelerators by beam energy and beam current.

Accomplishments of U.S. science in teaching and research reflect favorably on the culture that supports them. Visitors attracted to the United States by its scientific activities inevitably see the U.S. social scene in the light of cooperative endeavors of communities of dedicated scholars. Friendships built at the international level by collaboration in science permit meaningful discussion of politically more sensitive topics such as nuclear weapons and their control, cooperative efforts to protect the environment, and efforts to apply science and technology to the problems of developing countries.

Interaction and Scholar Exchanges

The most obvious of the collaborative interactions of physicists at the international level are probably the large international meetings that occur with a frequency of three or four per year. An example is the famous Rochester Conferences on elementary-particle physics. Most of the international meetings in physics are sponsored by the International

Union of Pure and Applied Physics (discussed later in the section on International Organizations and Cooperative Programs). At such meetings physicists make formal presentations of new experimental data or new theories; equally important are discussions that occur informally. Although such meetings are invaluable for the exchange of information and as a means of establishing associations that may lead to continuing communication and possible collaboration, longer-term exchanges of scholars among nations are even more productive.

More lengthy exchanges can be subdivided into three types. The first is the short-term visit in which a physicist from another country visits universities and laboratories of interest to him, often as an adjunct to attendance at an international meeting. This sort of visit, which is relatively common, adds depth and more leisurely discussion to the kind of interaction that occurs in a more intensive way at a meeting. A prominent university, such as Cambridge in England or Columbia in the United States, welcomes several score of foreign visitors each year. Major research laboratories also receive many visitors from abroad. Brookhaven National Laboratory, for example, has kept records that show that in the six-month period ending June 30, 1971, there were 124 short-term visits (one to three days) from abroad. The four countries chiefly represented by these visitors were Japan, France, the United Kingdom, and the Soviet Union; the number of visitors from each was approximately equal—about 20.

A second category of visitor is the senior scholar who visits a particular laboratory for a period of some months, often to perform a particular experiment using equipment not available in his own laboratory. Thus, Brookhaven National Laboratory will customarily have a dozen or more foreign scholars in residence for periods of from three months to one year performing experiments involving special equipment. During the first half of 1971, this laboratory had 95 foreign visitors with appointments of one month or longer; 13 of these held postdoctoral appointments. Similarly, there were a number of U.S. physicists with such appointments in temporary residence at CERN in Geneva during 1971.

Many unique facilities exist in different countries that could be used to a much larger degree not only by the physicists of the home country but by groups from other countries. Such international collaboration should be strongly encouraged. Its value lies not only in the pooling of ideas and experience but also in the cooperative use of instrumentation. Many groups of physicists have developed instruments that could be exploited advantageously at facilities in other countries. There are, for example, three American groups making use of the storage rings at CERN with their own instruments. There also are several European and U.S. groups using the Serpukhov accelerator in the Soviet Union. A striking example is the construction of a giant bubble chamber by French scientists at the Serpukhov laboratory. Other examples are collaboration among nations in the use of radio telescopes for long-baseline radio interferometry.

However, the frequency and significance of such visits increase rapidly with the cost and sophistication of the equipment involved. It is no accident that a particularly large fraction of such visits are to the great accelerators and the large optical and radio-astronomical telescopes; thus, one can predict that the number of applications from foreign particle physicists to work at the new National Accelerator Laboratory at Batavia, Illinois, will be large. Relatively long-term exchanges are important among theoretical physicists also because of the opportunity to do collaborative work with leading scholars.

The third category of international exchange is that of students—both foreign citizens studying in the United States and U.S. students who go abroad for postdoctoral training. Some data on these interchanges appear below; however, the implications and significance of such exchanges are clear even without detailed data. They represent a large-scale effort toward improving international understanding and appreciation through research, teaching, and interaction in the academic environment. Students returning to their home countries after a year or two of study abroad bring back not only new scientific perspectives but a better understanding of a different culture and often close and lasting ties with many of its representatives.

Other than these three main categories of international contact, there is a fourth composed of scholars who come from other countries to, for example, the United States, and remain here more or less permanently. Nearly one fourth (23 percent) of the respondents to the most recent National Register Survey of U.S. PhD physicists were foreign born. About half are now U.S. citizens. Some came to this country as young children; some are scholars who immigrated to the United States many years ago. But a large proportion falls into the category of young postdoctoral students who were attracted by the research opportunities in the United States and found permanent employment in industry or academic institutions. In some instances this so-called brain drain robs small countries of able and much needed scholars. The U.S. Government tends to discourage immigration during periods in which there are job shortages, but for the most part the influx of foreign scholars is regarded as strengthening and enriching our national capabilities.

Table 8.4 shows the extent to which citizens of other countries accept postdoctoral appointments in the United States. In practically all fields, including physics, at least half of the postdoctorals are foreign citizens. The principal countries of origin of the physical sciences postdoctorals in the United States are shown in Table 8.5. The position of the United States

TABLE 8.4   Citizenship of Postdoctoral Fellows in the United States [a]

Field	U.S. Citizen	Foreign Citizen
Physics	633	629
Mathematics	144	95
Chemistry	603	1048
Earth sciences	86	103
Engineering	99	174
Biosciences	503	400

[a] Source: *The Invisible University: Postdoctoral Education in the United States* (National Academy of Sciences, Washington, D.C., 1969), Table B-1.

in science is suggested by its ranking on the list of countries that are the principal producers of the world's physics literature (Table 8.3). The contributions of the United States to the education of citizens of countries with less well-developed facilities is suggested by the high positions of India and Taiwan on the list of countries in Table 8.5.

In view of the usefulness of postdoctoral training in the education of scientists and the broader utility of international exchanges of scholars, it is unfortunate that there has been a marked decline in recent years in the number of U.S. scientists studying abroad and also in the number of foreign scientists coming to the United States. A primary reason for this decline is found in budgetary constraints on the support of science. As an example,

TABLE 8.5   Principal Countries of Origin of Physical Sciences Postdoctoral Fellows

Country	Number
United States	1512
United Kingdom	399
India	284
Germany	193
Japan	190
Taiwan	102
Australia	86
Canada	82
Israel	63
Switzerland	62
Italy	60
France	51

the number of postdoctoral appointments at Brookhaven National Laboratory decreased from 83 in January 1967 to 52 in January 1971. As a second example, the number of postdoctoral fellows in a typical U.S. physics department, that of Cornell University, declined from 42 in 1967 to 32 in the spring of 1971.

There has been a comparable decline in the number of U.S. scholars studying abroad. This decline is partly explained by general financial stringency, but it is also a result of specific policy decisions in that several categories of postdoctoral fellowship program have been terminated by the National Science Foundation and the National Institutes of Health. The physics profession, representing a field in which international exchanges are of particular significance, should take the lead in working to bring about a reversal of this trend.

Some Data on Exchanges

The foreign student is a familiar participant in graduate education in the United States. In 1969, 22 percent of the 11,000 full-time graduate students in physics doctorate departments were foreign citizens, a proportion that is high for the fundamental sciences in general, in which the averages are 19 percent for the physical sciences, 18 percent for the mathematical sciences, and 15 percent for the life sciences. In engineering, however, 35 percent of the full-time graduate students in U.S. institutions were from foreign countries.

The substantial number of foreign graduate students must be viewed in part as an indication of a positive U.S. educational policy, not merely as receptivity to foreign demand for graduate education. Only 4 percent of the foreign students receive their principal support from foreign sources. The support that foreign physics students receive from U.S. sources further suggests their acceptance at the departmental level; compared with U.S. citizens, foreign students are only half as likely to have fellowships and traineeships, and they depend heavily on nonfederal sources of support.

The proportion of PhD's from U.S. institutions earned by foreign citizens corresponds to the level of foreign-student participation in these programs. The foreign component of total U.S. PhD's in physics has risen slowly from approximately 13 percent in 1958 to 17 percent in 1970, a period during which U.S. PhD production in physics increased from 500 per year to 1500 per year. The involvement of foreign students has more than kept pace with the rapid expansion in physics PhD programs.

The destinations of new PhD's of both U.S. and foreign citizenship present another aspect of international interaction. Typically, about 8 percent of the U.S. doctorates go abroad for their first employment or for

TABLE 8.6 Physics PhD's: Immediate Postdoctoral Plans, Fiscal Years 1958–1970

| Citizenship | Postdoctoral Study | | Employment | | | |
	U.S.	Foreign	U.S.	Foreign	Unknown	Total
United States	1695	420	6024	152	1979	10,270
Foreign	392	139	626	301	460	1,918
						12,188
Incomplete data						303
TOTAL U.S. DOCTORATES IN PHYSICS AND ASTRONOMY, FISCAL YEARS 1958–1970						12,491

postdoctoral study. In 1970, approximately 140 new physics doctorates went abroad for these purposes. Table 8.6 shows the pattern of these postdoctoral destinations for the 13 years 1958–1970.

Compared with new doctorates in other fields, physicists show a strong tendency to go abroad for further study or employment. The average of 5.7 percent for physicists can be compared with an average of 3.1 percent for all fields taken together. The importance of foreign postdoctoral experience for U.S. physicists is indicated by two comparisons: (1) 20 percent of the U.S. citizens who elect immediate postdoctoral study go abroad compared with 7 percent of all postdoctoral fellows, immediate and senior, in all fields; (2) postdoctoral study is the main attraction for young physicists going abroad, whereas foreign employment attracts the majority of those in other fields, as Table 8.7 shows.

The data indicate that postdoctoral study is an important correlate of foreign contact for U.S. citizens with new PhD's in physics. The number of young physicists going abroad has averaged 46 per year during the past five years. In addition to these young PhD's, senior postdoctoral recipients generally choose foreign host institutions. Although data comparable with those cited above for immediate postdoctoral students are not available, we

TABLE 8.7 Foreign Postdoctoral Destinations

U.S. Citizens	Study	Employment	Sample Size
Physics	73%	27%	572
All fields	46%	54%	1978

estimate that there were about 80 senior postdoctoral fellows abroad in 1967, perhaps half that number in 1970, the reduction being a consequence of cuts in NSF postdoctoral support. Taken together, U.S. citizens engaged in postdoctoral study in physics in foreign countries currently amount only to approximately 100 persons. As noted previously, the foundation for international interaction and cooperation in physics is sustained by an extremely modest commitment of human and fiscal resources.

Postdoctoral study provides not only foreign experience for U.S. citizens but enables foreign physicists to obtain substantial acquaintance with the U.S. community. Half of the 1200 postdoctoral appointments in physics are held by foreign citizens. Most (85 percent) have come to this country after receiving a PhD, so that postdoctoral study is their first exposure to the United States. Possessing a combination of scientific maturity and intellectual receptivity, these 500 or so foreign-trained PhD's who come to the United States for a year or two constitute an effective channel of interaction with world physics, for 88 percent of them return to their respective countries. The other component of the foreign postdoctoral group in the United States is made up of foreign citizens who earned a PhD in the United States; they number approximately 100. In contrast to those with a foreign PhD, the foreign postdoctorates with a U.S. PhD are more likely to remain in the United States; 60 percent of them do so.

Those who stay become part of the foreign-born or foreign-citizen component of the U.S. physics community. As Table 8.8 shows, these scientists amount to almost one fourth of the PhD physicists in the United States. The median ages, also shown in Table 8.8, give a clue to immigration characteristics. Those who become U.S. citizens typically are eight years older than the American-born group, and the noncitizens are younger, suggesting that naturalization is correlated with career development. A majority of the foreign-born physicists in the United States hold U.S. doc-

TABLE 8.8 Foreign and American Component of the U.S. Physics Community [a] (PhD's in 1970)

Citizenship/Birth	Number	Distribution	Median Age
U.S.-born citizens	15,500	75.5%	37.0
Foreign-born U.S. citizens	2,100	10.5%	43.7
Foreign citizens	2,400	12.0%	34.9
TOTAL	20,000		

[a] Based on 1970 National Register, assuming an 83 percent response from all groups. Foreign visitors and returning postdoctoral appointees are not included.

TABLE 8.9   Degree Background of Foreign-Born and Foreign-Citizen Physicists

Country of PhD	Number	Percent
United States	2800	65
Foreign country	1500	35

torates, but those with PhD's from institutions in other countries constitute a substantial number, as Table 8.9 shows.

One indication of the intellectual attraction of U.S. physics for foreign scientists is the distribution of foreign physicists among subfields. Table 8.10 presents data on the total number of PhD physicists in each of the sub-fields, together with the concentration of foreign born and the median age of the foreign-born groups. The subfields that have attracted the largest concentration of physicists from abroad are also those that have attracted the younger ones, as Figure 8.9 shows. The international character of the U.S. program in elementary-particle physics is especially promi-

TABLE 8.10   Foreign-Born Physicists in the United States: Concentration and Median Age by Subfield [a]

Physics Subfield	Total PhD's	Concentration of Foreign Born (%)	Median Age of Foreign Born
Acoustics	332	16.8	41.8
Astronomy	643	21.2	38.8
Astrophysics and relativity	255	25.2	36.6
Atomic, molecular, and electron	1102	23.4	36.8
Condensed-matter	4235	23.3	37.6
Earth and planetary	724	19.5	38.5
Elementary-particle	1440	28.3	35.4
Fluids	576	27.5	39.9
Nuclear	1824	21.7	38.2
Optics	1110	21.1	39.9
Physics in biology	277	20.9	41.3
Plasmas	542	27.2	38.0
Other	3552	15.1	42.6
All subfields	16,612	21.8	38.4

[a] Source: Data based on the 1970 National Register of Scientific and Technical Personnel.

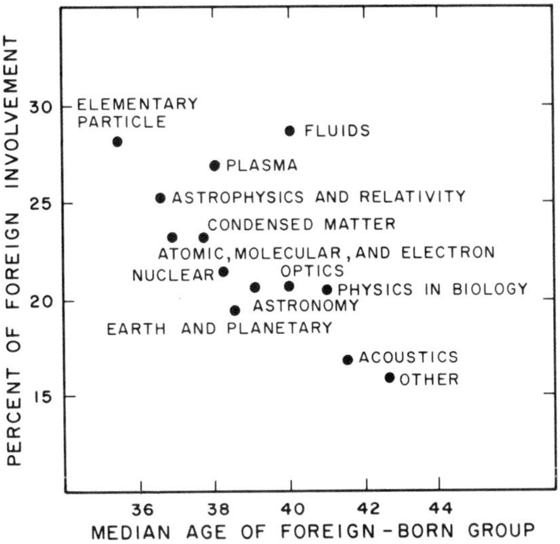

FIGURE 8.9 The attraction of physics subfields for foreign-born physicists as indicated by concentration and median age.

nent in these data. They also suggest that the motivation to work in the United States is related to this country's fundamental work in physics rather than to assumed rewards for technological creativity. In this respect the motivation of foreign physicists is clearly different from the scientific goals that motivate foreign graduate students who come to this country, for technologically oriented subjects such as engineering or agricultural economics are those that attract most foreign graduate students. The employment pattern of foreign-born physicists in the United States also reflects the emphasis on fundamental science rather than applied physics and technology, as suggested in Table 8.11.

The 200 or more PhD's who have become U.S. citizens can be considered as permanent residents; however, there is evidence for a substantial return

TABLE 8.11 U.S. Employer Destination of New Physics PhD's

Employment	U.S. Citizen $N=7118$ (%)	Foreign Citizen $N=824$ (%)
Academic institution	49	69
Industry	29	19
Government laboratory or federally funded research center	13	6
Other	9	6

migration among the remaining more than 200 foreign physicists who reside in the United States. The foreign contact provided by permanent residents can be viewed as an enrichment of the U.S. physics community; return migration is more truly in the spirit of an international exchange. Although data comparable with those cited for new PhD's or postdoctoral appointees are lacking, a longitudinal analysis of the 1968 and 1970 National Register Surveys affords an indication of this important export component of foreign contact. Between 1968 and 1970 about 450 established, foreign-citizen PhD's joined the U.S. physics community, while 250 others returned to foreign countries.

INTERNATIONAL ORGANIZATIONS AND COOPERATIVE PROGRAMS

Exchange arrangements, noted in the preceding section, are not the only aspect of international relations in science. There are also governmental and nongovernmental organizations concerned with cooperation in science and technology.

In the governmental category fall the activities of the United Nations and many of its specialized agencies—e.g., the World Meteorological Organization, the International Atomic Energy Agency, the U.N. Committee on Peaceful Uses of Outer Space, the International Telecommunications Union, and UNESCO. In addition, many bilateral and multilateral programs are undertaken by various nations. The United States, for example, has conducted large international programs, both bilateral and multilateral, in atomic energy and space research.

Within the last decade or so, operating organizations, financed by governments and conducting research in their own laboratories, have come into being. Perhaps the best known of these is CERN, the European Organization for Nuclear Research, which represents a collaboration among 12 Western European nations. Its laboratories at Geneva engage the services of some 400 scientists and engineers in elementary-particle and medium-energy physics research and, in addition, approximately 800 visiting workers from many countries. CERN's present facilities represent an investment in excess of $550 million. Construction has started on a 300-GeV accelerator that will cost approximately $250 million. Without cost-sharing by the 12 member nations, it is virtually certain that such facilities could not have been established in Western Europe.

The European Space Research Organization (ESRO) has a similar rationale and consists of a group of Western European nations. ESRO has launched some of its experiments, but it also has worked collaboratively with NASA, which has provided launch vehicles on several occasions. Two other examples of major research facilities similar to CERN and ESRO in their structure are the Joint Institute of Nuclear Research at Dubna, owned by a group of Eastern European countries, and the European Southern Sky Observatory (ESO), located in Chile but owned by six West European nations.

The need for an international facility would seem to be less for theoretical physics, but in fact an International Centre for Theoretical Physics was established about eight years ago at Trieste by the International Atomic Energy Agency (IAEA). The primary function of the Centre is to serve as a research institute where physicists from developing countries, who would ordinarily have little opportunity for interacting with other physicists, can get together with each other and with physicists from developed countries to pursue advanced theoretical research.

A collateral virtue of the Centre is that, because it is strongly supported by both the United States and the Soviet Union, it can serve as a meeting ground for international working groups. Thus the first major activity of the Centre (in 1964) was a month-long set of lectures on plasma theory by an international group, from which emerged what is probably still the best text in the field (*Plasma Physics* published by the IAEA). Following this, in 1965–1966, about 30 people worked for varying periods during the year with average stays of a few months. In contrast to the usual activities of the Centre, most of the participants were senior scientists from the developed countries with strong interests in controlled fusion. About 50 publications resulted from this working group (primarily on instability theory, transport theory, and nonlinear wave theory), and the research continues to have an impact on the field. Moreover, continuing collaboration of some U.S. scientists with those of the Soviet Union, England, France, Germany, and Italy date from this working group.

Despite the oft-cited scientific, political, and cultural advantages of international collaboration *in facilities* for scientific research, the primary motivation almost invariably is an economic one. This is especially true for countries that are technically well developed but are limited in their resources. Sensitive to economic competition and keenly aware that research is the basis for technology, they feel it essential to participate in research at the scientific frontiers but find that some of the facilities required are prohibitively costly. If a country cannot afford to be the sole owner of a facility, it must either resign itself to a noncompetitive position or enter into collaboration with other countries. Thus far, the United States has not really had to make such a choice, but budget pressures and increasingly costly special facilities for research are likely to force it to do so in the future.

Various nongovernmental organizations also contribute to international cooperation in the sciences. Thus societies in various countries cooperate as suggested by the program in *Physics Abstracts* conducted by the American Institute of Physics and the Institution of Electrical Engineers in England. It is expected that the newly established European Physical Society will have close ties to, for example, the American Physical Society. But the most extensively active and formal mechanism for cooperation among scientists is the International Council of Scientific Unions (ICSU).

The ICSU consists of 16 discipline-oriented unions and of almost a score of specialized committees. Many of the latter are interdisciplinary in character and represent joint efforts of two or more unions. Of the 16 unions, 10 are concerned with the physical and mathematical sciences, 5 with the life sciences, and 1 with the history and philosophy of science. Appendix A lists and provides information on the components of ICSU.

The functions of the unions are essentially similar even though the internal structures and approaches of unions vary. Thus, although all unions have formal national adherents, like the U.S. National Academy of Sciences, which maintain national committees to work with each union, the International Astronomical Union (IAU) has aspects of a membership organization, analogous to a domestic scientific society. Some unions, like the International Union of Geodesy and Geophysics (IUGG), hold congresses of an unrestricted nature, while the International Union of Radio Science (URSI), in an effort to keep its sessions from becoming unwieldy, establishes a quota-type system. Most unions conduct a major scientific and business congress once every three years, but the triennial assemblies of the International Union of Pure and Applied Physics (IUPAP) have been limited almost wholly to business matters—elections, commission reports, nominations to commissions, relationship with other units of ICSU—although the General Assembly of 1972, marking IUPAP's fiftieth anniversary, include scientific sessions where some 20 invited papers by scientists from various parts of the world survey major aspects of contemporary physics.

The primary functions of ICSU unions are threefold. First, they provide a means for communication of research results among scientists everywhere. This is done, in general, at the triennial or quadrennial assemblies of the unions and at conferences of subunits. IUPAP, in particular, relies on 16 commissions for this purpose (see below). This function is analogous to that of domestic scientific societies, whose meetings provide not only for the presentation of papers and their formal discussion but, equally important, for informal discussions. Only the latter, as a general rule, permit one to explore perplexing aspects of a piece of research and to learn about what has happened since the paper was prepared, as well as plans for continued research. The value of such interchanges at an international level is even greater than that domestically.

Another function served by the unions in their respective disciplines concerns scientific standards. Both science and technology benefit enormously from common standards of measurement and agreed-upon fundamental units. Such international standards are developed by the unions, and so too are methods of calibration. In addition, the unions establish common nomenclature, in terms and abbreviations, facilitating comprehension of published reports.

A third function is concerned with the issuance of some types of technical publication—for example, proceedings of significant symposia and triennial summaries of the major scientific work in each nation. The latter, which include literature citations, are most useful to workers everywhere as reference tools as well as summaries. Both the URSI and the IUGG have consistently issued such volumes at three- or four-year intervals.

The special committees of ICSU, while active in some of these ways, have somewhat more specific and programmatic functions possessing two characteristics: (a) an interdisciplinary character, thus embracing the interests of two or more unions, and (b) the actual or prospective existence of an international research program. Examples of the latter are the International Geophysical Year (1957–1959), whose antecedents go back to the International Polar Years of 1882–1883 and 1932–1933, the International Years of the Quiet Sun (1964–1965), and the Upper Mantle Program (1966–1970). Plans are under way for several similar international endeavors in the years ahead: the Global Atmospheric Research Program, which will use satellite systems to measure atmospheric circulation parameters; the International Geodynamics Project, concerned with the solid earth; and the International Magnetosphere Survey.

Appendix A indicates, of the 16 unions in ICSU, 10 are concerned with physical and mathematical sciences, in both their pure and applied aspects. The disciplines encompassed by these unions are astronomy, geodesy and geophysics, chemistry, radio science, physics, geography, crystallography, mechanics, mathematics, and geology. In varying degrees each of these deals with specific aspects of physics—for example, astronomy, crystallography, and radio science. Those concerned with geophysical sciences are discussed in the Report of the Panel on Earth and Planetary Physics. Here a further description of IUPAP may be of interest.

The International Union of Pure and Applied Physics (IUPAP) consists of physicists representing 38 countries.* These are usually appointed by national academies. An executive committee, including a president, several vice presidents, secretary-general, and associate secretary-general is selected at each triennial assembly to administer the union's affairs. IUPAP's annual budget is approximately $60,000, of which less than 25 percent

---

* See Appendix B for information on IUPAP adherents and commissions.

represents a subvention from ICSU (made possible by a UNESCO grant), while the bulk comes from subscriptions of the adhering academies. Except for some travel costs of the executive committee and minor office expenses, the funds are used by IUPAP to assist its commissions in the conduct of their scientific conferences. Hosts, in whose countries IUPAP or its commissions meet, seek public and private funds to defray most of the costs of meetings.

In addition to its finance, symbols, units and nomenclature, and publications commissions, IUPAP has 14 commissions concerned with subfields of physics: thermodynamics and statistical mechanics, cosmic rays, very-low-temperature physics, acoustics, magnetism, solid state, particles and fields, nuclear physics, atomic masses and related constants, physics education, atomic and molecular physics (including spectroscopy), plasma physics, and optics. Each of these consists of a chairman, a secretary, and some three to seven members appointed by the triennial assemblies of IUPAP on the basis of nominations from the delegates. Each conducts a major international conference, once every two or three years or so, where research results are presented in invited and contributed papers, thus providing each field with a periodic synthesis and assessment of the subfield. The proceedings are generally published.

Three types of conference are sponsored by IUPAP through the commissions. General conferences are intended to provide "an overview of the entire fields of interest" to a Commission. These usually occur at intervals of about three years "if advances in the field warrant." Attendance can range from 750 to 1500 participants. Topical conferences may be held "on broad subfields" of a Commission's interest, usually in years between any general conferences and with participation of between 300 and 600. Finally, special conferences on more specialized topics are possible, with attendance in the range of 50 to 200.* Some indication of the number and nature of such meetings is provided in Appendix C.

To describe the work of the commissions of IUPAP, let alone of other ICSU unions and committees that are related to physics,† is beyond the scope of this section. However, it may be of interest to sketch the origins and activity of one commission as an example: the Commission on Particles and Fields.‡ Although it was not established until 1957, its roots go back to the Rochester Conferences, the first of which was held

---

* The above information, as well as criteria for IUPAP meetings, may be found in *I.U.P.A.P. General Report 1970,* pp. 81–83.
† The Report of the Panel on Earth and Planetary Physics does, however, describe some of the activities of unions and committees concerned with the geological and geophysical sciences.
‡ Formerly called the High Energy Commission or the High Energy Nuclear Physics Commission.

in 1950. In that year, when the Berkeley and Rochester synchrocyclotrons and the Berkeley and Cornell electron synchrotrons had started operations, U.S. scientists took initiative to organize the first Rochester Conference in which "experimentalists would be given 'equality' with the theorists." *

The first two Rochester Conferences, 1950 and 1952, "were modest in scope, of short duration, supported locally and still not truly international." * The third included scientists from England, France, Italy, Australia, Holland, Japan, and other countries. The fourth conference, essentially similar to the third in subjects and format and the last attended by Fermi, heard results on pion–nucleon scattering experiments at Brookhaven, nucleon–nucleon polarization effects in the several hundred MeV region observed at four U.S. synchrocyclotrons and at Harwell. The fifth conference in 1955 showed progress in its internationalization, for scientists from 15 countries participated, and, for the first time, the sponsorship of IUPAP was obtained. Not until the sixth conference of early 1956 were Soviet scientists present. At the seventh conference in 1957 it was proposed to the IUPAP General Assembly that a high-energy commission be established. The Commission was established, and its membership was chosen to reflect work in high-energy physics in the United States, the Soviet Union, Western Europe, and, somewhat later, elsewhere. The Commission decided to pursue the model of the Rochester Conferences but set up a rotation of site—Geneva in 1958, Kiev in 1959, Rochester in 1960. Moreover, a second series of conferences, to be held biennially, was organized to deal with accelerators and instrumentation. In the triennium preceding IUPAP's 1969 General Assembly, the Commission summarized its work in that interval as follows:

The main business of the Commission was to discuss IUPAP sponsorship and financial support of conferences in the field of high energy physics. The regular International Conferences on High Energy Physics, held every even year, and the regular International Conferences on High Energy Accelerators, held every odd year, were prepared in detail. In addition, various more specialised meetings were discussed and IUPAP sponsorship was recommended for four conferences held in 1967, for three conferences in 1968, and for five conferences in 1969.

The conferences in the two main fields noted above sponsored by IUPAP were the following: International Conference on High Energy Physics (Austria, 1968) and two on High Energy Accelerators (United States, 1967; Soviet Union, 1969). In addition, several specialized conferences were held:

* R. E. Marshak, "The Rochester Conferences: The Rise of International Cooperation in High Energy Physics," *Bulletin of the Atomic Scientists,* June 1970, p. 93. (The narrative above is based largely on this article, with some information drawn from IUPAP's *Yearbooks* and *News Bulletins.*)

International Conference on High Energy Physics and Nuclear Structure (Rehovoth, Israel), February/March 1967.

International Theoretical Physics Conference on Particles and Fields (Rochester, N.Y., U.S.A.), June 1967.

International Symposium on Electron and Photon Interactions at High Energies (Stanford, Calif., U.S.A.), September 1967. Topical Conference on High Energy Collisions of Hadrons (CERN, Geneva, Switzerland), January 1968.

International Colloquium on Nuclear Electronics (Versailles, France), September 1968.

Topical Conference on Weak Interactions (CERN, Geneva, Switzerland), January 1969.

Third International Conference on High Energy Reactions (Stony Brook, N.Y., U.S.A.), September 1969.

International Symposium on Electron and Photon Interactions at High Energies (Liverpool, U.K.), September 1969.

International Conference on High Energy Physics and Nuclear Structure (Columbia University, N.Y., U.S.A.), September 1969.

APPENDIX 8.A. ICSU, THE UNIONS, SPECIALIZED COMMITTEES, SERVICES, AND INTER-UNION COMMISSIONS

The Unions

The 16 unions, as well as the technical commissions or associations of some unions, are as follows (date of establishment follows the union names in parentheses).

International Astronomical Union (IAU) (1919) Ephemerides. Documentation. Astronomical Telegrams. Celestial Mechanics. Positional Astronomy. Astronomical Instruments. Solar Activity. Radiation and Structure of the Solar Atmosphere. Fundamental Spectroscopic Data. The Physical Study of Comets. The Physical Study of Planets and Satellites. The Moon. The Rotation of the Earth. Position and Motions of Minor Planets, Comets and Satellites. The Light of the Night-Sky. Meteors and Meteorites. Photographic Astrometry. Stellar Photometry. Double Stars. Variable Stars. Galaxies. Stellar Spectra. Radial Velocities. Time. Structure and Dynamics of the Galactic System. Interstellar Matter and Planetary Nebulae. Stellar Constitution. Theory of Stellar Atmospheres. Star Clusters and Associations. Exchange of Astronomers. Radio Astronomy. History of Astronomy. Photometric Double Stars. Astrophysical Plasmas and Magneto-Hydrodynamics. Astronomical Observations from Outside the Terrestrial Atmosphere. Special Classifications and Multi-Band Colour Indices. Teaching of Astronomy. Cosmology. High-Energy Astrophysics.

International Union of Geodesy and Geophysics (IUGG) (1919) Geodesy. Seismology and Physics of the Earth's Interior. Meteorology and Atmospheric Physics. Geomagnetism and Aeronomy. Physical Sciences of the Ocean. Scientific Hydrology. Volcanology and Chemistry of the Earth's Interior.

International Union of Pure and Applied Chemistry (IUPAC) (1919)

International Union of Radio Science (URSI) (1919) Radio Measurements and Standards. Radio and Non-Ionized Media. Ionosphere. Magnetosphere. Radio Astronomy. Radio Waves and Circuits. Radio Electronics. Radio Noise of Terrestrial Origin.

International Union of Pure and Applied Physics (IUPAP) (1922) See Appendix 8.B for further information on IUPAP and its commissions.

International Union of Biological Sciences (IUBS) (1923)

International Geographical Union (IGU) (1923)

International Union of Crystallography (IUCr) (1947) Journals. Structure Reports. International Tables. Crystal Growth. Crystallographic Apparatus. Crystallographic Computing. Crystallographic Data. Crystallographic Nomenclature. Crystallographic Studies at Controlled Pressure and Temperature. Crystallographic Teaching. Electron Diffraction. Neutron Diffraction.

International Union of Theoretical and Applied Mechanics (IUTAM) (1947)

International Union of the History and Philosophy of Science (IUHPS) (1947–1956)

International Mathematical Union (IMU) (1952)

International Union of Physiological Sciences (IUPS) (1955)

International Union of Biochemistry (IUB) (1955)

International Union of Geological Sciences (IUGS) (1961) Stratigraphy. Petrology. Structural Geology. Marine Geology. Meteorites.

International Union of Pure and Applied Biophysics (IUPAB) (1966)

International Union of Nutritional Sciences (IUNS) (1968)

The Committees

Scientific Committee on Oceanic Research (SCOR) (1957)
Scientific Committee on Antarctic Research (SCAR) (1958)
Committee on Space Research (COSPAR) (1958)

Special Committee for the International Biological Programme (SCIBP) (1963)
Scientific Committee on Water Research (COWAR) (1964)
Committee on Science and Technology in Developing Countries (COSTED) (1966)
Committee on Data for Science and Technology (CODATA) (1966)
Committee on the Teaching of Science (1968)
Special Committee on Problems of the Environment (SCOPE) (1969)

The Permanent Services

ICSU Abstracting Board (IAB) (1953)
Federation of Astronomical and Geophysical Services (FAGS) (1956)

The Inter-Union Commissions and Panel

Inter-Union Commission on Frequency Allocations for Radio Astronomy and Space Science (IUCAF)
Inter-Union Commission on Radio Meteorology (IUCRM)
Inter-Union Commission on Solar Terrestrial Physics (IUCSTP)
Inter-Union Commission on Spectroscopy (IUCS)
Inter-Union Commission on Geodynamics (ICG)
Inter-Union Commission on Lunar Studies
ICSU Panel on World Data Centres (Geophysical and Solar) (WDC)

Adherents to ICSU and the Unions

	ICSU	IAU	IUGG	IUPAC	URSI	IUPAP	IUBS	IGU	IUCr	IUTAM	IUHPS	IMU	IUPS	IUB	IUGS	IUPAB	IUNS
Argentina	x	x	x	x	x	x	x	x	x	x	x	x	*	x	x	—	x
Australia	x	x	x	x	x	x	x	x	x	x	x	x	x	x	x	x	—
Austria	x	x	x	x	x	x	x	x	x	x	x	x	—	x	x	x	x
Belgium	x	x	x	x	x	x	x	x	x	x	x	x	x	x	x	x	x
Bolivia	—	—	x	—	—	x	—	—	—	—	—	—	—	—	x	—	—
Botswana	—	—	—	—	—	—	—	—	—	—	—	—	—	—	x	—	—
Brazil	x	x	x	x	x	x	x	x	x	—	—	x	*	x	—	—	x
Bulgaria	x	x	x	x	—	x	x	x	x	x	x	x	x	—	—	—	—
Burma	—	—	x	—	—	—	—	—	—	—	—	—	—	—	—	—	—
Canada	x	x	x	x	x	x	x	x	x	x	x	x	x	x	x	x	x
Ceylon	x	—	—	—	x	—	—	—	—	—	—	—	—	—	x	—	—
Chile	x	x	x	—	—	—	—	x	x	—	x	—	*	x	x	x	x
China, Rep. of	x	x	x	x	x	x	x	x	—	—	x	x	—	x	x	x	—
Colombia	x	x	x	x	—	—	—	—	—	·—	—	—	x	—	—	—	x
Cuba	x	—	x	x	—	x	x	x	—	—	—	x	x	—	x	—	—
Cyprus	·—	—	—	—	—	—	—	(†)	—	—	—	—	—	—	x	—	—
Czechoslovakia	x	x	x	x	x	x	x	x	x	x	x	x	x	x	x	x	x
Denmark	x	x	x	x	x	x	x	x	x	x	x	x	x	x	x	x	x
Dominican Rep.	—	—	x	—	—	—	—	x	—	—	—	—	—	—	—	—	—

Adherents to ICSU and the Unions—Continued

	ICSU	IAU	IUGG	IUPAC	URSI	IUPAP	IUBS	IGU	IUCr	IUTAM	IUHPS	IMU	IUPS	IUB	IUGS	IUPAB	IUNS
Ethiopia	—	—	X	—	—	—	—	X	—	—	—	—	—	—	—	—	—
Finland	X	X	X	X	X	X	X	X	X	X	X	X	X	X	X	—	X
France	X	X	X	X	X	X	X	X	X	X	X	X	X	X	X	X	X
Germany, Fed. Rep. (B.R.D.)	X	X	X	X	X	X	X	X	X	X	X	X	X	X	X	X	X
Germany, Dem. Rep. (D.D.R.)	X	X	X	—	X	X	X	X	X	X	X	X	X	X	X	X	X
Ghana	X	—	X	—	X	—	—	X	—	—	—	—	—	—	X	—	—
Greece	X	X	X	X	—	—	—	X	—	—	—	X	—	—	—	X	—
Guatemala	—	—	X	—	—	—	—	X	—	—	—	—	—	—	X	—	—
Guinea	—	—	—	—	—	—	—	X	—	—	—	—	—	—	—	—	—
Haiti	—	—	X	—	—	—	—	—	—	—	—	—	—	—	—	—	—
Hong Kong	—	—	—	—	—	—	—	X	—	—	—	—	—	—	—	—	—
Hungary	X	X	X	X	X	X	X	X	X	X	X	X	X	X	X	X	X
Iceland	—	—	X	—	—	—	—	X	—	—	—	X	—	—	X	—	—
India	X	X	X	X	X	X	X	X	X	X	X	X	X	X	X	X	X
Indonesia	X	—	X	—	—	—	—	X	—	—	—	—	—	—	X	—	—
Iran	X	X	X	—	—	—	—	X	—	—	—	—	—	X	X	—	X
Iraq	—	—	—	—	—	—	—	X	—	—	—	—	—	—	—	—	—
Ireland	—	X	X	X	—	X	X	X	—	—	—	X	—	X	X	X	—
Israel	X	X	X	X	X	X	X	X	X	X	X	X	X	X	X	X	X
Italy	X	X	X	X	X	X	X	X	X	X	X	X	X	X	X	X	X
Jamaica	(†)	—	—	—	—	—	—	X	—	—	—	—	—	X	X	X	X
Japan	X	X	X	X	X	X	X	X	X	X	X	X	X	X	X	X	X
Kenya	(n)	—	—	—	—	—	—	—	(†)	—	—	—	—	—	—	—	—
Korea, Dem. People's Rep. of	X	X	X	—	—	—	—	—	—	—	—	—	X	—	X	—	—
Korea, Rep. of	X	—	X	X	—	X	X	X	—	—	—	—	X	—	—	—	—
Lebanon	—	—	X	—	—	—	—	—	—	—	—	—	—	—	—	—	—
Luxembourg	—	—	—	—	—	—	—	—	—	—	—	—	—	—	X	—	—
Malagasy Republic	X	—	X	—	—	—	—	X	—	—	—	—	(†)	—	X	—	—
Malawi	—	—	—	—	—	—	—	X	—	—	—	—	—	—	X	—	—
Malaysia	—	—	X	—	—	—	—	X	—	—	—	X	—	—	X	—	—
Mexico	X	X	X	X	X	X	—	X	—	—	—	X	*	X	X	X	X
Monaco	X	—	X	—	—	—	X	—	—	—	—	X	—	—	—	—	—
Morocco	X	—	X	—	X	—	X	X	—	—	—	—	—	—	X	—	—
Netherlands	X	X	X	X	X	X	X	X	X	X	X	X	X	X	X	X	X
New Zealand	X	X	X	X	X	—	X	X	X	—	—	—	X	—	X	X	X
Nigeria	X	—	—	X	X	—	X	X	—	—	—	—	—	—	X	—	X

Adherents to ICSU and the Unions—Continued

	ICSU	IAU	IUGG	IUPAC	URSI	IUPAP	IUBS	IGU	IUCr	IUTAM	IUHPS	IMU	IUPS	IUB	IUGS	IUPAB	IUNS
Norway	X	X	X	X	X	X	X	X	X	X	X	X	*	X	X	X	X
Pakistan	X	—	X	—	—	X	—	X	X	—	—	X	—	—	—	—	X
Panama	—	—	—	—	—	—	—	X	—	—	—	—	—	—	—	—	X
Peru	X	—	X	—	X	—	—	—	—	—	—	—	X	X	—	—	—
Philippines	X	—	X	—	—	—	—	X	—	—	—	—	—	—	—	—	X
Poland	X	X	X	X	X	X	X	X	X	X	X	X	X	X	X	X	X
Portugal	X	X	X	X	X	—	—	X	—	X	X	X	—	X	X	—	X
Rhodesia	—	—	X	—	—	—	X	—	—	—	—	—	—	—	—	—	—
Romania	X	X	X	X	—	X	X	X	—	X	X	X	X	—	X	X	—
Senegal	—	—	X	—	—	—	—	X	—	—	—	—	—	—	—	—	—
Sierra Leone	—	—	X	—	—	—	—	(†)	—	—	—	—	—	—	—	—	—
Singapore	X	—	—	—	—	—	—	(†)	—	—	—	X	—	—	—	—	—
South Africa	X	X	X	X	X	X	X	X	X	—	—	X	—	—	X	—	—
Spain	X	X	X	X	X	X	X	X	X	X	X	X	X	X	X	—	X
Sudan	—	—	X	—	—	—	—	X	—	—	—	—	—	—	—	—	—
Swaziland	—	—	—	—	—	—	—	—	—	—	—	—	—	X	—	—	—
Sweden	X	X	X	X	X	X	X	X	X	X	X	X	X	X	X	X	X
Switzerland	X	X	X	X	X	X	X	X	X	X	X	X	X	X	X	X	X
Syria	—	—	X	—	—	—	—	—	—	—	—	—	—	—	—	—	X
Tanzania	(n)	—	—	—	—	—	—	X	—	—	—	—	—	—	X	—	—
Thailand	X	—	X	—	—	—	—	X	—	—	—	—	—	—	X	—	—
Tunisia	X	—	X	—	—	—	—	X	—	—	—	—	—	—	X	—	—
Turkey	X	X	X	X	—	—	—	X	—	X	—	X	X	—	X	—	—
Uganda	(n)	—	—	—	—	—	—	X	—	—	—	—	—	—	X	—	—
U.S.S.R.	X	X	X	X	X	X	X	X	X	X	X	X	X	X	X	X	X
United Arab Rep.	X	X	X	X	—	X	X	X	—	—	—	—	X	—	X	—	—
United Kingdom	X	X	X	X	X	X	X	X	X	X	X	X	X	X	X	X	X
U.S.A.	X	X	X	X	X	X	X	X	X	X	X	X	X	X	X	X	X
Uruguay	X	X	—	—	—	—	—	X	—	—	—	—	X	—	—	—	—
Vatican City State	X	X	—	—	—	—	—	X	—	—	—	—	—	—	—	—	—
Venezuela	X	X	—	X	—	—	—	—	—	—	—	—	X	X	X	X	—
Vietnam, Dem. Rep.	X	—	—	—	—	—	—	—	—	—	—	—	—	—	—	—	—
Vietnam, Rep.	X	—	X	X	—	—	—	—	—	—	—	—	—	—	—	—	—
Yugoslavia	X	X	X	X	X	X	X	X	X	X	X	X	X	—	X	—	X
Zaire	—	—	—	—	—	X	—	—	—	—	—	—	—	—	X	—	X
Zambia	—	—	—	—	—	—	—	X	—	—	—	—	—	—	X	—	—
	62	44	69	43	37	38	41	69	31	29	32	41	39	35	60	27	34

(ⁿ) Scientists from Kenya, Uganda, Tanzania have constituted a joint Academy in Nairobi called East African Academy, which is a member of ICSU.

(*) Norway adheres to the IUPS through the Scandinavian Physiological Society; Argentina, Brazil, Chile, Mexico, Peru and Venezuela adhere as a group through the Latin-American Physiological Association. This Association also adheres to IUB.

(†) National Associate.

APPENDIX 8.B. THE INTERNATIONAL UNION OF PURE AND APPLIED PHYSICS

Objectives

The statutes define the objectives and characterize the national committees in Articles I and II as follows:

I. Aims of the Union and Conditions of Membership

   1. The aims of the Union are:

      (i) the stimulation and promotion of international cooperation in Physics;

      (ii) the co-ordination of the work of preparing and publishing abstracts of papers and tables of physical constants;

      (iii) the promotion of international agreements on the use of symbols, units nomenclature, and standards;

      (iv) the encouragement of interesting research.

    The Union may organize international meetings.

    Individual nations may join the Union through their National Academies, their National Research Councils, equivalent national societies or groups of societies, or, if suitable ones do not exist, through their Governments.

    Membership of a given nation through several distinct organizations is not permitted, unless these several organizations have previously agreed to share Union dues and voting rights.

    The word "nation" includes dominions, diplomatic protectorates, or other territories which have an independent scientific community.

II. National Committees

   2. The body responsible for initiating its country's membership in the Union will set up a National Committee, which will maintain liaison with the Union.

   3. The National Committees will, in their respective countries, encourage and co-ordinate study in various fields of Physics, with emphasis on international aspects. Every National Committee may, of itself or in collaboration with other National Committees, submit to the Union for discussion problems which are within its competence.

    The National Committee elect their delegates to Union assemblies. They also elect a Delegation Head who votes on the delegation's behalf on questions of administration as laid down in Articles 14 and 16.

Commissions

Financial Commission
Commission for Symbols, Units and Nomenclature (1931)
Commission on Thermodynamics and Statistical Mechanics (1945)
Commission on Cosmic Rays (1947)
Commission on Very Low Temperature (1949)
Commission on Publications (1949)
Acoustics Commission (1951)
Semiconductors Commission (1957)
Magnetism Commission (1957)
Solid State Commission (1960)
Commission on Particles and Fields (1957)
Nuclear Physics Commission (1960)
Atomic Masses and Related Constants Commission (1960)
Commission on Physics Education (1960)
Atomic and Molecular Physics and Spectroscopy Commission (1966)
Plasma Physics Commission (1969)
International Commission for Optics (1948)

Adherents

Argentina, Australia, Austria, Belgium, Bolivia, Brazil, Bulgaria, Canada, China (Taiwan), Cuba, Czechoslovakia, Denmark, Finland, France, German Democratic Republic, Federal Republic of Germany, Great Britain, Holland, Hungary, India, Ireland, Israel, Italy, Japan, Mexico, Norway, Pakistan, Poland, Romania, South Africa, South Korea, Spain, Sweden, Switzerland, United Arab Republic, United States of America, Union of Soviet Socialist Republics, and Yugoslavia.

APPENDIX 8.C. IUPAP CONFERENCES, 1970–1971

1970

Second International Conference on Atomic Physics, England, July 21–24
Symposium on Polarization Phenomena in Nuclear Reactions, U.S.A., August 31–September 4
Angular Correlations in Nuclear Physics, Holland, 1st week September
International Conference on Magnetism, France, September 14–19
The Density of Electronic Charge Spin and Momenta, France, September 9–12
Magnetic Thin Films, Czechoslovakia, September 21–23
International Working Seminar on the Role of History of Physics in Education, U.S.A., July 13–17
International Congress on the Education of Teachers of Physics in Secondary Schools, Hungary, September 11–17
XVth International Conference on High Energy Physics, U.S.S.R., August 26–September 4
Statistical Mechanics, Mexico, October 19–24
12th International Conference on Low Temperature Physics LT–12, Japan, September 4–10

Transport Properties of Solids, Australia, August 27–29

Décharges et Isolement Électriques dans le Vide, Canada, September

IVth International Congress on Radiation Research, France, June 28–July 4

Semiconductors, U.S.A., August 17–21

Radiation Effects in Semiconductors, U.S.A., August 24–26

Precision Measurements and Fundamental Constants, U.S.A., August 3–7

The Physics and Chemistry of Semiconductor Heterojunctions, Hungary, October 11–17

Physics of Metastable Alloys, Yugoslavia, September

International Conference on Thermodynamics, Wales, April 1–4

Holography, France, July

Instrumentation, U.S.S.R., September 8–12

1971

International Conference on Statistical Mechanics, U.S.A., March 29–April 2

Conference on Theoretical Physics and Biology, France, June 21–26

12th International Conference on Cosmic Rays, Tasmania, August 16–25

7th International Conference on Acoustics, Hungary, August 18–26

International Conference on Amorphous and Liquid Semiconductors, U.S.A., August 8–13

International Conference on Crystal Growth, France, July 5–9

International Conference for Solid Surfaces, U.S.A., October 11–18

Conference on Colour Centres in Ionic Crystals, England, September 13–17

International Conference on Light Scattering in Solids, France, July 19–23

VIIIth International Conference on High Energy Accelerators, Switzerland, September 20–24

International Conference on Electron and Photon Interactions at High Energies, U.S.A., August 23–28

International Conference on Duality and Symmetry in Hadron Physics, Israel, April 5–7

International Conference on High Energy Physics and Nuclear Structure, U.S.S.R., September 7–12

Conference on Statistical Properties of Nuclei, U.S.A., August 23–27

IVth International Conference on Atomic Masses and Related Fundamental Physical Constants, England, September 6–10

3rd International Conference on Vacuum UV Radiation Physics, Japan, August 30–September 2

VIIth International Conference on the Physics of Electronic and Atomic Collisions, Netherlands, July 26–31

Xth International Conference on Phenomena in Ionized Gases, England, September 13–18

International Symposium on Plasma Physics, Canada, July 6–7

7th International Conference on General Relativity, Denmark, July

9

THE
INSTITUTIONS
OF
PHYSICS

It was the erection and institution of—Salomon's
House—that house for the finding out of the true
nature of all things (whereby God mought have
the more glory in the workmanship of them and
men the more fruit in the use of them)—and
withal to bring unto us books, instruments and
patterns in every kind.

FRANCIS BACON (1561–1626)
The New Atlantis

INTRODUCTION

Much physics has become big physics—big equipment, big institutions, big
money, and correspondingly big involvement with the federal government.
In this chapter we shall try to show how the transition from little physics
to big physics grew almost inevitably from the fundamental character of
physics, and how this transition has affected the institutions—national
laboratories, industrial laboratories, and universities—in which physics is
pursued.

When physics, as it is now defined, first started, say at the time of
Galileo, it was primarily concerned with discerning the simplest laws of
nature—the laws of motion, the laws of gravitational attraction. But
once laws of nature were discovered and proved, physicists began to apply
them to increasingly complex situations. The complexity of the individual

580

situation could be described by what physicists call, broadly, initial or boundary conditions. The distinction between laws of nature and initial conditions is illustrated by Newton's discovery of the laws of gravitational attraction and the laws of motion. These were general laws, applicable to a large class of situations. To apply these laws to predict the motion of two interacting planets, Newton had to know the initial conditions that characterized the specific situation he was studying. In this case he needed to know the initial distances and relative velocities between the two planets and, if the planets were not spherical and, if very great accuracy were demanded, their shapes and mass distributions. The importance of initial or boundary conditions is strikingly apparent in the application of Maxwell's equations governing electromagnetic radiation to the prediction of the behavior of a microwave cavity or waveguide. Their importance is equally obvious in the prediction of the behavior of open and closed organ pipes, or violin strings, from the simple laws of mechanical vibrations.

The distinction between laws of nature and initial conditions is fundamental in physics. For physics tends to develop along two rather different, though interrelated, lines. In Chapter 3 we alluded to this distinction in the discussion of fundamental physical laws, which govern the interactions of the simplest physical systems, and *organized* or *disorganized* complexity, which comprises those physical phenomena involving large numbers of interacting simpler systems, such as atoms in a crystal lattice or molecules in a gas. In this chapter we shall preserve this distinction but shall use a rather different terminology, Weisskopf's * "extensive physics" and "intensive physics."

Extensive physics is the application of known laws of nature to the understanding of complex natural phenomena. These natural phenomena always involve atoms or nuclei in a variety of different configurations. Here one is concerned with vast numbers of complex situations, each of which is distinguished by specific and unique initial conditions: the particular configuration of atoms and nuclei that nature has presented at a particular time and place. The fundamental laws of nature in these situations are not at issue. Physicists believe that they understand or can fully describe the electrical and gravitational interactions of matter and that they know in principle what determines the motions, therefore the energies, of interacting bits of matter. Quantum mechanics, plus a knowledge of the laws of electrical attraction, implies all of chemistry, at least from the point of view of the physicist. Four of the main branches of modern physics—atomic and molecular, condensed matter, plasma and

---

* V. F. Weisskopf, "A Defense of Science," *Science, 147,* 1552–1554 (March 26, 1965). See also Chapter 5.

fluid dynamics' and, to a lesser degree, nuclear—are concerned with the application of known laws of nature (meaning known individual interactions) to the prediction of phenomena associated with different aggregations of elementary particles. In three of these, atomic and molecular, condensed matter, and plasma and fluid dynamics, the underlying interactions are electromagnetic. In the fourth, nuclear, there are in addition specifically nuclear interactions. But by and large what these branches of physics are principally about is the explanation and prediction of phenomena arising from complex configurations of elementary atomic and nuclear entities. It is the physics of initial conditions or extensive physics; it is the physics of complexity.

Much physics, in fact most science, is of this character. Certainly chemistry, the environmental sciences, and biology (unless one admits the existence of completely biotonic laws) are concerned with initial conditions: the application of known laws of nature to complex situations, each characterized by certain initial conditions. The marvelously ingenious phenomena such as the Mössbauer effect or the laser or even superconductivity were, in retrospect, always implicit in the conceptual framework of physics as it existed at the time of their discovery—or at least before the phenomena were explained (as in the case of superconductivity). The essential word here is complexity: New phenomena emerge by virtue of the inherent complexity of the physical system, a complexity that ordinarily defies attempts to unravel it.

There are in physics, however, two additional threads that transcend the initial conditions. There are the deeply fundamental questions, almost philosophic in flavor that ask: Are there any more fundamental, simpler principles that underlie the laws of nature? In modern physics the answers to this question are usually sought in the underlying structure and character of space–time itself. Thus general relativity teaches that gravitation is a consequence of the geometry of space; or, more generally, modern physics is dominated by the view that the laws of conservation can be derived from the symmetry properties of the surrounding world. For example, to speak of the simplest, conservation of energy derives from the invariance of the laws of nature with respect to time displacement, and conservation of momentum derives from the invariance of the laws of nature with respect to translations in space–time.

These almost philosophic considerations, which underlie and tend to dominate much of the deepest thinking in modern physics, are not directly relevant as far as the relation between physics and its public patrons is concerned. Such generalizations are very much the products of single minds of great genius; though the impact of such work on physics is enormous, it is very inexpensive.

The other thread in modern physics that transcends the initial conditions is concerned with discovering new laws of nature, not simply applying known laws of nature to ever more complex situations. The areas of physics in which completely new and unfathomed laws of nature—that is, laws of interaction—may exist are those concerned with extremely small and extremely large distances (elementary-particle physics and astrophysics, respectively). In the case of small distances, the physicist sought originally to understand, or to describe, the so-called nuclear force, the elementary, intrinsically nuclear interaction among nuclear particles. The general idea was to reduce nuclear physics to an initial condition science in the same way that chemistry has been so reduced, but with the nuclear forces playing the role that electromagnetic force plays in chemistry. However, in the course of seeking to elucidate the nuclear force, a rich and totally unsuspected world of entirely new entities was discovered. The relation of these entities to the rest of physics is not completely clear. Yet, from a more thorough understanding of these may come insights into what philosophers would call ontological questions: How do protons and neutrons happen to exist? And why is the elementary charge just 4.8×10^{-10} esu? And why is the range of the nuclear force approximately 10^{-13} cm?

In the case of astrophysics, which offhand one might classify as initial condition or extensive science, completely new and unsuspected laws of nature could be at work. Here the point is that when one is dealing with sufficiently great distances and sufficiently large masses, one enters regions of physical stress that are so enormous as to affect the very substrate, namely, space–time, itself. Whether totally new theories of matter and space will come from such investigations is difficult to predict. At the very least, there is a connection between the phenomena that occur in extremely dense bodies like neutron stars and the phenomena that one finds as one probes to smaller and smaller distances in the hearts of nuclei. In addition, a problem arises at the origin of the universe when the laws of nature, as physicists understand them, simply cannot apply before a certain critical time.

The search for new fundamental laws of nature among the elementary particles, and possibly among some stars, and the search for more fundamental principles underlying the laws of nature constitute what Weisskopf * calls intensive physics. It includes what are in many respects the most challenging subfields in modern physics—challenging because of their subtlety and novelty—but also the most difficult and expensive to

* V. F. Weisskopf, "A Defense of Science," *Science, 147,* 1552–1554 (March 26, 1965).

explore. The phenomena of elementary-particle physics are extremely rare in the terrestrial environment. To create the tiny, short-lived particles of elementary-particle physics requires enormous energies, and enormous energies in the laboratory mean huge accelerators. These accelerators are among the most expensive pieces of scientific equipment man has ever devised—matched in expense only by the satellite systems with which, together with the huge optical and radio telescopes, the astrophysicist probes the dim outer reaches of the universe.

The quest for new laws of nature at the very limits of distances both small and large has placed these subfields at the mercy of the public; either the quest is undertaken at public expense, or it is not undertaken at all. The seemingly inexorable urge of physicists to discover ever more subtle laws of nature has made the pursuit of physics completely dependent on the society in which it is embedded.

But most modern physics, and indeed most modern science, relates to the physics of initial conditions or extensive physics, the physics of complexity, which entails the investigation of the almost endless variety of phenomena that occur in nuclei, atoms and molecules, solids, and plasmas. These phenomena are often of great subtlety and beauty: for example, the action of masers and lasers, the fission of a uranium nucleus, or the complex modes of vibration of a magnetically confined plasma. Their experimental investigations and theoretical elucidation require the highest order of intellectual acuity and insight. Often some seemingly simple phenomena defy explanation even after many years of concentrated effort by great minds. For example, turbulence, whether in a plasma confined magnetically or in water flowing in a pipe, is still not fully understood.

Since initial-condition or extensive physics deals with objects and phenomena that are close to man's experience, it is natural that this kind of physics usually lends itself to application. Many modern technologies have sprung or will spring from initial-condition physics—nuclear power from nuclear physics, the modern transistorized computer from condensed-matter physics, and, it is hoped, controlled fusion from plasma physics. For the same reasons, the various parts of initial-condition physics impinge strongly on other sciences. Much of current chemistry is yesterday's initial-condition physics; hardly any of the environmental sciences could proceed without analytical tools contributed by initial-condition physics; and modern biology without x-ray crystallography would be difficult to imagine.

Thus, the branches of initial-condition physics relate strongly to each other, to other sciences, and to technology. But, in addition, each possesses an internal logic and structure that provide the foundation for their study as basic sciences. By basic science is meant science that is pursued simply to extend knowledge without regard to its application and that receives its motivation from the inner logic of the discipline or subfield under study. Many of the best minds in physics address the basic problems of initial-condition physics. (See also Chapter 4.)

THE INSTITUTIONS OF PHYSICS

The modern institutions of physics have developed to accommodate its various branches. Perhaps the most striking development has been the rise of large centralized laboratories for the study of many aspects of physics and astronomy.

Physics in National Facilities

The movement toward centralized, expensive facilities was foreshadowed in the development of nuclear physics during the immediate post-World War II days. However, the magnitude and complexity of the facilities required for high-energy physics overshadowed anything needed for other branches of physics with the exception of astrophysics and space astronomy. In 1952, the most expensive piece of physical equipment was the Atomic Energy Commission's (AEC's) Materials Testing Reactor (MTR) in Idaho, which cost about $20 million. By 1964, the Stanford Linear Accelerator (SLAC), costing $114 million, had been built. Currently, the 200-GeV accelerator at the National Accelerator Laboratory (NAL) is under construction at a cost of $240 million. These figures can be compared with the $6.5 million spent for the 200-in. optical telescope at Mount Palomar California, the $14 million spent in 1965 for the 140-ft radio telescope at Greenbank, West Virginia, or the $75 million each for orbiting astronomical observatories.

There has also been a trend toward large, centralized laboratories in initial-condition or extensive physics. World War II showed the power and usefulness of this kind of physics, especially when it was conducted in large, interdisciplinary institutions. Before World War II, rather few large corporation laboratories employed physicists (in contrast to chemists who often worked in industrial laboratories); today physicists work in many industrial laboratories. Research on condensed matter is conducted at many interdisciplinary laboratories on university campuses, notably those sponsored by the Advanced Research Projects Agency (ARPA) of the Department of Defense (DOD). And many government and national laboratories are powerful bases for the study of nuclear, atomic, molecular, plasma, and condensed-matter physics.

These laboratories are supported by various agencies, some of which

TABLE 9.1 Selected Examples of Laboratories Having Major National Facilities for Physics

Agency and Laboratory	Date Established	Location	Parent Organization	Major National Facilities
Atomic Energy Commission				
Ames Laboratory	1942	Ames, Iowa	Iowa State U.	Research reactor metals process development plant
Argonne Nat. Lab.	1942	Argonne, Ill.	Argonne U. Assoc. and U. of Chicago	Zero Gradient Synchrotron, CP5 Reactor, low-energy accelerators
Brookhaven Nat. Lab.	1947	Upton, N.Y.	Associated U.'s, Inc.	Alternating Gradient Synchrotron, high-flux beam reactor, Tandem Van de Graaff, medical research reactor, cyclotron
Cambridge Electron Accelerator	1956	Cambridge, Mass.	MIT and Harvard U.	Electron synchrotron
Lawrence Radiation Lab.	1936	Berkeley, Calif.	U. of California	Bevatron, 180-in. cyclotron, Hilac, other accelerators
Los Alamos Scientific Lab.	1943	Los Alamos, N.M.	U. of California	Los Alamos Meson Physics Facility, low-energy accelerators
Nat. Accelerator Lab.	1967	Batavia, Ill.	U. Research Assoc.	Proton synchrotron, energy 200 GeV or higher[a]

	Year	Location	Operator	Description
Oak Ridge Nat. Lab.	1943	Oak Ridge, Tenn.	Union Carbide Nuclear Co.	Oak Ridge research reactor, high-flux reactor, isochronous cyclotron, tandem accelerator, electron linear accelerators
Princeton–Pennsylvania Accelerator[b]	1956	Princeton, N.J.	Princeton U. and U. of Penn.	3-GeV proton synchrotron
Stanford Linear Accelerator Center	1961	Stanford, Calif.	Stanford U.	20-GeV electron Linac
U.S. Air Force National Magnet Lab.	1960	Cambridge, Mass.	MIT	Very-high-field magnets, dc and pulsed
National Aeronautics and Space Administration Lunar Receiving Lab.	1968	Houston, Tex.	Manned Spacecraft Center	Facilities for quarantine and examination of lunar samples
Virginia Assoc. Research Center	1960	Langley Field, Va.	Virginia Polytechnic Inst.	600-MeV cyclotron, Linac
Atomic Energy Commission and National Science Foundation				
MURA	1954	Stoughton, Wis.	U. of Wisconsin	Electron synchrotron, used as intense source for ultraviolet radiation

[a] Under construction. Completion expected about July 1, 1972.
[b] Federal support for elementary-particle physics research terminated June 30, 1971.

587

were created primarily to build and operate major facilities.* Table 9.1 lists those laboratories having major national facilities for physics; only those that make facilities available to a substantial extent to outside users are included. Facilities for astronomy are not listed in this table.† Among the best known are the National Astronomy and Ionospheric Center at Arecibo, the National Radio Astronomy Observatory, the Kitt Peak National Observatory, the Cerro Tololo Inter-American Observatory, and parts of the National Center for Atmospheric Research. The Goddard Space Flight Center also is a major laboratory for astronomy. Laboratories that accept visitors for cooperative work but in which the work is not centered around a major experimental facility also are not included. Further, the activities of the National Aeronautics and Space Administration (NASA) in providing satellites for experiments in space physics and in such projected activities as the high-energy satellites do not fit into this particular table; some of these cost as much as, or more than, the big accelerators. Many large facilities at universities also are omitted here because these are primarily intended for the use of local scientists; however, some of these appear in later tables.

The oldest laboratory in Table 9.1, the Lawrence Berkeley Laboratory, originated in 1936 as part of the Berkeley campus of the University of California; at that time the concept of a national facility was not involved. After the war, the construction of major new accelerators at this laboratory and the needs of experimenters in less-favored locations led to enlargement of the scope of the laboratory. The laboratories at Ames, Argonne, Oak Ridge, and Los Alamos were established during the war as part of the effort to develop nuclear reactors and nuclear weapons. In the early postwar period, their functions were enlarged to provide special facilities for research by a wider community, and they achieved the status of national laboratories. Brookhaven was the first of these laboratories founded expressly for the purpose of making special facilities available to outside scientists. The AEC is the sponsor of the largest array of laboratories with facilities for physics, largely because of the

---

* A careful examination of the national laboratories and multiprogram laboratories financed by the AEC is given by Harold Orlans, *Contracting for Atoms* (The Brookings Institution, Washington, D.C., 1967). See also *A Statistical Summary of the Physical Research Program,* issued annually by the Division of Research of the AEC. Forty large laboratories sponsored by various federal agencies are discussed in the report *Contract Research and Development Adjuncts of Federal Agencies,* by J. G. Welles *et al.* (Denver Research Institute, Denver, Colo., 1969). For a condensation of much of this report, see D. C. Coddington and J. G. Millikin, "Future of Federal Contract Research Centers," *Harvard Business Review* (March-April 1970).

† The Astronomy Survey Committee provides detailed information on facilities in this field in their report, *Astronomy and Astrophysics for the 1970's* (National Academy of Sciences, Washington, D.C., 1972).

historical fact that its years of emergent vigor coincided with the rising need for such installations and because of the match that developed between this need and the needs of the AEC.

In astronomy and certain other fields, the National Science Foundation (NSF) now plays a similar role. Large in-house laboratories with strong programs in physics research also exist in other agencies. For example, the Department of Commerce operates the National Bureau of Standards; NASA has The Lewis Laboratory and Langley Field; and the DOD, or its constituents, operates Ft. Monmouth, Picatinny Arsenal, Frankford Arsenal, Edgewood Arsenal, Ft. Belvoir, the Naval Research Laboratory, Wright-Patterson AFB, and a number of others. Laboratories such as these are not concerned in a major way with the operation of large facilities for the benefit of outsiders and are not considered here.

It may be useful to discuss briefly the typical procedures followed in the national laboratories listed in Table 9.1, for there is widespread misinformation and misunderstanding about laboratory operating procedures. Each of the laboratories has been endowed with a large measure of independence. However, in each, programmatic needs of the sponsor are made known to the laboratory management, and responsiveness to these needs is ensured by a combination of methods. One is the annual budget cycle in which the laboratory requests funds for various specific purposes and the agency awards support in accordance with the compatibility of these requests with its goals, its estimate of the ability of the laboratory to fulfill its commitments, and the availability of funds to the agency. In addition, the agencies have technical staffs who monitor the work in the laboratories and influence the directions such work takes. Different programs within an agency exert this control in varying degrees. Generally, the more applied the effort and the more directly it relates to a programmatic mission, the closer is the supervision. In basic science the control usually is indirect—exercised through informal interaction and through control of annual budgetary levels—but no less real. Large facilities are constructed only after specific approval, extending beyond the agency to the Congress.

Management and Review Procedures   The complex managerial responsibilities of a large, multipurpose laboratory require corresponding authority resident in the laboratory. Nevertheless, many forms of outside monitoring and influence exist in addition to those coming from the sponsoring agency. Each of the laboratories listed in Table 9.1 is operated by a contractor, and the contractor is a layer of authority superimposed on the laboratory director. In the case of the Lawrence Radiation Laboratory, the University of California is the contractor; for Argonne National Laboratory, it is the University of Chicago together with an interuniversity consortium set up specifically for this purpose and known as Argonne University Associates; and for Brookhaven National Laboratory it is Associated Universities, Inc. (AUI).

The Brookhaven–AUI example offers a good illustration of this kind of laboratory management arrangement. The AUI is a nonprofit corporation chartered under the education laws of the State of New York. It has a president, who is the chief executive officer, a small corporate staff, and a governing Board of Trustees. The trustees consist of two members from each of nine eastern universities and as many as six trustees-at-large. Nominations for the university representatives are made by the universities, and these and the remaining trustees are elected by the Board for specified terms. University representation is dominant on the Board (in practice it has been virtually 100 percent), and nationwide, rather than just regional, interest is afforded in practice by the trustees-at-large, who generally come from all parts of the United States. The Board sets policy for operation of the laboratory and has kept close watch over its personnel and programs. A comprehensive set of visiting committees chosen by the Board reviews the scientific programs of the laboratory in detail each year and reports directly to the Board. The high-energy physics program has additional guidance from two major groups: the High Energy Advisory Committee, which reviews all proposed experiments and assigns them priority, and the High Energy Discussion Group, which comprises all the outside users of high-energy facilities together with the senior high-energy staff of the laboratory. This group reviews the plans for improved facilities (beams, detectors, bubble chambers, and the like) and offers guidance on priorities to be assigned in this area. Many other fields of research have advisory groups of various kinds: the double MP Tandem Van de Graaff has a Users Group Committee and an Advisory Committee similar to those in high energy. The National Neutron Cross Section Center has its own advisory committee of outside experts and receives further guidance from the U.S. Nuclear Data Committee. The Physics Department has a regular board of outside consultants who review problem areas.

The general trend in all the laboratories in Table 9.1 has been toward the formation of more advisory committees and groups that provide outside wisdom and influence in their scientific programs. Each of the high-energy facilities in the United States has at least one such committee, and the various medium- and low-energy facilities are acquiring them. The National Magnet Laboratory has a technical advisory committee that reviews its programs annually, as does the Virginia Associated Research Center. The sponsoring agencies also have such committees to review affairs regularly from the viewpoint of the sponsor. The High-Energy

Physics Advisory Panel to the AEC (HEPAP), which advises on certain national high-energy programs in the AEC, is one of the best-known examples. In addition, the General Accounting Office, which reports to the Congress, has been given increasingly frequent assignments for *ad hoc* investigation of scientific programs; these investigations extend far beyond the fields traditionally assigned to auditors. Recently, for example, the AEC's management of its high-energy laboratories and the manner in which the laboratories make decisions and conduct their research programs have been the subjects of two separate General Accounting Office audits. The dissemination of scientific information in AEC laboratories was the subject of a third such review.

But of all the review mechanisms, the informal but critical opinions of the scientific community are the most important in maintaining high standards in various research programs. How is this opinion marshaled? The supporting agency's programs are administered by scientists who keep in close touch with developments in a field and the opinions in the community. For the largest laboratories, the work of which is prominent, this evaluation poses no difficulties. The small programs, which are more numerous and less clearly visible, require additional evaluation, which, customarily, is supplied through the mechanism of written proposals that are reviewed by referees chosen by the agencies.

The Physics Survey Committee believes that the system of reviews of programs has worked well, with no more overlap than any science needs for normal verification. This system has fostered vigorous scientific research for the national benefit and has stimulated the work of both national laboratory and university-based scientists.

Physics in Industrial Laboratories

Most research in industry is by its very nature applied. Few companies are sufficiently large, as, for example, are Bell Telephone Laboratories, IBM, or General Electric, to be able to support a significant amount of basic research, and few have found basic research sufficiently profitable to justify the substantial investment required for this type of work.

The principal opportunities for physicists in industry relate to applied research, certain interdisciplinary areas of research, and of course, design and development. However, there is growing competition for such jobs from the engineering community, as discussed in Chapter 12. That chapter also describes trends in the employment of physicists by industry and offers some recommendations for improving the interaction between university-based and industrial research efforts. (In addition, Chapter 7 briefly discusses the role of the physicist in industry.)

Table 9.2 presents a list of corporations belonging to the Industrial

TABLE 9.2   Physicists Employed in Research (and Development) in Major Industries

Company	Number of Research Personnel [a]	Number of Physicists
AMF Incorporated	94	3
Air Products & Chemicals	293	2
Alcan Research & Development	690	16
Allis-Chalmers	100	7
American Can	1,139	10
American Cyanamid	2,467	10
American Metal Climax	165	1
American Standard	2,245	29
American Tobacco	222	1
Amoco Production	341	8
Armstrong Cork	642	27
Atlantic Richfield	908	20
BASF Wyandotte	263	2
Bell & Howell	80	7
Bell Telephone Laboratories	13,841	614
Borg-Warner	1,273	11
Brown & Williamson Tobacco	158	2
Burlington Industries	225	1
Cabot Corporation	355	3
Canadian Industries	460	2
Carborundum Company	432	8
Carpenter Technology	236	4
Carrier Corporation	450	4
Celanese Corporation	1,776	19
Chevron Research	1,620	19
Consolidated-Bathurst	92	4
Continental Can	1,010	25
Continental Oil	583	12
Corning Glass	1,361	52
Dunlop Research Centre	40	2
Dupont	4,300	160
Eastman Kodak	1,420	97
Esso Production Research	669	37
FMC Corporation	2,633	9
Farbwerke Hoechst	677	31
Firestone	245	9
Ford	780	40
Foxboro	76	7
GAF Corporation	763	10
GTE Laboratories	449	39
Gates Rubber Company	88	4
General Electric	1,800	150
General Mills	821	1

TABLE 9.2—*Continued*

Company	Number of Research Personnel [a]	Number of Physicists
General Motors	1,425	40
Gillette	642	21
Goodyear Tire and Rubber	556	15
Grinnell	81	1
Gulf Research and Development	2,146	31
Harbison-Walker Refractories	97	2
Hercules	1,305	45
IBM	1,350	194
International Nickel	398	4
ITT	14,900	60
Johns-Manville	614	9
Jones & Laughlin Steel	333	8
Kaiser Aluminum & Chemical	349	4
Kennecott Copper	477	16
Keuffel & Esser	103	3
Kimberly Clark	277	5
Leeds & Northrup	156	18
Lubrizol	1,054	1
MacMillan Bloedel	100	4
Marathon Oil	295	14
Mead	221	3
Mine Safety Appliances	245	8
Mobil Oil	1,974	45
Moore Business Forms	197	2
NL Industries	1,127	18
National Cash Register	2,860	21
National Distillers & Chemical	350	3
National Steel	113	2
North American Philips	692	37
Norton	461	14
Ing. C. Olivetti	463	25
Owens-Corning Fiberglas	350	11
Owens-Illinois	1,625	37
Pennzoil United	52	1
Pfizer	813	1
Phillips Petroleum	1,169	17
Pillsbury	273	2
RCA Laboratories	1,180	127
Raytheon	170	41
Research Corporation	26	3
SCM Corporation	354	14
Scott Paper	773	6
International Paper	502	6
Selas Corporation	60	2

TABLE 9.2—*Continued*

Company	Number of Research Personnel [a]	Number of Physicists
Shell Development	2,476	51
Tennessee Eastman	525	8
Texaco	1,901	38
3M Company	3,356	66
Timken	168	2
USM Corporation	534	5
Union Carbide	4,520	100
Uniroyal	1,431	20
U.S. Steel	1,706	31
Universal Oil Products	717	4
Westinghouse	1,513	138
Xerox	553	35
Youngstown Sheet and Tube	158	3

[a] Includes professional, administrative, and technical personnel and often includes those employed in development as well as research, for many companies do not distinguish between these two activities.

Research Institute that employ physicists. The total number of research personnel indicated for each company in Table 9.2 includes professional, administrative, and technical manpower and frequently includes both research and development personnel, for many companies do not distinguish between these two activities in their records. Figures 9.1 presents data on the annual costs per scientist or engineer to large U.S. industries that are significantly engaged in research and development.

Some indication of the production of industrially funded research that enriches the publicly available store of knowledge in basic and applied physics was obtained by counting the entries in the Corporate Index section of the 1969 *Science Citation Index*. Table 9.3 presents the findings of this search. The data do not include any work of federally funded research and development centers even when these are managed by an industrial organization.

Table 9.4 compares the production of published papers in different types of industries, in regard to both the ratio of physics papers to total papers published and the ratio of papers appearing in American Institute of Physics (AIP) journals to the total number of physics papers. The optical industries had the highest ratio of physics papers to total published papers, and optical, instrumentation, and aerospace organizations had the highest ratios of papers appearing in AIP journals to total physics papers. The types of industries represented in this table also give some

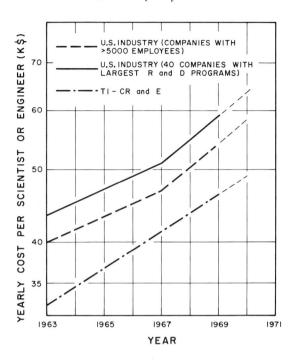

FIGURE 9.1 Annual cost per scientist or engineer to large U.S. industries that are significantly engaged in research and development.

indication of the nature of the work in which physicists in industry are chiefly engaged.

In the future, however, it will be important for trained physicists to become engaged in many kinds of industrial and related research in which they have not formerly participated. They should be engaged in basic studies of industrial processes, which previously have been the exclusive province of the engineer. Of course, physicists should be but a small minority in such studies, but they have a unique and distinctive contribution to make. Their value will lie in their capacity to take a fresh look at many industrial problems, for they will be less influenced by the traditions and detailed knowledge stemming from long, successful practice of a particular way of doing things. This kind of new look at well-established technologies will become increasingly necessary as the public puts pressure on industry to alter its practices to meet new environmental standards and new aspirations in regard to safety standards and psychologically satisfying work environments.

This need can be illustrated by the physics of the internal combustion engine. This technology evolved successfully without the application of much science. There was not much more that could be done to increase efficiency and improve performance; therefore, further detailed

TABLE 9.3   1969 Publications of U.S. Industrial Research Organizations

Organizations	1969 Papers in Physics Journals	1969 Total Papers	Ratio Physics to Total	Cumulative Total of Physics Papers
Bell Telephone Laboratories	614	1579	0.39	614
IBM	253	786	0.32	867
General Electric	158	1094	0.14	1025
RCA	137	335	0.41	1162
North American Rockwell	109	287	0.38	1271
Westinghouse	89	598	0.16	1360
Ford	68	240	0.28	1428
Gulf General Atomics	61	198	0.31	1489
McDonnell Douglas	58	215	0.27	1547
Lockheed Aircraft	56	330	0.17	1603
Boeing	53	306	0.17	1656
Texas Instruments	49	148	0.33	1705
Perkin-Elmer	46	78	0.59	1751
Monsanto	42	290	0.14	1793
Bolt Beranek & Newman	41	75	0.55	1834
Eastman Kodak	37	170	0.22	1871
AVCO-Everett Research Lab.	37	141	0.26	1908
Sperry Rand	36	111	0.32	1944
Hughes Aircraft	36	149	0.24	1980
Union Carbide	35	225	0.16	2015
Xerox Corp.	35	90	0.39	2050
Raytheon	34	125	0.27	2084
Dupont	32	324	0.10	2116
United Aircraft	32	133	0.24	2148
Varian Associates	31	89	0.35	2179

understanding of the physics of the engine did not promise much in the way of economic rewards. But suddenly the imposition of emission standards demanded a fresh look at the internal combustion engine from a more fundamental standpoint than previously. It is in this kind of situation that the research style and mode of thinking of the physicist can be most fruitful. Unfortunately, industry for the most part has not learned how to use physicists in this way, and the present academic training of physicists does not encourage them to transfer the modes of thinking characteristic of physics to industrial problems. In most cases industry uses physicists and engineers in roughly equivalent roles; occasionally physicists also are used for very-long-range basic research efforts. But they are not used, as they well could be, to examine old applied

TABLE 9.4   Characteristics of Publications of Different Types of Industrial Organizations

Type	Ratio of Physics Papers to Total Papers [a]	Ratio of Papers in AIP Journals to Total Physics Papers
Large electronic	0.29	0.77
Small electronic	0.32	0.74
Optical	0.35	0.86
Aerospace	0.29	0.85
Manufacturing	0.18	0.73
Instrumentation	0.24	0.87
Metallurgical	0.16	0.41
Chemical	0.11	0.73

[a] See Table 9.3.

problems from a new and innovative point of view and, particularly, to examine the technological problems of an industry from a holistic standpoint rather than attacking, piecemeal, discrete technical problems defined by existing technology. The unique contribution of physicists in World War II rested on their new approach to military problems, often beginning with a complete redefinition of the problem to be solved. As the various environmental crises produced by the growth of industrial civilization multiply, such an approach to industrial problems would appear to be increasingly necessary, and physicists should be better prepared than most scientists to meet this new kind of challenge, much as they did in the early stages of the development of the electrical and communications industries.

Physics in Universities

Approximately half of the physics community is located in the universities and is chiefly engaged in research and teaching. These physicists perform much research for the government, and, through undergraduate, graduate, and postdoctoral training programs, they produce the physicists needed by government and industry. (The training and characteristics of physicists in the universities are discussed in detail in Chapter 12.)

The changing mode of research in physics has had particular impact on the heavily research-oriented, university-based physicist. For example, for some years astrophysicists and high-energy physicists working in the universities have found that successful work in these subfields necessitates the use of major facilities located in large regional or national laboratories. This type of user activity at a place remote from the home campus is now spreading to other subfields of physics as they, too, begin to depend more heavily on facilities too large to be acquired and adequately maintained by a single university. Some of the best opportunities for frontier research in nuclear physics soon will be afforded by heavy-ion accelerators and meson factories, and high-magnetic-field work in the physics of condensed matter may require the facilities of the National Magnet Laboratory, to mention but two examples.

The shift of activity away from the campus to some central facility presents a number of problems to those who wish to remain active at the frontier of a particular subfield. The success of university user groups working in high-energy physics is an illustration of the great benefits that can derive from such cooperative arrangements in spite of the many difficulties they present.

The problems are of two general types—those relating to faculty and those involving graduate students. In the case of faculty, the absence from the campus that user-group activity enforces inevitably interferes with the normal faculty responsibilities of teaching, committee work, supervision of graduate research, and the like unless special arrangements are made. These arrangements include, for example, the assignment to two faculty members engaged in a particular off-campus user activity of otherwise individual responsibilities so that by prearrangement one member of the pair is available on the home campus at any given time. Another such arrangement is to confine local obligations to a single semester each year, leaving the remaining semester free for off-campus activity. This practice is less satisfactory than the previous example, for, during several months' absence from the campus, the scientist inevitably loses contact with the general life and problems of his home department and institution and becomes more involved in those of the institution at which he is a user. Not the least of the problems is that of persuading colleagues of the user in his own institution that they are not victims of uneven privilege, and that maintaining contact with frontier research in big science is an essential aspect of ensuring the vitality of overall departmental activity.

Problems of a different kind face the graduate student working in a user group off campus. Clearly, enforced absence from the campus causes some disruption and inconvenience. If the absences are infrequent and brief, at least until the student has had an opportunity to complete his formal course requirements, these problems are easily managed. In effect, the student trades his full participation in the intellectual life of the home institution—its seminars, colloquia, and the like—for those of the host institution. To the extent that the latter is a national or international center, he could derive great benefit from this tradeoff.

But more important is the style and content of his work. As experimental work shifts from smaller local facilities to larger ones of national or international significance, it is clear that the pressures for the greatest possible efficiency in the use of the facility mount rapidly, as does competition for access to it. Under these circumstances it is difficult to allow a student a major role in the overall design of an experiment, its execution, and its analysis. It also is difficult, when scheduling is tight and long-term, to offer sufficient time and flexibility to allow a student to follow his curiosity into whatever new channels unfold during the course of an experiment. What is involved here is a different style of experimental educational experience and one that is intrinsically more specialized than has been traditional in physics. But there is little or no choice; physics departments must accept such changes if they are to remain active at the research frontiers that attract some of the most able students.

Therefore, the Physics Survey Committee urges universities to be flexible in their demands on faculty and students, who, because of their field of research, must spend a significant amount of time at sites remote from the campus using major facilities not otherwise accessible. Universities should recognize that this type of research endeavor is necessary to retain personnel active at the frontiers of their respective fields, and that the returns to the university in intellectual vitality and excitement will far outweigh whatever inconveniences and additional expenses are entailed.

The Committee also notes that in a number of subfields of physics, especially the more applied ones, the research frontiers are not at the universities but at industrial or more specialized government laboratories. Procedures should be developed that will facilitate dissertation research in such facilities under the joint supervision of a university faculty member (full-time or adjunct), who must protect the student's interests and maintain educational standards, and a representative of the facility at which the research is taking place, who must be responsible for the relevance of the project to the overall mission of the facility.

University personnel also should recognize that they can make a major contribution to the effectiveness of work in national facilities through, for example, serving on review committees or accepting joint appointments that afford fuller participation in and responsibility for the programs and work of a facility.

The importance of close ties between the universities and industry and the potential contribution of the universities to industrial research were emphasized in a recent article in *The Bulletin of the California Institute o* *Technology* *:

> In the early decades of its existence, Caltech played a remarkable role in providing leadership for the introduction of science and research into industry. Under Theodore von Kármán, Caltech became the scientific mecca for the aircraft industry. It similarly led the way in the development of high-voltage transmission lines, of pumps for the new water aqueducts, of standards for the construction of earthquake-resistant structures, and a host of similar activities.
>
> After World War II, when the federal government began to support research in universities on a major scale, the ties with industry began to drop away, both here at Caltech and across the nation.
>
> Now there is a growing realization that much was lost in this disengagement. Modern industry is a fertile source of ideas for new areas of research. The faculty needs a close association with industry if it is to educate properly young engineers for industrial positions. Industry must be on more intimate terms with students if it is to attract the bright students into industry, and industry can well afford to become acquainted with the advanced research taking place in universities which will hit the industrial frontiers in five to ten years.
>
> In order to reestablish closer ties with industry we have begun to invite distinguished experts from industry to spend a year at Caltech as visiting faculty members to teach and to participate fully in the life of the campus. . . .
>
> As part of our program of strengthening ties with industry, we have adopted a new policy which will permit students to enroll for graduate work on a part-time basis. This should permit some of the excellent young men in industry to further their graduate education.

COMPARISONS AMONG INSTITUTIONS

General

The institutions of physics can be distinguished by their particular institutional purposes and goals as well as by their scientific objectives. The industrial laboratory seeks to develop products that will prosper in the marketplace; the government laboratory, to conduct research in support of the missions of the agency to which it belongs; the university, to promote education and develop new knowledge; and the national laboratory, partly to conduct applied work in support of its parent agency and partly to provide to the scientific community extremely expensive facilities and to conduct research with these extramural scientists.

The role of the AEC national laboratories deserves some elaboration since these are *par excellence* homes of big physics. Moreover, the AEC is the principal source of support for three subfields of physics: high-

* *The Bulletin of the California Institute of Technology* (November 1971), pp. 45–46.

energy, nuclear, and plasma. The national laboratories of the AEC differ widely in purpose and character. At one extreme are institutions like the Argonne and Oak Ridge National Laboratories that have heavy applied commitments in reactor development, flourishing research enterprises in elementary-particle and nuclear physics, and broad activities in many other branches of physics, chemistry, and engineering. At the other extreme are the Stanford Linear Accelerator and the National Accelerator Laboratory, which are devoted entirely to high-energy physics. To understand this wide diversity in purpose of the AEC laboratories, one must recall that the Atomic Energy Act of 1954 directs the AEC not only to develop nuclear energy but also to conduct research in "nuclear processes." * In other words, the AEC is directed by law to serve both as a "National Science Foundation" for the exploration of nuclear phenomena and as an agency to develop nuclear energy and its applications. The latest amendment to the Act that established the AEC also removes the restrictions to nuclear phenomena and to nuclear energy insofar as these relate to the overall energy questions facing the nation.

The differences in institutional purposes and in the character of the physics conducted to achieve these purposes naturally imply variations among the different institutions. Both the national laboratory and the industrial laboratory are much more hierarchical and interdisciplinary than is typical of the university physics laboratory, and they both place greater emphasis on applied research than does the university. Because of the complexity of some of the big-physics undertakings at the national laboratories, the labor must be divided among interacting specialists. There are computer experts, instrument technicians, engineers, and many parascientific specialists who add greatly to the logistic power of these institutions. Because the work is so much a team activity, it must be organized and orchestrated to a degree that is less common and, indeed, less appropriate in university settings. Time on a big accelerator is at such a premium that it must be rather rigidly scheduled and rationed.

In the large government and industrial laboratories, the research aspirations of the individual scientist are usually more closely identified with the purposes of the institution than is the case in the university. This situation is natural since the most easily articulated purpose of a university is education rather than research. But the degree of such identification between individual and institution varies greatly in the large laboratories; it ranges from strong identity of purpose at the industrial laboratory to

---

* Joint Committee on Atomic Energy, *Atomic Energy Legislation through 91st Congress, 1st Session,* Atomic Energy Act of 1954, Chapter 4, Section 31, page 16 (U.S. Government Printing Office, Washington, D.C., 1970).

relative independence in some national laboratories in which the atmosphere is a cross between a university and an industrial laboratory.

In the industrial laboratory, the physicist's work usually fits into a wider applied context; therefore, the audience for his research is his fellow workers, including management, who are interested not so much in the implication of his work for physics as in its implications for the future of the organization and the job at hand. The industrial physicist derives much of his reward and job satisfaction outside the community of physics.

The academic physicist is at the other extreme; his professional aspiration is approval and regard from colleagues in his discipline. How the university management regards him is not unimportant, but his primary identification is more often with colleagues in the same discipline at other institutions than with colleagues in different disciplines at his own institution.

The physicist in the governmental, or more especially, the major national laboratory occupies an intermediate position. All these laboratories fulfill multiple purposes, in varying mixes. These purposes are

1. Providing special facilities for the use of the scientific community at large;

2. Conducting in-house and collaborative research and development programs with their own facilities to ensure a core of experts resident at the facilities for their nurture and advancement and to maximize their use, including full exploitation, which would not be readily achieved by visiting users alone;

3. Conducting research and development programs of direct mission interest to the sponsoring agency, and sometimes to other agencies, which may or may not be directly related to the special facilities for regional or national use;

4. Providing a core of experts for diverse technical advice and assistance to the sponsoring agency and sometimes to other agencies;

5. Conducting training and information dissemination programs of interest to the sponsoring agency or of importance to the health of science more generally;

6. Conceiving, designing, and building new and more advanced facilities for regional or national use as needs and opportunities arise.

At a laboratory, such as the Stanford Linear Accelerator, with one major facility, activities 1 and 2 constitute the principal program. At multidisciplinary laboratories, such as Argonne and Oak Ridge National Laboratories, all six activities are present and play strong roles.

The six kinds of activity are symbiotic and synergistic to a degree that is not always recognized. For the complex task of building and maintaining these advanced facilities, technical staffs of the highest quality are needed. Such people are attracted by an atmosphere of excellence and by the possibility of being more than technicians who merely provide the tools with which other persons win scientific laurels. The devices at the forefront of research are not static. They are undergoing constant evolution, modification, and upgrading. In spite of this effort, their lifetimes are finite and eventually they must be supplanted. Thus all the laboratories that have built large accelerators have also included scientists, who, along with the outside users, have conducted experimental programs with the machines. In some cases there were no outside users when the machine was brought into operation, or such users have not been sufficiently numerous to exploit it fully. The same situation that characterizes the large accelerators also holds true for the reactors used for research in the physics of condensed matter and for facilities used in plasma physics and physics in biology.

Age Distributions

Figure 9.2, parts (a) through (d), presents comparisons of the age distribution of PhD physicists in academic institutions, government laboratories, industry, and federally funded research and development centers with the overall age distribution for all physicists as reported in the 1970 National Register of Scientific and Technical Personnel. Contrary to prevalent folklore, there is marked similarity in the distributions in all four employment categories.

But more surprising are the corresponding data shown in Figure 9.3, parts (a) and (b), for the Brookhaven and Oak Ridge National Laboratories. At Brookhaven a well-developed policy of short-term temporary appointments similar to academic research associateships and term assistant professorships has been in force for many years. However, at Oak Ridge, with the exception of a very limited number of temporary fellowships, appointment to the Laboratory staff traditionally has been for an indefinite period. Yet the data in Figure 9.3 do not indicate the development of a relatively aging staff at Oak Ridge, and the normal patterns of mobility provide a balanced age profile at both institutions. Figure 9.4, in which the age profiles of all physicists in federally funded research and development centers in 1970 are compared with those in 1964, illustrates this finding more dramatically. The total PhD manpower involved has remained essentially constant at 2000 during this interval; the turnover policy illustrated by the net gain and net loss data of Figure 9.4 involves the hiring of some 600 young (≤ 37 years of age) PhD's to replace 600

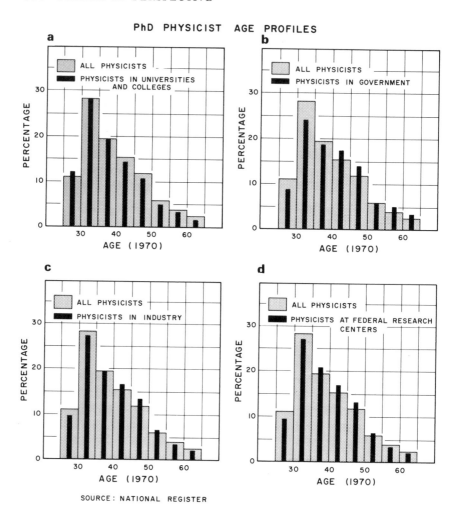

PhD PHYSICIST AGE PROFILES

SOURCE: NATIONAL REGISTER

older (>37 years of age) PhD's who have left the research center staffs through resignations, retirements, or death.

In contrast to the national laboratories in which a balanced age profile has been maintained by a highly unbalanced in-and-out mobility, the remarkable growth of the university and college sector, which has drawn physicists from all age cohorts, has maintained the balanced age profile shown in Figure 9.1(a). From an institutional point of view, the growth of university physics has been in both number of institutions and size of faculties. Figure 9.5 presents the numbers of U.S. educational institutions engaged in PhD-level graduate research as a function of time from 1959 to 1971. During the period of general scientific expansion

◄ FIGURE 9.2 (a) Age distribution of PhD physicists in universities and colleges compared with the age distribution for all PhD physicists.[a]

	25–29	30–34	35–39	40–44	45–49	50–54	55–59	60–64
Academic PhD's N = 8208	12.2%	28.3%	19.6%	14.4%	10.8%	5.1%	3.8%	1.7%
All PhD's N = 16,229	11.0%	28.2%	19.6%	15.4%	11.9%	6.0%	3.9%	2.5%

(b) Age distribution of PhD physicists in government laboratories compared with the age distribution for all PhD physicists.[a]

	25–29	30–34	35–39	40–44	45–49	50–54	55–59	60–64
Government PhD's N = 1467	8.8%	24.2%	18.9%	17.5%	14.0%	6.0%	5.2%	3.4%
All PhD's N = 16,229	11.0%	28.2%	19.6%	15.4%	11.9%	6.0%	3.9%	2.5%

(c) Age distribution of PhD physicists in industry compared with the age distribution for all PhD physicists.[a]

	25–29	30–34	35–39	40–44	45–49	50–54	55–59	60–64
Industry PhD's N = 3802	9.8%	27.4%	19.9%	16.6%	13.5%	6.4%	3.4%	2.1%
All PhD's N = 16,229	11.0%	28.2%	19.6%	15.4%	11.9%	6.0%	3.9%	2.5%

(d) Age distribution of PhD physicists in federally funded research and development centers compared with the age distribution for all PhD physicists.[a]

	25–29	30–34	35–39	40–44	45–49	50–54	55–59	60–64
Research Center PhD's N = 1912	9.5%	26.9%	20.9%	17.1%	13.4%	6.3%	3.5%	2.0%
All PhD's N = 16,229	11.0%	28.2%	19.6%	15.4%	11.9%	6.0%	3.9%	2.5%

[a] Percentages do not add to 100 because of the omission of the 65 and over age group.

in the 1960's, the number of institutions offering the PhD in physics grew at a rate of 7.6 new institutions per year, a rate considerably in excess of that characterizing the number of PhD-granting institutions, which was 4.5 per year.* Appendix 9.A presents a table in which the institutions granting PhD's in physics and astronomy are ranked according to the number of doctorates awarded in these fields in two five-year intervals, 1961–1965 and 1966–1970. This Appendix also includes a brief discussion of these data and the trends they depict.

Structure of Physics Departments

An obvious trend during the 1960's was toward the formation of larger

---

* National Science Board, *Graduate Education. Parameters for Public Policy* (U.S. Government Printing Office, Washington, D.C., 1969).

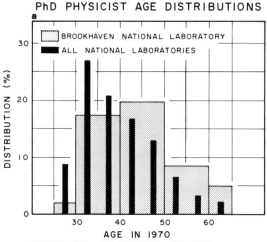

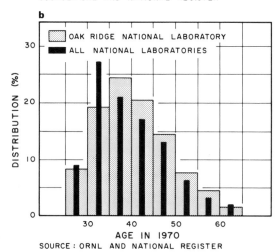

FIGURE 9.3 (a) Age distribution of Brookhaven National Laboratory PhD research staff members compared with the age distribution for all PhD's employed in federally funded research and development laboratories (as reported in the 1970 National Register of Scientific and Technical Personnel). (b) Age distribution of Oak Ridge National Laboratory PhD research staff members compared with the age distribution for all PhD's employed in federally funded research and development laboratories (as reported in the 1970 National Register of Scientific and Technical Personnel).

physics departments, as Figure 9.6 indicates. This figure compares the number of PhD-granting departments according to the size of the faculties in 1960 and 1968. The ever-increasing sophistication of physics research clearly has favored the formation of larger groups using central facilities. This conclusion is further substantiated by the total faculty size distribution of 18 physics departments in New York and California that have begun offering the PhD only within the last ten years; Figure 9.7 presents these data. The median faculty size for these departments in 1970 was 23.

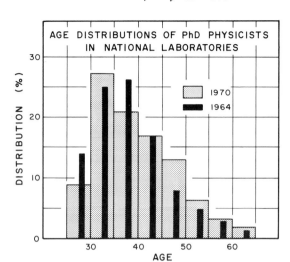

FIGURE 9.4 Age distribution of PhD's in federally funded research and development laboratories in 1964 and 1970 (as reported in the National Register Surveys for those years).

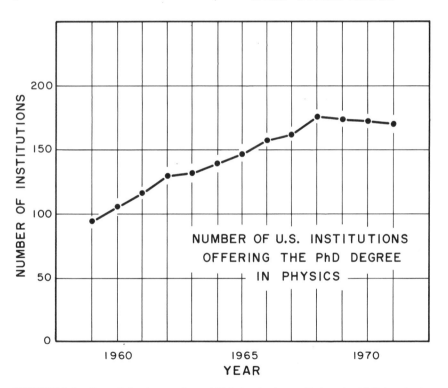

FIGURE 9.5 Growth in the number of U.S. institutions offering the PhD in physics in 1959–1971.

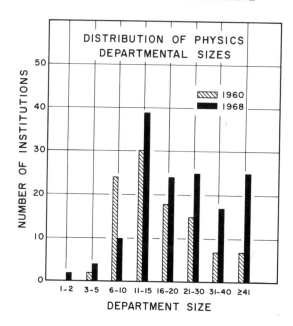

FIGURE 9.6 Comparison of departments granting the PhD in physics in regard to size of faculty in 1960 and 1968.

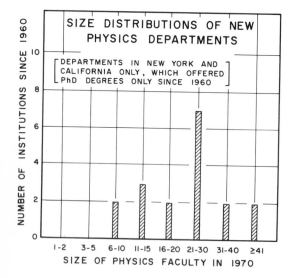

FIGURE 9.7 Physics faculty size distribution of 18 departments in New York and California that began offering the PhD in physics after 1960.

Seventy percent of the faculty was in departments larger than the median; a physics PhD program does not start small!

With the continued balance in the age profile, the age-rank structure in U.S. academic institutions, depicted in Figure 9.8, has been generally stable between 1964 and 1970, the time during which the PhD physics faculty nearly doubled. Rather large variation in rank structure occurs among academic institutions, as Table 9.5 shows. As an example, Figure 9.9 presents a comparison of the rank structure of two large private institutions, Harvard University and Massachusetts Institute of Technology, having an aggregate faculty of 163, and two large public institutions, Berkeley and Illinois, having an aggregate faculty of 138, with a sample of PhD's in all U.S. institutions, an aggregate of 6432 PhD's.

Detailed data pertinent to each department are not readily available, but for one institution, Massachusetts Institute of Technology, the median age by faculty rank corresponds closely to the national pattern. Therefore, the pronounced shift to higher faculty rank probably is accompanied by an overall aging of the faculty (see also Chapter 12).

Several features of the growth patterns of U.S. PhD-level institutions in physics are apparent from Table 9.5. During the last decade, the overall growth in number of faculty members has been much smaller in established institutions than in emerging institutions. The aggregate

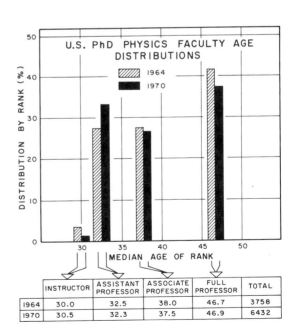

FIGURE 9.8   Age-rank structure of PhD physics faculty in U.S. academic institutions.

	INSTRUCTOR	ASSISTANT PROFESSOR	ASSOCIATE PROFESSOR	FULL PROFESSOR	TOTAL
1964	30.0	32.5	38.0	46.7	3758
1970	30.5	32.3	37.5	46.9	6432

TABLE 9.5 The Rank Structure and Tenure Latitude of Representative Aggregates of U.S. Physics Departments [a]

Academic Year	Institution Aggregate	No.	Number of Faculty	Academic Rank Structure (%)				Tenure Latitude (%)	
				Lecturer/ Instructor	Assistant Professor	Associate Professor	Full Professor	Untenured	Tenured
1970–71	Harvard,[b] MIT, Berkeley, and Illinois	4	301	7.0	16.3	14.3	62.5	23.2	76.8
1959–60	Same as above	4	215	13.5	24.6	16.7	45.2	38.1	61.9
1970–71	"New"[c] New York PhD departments	11	281	7.5	31.7	23.2	37.8	39.0	61.0
1970–71	"Established"[c] New York PhD departments	11	311	4.8	22.8	28.4	44.0	27.6	72.4
1959–60	Same as above	11	227	14.5	26.4	21.6	37.6	40.8	59.2
1969–70	PhD's in all institutions, granting all degrees 1970	848	6421	3.1	32.8	26.4	37.8	35.8	64.2
1959–60	PhD's in all institutions, granting all degrees 1960	621	2581	6.6	25.5	26.4	41.6	32.0	68.0

[a] Source: AIP Faculty Directories and National Register.
[b] Includes the Division of Applied Physics, which was formed since 1959.
[c] "New" means having established a PhD program after 1959, and "Established" means already having a program in 1959.

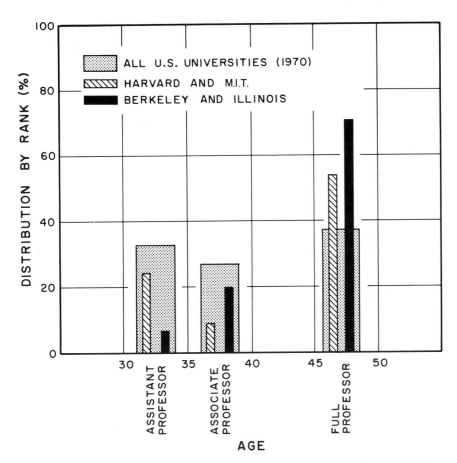

FIGURE 9.9 Comparison of age-rank structure of two large private institutions (Harvard University and Massachusetts Institute of Technology) with that of two large public institutions (University of California at Berkeley and University of Illinois).

faculty of Harvard, Massachusetts Institute of Technology, the University of California at Berkeley, the University of Illinois, and the established New York State institutions grew by only 38 percent, while the aggregate faculty of newly established New York institutions grew by 190 percent.* (See Appendix 9.A.) In the nation as a whole, for institutions at all levels of degree production, the growth in PhD faculty was

---

* The 1959–1960 aggregate faculty of the New York institutions that subsequently began granting the PhD was 97.

149 percent, 9 percent of which was the result of upgrading; that is to say, the fraction of non-PhD faculty declined from 28 percent in 1959–1960 to 24 percent in 1969–1970. At the same time, the number of institutions offering the bachelor's degree or higher degrees in physics increased by 43 percent, from 621 in 1959–1960 to 848 in 1969–1970, an overall growth that may be compared with the 84 percent growth of the PhD-granting components of this number, as shown in Figure 9.5, which has resulted largely from the transformation of bachelor's- and master's-granting institutions into ones offering the PhD.

The faculty rank structure has shifted toward a heavier concentration of full professors in those PhD-granting institutions that were already established in 1959–1960, as Table 9.5 shows, and the fraction of untenured faculty positions has tended to decrease. The newer institutions have a lower percentage of full professors, and a smaller fraction of their faculties are in tenured positions. In the United States as a whole there is a somewhat smaller fraction of tenured faculty than is found in the two aggregates of established institutions. In the process of institutional maturation and consolidation that is likely in the present decade, somewhat more latitude in faculty development is available to newly expanded departments and to U.S. academic institutions in general than to the already well-established institutions.

A different aspect of the faculty-rank structure of graduate institutions was explored by the National Science Board.* The Board found that "the quality of the institutions reflects the (relative and absolute) numbers of full professors," and that little, if any, quality significance can be associated with other features of faculty-rank structure. Their findings in terms of their quality class categories appear in Table 9.6, in which A corresponds to distinguished and F and G are marginal to inadequate.

Although institutional lists corresponding to the National Science Board categories are not available, the Cartter ranking,† as updated by Roose and Andersen,‡ and applied by Elton and Rodgers,§ provide a convenient quality ranking for physics graduate institutions, which, as Table 9.7 shows, can then be applied to the aggregates indicated in Table 9.5 to afford a comparison with all U.S. physics PhD-granting institutions.

* National Science Board, *Graduate Education. Parameters for Public Policy* (U.S. Government Printing Office, Washington, D.C., 1969).

† A. M. Cartter, *An Assessment of Quality in Graduate Education* (American Council on Education, Washington, D.C., 1966).

‡ K. D. Roose and C. J. Andersen, *A Rating of Graduate Programs* (American Council on Education, Washington, D.C., 1971).

§ C. F. Elton and S. A. Rodgers, "Physics Department Ratings: Another Evaluation," *Science, 174*(4009), 565–568 (Nov. 5, 1971). (Elton and Rodgers have found the Cartter ratings for physics institutions to be essentially supported by objective data.)

TABLE 9.6   Full Professors as Percentage
of Total Faculty—All Disciplines [a]

Quality Class	Median Percentage
A	42
B	38
C	32
D	30
E	27
F	28
G	28

[a] Source: National Science Board (NSB 69-2).

These data show that the highest quality large physics departments have a concentration of full professors well above the median for the National Science Board "distinguished" category, and that even the established New York State institutions, spread as they are over the Roose-Andersen categories, are more heavily staffed with full professors than the overall median of distinguished U.S. graduate schools. Finally, educational institutions in physics in general have a concentration of full professors that corresponds to the second rank in the National Science Board study, "strong." The newer institutions apparently are relatively overlooked in the Roose-Andersen ratings, as one of those in New York that is omitted in this ranking has a faculty of 60, 50 percent of whom are full professors, including two Nobel Laureates.

TABLE 9.7   Number of Graduate Institutions by Quality Category:
U.S. Physics Total and Selected Aggregates [a]

Institution Aggregate	Quality Category			
	I	II	III	0 [b]
Harvard, MIT, Berkeley, and Illinois	4	0	0	0
New York State "New" PhD Departments	1	0	0	10
Established New York State PhD Departments	3	3	3	2
All U.S. PhD Departments	30	21	18	101

[a] Source: Roose and Andersen, 1971.
[b] Not listed.

Salary Distributions

The salaries of physicists vary with the institutions in which they work. In Figures 9.10 and 9.11 the median salary of full-time PhD's employed by academic and nonacademic institutions is plotted as a function of age. (Data on the academic group are based on 12 months.) The range indicated in these figures is between the twenty-fifth and seventy-fifth percentiles. The changing activity pattern with age also is evident in these figures. Characteristic institutional variation independent of the age factor is shown in Figure 9.12 for PhD's in the 35 to 40 age range, that is to say, PhD's with about ten years of professional experience. The major difference is between academic and nonacademic sectors, each of which em-

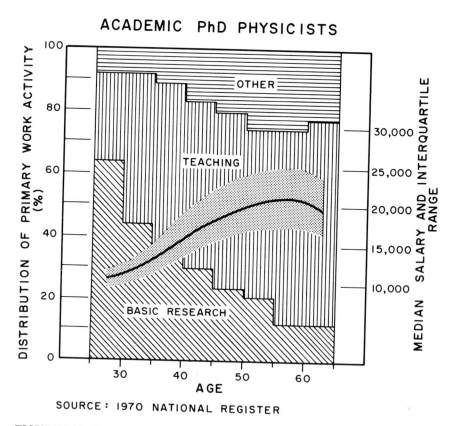

FIGURE 9.10 Primary work activity and salary by age of PhD physicists employed in academic institutions (based on 12 months).

NONACADEMIC PhD PHYSICISTS

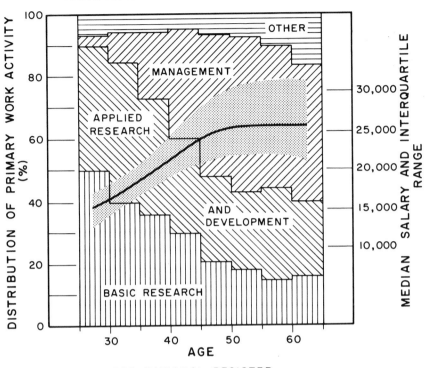

SOURCE: 1970 NATIONAL REGISTER

FIGURE 9.11   Primary work activity and salary by age of PhD physicists employed in nonacademic institutions.

ploys about 50 percent of the PhD physicists, with industrial salaries being slightly higher than those in government-related institutions. Another characteristic variation is reflected in salaries according to work activity, as shown in Figures 9.13 and 9.14, for academic and nonacademic PhD employees, again in the 35–40 age bracket. The data show clearly the economic premium that U.S. society places on management responsibilities.

Institutional Mobility

There has long been discussion concerning patterns of mobility for physicists in the three employment categories—academic, governmental, and industrial. A rather well-developed folklore exists at the student level

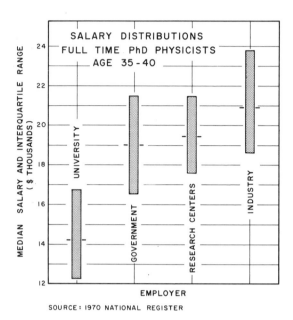

FIGURE 9.12 Variations in salary among employing institutions for PhD physicists in the 35–40 age range who are employed full time. (Data on the academic group are based on 12 months.)

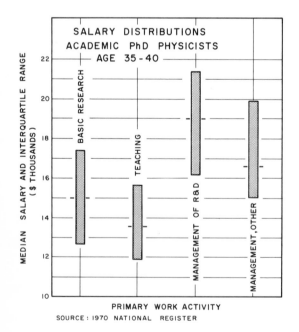

FIGURE 9.13 Median salary (12 months) and interquartile range by primary work activity for PhD's aged 35–40 and employed in colleges and universities.

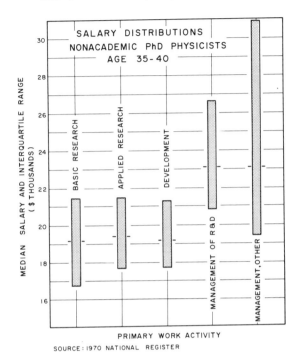

FIGURE 9.14 Median salary and interquartile range by primary work activity for PhD's aged 35–40 and employed by government and industry.

that suggests that the system is diode coupled; that is to say, at least during periods of expansion, such as the early 1960's, it is relatively easy to move to the right in this listing (from academia to government to industry) but relatively difficult to move countercurrent from the industrial laboratory to the government laboratory to the university.

Such a picture is much too simplistic. In part the effect is a real one reflecting a natural reluctance to move against a salary gradient. But, although it is rather rare for young PhD's with only a few years of industrial or national laboratory experience to return to university faculties, a large number have done so after a longer period during which their research activities, relatively uninterrupted by teaching and academic committee activities, gained national, or international, reputations for them. In fact, there is growing evidence to suggest that the career pattern of the most successful academic physicists involves just such an interim period in governmental or industrial laboratory research.

Some concern has been expressed that national laboratories in responding to short-term financial constraints may have sacrificed young personnel in favor of older and more established members of laboratory staffs. Some of the problems inherent in such a reaction are discussed in Chapter 6.

The Committee suggests that any national laboratories that have not already done so should develop and implement an employment policy for professional staff to ensure an adequate mix of both old and young, temporary and permanent staff. The usual arrangements employed by universities to achieve some measure of balance in staffing are likely to be inappropriate to some national laboratories in view of the continuity required in certain applied programs. However, other types of arrangements can and should be explored.

DISTRIBUTION OF EFFORT AND INTERRELATIONSHIPS AMONG INSTITUTIONS

The commitment of physics to very large facilities is evident in Table 9.8, parts A through F. This table lists by subfield the initial capital investment in large facilities. The most expensive facilities are those in high-energy physics, with a total capital expenditure of $540 million. Nuclear physics, including both medium-energy and nuclear structure physics, is next, with a total of $350 million. The condensed-matter laboratories represent a capital expenditure of about $33 million; however, the total capital investment in condensed-matter physics is much larger, for it

TABLE 9.8A Initial Capital Investment in Big-Physics Facilities in the United States: High-Energy Particle Physics (>1 GeV)

Laboratory and Facility	Type [a]	Energy	Year	Capital Cost [b] ($Millions)
Argonne (ZGS)	PS	12.7 GeV-p	1963	50.0
Brookhaven (AGS)	PS	33.0 GeV-p	1960	30.7
Caltech	ES	1.5 GeV-e	1952 [c]	1.6
U. of California, Berkeley (Bevatron)	PS	6.2 GeV-p	1954	30.0
Cornell	ES	10.0 GeV-e	1967	11.5
MIT–Harvard (CEA)	ES	6.3 GeV-e	1962	25.0
National Accelerator (NAL)	PS	200.0 GeV-p	(1972)	240.0
Princeton–Pennsylvania (PPA)	PS	3.0 GeV-p	1963 [c]	30.0
Stanford Linear Accelerator (SLAC)	EL	20.0 GeV-e	1966	114.0
Stanford (Mark III)	EL	1.2 GeV-e	1960	7.0
TOTAL				539.8

[a] PS = Proton synchrotron; ES = electron synchrotron; EL = electron linear accelerator.
[b] Initial costs; most have had substantial additions.
[c] Now closed.

TABLE 9.8B Initial Capital Investment in Big-Physics Facilities in the United States: Low- and Medium-Energy Nuclear Physics (<1 GeV)

Type	Energy (MeV)[a]	Number[b]	Est. Total Investment[c] ($Millions)	Recent Examples — Facility with	Approx. Cost ($Millions)
High-Voltage Machines					
Single stage	≤ 5	40[d]	10	4-MeV Dynamitron	0.5
Single stage	≥ 5	15[d]	15	6-MeV CN VdG	1.0
Tandem, two stage	≤ 22	27	80	15-MeV FN VdG	4.0
Tandem, three stage	≤ 32	7	35	30-MeV MP–MP VdG (Brookhaven)	12.0
Cyclotrons					
Lawrence (FF)	≤ 23[d]	8	8	(No current construction)	—
Isochronous (AVF)	≤ 200	18[e]	60	200-MeV AVF Cyclotron (Indiana)	10.0
Synchrocyclotron (FM)	≤ 730	5[f]	25	600-MeV FM Cyclotron (NASA 1967)	15.0
Linear Accelerators					
Positive ion	≤ 800	3[g]	60	800-MeV Proton Linac (LASL)	55.0
Electron	≤ 150[e]	15[g]	40	140-MeV Electron Linac (ORNL)	6.0
Electron	> 150[e]	3	17	400-MeV Electron Linac (MIT)	8.0
Betatrons	≤ 25[e]	2[h]	—	(No current construction)	—
Synchrotrons	≤ 180[e]	3[h]	—	(No current construction)	—
TOTAL		146	350		

[a] Rated proton energy, unless otherwise indicated.
[b] Some 900 additional accelerators, mostly small, are now in use.
[c] Initial costs; most have had additions.
[d] Many other single-stage HV machines are used in other fields of science and industry.
[e] Several smaller AVF cyclotrons are applied to the biomedical and isotope fields.
[f] The Harvard FM cyclotron is now used in biomedical work.
[g] Many other small linear accelerators are now used in other fields of science and industry.
[h] Several other betatrons and small electron synchrotrons are still used in other sciences and industry.

TABLE 9.8C Initial Capital Investment in Big-Physics Facilities in the United States: High-Powered Research Reactors ($\geq$ 10 MW) [a]

Facility	Power (MW)	Year (Start-up)	Cost [b] ($Millions)
Materials Testing Reactor (Idaho)	40	1952	20 [c]
Oak Ridge Research Reactor	30	1958	5
Plum Brook Reactor (NASA)	60	1961	15
High-Flux Isotope Reactor (Oak Ridge)	100	1965	15
Brookhaven High Flux Beam Reactor	40	1965	10
National Bureau of Standards	10	1967	6
TOTAL			71

[a] Reactors used substantially for physical research; large engineering test reactors and many research reactors of lower power are not listed.
[b] Initial costs; some have had additions.
[c] Now closed.

TABLE 9.8D Capital Investment in Big-Physics Facilities in the United States: National Plasma Laboratories

Number of Laboratories	FY 1954– FY 1971 ($Millions)	FY 1972 ($Millions)	FY 1954– FY 1972 ($Millions)
Four (Los Alamos, Princeton, Oak Ridge, and Livermore)	334.4	26.2	360.6

[a] For FY 1951–FY 1953, total was approximately $1 million.

includes many smaller facilities not covered by this table. The large research reactors, which support work in condensed-matter and nuclear physics, as well as in many other branches of science and engineering, represent an investment of $71 million. Finally, there are large facilities on the borderline between physics and astronomy, such as radio telescopes (some $14 million), optical telescopes (about $6.5 million), and the Orbiting Astronomical Observatory ($240 million).*

* It is difficult to draw the line between devices built for basic research and those built for applied research and development. Thus the Materials Testing Reactor, when it was built, was intended for basic as well as applied research and consequently was provided with beam holes. The much more expensive Advanced Test Reactor and the Fast Flux Test Facility were dedicated from the beginning as engineering test devices. The choice of devices to be included in Table 9.8 is, therefore, somewhat arbitrary; for example, the Orbiting Astronomical Observatory is omitted on the grounds that it is devoted primarily to research in astronomy rather than physics.

TABLE 9.8E Initial Capital Investment in Big-Physics Facilities in the United States: Condensed Matter, Materials, and Metallurgy

Facilities and Research Groups	Year (Start-up)	Cost ($Millions)
National Magnet Laboratory, MIT	1963	5.5
Divisions of national laboratories (est. at $1.7 million × 5 years)	—	8.5
Federal support for university groups (est. at $3.8 million × 5 years)	—	19.0
TOTAL		33.0 [a]

[a] This amount includes only large facilities. The total capital cost is much larger, but it is in general spread among many facilities.

These figures indicate that high-energy physics and space astronomy are among the biggest of the big sciences, but that, to an increasing degree many of the other extensive areas of physics also are becoming big science. Thus the Los Alamos Meson Physics Facility, which is to be used for medium-energy physics (a borderline between nuclear and elementary-particle physics) will cost $55 million. Some big plasma devices, such as the Los Alamos Scyllac, cost as much as $8 million. Even in the National Magnet Laboratory, there are single devices that cost as much as $1 million. Most such investments are made outside the university proper, though in several cases, notably the Stanford Linear Accelerator, the installation adjoins a university campus. Therefore, it is reasonable to expect physics, which in 1940 was predominantly university-based, to

TABLE 9.8F Initial Capital Investment in Big-Physics Facilities in the United States: Major Facilities for Research in Astronomy [a]

Support of Astronomy	As of	Cost ($Millions)
Federally supported optical astronomy	1963	17.0
Federally supported radio astronomy	1963	27.0
State and private university-supported astronomy	1963	12.0
TOTAL		56.0

[a] For more detailed and recent data on facilities and operating costs in astronomy, see the report of the Astronomy Survey Committee, *Astronomy and Astrophysics for the 1970's* (National Academy of Sciences, Washington, D.C., 1972).

move toward other institutions. Such a trend was clearly evident in the 1950's, and in 1960 the number of physicists employed in government and industry taken together was greater than the number working in the universities. However, in the 1960's, with a slowing in the rate of employment of physicists in government and industrial laboratories and growing national emphasis on the development of strong graduate education programs, a shift toward university employment again occurred. (See Chapter 12 for a detailed discussion of these trends.)

Figures 9.15 and 9.16 show, respectively, the growth of the AEC national laboratories during the period 1948–1970 and the distribution of funds in these laboratories among major research areas in the physical sciences and within physics.

Though the relative amount of money spent in the universities has diminished (principally because the amount spent in the large establishments has increased), the influence of university physics remains strong in all branches of physics. In experimental elementary-particle physics, much nuclear physics, many aspects of condensed-matter physics, plasma physics, and even atomic and molecular physics, although much work is

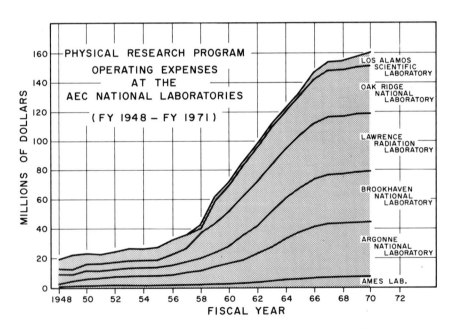

FIGURE 9.15 Growth in the research programs of national laboratories operated by the Atomic Energy Commission, 1948–1970.

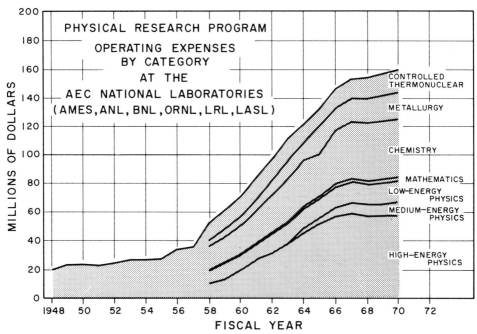

FIGURE 9.16   Distribution of funds among various physical sciences and subfields of physics in national laboratories operated by the Atomic Energy Commission, 1948–1970.

conducted outside the university, the interactions between the university and nonuniversity establishments still are vital. Most high-energy physicists are on university faculties, as are increasing numbers of medium- and low-energy physicists. They travel to national facilities to perform their experiments. This pattern is characteristic of other physics subfields as well. And, of course, in many cases, notably the Stanford Linear Accelerator Center and the Lawrence Berkeley Laboratory, the national facilities are close to or a part of a university.

In the Westheimer report on chemistry,* the panel on chemistry in the federal laboratories predicted that, because instrumentation for chemistry had become so elaborate and expensive and because the science had become so highly rationalized, professionals working in national facilities would outperform those working in universities in which, typically, a

---

* Committee for the Survey of Chemistry (F. H. Westheimer, Chairman), *Chemistry: Opportunities and Needs,* NAS–NRC Publ. 1292 (National Academy of Sciences–National Research Council, Washington, D.C., 1965).

professor employs students and postdoctoral fellows. Yet, university science traditionally has been the preponderant source of truly new intellectual breakthroughs. For example, the first xenon compounds were discovered by Neil Bartlett at the University of British Columbia *; the massive investigation of their properties was largely the work of the Argonne National Laboratory. An examination of the institutional affiliations of contributors to issues of the *Physical Review* provides further evidence of the leading role of university-based physicists, as Table 9.9 shows.

In high-energy physics most of the papers involve university authors. However, if one divides elementary-particle physics into theoretical and experimental, one finds that essentially all the experimental work is performed at a national facility and nearly all the theoretical work takes place at the university. University-affiliated physicists predominate even when the actual work is conducted at a national facility.

Obviously, physics would lose something essential if, for whatever reason, it were not conducted in close contact with the university. Throughout this Report we stress the important connection between education and research; despite the difficulties now imposed both by the massiveness of certain parts of physics and by the antiscientific attitudes that characterize some university campuses, we believe that this connection is a great source of strength in U.S. science. In countries where this connection is weaker than it is in the United States, the scientific enterprise appears less healthy and productive.

That the university has been the source of much of the most original thought in U.S. science is apparent in the finding that 19 of the 22 Nobel Prizes in physics awarded to Americans since World War II were won by university professors. It is difficult to visualize how U.S. physics could remain at the forefront of research in this discipline without preserving, insofar as is possible in this day of big physics, a close connection with the university. Therefore, a central task of science and university planners during this decade is to work out the relations between the universities and the central facilities that will enable the universities to participate at the forefront of the physics enterprise and, at the same time, will preserve the professor–student relation that has been fruitful in the past. This problem is one that elementary-particle physics and astrophysics have had to face and have attempted to meet through user groups, though these, too, present a number of problems. For example, any given facility can

---

* N. Bartlett and N. K. Jha, "The Xenon–Platinum Hexafluoride Reaction and Related Reactions," in *Noble-Gas Compounds*, H. H. Hyman, ed. (The University of Chicago Press, Chicago, Ill., 1963).

TABLE 9.9 Institutions Contributing Papers to the *Physical Review,* 1939, 1949, 1959, and 1969 [a]

	1939 Papers	1949 Papers	1959 Papers	1969 Papers
High-Energy Physics				
National Labs.	0	0	52	132
Other Govt. Labs.	0	0	8	108
University Labs.	30	54	156	936
Industrial Labs.	0	0	0	36
SUBTOTALS	30	54	216	1212
Nuclear Physics				
National Labs.	0	38	104	192
Other Govt. Labs.	0	0	20	36
University Labs.	102	182	124	504
Industrial Labs.	0	6	12	12
SUBTOTALS	102	226	260	744
Plasma and Fluid Mechanics				
National Labs.	0	0	4	12
Other Govt. Labs.	4	0	4	0
University Labs.	22	18	4	144
Industrial Labs.	6	4	8	0
SUBTOTALS	32	22	20	156
Atomic and Molecular Physics				
National Labs.	0	6	12	48
Other Govt. Labs.	2	2	4	72
University Labs.	82	54	36	360
Industrial Labs.	2	2	12	12
SUBTOTALS	86	64	64	492
Solid-State Physics				
National Labs.	0	12	24	144
Other Govt. Labs.	0	4	28	192
University Labs.	44	42	196	900
Industrial Labs.	10	22	132	288
SUBTOTALS	54	80	380	1524
TOTAL PAPERS [a]	304	446	940	4128

[a] 1939 estimated from six-month sample, 1622 pages. 1949 estimated from six-month sample, 1014 pages. 1959 estimated from three-month sample, 1695 pages. 1969 estimated from one-month sample, 2619 pages.

serve effectively only a limited number of research workers if any of them is to gain a sense of the personal participation and satisfaction that are essential to continuing high-quality work. In addition, a facility cannot be efficiently maintained or updated unless there is a sizable group of research users in-house who regard it with a much greater sense of involvement than would a typical outside user. Without such a high-quality, dedicated in-house group, constituting at least one fourth of the total number of users, the facility probably is doomed to early obsolescence and mediocrity. Consequently, this Committee urges that the sponsors and managers of all national facilities devote particular effort to maintaining the integrity and quality of the in-house research groups, and that these constitute at least 25 percent of the active users of a facility. At the same time, we must emphasize that these in-house groups should recognize and accept this responsibility to contribute significantly to maintaining the strength of a facility and to developing ancillary instrumentation for the benefit of all users. In addition, the management and the program committees associated with national facilities should develop and publish a statement concerning the research capacity of a facility. It is wasteful of both talent and money if extensive effort goes into planning by a prospective user-group when there is little probability of a project's survival in the competition for time and support at the needed facility.

Many national laboratories have evolved highly effective mechanisms for implementing user-group activity. Such mechanisms include user participation in the initial design of major facilities, provision of adequate technical and instrumentation support during use of the facility, and provision of adequate and attractive housing for visiting scientists and their families. Other national laboratories should consider these and other possible steps for facilitating user participation.

But the problem of planning for and making effective the work of outside users is not limited to elementary-particle physics and astrophysics. National facilities are becoming more numerous in many subfields. Nuclear physics, for example, will need to develop similar arrangements for the use of the Los Alamos Meson Physics Facility and the proposed National Heavy Ion Physics Facility. Further, in nuclear physics the activity in supertransuranic species will require a unique combination of people and facilities in physics, chemistry, large-scale instrumentation, and health physics. Similarly, major programs applying the concepts and technologies of physics to the central problems of biology, such as the MAN program at the Oak Ridge National Laboratory would be difficult to mount in any other type of institutional framework.* Consequently, national

_____

* There are, of course, other types of major problems of importance to society with regard to which the universities present a unique competence; an example would be

laboratories, many of which are established specifically to attack major physics-related problems in various subfields, should continue to be alert to possible areas of national concern to which their personnel and facilities might make a major contribution.

Recent reductions in support of research and development activities in various institutional settings have an impact on university and national laboratory interaction. The question arises: Can one adopt some general policy regarding the apportionment of reductions in support among the different kinds of institutions—national laboratories, other government laboratories, and universities—all of which derive either all or a major part of their support from the federal government?

It would be easy to say that, when such choices must be made, the installations at which the newest, and potentially most useful, discoveries are likely to occur should receive preference. Thus if the choice is between the 200-GeV accelerator and lower-energy accelerators, the tendency would be to favor the high-energy accelerator. But there are other considerations that preclude so simple a decision, considerations pertaining to educational, social, and economic effects. For example, the Oak Ridge National Laboratory is a major influence on a region of the country generally regarded as underdeveloped, both scientifically and economically.

A possible conflict arises in the present climate of financial constraints between basic researchers in the universities and in government or national laboratories. The university physicist who applies for support must place his proposal in competition with proposals from peers in his discipline. Whether the agency making the allocation of funds uses a review board to rate proposals or uses its own staff to make such ratings, the prospective university grantee is subjected to sharp and public scrutiny. However, physicists in government laboratories who are doing basic research are not subjected to the same kind of peer scrutiny, though their work is carefully reviewed by the laboratory management, visiting committees, and the scientific staff of the supporting agency. Their support is more or less assured as long as the institution receives support and as long as the management of the institution and that of the supporting agency have confidence in them. Consequently, it has been suggested that physicists at government laboratories should be subjected to the same kind of review as that given university faculty who seek federal grants.

To impose this kind of restriction and second-guessing on the management of a government laboratory would greatly weaken the management and, therefore, weaken the laboratory as well. Basic physics, though performed widely in government laboratories, generally is not the prime

---

one requiring the work of scientists, economists, political scientists, lawyers, and historians. Such a mix of recognized experts would be difficult to find in other than a university context.

purpose of these laboratories; this purpose usually is the achievement of some governmental objective through the use of science and technology. It is better achieved in almost all instances if the applied research is reinforced by basic physics; this is the underlying rationale for the support of basic physics in government laboratories. The quality of the detailed research activity in progress in these laboratories is ultimately the responsibility of the laboratory director. A variety of mechanisms exists for the discharge of this responsibility. To avoid unfortunate misunderstanding, individual laboratories probably should publicize more widely the nature of the ongoing review mechanisms that they employ. In addition, the management of the national laboratories should examine their basic physics research programs carefully to identify any that either do not contribute significantly to the mission of the laboratory or do not require the special facilities and technical support of the laboratory. Where such activities are found, their termination or transfer to a suitable university or industrial milieu should be seriously considered.

Through careful planning and the continuation of present efforts toward fuller cooperation, the various institutions of physics can fulfill ever more complementary and mutually reinforcing roles, thereby strengthening the overall physics research enterprise.

APPENDIX 9.A INSTITUTIONS GRANTING DOCTORATES IN PHYSICS AND ASTRONOMY RANKED ACCORDING TO NUMBER OF DOCTORATES AWARDED IN THESE FIELDS IN TWO FIVE-YEAR PERIODS, 1961–1965 AND 1966–1970 [a]

Institution	1961–1965		1966–1970		Percent Increase (or Decrease)
	Rank	PhD's	Rank	PhD's	
U. of California, Berkeley	1	252	1	299	17.6
Harvard	2	192	4	201	4.7
MIT	3	174	2	262	50.6
Columbia (N.Y.)	4	122	8	152	24.6
Caltech	5	118	13	135	14.4
Princeton	6	109	9	149	36.7
U. of Illinois	7	108	3	202	87.0
Cornell	7	108	5	177	63.9
U. of Wisconsin	9	98	10	148	51.0
Yale	9	98	7	162	65.3
Stanford	11	97	11	144	69.6

APPENDIX 9.A—*Continued*

Institution	1961–1965		1966–1970		Percent Increase (or Decrease)
	Rank	PhD's	Rank	PhD's	
U. of Michigan	12	95	12	142	49.4
U. of Chicago	13	88	16	118	34.1
Penn State	14	79	19	108	36.7
U. of Maryland	15	75	6	170	126.4
U. of Colorado	15	75	17	113	50.5
U. of Washington	17	70	26	85	21.4
Ohio State	18	69	20	106	53.6
UCLA	19	68	18	112	54.7
U. of Rochester (N.Y.)	20	65	14	120	84.6
Johns Hopkins	21	64	26	85	32.8
Case Western Reserve	22	61	20	106	73.8
Carnegie-Mellon	23	58	40	61	5.2
New York U.	24	57	29	80	40.3
Purdue	25	54	22	102	88.8
U. of Pennsylvania	26	52	23	93	78.8
U. of Texas	26	52	14	120	130.8
Catholic U. of America	28	50	30	72	44.0
Indiana U.	29	49	42	59	20.4
Iowa State	29	49	25	87	77.5
Brown U.	31	48	34	67	39.6
U. of Virginia	31	48	38	63	31.2
Duke	33	47	38	63	34.0
Rensselaer Polytech. Inst.	34	45	24	89	96.8
Rice U.	35	41	32	69	68.3
U. of Minnesota	36	38	35	66	73.7
Oregon State	36	38	107	11	(−71.0)
U. of Pittsburgh	38	37	37	64	73.0
Notre Dame	38	37	48	49	32.4
Syracuse	40	36	36	65	80.3
Michigan State	41	35	40	61	74.3
Washington U. (Missouri)	42	32	52	42	31.2
U. of Tennessee	43	31	31	71	128.9
U. of North Carolina	44	28	55	40	42.9
Florida State	44	28	44	54	92.8
U. of Florida	46	27	50	47	74.1
Vanderbilt	46	27	46	52	92.6
Louisiana State U. (and A&M Col.)	48	26	56	37	41.3
Northwestern	49	25	43	57	128.0
U. of Utah	49	25	56	37	48.0
U. of Cincinnati	51	23	91	18	(−21.7)

APPENDIX 9.A—*Continued*

Institution	1961–1965		1966–1970		Percent Increase (or Decrease)
	Rank	PhD's	Rank	PhD's	
Brandeis	51	23	51	46	100.0
Stevens Inst. of Technol.	51	23	62	34	47.8
Rutgers	54	22	33	68	209.1
Polytech. Inst. of Brooklyn	54	22	47	50	127.3
U. of Missouri (Columbia)	56	21	66	31	47.6
U. of Kansas	56	21	77	24	14.3
U. of Connecticut	56	21	69	29	38.1
U. of Iowa	59	20	52	42	110.0
Georgetown	59	20	63	33	65.0
Georgia Inst. of Technol.	59	20	58	36	60.0
St. Louis U.	62	19	91	18	(−5.3)
Temple U.	62	19	67	30	57.9
Texas A&M	64	18	49	48	165.0
Illinois Inst. of Technol.	64	18	73	28	55.0
Lehigh U.	64	18	58	36	100.0
New Mexico State	64	18	80	23	27.5
U. of Alabama	68	17	91	18	5.9
U. of Nebraska	69	15	98	16	6.6
U. of Oklahoma	70	14	65	32	128.6
North Carolina State	70	14	91	18	28.6
Wayne State U. (Michigan)	72	13	73	28	115.4
U. of Arizona	72	13	54	41	215.4
U. of New Mexico	72	13	91	18	38.5
Boston U.	75	12	63	33	174.9
SUNY, Buffalo	75	12	69	29	143.6
U. of Oregon	75	12	67	30	149.9
U. of California, San Diego	75	12	28	84	600.0
U. of Southern California	79	11	85	20	81.8
Kansas State (Col. of Agri. & Appl. Sci.)	79	11	85	20	81.8
Washington State	81	10	76	26	160.0
U. of Kentucky	81	10	80	23	130.0
Va. Polytech. Inst.	81	10	58	36	260.0
U. of California, Riverside	81	10	44	54	440.0
Fordham	85	9	91	18	100.0
Oklahoma State	86	7	61	35	400.0
Tufts U.	86	7	83	21	200.0
Ohio U.	86	7	73	28	300.0
U. of West Virginia	86	7	85	20	185.6

APPENDIX 9.A—*Continued*

Institution	1961–1965		1966–1970		Percent Increase (or Decrease)
	Rank	PhD's	Rank	PhD's	
U. of Arkansas	90	6	103	13	116.2
U. of Alaska	90	6	113	9	50.0
Brigham Young	90	6	88	19	215.8
Arizona State	93	5	88	19	280.0
Texas Christian	93	5	98	16	220.0
Howard U.	93	5	113	9	80.0
Tulane U. (La.)	96	4	83	21	280.0
Utah State (Col. of Agri. & Appl. Sci.)	96	4	107	11	175.0
U. of New Hampshire	98	3	105	12	300.0
Colorado State	98	3	98	16	429.0
U. of South Carolina	98	3	103	13	330.0
U. of Missouri (Rolla)	98	3	82	22	627.0
Naval Postgrad. School	98	3	113	9	200.0
U. of Denver	103	2	120	7	250.0
St. John's U. (N.Y.)	103	2	145	1	(−50.0)
U. of Delaware	103	2	97	17	750.0
Bryn Mawr	103	2	120	7	250.0
U. of Georgia	103	2	125	6	200.0
Inst. of Paper Chem. (Wis.)	103	2	133	4	100.0
U. of Nevada	103	2	129	5	150.0
U. of California, Davis	110	1	69	29	2800.0
Yeshiva U.	110	1	77	24	2300.0
Clark U. (Mass.)	110	1	133	4	300.0
George Washington	110	1	117	8	700.0
Colo. School of Mines	110	1	141	2	100.0
SUNY-Col. at Syracuse	110	1	—	0	(−100.0)
Boston College	110	1	107	11	1000.0
Alfred U.	110	1	145	1	0.0
Clemson	110	1	77	24	2300.0
Worcester Polytech. Inst.	110	1	107	11	1000.0
Emory U.	—	—	145	1	—
American U.	—	—	129	5	—
Auburn	—	—	117	8	—
U. of Massachusetts	—	—	125	6	—
U. of Wyoming	—	—	113	9	—
Adelphi U.	—	—	117	8	—
Baylor	—	—	137	3	—
Rockefeller U.	—	—	120	7	—
U. of Houston	—	—	107	11	—
Texas Tech.	—	—	125	6	—
U. of Akron	—	—	145	1	—

APPENDIX 9.A—*Continued*

Institution	1961–1965		1966–1970		Percent Increase (or Decrease)
	Rank	PhD's	Rank	PhD's	
U. of So. Illinois	—	—	145	1	—
U. of Hawaii	—	—	137	3	—
U. of Mississippi	—	—	133	4	—
U. of Miami	—	—	133	4	—
City U. of New York	—	—	102	14	—
U. of California, Santa Barbara	—	—	88	19	—
U. of Idaho	—	—	137	3	—
Kent State	—	—	137	3	—
SUNY, Albany	—	—	141	2	—
Northeastern (Mass.)	—	—	98	16	—
SUNY, Stony Brook	—	—	69	29	—
U. of Vermont and State Agri. Col.	—	—	129	5	—
Dartmouth	—	—	112	10	—
U. of California, Irvine	—	—	125	6	—
U. of Toledo	—	—	141	2	—
Clarkson Col. of Technol. (N.Y.)	—	—	105	12	—
Drexel U. (Phila.)	—	—	120	7	—
Lowell Tech. Institute	—	—	141	2	—
New Mexico Inst. of Mining & Technol.	—	—	145	1	—
William and Mary Col.	—	—	120	7	—
U. of Santa Clara	—	—	145	1	—
U. of California, Santa Cruz	—	—	145	1	—
Cooper Union	—	—	145	1	—
Montana State	—	—	129	5	—

a Source: National Research Council, Office of Scientific Personnel, Doctorate Records File, January 1972.

Discussion

The median number of PhD degrees awarded in physics and astronomy by the 119 institutions granting such degrees in the interval 1961–1965 was 20; and the mean number, 34. One third (that is, 40) of these 119 PhD-granting institutions awarded three fourths (76.5 percent) of the 4077 doctorates in physics and astronomy. The top ten institutions granted one third (33.8 percent) of the degrees.

In the interval 1966–1970, the number of institutions awarding the PhD in physics and astronomy increased by 35 to a total of 154. The median number of degrees awarded in these fields was 24, a slight increase over the median for 1961–1965, and the mean increased to 45, as compared with 34 in 1961–1965. Again one third (51) of the 154 institutions awarded three fourths (75.8 percent) of the 6889 PhD's granted in this five-year period. However, the percentage produced by the ten top-ranking institutions decreased from one third in 1961–1965 to somewhat more than one fourth (27.9 percent) in 1966–1970.

One institution (Caltech) that had ranked fifth in the 1961–1965 interval dropped to thirteenth in 1966–1970, with the University of Maryland rising in rank from fifteenth in the first half of the 1960's to sixth in the second half of that decade.

Only five institutions reported a decline in the number of doctoral degrees awarded in physics and astronomy in the 1966–1970 interval compared with 1961–1965: Oregon State University, University of Cincinnati, St. Louis University, St. John's University, and the State University of New York-College at Syracuse. One institution, Alfred University, produced only one PhD in these fields in each of the two intervals studied. The remaining 113 institutions that had PhD doctoral pro-

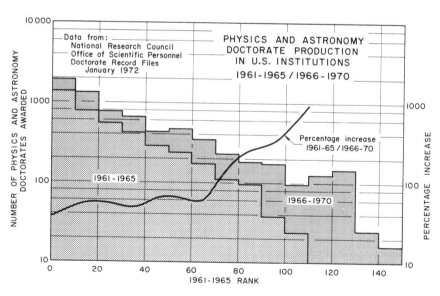

FIGURE 9.A.1 PhD degrees in physics and astronomy awarded by U.S. institutions in 1961–1965 and 1966–1970.

grams in physics and astronomy in 1961–1965 reported increases ranging from 4.7 percent to 2800.0 percent in the number of these degrees awarded in 1966–1970. In addition, the 35 institutions that began doctoral programs in physics and astronomy in 1966–1970 produced 223 doctorates (3 percent of the 6889 produced by all 154 institutions during this interval).

Figure 9.A.1 depicts the data presented in this appendix.

10

THE

SUPPORT

OF

U.S. PHYSICS

INTRODUCTION

In a third of a century, expenditures for physics research in the United States have risen nearly two orders of magnitude. From a small individualistic enterprise, supported largely by foundations and universities, physics has grown to a massive operation involving expenditures of more than $500 million per year and drawing substantial support from federal agencies.

Of the many federal sponsors of physics, the most conspicuous are the mission-oriented agencies, the Department of Defense (DOD), Atomic Energy Commission (AEC), and National Aeronautics and Space Administration (NASA). A smaller but vitally important part is contributed by the National Science Foundation (NSF). Each of these agencies has its own distinct mission, but each has been able to reap rich rewards from its support of physics while simultaneously contributing in an important way to the development of the science itself.

The evolution of this mutually beneficial relationship between the mission-oriented agencies and the science has not been a trivial task. On the part of the agencies, too great an emphasis on relevance could have destroyed the science; on the part of the physicists, refusal to consider the application of their work could have robbed it of its vitality as a component in a growing technological society. Compromises have been re-

635

quired on both sides, and, by and large, the patterns of support that emerged have been eminently successful in bringing about the realization of the manifold contributions that physics can make to society. Although we must view the diminishing funding levels of the past several years with the greatest concern, we cannot seriously fault the mechanisms by which the support has been allocated. Many agencies of government have been involved in the support and exploitation of physics, each with its own criteria and administrative pattern. This multiplicity of sources of support may lack a certain neatness, but it reflects clearly the fact that physics interacts with society in many different ways, through many different channels.

Society's problems are not limited to defense, energy production, or prestige in space. There are other pressing matters bearing on the quality of life: health, transportation, environmental control, and equitable distribution of goods and services. Physics and physicists have something to say about these problems, too, but just what they can contribute is difficult to specify, except at the most superficial level. It remains an important challenge for both the physics community and the federal agencies to develop mechanisms through which the necessary intellectual interactions can be promoted. If we have learned anything from the experience of the agencies that have used physics successfully, it is that such interactions must involve a simultaneous concern for the pursuit of the mission and the growth of the science. Even in the short range, the fruits of research must be replenished as they are harvested.

To these ends, we recommend:

1. That the present multiplicity of support sources for basic physics research be preserved and strengthened;
2. That such agencies as the Departments of Transportation (DOT), Health, Education and Welfare (HEW), and Housing and Urban Development (HUD) seek to evolve mechanisms by which the resources of physics can be exploited and expanded in the context of their missions.

PATTERNS OF SUPPORT

Support as It Appears to a Physicist

Physicists work in universities (50 percent), federally funded research and development centers (12 percent), government laboratories (9 percent), and industrial laboratories (25 percent).* In the latter two categories,

---

* Figures here refer to PhD physicists; research centers includes federally funded research and development centers operated by universities or industry. See Chapters 9 and 12.

the support problem is relatively simple because the laboratories involved have definite missions to fulfill. As long as the parent organization (government or industry) finds that its physicists are contributing to the mission, then support is forthcoming for most of the projects in the laboratory.

The physicist in the university, on the other hand, is committed to following his research wherever it leads him; his activities are not mission-oriented (be they in pure or applied physics). He has a special problem in seeking funds to support his research because he cannot promise a specific result to the sponsor; he can only agree to work toward certain goals, which he sets for himself.

Support of physics, even in universities, usually requires more than the physicist's salary (most of which is usually paid during the academic year by the university). In experimental work, he requires research instruments and often one or more technical assistants; if he works with a large accelerator or telescope, he requires a supporting organization of engineers, technicians, and others. His graduate students, who are learning physics while helping in his laboratory, must be paid for. In theoretical work, he needs computer time and, often, the services of graduate research assistants and associates. The expenses add up to about $50,000 per year for each physicist engaged in an average project, a sum that universities are simply unable to defray. The university physicist therefore seeks funds from one of three sources—the federal government, industry, or private foundations. Currently some 85 percent of all research funds for university physicists are provided by the federal government.

Although about two dozen agencies are engaged in the support of physics, four of them—the AEC, NASA, DOD, and NSF, in that order—together supply over 95 percent of the federal funds allocated to physics. Basically, the choice of agency is based on compatibility between the goals of the agency and the research project. Thus, to oversimplify, if the project is in nuclear physics, the AEC will be particularly interested; if in solid-state physics, with potential application to defense problems, the DOD will be interested. Of the four agencies, the NSF has the broadest goals—the support of basic research—so that if there is relatively little compatibility with the interests of other agencies, a physicist will submit his proposal to the NSF.

In formulating a project for submission to a federal agency, the physicist first writes a research proposal stating the goal, methods, and resources for the project. Estimates of costs are made, often with the help of university financial officers, and a budget for a 1- to 5-year period is prepared. Since the funds sought are to be transmitted to the university for administration, approval of the proposal by the university is required. This usually involves the departmental chairman, the provost or other officer of the university, and a representative of the university's financial offices.

The proposal is then transmitted formally by the university to the agency or agencies that may have an interest in the research area.

Agencies usually require about six months to process the proposal. It is typically examined by referees outside the agency, appropriate experts who submit written critiques to the agency. It is then considered by the agency staff. If, as in most cases, the agency is mission-oriented, questions will be asked concerning its relevance to the mission. In any case, the budget will be scrutinized to see where dollars can be saved and used for other worthwhile investigations. If funds are limited, as they usually are, discussions will occur between the university physicist and the agency staff as to where and by how much the budget can be trimmed. If the proposal is approved, a contract is negotiated between the agency and the university, and the work begins. If the proposal is rejected, the typical physicist remains hopeful and submits it again, perhaps in modified form, to the same or to other agencies.

Historical Background *

The congenial association of science with the federal establishment has a long history, characterized for the most part by a tacitly understood dichotomy between the motivations of scientists seeking federal support and those of the government in supporting science. The Lewis and Clark expedition of 1804–1806—certainly an enterprise with a strong scientific flavor—was justified to Congress as essential to the extension of foreign commerce. Like many modern scientific endeavors, it derived its main logistical support from the military. The U.S. Naval Observatory—again a basically scientific enterprise from the beginning—was first funded as a Depot of Charts and Instruments, a highly mission-oriented organization in the eyes of the Congress of the time. These instances illustrate a kind of ambivalence, which might seem hard to live with, save that scientists have lived with it for many centuries.

Since the time of Archimedes, scientists have been trying to explain to their governments that science should be supported for its own sake. For an equally long time, governments have persisted in basing their support of science on the belief that it is essential to the national security, national welfare, or national prestige. Both points of view have validity, of course, and the support of science has paid handsome dividends both because of

---

* See "Toward a Science Policy for the United States," Report of Subcommittee on Science, Research and Development, Committee on Science and Astronautics, U.S. House of Representatives, October 15, 1970, Part III; and H. Brooks, in "National Science Policy," Hearings of Subcommittee on Science and Astronautics, July–Sept. 1970, p. 931.

and despite a narrow outlook on the part of many of its patrons. Without a focus in the real world, without the occasional flexing of the scientific muscle for the common weal, science would soon become detached and sterile; but if scientists did only what they were directed to do, the golden eggs would soon cease to appear.

The great episodes in the romance of science and government have been mainly associated with wars, until quite recently, when NASA was created for a considerably different, though still mainly nonscientific, objective. The National Academy of Sciences, the first formal link between government and the scientific community—was created in the shadow of the Civil War. When World War I broke out, the National Research Council (NRC) was created as a constituent body of the Academy to mobilize scientific effort in pursuit of the national defense. The NRC did not itself become an important supporter of research during the war but rather acted to recruit and coordinate and channel scientific skills into the war effort, skills that were largely exercised by scientists in uniform. After the war, the NRC lost its intimate contact with the federal establishment, but, with support—generous for the time—from private foundations, was able to continue as a focus for the scientific community and as a channel of communication to the government.

World War II brought quite a different kind of science–government relation into being, with the establishment of the Office of Scientific Research and Development (OSRD). With its head, Vannevar Bush, reporting directly to the President, OSRD succeeded in maintaining an autonomy that permitted a large measure of scientific judgment in just how science was to serve the government's aim. For the first time in war the scientists functioned outside, and often independent of, the military. Although OSRD did not construct its own laboratories, it carried out research and development enterprises on a vast scale through contracts with universities and industries. The basic science component of these efforts was not very large, but its obvious contribution to the success of the mission served to emphasize the importance of a continuing basic research program in support of even a sharply focused development effort.

The OSRD was not the only means for channeling science into the war effort. Indeed its expenditures were exceeded by several other agencies, most notably the Manhattan Project, which took over the nuclear bomb development at an early stage. However, also in the brilliant success of this project, the crucial importance of a management that accorded a substantial role to basic science was clearly evident. In both OSRD and the Manhattan Project completely new techniques for the fruitful interaction between scientists and their public customers were developed; these were primarily administrative instruments that made possible a suitably loose coupling between the stated goal and the procedure for reaching the goal.

It should be observed that what is here referred to as the basic research component of the war effort was hardly basic in the most restrictive definition. Rather, it consisted in dipping into the well for the really fundamental knowledge and pressing further research only along lines that were expected to produce yields in the short term. With some interesting exceptions, contributions to fundamental science during the war period were not very great—as the near-vanishing thickness of the scientific journals of the period attests. In fact, the massive dislocations in nearly all university laboratories, and the near-complete cessation of graduate education, would have left the scientific establishment in complete collapse had the government not come to its rescue at the end of the war.

Chief among the rescuers was the Office of Naval Research (ONR), created in 1946 as the successor to the Navy's Office of Research and Inventions. From the outset, the ONR took the view that the vitality of basic research was important to the long-range mission of the Navy and set out to offer its support to all kinds of basic science, with little explicit regard for its relevance to immediate Navy problems.

Quite possibly the most important contribution ever made to the integration of government and science was this recognition by the ONR that advances in technology depend on the health and vitality of the whole body of science and not just that part in which near-term gains are visible. By thus accepting the scientists' assertion that "science for its own sake" is necessary for the integrity of the subject and that an integrated, whole science is the essential base of technology, the ONR came close to resolving the old dichotomy between the motivations of the scientists who sought federal support and those of the agencies that offered it.

There were many who saw dangers in the domination of basic science by federal money. Concerns were expressed about the hazards incident to centralized control of any kind, and most specifically those incident to control by the military. Moreover, if the buyer and seller have different understandings of the purpose of the contract, instabilities disadvantageous to both parties may develop. In the event, however, the "domination" by the ONR proved much less onerous than it might have been. Taking to heart the lessons taught by the OSRD, the ONR established a strong cadre of highly qualified scientists as the controlling agents and the necessary interface between the laboratory and the purse. Without these dedicated men, and their successors in every successful science-supporting agency since, science might indeed have languished in the postwar years.

Following the lead of the ONR, several other agencies were formed to capitalize on the scientific momentum generated by the war. Thus the AEC, a new civilian agency, took over the development of nuclear energy in 1946, and other agencies parallel to ONR were formed in the Air Force and in the Department of the Army. Again, these new agencies entertained a lively appreciation of the need for rejuvenation of basic science relevant to their missions.

Although the mission agencies have established an enviable record in their enlightened support of science, the very fact that they have well-defined objectives makes their support of basic science more vulnerable than it ought to be. As long ago as 1944, President Roosevelt commissioned Vannevar Bush to study how the OSRD experience might be carried over into peacetime. In his report, *Science, The Endless Frontier,* Bush emphasized the need for a permanent structure within the government, adapted to supplementing the support of basic research and education. He envisioned the new organization as largely patterned after the OSRD, with scientists providing the leadership and with missions to support basic science in universities, to support higher education through scholarships and fellowships, and to assume major responsibility for the support and coordination of research in the public interest, including medical research and long-range scientific research on military matters.

At the culmination of a long debate, an entirely new kind of agency, the NSF, was founded in 1950, to take over the broad support of basic scientific research. Its goal was defined so broadly ("to provide for the welfare and support of basic science in the United States") that it could fund types of research not easily justifiable under the missions of the DOD, AEC, or NASA. Thus the establishment for federal support of scientific research, as experienced by the university physicist, was completed. In one way or another, all had grown from the World War II experience. Implicit in this development was the feeling in both the executive and legislative branches that funds spent for physics could be justified as contributing to the national security of the United States. Of course, some of the research money was spent directly on new ideas for defense systems, but much more of it was not. The argument for the latter was that, by increasing the overall competence in and knowledge of the physical sciences, the nation was providing a pool on which to draw in an emergency. Indeed, this view is undoubtedly correct. For example, the constant support of solid-state physics made the transistor possible and, with it, the miniaturized circuits and computers that are the heart of missile guidance systems.

Statistical Data on the Support of Physics

Figure 10.1 shows the support of U.S. research and development and of basic science from all sources in the period 1923–1970. The following points are of interest:

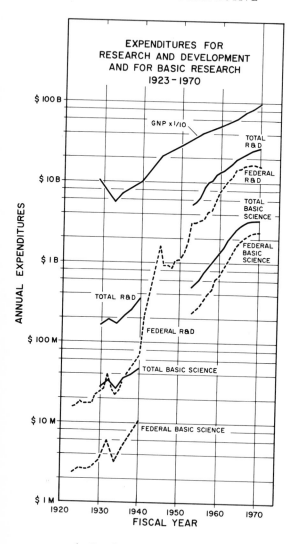

FIGURE 10.1 Expenditures for research and development and for basic research, 1923–1970. Except for the period, 1940–1953, figures do not include capital equipment or plant. Figures from *National Patterns . . .* derive from surveys by performers and differ somewhat from those in *Federal Funds. . . .* [Sources: Data on 1920–1940 from V. Bush, *Science, The Endless Frontier* (U.S. Government Printing Office, Washington, D.C., 1945). Data on 1940–1953 from *Federal Funds for Research, Development, and Other Scientific Activities,* Volume XX (U.S. Government Printing Office, Washington, D.C., 1971). Data on 1953–1970 from *National Patterns of R and D Resources,* NSF 70–46 (U.S. Government Printing Office, Washington, D.C., 1970).]

1. In the four decades from 1930 to 1970, total annual expenditures for research and development rose from $200 million to $27 billion, more than two orders of magnitude. The basic research component, amounting to typically 10 percent of the total, shows the same increase.

2. The fraction of research development and of basic research supported by the federal government rose from one tenth to nearly two thirds over this period.

3. The rate in increase of federal support of research and development and of basic science above shows an abrupt decline around 1965; in

dollars of constant purchasing power, the support in 1970 was significantly less than in 1965.

Figure 10.2 shows the trend of federal support of basic research in the *physical and environmental sciences,* in 1957–1970. Over the period 1958–1964, total federal support of these fields increased more than

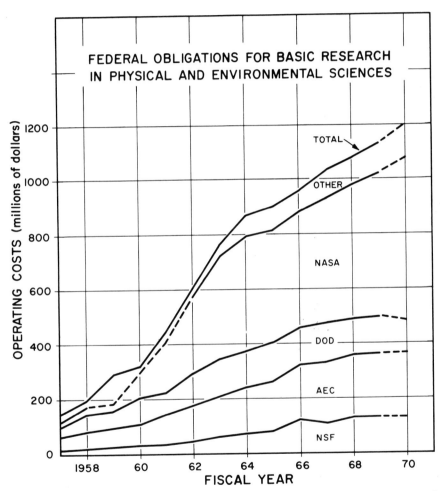

FIGURE 10.2   Federal obligations for basic research in physical and environmental sciences, 1957–1970. [Source: Data from *Federal Funds for Research, Development, and Other Scientific Activities,* Volumes VII–XX. Figures are not operating costs and do not include capital equipment and research and development plant. Figures for 1970 are estimates.]

fourfold; from 1964 to 1970, the increase has been only a factor of 1.3—
most of this modest increase has been offset by inflation.

Also shown in Figure 10.2 is the distribution of support for physical
and environmental sciences among the principal agencies. Here it may be
seen that a large part of the rise in the early 1960's is to be ascribed to
dramatic increases in NASA expenditures: Their leveling after 1964 is
partially responsible for the abrupt change in the trend of the total support
level.

The situation of federal support of basic *physics* in 1959–1970 is
portrayed in Figure 10.3. The trend shown here is quite similar to that

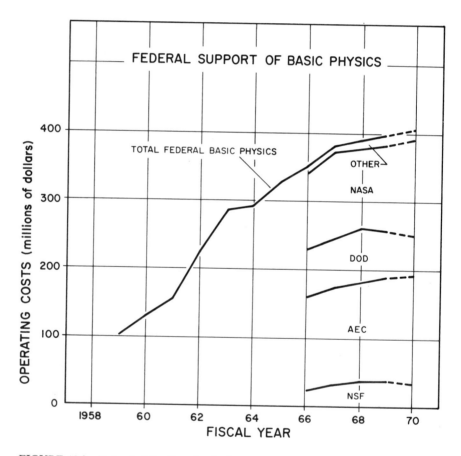

FIGURE 10.3   Federal obligations for basic research in physics, 1959–1970. [Source:
Federal Funds for Research, Development, and Other Scientific Activities, Volumes
IX–XX.]

of Figure 10.2. In 1959–1963, physics represented almost exactly 20 percent of the total of all basic science expenditures by the government. Since 1964, the fraction has dropped to about 18 percent. As shown by the indicated distribution among agencies for 1966–1970, the AEC and NASA are the principal supporters of physics. The once dominant DOD has retired to third place, with the NSF a poor fourth.

Figures 10.4, 10.5, and 10.6 show rough distributions of the funding supplied by the AEC, NSF, and DOD to various subfields of physics, as compared with total expenditures for basic physical and environmental science.

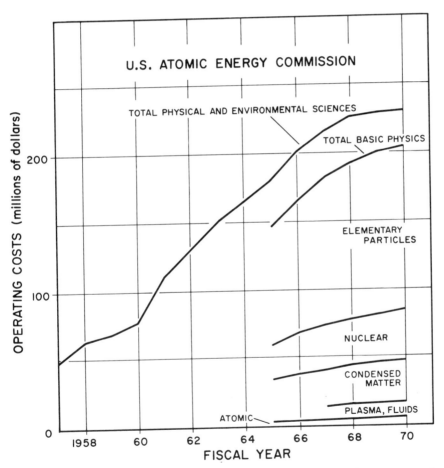

FIGURE 10.4 Distribution of AEC support among subfields of physics, as compared with total support of basic physics and the total for basic research in physical and environmental sciences.

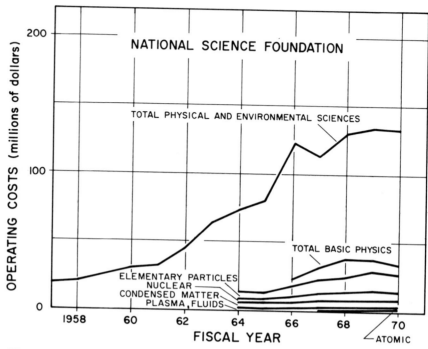

FIGURE 10.5   Distribution of NSF support among subfields of physics, as compared with total support of basic physics and the total for basic research in physical and environmental sciences.

Since the AEC is the lead agency for elementary-particle physics, it is not surprising that a substantial fraction of all the funding for this subfield comes from this agency. The DOD's former relatively strong position in elementary particles and nuclear physics has been drastically cut back since 1968. On the other hand, basic condensed-matter physics is about equally important in the AEC and DOD. The NSF, with a much smaller total physics program, has about the same relative emphasis on elementary particles, nuclear physics, and condensed matter as does the AEC.

Finally, Figure 10.7 shows the federal expenditures for basic physics research as a fraction of the total federal budget for 1959–1970. The fraction rose rapidly from 0.1 percent in 1959 to nearly 0.3 percent in 1963–1966 and has declined to about 0.2 percent since.

The Present Situation

The data in the previous section raise some questions. Why do the curves have the shape that they do? The rapid rise during World War II

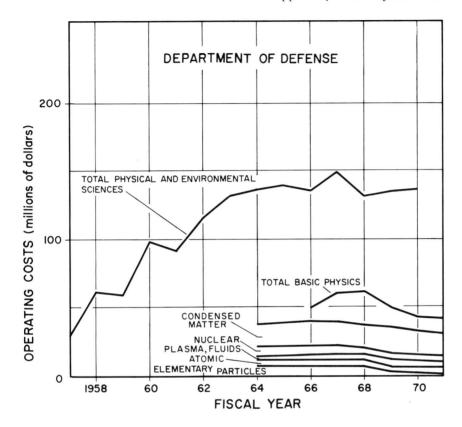

FIGURE 10.6 Distribution of DOD support among subfields of physics, as compared with total support of basic physics and the total for basic research in physical and environmental sciences.

is easy to understand. The reason for the continued rise after the war is not so obvious. The highly successful experience during World War II led to a large momentum for growth in the postwar period. The cadres of young men who had worked in the wartime laboratories were eager to get back to their own research interests, and many methods developed during the war were ripe for exploitation for many other purposes. Universities, aware of the successes of physics, rapidly expanded their facilities, and a major demand for research funds developed almost overnight.

On the federal side, administrators in key offices responded to this demand sympathetically, for the reasons stated above. Congress, impressed with the successes of wartime science and concerned about further military threats, voted substantial research funds into the agency budgets. But

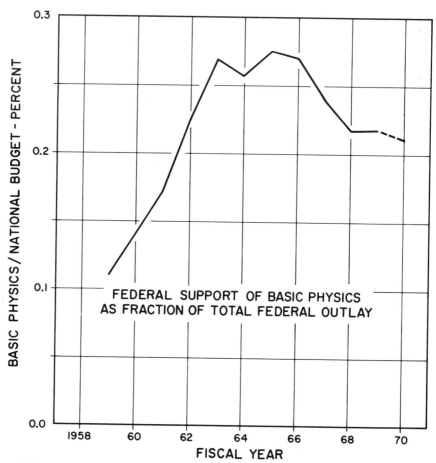

FIGURE 10.7   Federal obligations for basic research in physics as fraction of federal budget expenditures.

starting in about 1966, a number of forces came into play to decrease the rapid growth rate of 1950–1960.

First, the physics budget had been growing relative to the total federal budget, as shown in Figure 10.7. Moreover, several very large projects, such as the Stanford Linear Accelerator Center (SLAC), had been initiated. Suddenly the physics budget became more visible and, therefore, an identifiable target for economies.

Second, a high-priority program had been initiated by the President's Science Advisory Committee to respond to the clearly envisaged national needs for trained scientific and technical manpower in the military and space effort. In the late 1960's, not only had the goals of this program been achieved—earlier than planned as a result of enthusiastic university expansion—but manpower needs were rapidly dwindling as both the military and space programs were declining.

Third, during this period, the United States was becoming increasingly involved in an unpopular war in Vietnam. This generated several pressures and counterpressures. To carry out its Vietnam operations without raising taxes, the government could only cut the budgets of civilian agencies. Thus, the budgets of NASA and the NSF, significant supporters of physics, were reduced. The Vietnam war did not require large-scale advanced technology, so little new support for physics was generated by it. On the contrary, during this period, disagreements developed within the university community over the war issue, with the result that support for close DOD–university relations deteriorated. In addition, the Mansfield Amendment of 1969 forced the DOD drastically to reduce its support of basic research in some fields.

Fourth and finally, the period 1960–1970 was one of rapid social change in the United States, and this too had its impact on support of physics. At the beginning of the period, the majority of citizens, including students, were seeking a higher standard of living for themselves and for their families. By the end of the period, minority groups had become increasingly determined to achieve what they regarded as first-class citizenship and, in the process, had impressed on the public what miserable conditions they often endured. The result has been an increasing determination to reorient U.S. society so that all groups might enjoy a high standard of living; this objective had led to large federal programs concerned with poverty, health care, education, and the like, all of which have exerted pressure on the discretionary fraction of the federal budget.

A related factor is the intense concern about the physical environment and the quality of life. During the past decade an increasing number of cities were enveloped in smog, and an increasing number of streams and lakes became polluted. Recent large-scale publicity has made the public aware that uncontrolled pursuit of a higher standard of living may lead to increased pollution. Thus far, federal expenditures for improvement of the environment have not been substantial, so there has been no large effect, positive or negative, on the support of physics, but such effects may be anticipated in the near future. Perhaps just as important from the point of view of physics is a growing awareness that means must be found to conserve such irreplaceable resources as fossil fuels. Once the short-term economic obstacles are overcome, one can confidently expect increased emphasis on high-technology replacements for such resources.

Rationale for the Support of Physics

In the war and postwar periods, physics contributed effectively to national security. Industry supported physics because its findings increased profits. And private foundations supported physics for reasons consistent with their charters—usually for its educational and cultural value.

In short, in each case there were benefits rendered for funds conferred. Clearly, this is the principal reason that government, industry, or foundations can be expected to support physics in the years ahead. It is unrealistic to expect otherwise.

The solution to a physics problem begins when someone has a relevant idea and is completed when that idea is either proved or disproved and is reported to the public in the literature. There are profound satisfactions in this endeavor to those engaged in it, including intellectual challenge and excitement, aesthetic satisfaction, the sense of fellowship with collaborating and even competing colleagues, and the pleasure of discovery. As there are only 40,000 professional physicists, purely intellectual activity on the part of such a tiny fraction of the population cannot justify major public support, even though everyone has a sense of curiosity about the unknown and all can share in the exploration of the unknown.

Of more specific social interest is the role of physics in education, broadly conceived. A large fraction of all physicists is engaged in teaching, formally or informally, in universities, colleges, high schools, technical institutes, industrial seminars, and the like. There they attempt to communicate in detail what it means to solve a physics problem. In the most intensive example of education, a graduate course at a university, the professor may lecture in great detail about a very specialized problem, requiring his students to reproduce the steps or even to improve on them; while in a more informal setting, the physicist may lecture at a seminar of executives without technical background on how certain current developments in physics may be of interest to industry. Although the activity thus embraced under education is diverse, all of it is characterized by a more experienced person sharing his know-how and knowledge of physics with less experienced persons. Because physics is a highly developed and efficient way of coping with the physical world, this sharing of experience transmits a valuable resource. By acquiring this resource, a person's vision of the world is permanently changed and clarified. How can one estimate the benefits from this process? Millions of citizens are willing to pay directly or indirectly for the support of educational institutions that foster such activity. Moreover, both industry and government spend significant sums to improve the technical proficiency not only of technical personnel but of management through a variety of educational programs.

Every act of the physics enterprise *per se,* because it somewhat alters the understanding of the physical world, has implications for dealing with the physical world and thus with technology. A common example is research on semiconductors. These materials, often used in exceedingly minute quantities, have electrical properties that make them superbly useful in electronic circuits. A solid-state physicist now can calculate that a certain mixture of two elements will have specific properties currently in demand in the electronics industry. He can carry out the calculation, do an experiment to verify the calculation, and publish the result. The information will be quickly utilized by electronic engineers to improve, say, a circuit in a commercial stereo receiver. (See also Chapters 4 and 7.)

Physics—often through technology—also has impact on the overall national security and economic health of the nation. National security needs no additional comment beyond noting that any realistic program for disarmament or de-escalation of military pursuits inevitably will depend equally, if not even more, on the most advanced science and technology for detection, surveillance, and the like.

The question of the national economic health has been less discussed but is of even greater importance to the nation's future and to life in the United States. The position of the United States in international trade has become increasingly dependent on the high-technology components of the economy,* quite apart from the direct contribution of these components to the U.S. standard of living. The current international economic instabilities are an explicit recognition that the U.S. economy and, in particular, these vital high-technology components are becoming increasingly vulnerable to foreign competition. Although the direct connection cannot be demonstrated uniquely and unambiguously, there exists evidence that the development of these strong high-technology industries has reflected the strength of U.S. science—and frequently U.S. physics. Maintenance of U.S. strength in science is an important part of any coherent response to the present economic difficulties.

Can the federal government support the applied physics involved in these benefits without supporting pure physics, which yields no such obvious, immediate benefit? Virtually every experienced scientist is convinced that this would be impossible, because science is an organic unity; pure and applied physics merely describe poles of a continuum that includes every type of activity in physics from the mathematical theory of elementary particles to fairly practical electrical engineering. There is such extraordinary interaction and feedback among all the parts of this organic unity that the thought of supporting one part to the exclusion of others is a contradiction in terms.

It is not easy to say just how the benefits of applied physics performed

---

* See, for example, Chapter 7.

in universities are distributed. For an industrial concern with a product that it is trying to improve, applied physics research may sometimes be justified on a profit-and-loss basis. Even here, the most enlightened industrial laboratories permit a considerable degree of freedom for their physicists because the impact of any line of activity on future profits is so difficult to predict. In the case of university research, one can only say that, because the results are freely published, the whole industrial community benefits, including those parts of it devoted to making products required by the federal government.

In summary, the benefits of physics, as with all science, largely accrue to the public as a whole; thus the appropriate prime instrument of support is the federal government. The level of support will naturally reflect the degree to which the public, through its elected representatives, believes, or is convinced, that physics is relevant to current national problems. It is the responsibility of the physics community to find ways to be relevant. One example of an area in which deeper understanding of physics is clearly required for successful resolution of national problems, as we have indicated, is energy resources; achievement of controlled thermonuclear power would be a decisive advance.

The organization and the research and development budgets of various federal agencies that support basic research in physics in connection with their missions appear in Appendix 10.A. The Appendix also includes information on federal agencies that could become more actively involved in the support of physics in the future.

FEDERAL POLICY MAKING

Even to understand the formal structure of the machinery by which scientific policy is made and put into effect in the government is a nontrivial task. As with any important government function, actual decisions must be based on a combination of technical, budgetary, and political considerations, and the mechanism must permit some kind of pressure equilibration among these factors at all stages of the process. In principle, the detailed technical judgments are made within the agencies, and the political–economic input occurs at the Presidential level, through the Office of Management and Budget (OMB), and in the Congress. In fact, well-functioning mechanisms exist, both formal and informal, for political and economic intervention at the technical levels and, to a somewhat lesser extent, for technical input at the political levels.

The annual budget is the embodiment of the federal science policy. It is in the process of formulating the science components of the budget, and eventually of confronting them with other demands on the public purse, that policy takes its first essential step toward realization.

Evolution of a Budget

The Agencies In such agencies as the NSF, AEC, NASA, and DOD—agencies with comparatively large science components—construction of the science budget is a long and arduous procedure, involving many kinds of input. At one end is a set of overall agency target totals provided by the OMB; at the other is a set of desires expressed by managers of individual programs. At this level, the scientific community is heard through submission of proposals for support of specific projects; the voice of the political community is first heard as a rather generalized statement that next year's allocation for science should be higher or lower by so many percent. The process that intervenes before a final budget is adopted provides a number of further opportunities for intervention of both a technical and political nature and ultimately aims at some reconciliation of the various points of view.

Typically, any given year's budget will have a history extending back several years. Many agencies have outlined plans for five years or more ahead,* plans that initially may be rather generalized but that become progressively sharper as annual budget deadlines approach.

Specific budget plans for fiscal year Y (July 1, $Y-1$, to June 30, Y) may well begin within an agency as much as two winters earlier. At this stage, there is available only an educated guess as to the likely total to be allowed, and budgets of individual programs are only broadly constrained. As individual budget items pass through the administrative hierarchy, they are summed and confronted with the (usually much smaller) target total. At the same time the target may be adjusted to reflect new assessments of congressional opinion or the possible attractiveness of new programs. The process undergoes a number of iterations at intermediate management levels, involving largely technical considerations in arriving at an appropriate balance among various competing scientific activities. To a degree that varies markedly from one agency to another, consultative scientific advisory groups may have some influence on these determinations, either in detail or in formulation of long-range policy. Through the agency's fiscal arm, and through final consideration of the budget at the top administrative level, the science budget is matched to other components of the overall mission of the agency and prepared for submission to the OMB in the fall of $Y-2$.

---

* See, for example, Joint Committee on Atomic Energy Hearings, February and March 1971, p. 265.

The Office of Management and Budget and the Office of Science and Technology Concurrent with the formation of budgets within the agencies, the broad lines of policy are being established by the OMB, in consultation with the Office of Science and Technology (OST). In the spring of Y − 2, a general review is conducted by the OMB to estimate political and economic factors likely to influence the budget and to establish the general framework within which the agencies' needs must be accommodated. The OMB will at this time provide target figures while permitting the agencies to work up alternative plans for various levels above and below the targets. By about June or July the agencies will receive planning figures.

An important element of the spring previews is the identification of problem areas for special study during the summer. These may be generalized problems, such as transportation, involving the missions of several agencies, or they may be quite specific, such as the federal support of graduate training or the extent to which science support should be directed to the solution of social problems. The questions to be reviewed may be initiated by the President or may be generated by pressures from various sources inside and outside the government.

The actual studies arising from the spring review will be carried on by the agencies or by *ad hoc* groups of OMB and OST staff, who solicit detailed reports or responses to specific questions from the cognizant agencies. Often special consultant panels are used, and some problems may be referred to the President's Science Advisory Committee (PSAC). At the end of summer, summary reports are written, which may then form the basis for a more informed judgment in making decisions on the budget in the fall.

President's Budget The formal confluence of the agency budgets and the President's science policy occurs in the fall of Y − 2. Hearings or discussion with agency representatives are held by the OMB that embrace not only fiscal and managerial matters but, more importantly, a critical examination of an agency's program content and direction. It is in the process, for example, that the NSF's budget for nuclear physics or the AEC's allocation to a national laboratory must be defended, and it is through this process that the ultimate compromise is hammered out between the agency's technical evaluation of what it ought to do and the President's economic and political evaluation of what it can be allowed to do. Congress may later change these decisions, but the Executive position is solidified here, and the agencies are expected to support this position in their later presentations to Congress.

The final stage in the preparation of the President's budget—after successive levels of review and discussion—is review within the OMB before the Director or his deputy with responsibility for the total budget. The OST participates in this review, as may representatives from other Executive offices, such as the Domestic Council. Agency interests are represented by the OMB examiners, who now assume the role of presenters of the agencies' programs and defenders of decisions made at lower levels in the OMB. In addition to serving as the final review, the operation has the function of anticipating and resolving questions of a policy or political nature and examining individual agency programs in the larger context of the overall government.

The Congress The President's budget for fiscal year Y is submitted to Congress in January $Y - 1$, six months before the fiscal year starts. Generally, each agency's budget will be subjected to hearings before one appropriations committee in each House. In some cases, for example, the AEC and NSF, separate authorization committees are also involved. Preparation for the hearings is often made by the committee staff, who address specific questions to be answered by the agencies and who arrange for formal testimony as required. Again, even rather technical questions may come under discussion, and the outcome represents some kind of balance between the persuasive power of the agency and the political judgment of individual congressmen or staff members. It is possible for even rather solidly backed decisions to be reversed in this process; and the consequent rearrangements may upset many carefully contrived balances within the Executive Branch. In recent years, congressional action has been sufficiently time-consuming that the budget is not formally determined until fall, well into the fiscal year. As a final check on the whole process, the Executive Branch can in some cases withhold funds voted by the Congress, for example, if the situation has changed since the vote, if Congress has imposed a blanket expenditure ceiling, or if unforeseen technical problems arise that raise questions about the validity of continuing a project.

The entire process of establishing and implementing science policy is a multiply connected system, one with many inputs and cross-links. Even though one can identify several discrete points at which decisions appear to have been made in more or less formal actions, such actions are always preceded by extensive informal negotiations and are often followed by later stages at which the decisions may be reversed. There is really no place at which only technical or only economic considerations enter, and there is rarely a point at which one can say, "The decision was made here." The process is very much one of negotiation and consensus, with a strong awareness at every stage that what evolves must preferably be technically sound but certainly must be politically acceptable.

The Policy-Making Arms

The Office of Management and Budget The arbiter and monitor of the President's scientific policy is the OMB, in the Executive Office of the President (see Figure 10.8). Its functions include: preparation of the budget and formulation of the fiscal program of the government, supervision of the administration of the budget, and clearing and coordinating departmental advice on proposed legislation, among other functions.*

Under the Director, the OMB has six program divisions, with a staff of about 200 budget examiners. Each of the divisions has responsibility for monitoring the several agencies in its cognizance and for making recommendations to the OMB Director regarding program levels. The Divisions and their assigned agencies are

1. Economics, Science, Technology: NASA, AEC, NSF, Commerce, Transportation
2. Natural Resources: Interior, Agriculture
3. Human Resources: HEW, HUD, OEO, Labor
4. National Security: DOD
5. International: State, AID, Military Aid
6. General Government: Justice, Treasury

Three Assistant Directors head the six Divisions, reporting to the Deputy Director for the Budget.

By and large, budget examiners are assigned to individual agencies, typically two or three to each agency. Often they are persons with technical background who are expected to familiarize themselves in detail with the workings of the agency, sitting in on the discussions and visiting outside establishments. Their prime responsibility is program analysis, including considerations of cost effectiveness, and these factors presumably weigh strongly in their recommendations to the OMB.

The mechanism of policy making in the OMB has already been discussed above, in connection with the construction of the budget. As noted there, the spring preview and the following summer studies are crucial in defining which issues will be joined in a given year. In addition to these exercises, each Division's program is presented to the directors twice a year. What finally emerges as policy contains many components, including the examiners' recommendations, consultations with other Executive offices— the OST, Domestic Council, and others—and a pragmatic evaluation of congressional attitudes.

---

* *U.S. Government Organization Manual—1971/72* (U.S. Government Printing Office, Washington, D.C., 1971).

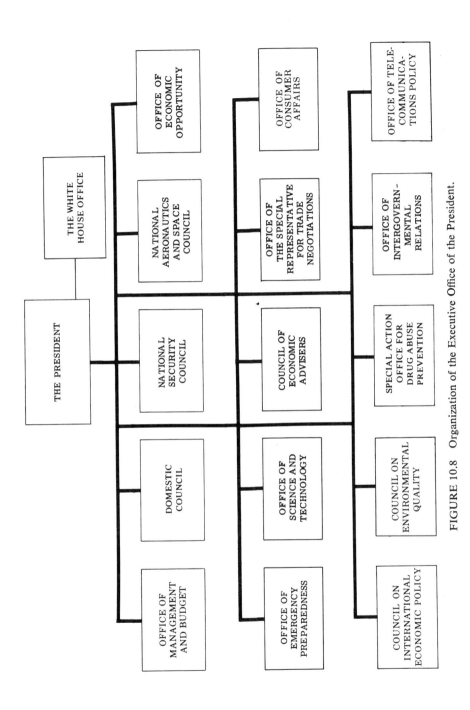

FIGURE 10.8 Organization of the Executive Office of the President.

657

Comment The role and power of the OMB in science policy have increased markedly in the last decade, and its influence is ubiquitous in the agencies, even at low administrative levels. Although the agencies seem to be able to cope with the constraints thus imposed, there is danger that rigid budgetary controls that inhibit transfers of funds as scientific needs change can result in a scientific establishment too ponderous to operate effectively. The same remark applies also to the Congress, which has, for example, chosen to divide the NSF budget into more than a dozen line items, thereby rendering it difficult for that agency to respond to changing conditions and new opportunities.

The Office of Science and Technology The OST is also located in the Executive Office of the President (see Figure 10.8). Its Director provides advice to the President on policies and programs "to assure that science and technology are used most effectively in the interests of national security and general welfare." * Among its functions are the review and coordination of major federal activities in science and technology and evaluation of major policies and programs of the various agencies.

Besides his obligations in the OST, the Director also serves as the President's Science Adviser, as chairman of the President's Science Advisory Committee, and as chairman of the Federal Council for Science and Technology. The OST provides staff support for all of these enterprises. It has a staff of about 24 scientists and makes use of some 200 consultants in OST or PSAC panels.

The OST has, in varying degrees, a close working relationship with the OMB, participating in the budget reviews and in the summer studies or other analyses of particular issues. Typical of recent studies in which the OST has been involved are examinations of future needs for scientists and formulation of policy for graduate student support, the proper role of the NSF in applied research (for example, RANN), and transfer of the Interdisciplinary Development Laboratories and the National Magnet Laboratory from the DOD. In all these questions the OST assists in formulating a policy and takes an active part in the negotiations required to bring it to fruition. The OST's power in the government lies not so much in explicit authority as in its ability to promote a consensus among the agencies and other Executive offices with line responsibilities.

The Federal Council for Science and Technology The FCST is mainly a coordinating organization, consisting of the Science Adviser as chairman and officers of policy rank from eleven departments and agencies, for

---

* *U.S. Government Organizational Manual—1971/72* (U.S. Government Printing Office, Washington, D.C., 1971).

example, assistant secretaries for research and development, Administrator of NASA, Chairman of the AEC, and Director of the NSF. The Council has committees for various specialties, including atmospheric sciences, materials research and development, high-energy physics, scientific and technical information, and others.

Science Policy in the Operating Agencies

Policy making within an operating agency is necessarily a specialized process concerned largely with the technical question of how best to fulfill an agency's mission. While political considerations are by no means absent, the principal requirement is for broad technical expertise.

In every agency, decisions on science policy are made on many levels, to an extent determined at each level by the need and the degree of delegated authority. It seems to be true in most agencies that the delegation is adequately broad and that communication channels work well, in the sense that policy can be effectively implemented, and at the same time, be influenced by recommendations coming up from the technical levels.

The external scientific community exerts its strongest influence on the agencies through submission of proposals for specific research activities. Especially after surviving the refereeing procedure, such proposals serve as clues to the program officers about possibly rewarding lines of investigation for which he should be seeking funds. Naturally, the program officer also follows the scientific literature and bases his determination of what lines to pursue partly upon his assessment of the promising trends thus indicated by the scientific community.

Most agencies rely heavily on scientific advisory bodies of one form or another to assist them in formulating policy. To the extent that such bodies have adequate representation outside the government, they may constitute an important control and input mechanism. Finally, the agencies may commission *ad hoc* groups to advise them on special problems.

The following paragraphs illustrate the mechanics of the decision-making process for a few agencies. Dramatic shifts in policy or budget levels are not generally to be expected; the time for fruition of a research project is usually four or five years, so good management dictates that only a fraction (of the order of one fifth) of a given program is really disposable in any year. Even for that fraction, commitments in facilities and people exert strong (and not necessarily undesirable) inertial forces. Consequently, program management must generally consist in adjusting the system on a several-year time scale, and only rarely can these adjustments be expected to have visible effect in one year. This intrinsic time lag has the salutary effect of requiring constant attention to the long-term prospects of an agency and of a science, but, concomitantly, it imposes real difficulties in an agency's ability to respond quickly to new opportunities.

Atomic Energy Commission

Planning   Internal inputs to the Commission on Research and Development Policy come from the General Manager, the AEC staff, and the General Advisory Committee, a statutory committee of outside scientists appointed by the President. An important element in the Commission's budget process is the internal Budget Review Committee, consisting of several of the Assistant General Managers and the Controller.

Within the Divisions of Physical Research and Controlled Thermonuclear Research (CTR), decisions are largely made by consultations among the staff. Two programs, high-energy physics and controlled thermonuclear research, have standing advisory committees of outside scientists.

The Division Directors participate in the presentations to the Commission, in the OMB budget hearings, and in the hearings of the congressional committees responsible for authorization and appropriations legislation. They have thus rather direct communication to all relevant centers of decision making. Through their direct participation in the budget process, they have considerable influence on the allocations to the several programs in their Divisions. The proposed budget is formulated in terms of the main disciplines of research in the physical sciences supported under the programs; execution of the approved budgets follows the same format.

Extramural Program   A part of the activities of the Divisions of Physical and CTR Research—about $90 million—is carried out off-site, that is, by contracts with universities, industry, or nonprofit institutions.

In selecting projects for support, the AEC places emphasis upon the scientific merit of the proposal, background and experience of the principal investigator, and the facilities and environment of the institution submitting the proposal for the performance of the proposed research. Proposals for AEC assistance are usually initiated by the scientist interested in doing the work and are submitted through an appropriate administrative official of his institution. Occasionally, the AEC may request investigators to undertake research of particular interest to the AEC; however, this is a rarity in the Commission's Physical Research Program.

Proposals on subjects of interest to AEC and submitted by universities normally are subjected to a "peer review process," ranging from relatively informal methods to the use of more formal review committees or boards, depending on the size or complexity of the proposal. Within the AEC the review process leans toward the informal, at least as far as the smaller (e.g., $100,000 or less) proposals are concerned. Whenever an unsolicited proposal from an educational institution is received in the AEC, it is assigned to an AEC staff scientist for review. He will solicit the opinions, usually by an exchange of letters with about three to five other scientists throughout the United States who he has reason to believe are knowledgeable in the field of the proposal. These other scientists may be at universities or may be associated with the large national laboratories, such as Argonne or Brookhaven. They themselves may well be, and in most cases are, supported by the AEC.

The AEC staff scientist considers the comments of the reviewers and then arrives at his judgment as to whether, in his view, the proposal is worthy of support. A tight budgetary situation tends to require a secondary judgment as to which few of the relatively large number of proposals found to be technically sound should be the ones to actually receive funding. This secondary judgment is based largely on consideration of AEC program interest and program balance.

When large or unusual proposals are involved, the AEC will use a more formal approach, i.e., specially appointed standing or *ad hoc* committees, panels, or boards that meet recurrently to discuss the merits of the proposal and arrive at a recommendation.

In all cases, the recommendation made by a responsible technical reviewer or review committee to approve a proposal must be reviewed further, concurred in, and approved by others within the Divisions of Physical Research and CTR before an AEC field office is authorized to negotiate an appropriate contract with the university. Scientific quality, program balance, and levels of funding for physical research are controlled by the Headquarters, Division of Physical Research and CTR. Negotiation, execution and administration of contracts are the responsibility of the managers of the various AEC operations offices, which are located in a number of different places throughout the country.*

National Science Foundation

Planning As originally conceived, the NSF was to be the principal coordinator for science within the federal government and thus might have been a strong force in formulating overall science policy. In the event, the formal coordinating responsibility has been transferred to the OST and the FCST, and the NSF's policy planning is largely restricted to its own mission.

Broad policy questions are determined by the National Science Board, which not only acts on proposals emanating from the Foundation officers but also conducts extensive studies of its own.† Among the pressures that act on the Board are the recommendations of the various Assistant Directors, shifts in complexion of programs of other agencies—as occasioned, for example, by the Mansfield Amendment—and societal pressures of one kind or another, transmitted by the President, the OMB, or the Congress.

Program Decisions The actual research program is strongly influenced by intra-agency pressures, by the zeal of Program Directors, Section Heads,

---

* G. Kolstad, AEC, private communication (1971).
† For example, National Science Foundation, National Science Board, *Toward a Public Policy for Graduate Education in the Sciences*, NSB 69-1 (U.S. Government Printing Office, Washington, D.C., 1969).

and Division Directors in putting forward attractive research proposals. Thus ultimately, in some measure, the program of the Foundation is directly sensitive to the interests of the scientists who come to it for funds.

The Foundation supports research in response to unsolicited proposals, initiated by the investigators and submitted by their universities on their behalf. Proposals are reviewed by four to six scientists in a field, selected by the appropriate Program Director; they are asked to comment on the scientific interest of the proposal, the competence of the investigator, and the promise of the proposed method of investigation. Over the past five years, some 2500 individuals have reviewed proposals for Physics Program Directors in the NSF, some of them many times. Because many projects are regularly renewed, only about 10 percent of the total funds are likely to be allocated for new proposals in any given year.

The Physics Advisory Panel, a body of ten outside physicists, meets twice a year to review actions of the Program Directors and to advise on policy.

Department of Defense

Planning Overall research and development policy for the DOD is made by the Director of Defense Research and Engineering (DDRE). He is assisted by a considerable technical staff, by the Advanced Research Projects Agency (ARPA), the Weapons Systems Evaluation Group (WSEG), and the Defense Science Board, a statutory body. The latter is the senior advisory group to DDRE and consists of 24 members-at-large, appointed by the Secretary of Defense, plus the chairmen of the Army Scientific Advisory Panel, the Naval Research Advisory Committee, and the Air Force Scientific Advisory Board. The Board meets three times each year and conducts its studies through *ad hoc* task groups.

Air Force

Planning The senior advisory group is the Scientific Advisory Board (SAB), advising the Secretary and the Chief of Staff. Like the Defense Science Board, SAB is statutory and is composed largely of civilian scientists. The Director of Laboratories (see Figure 10.A.4) has the Scientific Advisory Group (SAG) on the research side; SAG also acts as a Board of Visitors to monitor activities of the various research and development laboratory components, including the Air Force Office of Scientific Research (AFOSR). Within AFOSR, the Directorates for various disciplines have Research Evaluation Groups (REG's) available to them, for both policy advice and evaluation of proposals. In the Physics Directorate, the REG operation is contracted to a university.

Budgets   Budgets are fixed in principle by DDRE, but at each administrative level below, there is an understood authority to reprogram about 5 percent to meet changing situations.

Extramural Program   The AFOSR operates on the basis of unsolicited proposals. As with other agencies, continuity of program is important; since the average project takes four to five years to develop, only 20–25 percent of the budget is disposable in any year.   Criteria for selection are

1. Relevance to Air Force mission;
2. Scientific merit;
3. Capability of investigator;
4. Adequacy of facilities;
5. Cost considerations.

Proposals are mainly evaluated in-house with help from the REG's as needed.

Army

Planning   A civilian panel, the Army Scientific Advisory Panel (ASAP) advises the Assistant Secretary for Research and Development and the Chief of Research and Development.   The Army Research Office (Durham) [ARO(D)] has a separate advisory group consisting of scientists from Army laboratories plus a few civilian scientists.

Extramural Program—Proposals to the ARO (Durham) are generally reviewed by independent scientists nominated for each proposal by an NAS–NRC committee advisory to the ARO(D).   They are also referred to various of the in-house laboratories for evaluation of Army interest.   Of the some 650 projects in progress in 1969, 153 were dropped in that year and 120 new ones accepted.

Navy

Planning   Long-range planning of Navy mission requirements is carried out by the Chief of Naval Operations.   From these plans are generated research and development requirements (partially by the ONR) that are translated into research programs by the ONR.   From the opposite direction, the ONR scientific staff also has responsibility for recognizing new scientific developments that may bear on the Navy's mission at some time in the future.   As compared with the early days of the ONR, the test of relevance is now applied with greater stringency.

The Naval Research Advisory Committee (NRAC), which reports to the Secretary, is the statutory civilian advisory group.

Extramural Program   About two thirds of the ONR's budget is spent on the extramural program in universities and industry. The ONR has a comparatively large scientific staff—about 190 persons—and reviews most proposals in-house.

National Aeronautics and Space Administration

Planning   The Associate Administrator and his Deputy (Planning) (see Figure 10.A.7) are responsible for developing NASA's long-range plans. In this work, they have the guidance of the President's Science Advisory Committee (PSAC), the Space Science Board (SSB), numerous internal and external advisory committees, and the results of technical studies conducted both in-house and under contract to NASA.

The SSB is an organ of the National Academy of Sciences–National Research Council, advisory to NASA's Administrator. In practice, the principal point of contact is with the Office of Space Science and Applications (OSSA).* As its name implies, the SSB is concerned with research opportunities and priorities in the space program.†

The National Aeronautics and Space Council is the President's committee to advise him on policies, plans, and programs and to coordinate the activities of various agencies in the field. It is composed of the Vice President, Secretary of State, Secretary of Defense, Administrator of NASA, and Chairman of the AEC.

A major NASA in-house advisory body has recently been formed, the Space Program Advisory Council (SPAC), which has Committees on Applications, Physical Sciences, Life Sciences, and Space Systems. Formally, SPAC reports to the Deputy Administrator, but it and its committees will have working relationships with the Associate Administrators. Appointments are made by NASA, but not over one fourth of the members may be NASA employees.

NASA projects are typically long range, of five to ten years' duration. The planning process is therefore an especially demanding exercise, requiring numerous detailed feasibility studies and continued iteration as the technology develops. The scientific community has a number of input points, through proposal pressure and the various advisory committees,

---

* As of December 3, 1971, OSSA was split into the Office of Space Science (OSS) and the Office of Application (OA).
† See, for example, Space Science Board, *Priorities for Space Research 1971–1980* (National Academy of Sciences, Washington, D.C., 1971).

but, as remarked earlier, the ultimate decisions must have a strong political ingredient if spending money on this scale is to continue to command popular support.

Budgets   As with all federal agencies, the production of the annual budget represents a major effort, extending over half the year or more. In NASA, budget submissions are made by the various field installations in accordance with general guidelines issued by the program administrators. In subsequent treatment at headquarters, candidate new starts are narrowed down and various options reviewed at successively higher levels within NASA. Staff members of the OMB are briefed two or three times a year by Division Directors, and they participate in the later stages of the budget discussion.

Extramural Program   Of a total basic research budget * of $637 million, NASA spends $238 million in-house, $260 million in industry, $61 million in universities, and $69 million in federally funded research and development centers administered by universities.† The OSSA spends about $24 million of its Supporting Research and Technology budget at academic institutions. Division of the funds among various scientific areas depends on the pertinence of the work to NASA's future program, the quality of the work, possible applications, and continuity of the overall research program.

The agency announces to interested scientists "Opportunities for Participation in Space Flight" (AFO). Experiments for space missions proposed in response to an AFO are selected by the Space Science and Applications Steering Committee (SSASC), advisory to the Associate Administrator for OSSA. Proposals may be from the scientific community either within or outside of NASA. They are reviewed by the OSSA, with advice of a subcommittee of the SSASC containing both NASA and outside scientists. Final determination is made by the Associate Director, with advice from the SSASC. Approximately 30–40 percent of scientific flight experiments are performed by in-house groups, the remainder by industrial and academic groups.

A majority of research grants to universities result from unsolicited proposals for work on specific research problems. Proposals are largely reviewed in-house and may be funded either through the Headquarters or Field Offices. In the evaluation, considerable stress is placed on achieving

---

* NASA includes in its basic research budget apportioned charges for launch vehicles, launch operations, spacecraft, and tracking and data acquisition.
† Obligations, fiscal year 1970; *Federal Funds for Research, Development, and Other Scientific Activities,* Volume XX (U.S. Government Printing Office, Washington, D.C. 1971).

a balance among various areas of support and maximizing the relevance to the overall mission. University programs are monitored by the Office of University Affairs.

National Institutes of Health

Planning A Program and Budget Review Committee reporting to the Director of NIH holds periodic hearings on annual budgets and five-year projections. The actual appropriations may be strongly influenced by the external constituencies of individual institutes—for example, the American Heart Association or American Cancer Society—as well as by congressional pressures. By law, each Institute has an Advisory Council, consisting of twelve people, eight professionals and four lay members, appointed by the Secretary of Health, Education, and Welfare for four years. The Advisory Councils have particular importance in the extramural program—all grants must be approved by them.

Extramural Program Over 30,000 grant applications, including renewals, are addressed each year to the NIH. Such proposals are handled in the first instance by the Division of Research Grants, which determines which Institute is likely to support them. Technical evaluation is carried out by one of 50 study sections, each consisting of 5–15 outside experts, who meet three times a year to assign priorities on the basis of scientific quality. About half of the proposals are eliminated at this stage, and the remainder go to project directors at the respective Institutes for further evaluation—particularly as to mission priority—and final recommendation to the Advisory Council. Approved projects are generally funded for one year at a time, with an informal commitment to extend for several years. In 1969, there were some 12,000 active research grants in just under a thousand institutions, 200 of which were foreign. Because of the policy of automatic renewal, only approximately one fourth of the projects in any given year will be in competition.

APPENDIX 10.A. AGENCY MISSIONS

Both history and logic point to the federal government as the principal sponsor of scientific research at it applies to national hopes and needs that depend on the acquisition of new knowledge. To this end some agencies, notably the NSF, have a broad charter to support both basic and applied research, without the restraint of explicit relevance to a stated mission other than advancement of the common weal. Other agencies, having more restrictive missions, find nonetheless that research has an essential role in the pursuit of their objectives, and so these agencies, too, comprise a significant source of support, as indicated in Figure 10.3.

The historical development of science support by various agencies represents a pattern that is by no means either accidental or without logical foundation. Agencies concentrate their support in areas most likely to be relevant to their individual missions. At the same time, very-long-term benefits must be kept in mind, and each agency needs to maintain some involvement with every field of science that bears on its mission.

The following sections outline the current involvement in physics of various agencies that conduct physics research through in-house laboratories, by contracts and grants, or by a combination of these. The outline is far from exhaustive, but the examples chosen illustrate a considerable range in degree of involvement. Tables 10.A.1 through 10.A.4 list the various departments, agencies, and units reporting basic or applied research in physics, together with their operating funding levels (obligations, exclusive of plant) for fiscal year 1970. The following points should be noted:

1. Total federal obligations for basic research are about $2.1 billion, for basic plus applied research about $5.6 billion. (Total federal research and development operations amount to about $15.3 billion.)

2. Of the total research budget, physics comprises about 10 percent; of the basic research budget, 18 percent.

3. Three agencies, the AEC, DOD, and NASA, account for 91 percent of the basic plus applied physics research budget. Two of these, the AEC and NASA, make up 76 percent of the basic physics budget, with the DOD plus the NSF accounting for another 19 percent.

4. Almost all departments have *some* component of physics research, albeit in some cases a rather small one.

Table 10.A.5 is a condensed listing showing federal research and development obligations to universities and colleges for the period 1960–1971, by agency involved.

The Atomic Energy Commission

Mission The AEC is an independent agency responsible to the President and the Congress. It was established by the Atomic Energy Act of 1946 to assume the responsibility for the development, use, and control of atomic energy and for the production of nuclear weapons. In 1954, the functions and responsibilities of the AEC were expanded to provide for greater emphasis on developing and promoting peaceful uses of atomic energy.

TABLE 10.A.1 Federal Obligations for Research, by Agency and Field of Science, Fiscal Year 1970 ($Thousands)

Agency and Subdivision	Total	Life Sciences	Psychol-ogy	Physical Sciences	Environ-mental Sciences	Mathe-matics	Engi-neering	Social Sciences	Other Sciences
Total, all agencies	5,601,906	1,533,432	113,328	1,010,450	586,631	102,138	1,967,739	215,852	72,336
DEPARTMENTS									
Department of Agriculture, total	271,925	166,681	32	36,908	4,838	686	28,909	33,663	208
Agricultural Research Service	152,468	98,631	32	30,757	859	332	20,707	1,150	—
Cooperative State Research Service	61,052	41,983	—	3,053	350	—	2,092	13,574	—
Economic Research Service	14,904	—	—	—	—	—	—	14,904	—
Farmer Cooperative Service	884	—	—	—	—	—	—	884	—
Forest Service	41,933	26,067	—	3,098	3,629	145	6,110	2,884	208
National Agricultural Library	208	—	—	—	—	—	—	—	208
Statistical Reporting Service	476	—	—	—	—	209	—	267	—
Department of Commerce, total	94,114	24,157	1,096	17,156	22,090	1,347	15,689	10,826	1,753
Bureau of the Census	1,665	—	567	—	—	266	—	832	—
Economic Development Adminis-tration	3,809	—	—	—	—	—	—	3,809	—
Maritime Administration	4,878	—	—	—	—	89	3,268	1,521	—
National Bureau of Standards	27,055	—	—	15,779	—	992	9,456	—	828
National Oceanic and Atmospheric Administration	52,915	24,157	529	1,377	22,090	—	2,965	1,159	638
Office of Business Economics	3,505	—	—	—	—	—	—	3,505	—
Patent Office	287	—	—	—	—	—	—	—	287
Department of Defense, total	1,556,783	104,003	25,594	219,639	128,253	61,899	972,775	4,693	39,927

668

Department of the Army	291,536	52,863	6,942	64,021	13,941	5,620	129,057	2,767	16,325
Department of the Navy	272,922	15,595	6,257	62,509	46,895	29,026	112,640	—	—
Department of the Air Force	665,000	26,707	6,384	31,536	25,800	6,108	568,265	200	—
Defense agencies	323,234	8,218	5,728	60,892	41,617	20,715	161,760	1,400	22,904
Departmentwide funds	4,091	620	283	681	—	430	1,053	326	698
Department of Health, Education, & Welfare, total	1,129,271	897,108	60,484	41,283	630	8,170	22,667	84,368	14,561
Food and Drug Administration	21,805	15,492	—	6,313	—	—	—	—	—
Health Services & Mental Health Administration	160,358	99,050	33,971	734	—	349	87	25,526	641
National Institutes of Health	868,788	770,946	17,715	34,236	630	7,821	22,435	1,085	13,920
Office of Child Development	1,000	500	—	—	—	—	—	500	—
Office of Education	33,066	—	6,825	—	—	—	—	26,241	—
Social and Rehabilitation Service	31,885	11,120	1,973	—	—	—	145	18,647	—
Social Security Administration	12,369	—	—	—	—	—	—	12,369	—
Department of Housing & Urban Development	7,097	—	—	—	—	—	7,097	—	—
Department of the Interior, total	121,692	20,716	58	13,952	43,570	1,816	39,892	1,494	194
Bonneville Power Administration	524	—	—	25	31	20	448	—	—
Bureau of Land Management	710	650	—	—	—	—	—	60	—
Bureau of Mines	33,894	—	—	7,462	1,821	571	24,040	92	—
Bureau of Outdoor Recreation	92	—	—	—	—	—	—	73	—
Bureau of Reclamation	7,202	25	—	—	4,810	—	2,294	—	—
Bureau of Sport Fisheries & Wildlife	18,041	18,041	—	—	—	—	—	—	—
Geological Survey	40,963	—	—	3,891	34,748	950	1,374	—	—
Office of Coal Research	2,341	—	—	—	—	—	2,341	—	—
Office of Saline Water	7,175	—	—	2,174	—	—	5,001	—	—
Office of Water Resources Research	10,750	2,000	58	400	2,160	275	4,394	1,269	194

TABLE 10.A.1—*Continued*

Agency and Subdivision	Total	Life Sciences	Psychology	Physical Sciences	Environmental Sciences	Mathematics	Engineering	Social Sciences	Other Sciences
Department of Justice, total	5,334	326	1,239	630	—	240	363	2,536	—
Bureau of Narcotics and Dangerous Drugs	839	326	114	—	—	—	—	399	—
Law Enforcement Assistance Administration	4,495	—	1,125	630	—	240	363	2,137	—
Department of Labor, total	11,286	—	—	—	—	—	—	11,286	—
Bureau of Labor Statistics	2,566	—	—	—	—	—	—	2,566	—
Labor-Management Services Administration	290	—	—	—	—	—	—	290	—
Manpower Administration	6,214	—	—	—	—	—	—	6,214	—
Workplace Standards Administration	2,216	—	—	—	—	—	—	2,216	—
Postal Service	4,577	—	—	—	—	—	—	—	4,577
Department of State, total	18,895	14,421	143	—	—	—	1,392	2,939	—
Departmental funds	125	—	—	—	—	—	—	125	—
Agency for International Development	18,627	14,421	—	—	—	—	1,392	2,914	—
Peace Corps	143	—	143	—	—	—	—	—	—
Department of Transportation, total	88,263	3,145	3,701	32	15,379	2,419	51,313	11,549	725

Agency	1	2	3	4	5	6	7	8	9
Federal Aviation Administration	9,960	1,484	395	—	400	350	7,331	—	—
Federal Highway Administration	28,608	—	—	—	—	—	27,177	1,431	—
Federal Railroad Administration	4,743	—	—	—	—	400	4,020	323	—
National Highway Traffic Safety Administration	15,892	1,651	3,286	—	—	1,554	8,293	1,108	—
Office of the Secretary	8,956	10	—	—	249	—	1,902	6,805	—
United States Coast Guard	13,881	—	—	32	13,244	—	595	—	—
Urban Mass Transportation Administration	6,223	—	20	—	1,486	115	1,995	1,582	725
Department of the Treasury, total	286	—	—	270	—	—	16	—	—
Bureau of Engraving and Printing	286	—	—	270	—	—	16	—	—
OTHER AGENCIES									
Advisory Comm. on Intergovt. Relations	553	—	—	—	—	—	—	553	553
Atomic Energy Commission	433,117	80,719	—	300,278	9,462	5,779	36,879	—	—
Civil Aeronautics Board	263	—	—	—	—	—	—	263	263
Civil Service Commission	279	279	—	—	—	—	—	—	—
Environmental Protection Agency	52,984	19,311	8	8,430	4,801	193	19,718	523	523
Federal Communications Commission	1,015	—	—	—	—	—	1,015	390	390
Federal Home Loan Bank Board	390	—	—	—	—	—	—	388	388
Federal Trade Commission	388	—	—	—	—	—	—	—	21
General Services Administration	215	—	—	—	—	—	194	—	968
Library of Congress	968	—	—	—	—	—	—	968	968
National Aeronautics & Space Administration	1,409,732	78,229	3,664	290,271	297,629	2,350	736,359	1,230	8,734
National Science Foundation	274,789	60,119	7,720	76,276	59,270	17,161	25,927	19,582	19,582
Office of Economic Opportunity	29,400	2,600	5,000	—	—	—	531	21,800	21,800
Office of Emergency Preparedness	1,016	—	—	237	83	18	279	18	485
Office of Science and Technology	929	284	10	—	—	—	279	18	—

TABLE 10.A.1—*Continued*

Agency and Subdivision	Total	Life Sciences	Psychology	Physical Sciences	Environmental Sciences	Mathematics	Engineering	Social Sciences	Other Sciences
Small Business Administration	101	—	—	—	—	—	—	101	—
Smithsonian Institution	18,217	7,709	—	3,195	586	—	—	6,727	—
Tennessee Valley Authority	6,376	1,727	—	890	—	—	3,406	353	—
U.S. Arms Control & Disarmament Agency	3,732	—	—	303	40	—	2,811	395	183
United States Information Agency	157	—	—	—	—	—	157	—	—
Veterans Administration	57,762	52,177	4,300	700	—	60	350	175	—

672

TABLE 10.A.2 Federal Obligations for Research in Psychology and Physical Sciences, by Agency and Detailed Field of Science, Fiscal Year 1970 ($Thousands)

Agency and Subdivision	Total	Psychology			Total	Physical Sciences			
		Biological Aspects	Social Aspects	Psychological Sciences		Astronomy	Chemistry	Physics	Physical Sciences
Total, all agencies	113,328	45,141	64,175	4,012	1,010,450	210,950	243,894	538,333	17,273
DEPARTMENTS									
Department of Agriculture, total	32	—	32	—	36,908	—	30,186	6,722	—
Agricultural Research Service	32	—	32	—	30,757	—	25,439	5,318	—
Cooperative State Research Service	—	—	—	—	3,053	—	3,053	—	—
Forest Service	—	—	—	—	3,098	—	1,694	1,404	—
Department of Commerce, total	1,096	529	567	—	17,156	823	7,402	8,598	333
Bureau of the Census	567	—	567	—	—	—	—	—	—
National Bureau of Standards	—	—	—	—	15,779	823	6,025	8,598	333
National Oceanic and Atmospheric Administration	529	529	—	—	1,377	—	1,377	—	—
Department of Defense, total	25,594	5,885	18,056	1,653	219,639	10,591	39,962	154,034	15,052
Department of the Army	6,942	3,209	3,276	457	64,021	8	26,897	30,259	6,857
Department of the Navy	6,257	1,040	4,850	367	62,509	4,040	5,779	49,651	3,039
Department of the Air Force	6,384	538	5,846	—	31,536	6,397	6,109	19,030	—
Defense Agencies	5,728	1,063	3,836	829	60,892	146	1,157	54,433	5,156
Departmentwide funds	283	35	248	—	681	—	20	661	—
Department of Health, Education & Welfare, total	60,484	26,295	33,790	399	41,283	—	40,725	94	464

TABLE 10.A.2—Continued

Agency and Subdivision	Psychology				Physical Sciences				
	Total	Biological Aspects	Social Aspects	Psychological Sciences Total	Total	Astronomy	Chemistry	Physics	Physical Sciences
Food and Drug Administration	—	—	—	—	6,313	—	6,313	—	—
Health Services & Mental Health Administration	33,971	15,727	18,146	98	734	—	734	—	—
National Institutes of Health	17,715	10,568	7,147	—	34,236	—	33,678	94	464
Office of Education	6,825	—	6,524	301	—	—	—	—	—
Social and Rehabilitation Service	1,973	—	1,973	—	—	—	—	—	—
Department of the Interior, total	58	—	58	—	13,952	—	12,148	1,804	—
Bonneville Power Administration	—	—	—	—	25	—	—	25	—
Bureau of Mines	—	—	—	—	7,462	—	6,642	820	—
Geological Survey	—	—	—	—	3,891	—	2,941	950	—
Office of Saline Water	—	—	—	—	2,174	—	2,174	—	—
Office of Water Resources Research	58	—	58	—	400	—	391	9	—
Department of Justice, total	1,239	—	1,239	—	630	—	—	630	—
Bureau of Narcotics and Dangerous Drugs	114	—	114	—	—	—	—	—	—
Law Enforcement Assistance Administration	1,125	—	1,125	—	630	—	—	630	—
Department of State, total	143	—	143	—	—	—	—	—	—
Peace Corps	143	—	143	—	—	—	—	—	—

Department of Transportation, total	3,701	395	3,306	—	32	32	—	—	—
Federal Aviation Administration	395	395	—	—	—	—	—	—	—
National Highway Traffic Safety Administration	—	—	—	—	32	32	—	—	—
United States Coast Guard	3,286	—	3,286	—	—	—	—	—	—
Urban Mass Transportation Administration	20	—	20	—	—	—	—	—	—
Department of the Treasury, total	—	—	—	—	270	—	70	40	160
Bureau of Engraving and Printing	—	—	—	—	270	—	70	40	160
OTHER AGENCIES									
Atomic Energy Commission	—	—	—	—	300,278	—	64,218	236,060	—
Civil Service Commission	279	—	279	—	—	—	—	—	—
Environmental Protection Agency	8	8	—	—	8,430	—	5,421	2,073	936
National Aeronautics & Space Administration	3,664	3,664	—	—	290,271	176,404	19,843	94,024	—
National Science Foundation	7,720	4,300	1,460	1,960	76,276	19,922	22,241	33,785	328
Office of Economic Opportunity	5,000	—	5,000	—	237	15	75	147	—
Office of Science and Technology	10	5	5	—	3,195	3,195	—	—	—
Smithsonian Institution	—	—	—	—	890	—	890	—	—
Tennessee Valley Authority	—	—	—	—	—	—	—	—	—
U.S. Arms Control & Disarmament Agency	—	—	—	—	303	—	303	—	—
Veterans Administration	4,300	4,060	240	—	700	—	410	290	—

TABLE 10.A.3 Federal Obligations for Basic Research, by Agency and Field of Science, Fiscal Year 1970 ($Thousands)

Agency and Subdivision	Total	Life Sciences	Psychology	Physical Sciences	Environmental Sciences	Mathematics	Engineering	Social Sciences	Other Sciences
Total, all agencies	2,062,256	576,041	57,237	704,453	352,380	58,906	242,641	66,118	4,480
DEPARTMENTS									
Department of Agriculture, total	115,709	75,667	5	17,796	1,573	307	10,292	10,069	—
Agricultural Research Service	73,342	49,529	5	15,181	276	99	8,042	210	—
Cooperative State Research Service	23,200	15,954	—	1,160	133	—	795	5,158	—
Economic Research Service	4,322	—	—	—	—	—	—	4,322	—
Farmer Cooperative Service	221	—	—	—	—	—	—	221	—
Forest Service	14,548	10,184	—	1,455	1,164	145	1,455	145	—
Statistical Reporting Service	76	—	—	—	—	63	—	13	—
Department of Commerce, total	39,201	6,943	319	11,510	11,571	420	3,887	4,367	184
Bureau of the Census	308	—	109	—	—	178	—	21	—
National Bureau of Standards	14,350	—	—	11,089	—	242	2,835	—	184
National Oceanic and Atmospheric Administration	21,038	6,943	210	421	11,571	—	1,052	841	—
Office of Business Economics	3,505	—	—	—	—	—	—	3,505	—
Department of Defense, total	246,670	28,886	10,482	68,411	49,329	28,994	58,753	1,815	—
Department of the Army	45,451	16,923	878	11,181	4,379	1,925	9,552	613	—
Department of the Navy	81,944	9,763	3,946	24,018	27,081	5,728	11,408	—	—
Department of the Air Force	81,000	2,200	759	28,602	17,869	5,634	25,736	200	—
Defense agencies	38,275	—	4,899	4,610	—	15,707	12,057	1,002	—

676

Department of Health, Education, & Welfare, total	387,940	300,257	35,380	23,957	165	3,182	2,287	22,528	184
Health Services & Mental Health Administration	60,340	23,142	20,674	734	—	—	—	15,790	—
National Institutes of Health	319,385	277,115	12,406	23,223	165	3,182	2,287	823	184
Office of Education	7,812	—	2,300	—	—	—	—	5,512	—
Social Security Administration	403	—	—	—	—	—	—	403	—
Department of the Interior, total	49,923	8,886	—	8,399	29,183	751	2,313	350	41
Bureau of Land Management	15	15	—	—	—	—	—	—	—
Bureau of Mines	6,248	—	—	4,120	173	120	1,835	—	—
Bureau of Reclamation	82	12	—	—	—	—	39	31	—
Bureau of Sport Fisheries & Wildlife	8,459	8,459	—	—	—	—	—	—	—
Geological Survey	30,667	—	—	1,945	28,146	576	—	—	—
Office of Saline Water	2,174	—	—	2,174	—	—	—	—	—
Office of Water Resources Research	2,278	400	—	160	864	55	439	319	41
Department of Labor, total	2,652	—	—	—	—	—	—	2,052	—
Bureau of Labor Statistics	1,422	—	—	—	—	—	—	1,422	—
Labor-Management Services Administration	290	—	—	—	—	—	—	290	—
Manpower Administration	340	—	—	—	—	—	—	340	—
Department of Transportation, total	21,099	—	10	—	11,130	55	8,797	1,107	—
Federal Highway Administration	7,390	—	—	—	—	—	7,020	370	—
National Highway Traffic Safety Administration	1,127	—	—	—	—	—	1,127	—	—
United States Coast Guard	11,080	—	—	—	11,080	—	—	—	—

TABLE 10.A.3—*Continued*

Agency and Subdivision	Total	Life Sciences	Psychology	Physical Sciences	Environmental Sciences	Mathematics	Engineering	Social Sciences	Other Sciences
Urban Mass Transportation Administration	1,502	—	10	—	50	55	650	737	—
OTHER AGENCIES									
Atomic Energy Commission	286,669	37,599	—	230,966	—	5,779	12,325	—	—
Civil Service Commission	83	—	83	—	—	—	—	—	—
Environmental Protection Agency	5,450	2,254	—	2,501	561	—	132	2	—
National Aeronautics & Space Administration	637,034	50,509	2,854	262,874	195,218	2,350	121,999	1,230	—
National Science Foundation	244,977	52,662	7,600	74,649	53,031	17,061	21,669	14,234	4,071
Office of Economic Opportunity	1,600	—	—	—	—	—	—	1,600	—
Office of Science and Technology	372	114	4	95	33	7	112	7	—
Smithsonian Institution	18,217	7,709	—	3,195	586	—	—	6,727	—
Veterans Administration	5,260	4,555	500	100	—	—	75	30	—

TABLE 10.A.4 Federal Obligations for Basic Research in Psychology and Physical Sciences, by Agency and Detailed Field of Science, Fiscal Year 1970 ($Thousands)

Agency and Subdivision	Psychology				Physical Sciences				
	Total	Biological Aspects	Social Aspects	Psychological Sciences	Total	Astronomy	Chemistry	Physics	Physical Sciences
Total, all agencies	57,237	29,536	24,721	2,980	704,453	203,882	143,106	352,954	4,511
DEPARTMENTS									
Department of Agriculture, total	5	—	5	—	17,796	—	15,306	2,490	—
Agricultural Research Service	5	—	5	—	15,181	—	13,273	1,908	—
Cooperative State Research Service	—	—	—	—	1,160	—	1,160	—	—
Forest Service	—	—	—	—	1,455	—	873	582	—
Department of Commerce, total	319	210	109	—	11,510	823	4,114	6,246	327
Bureau of the Census	109	—	109	—	—	—	—	—	—
National Bureau of Standards	—	—	—	—	11,089	823	3,693	6,246	327
National Oceanic and Atmospheric Administration	210	210	—	—	421	—	421	—	—
Department of Defense, total	10,482	2,461	7,197	824	68,411	9,578	11,884	42,891	4,058
Department of the Army	878	295	126	457	11,181	—	4,012	6,972	197
Department of the Navy	3,946	1,040	2,539	367	24,018	3,796	2,235	17,987	—
Department of the Air Force	759	63	696	—	28,602	5,782	4,888	17,932	—
Defense agencies	4,899	1,063	3,836	—	4,610	—	749	—	3,861
Department of Health, Education, & Welfare, total	35,380	19,339	15,845	196	23,957	—	23,876	49	32

TABLE 10.A.4—Continued

Agency and Subdivision	Psychology				Physical Sciences				
	Total	Biological Aspects	Social Aspects	Psychological Sciences	Total	Astronomy	Chemistry	Physics	Physical Sciences
Health Services & Mental Health Administration	20,674	11,607	9,067	—	734	—	734	—	—
National Institutes of Health	12,406	7,732	4,674	—	23,223	—	23,142	49	32
Office of Education	2,300	—	2,104	196	—	—	—	—	—
Department of the Interior, total	—	—	—	—	8,399	—	7,220	1,179	—
Bureau of Mines	—	—	—	—	4,120	—	3,521	599	—
Geological Survey	—	—	—	—	1,945	—	1,369	576	—
Office of Saline Water	—	—	—	—	2,174	—	2,174	—	—
Office of Water Resources Research	—	—	—	—	160	—	156	4	—
Department of Transportation, total	10	—	10	—	—	—	—	—	—
Urban Mass Transportation Administration	10	—	10	—	—	—	—	—	—
OTHER AGENCIES									
Atomic Energy Commission	—	—	—	—	230,966	—	39,016	191,950	—
Civil Service Commission	83	—	83	—	—	—	—	—	—
Environmental Protection Agency	—	—	—	—	2,501	—	1,572	929	—
National Aeronautics & Space Administration	2,854	2,854	—	—	262,874	170,358	18,287	74,229	—
National Science Foundation	7,600	4,210	1,430	1,960	74,649	19,922	21,741	32,892	94
Office of Science and Technology	4	2	2	—	95	6	30	59	—
Smithsonian Institution	—	—	—	—	3,195	3,195	—	—	—
Veterans Administration	500	460	40	—	100	—	60	40	—

TABLE 10.A.5 Federal Research and Development Obligations to Universities and Colleges 1960–1971, by Agency ($Millions) [a]

Agency	1960	1961	1962	1963	1964	1965	1966	1967	1968	1969	1970	1971 (est.)
TOTAL, ALL AGENCIES	458.6	584.9	755.0	899.5	1077.3	1193.8	1350.5	1454.5	1490.3	1525.8	1472.9	1616.6
HEW	157.8	220.7	309.8	350.4	418.6	472.6	534.4	619.8	671.3	695.0	646.6	768.6
DOD	154.5	191.1	200.2	237.5	292.0	291.0	295.3	279.9	244.4	252.8	213.5	195.1
NSF	59.9	68.4	89.3	115.3	127.4	142.1	187.4	207.8	221.0	212.6	228.0	251.8
NASA	10.4	17.7	53.5	78.2	105.6	124.1	133.2	124.1	130.6	125.1	131.2	125.0
AEC	39.3	49.4	54.9	67.4	69.4	74.4	82.2	89.5	92.6	101.4	100.3	94.9
Dept. of Agriculture	31.6	32.9	38.6	40.6	48.6	58.4	61.6	64.2	61.1	61.5	64.8	68.0
OEO	—	—				7.1	20.9	15.5	13.8	25.4	20.0	22.4
Dept. of the Interior	1.5	2.2	2.8	3.7	5.5	9.8	19.7	23.4	25.7	23.8	18.5	23.1
DOT	0.3	0.7	0.7	0.8	0.6	0.4	1.4	11.1	11.9	12.8	10.9	10.5
AID	0.1	0.3	2.7	1.3	3.9	6.0	3.7	3.2	3.5	5.0	8.2	17.4
Dept. of Labor			0.1	0.5	1.1	2.2	3.1	3.0	3.2	3.0	3.6	4.4
HUD	—	—					0.4	2.5	1.1	1.0	0.5	1.2
Dept. of Commerce	0.9	0.8	1.6	2.8	2.9	3.7	3.7	6.0	7.4	1.8	4.5	6.3
All other	2.3	0.7	0.8	1.0	1.7	2.0	3.5	3.5	2.7	4.6	22.3 [b]	21.9 [b]

[a] Exclusive of federally funded centers. See NSF series on *Federal Funds for Research, Development, and Other Scientific Activities.*
[b] Environmental Protection Agency, $18.2 million in 1970, $17.9 million in 1971.

Chapter 4, "Research," of the Atomic Energy Act states in part:

Sec. 31. RESEARCH ASSISTANCE.

a. The Commission is directed to exercise its powers in such manner as to insure the continued conduct of research and development and training activities in the fields specified below, by private or public institutions or persons, and to assist in the acquisition of an ever-expanding fund of theoretical and practical knowledge in such fields. To this end the Commission is authorized and directed to make arrangements (including contracts, agreements, and loans) for the conduct of research and development activities relating to

(1) nuclear processes;

(2) the theory and production of atomic energy, including processes, materials, and devices related to such production;

(3) utilization of special nuclear material and radioactive material for medical, biological, agricultural, health, or military purposes;

(4) utilization of special nuclear material, atomic energy, and radioactive material and processes entailed in the utilization or production of atomic energy or such material for all other purposes, including industrial uses, the generation of usable energy, and the demonstration of the practical value of utilization or production facilities for industrial or commercial purposes; and

(5) the protection of health and the promotion of safety during research and production activities.

b. The Commission is further authorized to make grants and contributions to the cost of construction and operation of reactors and other facilities and other equipment to colleges, universities, hospitals, and eleemosynary or charitable institutions for the conduct of educational and training activities relating to the fields in subsection a.

Sec. 32. RESEARCH BY THE COMMISSION—The Commission is authorized and directed to conduct, through its own facilities, activities and studies of the type specified in Section 31.

Sec. 33. RESEARCH FOR OTHERS—Where the Commission finds private facilities or laboratories are inadequate to the purpose, it is authorized to conduct for other persons, through its own facilities, such of those activities and studies of the types specified in Section 31 as it deems appropriate to the development of atomic energy. The Commission is authorized to determine and make such charges as in its discretion may be desirable for the conduct of such activities and studies.

A significant enlargement of the AEC's scope was introduced in 1971 by amendment to the Atomic Energy Act. To Section 31, partially quoted above, is added *:

(6) the preservation and enhancement of a viable environment by developing more efficient methods to meet the Nation's energy needs.

In Section 33, the word "atomic" is deleted in the reference to "development of atomic energy."

These changes would appear to give the AEC a broad mandate for

---

* Authorizing Appropriations for the AEC for FY 1972, Report of the Joint Committee on Atomic Energy, June 3, 1971.

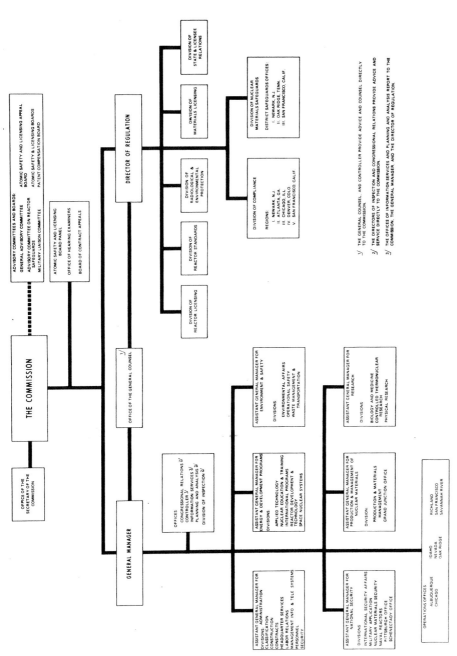

FIGURE 10.A.1 Organization chart for the AEC.

683

TABLE 10.A.6   Budgets of the AEC Divisions of Physical Research and Controlled Thermonuclear Research

Item	Amount Budgeted ($Millions)			
	1969 [a]	1970 [b]	1971 [b]	1972 [c]
High-energy physics	118.6	120.5	118.6	116.4
Medium-energy physics	11.3	12.8	13.0	13.1
Low-energy physics	29.6	29.4	27.6	25.3
Mathematics, computer research	5.7	5.8	5.4	4.8
Chemistry	54.5	53.9	51.6	49.0
Metallurgy, materials	27.5	27.7	26.8	25.2
Controlled thermonuclear	26.5	27.7	28.4	29.8
TOTAL PHYSICAL RESEARCH	273.9	277.8	271.4	263.6

[a] Costs, fiscal year 1969.
[b] Hearings, Joint Committee on Atomic Energy, AEC Authorizing Legislation, fiscal year 1972, February-March 1971, p. 260.
[c] Report, Joint Committee on Atomic Energy, Authorizing Appropriations for AEC, FY 1972, July 8, 1971, p. 3.

research in all matters relating to energy production, whether nuclear or not.

Structure   The AEC is organized with five members serving as Commissioners and having equal responsibility and authority in all decisions and actions of the Commission. The administrative and executive functions of the Commission are discharged by the General Manager and Deputy and Assistant General Managers. Of the six Assistant General Managers, three are concerned with research: Research and Development, Military Applications, and Reactors. Within these directorates, physics research is carried on in the Division of Research (RES), Division of Biology and Medicine, Division of Military Applications (DMA), and Reactor Development and Technology (RDT), among others * (see Figure 10.A.1).

Research Budget   Table 10.A.6 exhibits the budgets of the Divisions of Physical Research and Controlled Thermonuclear Research for fiscal years 1969–1972, as distributed among major fields. For all of AEC, the total of applied and basic research in physics was reported as $236.1 million for fiscal year 1970 (Table 10.A.2).

* *U.S. Government Organization Manual—1971/72* (U.S. Government Printing Office, D.C., 1971). A recent reorganization (Dec. 1971) places controlled thermonuclear research in a separate division.

Program in Physics   DMA and RDT are oriented to the specific missions indicated by their titles. Research programs include nuclear-structure physics (Los Alamos and Livermore), neutron physics, neutron cross-section measurements, and presumably, a considerable amount of atomic, condensed-matter, and plasma physics. Some physics research is also carried out in the Division of Biology and Medicine.

The program of the Division of Research is indicated by the budget categories in Table 10.A.6. The AEC has prime responsibility in the government for high-energy physics and furnishes over 90 percent of the support in this subfield. In nuclear and plasma physics (controlled thermonuclear research), again the AEC support is dominant. The AEC conducts most of its research in federally funded research and development centers (FFRDC) administered by industry ($66 million) or universities ($245 million).* Individual university contracts amount to about $97 million.

The Division of Research spends about $207 million in FFRDC's, and $71 million in off-site contracts, mainly at educational institutions.† In 1970, there were 566 off-site contracts, in 153 institutions. Of these, about 150 contracts, aggregating some $45 million, were in physics. Argonne, Brookhaven, SLAC, and the Lawrence Berkeley Laboratory (LBL) all have physics budgets of more than $20 million; they will soon share this distinction with the National Accelerator Laboratory at Batavia.

Comment   Although the mission of the AEC is a strictly practically motivated one of development of energy resources for peaceful and other purposes, the legislation and practice encourage a broad support of basic science, at least in those fields that have a discernible relevance. In basic physics, the AEC accounts for nearly half of the total federal support. Of its total budget of $2.3 billion, the AEC spends $433 million on basic and applied research and $287 million on basic research.‡

The AEC is very much the purveyor of big science, as is indicated by its emphasis on national laboratories. To some extent this is an outgrowth of the Manhattan Project, which attacked a big problem in a big way, at the time unique in the history of science. Continuation of this mission-

---

* In fiscal year 1970; see *Federal Funds for Research, Development, and Other Scientific Activities,* Volume XX (U.S. Government Printing Office, Washington, D.C., 1971), p. 127.
† *Research Contracts in the Physical Sciences,* July 1, 1970.
‡ In fiscal year 1970; see *Federal Funds for Research, Development, and Other Scientific Activities,* Volume XX (U.S. Government Printing Office, Washington, D.C., 1971), p. 127.

directed effort after the war clearly required operation of large-scale enterprises, and these have changed the course of physics. Big physics, big chemistry, big astronomy—these have become a way of life for the scientist. Whether little science, and little institutions, can long survive will depend on whether their contributions to science in the larger sense, including not only progress but also education, culture, and aesthetic pleasures, continue to be recognized and respected.

The National Science Foundation

As originally conceived, the NSF was to be the principal agency for support of science and for coordination of the scientific efforts of all federal agencies. Specifically included in its domain was to be support of basic scientific research in medicine and the natural sciences as well as defense-related research.

As matters worked out, however, the NSF legislation was sufficiently delayed that support of some of the fields of science became well established in other agencies: For example, the National Institutes of Health (NIH) assumed the support of most medical research, and defense-related work was vigorously supported by the DOD. Even so, the NSF has a broad charter and would be a more dominant influence if it were adequately funded.

The stated objective of the NSF is "to strengthen research and education in the United States." Its activities include: (a) development of information on scientific resources; (b) award of grants to universities and other nonprofit institutions in support of scientific research; (c) support of national research centers; (d) support of education through graduate fellowships, teachers' institutes, and development of awards; and (e) coordination of scientific information activities. The role of coordinating federal science was formally transferred to the Office of Science and Technology (OST) in 1964.

Structure The Foundation consists of the National Science Board (NSB) of 24 members, a Director, Deputy Director, and six Assistant Directors. The Board, the Director, his Deputy, and four of the Assistant Directors are appointed by the President, with the advice and consent of the Senate.

The six Directorates are Institutional Programs, Education, National and International Programs, Research, Research Applications, and Administration.

Among other things, the Assistant Director for National and International Programs is responsible for the support of certain national research centers, specifically the Kitt Peak National Radio Astronomy Observatory, the Cerro Tololo Inter-American Observatory, the Arecibo Radio Astronomy Observatory, and the National Center for Atmospheric Research. Responsibility for Arecibo was transferred from the DOD in 1970. These installations are operated by university associations under contract to the NSF.

Physics research is mainly supported in the Directorate of Research, which has divisions for environmental sciences, biomedical sciences, engineering, social sciences, and mathematical and physical sciences. Within the last-named division are sections for astronomy, chemistry, mathematics, and physics.

Budget After a distinctly modest beginning at a level of only $0.5 million in 1950, the total NSF budget passed the $500 million mark in 1971; the 1972 budget is $622 million. Projections for 1973–1977 envisage an increase from $850 million to $1,900 million.* Whether these happy prognostications are realized remains to be seen. Some encouragement for an optimistic view is to be found in the passage by the Senate of an authorization bill calling for $766.5 million in fiscal year 1972 and $907 million in fiscal year 1973. The President's recommended NSF budget for fiscal year 1973 is $653 million.

The main items in the budgets for 1970, 1971, 1972, and 1973 are shown in Table 10.A.7. A breakdown of the physics program is given in Table 10.A.8. The NSF reports total obligations for basic physics research as $33.8 million, and for applied physics research as $0.9 million in fiscal year 1970 (see Tables 10.2 and 10.4).

Program in Physics Most of the NSF's support of basic physics research is channeled through the Physics Section budget given in Table 10.A.8. The annual publication *NSF Grants and Awards* lists the individual grants, with titles, names of principal investigators, and amounts. In 1970, there were 245 grants to 110 institutions; the average award was $115,000. Nuclear and elementary-particle physics account for nearly two thirds of the budget.

As may be seen from Table 10.A.6, the AEC support of physics suffered a decline of about $4.5 million between 1971 and 1972. Part of the increase in the NSF budget in the same period was used to pick up university operations dropped by the AEC. Further increases in the obligations assumed by the NSF in 1972 arise from projects dropped by the DOD and outright transfers, for example, the National Magnet Laboratory. The Materials Research Interdisciplinary Laboratories, formerly financed

---

* Report, Committee on Science and Astronautics, May 17, 1971, pp. 92–204.

TABLE 10.A.7   The Budget of the National Science Foundation

Category	Amount ($Millions)			
	FY 1970 [a] (Actual)	FY 1971 (Actual)	FY 1972 (Est.)	FY 1973 (Est.)
Scientific research project support	161.7	180.4	246.6	275.3
Specialized research facilities	6.5	—	—	—
National and special programs	39.1	49.8	85.6	109.1
Research applied to national needs	—	34.0	56.0	80.0
National research centers	27.2	36.9	40.2	42.3
National sea grant program	9.0	6.1	—	—
Computing activities	16.9	15.0	21.0	20.5
Science information	11.4	10.7	9.8	9.5
International cooperative science activities	1.7	2.2	4.0	4.7
International government science program	0.5	0.8	1.0	1.0
Institutional support for science	44.7	34.4	21.0	12.0
Science education improvement	80.9	68.3	66.1	70.0
Graduate-student support	39.3	30.5	20.0	14.0
Planning, policy studies	2.1	3.2	2.7	2.5
Management	19.7	21.8	24.1	26.8
Special foreign currency program	2.0	2.0	3.0	7.0
APPROPRIATION	440.0	513.0	622.0	653.0

[a] Housing and Urban Development—Space-Science Appropriations, Hearings April 1971, Part 3, p. 100. Fiscal year 1971, 1972, 1973 figures from President's budget, fiscal year 1973.

by the DOD Advanced Research Projects Agency (ARPA), have also been transferred to the NSF.

Problem-oriented research was not included in the NSF Act of 1950 but is permitted by amendments in 1968. A major push in this direction was undertaken in 1970 and 1971 under the title of IRRPOS (Interdisciplinary Research Relevant to Problems of Our Society). In 1972, this program was absorbed and expanded under the designation RANN (Research Applied to National Needs). Budgets in 1970 and 1971 were $12 million and $34 million, respectively; the President's budget for fiscal year 1972 included $81 million for RANN. The program includes advanced technology applications, environmental systems and resources, social systems, and human resources. About 60 percent will be applied research.*

* Hearings, Subcommittee on Appropriations: HUD—Space Science, April 1971, p. 223ff.

TABLE 10.A.8 National Science Foundation Specialized Research Project Support, Physics and Materials Science

Item	Amount ($Millions)		
	FY 1971	FY 1972	FY 1973 [a]
Physics			
Atomic, molecular, and plasma	2.75	3.55	
Elementary-particle [b]	10.31	14.50	
Nuclear [c]	9.26	11.75	
Theoretical	2.72	3.65	
Solid-state and low temperature [d]	0	0	
TOTAL PHYSICS	25.04	33.45	36.0
Materials Science			
Engineering materials	3.88	5.30	
Solid-state and low-temperature physics	5.66	7.90	
Solid-state chemistry	0.93	1.30	
National Magnet Laboratory	0.40	2.10	
Materials research laboratories	0	12.80	
TOTAL MATERIALS	10.88	29.40	37.4

[a] President's budget, fiscal year 1973.
[b] Includes HEPL (Stanford) and Chicago.
[c] Includes Nevis (Columbia) and LAMPF users; does not include nuclear theory.
[d] Transferred to Materials Science in 1971.

The RANN program is a new venture in several respects: First, it puts the NSF squarely in the problem-solving business. The expressed motivation is the application of research and research techniques to identified problems, such as energy sources, urban engineering, and weather modification. These efforts, while clearly laudable in themselves, will require the NSF to align itself closely with other agencies pursuing the same aims. Another departure from the NSF practice lies in the expressed intent to seek out or assemble teams to solve the identified problems, as opposed to offering support to people who come with ideas.

The House Committee on Science and Astronautics, in its report * authorizing the NSF budget for fiscal year 1972, expressed some reservations on the projected meteoric growth of RANN, questioning whether problem-oriented research ought not to be supported by the agency having the primary mission for the application. Although the question was gently put, it was accompanied by a reduction of $31 million in the authorized

* Report 92–204, May 17, 1971.

budget for RANN. In the final Act, a maximum of $59 million was authorized for this purpose.*

Comment As previously indicated, the NSF is the only agency whose primary mission is the support of basic science. It is the agency that can and must promote science for what it is—an integrated whole, serving culture, curiosity, and the national economy alike. To serve these purposes effectively, the NSF needs to be able to exercise more influence, not at the expense of other agencies supporting science for perfectly legitimate and persuasive reasons of their own, but more influence in the sense of having a commanding position in every important field of science. To achieve such a position requires not only a broad base of projects widely distributed to draw on the whole educational establishment but also a stronger participation in the larger scientific centers.

The evolution of the RANN program will doubtless enhance the influence of the NSF if adequate incremental funding is made available, but there is room for serious concern that too great an emphasis on RANN might subvert the main function of the NSF.

The Department of Defense

Mission The DOD provides for the military security of the United States through coordination and control of the Departments of the Army, Navy, and Air Force and through strategic direction of the Armed Forces. As an operation that depends heavily on technological innovation, the DOD makes the largest federal expenditure on research and development, some $7.4 billion in 1970, about half the total for all federal agencies.

Structure The Secretary of Defense is assisted by the Armed Forces Policy Council, the Deputy Secretary, various Assistant Secretaries, the military departments represented by their Secretaries, and the Joint Chiefs of Staff (see Figure 10.A.2).

The focus of research and development is the Director of Defense Research and Engineering (DDRE), in the Secretary's office, who supervises all such activities in the DOD and is advised by the statutory Defense Science Board. All the work is administered in the military departments except for a few special agencies:

1. The Advanced Research Projects Agency (ARPA) is a separately organized research and development agency directly under DDRE; it may

---

* House Reports 92–412, 92–337.

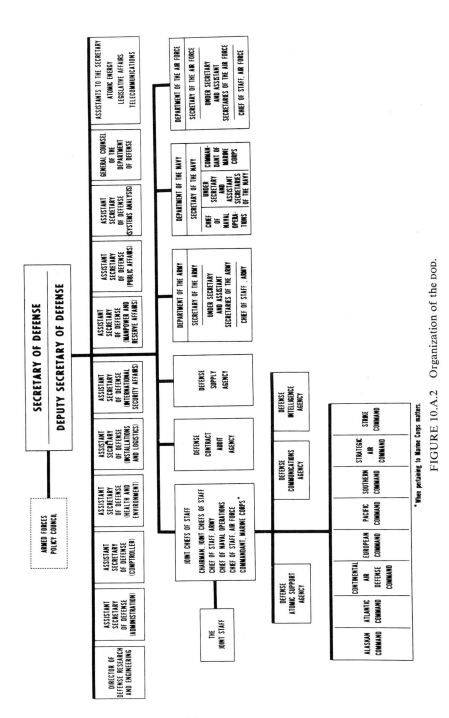

FIGURE 10.A.2 Organization of the DOD.

691

use the services of the military departments or contract with outside organizations.

2. The Weapons Systems Evaluation Group (WSEG) is a creation of the Joint Chiefs of Staff (JCS), but is administratively under DDRE. Its work is done by civilian analysts, employed by the Institute for Defense Analysis, a nonprofit organization under contract to the DOD.

3. The Defense Nuclear Agency (DNA), successor to the Defense Atomic Support Agency, and to the Armed Forces Special Weapons Project, organized in 1947 to take cognizance of nuclear weaponry, is responsible to the Secretary through the JCS. The DNA is the central coordinating agency, with the AEC, on research and development, production, stockpiling, and tests of nuclear weapons.

Budget Of the total of approximately $7 billion for research, development, test, and evaluation, the DOD spends about 5 percent on research and about 12 percent on exploratory development. These categories, designated by DOD as fiscal "sub-elements 6.1 and 6.2," respectively, correspond only roughly to the NSF classification of basic and applied research. The two sets of numbers are shown in Table 10.A.9 for 1966–1971. When inflation is taken into account, the numbers indicate a general decrease of technical effort of about 30 percent in the past five years.

Of the basic research budget, about 42 percent is spent in-house and about 45 percent is contracted to universities; less then 10 percent goes

TABLE 10.A.9 Comparison between the Department of Defense and National Science Foundation Obligations for Research

Category	Obligations ($Millions)					
	FY 66	FY 67	FY 68	FY 69	FY 70	FY 71
Research 6.1 [a] (DOD)	380	398	368	404	368	(370)
Exploratory development 6.2[a] (DOD)	1,097	1,017	906	875	857	(897)
TOTALS	1,477	1,415	1,274	1,279	1,225	(1,267)
Basic research (NSF) [b]	262	284	263	276	247	(243)
Basic and applied research (NSF) [b]	1,849	1,591	1,577	1,411	1,557	(1,479)

[a] Hearings, DOD Appropriations for 1971, part 6, p. 15.
[b] *Federal Funds for Research, Development, and Other Scientific Activities.*

TABLE 10.A.10 Department of Defense Research Obligations by Discipline

Category	FY 69 ($Million)	%	FY 70 ($Million)	%	FY 71 ($Million)	%
General physics	39.7	9.8	33.6	9.1	31.8	8.6
Nuclear physics [a]	34.7	8.6	33.4	9.1	28.5	7.7
Chemistry	14.7	3.7	13.2	3.6	13.5	3.6
Mathematical sciences	43.7	10.8	42.6	11.6	35.9	9.7
Electronics	31.6	7.8	27.7	7.5	29.5	8.0
Materials research	42.8	10.6	38.4	10.4	49.2	13.3
Mechanics	31.7	7.9	30.3	8.2	30.9	8.4
Energy conversion	16.7	4.1	14.1	3.8	15.5	4.2
Terrestrial sciences	13.7	3.4	12.6	3.4	14.0	3.8
Atmospheric sciences	31.2	7.7	31.1	8.4	32.6	8.8
Biological and medical	44.2	10.9	38.3	10.4	39.2	10.6
Behavioral and social	16.3	4.0	12.9	3.5	13.1	3.5
Oceanography	32.1	7.9	30.0	8.2	28.7	7.8
Astronomy and astro-physics	11.1	2.8	10.3	2.8	7.2	2.0
TOTALS	404.2	100	368.5	100	369.6	100

[a] Includes weapons effects.

to industry.* This distribution is in marked, and understandable, contrast to the situation for total research and development: 27 percent in-house; 64 percent industry; and 3 percent universities.

A breakdown of the 6.1 budget category into scientific disciplines is presented in Table 10.A.10. Again it should be noted that this budget includes some applied research under the NSF definition. Thus, for example, NSF reports an estimated total outlay for basic physics research in DOD as $42.9 million in fiscal year 1970 (Table 10.4); components of this total are presumably to be identified in several of the items in Table 10.A.10.

Project Themis This operation was originally conceived as a means of broadening the institutional base for federally funded research. Distribu-

* In fiscal year 1970; see *Federal Funds for Research, Development, and Other Scientific Activities,* Volume XX (U.S. Government Printing Office, Washington, D.C., 1971).

tion of Themis funds was to serve, as far as practicable, to encourage new institutions not already established in the research field. In 1967, DOD funding was $20 million; it rose to nearly $30 million in 1969. In 1970, the Senate Armed Services Committee determined that Themis was inappropriate for DOD funding, and it has since disappeared as a line item. In 1971, about $6 million of DOD research funds were allocated to continuation of 30 projects that started under Themis. In all, Themis funding since 1967 provided start-up for 118 interdisciplinary programs in 76 universities.*

Mansfield Amendment † Section 203 of the fiscal year 1970 Military Procurement Act prohibited support by the DOD of any research not directly and apparently related to a specific military mission or function. Although the Section was in force only in fiscal year 1970, it has led to a massive review of DOD projects and a strong effort to relate them to specific military needs. In practice, the review has resulted in discontinuation of only a relatively few projects ‡ that would not have been dropped for other policy reasons—for example, a decision that the DOD should depend on other agencies for nuclear-physics research—but it has probably changed the attitude of some program managers toward basic research proposals and has undoubtedly discouraged many scientists from submitting proposals to the DOD.

Comment That the DOD has been, for 25 years, the largest source of research and development funds—accounting for half of the federal component—is a source of concern that U.S. research and development, and science, may be too strongly oriented toward the military.§ As far as basic science is concerned, the situation is distinctly ameliorated by the DOD's singularly enlightened science management and by the fact that a number of other agencies—NASA, HEW, AEC, and NSF—also have strong commitments to the support of basic science (Table 10.A.3). Nonetheless, the concern is real, and the question of the DOD's continued participation in scientific endeavors was brought to a head by the Mansfield Amendment.

---

* Hearings, DOD Appropriations, 1971, part 6, p. 110.
† See a discussion by R. W. Nichols, "Mission-Oriented R & D," *Science, 172,* 29 (1971).
‡ Of 6600 DOD projects reviewed in fiscal year 1970, 220—representing $8.8 million— were affected by Sec. 203. In the same year, tightening of the budget led to a reduction of $64 million in funds for research and exploratory development. (Sen. Mansfield in hearings on National Science Policy, Committee on Science and Astronautics, July–Sept. 1970, p. 608.)
§ See H. Brooks, in Hearings on National Science Policy, Committee on Science and Astronautics, July–Sept. 1970, p. 931.

According to Senator Mansfield,* the intent of Section 203 was not to put the DOD out of science but rather to reduce the research community's dependence on DOD in areas in which other agency sponsorship might be more reasonable. As it is, he points out, the DOD has been a kind of back-door NSF, and the NSF has been the orphan child.

Although there is much to be said in favor of strengthening the NSF and other civilian agencies, there is danger in too narrowly defining the role of the DOD in science. First, from the point of view of the health of science itself, it is important to keep open a multiplicity of support agencies, with different motivations and different managerial outlooks. Since scientists cannot say exactly what basic science is for, they have to rely on a continuing dialogue involving all the users of science, including the DOD.

From the point of view of the DOD as an agency with an important mission, withdrawal from contact with some fields of science may be costly. The DOD's dependence on technology is so vast, and so critical, that it needs, probably more than any other agency, to be sure that it has a complete establishment. Completeness in a technological enterprise necessarily includes access to the systematic exploration of the future, and that is what science is for.

Department of the Air Force

Structure for Research and Development Within the Air Force, top management is in the hands of the Assistant Secretary for Research and Development, reporting to the Secretary, and the Deputy Chief of Staff for Research and Development, reporting to the Air Staff (Figure 10.A.3). Most of the scientific operations are carried out by the Air Force Systems Command through the Director of Laboratories. Under his cognizance, on the research side, are four in-house laboratories and the Air Force Office of Scientific Research (AFOSR) (Figure 10.A.4).† Although the AFOSR is the principal outside contracting agency, most Air Force laboratories support some external research and development effort.

Budget The Air Force spent approximately $80 million annually for basic research in fiscal years 1970 and 1971, distributed as shown in Table 10.A.11. In 1971, $31 million was allocated to the AFOSR, shown in the last column of the table. In 1972, $5.2 million of the AFOSR budget for solid-state physics was recategorized as materials research and is now carried in that category, with a corresponding decrease in general physics.

* Hearings on National Science Policy, Committee on Science and Astronautics, July–Sept. 1970, p. 608.
† *Air Force Research Objectives,* 1971, AF Systems Command.

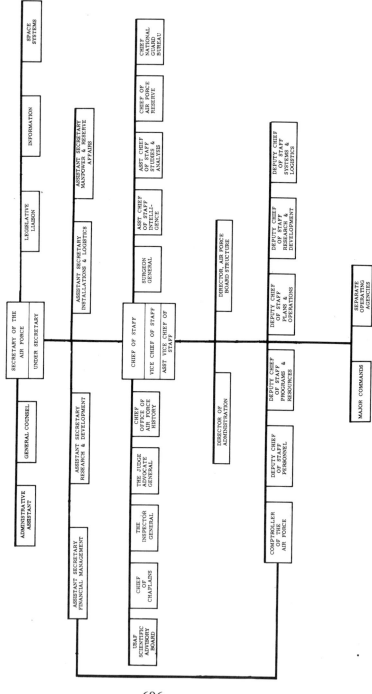

FIGURE 10.A.3 Organization of the Department of the Air Force.

696

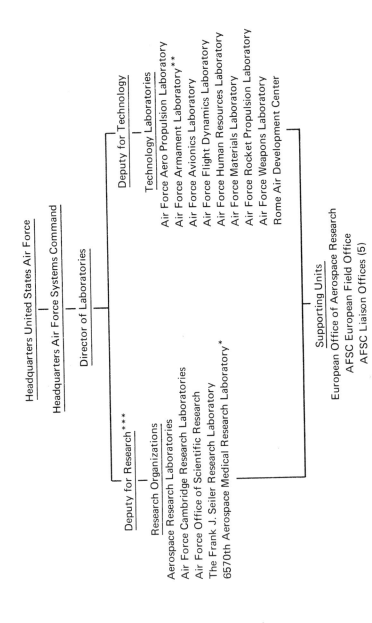

Headquarters United States Air Force

Headquarters Air Force Systems Command

Director of Laboratories

Deputy for Research***

Research Organizations
Aerospace Research Laboratories
Air Force Cambridge Research Laboratories
Air Force Office of Scientific Research
The Frank J. Seiler Research Laboratory
6570th Aerospace Medical Research Laboratory*

Deputy for Technology

Technology Laboratories
Air Force Aero Propulsion Laboratory
Air Force Armament Laboratory**
Air Force Avionics Laboratory
Air Force Flight Dynamics Laboratory
Air Force Human Resources Laboratory
Air Force Materials Laboratory
Air Force Rocket Propulsion Laboratory
Air Force Weapons Laboratory
Rome Air Development Center

Supporting Units
European Office of Aerospace Research
AFSC European Field Office
AFSC Liaison Offices (5)

*R&D program is under the technical direction of DL.
**Development program is under the technical direction of DL.
***This position was deleted in December 1971.

FIGURE 10.A.4   Organization chart for the Director of Laboratories, Air Force Systems Command.

697

TABLE 10.A.11   Department of the Air Force

Category	Obligations for Basic Research [a] ($Millions)		
	Total AF FY 1970	Total AF FY 1971	AFOSR FY 1971
01 General Physics	16.4	15.2	10.3
02 Nuclear Physics	1.5	0	0
03 Chemistry	4.9	4.5	3.3
04 Mathematics	5.6	6.5	4.3
05 Electronics	7.9	8.4	2.9
06 Materials	2.6	3.8	0
07 Mechanics	8.9	8.3	4.2
08 Energy Conversion	6.3	6.2	3.9
09 Terrestrial Science	1.8	2.0	0.1
10 Atmospheric Science	16.7	18.8	0
11 Astronomy	5.4	3.6	0
12 Biomedical	2.2	1.7	1.4
13 Human Resources	0.8	1.4	1.1
TOTALS	81.0	79.6	31.4

[a] Hearings, DOD Appropriations for 1971, part 6, p. 593.

Program in Physics   As is generally the case in the DOD, there is a strong effort to connect basic research objectives to short-range or long-range problems of the overall mission. Frequent meetings are held with the product divisions—those concerned with hardware—to define research needs as generated in the developmental laboratories, and these, together with interpretations of the overall Air Force plan, serve to guide the general direction of the basic research program. Thus, problems of reconnaissance and communication encourage an interest in electronics and solid state, and problems of energy conversion and power generation encourage work in plasma physics. Some subjects, for example, nuclear physics, which would seem to have relevance for the Air Force, are not supported on the basis that the Air Force can maintain adequate contact through research carried on by other agencies. In the category of general physics, the AFOSR supports about 150 projects grouped under a dozen general headings. Other physics-related programs are included in Electronics, Mechanics, and Energy Conversion.

Department of the Army

Structure for Research and Development   The Army has a dual civilian–military top management: The Assistant Secretary for Research and Development reports to the Secretary of the Army; the Chief of Research and Development reports to the Chief of Staff of the Army (Figure 10.A.5.) There is, however, close liaison between the Assistant Secretary for Research and Development and the Chief of Research and Development. Much of the Army's research and development is carried out by the Army Materiel Command and its subordinate Commodity Commands, laboratories, and centers; the Corps of Engineers; and the Army Medical Department. These agencies operate a number of scientific establishments located in various Army arsenals, laboratories, proving grounds, and training centers. They conduct fundamental research, primarily in-house.

Programs of broad Army-wide interest are managed by the Director of Army Research, in the Office of the Chief of Research and Development. Within this directorate are divisions for life sciences, environmental sciences, and physical and engineering sciences, among others. The prime operating arm for support of such Army-wide basic research under contracts or grants, and in mathematics and the engineering, physical, and environmental sciences, is the Army Research Office, Durham [ARO(D)]. In addition to projects funded by the Chief of Research and Development, the ARO-D monitors certain projects sponsored by ARPA, the Army Materiel Command, and the DNA.

Budget, Program in Physics   The Army spends about $290 million per year on basic and applied research (fiscal year 1970: Table 10.A.1). The Division of Physical and Engineering Sciences spends $110 million on physical sciences, engineering, and mathematics. Approximately half of the basic research in this category—about $12 million—is spent by the ARO-D in its extramural program. A total of about 650 projects is supported by this office. Table 10.A.12 lists the subcategories of physics, with the numbers of projects in each in 1969. According to the NSF report, *Federal Funds for Research, Development, and Other Scientific Activities,* the Army spent $8.0 million for basic physics research in 1969, about half in-house and half extramural; most of this is presumably represented in the categories of Table 10.A.12.

Primary emphasis in Army research is not so much on disciplines but rather on problem areas. Thus, for example, an understandable preoccupation with material deterioration leads to strong support of fields of physics and chemistry that might have relevance for corrosion, whereas

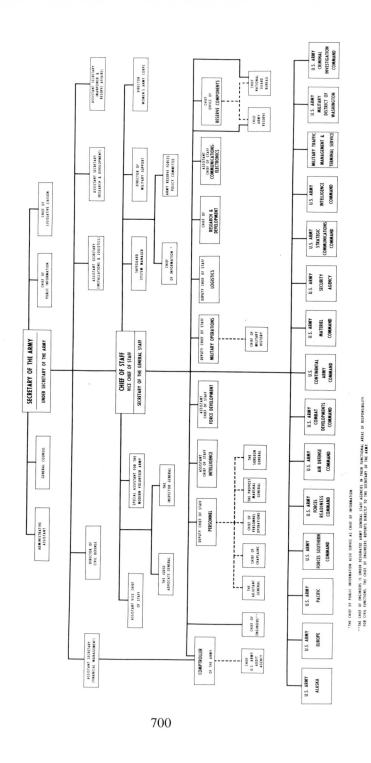

FIGURE 10.A.5   Organization of the Department of the Army.

700

TABLE 10.A.12 Physics Program, ARO (Durham), 1969 [a]

Subject	Number of Projects
Radiation physics	9
Atomic, molecular, plasma physics	50
Instrumentation, general physics	7
Cryophysics, liquids	17
Electromagnetic radiation, optics	34
Structure of solids	30
	147

[a] *Research in Progress, 1969,* U.S. ARO(Durham).

areas with less identifiable relevance are left to other agencies. Problem areas of special interest to the Army are detailed in the publication *Military Themes for Oriented Research of High Scientific Merit,* ARO(Durham).

Department of the Navy

Structure for Research and Development The Assistant Secretary for Research and Development is responsible for all matters relating to research, development, test, and evaluation within the Navy. Although a good deal of applied research and development is conducted in components of the Naval Material Command (Figure 10.A.6), all basic research (budgetary category supplement 6.1) is controlled by the Office of Naval Research (ONR), reporting to the Assistant Secretary for Research and Development. The ONR maintains administrative offices in London, Chicago, Boston, and Pasadena. It operates the Naval Research Laboratory (NRL) and the Naval Biomedical Laboratory (NBL).

Budget, Program in Physics Navy budgets for research, development, test, and evaluation are given in Table 10.A.13. Again, the categories listed do not coincide entirely with the classification used by the NSF: for basic and applied research, Table 10.A.1 lists $273 million (fiscal year 1970); for basic research only, $82 million (Table 10.A.3).

Of the research funds listed in Table 10.A.13, about $12 million, or 10 percent, goes to the directors of some 28 naval laboratories for in-house independent research. The remainder is called "Defense Research Sciences" and is used to support research in physical, engineering, environmental, biomedical, and behavioral sciences, part at in-house laboratories and part through the extramural program. About 30 percent of the Defense

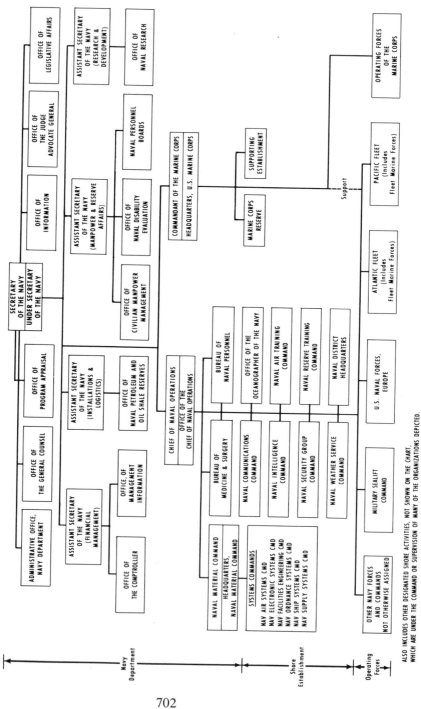

FIGURE 10.A.6   Organization of the Department of the Navy.

702

TABLE 10.A.13 Department of the Navy, Obligations for Research, Development, Testing, and Evaluation [a]

Category	Obligations ($Millions)		
	FY 69	FY 70	FY 71
1. Research	130	118	119
2. Exploratory development	262	236	243
3. Advanced development	297	281	347
4. Engineering development	346	393	532
5. Management and support	228	230	226
6. Operational systems development	928	943	730
TOTALS	2,192	2,200	2,197

[a] Hearings on DOD Appropriations for 1970, part 1, p. 337.

Research Sciences program is in oceanography. Total Navy support of basic and applied physics research is reported by the NSF as $50 million (fiscal year 1970, Table 10.A.2).

The ONR has divisions for earth sciences, physical sciences, mathematics and information, biology, psychology, materials, ocean sciences, and others. In physics, emphasis is given to acoustics, lasers, superconductivity—including refrigeration—electronics, and communications. Programs in high-energy physics, nuclear physics, and radio astronomy, in which the ONR once had a strong position, are being phased out by decision of the Director of Defense Research and Engineering.

The National Aeronautics and Space Administration

Mission In 1958, in the wake of Sputnik, NASA was established. The statutory functions of NASA were outlined in the National Aeronautics and Space Act, of which the relevant section is cited here:

Sec. 102.

(a) The Congress hereby declares that it is the policy of the United States that activities in space should be devoted to peaceful purposes for the benefit of all mankind.

(b) The Congress declares that the general welfare and security of the United States require that adequate provision be made for aeronautical and space activities. The Congress further declares that such activities shall be the responsibility of, and shall be directed by, a civilian agency exercising control over aeronautical and space activities sponsored by the United States, except that activities peculiar to or primarily associated with the development of weapons systems, military operations, or the defense of the United States (including the research and development necessary to make effective provision for the defense of the United States) shall be the responsibility of, and shall be directed by, the Department of Defense; and that determination as to which such agency has responsibility for and direction of any such activity shall be made by the President in conformity with section 201 (e).

(c) The aeronautical and space activities of the United States shall be conducted so as to contribute materially to one or more of the following objectives:

(1) The expansion of human knowledge of phenomena in the atmosphere and space;

(2) The improvement of the usefulness, performance, speed, safety, and efficiency of aeronautical and space vehicles;

(3) The development and operation of vehicles capable of carrying instruments, equipment, supplies, and living organisms through space;

(4) The establishment of long-range studies of the potential benefits to be gained from the opportunities for, and the problems involved in the utilization of aeronautical and space activities for peaceful and scientific purposes;

(5) The preservation of the role of the United States as a leader in aeronautical and space science and technology and in the application thereof to the conduct of peaceful activities within and outside the atmosphere;

(6) The making available to agencies directly concerned with national defense of discoveries that have military value or significance, and the furnishing by such agencies, to the civilian agency established to direct and control nonmilitary aeronautical and space activities, of information as to discoveries which have value or significance to that agency;

(7) Cooperation by the United States with other nations and groups of nations in work done pursuant to this Act and in the peaceful application of the results thereof; and

(8) The most effective utilization of the scientific and engineering resources of the United States, with close cooperation among all interested agencies of the United States in order to avoid unnecessary duplication of effort, facilities, and equipment.

In his special message to the Congress on May 25, 1961, President Kennedy enunciated the thrust of the U.S. space program in the following terms:

Finally, if we are to win the battle that is now going on around the world between freedom and tyranny, the dramatic achievements in space which occurred in recent weeks should have made clear to us all, as did the Sputnik in 1957, the impact of this adventure on the minds of men everywhere, who are attempting to make a determination of which road they should take. . . . First, I believe that this nation should commit itself to achieving the goal, before this decade is out, of landing a man on the moon and returning him safely to the earth. No single space project in this period will be more impressive to mankind, or more important for the long-range exploration of space; and none will be so difficult or expensive to accomplish.*

Shortly before the successful accomplishment of the manned lunar landing, President Nixon formed a Space Task Group chaired by the Vice President to study the scope and pace of the space program during the

* Special Message to Congress on Urgent National Needs, May 25, 1961 (Sec. IX).

1970's. They recommended that: "The national program for the next decade in space should focus on utilizing space capabilities for the welfare, security, and enlightenment of all people." *

Structure This agency is headed by an Administrator and his Deputy, an Associate Administrator for organization and management, and five Associate Administrators in charge of the offices of Manned Space Flight (OMSF), Space Science and Applications (OSSA), Tracking and Data Acquisition (OTDA), and Advanced Research and Technology (OART) (see Figure 10.A.7). The research program is coordinated and controlled by the Headquarters offices. It is executed or contracted for by the several field offices. Eighty percent of NASA's research and development is extramural.

The OSSA is operated through three field stations: Goddard Space Flight Center, the Jet Propulsion Laboratory, and Wallops Station. It has responsibility for selecting all scientific experiments for NASA space missions, for the development of instruments to be flown on future missions, and for ground-based work needed in support of NASA's scientific objectives. The OSSA also manages the flight program of automated spacecraft in support of both scientific and applications missions.

The OMSF operates three NASA centers and is responsible for the Apollo and Skylab programs and for the development of the Space Shuttle.

The OART directs ground-based research and development on aircraft, spacecraft, launch vehicles, nuclear and other propulsion systems, electronics, and other advanced technology of importance to NASA.

Program Under the general heading of Supporting Research and Technology (SR&T), NASA funds the development of instruments, spacecraft systems, theoretical investigations, and laboratory research. A major part of this program is basic research that is relevant to the scientific investigations being conducted in the flight program. Areas of interest to physicists include the earth environment above 80 km (neutral atmosphere, ionosphere, radiation belts, magnetopause), the interplanetary medium (solar wind, solar magnetic fields, micrometeorites), lunar and planetary investigations (sample analysis, remote sensing of planetary atmospheres and surfaces), astrophysics (cosmic rays, x-ray and gamma-ray astronomy, relativity), and environmental sciences (remote sensing of the earth's atmosphere, surface, and subsurface).

The flight program of NASA uses instrumented aircraft, sounding rockets,

---

* *The Next Decade in Space,* a report of the Space Science and Technology Panel of the President's Science Advisory Committee, March 1970.

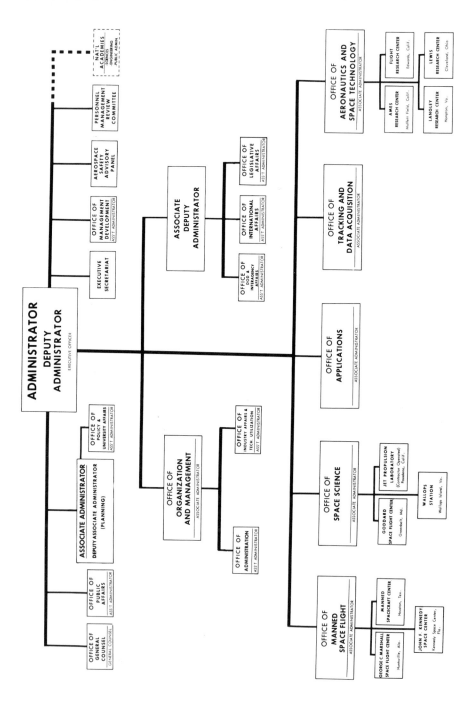

FIGURE 10.A.7 Organization of the National Aeronautics and Space Administration.

706

satellites, and deep-space probes. The space missions include the Apollo lunar landing, Skylab, Orbiting Geophysical Observatory, Explorer (generic name for relatively small satellites), Pioneer, Mariner, Viking, and the Applications Technology Satellite.

Budget For fiscal year 1970, NASA reported obligations for basic research in physics as $74 million; astronomy as $170 million; and environmental sciences (atmospheric, oceanographic, and earth sciences) as $195 million.*
A breakdown of the physics budget is shown in Table 10.A.14. These figures include substantial charges for work not normally considered

---

* These estimates are more recent than those cited in Tables 10.A.3 and 10.A.4.

TABLE 10.A.14 National Aeronautics and Space Administration Estimated Obligations for Basic Physics Research in Fiscal Year 1970 ($Thousands)

Program/Project	Amount
OMSF	
Apollo experiment	1,860
Spaceflight operations	399
OSSA	
Physics and astronomy programs—physics	
supporting research and tech/advanced	
studies	6,797
Explorers	7,277
Sounding rockets	5,427
Data analysis	2,032
Lunar and planetary program	
Pioneer	12,764
Helios	192
OART	
Basic research program—SRT	359
Space power and electric propulsion: space	
power SRT	2,208
OUA	
Training and research grants	860
Total, Direct Program	40,175
Tracking and data acquisition	17,630
NASA Center operation and other costs	16,424
Total, Other	34,054
NASA TOTAL (as reported to NSF)	74,229

basic research. They include apportioned charges for launch vehicles, launch operations, spacecraft that provide the laboratory bench for the experiments, and a worldwide tracking network to recover the data. Even the basic research instruments have to be developed far beyond the laboratory stage, because they are generally expected to operate unattended for a year or longer.

Comment   As is the case with the DOD, the relationship between NASA and the scientific community involves a certain ambivalence, a feeling that is perhaps enhanced by the fact that NASA's mission is far more specifically identified with science than is that of the DOD. Again we have the problem of motivation: if a scientific program is aimed too directly at a specific goal, the opportunities to explore paths to new knowledge will be limited by that goal rather than by the paths that emerge in the course of work. At the same time, an agreed-on goal may often be a powerful stimulant to productive scientific activity.

Although both the legislative history and the wording of the Space Act give some emphasis to the intended role of NASA in promoting the expansion of basic science, it is perhaps fair to say that, at least until the present time, NASA has been primarily a mission-oriented agency. The overwhelming priority established by President Kennedy for the Apollo mission served to focus strongly a major part of NASA's effort on accomplishing the prodigious engineering task, necessarily relegating the purely scientific effort to a somewhat secondary role.

It cannot be denied that the successful accomplishment of this mission, marked by Neil Armstrong's first step on the lunar surface, launched mankind on the most magnificent adventure in his history on this planet— the journey to the stars. This is an adventure that the meanest of minds cannot ignore.

At the same time, it is necessary to recall that much needs to be learned in fields of science that would appear to be irrelevant to this inspiring goal, and if they are allowed to languish, other equally important goals may fail of achievement or even of identification as such. The single-minded devotion of NASA to its primary mission has had salubrious effect on our national pride, the prestige of science, and on science itself. These successes should not, however, blind those who shape science policy to the continuing need for a much broader base of support for science than can be mustered around any single mission.

The Department of Agriculture

Mission   "The Department of Agriculture is directed by law to acquire and diffuse information on agricultural subjects in the most general and

TABLE 10.A.15 Department of Agriculture, Physical Science Research

Service	Budget ($Millions)		
	All Research [a]	Physical Sciences [a]	Physics [a]
Agricultural research	152	30.8	5.3
Cooperative state research	62	3.0	0.0
Forest	42	3.1	1.4
TOTALS	272 [b]	36.9	6.7

[a] Operations, basic and applied, fiscal year 1970; see Tables 10.A.1 and 10.A.2.
[b] Includes non-physical-sciences research budgets of divisions of the Department not listed in this table.

comprehensive sense. The Department performs functions relating to re-search, education, conservation, regulatory work, agricultural adjustment, surplus disposal, and rural development." *

Structure of Research Organization Reporting to the Secretary is the Science and Education Director, who is responsible for five Services: Agricultural Research (ARS), Cooperative State Research (CSRS), Extension, National Agricultural Library, and the Animal and Plant Health (APHS). He also coordinates the research activities of the Forest Service, Economic Research Service, Statistical Reporting Service, and the Farmer Cooperative Service (see Figure 10.A.8).

Research Budget Budgets for the various research services appear in Tables 10.A.1 and 10.A.2. Of the seven mentioned there, only ARS, CSRS, and the Forest Service have appreciable budgets in the physical sciences (see Table 10.A.15).

Physics Program The ARS is a large organization, with 8759 full-time employees, spending about $192 million (in fiscal year 1972). It operates some 133 federal facilities and has work in progress in 196 other locations, such as state experimental stations and land-grant colleges. As is evident from the budget, physics as such plays a relatively small role in the program. Some of the problems are soil and water resources, soil physics, hydrology, salination, mechanics of sewage treatment, fibers, food texture, structure, aroma, defect detection, and sterilization. Most of this work is done in-house, some in cooperation with other agencies, for example, the DOD.

* *U.S. Government Organization Manual—1971/72* (U.S. Government Printing Office, Washington, D.C., 1971).

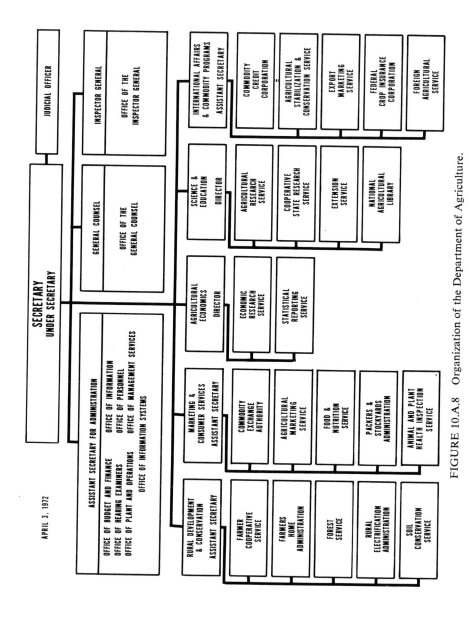

FIGURE 10.A.8 Organization of the Department of Agriculture.

710

The CSRS is largely a granting agency, administering funds that go to states under the Hatch Act. The states in turn distribute the funds to land-grant colleges to support agricultural research.

Comment The ARS recognizes the need for closer contact with basic physical sciences. Part of this need is being met through the operation of 15–20 small group laboratory operations in the ARS with unrestricted missions and part through the appointment of specialized scientific aides to the Director.

Department of Housing and Urban Development

Mission This Department was established in 1965 for the purpose of administering federal programs in housing and community development and providing assistance in achieving the maximum coordination of federal programs relating to housing, urban development, and mass transportation through state, local, or private action.*

Organization for Research Mainly a management and administrative department, HUD has little in-house research capability. Under the Secretary are seven Assistant Secretaries, including the Assistant Secretary for Research and Technology, who is the principal focal point for all research, development, and demonstration activities (see Figure 10.A.9). Under his cognizance is the Office of Research and Technology, with about 60 professionals, including architects, engineers, sociologists, public administrators, and a few physicists and chemists. The Office does not conduct research but rather administers the work carried on through contracts. Much of the work is contracted to industry, nonprofit organizations, and public bodies. Such contracts usually result from identification of a specific problem by HUD, followed by a request for proposals. Rather few contracts originate from unsolicited proposals. The Urban Institute conducts studies for HUD and other agencies. Although it is a part of the Department of Commerce, the National Bureau of Standards also undertakes projects for HUD, as it does for other agencies, and renders advice on evaluation of proposals.

Research Budget In fiscal year 1970, HUD obligated about $30 million to research and development: $7 million to applied research and $23 million to development.† The applied research funds were in the category called engineering; no expenditure is listed in the physical sciences.

* *U.S. Government Organization Manual—1971/72* (U.S. Government Printing Office, Washington, D.C., 1971).
† *Federal Funds for Research, Development, and Other Scientific Activities,* Volume XX (U.S. Government Printing Office, Washington, D.C., 1971).

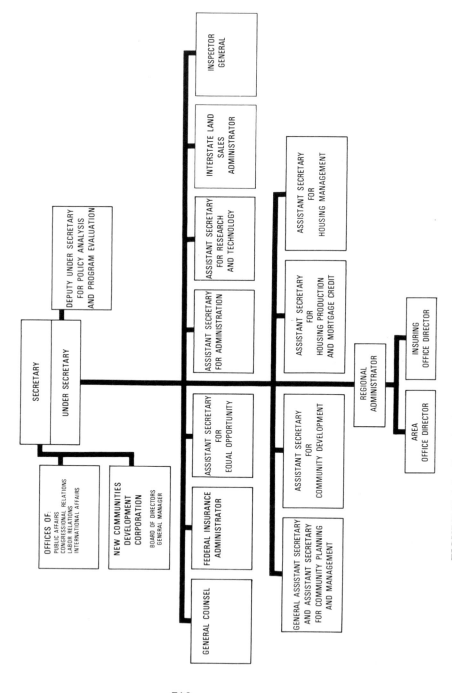

FIGURE 10.A.9   Organization of the Department of Housing and Urban Development.

712

Research Program The Office of Research and Technology administers programs in housing research, urban studies, low-income housing demonstration, open space, urban beautification, urban renewal, and technology of experimental housing. Although the programs are largely addressed to applications of existing technology, there is an acute recognition of the need for more basic research in both hardware and software. Central to many of the problems is the construction of a valid model of a city, including the various environmental, economic, and political elements, and the coupling among them. Even the classification of the fundamental ingredients is at the pre-Mendeleev stage. Attempts thus far to apply the techniques of the physical sciences have not been conspicuously successful. It is likely that an aggressive attack on such matters by HUD will require development of a strong in-house research group.

National Institutes of Health

Mission The Department of Health, Education, and Welfare (HEW) was established in 1953 to administer those agencies of government whose major responsibilities are to promote the general welfare in health, education, and social security. Among the organizations included is the Public Health Service, which has responsibility for all health and health-related programs in the Department.

Organization for Research The Public Health Service (PHS) consists of three operating agencies: the Food and Drug Administration, the Health Services and Mental Health Administration, and the National Institutes of Health (NIH) * (see Figure 10.A.10). Direction of these agencies is the responsibility of the Assistant Secretary (Health and Scientific Affairs). The National Advisory Health Council and advisory councils established for major program areas provide policy guidance and recommend research grant applications.

The NIH conducts and supports research in the causes, prevention, and cure of disease; administers programs to meet health manpower needs; administers federal standards for biological products; and operates the National Library of Medicine. The research organization consists mainly of ten categorical Institutes specializing in various diseases. Although the NIH is administered by a single Director, the individual Institutes have a large degree of autonomy in their operation.

---

* The Environmental Health Service, formerly in HEW, has been transferred to the Environmental Protection Agency.

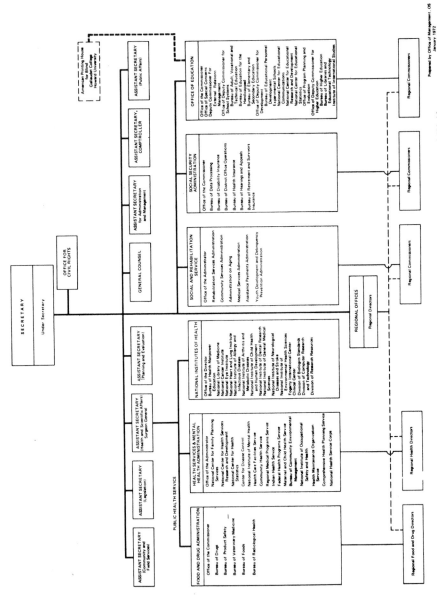

FIGURE 10.A.10 Organization of the Department of Health, Education, and Welfare.

TABLE 10.A.16 Consolidated Obligations, NIH Institutes and Research Divisions ($Thousands)

Activity	FY 1970	FY 1971
Research Programs	830,998	976,317
Regular grants	437,317	486,789
General research support grants	50,310	46,173
Centers, resources and other grants	102,649	131,309
Laboratories and clinics	91,996	112,460
Collaborative research and development	124,845	179,249
Other direct research	23,881	20,337
Research Training	179,101	181,110
Grants	127,899	129,156
Traineeships	2,665	2,336
Fellowships	20,089	21,361
Research career program awards	27,708	27,858
Direct training	740	399
Program Management	27,901	32,450
TOTALS	1,038,000	1,189,877

Research Program, Budget Of the total HEW obligations of $1.1 billion for basic and applied research in fiscal year 1970, the NIH accounted for $869 million (Table 10.A.1). Intramural obligations amounted to $165 million, $516 million went to universities, $121 million to other nonprofit institutions, and $34 million to industry. Some $771 million was devoted to the life sciences, and $34 million was spent in the physical sciences, almost all in chemistry; physics accounted for $94 thousand in 1970. Consolidated obligations for the Institutes and research divisions in 1970 are given in Table 10.A.16.

Each Institute operates both intramural and extramural programs. Within the intramural programs of the Institutes, there are about 200 major laboratories and clinics, involving some 2000 research scientists in all. Overall direction of the research program in each Institute is in the hands of its Scientific Director.

The extramural programs are operated through both contracts and grants. The former instrument is used largely for the procurement of specified services—synthesis of compounds, animal tests, and the like—while the grant more often supports independent research projects. Within the grant category are research grants, training grants, and general institutional support grants. Both operating and research grant funds are appropriated by Congress for the individual Institutes.

As is indicated by the legislative language and by the organization into categorical Institutes, the NIH is primarily a mission-oriented agency: Only about one third of the budget is characterized as basic, as opposed to applied, research. Even in the grant program, relevance to the mission is an important criterion. Such fields as organic chemistry and physics are regarded as peripheral, and the NIH has few programs in these areas.

Department of Commerce

Mission "The mission of the Department is to promote full development of the economic resources of the United States," * It does this through a variety of programs designed to encourage economic progress in communities and industries. Among its specific functions are collection of demographic, economic, business, scientific, and environmental data. Commerce administers the patent system and provides weather and other environmental services; it also conducts scientific research in physical measurement standards; in engineering, product, and commodity standards; in the oceans, earth, and atmosphere; and in advancing selected fields of technology.

Organization for Research Under the Secretary, there are two major positions concerned with research and development: the Administrator of the National Oceanic and Atmospheric Administration (NOAA) and the Assistant Secretary for Science and Technology (see Figure 10.A.11).

A comparatively new organization, NOAA was assembled partly from the former Environmental Science Services Administration (ESSA) (under the Assistant Secretary for Science and Technology) and from some agencies transferred from the Department of the Interior. NOAA's responsibilities include oceanography, study of the shape and size of the earth, and the upper and lower atmosphere.

Currently under the Assistant Secretary for Science and Technology are the Patent Office, National Technical Information Service, Office of Telecommunications, and National Bureau of Standards.

Budget, Research Program Research obligations for fiscal year 1970 (estimated) are reported in Table 10.A.17.

---

* *U.S. Government Organization Manual—1971/72* (U.S. Government Printing Office, Washington, D.C., 1971).

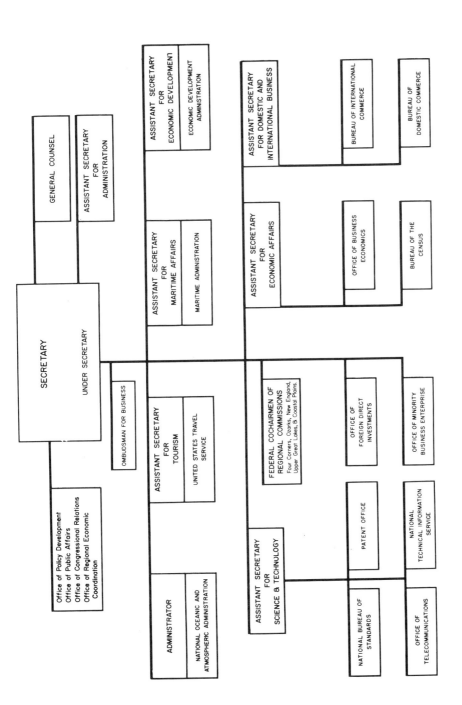

FIGURE 10.A.11 Organization of the Department of Commerce.

TABLE 10.A.17 Department of Commerce, Fiscal Year 1970 Obligations, Research [a] ($Millions)

Unit	Total Research	Physical Sciences	Environmental Sciences	Physics
NOAA	52.9	1.4	22.1	0.0
NBS	27.1	15.8	0.0	8.6
TOTAL, COMMERCE	94.1 [b]	17.2	22.1	8.6

[a] See also Tables 10.A.1–10.A.4.
[b] Includes non-physical-sciences research budgets of divisions of the Department not listed in this table.

As mentioned earlier, ESSA's operations have been in the environmental sciences; ESSA has had major programs in meteorology, hydrology, surveying (Coast and Geodetic Survey), oceanography, electromagnetic-wave propagation, and properties of the atmosphere.

The NBS conducts research in several broad program areas: basic standards, materials research, applied technology, radiation research, computer sciences, and information services. Roughly half of the work in the NBS is done for agencies other than Commerce. A breakdown of the NBS budget for physics research is given in Table 10.A.18.

TABLE 10.A.18 Physics Research Expenditures at the National Bureau of Standards, Fiscal Year 1970

Category	Amount ($Thousands)
Electron, atomic, molecular	2,530
Elementary-particle	425
Nuclear	1,425
Plasmas, fluids	1,075
Condensed-matter	2,796
Space, planetary	300
Biophysics	50
Optics	740
Acoustics	330
Miscellaneous physics	716
TOTAL	10,387

Because the Department of Commerce is strongly oriented to practical economic output, there are strong pressures on the research organization to undertake problems of immediate relevance. Thus, for example, the importance of the textile industry and the consequent congressional pressure led to large programs in fabric flammability. Similarly, programs for weather modification have a wide appeal and may receive higher priority than the underlying basic research. Despite these pressures, however, the scientific bureaus enjoy a degree of autonomy and are able to carry forward some long-range programs.

Department of the Interior

Mission The Department of the Interior is the custodian of the nation's natural resources. As such, it has responsibility for conservation and development of mineral, water, and wildlife resources; reclamation of lands; abatement of pollution; and management of hydroelectric power systems.

Organization for Research In the Office of the Secretary is a Science Adviser who works with the Secretary on scientific matters. The position of Science Adviser has been vacant since November 1970; his functions have been performed by the Under Secretary and his staff. The principal organizations concerned with physical and environmental sciences are the Bureau of Mines and the Geological Survey, both under the Assistant Secretary for Mineral Resources (see Figure 10.A.12). Research in the physical and environmental sciences is also supported by the Bureau of Reclamation, the Bonneville Power Administration, and the Offices of Coal Research, Saline Water, and Water Resources Research. Environmental programs that are largely biological are supported by the Bureau of Sport Fisheries and Wildlife and the National Park Service.

TABLE 10.A.19 Department of the Interior, Obligations for Research, Fiscal Year 1970

| Unit | Amount ($Millions) | | | |
	Total Research	Physical Sciences	Environmental Sciences	Physics
Bureau of Mines	34	7	2	0.8
Geological Survey	41	4	35	1.0
TOTAL, INTERIOR	122 [a]	14 [a]	44 [a]	1.8

[a] Includes research budgets of divisions of the Department not listed in this table.

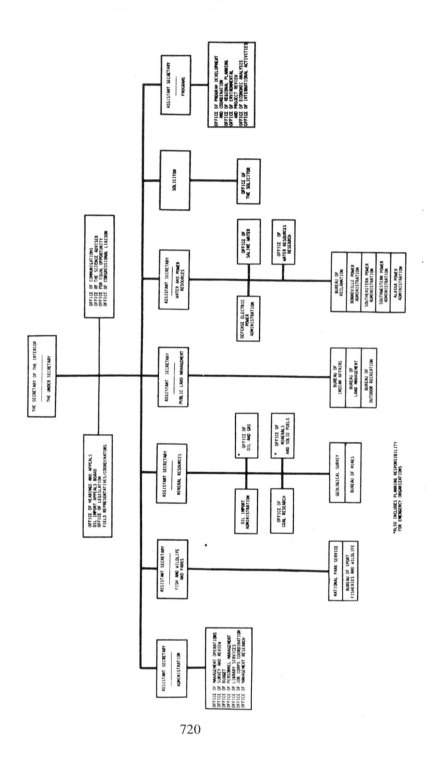

FIGURE 10.A.12   Organization of the Department of the Interior

Budget, Research Program   In fiscal year 1970, the total research budget of Interior was $122 million *; about $86 million of this research was in-house and $18 million was performed by universities. The Bureau of Mines obligated $34 million, of which $9 million was in physical and environmental sciences; and the Geological Survey $41 million, with $39 million in these two areas (see Table 10.A.19).

The Bureau of Mines conducts research on extraction, processing, use, and disposal of minerals, mineral fuels, and helium. Essentially all the work is performed in-house, in about a dozen laboratories; only recently has a small start been made on a university contract program.

The Geological Survey performs surveys covering topography, geology, and mineral and water resources. The basic-research program is mainly in geology, geochemistry, and geophysics; most of the work is rather directly oriented to solution of practical problems in these fields. As with the Bureau of Mines, most of the funds are spent in-house, although considerable use is made of temporary (summer) employees, particularly in field work. The Geology Division employs about 1800 scientists, about 1200 of whom are geologists. University participation has in the past been largely through the temporary employment program, but more recently there has been a trend toward formal contracts and grants. About $8 million was allocated for this purpose in 1970. Often such contracts result from solicitation of competitive proposals from a number of institutions by the Geological Survey. Since the Geological Survey has a large in-house effort, it has adequate expertise both to formulate problems and to evaluate proposals for research along its established lines. Partly because of its venerable age, the Survey enjoys close relations with its counterparts in the academic and industrial communities, through both former employees and faculty and student participation in the summer field programs.

General Comment

Our examination of the involvement of federal agencies in the support of physics confronts us with a rather impressive fact: Although physicists might wish it otherwise, basic physics is supported by people who expect some tangible return. Whether for better weapons, cheaper energy, or national prestige in the space race, 85 percent of the physics effort is supported by agencies with specific technological missions. On the one hand, this observation carries a certain reassurance that the agencies in question do recognize their need for, or obligation to, basic research; on the other hand, one cannot repress some feeling of vulnerability in this close link with

* Table 10.A.1.

specific missions. That so little as 10 percent of the support of basic physics should be dedicated to the advancement of the science *per se* is distinctly uncomfortable.

Quite evidently, the base of support for physics could be broadened in two ways. First, and most obvious, the NSF, as envisioned in the Bush report, should become the principal supporter of basic research. In the totality of federal support of basic science, the NSF now accounts for only 11 percent; as we have seen, it is even somewhat less in physics. At this level, the NSF can hardly be regarded as the bellwether of science; it cannot even serve very effectively as a flywheel to damp fluctuations arising in times of financial pressure. At a level twice or three times the present one, it might have a better chance.

The second direction that needs to be pursued is the awakening of other mission agencies, such as the Department of Transportation, HUD, and NIH, to the value of physics, and the awakening of physicists to the challenges of the new problems that concern such agencies. It is a commonly held belief—at least among some physicists—that physicists can do anything; however, whether they can bring their wits to bear on the kinds of muddy problems in city design, transportation systems, or human diseases that people in the outside world have to solve remains to be seen. It is almost certainly *not* true that the average physicist is better at community planning than the professional planner; but he does have something to offer, and it is up to him, and to the agency in question, to find out what it is. From the side of the agencies, what is needed is a way to discover which of their problems have a physics content, and what in physics has a bearing on their missions. To do this, they will need both to develop a substantial in-house capability in physics and to establish communication with the field through support of extramural programs in basic-physics research.

11

PHYSICS IN EDUCATION AND EDUCATION IN PHYSICS

I am much assured, there is not a learned man in all
the world who hath not found by experience, that
skill in any Faculty—is not to be attained, without a
timely beginning, a constancy and assiduity in study,
especially while they are young.

> SETH WARD (1617–1689)
> *Vindiciae Academiarum,* pp. 64–65 (1654)

A sound knowledge of at least the principles of general
physics is necessary to the cultivation of any
department.

> JOHN WILLIAM STRUTT (1842–1919)
> Third Baron Rayleigh

A theoretical science, unaware that those of its
constructs considered relevant and momentous are
destined eventually to be framed in concepts and
words that have a grip on the educated community
and become part and parcel of the general world
picture—a theoretical science, I say, where this is
forgotten, and where the initiated continue musing to
each other in terms that are, at best, understood by a
small group of close fellow travellers, will necessarily
be cut off from the rest of cultural mankind; in the
long run it is bound to atrophy and ossify. . . .

> ERWIN SCHRÖDINGER (1887–1961)
> British Journal for the Philosophy of Science
> Volume III, pp. 109–110 (1952)

723

The Meaning of Physics Education

In Chapter 3 we stated that science is knowing. As the term is generally used today it means knowing about the nature and structure of the physical universe and its inhabitants. The process of acquiring scientific knowledge is in essence education. It is the education of mankind, even if only a few participate directly, for the knowledge they gain is available to all. This Report tells elsewhere of the knowledge that the pursuit of physics brings to mankind and of the conditions necessary for continuing the pursuit.

Society ensures that some of its young are prepared to carry on the process of scientific inquiry. It asks that others be prepared to put scientific knowledge to use. But society manifestly believes that people, in general, should share some of what mankind knows about the universe and its inhabitants, inasmuch as it sanctions the inclusion of science among the subjects taught in the institutions it has created for education.

Thus, to write of science as man educating himself to the ways of nature is not to state that the whole purpose of science is to further man's control over nature for his own immediate benefit. To be sure, man educates himself to gain a livelihood but also to gain awareness from his own reflections and actions, from those of his fellows as expressed in the arts, and from what the social, behavioral, and biological disciplines tell him about himself and other living things. The inanimate universe in which he lives also influences his awareness; the view he holds of his place in this universe is a part of his awareness. Education that does not offer a view of science from this perspective, that does not stress this view, is indeed likely to lead to public misapprehension about and alienation from science.

The aspect of control over nature mentioned above is assuming an unaccustomed and urgent meaning different from the one commonly associated with it. Technology, emerging from an incubation period of several thousand years, now confronts society with the necessity of regulating its manners and further growth if a habitable steady-state earth is to be maintained. Guiding society and technology into this unaccustomed regime will be a delicate business, requiring balance between stifling restraint and reckless laissez faire. Science, which society has supported mainly as a lever for developing new technology, will now play a large role in providing the tools to manage it. Energy, matter, time, space, and even entropy (which is now written about in the popular press) will be with society in the decades ahead, in politics and economics and in national policies and international negotiations. Society, the ultimate manager of technology, must comprehend the several roles of science, as well as some universally applicable scientific principles.

Although the foregoing comments, and much that follows, refer to science in education generally, the Committee's principal objective has been to examine the role of physics in education. Whatever we may conclude about the purposes, approach, and content of physics education, we should take into account its coupling to the entire educational endeavor, a complex structure and one on which society relies heavily. Therefore, it is useful to construct simplified models such as that in Figure 11.1. Physics can be found in every element of it.

There is a dimension, however, that Figure 11.1 lacks: the diversity of education in the United States. An instructor from almost any other nation, by outlining the curriculum and methods of his own institution, whatever its type, will be giving a fairly accurate description of all such institutions in his country. This is far from true in the United States, in which there are 22,000 independent school districts and 2400 independent institutions of higher education.* A degree of conformity exists, of course, among institutions of a given class, but it is a product of informal exchanges of views about education, competitive spirit, and whatever nationally held mores and goals there may be, rather than of central planning. The provision of federal support on an increasing scale has brought with it an obligation to adhere to norms, but the obligation has resulted in convergence more of form than of educational substance. Serious concern is often expressed today over the tendency toward homogenization,† but diversity persists as an outstanding feature that renders difficult the task of describing U.S. education either by gathering a body of meaningful statistics, comparing data from year to year, or merely attempting to characterize such a relatively small component as physics education.

Figure 11.1 gives no measure of the dimensions of the enterprise it illustrates. In 1969–1970, about 30 percent of the population of the United States was numbered among the full-time students, teachers, and administrators,* and $70 billion, or about 7 percent of the Gross National Product, was expended on all phases of education.‡

In this same year, the federal government obligated about $70 million, or 0.1 percent of the cost of education, for the improvement of precollege and undergraduate science education. The total National Science Foundation (NSF) obligation for science education was $120 million.§

---

* U.S. Department of Health, Education, and Welfare, Office of Education, National Center for Educational Statistics, *Digest of Educational Statistics, 1970*, Publ. OE 10024–69 (U.S. Government Printing Office, Washington, D.C., 1970).

† U.S. Department of Health, Education, and Welfare, Office of Education, *Report on Higher Education* (U.S. Government Printing Office, Washington, D.C., 1971); H. L. Hodgkinson, *Institutions in Transition* (McGraw-Hill, New York, 1971).

‡ U.S. Department of Health, Education, and Welfare, *Digest of Educational Statistics, 1970*, p. 20.

§ *National Science Foundation Annual Report, 1970* (U.S. Government Printing Office, Washington, D.C., 1971), pp. 57 and 122.

STRUCTURE OF U.S. EDUCATION

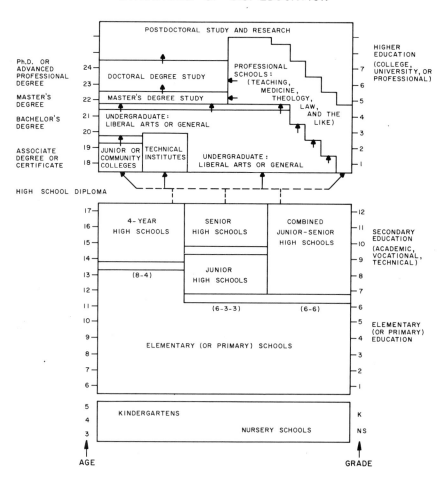

FIGURE 11.1   The structure of education in the United States. This figure presents a general picture of the structure of education in the United States. At the right side of the chart, three levels of education are indicated: elementary, secondary (high schools), and higher (colleges, universities, and professional schools). The approximate age of persons in each level is given at the left side. Three structural patterns below the college level are in common use. The pattern shown at the left is commonly called the 8–4 plan, meaning that after nursery school and kindergarten the pupils spend 8 years in the elementary school and 4 in the high school. The pattern in the center is generally called the 6–3–3 plan, indicating that after kindergarten the pupils spend 6 years in the elementary school, 3 in the junior high school, and 3 in the senior high school. The pattern at the right, called the 6–6 plan, means that pupils spend 6 years in the elementary school and 6 in the high school. All three plans lead

As is well known, the rate of growth of the enterprise has for two decades been a major determinant of U.S. education. A few examples suggest the scale of this growth. Between 1955 and 1965, the number of high school graduates increased more than 85 percent; the number of these entering college increased 110 percent. Today, more than half of the young people in the United States enter college; 20 years ago less than 25 percent did so. Currently 31 percent of the 18- to 24-year-old group is enrolled in college; two decades ago the fraction was 14 percent. In the same period, the number of institutions of higher education has increased from 1850 to 2400, and average enrollment has doubled. The number of public two-year colleges has nearly doubled, and they now enroll about one third of all college freshmen and sophomores.* Figure 11.2 shows the growth in the number of physics faculties in four categories of institutions.

* U.S. Department of Health, Education and Welfare, Office of Education, National Center for Educational Statistics, *Digest of Educational Statistics, 1970,* Publ. OE 10024–69 (U.S. Government Printing Office, Washington, D.C., 1970), pp. 49 and 66; Tables 102, 88, and 98.

---

to high school graduation at the age of 17 or 18 years. High schools generally can be classified as comprehensive or specialized. The comprehensive high school provides two or more programs in academic, vocational, technical, or general education in the same school. The specialized high school concentrates on one type of program. Large city school systems tend to specialize in the high schools, providing separate schools for vocational and technical programs. Vocational and technical high schools, however, sometimes offer the general subjects usually required for college entrance, so that a student who selects these courses can enter a college or university. Graduates of the high school may enter a junior college, a technical institute, a 4-year college or university, or a professional school. The junior college normally offers the first 2 years of a standard 4-year college program and a broad selection of terminal-vocational courses. Academic courses offered by the junior colleges are transferable for credit to 4-year colleges and universities. The technical institute offers postsecondary technical training not leading to professional degrees. Professional schools, as indicated at the upper right of the figure, begin at different levels and have programs of different lengths. For example, medical students must complete at least 3 years of premedical studies at a college or university before they can enter the 4-year course of a medical school; engineering sudents, on the other hand, can enter an engineering school immediately upon completion of a secondary school program. [Source: The figure is based on information contained in U.S. Department of Health, Education, and Welfare, Office of Education, National Center for Educational Statistics, *Digest of Educational Statistics, 1970,* Publ. OE 10024–69 (U.S. Government Printing Office, Washington, D.C., 1970).]

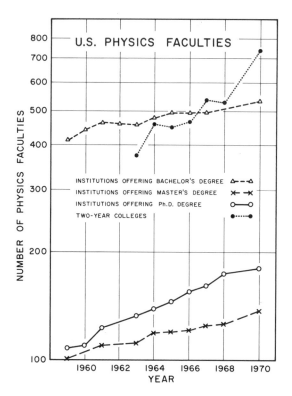

FIGURE 11.2 Number of physics faculties listed in American Institute of Physics directories, 1959–1970.

With unexpected abruptness, the period of rapid growth is ending. It now is widely recognized that, although the numbers of young people of secondary school and college age will continue to rise, the fractional growth will be much smaller than in the recent past. The population of elementary-school-age children will decrease during the next decade. Simultaneously, economic and political factors have produced a sudden leveling of state and federal support for education. For perhaps the first time in U.S. history, education has lost its special priority in the minds of both the general public and its elected officials. Effects of the stresses of rapid growth, followed by the shock of equally rapid decreases in support, are among the principal themes of this Report. Educationally, the effects on both present and future students, recent graduates at all levels, faculties, and institutions have been profound.

Of even deeper significance than some possibly transient phenomena are, of course, the changes taking place in the structure of society. Few doubt that these changes will affect the educational complex in vital ways. Thoughtful consideration and recommendations are being fostered by many agencies, public and private; the ongoing studies of the Carnegie Commission on Higher Education are outstanding. In contrast, warnings from wildly disparate directions suggest that the complex is an artifact that could, or should, be discarded or disassembled. Some state that the educational complex is decadent and pointless and that "greening" is about to take care of it; others, that constriction of freedom and choice are being brought about by managerial expertise—and, thankless child, the computer—acting in concert to promise the supremacy of accountability over education. In one case, the complex crumbles as the ivy takes over; in the other, the word "system" becomes an understatement. For either future, this chapter is totally irrelevant; it has meaning, even in concept, only for a more recognizable case.

Accordingly, we will speak to change and not to revolution. We assume, for the next decade or so, schools, colleges, universities, grade levels, and degrees. We expect loosening of structure and program but still recognizable form. We regard motivation to learn and understand as still a basic human trait, and scholarship and research still worthy activities that will be supported. The educational complex will retain its pyramidal shape; ability and drive will still be required to reach the highest levels as opportunity is more evenly distributed among the U.S. population.

Physicists, applying themselves to education, with rare exceptions have confined their attention to portions of the system that would seem to lie among the most readily isolated—the high school course for the talented and scientifically oriented student, the undergraduate physics major, and the graduate education of physicists. These interests are close to their hearts and are by far the easiest educational concerns for them to act on. Clearly, there is value in devoting such attention to the education of future physicists and future scientists, for among them are those who will help man in his further education of himself. However, for 20 years high school physics enrollment, unique among the sciences, has steadily declined in percentage * (see also Chapter 12). The times have called into question the propriety of the traditional mode for the education of physicists as well as the need for the capacity to educate them, which was built up during the 1960's through costly and cooperative federal and university effort. Despite some appearances, the education of a new generation of physicists is not a matter isolated from the rest of education.

Surely there is always good in examining the merit of any public policy;

---

* "Enrollment Increase in Science and Mathematics in Public Secondary Schools, 1948–49 to 1969–70," Science Resources Studies Highlights, NSF 71–30 (National Science Foundation, Washington, D.C., 1971).

such questioning of the value of science education should not invoke paranoid reactions from physicists. It is being accompanied, however, by a more general criticism. Some of this criticism can be interpreted as a reaction against exaggerated reliance on science as the source of cultural, ethical, and political norms. Some can be viewed as a reaction against the segregation of science as a special category of knowledge not accessible to the common person and of scientists as a special group of people having special powers. Not all of this criticism is unjust; it has been voiced in many guises by people of many backgrounds.* Schrödinger's warning, which is quoted at the beginning of this chapter, in the long run may be as useful to science as is the equation that bears his name. One of the principal problems of those who would improve science education, however, is how to compete effectively for public attention and support while still maintaining the intellectual integrity of science.

A survey of physics made in today's climate of opinion that confined itself to the examination of research and the education of professional physicists would risk enhancing whatever degree of alienation from science already exists in U.S. society. We believe that the times require a restatement of the values of physics in the education of man and of the need for emphasis of these values in the education of both scientists and non-scientists. Even more urgently they require that all physicists examine the extent to which they have been discharging their educational responsibilities. Physicists must review not only the quality and adequacy of their past efforts but also the implications of the omissions they may have countenanced.

To be specific, we find no educational need that compares in ultimate significance with the improvement of the general public's understanding of science. Some may be surprised or dismayed with the emphasis we give to this finding. Many scientists regard the achievement of improved public awareness as a problem soluble only in fantasy. The diffident attitude of society toward science (physics often is singled out) has been matched by a complementary diffidence respecting school science on the part of the scientists, the two attitudes tending to reinforce each other. Because physicists have a place, in fact a strategic place, in this circular situation that has been created, they could (and should) take actions intended to arrest the recycling of diffidence. Moreover, those who now are

---

* See, for example, "Non-scientists Dissect Science," in *Impact of Science on Society,* XIX(4) (Oct.-Dec. 1969).

members of physics faculties will realize that few new faces will join them during the next two decades; they cannot expect new entrants into the profession to relieve them of these responsibilities.

EDUCATION FOR PUBLIC UNDERSTANDING

The main task for physics in educating the public is to make real to it the way in which past generations gradually wrested a deeper view from nature, a better representation of the way things work.

In an earlier day, the phenomena of natural philosophy—mechanics, heat, sound, light, electricity, magnetism, and astronomy—all seemed distinct. Physics now encompasses and organizes them with only a few concepts and mathematical relationships. At one time the sun, moon, and stars were thought to be made of some ineffable stuff, unlike anything on earth. Galileo's first observation of the moon through his telescope and Newton's demonstration that the matter of which it is made behaves in the same way as the familiar matter of the earth were the precursors of man's recent lunar visits, each made with full confidence that man would neither be eaten by dragons nor vanish in a puff of gamma rays. Both the extension of earthly physics to the most remote galaxies that scientists have detected and the application of laboratory knowledge of atoms, nuclei, and subnuclear particles to the understanding of such phenomena pose problems. But neither effort has encountered any real contradiction; the same fundamental physical laws that apply to the earth hold also, as far as physicists can determine, for the universe.

Today the phenomena of chemistry and biology are described with greatest depth and clarity in the terms of physics and through the application of the instruments of physics, although their more complex aspects, especially those of biology, also require the specialized concepts and techniques of the particular discipline. The practical arts, which depend heavily on familiarity with materials, grow increasingly less dependent on empirical knowledge and more dependent on theoretical understanding gained through physics. It seems fair to say that one of the most far-reaching consequences of the pursuit of physics has been a continuing unification of knowledge, which makes it possible for man to retain the concept of the wholeness of nature while exploring it in all its rich detail. This consequence must be counted a humanistic achievement of high order. (See also Chapter 3.)

Unfortunately, in the United States the physicist's model of the way things work is now found in the minds of only a small fraction—say, one tenth—of the population. To accept this situation is to admit that most U.S. citizens are living in the shadow of a philosophy of natural events that was shown to be inadequate centuries ago. Yet the world they inhabit is embedded in space and time, is fashioned of matter and energy, and changes by the interplay of energy and entropy. To live without knowing these facts and understanding at least some of their implications means, for most of modern society, living with power it cannot master.

The kind of public understanding that is needed goes beyond appreciation of the significance of a speed of 25,000 miles per hour or beyond scanning the *Scientific American*. There are aspects of contemporary science that have to do with the ideas and the questions about nature that man has asked throughout history—for example, is all matter the same, ultimately? Is the universe finite or infinite? The public should realize that these persistent questions have their modern interpretations, that they remain the subject of investigation and debate, and that the names of contemporary scientists can be attached to their pros and cons, just as can those of the ancient Greek and medieval natural philosophers.

Martin Green, a literary critic and teacher of English, who, challenged by the "Two Cultures" debates a few years ago, completed courses in mathematics, physics, and biochemistry, wrote subsequently *:

We need a new humanism, to show how a man can know what Rutherford knew and what Eliot knew, and can become, not despite but because of that knowledge, fully a citizen of our world. We need to break the tabu of incommunicability that has been laid on that knowledge; to learn to transmit it to those with equal though different intellectual experience. The belief that a non-scientist cannot achieve any significant understanding of science must be dismissed as a delusion, a symptom of the disease itself. The difficulties are of course great. But they cannot be insoluble. For the problem is not to acquire a certain amount of information, or even understanding, but to employ a certain amount of serious attention. It is *our* activity, the way we act on our tiny section of physics or chemistry, the depth in us to which we are engaged, not the depth of the science we know, that is important. What we need is intellectual engagement, imaginative participation at the same level of intellect and imagination as we know in our own subject; though nine-tenths of this material must be taken on trust. A good teacher of science can do that in a year starting literally from scratch, just as a good teacher of literature can.

To teach physics the way Green asks, which of course is the way all physicists would ask, poses a fundamental difficulty that seldom can be overcome in a year of study, at least in the United States. The difficulty arises in part because physics is inherently a quantitative discipline and an

---

* M. Green, *Science and the Shabby Curate of Poetry* (W. W. Norton & Company, New York, 1965), p. 74.

experimental one that requires for the study of its contemporary state the use of mathematics often foreign to the student and, to a lesser extent, apparatus that is advanced and costly. Moreover, at any given time the concepts used at the working level are abstract and not a part of common experience. Until the present time, U.S. educational practices have been prompt and effective in sorting out (or creating?) two populations, roughly speaking, one of which has a proclivity for quantitative disciplines and another larger one that tends to shun the quantitative and instrumented. Presented in a college course with this second group, an instructor in physics must weigh two strategies. He can either take sufficient time to develop carefully the most basic ideas, making certain each is assimilated before progressing, thus forfeiting most of contemporary physics, or he can undertake a more superficial survey of all of physics.

There is another totally different approach, that is, to attempt to alter the conditions that bring about this division in the population.

In this section we treat several approaches, because we feel compelled to attack the problem of scientific literacy for the public in any promising way. However, success, if possible at all, would require a time measured in generations and still might never imply that "a man can know what Rutherford knew and what Eliot knew" in any literal sense. It is most unlikely that any one person in the seventeenth century knew what Galileo knew and what Donne knew, or in the nineteenth century, what Maxwell knew and Henry Adams knew. Possibly, neither the arts nor the sciences would benefit if all tension between them were to vanish. What is important for all society is that there be a general realization that both science and the arts are the creations of man and that both extend his vision.

Physics, as aesthetic and fascinating as its intellectual aspects are, appeals to other facets of human interest because it integrates the ingenious and the practical with the abstract and philosophical. It is as much the shaving of a metal surface in a vacuum and the use of a phonograph turntable as it is tensor analysis and debate over chance versus determinism. There is an appropriate distribution of emphasis for differing educational contexts, and all should be exploited.

The Physicist's Stake in the Schools

Models of the world enter the mind in the earliest years. Well founded, based on tentative and testable schemes, they can be developed and adapted throughout a lifetime. Insufficiently anchored, ignored, or derived from lore or unprocessed perceptions, they can limit man's outlook and activities throughout his life, even make him unsure or afraid, and distort a whole society. This is the stake that society, and the physics community in particular, has in good physics in the schools.

Physicists often sense deeply the powerful fascination that some young people feel for the world of physical phenomena. A recent study describes this power and its effect on those who are susceptible to it.* Prior to the ninth grade, half of those children who will concentrate on physics as college students already have selected science as their major field of study; the figure reaches 95 percent prior to graduation from high school. For the sake of their own profession, if for no other reason, physicists have cause to be concerned about the teaching of their subject in the schools.

But good physics is not merely verbal formulas, nor is it the memorization of facts, dexterity with this or that instrument, or the solution of test problems. These are all surface manifestations. Greater depth does not imply only cognition, although it includes cognition. It also encompasses an emotional component, made manifest in attitude, interest, even delight. An example of understanding in depth is a student making what is for him a new sentence—not a learned one, somewhat rephrased, or one chosen from a set of alternatives, but one made from words he knows, to voice his own observation or deduction.

There was a time when the task of the schools was to teach symbols and their manipulations. The Yankee farmer and his family lived in a rich flow of experience, but symbols were in short supply, and abstract knowledge was unneeded and virtually unknown. The Bible was the TV set. Now the situation is reversed, and the school has a new task. The factory worker and his family seldom see the sky, and when they do, it is a static source of wonder, not a dynamic one of vital information about direction, wind, and weather.

Experience today is social interaction with others and observation of a TV screen. Phenomena at a simple level are rare, but society is inundated with symbols. What must the schools do? The answer seems clear: Provide guided, inquiry-oriented experience with phenomena.

A start on improving school science education was made more than 15 years ago. To active physicists, the old high school curriculum, textbooks, and laboratory experiments spoke trivially and dully to the physics of an earlier century. The task was to create a revolutionary new curriculum for the entire nation or at least a model that would stimulate change in many locales. The agency for doing so was the Physical Science Study Committee (PSSC), a voluntary group of research physicists, college and

---

* W. R. Snelling and R. Boruch, "Factors Influencing Student Choice of College and Course of Study," *J. Chem. Educ., 47,* 326 (1970); see also J. R. Butler, *Occupational Choice* (Her Majesty's Stationery Office, London, England, 1968).

university instructors, and high school teachers supported by the National Science Foundation (NSF). Its product now is offered to a substantial fraction of those who take high school physics. Most physicists know about this course and recognize it as a landmark, although its effectiveness has been difficult to quantify. (Further discussion of it occurs in a subsequent section.)

No surprise can be attached to the primacy given by the critics of that time to the high schools. The students enrolled in college courses come almost without exception from high school physics courses; as we have just noted, 95 percent of the eventual physics majors have had this experience.* The high school students of physics, generally seniors, were largely self-selected; of considerably above-average attainment (see Chapter 12), they had made a commitment of some sort to intellectual development. The teachers of high school physics were in some way related professionally to the college and university instructors.

The mathematicians, embarking on a global transformation of all of school mathematics, and the PSSC, with its panoply of texts, supplemental monographs, films, and new laboratory apparatus, opened an age of massive curricular design. The high schools soon had available multiple new curricula in biology and chemistry. An alternative approach to physics was constructed. A course based on technology, with emphasis on computers, was introduced. Simultaneously it became apparent that these courses demanded better preparation in the lower grades. The ardor of a few talented teachers from the research ranks and the schools, combined with the communal experience gained from the high school episode, gave impetus to several major projects to enliven science in the elementary grades. The traditional ninth grade course in general science was regarded as occupying a key position. It is, in fact, the last time that a large fraction of the pupils (67 percent in the public junior high schools †) have any contact with the physical sciences. This course is receiving attention from earth scientists as well as from physicists and chemists.

The spirit of this educational movement enabled it to leap the barrier into the colleges with the design and introduction of a physical science course for students who would not major in one of the sciences or technologically based professions. The PSSC course, which has been regarded by some as too demanding for high school students in its undiluted form,

---

* *Summary Report. Survey of Physics Bachelor's Degree Recipients* (American Institute of Physics, Division of Education and Manpower, New York, 1971), Table 1.
† L. E. Rogers, *Science Teaching in the Public Junior High School* (U.S. Department of Health, Education, and Welfare, Office of Education, Bureau of Research, Washington, D.C., 1967), Table A.

appeared to have potential for the education of the college nonscience major, so an appropriate version of it was prepared. In turn, the physical science course for college students also presented possibilities for useful application to secondary school general science courses; consequently, a suitable modification of it now is being used in many of these classes.

The age of massive curricular design has not ended, at least for the schools. A recent count * shows 20 curricular projects in elementary science, 13 others in general science, 12 in integrated physical sciences, and 14 specifically in physics. We cannot describe all of these but will mention certain ones to suggest their flavor.

Grade School Physics

Three new curricula for the elementary schools are known as "Science, A Process Approach," † "Science Curriculum Improvement Study," ‡ and "Elementary Science Study." § All three merge sciences rather than split them, but all, inspection readily shows, rely heavily on physical sciences. In outlook and detail they differ widely. The most nearly conventional emphasizes the conceptual growth of the student and directs each experiment at some specific objective, be it an understanding of measurement or of animal feeding. Another, stimulated by a theoretical physicist, directs more of its effort, though by no means all, at an explicit introduction to some general concepts such as systems and interactions. Its materials often become symbols: arrows, little men for observers, or diagrammatic planes. The third, unabashedly opportunistic, presents real materials, almost never symbolic, and demands much more of a research atmosphere. A paper by a leader of the team originating this curriculum speaks of "messing about in science" (quoting Rat in *Wind in the Willows*) as an indispensable precondition to more formal and compact learning.

Studies of the efficacy of the new curricula are beginning to appear in the literature. For example, Stafford ‖ tested 60 beginning-first-grade children who had been exposed to inquiry-oriented science and an equal number who had been exposed to a conventional science program; the

---

* R. F. Tinker, Commission on College Physics, University of Maryland, private communication.
† American Association for the Advancement of Science; Xerox Corporation.
‡ Science Curriculum Improvement Study, University of California, Berkeley; Rand McNally.
§ Education Development Corporation, Newton, Mass.
‖ J. W. Renner and D. D. Stafford, "Inquiry, Children and Teachers," *Sci. Teacher, 37,* 55 (1970).

remainder of their program was the same. An attempt was made to match the groups in regard to readiness for first-grade work; however, the *control* group had a definite, although small, advantage. Tests based on Piaget's study of the development of the notion of conservation were given at the beginning of the term and again after four months of schooling. The main feature of the result was that the test group not only outscored the control group, but its gain in score was 50 percent larger than that of the other.

More recently, Weber and Renner * have studied the performance of elementary students after nearly five years of schooling. Again, two groups of 30 students each were identified, one of which had experienced science only through inquiry-oriented programs, and the other, through a conventional one. The two groups were matched in all other criteria deemed relevant to this study. They were tested in six areas of performance: observing, measuring, classifying, experimenting, interpreting, and predicting. In each of these categories the test group gave a substantially larger number of acceptable responses. Statistically, the performance of this group was shown to be superior, with an extremely high degree of confidence. The authors include observations not represented in the quantified data: the inquiry-trained pupil seemed to be more creative in designing possible solutions to the tasks set and was less willing to give up than his conventionally taught counterpart.

Although these studies cannot yet be considered definitive, they strongly underscore the value of asking what will happen to 20 newly minted elementary science courses of this kind. Do the schools have a place for them? Is it realistic to imagine physics as a component of the elementary curriculum? Who will teach the courses? A massive compilation of statistical information about science teaching in the school system can only suggest answers to these questions.

A report, *Science Teaching in the Elementary Schools,*† based on a survey made a decade ago, remains the most comprehensive source of information about, and probably a fair description of, the subject. From it one learns that U.S. public elementary schools regard the teaching of science as an important responsibility in all grades from kindergarten on. It is most frequently taught as a separate subject, increasingly so with grade progression. About 50 minutes per week are given to science in the kindergartens; well over 200 minutes per week of science are taught in the

---

* M. C. Weber and J. W. Renner, "How Effective is the scis Science Program?" Unpublished manuscript.
† By P. E. Blackwood, U.S. Department of Health, Education and Welfare, Office of Education, 1965.

seventh and eighth grades of large schools. The national average for all grades, weighted by student numbers, is about 100 minutes per week. No direct information is available on the fraction of this time given to topics that could be called physics. An indirect measure is gained from a section of the survey seeking information on the availability of certain kinds of science equipment: About 40 items were listed, and 15 of these were distinctively physical in nature; others, although also physical, were more generally applicable. Three main divisions of science seem to be covered— physical sciences, earth sciences, and life sciences. The assumption that one third of the time is available for physical sciences, with a large component of this and some from earth sciences being physics-based, does not appear unreasonable. Accordingly, we adopt one third as a reasonable fraction for time given to physics. Each student experiences an average of 350 hours of science instruction in the years from kindergarten through grade 6, of which 120 hours is estimated to be physics.

Let it be clearly understood, however, that, by pointing out that a certain number of hours are available for physics, we are neither advocating that school children be taught that science is a set of exclusively defined disciplines nor sanctioning the present distribution of effort, either among the broad science areas or over the entire range of studies. Rather, we are calling attention to the present allocation of time in which topics from physics can and should be an important component of the integrated general science appropriate for students at this stage of their development.

In more than 80 percent of the schools, the classroom teacher alone handles science in grades up to and including the fifth; for higher grades, this is true in 70 percent of the schools. Consultant help is available to more than 40 percent of the schools, but, of these, only 15 percent have an elementary science consultant and 29 percent a classroom teacher with special competence in science. About half of the schools for which such help is available report using it rarely or never; when such consultants are used, it most often is to provide materials. Only 15 percent of all schools spend as much as $1.50 per pupil per year on science supplies and equipment; half spend less than 40 cents.

It is encouraging to learn that about 3.33 million grade school pupils, about 10 percent of the total, are engaged at present with the three largest new curricula.* Their use requires two to three hours per week averaged over six grades. However, serious obstacles stand in the way of their widespread adoption. Among the obstacles, two are prominent. One of these is cost; the new curricula demand an expenditure of about $10 per

---

* R. F. Tinker, Commission on College Physics, University of Maryland, private communication.

pupil per year (the national average total expenditure per student-year in the elementary grades is about $800 *). The other is teacher preparation, which we discuss later in this chapter.

Intermediate Science

Not long after the introduction of the PSSC courses and those in chemistry and biology, reasons for operating on their precursor, the conventional intermediate-grade general science course (or the somewhat more diffuse science curricula of the junior high schools) became widely apparent. The challenge of the new high school courses was great, and stronger preparation for them was necessary. The notion of an integrated class-room–laboratory, with simple instrumentation, appeared readily transferable from the elementary school context; some of the new high school materials were perceived as conceptually and practically adaptable, with little modification, to the general science curriculum. Additionally, and perhaps most persuasively, the general science course was (and still is) the last opportunity to reach a large proportion of the population. Seventy percent of all students take this or a physical science course; it enrolls five times the number of students that high school physics does, and girls populate it in equal proportion to boys, whereas the ratio of girls to boys in physics is only 1:3.†

About 25 general science and physical science courses have been developed as a result of these considerations. One of them, "Introductory Physical Science," ‡ was developed by a group continuing the work of the PSSC. Another, "Probing the Natural World," § comes from the Intermediate Science Curriculum Study, a group centered in Florida State University. At least one additional year-long course in physical sciences || comes from a team in which a former PSSC consultant played a leading role. As one of its features, this course introduces the caloric theory of heat to enable pupils to discover that a reasonably satisfactory theory can be discarded when a better explanation for an observation is found.

---

* U.S. Department of Health, Education, and Welfare, Office of Education, National Center for Educational Statistics, *Projections of Educational Statistics to 1979–80* (U.S. Government Printing Office, Washington, D.C., 1971), p. 82.
† U.S. Department of Health, Education, and Welfare, Office of Education, National Center for Educational Statistics. *Digest of Educational Statistics, 1970,* Publ. OE 10024–69 (U.S. Government Printing Office, Washington, D.C., 1970).
‡ Introductory Physical Sciences (IPS) Group, Educational Services, Inc., Prentice-Hall, Inc., Englewood Cliffs, New Jersey.
§ Intermediate Science Curriculum Study, Florida State University; Silver Burdett Co., Morristown, New Jersey.
|| *Physical Science, Laboratory Approach,* J. H. Mareau and E. W. Ledbetter, Addison-Wesley Publishing Company, Reading, Mass.

Most of these efforts share the view that a systematic approach to the understanding of matter and energy is basic for all science. They attempt to build the pupil's confidence in his own ability to answer questions by experimentation and observation. They intermix manipulation, activity, and reading and face honestly the practical difficulties of even the simplest experimentation, which produces ambiguous results at times. Questions often are deliberately ambiguous, and some, also deliberately, require more background than the course offers. One common theme is to make evident that "Science does not deal with absolute truth." * Another states: "When we 'explain' something, we build a bridge between the familiar and the unfamiliar by using mental pictures and theories and by making comparisons." † The mode is one of combining student initiative and discovery with guided progress, the teacher being urged to remain in the background.

The High School Courses

Two well-known courses ‡ dominate high school physics today. One of them, that of the PSSC, has been in use for a decade or more and has gained wide national, and even international, acceptance. "The Project Physics Course," having completed its incubation period, is just now beginning to reach a substantial number of students. Each of these courses has been shaped by a main idea; PSSC strives to present a rich, tight line of inductive argument, whereas Project Physics unfolds its subject in the context of the intellectual history of man. Both were produced through the collaboration of college level and high school instructors. Both accept the structural assumptions of the conventional high school—classes, books, tests, hours—although the newer course provides a set of options, as a step toward a more relaxed format. The target of both is the student who has academic life well in hand, although portions of Project Physics, which makes a genuine effort to be attractive to students who are not so strongly science-oriented, have been used with disadvantaged students. For those more inclined toward practical concerns and applications, there is a third course, "The Man-Made World." § This course is based on engineering concepts such as feedback, stability, and optimization; it introduces students to mathematical modeling and to a simplified computer. Yet it has not met with a popular response.

---

* "Introductory Physical Science."
† "Physical Science, Laboratory Approach."
‡ *Physics* (The Physical Science Study Committee, Educational Services, Inc.; D.C. Heath and Company, Boston, Mass.); and *The Project Physics Course* (Harvard University; Holt, Rinehart and Winston, New York, 1971).
§ Engineering Concepts Curriculum Project.

In the face of the effort expended on these projects, the statistics on high school physics enrollment are distressing. The NSF reports * that "in every [science] course except physics a greater proportion of the students in the appropriate grade enrolled in the course in 1969–70 than in 1948– 49. . . . [T]he proportion in physics declined steadily." The figures for physics in U.S. public high schools are: 1948–1949, 28 percent; 1960– 1961, 23 percent; 1969–1970, 18 percent. These data are compared with those for other subjects and with total high school enrollment in Figure 11.3.

The impact of Project Physics has yet to be registered in the statistics. Its proponents believe that, through appeal to a distinctly broader segment of the student body, including girls, it will alter the trend in physics enrollments. However, if the lesson of the statistics is in fact the one we deduce, optimism in regard to the impact of Project Physics should be restrained.

A more striking example of the subtleties involved in producing educational change by design is difficult to imagine. The care that was exercised in the development of these new courses was immense. In them concepts are developed on a sound basis of experiment and analysis. The unfolding of the material follows a clear line. The physics is as nearly impeccable as one could hope. Several versions were tried before final publication. They were received enthusiastically. What is the problem?

Although no dogmatic answer will suffice, it is difficult to escape the conclusion that the desire for a tightly organized, logically constructed presentation tends to assume precedence over the realities of the classroom, that is to say, the nature of the learning process, the student's need for repetition, and the teacher's need for breadth to enable him to handle the unexpected (for not even a committee can anticipate all the confusions and misinterpretations of a student).

The material of PSSC, Project Physics, and Man-Made World will retain its validity and will be used for a long time. The issue is the manner in which it is used. Goethe has written, "Most thoughts have been thought already thousands of times; but to make them truly ours, we must think them over again honestly, till they take firm root in our personal experience." A model that is well adapted to learning in this spirit is that of the new elementary curricula.

In fact, in only a few years high schools will be populated with students accustomed to the inquiry mode of learning. They will need a less linear, verbal, and formal course structure than is currently incorporated in any

---

* "Enrollment Increase in Science and Mathematics in Public Secondary Schools, 1948–49 to 1969–70," Science Resource Studies *Highlights,* NSF 71–30 (National Science Foundation, Washington, D.C., 1971).

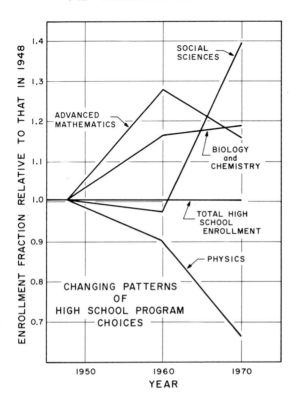

FIGURE 11.3   Fractional enrollment in certain high school science courses relative to the fractional enrollment in those courses in 1949. Note that the drop in physics and the lower slope in biology and chemistry took place during the decade following the introduction of the new high-powered curricula. [Source: The National Science Foundation.]

of the major high school curricula. The task for curriculum planners or teachers of high school physics is to adapt the excellent material now available to a freer, more flexible style of presentation and use.

This conclusion, that a change in teaching procedures is needed, is found also in a recent report by Elizabeth A. Wood * that was prepared for the American Institute of Physics. Her survey of the problems of science teaching in the schools was based on interviews with students, teachers, and administrators. Anecdotal rather than statistical, it is not readily summarized.

The author offers specific recommendations, all of which are pertinent to the present report. Many are as valid for the lower grades as for high schools. We quote a few that reinforce the attitudes expressed here:

We must begin to plan now for part-of-the-day teacherless teaching in the schools. This involves not only better teacherless materials, but also attitude reorientation on the part of school officials, parents, teachers and especially pupils.

* E. A. Wood, *Pressing Needs in School Science* (American Institute of Physics, New York, 1969).

Since projects undertaken by bright ambitious students soon get beyond the stage at which the teacher is able to give needed advice, it would be desirable to have a panel of local scientists in colleges or industrial laboratories who were willing to act as advisors to high school students.

Materials are needed for the students for whom reading and writing constitute a formidable barrier to the learning experience. In science we have an opportunity to "hook" them on the fun of learning through experimental work, with the exciting possibility of motivating them to read as a result.

Information covering science requirements for college entrance for a representative sample of colleges should be printed in concise form and made available to the schools. Misconceptions about college requirements inhibit innovation.

Opportunities should be sought for sympathetic cooperation between the education community and the science community.

Finally, Dr. Wood sees "an increasing divergence between the level of science in the text books (in terms of degree of abstraction, expected background, vocabulary and general sophistication of approach) and the level of competence of the *average* student and the *average* teacher to cope with a science course."

If the intent of high school science is general education rather than the generation of specialists, many ask, why should not chemistry and physics be taught in combination? Is there not a core of subject matter common to both, and is not this core the principal object of attention in both of the present subjects? Is it not because of tradition and vested interest that a pattern a century or more old still continues? And if a suitable combined course were to be introduced, might it not arrest the disturbing trend in physics enrollment?

These are cogent questions. Temperature, pressure, atoms, molecules, energy—the list of common topics is indeed lengthy. Both subjects are quantitative, both are laboratory sciences. Differences in research conducted today in physics or chemistry departments often depends more on local history than logical distinctions.

However, the two subjects are not identical. A report on education in chemistry * states that "a reasonable short definition of the scope of chemistry is: Chemistry is the integrated study of the preparation, properties, structure, and reactions of the chemical elements and their compounds, and of the systems which they form." A high school textbook † explains, "physics . . . deals with such features of the world as time, space, motion, matter, electricity, light and radiation; and some features of every event

---

* International Conference on Education in Chemistry, Division of Chemical Education, American Chemical Society, *Preliminary Report* (ACS, Washington, D.C., 1970), p. 11.
† The Physical Science Study Committee and Educational Services, Inc., *Physics* (D.C. Heath and Co., Boston, Mass.).

that occurs in the natural world can be seen in these terms." Incomplete as any such brief statements must be, they do convey the impression of distinctly different interests and approaches. The questions then reduce to whether there is pedagogic merit in bringing together two disciplines that in a certain basic sense are different, though they have much common subject matter. For a course designer it becomes a matter of finding a natural progression or alternation between the basic material of physics, that is, space, time, motion, force, energy, light, electricity, and magnetism, and that of chemistry, which is elements, compounds, reactions, synthesis, and analysis. From the standpoint of conventional physics these subjects of major concern to the chemist occur very late in the progression of manifestations of basic physical notions and laws. If they are not to be so treated, then a carefully designed sequence of topics is necessary so that the course will not consist of a collection of inhomogeneous lumps of subject matter. Until recently, the principal efforts at a combined approach have been directed toward the college level; at least one of these * shows promise of adaptability for use in secondary schools, if teachers are prepared to move back and forth easily between the two disciplines.

Those Who Teach Physics in the Schools

In 1970, the elementary classrooms contained 1.2 million teachers, the secondary schools about one million.† Apart from growth and changes in the pupil–teacher ratio, the turnover rate of teachers is about 8 percent per year. The U.S. Office of Education estimates ‡ that between now and 1980 a small decrease in elementary enrollment and teaching staff will compensate for increases in the secondary grades, the consequence being that the total number of pupils and teachers will remain nearly constant, as will the turnover rate. The total expenditures by elementary and secondary schools in 1970 amounted to nearly $50 billion.§ The salaries of classroom teachers constitute about $20 billion.

Despite the existence of much detailed information about teaching practices in elementary schools, the subject-matter preparation of the teachers does not appear to have been a prominent item for study. The

---

* PSNS Project, Rensselaer Polytechnic Institute, *An Approach to Physical Science* (Wiley, New York).
† U.S. Department of Health, Education, and Welfare, Office of Education, National Center for Educational Statistics, *Digest of Educational Statistics, 1970,* Publ. OE 10024–69 (U.S. Government Printing Office, Washington, D.C., 1970).
‡ U.S. Department of Health, Education, and Welfare, Office of Education, National Center for Educational Statistics, *Projections of Educational Statistics to 1979–80* (U.S. Government Printing Office, Washington, D.C., 1971), p. 82.
§ U.S. Department of Health, Education and Welfare, Office of Education, National Center for Educational Statistics, *Digest of Educational Statistics, 1970,* Publ. OE 10024–69 (U.S. Government Printing Office, Washington, D.C., 1970).

most recent report from the U.S. Office of Education on science teaching in the elementary schools does not take up the matter of teacher preparation. In the absence of direct information, we turned to two indirect sources. The American Institute of Physics * estimated the average requirement in physical science imposed by 50 state universities on elementary education majors in 1961 as 2.9 semester hours. A second source was the records of 247 students in one university who completed requirements for an elementary provisional teaching certificate in June 1971. Seven percent listed a physics course and 8.5 percent, a chemistry course. The remainder satisfied the science requirement through descriptive courses in biology and the earth sciences.†

The subject-matter preparation of secondary schoolteachers of general science is presented in Figure 11.4, which summarizes data obtained in a recent survey by the NSF.‡ Some uncertainty exists because the NSF did not collect data specifically on general science; instead, the diffuse category, "General Science and Other," was used. In addition, there was a large fraction of nonrespondents to the survey. The surmise that the data do not mask a substantial background in physical science is supported by the similarity between the distributions shown by the two extreme sets of bars in Figure 11.4. In an earlier survey,§ the NSF ignored the physical science preparation of general science teachers. The conclusion that general science teachers are typically deficient in physics, and to a lesser extent in chemistry, at least as indicated by their college training, is clearly borne out by these data. About two thirds of the U.S. public junior high schools offer in-service training of some type for their science teachers, and about 80 percent provide some kind of specialist assistance. However, the subject-matter distribution of these modes of assistance was not examined in the survey.

The 1969 NSF survey shows that no more than 1 percent of those teachers whose major teaching assignment was general science reported a bachelor's degree in physics, and a like fraction reported a master's degree in this subject; significant fractions were found in biology and general science, and small fractions in chemistry and mathematics. A somewhat paradoxical finding, albeit hardly a comforting one, was that 19 percent of these teachers reported graduate credits in physics (the same percentage

---

* *Physics Manpower and Educational Statistics* (American Institute of Physics, New York, 1962).
† R. C. Salyer, College of Education, University of Washington, private communication.
‡ *Secondary School Science Teachers, 1969* (National Science Foundation, Washington, D.C.). We are grateful to J. C. Lewis and J. H. Andrews for additional data not included in their report.
§ *Secondary School Science and Mathematics Teachers* (National Science Foundation, Washington, D.C., 1963).

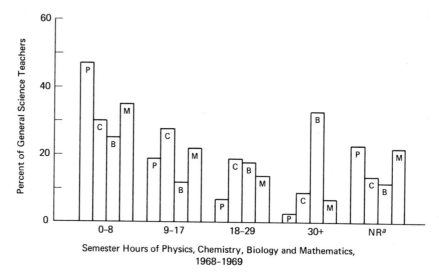

FIGURE 11.4 Preparation of general science teachers in basic sciences and mathematics. *NR=no response.

as for mathematics; 21 percent had graduate credits in chemistry; and 33 percent, in biology). Finally, a closer inspection of the data shows that 8 percent of the teachers had *no* college physics; for these teachers, their own exposure as pupils to general science probably constituted their only preparation for teaching the physics component of this course.

In view of the strategic position occupied by the specific teaching discipline, general science—and its subsidiaries, physical sciences, biological sciences, and earth sciences—we suggest that in future surveys the NSF report unambiguously on the preparation of those who teach these subjects and on the enrollments in them.

The preparation of those who teach physics in secondary schools, measured by physics courses taken, apparently has improved, as can be seen in Figure 11.5A. A detail omitted, because the comparison was not published, is that 2 percent of the 1968–1969 sample reported no credits in physics. Virtually all the teachers hold a bachelor's degree (not necessarily as highest degree), of which about 20 percent are in mathematics, 17 percent in physics, 14 percent in chemistry, and the remainder largely in general science and the other natural sciences. Only 4 percent have a bachelor's degree in education. About 10 percent hold a master's degree in physics and about 13 percent in other physical sciences or in general science.

Undoubtedly, because of the relatively small enrollment in physics, the

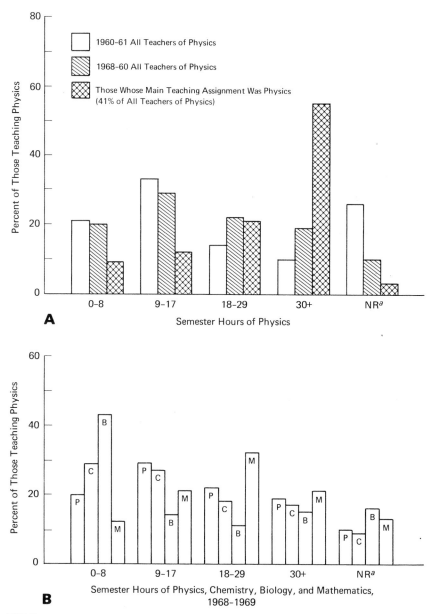

FIGURE 11.5 (A) Physics preparation of all teachers of physics in 1960–1961, all teachers of physics in 1968–1969, and those, in 1968–1969, whose main teaching assignment was physics. (B) Preparation of high school teachers of physics in basic sciences and mathematics in 1968–1969. "NR=no response.

assignment of teachers to this subject remains more haphazard than in the case of the other sciences; 59 percent of those teaching physics reported that their major teaching assignments were in other fields. This figure has not changed appreciably since the 1961 survey. The corresponding figures for other fields are: mathematics, 20 percent; biology, 25 percent; chemistry, 40 percent; earth sciences, 40 percent; general science, 40 percent. The 41 percent of those teaching physics who reported that physics was their main teaching assignment have a substantially stronger background than the average physics teacher. However, 9 percent of this group had no more than a year of college physics, and this subject, the report indicates, was their weakest science field.

We note that in conformity with earlier surveys and enrollments in the different science courses, the teaching of physics as a major assignment is confined to a small fraction—3 percent of all secondary school science teachers. This fraction should be contrasted with those for biology, 13 percent; chemistry, 6 percent; and mathematics, 38 percent. Like physics, earth sciences teaching is the major assignment of 3 percent. These figures have a special significance for those who plan and conduct the preparation of secondary school science teachers: No matter how well thought out, a teaching *major* in physics can be a productive enterprise in only a few special locations.

For teachers in service there has long been the expectation that a variety of supplemental training programs would be effective measures for refreshing their acquaintance with the subject matter they teach. These programs take the form of summer institutes, in-service courses, academic-year institutes, and the like. Many agencies, private and public, have provided support for all such activities. The NSF has been by far the most active (often under the prodding of the Congress).

Each year since the mid-1950's about 15 percent of the nation's high school science teachers have participated in an institute; the NSF has supplied more than half a million stipends for this purpose.* The Foundation estimates that 40 percent of the present secondary school teachers have attended an NSF-sponsored formal program, and that another 20 percent have participated in other programs.† Very little evaluation of the effectiveness of supplemental programs has been published except for anecdotal accounts collected by questioning participants. One published in-

---

* *National Science Foundation Annual Report, 1970.* [It is stated here (p. 68) that nearly 50,000 individuals were expected to participate in institute programs during 1970–1971.]

† *Secondary School Science Teachers, 1969* (National Science Foundation, Washington, D.C., 1971).

vestigation * concluded, on the basis of three standard tests given prior to and on the last day of four summer institutes for high school teachers, that participation had produced a gain in subject-matter mastery and in understanding science as a process. Quantitatively, however, the gains in mean scores for the three tests were 13 percent, 6 percent, and 2 percent—so slight that it is doubtful whether any long-term effects exist. There also is considerable anecdotal evidence to support the view that summer institutes are often presented at the same breakneck speed that contributes to the necessity for them in the first place. It is not surprising, therefore, that the Executive branch of the federal government has sought for several years to curtail expenditures of this kind.

Judgments about the future of the summer institute programs must be made carefully, however. It will always be necessary for science teachers at whatever level to update their information and teaching procedures from time to time. Moreover, summer institutes have been the principal way to disseminate the new curricula and methods among in-service teachers; in fact, currently no other means exist. The interest of the federal government in science education, which led to the development of these new curricula, must extend to their dissemination. To be effective, institutes for this purpose must be monitored carefully to make certain that they follow and accurately transmit the new approaches and procedures. The support program, therefore, must contain provision for some kind of control and evaluation.

Undoubtedly, supplemental programs have been of value; they could be more fruitful if conceived and carried out with greater comprehension of the problem they are supposed to alleviate. However, they do not constitute a facile solution to the problem.

Beneath the Surface

Relevant and intriguing though they be, numerical statements about the courses teachers have taken, such as those quoted in the preceding section, do not assess the quality of either preparation or performance. Over a period of years, many surveys of teacher characteristics, in addition to asking about course credits, have included questions on attitude in an attempt to elicit factors important to successful teaching, but these questions often suggest desirable responses. The staffs of summer institutes and others actively engaged in attempting to influence the attitudes and behavior of

---

* W. Welch and H. I. Walberg, "An Evaluation of Summer Institute Programs for Physics Teachers," *J. Res. Sci. Teaching, 5,* 105 (1967).

the teachers gain a different insight into the knowledge and attitudes that the teachers actually carry with them in daily life. It is interesting to juxtapose two conclusions, one from each source:

> The science teachers have a strong desire for self-improvement; they have participated extensively in NSF and other educational training programs; they have a good subject-matter foundation in their backgrounds which they value; and generally they felt well-prepared to teach their subjects and were committed to continuing their careers in education.*

> We at Harvard Project Physics are already encountering great difficulty with teachers who have no background in the history of science, in the philosophy of science, and have a confused idea about the nature of the laws of technology and science. The teachers we now have in the high schools are the product of the educational system which we have been running.†

The second of these quotations is consistent with an observation made a decade ago by Gruber,‡ who tested a group of 202 high school science teachers by asking each to make an outline, on some science subject he knew well, that would be suitable for a 20- to 30-minute talk by a high school senior. Gruber analyzed these outlines to determine the extent to which they showed perception of science—as a way of studying, of investigating and of viewing things; gave attention to theory, uncertainty, development, leading to information and comparison of ideas or approaches; and incorporated informational aspects into the process-of-science framework. He found that more than 60 percent of them gave negligible emphasis, and another 10 percent moderate emphasis, to this attitude toward science. His finding is the more striking because each teacher had been selected as a Fellow to participate in one of a number of academic-year institutes intended to focus on the subject matter of science and mathematics. On the basis of this and other information, he concluded that:

> High school teachers generally approach science teaching as a matter of conveying science as established facts and doctrines.

Such frankness is rare in the literature, but summer-institute directors frequently are willing sources of unpublished opinion on the disparity between the transcripts of the institute participants and their attitudes and performance.

---

* *Secondary School Science Teachers, 1969* (National Science Foundation, Washington, D.C., 1971); see Preface.
† Fletcher Watson *in* S. C. Brown, F. J. Kedves, and E. J. Winham, *Teaching Physics —An Insoluble Task?* (The MIT Press, Cambridge, Mass., 1971), p. 11.
‡ H. E. Gruber, "Science as Doctrine or Thought? A Critical Study of Nine Academic Year Institutes," *J. Res. Sci. Teaching, 1,* 124 (1963).

It comes as no surprise that elementary school teachers have even less understanding of science as an investigative and rational process than have high school teachers. Torrance * found in a survey of a group of 1000 elementary school teachers that only 1.4 percent regarded independent and critical thinking as the most important educational objectives.

We conclude that these somewhat fragmentary pieces of evidence indicate teacher attitudes toward science and preparation for teaching science that are seriously short of what is desirable. This statement is not a criticism of the teachers; it is the system that is at fault. Almost invariably, when asked what kind of additional study is most valuable, teachers ask for subject-matter courses. It might be said that they are seeking courses in methods, not the usual kind of course offered in departments of education but courses that give them experience with the methods of doing science instead of aiming solely at efficient transfer of subject matter.

A fair question is whether, through courses of any kind, teachers can be induced to improve their understanding of science and alter their performance. Results of studies are beginning to appear, suggesting that significant changes in teaching performance occur after the teacher has been in an inquiry-centered course. Work at the Science Education Center of the University of Oklahoma exemplifies the evidence for this conclusion. In one study,† 30 classes of elementary school children were observed. Half of these classes were taught by teachers who had been trained for the use of inquiry-oriented science material, the remainder by teachers who had had no contact with any of the new science curricula. The kinds of science experiences and questions the two sets of teachers asked were compared statistically. In brief, the results showed that the inquiry-trained teachers gave their pupils more than twice as many science experiences as the others and asked 50 percent more questions. Moreover, the favorable ratios were much greater in the categories of experience involving measurement and prediction and the categories of questions requiring skill, analysis, and synthesis. The noninquiry-trained teachers asked more questions in the categories of recognition and recall. Another study ‡ investigated the extent to which elementary school teachers altered their patterns of reading instruction as a consequence of a summer workshop in "new science." This investigation also was based on an analysis of types of questions about reading material asked by a group of inquiry-trained and

---

* E. P. Torrance, *Education and Creative Potential* (University of Minnesota, Minneapolis, Minn., 1963), Table 1.
† J. H. Wilson and J. W. Renner, "The 'New' Science and the Rational Powers: A Research Study," *J. Res. Sci. Teaching, 6,* 303 (1969).
‡ J. W. Renner and D. J. Stafford, "Inquiry, Children and Teachers." *Sci. Teacher, 37,* 55 (1970).

a matched group of conventionally trained elementary teachers in the second and fourth grades. For each grade, the two groups used the same reading material. The investigation showed a significant shift to questioning aimed at levels of thought above recognition and recall, requiring translation into other situations, analysis, and synthesis. Questions eliciting opinion and attitude were significantly more frequent among the inquiry-trained teachers. A study of before–after observations of questioning in social studies classes gave similar results.

A Value on School Physics

Can we assign a value to the importance of school physics to society—or to physicists for that matter? Importance is a matter of individual qualitative judgment, but in attempting to assess it, we often seek quantitative comparisons for help. A suggested comparison sets the yearly cost of school physics against other costs that are more commonly associated with physics.

We begin by estimating the amount expended each year in the United States for teaching physics in the schools. We restrict our estimate to the salaries of the teachers, the major component of this expenditure. It is necessary to estimate the fraction of time a teacher devotes to topics that can be reasonably considered physics. Especially in the lower grades, this fraction is likely to depend more on the background of the teacher

TABLE 11.1 Annual Expenditures by Schools for Teaching of Physics [a]

School Level	No. of Teachers (Thousands)	Percentage of Time for Physics	Average Salary, 1970 ($Thousands)	Expenditure ($Millions)
Elementary (Kindergarten–Grade 6)	1030	1.5	9.0	130
Intermediate (general science)	52	25.0	9.5	120
High school physics	21	32.0	9.5	64
ESTIMATED TOTAL SALARIES OF SCHOOLTEACHERS ATTRIBUTABLE TO PHYSICS				310

[a] Sources: *Science Teaching in Public Junior High School; Science Teaching in Elementary Schools; Secondary School Science Teachers, 1969; Estimates of School Statistics, 1970–1971* (Research Division, National Education Association, Washington, D.C., 1970); and *U.S. Registry of Junior and Senior High School Teaching Personnel in Science, Mathematics, and the Social Sciences* (National Science Teachers Association, Washington, D.C. 1971).

TABLE 11.2 Salaries of Physicists, 1970 [a]

Physicists	Number (Thousands)	Median Salary ($Thousands)	Expenditure ($Millions)
College and university faculty physicists	8.7	14.0	123
PhD physicists	16.0	17.4	278
All physicists	36.3	16.0	581

[a] Sources: Statistical Data Panel of the Physics Survey Committee; *Physics Manpower and Educational Statistics* (AIP, 1969).

and on the facilities actually at hand than on any adopted curriculum or syllabus prepared by a central school administration. Accordingly, little information about the average fraction of time actually used for science, much less physics, is available. However, using published data and spot checking against situations in individual locales, one can produce a conservative estimate, which is the basis for Table 11.1. The total given in this table may be inaccurate by as much as 25 percent, but only a rough approximation is needed.

The question before us is whether this total is commensurate with other sums for which the physicist customarily acknowledges responsibility. One such amount surely ought to be his own salary. Table 11.2 gives some estimates for comparison.

Another kind of expenditure, and surely one for which the physicist eagerly exercises responsibility, is that for his research. Table 11.3 gives current federal estimates of obligations for various components of research in 1971.

Comparing the various entries in Tables 11.2 and 11.3 with the total given in Table 11.1 suggests that the nation is demonstrating as much

TABLE 11.3 Federal Obligations for Research in Physics, 1971 [a]

Source of Support	Estimate ($Millions)
NSF total support for basic research	38
AEC total support for basic research	190
Federal obligations for basic research	420

[a] Source: *Federal Funds for Research, Development, and Other Scientific Activities*, NSF 70–38.

willingness to support physics in the classroom as in the research laboratory. Let us suppose for a moment that U.S. physicists were offered an additional $300 million per year. In a sense, they *have* been offered this sum, and it *is* now, and *will continue to be,* offered. The only string attached is that it must be used for *educating* the public, whose political actions will reinforce or hinder an enlightened policy for basic physics in the future. Thus far, physicists have shown little interest in the use of this large resource. A major challenge facing the physics community is to recognize this responsibility and develop means of exercising it meaningfully and wisely.

For Want of a Nail

The old allegory, it has been said recently,* describes the dilemma in regard to education. The new curricula offer promise of an eventual improvement in public understanding. Some believe they can help to motivate the culturally deprived. They are modest in terms of equipment and demands on physical facilities. Perhaps not ideal as they stand, they are loosening the tight conception of school science held in an earlier day and are stimulating further modification, adaptation, and evolution. Yet, they are unlikely to have the effects wanted by their originators, their financial backers, and those who sat in the stands and cheered. This bleak opinion results from evidence that the *way* in which the teachers are learning science, more than the *amount* of subject matter to which they are exposed, determines how they use the new materials. McKinnon and Renner,† in their study and analysis of intellectual development, conclude that "secondary and elementary teachers do not take advantage of inquiry-oriented techniques so necessary to the development of logical thought, because college professors do not provide examples of inquiry-oriented courses."

We must acknowledge that this conclusion is essentially correct. The example set for most students in physics courses (those students with strong *a priori* orientation toward physics being perhaps less susceptible) suggests strongly that this subject can be conveyed by lecturing, with occasional demonstrations (prearranged), and, when a laboratory accompanies the introductory course, by predigested experiments. The students are only learning to repeat old sentences, which they soon forget.

There is no point to forcing more new teachers through a system so demonstrably mistuned. To set it right will require some profound changes in the attitudes of a variety of sectors involved with education. In the

* J. D. Harren, *Science Education News* (AAAS) (July 1971).
† J. W. McKinnon and J. W. Renner, "Are Colleges Concerned with Intellectual Development?" *Am. J. Phys.,* **39**, 1047 (1971).

sciences, basically new concepts embracing re-evaluations of needs, opportunities, and potentialities are needed. Such changes will neither take place in accord with a grand scheme nor occur simultaneously at any given time. Where initiative is possible, it must be exercised. There is a point of entry for physics educators; they require no one's permission to use it. They need only to comprehend the problem, to realize what is at stake, and to muster their determination and their patience. They must undertake to teach the teachers.

There are traditional ways for science faculties to rationalize their historic indifference: future teachers are in the hands of educationists who load their programs with trivia; they cannot think quantitatively; they are not up to "honest" physics courses; the problems of elementary school science are so large that they can be managed only through specialist teachers; the faculties of institutions without teacher education programs have no responsibility. All these are stated as reasons for perpetuating the existing situation.

Let us face the matter more squarely. Nearly all future teachers are required to study science in a subject-oriented department; an honest course is one that induces students to learn for themselves; even granted a large increase in the availability of science consultants, the classroom teacher will continue to be the dominant influence in the classroom; no one is making a strenuous effort to produce science consultants, or even to use those who could become available; physics courses desirable for the purpose of teacher education can serve admirably as liberal arts courses, thus virtually all physics faculties could usefully engage in their development; given an 8 percent turnover rate of teachers,* a substantial impact can be made in a decade.

The behavior of in-service teachers is changed appreciably by contact with science taught in a manner that stresses inquiry and logical thought. Therefore, this mode of learning must be introduced in preservice teacher education if continued reliance on remedial measures is to be avoided. The course, "Physical Science for Nonscience Students," was developed with this intent. A few physicists are independently experimenting with courses taught in subject-matter departments and often with strong cooperation from professional educators.† These courses draw freely on the new curricula for their content and approach. It is becoming apparent

---

* U.S. Department of Health, Education, and Welfare, Office of Education, National Center for Educational Statistics, *Projections of Educational Statistics to 1979–80* (U.S. Government Printing Office, Washington, D.C., 1971), p. 82.
† See, for example, A. B. Arons, "Teaching the Sciences," in *The Challenges of College Teaching,* S. Cahn, ed. (Dr. Arons kindly supplied a copy of this work in advance of publication.)

to instructors that college students who intend to become teachers can be led to explore, measure, compute, err and recover, and draw unanticipated conclusions. In going through this process they recapitulate the steps and require the same time characteristic of schoolchildren, clearly not because of any fault of their own. In essence, here is the problem, and also a promising step toward solution. There are precedents for any physicist who chooses to look and wishes to act.

Ultimately the responsibility for the quality of physics teaching in the schools returns to the college and university faculties. They are the custodians of physics and they teach—or do not teach—those who emerge from their halls to teach the young.

College Physics for Nonscientists

Why is it that physicists consider their subject an exciting and rewarding intellectual enterprise, worthy of pursuit with fervor, whereas the world abounds with highly intelligent, well-educated persons who find physics dull, mysterious, or threatening? The answer to this question is related, of course, to the foregoing discussions. At least two attributes of the traditional teaching of physics contribute significantly to the problem: Physicists cater, in teaching, to their image of physics, which is pure and rigorous and is reinforced through their contacts with colleagues and their success with those few students who tend to see physics the way they do; for all but a few students the pace is too rapid and the instructor's desire to reach the modern physics at the end of the course too great. A student comment, frequently quoted, is apt: "Taking physics is like trying to take a drink from a fire hose."

These problems have many obvious parallels with the teacher education problem and their solutions are compatible. "Physical Science for Nonscience Students" is an example, as would be almost any course for future teachers designed with like objectives. As is the case with future teachers, the general student especially needs time to sense the significance of a physical law as a summary of experience to date, subject to evolving shifts in semantics and in extent of generality. These students benefit particularly from a realization of the provisional nature of (scientific) knowledge and from demonstrations of predictable and verifiable limits of validity, a facet of knowledge not always so readily shown in other fields.

Society faces environmental and social problems, many of which will be alleviated through the application of fundamental knowledge that is not yet available. In the political arena there is too little understanding of the structure on which technology, including the technology needed for control purposes, rests. One often wishes that more public servants could have had some contact with physics at the college level, contact that would have included an opportunity to gain insight into this structure. The so-called "informed" public also should be better educated in this subject. It is tempting, therefore, to orient courses for nonscience students toward current problems of major concern. This approach risks a superficial sampling of the basic material. It is more appropriate to teach within the guidelines suggested earlier, providing some ability to understand natural phenomena. For example, why are there aurorae; why do Polaroids help one see fish in water; why are sunsets red; why are no stars visible in the daytime; why does stopping a car traveling 60 miles per hour require four times the distance needed to stop one going 30 miles per hour; how can a sailboat go upwind; or even why is there thermal pollution from any type of power plant? It is also important to help these students to understand the motivations and social structure of science as well as to recognize that many types of order-of-magnitude calculation are not beyond them.

The apparent affinities between physics and the humanities have been the subject of much thought and writing. Changing modes in literature, music, and the fine arts during man's history frequently have coincided with the introduction into physics of radical new modes of thought. In the case of music and art, physics offers techniques and explanations to enrich the understanding of those who wish to practice or appreciate them. Physicists often feel deep personal affinity with the arts. Such affinities can be exploited in joint courses (for example, physics and literature, physics and music, physics and art), with the potential to attract students who ordinarily would never enroll in a "hard" science. In these courses, dialogue is best assured if instructors from the two disciplines are present simultaneously throughout the term. Although such cannot bring students more than a flavor of physics, they can provide a new perspective on it by exposing the conscious and unconscious use of metaphor deriving from it.

The history of science has become a recognized field, and many historians of science have strong backgrounds in physics. Nevertheless, a course conducted jointly by a physicist and a historian offers the possibilities of cultivating awareness of the thinking of some of the world's most creative persons and of an enriched exploration of the influence of physical ideas on man's intellectual development. The roles of physics and of physicists in affairs of the recent past offer an additional basis for approaching students who have a diversity of interests among the social sciences.

Physics faculty members often express the wish that students would study physics much as they do history or English—as a component of a liberal education. Students in the courses offered for nonscientists are the best existing approximation to this ideal. Because some of them, at least, have chosen freely, we must give their interests and needs the same quality of attention we give to those of our majors.

Physics Outside the Classroom

Today's world, with its bewildering array of materials, both natural and synthetic, its sophisticated manufacturing techniques, its wealth of display media, and its mobility, offers rich opportunity to learn in less formal contexts than the classroom. There are many motives for exploiting these possibilities. They provide ways for the public to keep in touch with discovery and new developments; a complementary aspect is the occasion they offer the scientist for describing his work. They provide additional stimuli for pupils in the schools. They can release science from its usually serious mode and show it to people in a lighter—even a playful—one.

Television is an obvious key to informal public education. In some parts of the world it is also an effective means of offering science information to the public. The British Broadcasting Corporation's (BBC) second channel presents an hour-long scientific documentary program in prime time each week, 40 weeks a year (one director, Alec Nisbett, is a theoretical physicist; several others have a scientific background), which is directed to an audience similar in level of information and education to the readership of the *Scientific American* or *The American Scientist*. The audience has reached 5 million, or 11 percent of the nation. The competition faced by this program is greater than that met in many parts of the United States, excluding the larger cities; the commercial and highly competitive BBC-1 channel schedules some of the leading drama and comedy series against it. In comparison with the United States, the response to scientific entertainment in the United Kingdom is immense. It is generally believed in the television industry that such a program would fail in the United States. Yet, recently in this country, two other British productions, one on science and one on culture more generally, have been shown. The two-and-one-half-hour presentation, "The Violent Universe," on recent discoveries in astrophysics and cosmology, was shown on National Educational Television (NET) and later by a major commercial network. The series, "Civilisation," made by the BBC and Kenneth Clark, was sponsored for NET presentation by a commercial firm. The time is ripe to apply existing expertise at the level of support and competence used by the BBC in these examples to produce a series on all science, with appropriate emphasis on physics and astronomy, for wide distribution in the United States. It should cover science at the level of the educated layman. The

American Institute of Physics may be an appropriate agency to initiate and coordinate this effort.

To reach critical mass in this activity as soon as possible the Committee recommends that the National Academy of Sciences exercise leadership in coordinating cooperative activities in this area by the major scientific societies. In the case of physics, the Committee, moreover, recommends that the American Physical Society levy a per capita assessment on its members; the proceeds of this assessment should be used by the American Institute of Physics as seed money in developing sources of significant support for its contribution to such joint scientific educational activities. It is long past time when physicists should begin paying for such activity themselves in addition to discussing its desirability.

In one western city of the United States there is an Exploratorium, the manifestation of a physicist's idea for inviting people to learn for themselves. Some would call it a science museum, but its originator dislikes the connotations of this term. The notion is to open a place for activity and have materials and tools available and also many examples of scientific and technological processes, equipment, and demonstration experiments. Schoolchildren, college students, and passers-by come in and pick up objects that interest them, finding out how they work or constructing something from materials kept at hand there. In another city, there is a Science Center, frequented by schoolchildren, for whom it is intended, that also has a demonstrated fascination for adults. In this Center is a "math" room, perpetually crowded with children drawing Lissajou's figures with a pendulum, making the Random Walker obey the directions of dice, creating soap-film surfaces, playing mathematical games, and, generally speaking, participating rather than merely observing.

A number of cities have well-established centralized science teaching resources. Nothing is new about these ideas, but, as with the examples cited, they can be modified to become creative responses to the contemporary scene. It is hardly necessary, much less possible, for many physicists to give the full effort required to initiate and carry through the development of an Exploratorium, although the field is far from saturated. However, the successful components, or even whole operations, can be copied with only moderate outlays of a scientist's time. Further, physicists can support and contribute to the spread of this kind of education at a variety of levels of commitment.

It is now easy to have one's own Exploratorium at home. Toys, notions, and ornaments explicitly illustrating physical phenomena have always enjoyed a certain amount of popularity among adults as well as children and continue to do so. In terms of economics, there would seem to be considerable leverage; of the $2.3 billion toy market in 1964, it was

found that 5 percent comprised educational and scientifically based toys.*
The extent to which physicists have contributed to the ideas for toys is not
known, but many are recognizable as variants of devices commonly
used in physics classes. Because the objective of a toy is to elicit free
response, physicists should welcome those that generate spontaneity and
suggest questions in contrast to those that lay out a structured program and
thereby mimic the classroom.

A rich variety of materials is readily available today, and at very low
cost, that make play of a kind unthinkable a few years ago quite inexpensive
now. Hand-held, manually operated vacuum pumps, lenses, and the like
—catalogs abound with such objects. Elizabeth Wood has discovered
how to use the television set as a stroboscope to show surface tension
ripples on the surface of a pie tin containing water. Her small book†
is written in this vein. To design a program for the exploitation of today's
commonplace is not the task of this Report, but to point out a happy way
of providing informal contact with phenomena and concepts that aid in
developing what is called physical insight cannot be amiss.

EDUCATION FOR THE PURSUIT AND USE OF PHYSICS

Through physics man comes to understand the most fundamental and
universal features of his world. Even at best this understanding is still
quite incomplete, however. Many obvious questions about matter, such as
why there is a natural unit of electrical charge or why the mass of an
electron is so small in comparison with that of a proton, cannot be answered
today in any reasonable manner. The search for understanding, a task
set by man for himself centuries ago, continues. As long as there are
young people interested in pursuing it further, there will be an older generation that attributes importance to educating them for the task.

In the past, physicists typically have had a view of their science that
stressed its universality and led them to deal with problems, whatever their
origins, that seemed amenable to attack through use of the outlook, methods, and knowledge of physics. After a period of growth of physics so
rapid that this attitude became obscured, physicists now face the task of
restoring the educational conditions that fostered it. The following analysis

---

* A. J. Wood Research Corporation, *Toy Buying in America. A One-Year Study Pre-
pared for the Toy Manufacturers of America, Inc.* (A. J. Wood Research Corporation,
Philadelphia, Pa., 1965).
† E. A. Wood, *Science for the Airplane Passenger* (Houghton Mifflin Co., Boston,
Mass., 1968).

of undergraduate and graduate education in physics has this objective as an underlying theme.

The pursuit of fundamental knowledge in physics has obvious usefulness (see Chapters 3 and 7). The individual bits of information gained from basic study, the instruments and techniques that evolve from it, and, above all, the great generalizations that allow much knowledge to be embraced as logical consequences of a few basic concepts and the point of view that extracts the essential core from a complex situation and sees the workings of universal principles seem to find their way into other sciences and technology. An acquaintance with physics has been considered important by those responsible for the education of other scientists, engineers, and technicians. The needs of these various professions differ and change from time to time. Accommodating them is one of the major responsibilities of physics faculties.

The Undergraduate Physics Major

Undergraduate education in the United States is expected to shape the young adult intellectually so that he is broadly equipped for a career, but at the same time it is supposed to help him to orient himself through social and humanistic modes of experience and understanding. In the next era, which promises to be one of unprecedented educational flexibility, institutions and their educational patterns may well offer a range of programs of varying length, depending on the particular objectives of a curriculum or the particular intentions of an individual student. With students seeking, and finding, ways to break the fixed mold of an earlier conception of undergraduate education, and with observers and critics of the educational system (such as the Carnegie Commission on Higher Education) advocating more flexible degree programs,* it is impossible at this moment to see clearly how educational patterns will change, although few would dispute that change is on the way.

The structure of a physics curriculum in the United States is governed by both a perception of the structure of the field and pedagogical considerations. It can be understood, then, that there is a general pattern throughout the nation. This pattern has already survived drastic change in the content of physics, though not without some substitutions of quantum for classical topics stimulated by changes in the direction of research to set new goals for students. Whether it is sufficiently adaptable to the changing role of physics is the question. This Committee believes that it will continue to

---

* S. H. Spurr, *Academic Degree Structures: Innovative Approaches* (McGraw-Hill, New York, 1970).

adapt, while retaining its vigor; we speak of a physics curriculum, anticipating that this term will have a roughly constant meaning for a number of years.

A common basis for the undergraduate major is first established by an introductory course covering the full range of physics, now always including calculus and extending over a period of one to two years. A course of this kind is usually characterized by reference to one of the standard textbooks.

All too frequently it is at this point that some of the most able students are turned away from physics. In their high school physics courses they had once before surveyed much of physics, albeit at a less demanding level. But they know many of the words and have an acquaintance with many of the concepts. Unless great care is taken, the freshman college physics appears as simply more of the same, and the most able students, frequently having their first exposure to other areas completely new to them, such as sociology, philosophy, and the like, concurrent with this freshman physics, find these other areas fresh, challenging, and relevant and withdraw from further contact with physics.

This is a danger that faces designers of undergraduate physics curricula at all levels. By its very nature physics is more hierarchical in its course structure than the typical humanities or social science discipline; with the burgeoning student demand for catholicity of overall undergraduate curricula, the physical sciences are at a very real disadvantage.

During the past 15 years—following the flight of the first Sputnik and the development of the new high school programs discussed above—a host of innovative projects and programs has brought a fresh approach to the study of elementary undergraduate physics, the character of which had remained surprisingly stable since the early nineteenth century in spite of many changes in substance. The new curricula and courses can contribute in different ways toward making college physics more interesting, contemporary, and challenging and the teaching of physics more effective. Largely in response to the notable improvements in secondary education in America during the same period, especially in mathematics, all the new physics courses for science majors aim at beefing up the material presented at the introductory level.

The pedagogy of physics has benefited enormously from what may be termed the "new physics" movement, which led, for example, to the Feynman * and Berkeley † courses. It seems unlikely that the next few

---

* R. P. Feynman, R. B. Leighton, and M. Sands, *Lectures on Physics* (Addison-Wesley, Reading, Mass., 1965).

† A. C. Helmholtz and E. M. Purcell, *The Berkeley Physics Course* (McGraw-Hill, New York, 1967).

years will call for the design of even more exacting introductory material. Rather, there is a need to separate the good from the not so good by the test of experience, to deploy the worthwhile innovations throughout the educational system, and to attend to the needs of those would-be physics students who are not prepared for the pace and rigor typical of the new programs. The beginning of a trend in this direction can be seen in the recent publication, by reader-sensitive commercial publishers, of shortened and less-ambitious versions of some leading introductory textbooks. In a counter trend, the designers of some new physics courses for nonscience majors suggest that their style may be appropriate for that proportion of the science majors who find the pace too demanding in the high-pressure physics courses.

Canonical Sequence   Building on the foundation laid by the introductory physics sequence, the dominant curricular progression in the junior and senior years embodies a more or less canonical sequence of courses that deal in greater depth with special areas of physics. Thus, most of the topics are repeated, but this time with the use of differential equations, computer programs, and some reference to current research. Such a sequence of courses typically includes classical mechanics, electricity and magnetism, optics, thermal physics, electronics, and quantum physics. The main variations among patterns concern the length of a particular course (for example, the choice between one and two semesters of classical physics) and the amount of modern physics and its various branches that is injected. A few schools have succeeded in establishing alternate routes through the junior and senior years toward the degree in physics. Since the standard curriculum has proved to be quite effective in preparing students for graduate school, there has been little incentive for change.

As physicists grow more aware of the importance of opening, to a major in this subject, career directions other than the conventional ones leading to the PhD in physics, they are beginning to re-examine the course options at the advanced undergraduate level. But even students bound for graduate school may profit from some rethinking of the traditional patterns of the junior–senior course structure—not because the traditional is necessarily suspect, but because it can embody too detailed a subdivision of the field, thereby tending to obscure the unity of physics, which we discuss in Chapter 4 and which should remain a central theme in all physics education. For example, the publication of several textbooks signals the appearance of thermal physics as a standard course, replacing separate courses in thermodynamics, kinetic theory, and statistical mechanics and coupling the subject strongly with the physics of condensed matter. The time may well be ripe for a review of other conventional courses, such as mechanics (should and could classical and quantum mechanics be combined?), electricity and magnetism (should more account be taken of plasma physics?), and optics, the renaissance of which is only beginning to be acknowledged in physics teaching.

The Place of Quantum Physics An important unresolved question facing physics departments in their educational work concerns the respective roles that classical and quantum physics ought to play in the undergraduate curriculum. It is often insisted that a good foundation in classical physics must be laid during the sophomore and junior years; on this foundation is placed a "frosting" of modern physics, capping the undergraduate curriculum with a course or two in quantum physics, atomic physics, and, perhaps, nuclear and solid-state physics. There is much to be said for this curricular structure. Physical measurements are made largely with the use of tools that obey classical physics; the physical intuition of young persons is developed by observation of the surrounding world, which is largely governed by the laws of classical physics; and many of the technological applications, for which the study of physics prepares students, deal exclusively with classical systems.

On the other hand, it is difficult to believe that anyone sufficiently interested in physics to declare himself a major will not be aware of quantum notions either through high school study or through his own reading. A large fraction of current work in physics is inaccessible to anyone who has no grounding in quantum physics; the interaction between physics and many neighboring disciplines (notably chemistry) is dominated by quantum ideas and atomic concepts; and some of the most significant contributions of physics to man's entire philosophical outlook during the past 70 years have developed from quantum physics. Textbooks treating quantum physics at the sophomore level in science or engineering are just becoming available.

It is obvious, although not surprising, that the physics curriculum at the college level has not yet fully adjusted to the impact of quantum physics. Analogously, it took many years after Newtonian mechanics was established to develop a modern·curriculum in which the new mechanics occupied its proper place. The Newtonian revolution, like the quantum revolution, had a strong theoretical and mathematical flavor. As a result, the education based on it tended to be excessively formal and abstract. Similarly, the great success of quantum physics has placed extraordinary, and perhaps excessive, emphasis on theory and formalism, not infrequently to the detriment of the development of physical intuition and structural understanding.

Back to the Phenomena   The question raised above is indicative of concern with the general style of physics teaching at the intermediate (and also the introductory) level. Obviously, as we have noted above, the most striking characteristic of the physics curriculum is its hierarchical structure, in which a chain of prerequisites guides the student from the bottom to the top, with even minor detours taken only at the student's peril.

Not surprisingly, the increasing tendency of undergraduate physics education toward emphasis on and reward of manipulative skill in handling the formalism of physics has brought forth loud calls for a return to nature. Many teachers are insisting that greater attention be paid to intelligent observation of natural phenomena, to the achievement of interpretative understanding, and to the acquisition of a modicum of physical insight while tackling physical problems with increasingly advanced mathematical techniques. Indeed, the charge has been made that the prevailing physics curriculum has the effect of turning students into mindless equation solvers instead of developing their intuitive faculties as observers and interpreters of the world. This is an extreme view.

One reason for this preoccupation with formalism is the fantastic growth of quantum mechanics and its impact on most of physics. But there is at least one other cause. To look at a problem from the proverbial, and somewhat subjectively defined, physical point of view may seem like the natural, innate, human approach, but actually it requires time, patience, and experience to develop physical intuition. Although good teaching can help a student substantially in this process of maturing, there is no reason to discard the learning crutch provided by easy familiarity with formal manipulation. Excessive mathematization of the curriculum for its own sake is certainly undesirable, but the use of formalism is obviously to be encouraged whenever it can eliminate impediments to physical understanding.

There is, however, every reason to advocate closer contact with the phenomena at all levels of physics education; this raises the question of the role of the laboratory in the physics curriculum. The variety of attitudes taken on this question throughout the United States indicates not so much a commendable endorsement of diversity as a basic state of confusion and indecision. The question is familiar: Should the laboratory serve primarily as an instructional device supporting and illustrating the lectures and recitation; should it be regarded fundamentally as a series of minute research projects, confronting the student with an opportunity to sharpen his manual skills and showing him what doing "real" physics is like; or should it be guided inquiry leading to concept formation and inductive reasoning? These questions will never be answered unequivocally or finally, even when such modifications as the "corridor laboratory" are accepted.

Most of the problems posed by the teaching laboratory are economic. Laboratory instruction is expensive, not only because of the use of equipment and supplies but even more so because of the requirement of supervision. This requirement means heavy teaching loads in small colleges and the use of large numbers of graduate teaching assistants in large universities. In recent years, laboratories accompanying large introductory courses frequently have been curtailed or entirely dropped for budgetary reasons. Although in most cases other rationales were found to justify the change, the educational merit of the usual introductory laboratory is, admittedly, difficult to demonstrate.

For a subject that has its roots in the study of phenomena, this admission comes with chagrin and should be a stimulus for continual efforts toward improvement. The main source of dissatisfaction has been the conventional, tightly structured, one-experiment-a-week, preprogrammed, apparatus version of a laboratory. Many efforts have been made in recent years to develop other formats called, usually, open-end laboratories. In these, students are given access to equipment after a degree of preparation that differs widely from one version to another.

In 1968, the Commission on College Physics held a workshop on open-end laboratories, and working models have been described in the literature.* They appear to offer a path to a more satisfying laboratory experience even in large introductory classes. Shonle concludes: "It is not necessary to be at a large or wealthy institution, or one with vast amounts of research equipment, to be able to institute the open-end laboratory. The necessity is, rather, a faculty willing not only to seek a new approach but also to participate actively in its realization."

A modification that seems to be gaining in popularity is the separation of laboratories from the junior–senior lecture courses and the establishment of separate intermediate and advanced laboratory courses. The divorce of the laboratory from the lecture courses can be a healthy development if it exploits the diversity of areas that most experiments draw on and if it does not undermine the phenomenological content of the lectures. The separation must be consciously viewed by students and faculty as a measure intended to strengthen the laboratory component of the physics curriculum and not—as may too easily happen—as a relegation of the laboratory to secondary status, perhaps even preparatory to dropping it altogether.

In examining the merits and drawbacks of laboratory courses, one should

---

* J. I. Shonle, "A Progress Report on Open End Laboratories," *Am. J. Phys., 38,* 450 (1970).

be aware of alternative, and possibly complementary, ways of achieving the prime objective of such instruction. Computer simulation might serve in some instances. In dealing with the role of physics experiments in the undergraduate curriculum, one might think again of three distinct levels of activity characterized by the magnitude of effort as 1-hour, 10-hour, and 100-hour investments of time, each answering different educational needs of the student. At the 1-hour level, the student would witness an experiment as a demonstration and illustration of the lecture material; at the 10-hour level, carefully planned laboratory experiments would be actively carried out by the student, following a more or less prescribed procedure; at the 100-hour level, the student would give a substantial fraction of his time for an extended period to a major experimental project, possibly involving him in a research enterprise under faculty supervision. Ideally, the total undergraduate experience of a physics major should include a blend of all of these modes of encountering observation and measurement in physics. In particular, we advocate that the use of appropriate lecture demonstrations not be confined to the introductory courses but be extended to the junior–senior level wherever and whenever possible.

Participation of undergraduate physics majors in original current research should be strongly encouraged. There is a vast amount of interesting little physics that can be done at modest expense, and the possible educational returns can be dramatic. An advanced undergraduate who works on a piece of research is much more likely to acquire a clear notion of what a commitment to physics means. He will have far more opportunity than usual to learn from direct contact with his professor; he will share in the excitement of exploration and in the frustrations that arise when things are not going as well as they might. Above all, at an early stage he will begin to see physics as a growing science, an open-ended enterprise, and one in which many issues and problems are not yet resolved. For such experience at the undergraduate level to be genuinely profitable, it is essential that the research problems be intelligently chosen. They must be of limited scope and, although they could certainly be collaborative in nature, should not usually consist of minute disconnected portions of large research projects. One should be under no illusion that the selection of suitable research problems for seniors is an easy task, but we regard as unduly pessimistic and restrictive the frequent assertion that undergraduate research is wasteful of student and faculty time because "the student does not know enough to do anything worthwhile." The current period of declining enrollments in undergraduate physics programs seems ideally suited for the establishment of the senior research project, at least as an elective component of the physics major or as an incentive for honors students.

Graduate School or Else?   Until quite recently, most undergraduate physics curricula were designed for one primary objective—preparation of
students for graduate work in physics. The number of graduating seniors
continuing their studies in graduate schools and the quality of the graduate
schools to which they were going were used as an easy measure of the
quality of an undergraduate program. Now it has become apparent that
the role of undergraduate programs in physics must be much more
broadly conceived. Such programs must be made attractive to students
who do not plan to pursue graduate work in physics, who are interested
in fields for which physics is a good background, or who may not take
any graduate work, seeking nonacademic employment instead after graduating from college.

When undergraduate physics curricula were entirely oriented toward
preparation for graduate school and most physics departments were striving
to become ever more involved in research, there was pressure to engage
in a continual escalation of academic demands placed on students. In
those schools in which two or more alternate major programs in physics
have been available, the less demanding and less professional BA programs
often have been allowed to atrophy, because it was less fashionable to
aspire to anything other than the pregraduate study bachelor's degree. The
same attitude also began to affect entire educational institutions. In an
atmosphere that placed a premium on training the heirs to science, universities with strong research activities had a natural advantage over small
liberal arts colleges, and it is not surprising that the gap grew rather than
diminished.

Today, it is important to recognize that an undergraduate degree in
physics provides a solid foundation for a great number of different professional careers and types of advanced study. This fact is easily demonstrated by perusal of any modern textbook in chemistry, biology, geophysics, or the medical sciences. Undergraduate physics also provides
a solid foundation for advanced study or work in all technical fields and
engineering. Students with a thorough knowledge of physics will be prepared to gain a depth of understanding of the principles underlying all of
these disciplines that cannot be matched otherwise. In any given school
or institution, students should have the opportunity to prepare for a
variety of interests and careers. Faculties and administrators should
realize that the nation needs diversity in physics education and should not
feel compelled to seek uniformity in their programs.

A special opportunity exists in those disciplines that were not long ago
regarded as distinct and almost independent but now are merging into a
more integrated field often called earth and planetary physics. This broad
and vigorously pursued area of investigation requires an understanding

of all of physics. There is a pressing demand that advances in physics and in technology be applied to observational problems. The researcher must rely continually on basic physics to interpret his data. With little or no distortion of the realities, it may be said that the entire field is, in principle, an application of physics and is recognizing this fact. Physics baccalaureates are welcomed into its graduate programs. Physics department faculties have good reason to respect the intellectual connections between the core subfields of physics and such important applications and should give careful attention to the implications of these connections in courses, seminars, informal activities of their students, and, whenever feasible, joint or exchange teaching programs.

An undergraduate major program in physics should be so constructed that it can be regarded as a reasonable and desirable background for students with interests in any of the liberal arts, not just in fields to which a knowledge of physics is directly relevant. It would seem most desirable if a fair fraction of college graduates in all professional areas had the foundation provided by the study of physics at the BA level.

There is good reason to hope that many more institutions will give increased emphasis to training teachers for secondary and elementary schools, that some will provide more opportunities for occupational preparation in practical areas, and that others will specialize in historical, political, or social problems to which the study of physics provides one avenue of access.

One approach, which we recommend, is that of having biophysicists, geophysicists, medical physicists, and the like participate actively in the undergraduate physics teaching programs. While teaching standard physics, use of examples and illustrations drawn from actual problems in such areas, rather than the all-too-frequently idealized and rather sterile examples that lurk in the problem sections of many physics textbooks, can awaken students to an appreciation of the potential of a physics education, or of physics itself, in a wide variety of career fields, without sacrificing the underlying rigor of the discipline.

The presence of outstanding individual teachers will greatly influence the contribution to physics education that a particular school can make. Probably no extensively funded collective educational improvement project, no matter how worthwhile and deserving of strong support, is likely to be as effective in making a good and competent and committed physicist of an undergraduate student as one of those rare persons—not necessarily the most charismatic or the best lecturer—who possesses the ability to stimulate and truly educate his pupils.

Finally, although we advocate that physicists and the managers of the institutions in which they teach initiate critical reviews of past practices and established methods in the educational system, we urge also the protection of the demonstrated capacity of many undergraduate physics programs to provide students efficiently with a solid background for graduate work in physics. It is essential that modifications and improvements in physics education not be allowed to weaken the traditional and strong pregraduate preparation needed for the continued health of physics.

We favor widespread open discussion of undergraduate curricula and practices. We urge attention to the discussions on these topics that are held under the auspices of the American Association of Physics Teachers and the articles that appear in its publications. We urge that Association to publish at periodic intervals surveys of undergraduate curricula and practices so that faculties and students, both prospective and those in residence, can inform themselves about the state of undergraduate physics education. It is not our intent to promote conformity; we have stated often in this Report that diversity is one of the great strengths of education in the United States. High quality is the ideal to be vigorously pursued.

Physics for the Technician

Recently, the Bureau of Labor Statistics issued a report projecting technical manpower needs to 1980.* It states that "Physics technicians and Mathematics technicians are expected to show the fastest growth rates among the technician occupational specialties, 95 percent and 91 percent, respectively." The projected demand is for 20,700 physics technicians by 1980, an increase of about 100 percent over the number in 1966. For the more inclusive category "Engineering and Physical Sciences Technicians," the need is projected to exceed 200,000. The Bureau of Labor Statistics projections, as we discuss in Chapter 12, tend to be controversial, and it is not our intention to comment on the degree of confidence to be accorded that report. The qualitative assumptions underlying the projections seem soundly related to well-established industrial trends: increasing utilization of technicians relative to total employment, due to the expansion of research and development activities; increasing complexity of industrial processes; and the growth of industries employing larger numbers of technicians. Recent cutbacks in industrial research programs raise some doubts about the soundness of these assumptions. Nevertheless, the employment of technicians probably will reach and maintain a substantially higher level than the present one.

* U.S. Department of Labor, Bureau of Labor Statistics, *Technician Manpower, 1966-80* (U.S. Government Printing Office, Washington, D.C., 1970).

In 1965, the largest source of new technicians was the upgrading of existing employees; the next largest source was postsecondary school training of a kind offered by technical institutes, junior and community colleges, vocational–technical schools, and the extension divisions of engineering schools. The rapid introduction of new two-year colleges guarantees that they will make an appreciably larger contribution to the technician work force than was the case in 1965.

A report by the NSF to Congress states *:

> The modern engineering technician occupies a position between the engineer and the skilled worker. His job is to translate the ideas of the engineer into working plans to be followed by the shopman in producing a product or carrying out a testing procedure. He must be acquainted with the associated engineering field and also with the detailed work procedures involved.
>
> The engineering technician curriculum is post-secondary, is most generally terminal and provides instruction in theory and applications related to science and technology.

The physics component of the curriculum for technicians still lacks full definition. There is a need for more suitable materials and for instructors who are well grounded in physics. These instructors must have a good understanding of the nature of the students who enter the program and the kinds of responsibilities these students will have when they enter employment.

The physics component of technical training has been studied by the Panel on Physics in the Two-Year Colleges of the Commission on College Physics. The report of a conference on this topic held in May 1969 has been published.†

According to this report, the student technician group includes both those with college preparatory and those with noncollege preparatory high school training. The typical entering student has a mathematics proficiency level below that of trigonometry and sometimes below algebra. He rarely has studied physics in high school. He is not likely to challenge a concept that he does not understand. He lacks self-confidence. Student technicians generally come from lower socioeconomic levels and find less family and peer-group support for their studies than do college students. Tests indicate a wide gap between the quantitative and the verbal abilities of the technical

---

* *The Junior College and Education in the Sciences,* Report of the National Science Foundation to the Subcommittee on Science, Research and Development of the Committee on Science and Astronautics, U.S. House of Representatives, 90th Congress, First Session (1967), p. 101.

† B. G. Aldridge, "Physics in Two-Year Technical Curricula," *The Physics Teacher,* *8,* 302 (1970).

student, the verbal being much the lower. One participant in the conference, a technician, made the following comment, which characterizes the aspirations of many potential technicians:

I found that my high school diploma did not qualify me for a good job, and I thought that two years at .... would qualify me for a better job. I chose the technician program rather than the liberal arts because I enjoyed fooling around with my hands, building things, rather than reading books.

It should be noted that such strong motivation can produce striking results.

There is a dichotomy of viewpoint that intrudes on discussions of physics education: Should the emphasis be on basic concepts, on the assumption that, when understood, their application can be left to the student (or in this case, learned on 'the job), or is it necessary to stress application to teach physics effectively? Technician education asks that both aspects be included, as can be seen in this portion of a report issued by the U.S. Office of Education *:

The technician must have sufficient knowledge of the basic principles and phenomena of the science underlying his specialty to be an effective, comprehending, and perceptive worker with his or her professional counterpart and to be able to master the inevitable (and often rapid) changes brought about by technological developments.

It should be assumed that the professionals in the field supply the deep theoretical components of the task, but the technician must be sufficiently grounded in the fundamental principles to permit some interpretation of phenomena he encounters, to have a sound understanding of the theory as it is applied in the field, and to learn of technological changes in his specialty by independent study of reports of developments as they occur. The basic science courses in his curriculum must provide the knowledge of the scientific principles and their application needed by the technician.

The basic science courses for the physical science and related engineering technologies are fundamental and applied physics, and usually some study of chemistry; these form the base for specialized courses in mechanics, statics, strength of materials, electronic circuitry, instruments and measurements, and other specialized applied physics as required by the particular technology.

Because of the characteristics of the student in a technical curriculum and his expectations from physics, the usual introductory college courses are unsuitable. Most technical physics instructors agree that there is a serious dearth of useful instructional material. They agree, also, that their courses need to focus initially on things rather than on abstractions: on machinery, equipment, and instruments with which technical students already are acquainted or are to become acquainted. The mathematics they study should be tied closely to physics and other technical studies. Laboratory work should be emphasized, with a written report as an important

---

* *Criteria for Technician Education—A Suggested Guide,* OE 80056 (U.S. Department of Health, Education and Welfare, Office of Education, Washington, D.C., 1968).

element of each exercise. Problem solving, involving practical problems related to industry and engineering also needs emphasis. The move toward generalization and abstraction should be deferred, although it must be made eventually. This progression of emphasis is similar to that advocated for essentially all physics students, although for the technical student it needs to be honored more strictly.

The desirability of breaking away from the traditional time quanta has appeared again in discussions of this curriculum. Here it is explicitly manifest in a proposal to offer laboratory work in modules involving from one to three weeks of instruction centered around a physical system of some sort, for example, a carburetor or an engine test bed. A careful distinction should be made between the use of such a system in a physics course and a technology course. In physics, one would go from the specific example to a general principle and then back to the example, seeking the manifestation of yet another general principle. In the technical course, the system is the principal object of attention. Sample modules have been prepared and are being tested.

As is the case with physics curricula more generally, the objectives of technical training curricula vary from institution to institution, and in any given location there will be a wide range of student background and ability. No single prescription can meet all the needs, so those who gather to discuss and develop technical physics courses must prepare materials that can be adapted to varied circumstances. In the two-year colleges, where most of the training of technicians will be accomplished, the problem becomes intensified, for physics instructors must also offer courses that prepare students for transfer into four-year colleges. The often inadequate working conditions and the heavy teaching assignments in two-year colleges exacerbate the problem.

One outcome of the 1967 meeting was the establishment of a National Steering Committee to stimulate and coordinate the production of physics instructional material for technicians. At present, it is supported by a grant made by the Esso Foundation to the American Institute of Physics. Also, The American Association of Physics Teachers has created a Committee on Physics in the Two-Year Colleges to carry on the work initiated by the Commission on College Physics. The interests of this Committee go beyond, but certainly include, the improvement of the education of technicians.

Others Who Use Physics

Students majoring in sciences other than physics are entering fields that have been influenced by physics in striking ways. Because it is the most quantitative and structured of all the sciences, physics serves as a model. Through its explications of atomic structure and interactions among atoms, it provides the theoretical basis for chemistry and those disciplines that grow from or depend on chemistry. Several recognized disciplines, for example, geophysics, atmospheric sciences, and health physics, can be considered in many respects as applied physics. Indeed, most measurement in science and medicine today is measurement of physical properties of matter and depends either on adaptation of techniques and apparatus originating in physics research laboratories or on electronic components, transducers, circuitry, and other devices that originated from the demands of physics research. The techniques of radioactive tracer labeling and control, ion kinetic energy spectroscopy, photoelectron spectroscopy, pion therapy, and semiautomatic scanning and pattern recognition, to mention only a few, are today undergoing adaptation to the particular needs of other sciences. There are many cogent reasons for including physics in the education of scientists and technologically oriented workers of all kinds.

The students in these various disciplines are guided and constrained by recommendations and requirements established by their particular faculties. These faculties confront the problem of fulfilling the practical need to provide their students with a sound and comprehensive background and, at the same time, responding to the demand for free choice of electives. In addition, they must weigh the willingness of the physics department to offer courses that match the backgrounds and needs of their students. Rarely do physics departments take the initiative in consulting with other science faculties to gain insight into other points of view on, and possible applications of, their courses; well might they do so periodically, recalling that possibly a majority of the students they teach, at least in elementary courses, are directed to them because of certain expectations.

Biology Students   For biologically oriented students the question of training in physics has considerable current significance. The words of biologist Dana L. Abell, written at a time when he was Associate Director of the Commission on Undergraduate Education in the Biological Sciences, are particularly compelling.*

Abell stated:

   An important inconsistency in planning curricula for undergraduates in biology, agriculture and natural resources has always been that training in physics is required but almost never used. The conviction that physics is an essential part of these programs is firm, however, even to the point that some people are already debating the content

---

* "A Working Conference on Source Material in Physics–Biology–Agriculture and Natural Resources," CUEBS Working Paper No. 2 (Commission on Undergraduate Education in the Biological Sciences, Washington, D.C., 1970).

of a "second course," i.e. physics beyond the introductory year. These people would seem to be maintaining that physics isn't used because the students with only one year of work aren't sophisticated enough to understand the material through which an interdisciplinary exchange between biology and physics has been accomplished.

CEANAR and CUEBS share the conviction that introductory physics courses belong in most of the major programs with which they are concerned, but we are convinced that elementary physical concepts can be introduced at many points in these programs. As we see it, interfaces do exist between these fields at all levels, but opportunities for interchange short of the highly sophisticated interdisciplinary field of biophysics or in more specialized fields, such as agricultural engineering have simply not been created.

Abell also made the following points, which suggest that physics faculties should question how well they are meeting the needs of these rapidly developing fields:

While faculties of physics and the biological sciences both decry the watering down of courses taken by biology students, circumstances conspire to make it a consistently less rigorous course than the one for majors. Yet, it is made to seem excessively formal and even without prospective physicists in it, it appears designed to fit the tastes of physicists alone. Undoubtedly, physics is best taught as a subject in its own right and the instructor *should* be allowed to choose illustrative material with which he will feel comfortable, but physical concepts do pervade all of science, however, and perfectly good examples of physical principles can be drawn from almost any aspect of biology. The fact that this is not done relates more to the training of physics faculties and to traditions in the teaching of physics than it does to any real boundaries between physics and the biological sciences. Greater use of illustrative materials from outside the traditional bounds of physics could have the effect not only of building a substantial bridge to other fields, but also of raising the level of sophistication as increased student interest brings greater incentive and commitment.

Students of Other Physical Sciences Students of the physical sciences, other than physics, do not encounter problems of the kind faced by the biologically oriented. Often, and with increasing frequency, students of other physical sciences find themselves in an introductory course offered for physics majors and engineering students. Few would argue that this arrangement is inappropriate. Beyond the introductory year, problems arise because subsequent courses, which might be suitable for other physical scientists, often are directed more to the interests of the physics major than is necessary. At the same time, the faculties of the other physical sciences tend to emphasize their own professional outlooks by offering courses that cover essentially the same ground but from the idiosyncratic viewpoints of their various disciplines. The validity of the contention that physics is basic to other physical sciences argues for intermediate physics courses congenial to students of these other fields who wish to explore the foundations of their major subjects. The use of qualified instructors from these other fields in physics departmental teaching activities has been noted above and has much to recommend it in bridging these interdisciplinary gaps.

Premedical Students   The long-standing requirement that premedical students complete a year of physics might be a residue of ancient modes of thought and education, for the nouns physician and physicist have a common root in the Greek word for nature. Until a century or so ago, physicians were, in fact, among the most learned men in science generally and contributed importantly to fundamental advances in physics (for example, Mayer and Helmholtz). One effect of the subsequent rapid differentiation between scientists and medical practitioners was to redefine the role of physics in medical studies. Ostensibly, it has been supposed to aid in inculcating a small amount of general scientific background and to induce a degree of familiarity with topics (for example, optics, acoustics, and fluid mechanics) that could be of direct professional value. Premedical students, and occasionally members of the physics faculty, commonly suspect that introductory physics is used as a screen to sort out aspirants having a certain degree and kind of academic capability.

Surely, if this last were the reason for the physics requirement, we would not concern ourselves over premedical students. The earlier, more cogent reasons must be reinterpreted in the light of the changing modes of medical education today. No longer will there be a single pathway into or through medical school, for it is becoming widely recognized by medical educators that the variety of forms of medical practice today requires greater variety in student background, interest, and talent than has been customary. Routes will be provided for some students whose undergraduate preparation has concentrated on the social sciences; many of them may need remedial work in the natural sciences. At the same time, medical schools are also encouraging stronger preparation in the basic sciences and mathematics as preparation for practice in the future. This latter tendency is revealed by recent data * showing that half of the nation's premedical students enroll in the introductory physics course taken by physics majors.

Apparently, the premedical physics requirement will continue to stand for some time as a kind of paradox. Still deemed important, as well it should be, given the strong trend in biology that leads in the direction of reductionism and the constantly increasing sophistication of medical instrumentation, the physics course is, nevertheless, only a small segment in the long road to either medical practice or medical research. Only

---

* R. G. Page and M. H. Littlemeyer, eds., *Preparation for the Study of Medicine* (University of Chicago Press, Chicago, Ill., 1969), Table 8.

about one percent of medical students have undergraduate majors in a physical science other than chemistry, and few medical students take more than the minimal requirement in physics. Faced with major problems that promise to alter medical practice and its organization in fundamental ways, it seems unlikely that medical educators will attempt to define their interest in a physics requirement more clearly than at present, if, indeed, they ever could.

In the meantime, physics faculties can recognize aspects of the course for nonscience majors that future physicians should encounter. Certainly, as urged by Abell, they can identify topics (for example, radiation, waves and pulses, fluid flow, and feedback systems) that promise to be significant in future biology and can add interest to lectures and laboratories. They can also use examples chosen from biomedical applications in their presentation of physics.

Engineering Students   Physics has, traditionally, been an important part of the curriculum for engineers, and necessarily so. The distinction between engineering and applied physics becomes more and more an artificial one. It is an experience shared by many instructors in some upperclass courses designed primarily for physics majors that engineers have been among the best (and worst) of the students. These remarks are not intended to imply that physics is a sufficient basis for engineering, but there is a strong coupling. In fact, many engineering faculties have evolved in the past 10–15 years into applied-science faculties, frequently staffed in part by physicists and capable of teaching aspects of physics that they feel their students need. They have grown all the more anxious to do so, as physics teachers during the same period have increasingly turned to abstract formulations of physical law at the expense of concrete illustration. The situation is further complicated by the proliferation of engineering specialties (such as aerospace, nuclear, chemical, civil, communication, electrical, geological, mechanical, and sanitary; all these appear as departments listed in the catalogue of a moderate-sized engineering school, and each may have its own attitude toward a physics requirement).

The important distinction between engineers and physicists, a distinction that must be kept in mind in assessing their special needs, is that of outlook. The engineering tradition is more pragmatic and less delighted with order-of-magnitude reasoning and with looking at various ways in which a problem can be viewed and solved than with learning the most expedient way of solving it. For engineering students, physics is useful insofar as it provides the foundation in fundamentals that allows a systematic approach to be used intelligently. In this sense the ability of physics to demonstrate that many complexities derive from great principles is important. Specifically, in physics courses intended for engineers, the conservation laws should be emphasized, as should fields from static and moving sources, velocity fields, and fundamental wave theory. Engineers can understand how things work without much physics, but the reinforcement stemming from the knowledge of why things work comes from physics and represents an investment in the future of engineering.

Finally, the interrelationships between physics and engineering will be fostered if the courses shared by students in the two fields begin to recapture some of the flavor of application. The once-popular engineering physics baccalaureate degree lost some of its appeal, possibly because it left no time for the leavening influence of the humanities. If this deficiency can be repaired in a satisfactory manner, this degree may well be very useful in a future that will be greatly concerned with applications of science.

Graduate Education

At best it is risky to attempt an assessment of graduate education in physics at this particular time. Graduate study is intensive professional training and thus is directly coupled to the economic prospects in the profession. Whether one speaks of a crisis in employment or merely of readjustment, it is agreed that the employment picture for the average scientist is undergoing striking changes, and these cannot help but have a profound influence on practices in graduate education.

As a working hypothesis (see Chapter 12) it seems reasonable to assume that in the 1970's, and perhaps well into the 1980's, no more than one half of those receiving the doctorate in physics will be able to find jobs in physics teaching at or above the college level and in research. Therefore, a large number of young PhD's will not be able to look forward to professional careers in which doing physics, in the sense of advancing knowledge, together with teaching and performing other academic duties, is the main part of one's activities. Those who plan graduate educational curricula must consider the needs of this growing fraction of graduate students. It is reasonable to expect that the shift into an era in which most professional physicists are no longer found in academic institutions will have a substantial influence on the education of professional physicists. The new demands placed on graduate education are hardly recognized as yet; it is far too early to say what new patterns will evolve and which of the old patterns will survive, but it is not inappropriate to summarize the present state of graduate education in physics and to attempt to identify and perhaps evaluate current trends.

At the end of the 1960's, graduate education in physics in the United States showed great basic strength. Measured by almost any conceivable yardstick, physics training had reached a level of remarkable effectiveness and the system was producing large numbers of competent—and sometimes brilliant—professional physicists. Eighty new graduate programs sprang up in emerging schools, and older programs were significantly expanded (see also the Appendix of Chapter 9). Almost without exception, the traditional model of the graduate department was followed by young faculty members. This development has contributed to the maintenance of sound academic standards throughout the university system, but it also has tended to inhibit innovation and experimentation at the graduate level and to constrict research directions. The extent of structural uniformity among graduate programs in physics is strikingly evident from an examination of the handbook on *Graduate Programs in Physics, Astronomy and Related Fields,* published in 1971 by the American Institute of Physics.

Entering the 1970's The coming decade offers an opportunity to reexamine admission criteria, because at the same time that graduate enrollments in physics departments are decreasing, these departments will be attempting to prepare students better for a far greater diversity in careers. The present screening methods favor those students who learn well from formal courses, who are conscientious and hardworking, and who are not likely to flounder in graduate school. Although it is not easy to document the contention that some truly creative individuals may have been stifled in their development toward becoming successful physicists, admissions policies and the rigidly structured program of the first two years may tend to discriminate against the inventive experimenter, with a keen intuition but a disinclination toward mathematical formalism, and the thoughtful theorist, with a philosophical bent but a distaste for the routine of problem-solving. Einstein's indictment of conventional formal education should serve as a constant reminder of the damage that might be unwittingly inflicted.

A generation ago, some of the most creative physicists came from the ranks of engineering, mathematics, or chemistry. Today, students in cognate areas find more difficulty in switching back and forth. Efforts should and could be made to encourage the crossing of fields, and in any event physics departments should welcome as graduate students talented applicants with undergraduate majors in allied subjects.

All phases of graduate education, including admissions, are seriously affected by the changing policy of the federal government with regard to the financial support of graduate students. The familiar attitude that it is in the national interest to provide substantial and direct federal support for the education of graduate students in physics and other sciences is quickly losing ground. Instead, it is being argued that the individual student, rather than society as a whole, is the chief beneficiary of graduate training, and that direct fellowship support for the majority of physics graduate students is no more appropriate than support for law or medical students. These changes are likely to alter the composition of the graduate student body by introducing elements of motivation and inhibition that have not been felt for many years. The effects of the changing federal economic policies on graduate training are discussed in Chapter 12.

The Core   The first two years of graduate work constitute the educational backbone of the PhD program at U.S. universities. It is during the rigorous and comprehensive course work that the U.S. student catches up with, and often surpasses, his foreign contemporaries, who usually receive more specialized education at an earlier age.

The core courses that are central to the graduate physics program appear to have their origins in the great pedagogic tradition of theoretical physics from Kirchhoff and Tait to Sommerfeld. Once again, this time working at greater depth and learning more powerful techniques, the student ranges over the central subjects of physics. This final cycle has been of greatest importance for physics, not alone because it provides a foundation upon which the student's advanced work in his specialized field can be built, but because each generation of students gains a common language and background, of which not only the subject matter but also the anecdotes and lore form a legitimate part. In the face of increasing specialization, physicists should be conscious of having maintained an underlying core of knowledge and an attitude that constitute two of the most characteristic features of physics as a discipline.

The core courses have not remained completely static, however. At an earlier time, considerable emphasis on continuum mechanics, including elasticity, acoustics, and fluid dynamics, and on optics was customary. After World War II, the emphasis shifted to give students background to pursue the new directions in research. Because the main thread of physics no longer seemed to require courses in continuum mechanics, and to avoid overly extending the period of classroom study, those courses gave way to quantum mechanics and its applications and extensions; that is, the core was altered. But the intent of a core program is to ensure study of those topics that are basic and broadly powerful in application and that underlie major unsolved scientific problems. Since continuum mechanics still contains such topics and problems, it is not surprising that, from time to time, they surface unexpectedly during the study of complex phenomena or in an effort to develop advanced technological applications. Currently, continuum mechanics is beginning to be viewed in a different light, and many are calling for its return to the core.

There can be little argument about the advantages of preserving a core course of study to stress the unity and common elements of all physics. Physicists must not be too rigid in defining it, but they also cannot allow it to respond lightly to winds of fashion in physics or to immediate demands of the marketplace. Because physics faculties, with all their diversity of interests and circumstances, strongly desire to provide basic, broadly powerful study for their students, their tendency to be conservative when re-examining the core subjects of physics is not inappropriate.

A Call for Restructuring   The normal pattern of graduate education in physics during the past few decades has embraced, by and large, a single model. Two years of course work are followed by three or four years of research leading to a PhD dissertation. The master's degree, which could be acquired after about two years of graduate-level work, has been severely reduced in prestige and has become either merely a way station in the progress toward the PhD or a consolation prize for those who are to be discouraged from going on. Students in the latter category often transfer to other schools and proceed eventually to the PhD.

It is difficult to change such patterns, ingrained and involving institutional attitudes and tradition, but there seems little doubt that the present period of ferment in higher education calls for serious re-evaluation. Under the slogan, "Less Time, More Options," * the Carnegie Commission on Higher Education has recommended that colleges and universities restructure their programs so that "a degree (or other form of credit) be made available to students at least every two years in their careers (and in some cases every year)." A refurbished (and possibly renamed) master's degree, based primarily on the course work of the first two years in graduate school, could produce professional physicists who might enter a career in education, industry, or government without the lengthy research experience embodied in a PhD dissertation.

A number of universities have instituted the Master of Philosophy degree. This is typically awarded on successful completion of all the normal requirements for the PhD degree *except* the dissertation and the original research on which it is based. It has not achieved the successful acceptance hoped

---

* Carnegie Commission on Higher Education, *Less Time, More Options: Education Beyond the High School* (McGraw-Hill, New York, 1971).

for it, at least as yet. Part of the failure stems from a form of academic snobbery that is reflected in greater interest in the number of doctorates available for listing in an institutional catalog than in the competence of the individuals involved relative to the discharge of their particular institutional duties. From an objective viewpoint, such a degree, if considered on its merits, would appear to be well suited to teaching and other careers in which a comprehensive knowledge of the field should be essential but in which the opportunity, or even the desire, to engage in original research is lacking.

We urge more physics departments and graduate schools to examine carefully the extent to which the introduction of such an M. Phil. degree might fulfill their objectives, and we urge colleges in particular to consider the possible advantages that might accrue to them through active recruiting of such M. Phil. graduates to their teaching faculties.

If the master's degree does not prove to be an appropriate vehicle for producing professionals for nonresearch careers in education, industry, or government, the introduction of new forms for graduate degrees, for example, the Doctor of Arts, should be considered by the physics community, which should also watch carefully, and critically, all initial attempts to establish new degree programs.

Preparation for careers in teaching at various levels should be an accepted part of graduate education in physics. It is appropriate here to applaud the emerging tendency for graduate departments to take an active interest in the preparation of graduate students for careers in teaching and science education.* It seems particularly desirable, as preparation for an era in which faculty openings will be scarce, that graduate students who desire academic careers have the opportunity to gain practical experience in teaching, not only as assistants in laboratory and recitation sections of the large elementary courses but also at the more advanced undergraduate, and even the graduate, levels as interns under the careful guidance of experienced teachers. Some leading departments have already instituted formal programs designed to involve faculty in helping graduate students to improve their teaching capabilities.

A Critique of the Research Component   It would seem desirable to take a fresh look at even the educational function of research training as it has evolved in graduate schools. It should be kept in mind that much of the present strength of physics in the United States has derived from an

---

* Commission on College Physics, *Graduate Preparation for Teaching—The Missing Component* (CCP, University of Maryland, College Park, Md., 1971).

insistence on intense specialization in graduate school. Opinions differ on how early a graduate student can and should be introduced to research —and the inhomogeneous preparation of graduate students precludes uniform rules about this—but it seems a sound principle that a period of apprenticeship in doing physics (as opposed to merely learning to know it) is beneficial to all advanced students. Even a student who intends to devote himself to a career in physics that will involve no research is well served by spending a modest amount of time participating in ongoing research in some particular area. The balance between general studies and specialization in research must be a strong function of the student's ultimate goals and his ability, but only those who work for at least a short period with new problems are going to be fully aware of the basic nature of physics as a science in perpetual self-renewal rather than as a fixed body of knowledge. A person who has acquired a reasonable degree of proficiency in a special area might not pursue his special field in his professional life, but he will find the possession of expertise in some area a valuable accessory to his personal development. His perspective in one specialty gives him a permanent advantage over the total generalist who knows a little bit about a great many things.

One may accept the foregoing premise without drawing the further conclusion that every graduate student must necessarily spend a large fraction of his efforts in research. There are many career opportunities in physics in which advanced study is most desirable but in which a relatively short exposure to specialized research experience suffices and the bulk of the student's time is better spent in other modes of learning.

The possibility of changing the role of the traditional MS degree and the wider introduction of the M. Phil. degree have already been mentioned. But even the work toward the PhD, with its strong and essential research component, could perhaps be made more rewarding and effective. Although the concept of apprenticeship for the research student has obvious validity, PhD students should not be allowed to spend such protracted periods of time as members of an academic research group that they develop a narrow and overly specialized outlook. The educational planning we envision would consciously expedite the student's progress toward becoming a self-reliant scientist. At the same time, such planning would provide the opportunity to inculcate in the PhD student attitudes that would prepare him better for a rewarding career as a physicist in the 1970's. Ways and means must be developed to ensure that the solid research experience of the graduate student, extending over a period of years, does not lead him to think solely in terms of a life in research and teaching, especially in his thesis field. Faculties should consider whether a more flexible attitude toward the PhD dissertation is possible, without compromising its traditionally high standards.

Above all, great care must be exercised by physics faculties to ensure that, whether by inadvertence or otherwise, the attitude that all careers other than academic are somewhat less prestigious, less acceptable, and the like not be allowed to propagate among their students. As we emphasize throughout this Report, such unfortunate attitudes are far more root causes of present difficulties in obtaining employment than any specific content or the structure of contemporary physics education. Only physics faculties can eradicate such attitudes; they bear a heavy responsibility to do so.

Breadth of Research Activity

It is a well-documented fact that teaching that is completely divorced from its underlying research base rapidly becomes sterile. During the 1960's, this recognition, coupled with the availability of the necessary supporting funds for both faculties and facilities, led to a general effort in physics departments toward establishing research activity in each of the major subfields of physics. Indeed, it has developed that such breadth of activity has become a rather widely used measure of departmental strength or attractiveness. To the extent that each of these activities can be maintained at a level above the critical one at which high-quality research and teaching are the result, these trends are clearly laudable. But relatively few universities have the resources to make even this attempt in realistic fashion.

Under conditions of limited funding and growth, the Committee argues strongly for selectivity, with each physics department concentrating its resources in those areas in which it can realistically hope to achieve or maintain excellence. We applaud the initiative that a number of regional groups have taken in arranging, where feasible, to mount complementary research activities such that students in the region have available to them a much broader range of expertise and opportunity than could be possible for any given one of the cooperating institutions. We urge that this concept receive serious consideration on a much broader scale throughout the nation.

Education for Industry

During a brief period around 1960, more physicists were employed by industry and the federal government than by academic institutions (see

Chapter 12). By the late 1960's, the dynamic factors governing employment in the three sectors produced a reversion to the more traditional balance, but not through the traditional kind of employment, that is, as regular academic faculty. All projections of employment patterns indicate that the nonacademic sector will offer the most opportunities for physicists during the next decade.

Because the decision to enter graduate education leading to an advanced degree is a career choice, though not necessarily a lifetime commitment, the student has a right to expect that the education and the degree that he seeks can offer him a realistic preparation for his chosen career. Because that education and degree are heavily subsidized with public funds, the public also has an interest in the nature of the education being provided.

In recent years, spokesmen for industry have indicated their concern with the graduate education of scientists and engineers and, in some instances, have spoken directly about physics education.* Reports of their opinions have been both published † and circulated informally.‡

By far the most universally endorsed theme is stated by Ascher:

> I believe that the demand for the general is the most important single demand that industry can put to the University; it is also an important contribution on the part of industry to the solution of the crisis that the University undergoes.

A second theme, clearly of equal urgency and not independent of the first, as we have noted previously, is a deep concern over the mental attitude that often accompanies a mismatch between career aspirations and actual careers open to individuals. Again concurrent with Ascher's statement is the view of the PhD degree as signifying predominantly training in research and specialization. This view attaches relatively little importance to the subject of the dissertation, pointing out that even if the physicist, no matter how employed, refuses to change his specialization every decade, his specialization will change. Spokesmen § point out that industry has a strong interest in people with the flexibility to shift from one area of work to another, and also to careers in which they will draw

---

* See, for example, S. D. Ellis, *Work Complex Study* (American Institute of Physics, New York, 1969); and A. M. Bueche, "Issues in the Changing Relationship between Industry and Academic Science" (General Electric Co., Corporate Research and Development, Schenectady, N.Y., 1970), unpublished paper.

† G. Diemer and I. H. Einch, "The Education of Physicists for Work in Industry," in *Proceedings of an International Seminar on Education of Physicists for Work in Industry,* Eindhoven, The Netherlands, December 2–6, 1968 (Centrex Publishing Co., Eindhoven, The Netherlands, 1968).

‡ E. Ascher, "A Contribution to the International Seminar on Education of Physicists for Work in Industry," unpublished manuscript.

§ For example, A. M. Bueche, General Electric Co., Schenectady, N.Y.

on their scientific backgrounds and the knowledge and perspective gained from fields quite different from those in which they may be working.

These general beliefs lead to more detailed opinions about the nature of graduate education. For instance, the student needs greater awareness of the role of physics in society. He may gain such awareness more through courses in which case studies are a principal part of the content than through introductory courses in various social sciences or humanities. Students also need to know something of the style of industrial research and the complexity of most industrial problems—problems that require integration of the knowledge and techniques from a number of disciplines, often extending beyond the sciences. The long period demanded today for PhD training (6.3 years is the median in the United States for a doctorate in physical sciences and engineering) is of concern because it consumes a substantial interval during the highly productive years of a young scientist, thus increasing the cost to both the nation and the individual in terms of productivity.

There is a strong desire to strengthen the bonds between industry and the academic institutions through temporary exchanges of personnel, internships, adjunct professorships, and the like. Collaboration between universities and industry could bring about successful programs for mid-career education to keep scientists abreast of their fields, offer an opportunity for those who are changing fields to acquire a new base of knowledge and skill, and provide broader knowledge of the overall state of science, as well as of the social context, for those whose careers are taking them out of the laboratory.

Industry's complaint of narrowness in PhD training is accepted by the academic community as directed more toward attitude than substance. Even elementary-particle theorists can employ their highly developed computational and programming skills to industrial problems, applying also the broad education in physics that they have received. In the United States, a relatively close coupling (compared with most other nations) between industrial and academic physics is apparent at almost any research conference. In recent years, the demand for rapid expansion of the educational system inevitably produced strains. However, the absence of intellectual barriers assures that in a steady-state climate differences in view are unlikely to persist.

AGENCIES AND ORGANIZATIONS

Ultimately determined by individual ability and performance, teaching nevertheless is influenced by the thoughts and practices of the profession generally and the resources provided by society. The teaching of physics is sensitive to these influences, most particularly through professional organizations of physicists and agencies of the federal government that have the responsibility for educational improvement. Over the past decade, changes have occurred in the nature and policies of these organizations and agencies. Indications of some forthcoming changes are now becoming apparent. Many organizations of a more general character have potential significance for physics education, but a description of them and of their existing and possible modes of interaction are beyond the scope of this Report. We will treat only those few that have the most immediate impact on physics education.

Whatever changes may be forthcoming in the next few years will occur during a period that promises generally constricting funds for the improvement of science education. Figure 11.6 illustrates the overall constraints that govern programs in physics. The one apparently encouraging trend, which is found in expenditures of the American Association of Physics Teachers (AAPT), is largely a consequence of increased journal costs and the willingness of members to underwrite at their expense the assumption of added responsibilities. In the face of a pressing need for more experimentation in education, greater opportunity for teachers to communicate with each other and to extend their range of communication, and the development of more stringent methods for evaluation of the results of experimentation, as well as for the means to disseminate widely those new methods that stand the test of more critical evaluation, the trends illustrated here are a matter of grave concern.

The Federal Presence in Science Education

Since its beginning in 1950, the NSF has been the principal agent for conducting federal programs in science education. Its many programs and activities throughout this period can be grouped roughly into three categories: encouragement and assistance for students, strengthening the qualifications of teachers, and the development of improved courses and curricula. Most of this effort has been allocated to the education of professional scientists, but increasingly the NSF has assumed responsibility for the improvement of science education at all levels and for a broader range of educational objectives. The resources given this agency by Congress and the Executive branch have sufficed to enable it to have a large quantitative impact on the professional education of scientists and even to exert a significant qualitative effect on high school physics. However, its broader role in education must be limited to pilot efforts at innovation, for it does not have the resources to mount, for example, a major national

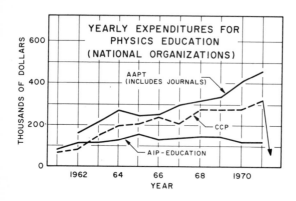

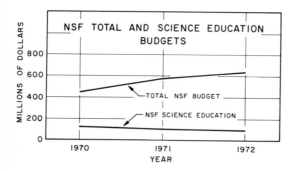

FIGURE 11.6 *Top:* Yearly expenditures for education by the three most active national physics organizations. *Bottom:* Total yearly NSF budget and yearly NSF budget for education.

program to improve elementary school education by re-educating one million teachers.

Although local school systems, state departments of education, and a number of federal agencies share responsibility for science education, the NSF is unique among them in its close link with the scientific community. When educational improvement can be regarded an adjunct or an outgrowth of their other activities, scientists find it the easier to participate. Elsewhere in this chapter we have stressed the importance of their doing so. The inclusion of the improvement of science education as an explicit and strong function of the NSF is the best current assurance of accomplishing this objective.

Further conditions are necessary if the NSF's educational activity is to have a quality matching its importance. Because science education is part of science, much of its style and even its vocabulary are those of science. Communication between the education staff of the Foundation and the outside scientific community is largely scientific discourse. It would take place haltingly and uneasily were not the personnel of the Foundation scientists. But there is a further, more stringent condition. The programs formulated by the NSF and their implementation and evaluation are the responsibility of the staff of its Education Directorate. Obviously, the quality of these activities is dependent on the quality of the staff. The staff, however, does not act in isolation; it has continual interaction with the scientific community through advisory panels and committees, reviewers of proposals submitted to the NSF, and informal contacts at meetings, to mention only the most prevalent mechanisms for exchanges of views. The quality of the advice and suggestions depends on the caliber of the persons with whom the staff interacts, and the nature of the staff is a major determinant of high-quality, productive interaction with the scientific community.

One can state with justice, therefore, that the argument for locating federal programs for science education in the NSF depends heavily on the ease with which its personnel can draw on the assistance of the best of the nation's scientist-educators. The caliber of the staff is the key. Though mindful of the NSF's many successes, we believe complacency to be unwarranted and urge that a continuing effort be made to attract ever more able persons to the education staff.

The task of surveying the NSF's education program is made easier by a report in 1970 from its Advisory Committee for Science Education.* This report reviews the Foundation's educational programs, especially the course content improvement activities, evaluates them, and recommends that emphasis be shifted toward scientific education of the public. In general, and in most of its particulars, our Report is consistent with this earlier one; it, too, recommends greater effort toward further precollege curriculum development, preservice teacher education, the improvement of courses, and the teaching of science for the nonscience undergraduate. It also advocates greater perspective on social needs in the education of scientists, improvement in the training of technical personnel, and reconsideration of the nature of advanced degrees. In some respects there are differing views, for example, the NSF report calls for greater emphasis on interdisciplinary, problem-oriented education for scientists than we do.

---

* Advisory Committee for Science Education, *Science Education—The Task Ahead for the National Science Foundation* (U.S. Government Printing Office, Washington, D.C., 1970).

The following excerpt from an address given by Roger Tory Peterson on acceptance of the Audubon Medal at the annual Audubon dinner, New York, November 1971, is cogent:

In the wake of the recent ecological awareness, any number of universities have set up departments and courses dealing with the environment. Even whole new colleges devoted to this theme have been launched—Hampshire College in Massachusetts, Green Bay in Wisconsin, and several others. Their avowed purpose is to examine the total environment on an interdisciplinary level.

It is a grand concept but a weakness is that on the undergraduate level few students have even one discipline at their command. Many of these nice young people like the idea—they like the word ecology, but they are completely lacking in background; they haven't done their homework. As my young son puts it: "They soon find that the subject is a very real one—perhaps more real than they are."

It has been my observation that by far the most sophisticated ecologists and environmentalists are those who did not start as generalists. They threw themelves into a specific interest—birds, botany, marine biology, or one of the other earth sciences. They lived it and breathed it. Eventually, through a sort of intellectual adaptive radiation they expanded their horizon to a number of peripheral things. Starting from a focal point their eventual interest in the wider aspects of ecology acquired relevance.

As the NSF begins the task of improving science education for the nonscientist, it should not have to phase out programs of proved merit that help to maintain the strength of science in the United States. Predoctoral fellowships offered through national competition would be in this category. The winning of an NSF Graduate Fellowship has become an undisputed mark of promise. A student holding one is welcome anywhere; he is able to develop himself as a scientist in the surroundings he thinks best suited to his purpose. Surely society gains from providing this kind of assistance for its most promising young scientists.

The U.S. Office of Education is entrusted with the expenditure of about $5 billion per year. In contrast, the NSF obligation for education reached its highest point, approximately $125 million, in 1968 and has declined to about $70 million per year since that time. The budgets of the two agencies reflect difference in the specialized concerns of NSF and the quite different congressional and executive intent behind the appropriations. The Office of Education has a broad mandate to support education at all levels, but most of its funds are distributed to the states and local school districts on a formula basis. It supports research and instruction in educational methodology. The improvement of subject-matter instruction is a less direct concern. However, if new science curricula are to be disseminated widely, resources more on the scale of those available to the Office of Education than those of the NSF will be needed.

We recognize a strong need for close cooperation between the Office of Education and the NSF if the objective of strengthening science education for the general public is ever to be achieved. Given the tested channels of communication between the science community and the NSF, and the dearth of such channels to the Office of Education, it is reasonable for most innovative and developmental efforts to be conducted through the NSF, at least for the foreseeable future. Yet to offer a substantial fraction of the in-service teachers improved preparatory study will require a better understanding of the learning process. In addition, to facilitate the distribution of science materials will require expenditures and channels to the schools that at present only the Office of Education is equipped to manage. To attack the problem by, for example, dividing the responsibility for science education according to the educational level, assigning the lower levels to one agency and the higher levels to the other, would present disadvantages, for this policy would tend to remove a part of science education from the habitat of the scientists and thereby weaken their interest and involvement. It is gratifying, therefore, to find in the 1970 Report of the NSF Advisory Committee on Education a call for strengthening the liaison between the two agencies.

The drastically diminished level of support for educational activities in the NSF has been a matter of great concern, although it is recognized that much of the decrease is a consequence of a deliberate reversal of a policy instituted by the federal government a decade ago to accelerate the production of scientists and engineers. Accompanied as it is by mandates to expand programs for the improvement of general science education and to emphasize the applicability of science to the achievement of socially desirable goals, this budgetary pressure is felt intensely by programs intended to improve the curricula and facilities for educating future scientists and engineers. Society cannot ignore the reality of the costs of education for science; like the study of medicine, it always has been relatively expensive. But the nation as a whole gains from the productivity of scientists and engineers and should not shirk the task of educating them effectively.

Legislation has been introduced to establish two new educational agencies, the National Institute of Education and the National Foundation for Higher Education. In the words of the President of the United States, the National Institute of Education would be aimed at a "serious, systematic search for new knowledge needed to make the educational opportunity truly equal." * It would be located in the Department of

---

* "Education for the 1970's, Renewal and Reform," *Messages to the Congress,* by Richard M. Nixon, March 1970.

Health, Education, and Welfare. Again in the President's words, "It would have a National Advisory Council of distinguished scientists, educators and laymen to ensure that educational research in the Institute achieves a high level of sophistication, rigor and efficiency." No one who looks critically at the present output of educational research will question the need for a central means of coordinating and evaluating it in the way indicated by the President, whether it be directed explicitly at the goal of equalizing educational opportunity or at other goals. If carried out in a truly critical spirit, the evaluation of educational research could lead to the recasting of education to give all people a more realistic opportunity for satisfaction from education.

The National Foundation for Higher Education, which would be much like the NSF and the National Foundation for the Arts and Humanities, would have as its objectives the support of ideas for improvement and reform in higher education, the strengthening of colleges and universities that have special needs or fulfill special functions, and the development of a national policy on higher education. To ensure coordination, the Director of the NSF and the Commissioner of Education would serve *ex officio* on the Higher Education Board, which would govern the new Foundation.

Should one or both of these agencies come into being, the effect on education in the United States could be profound. In particular, of course, the education programs of the NSF would be affected.

Professional Organizations

Until 1971, the professional interests of physicists in education were widely represented by three organizations, the American Association of Physics Teachers (AAPT), the American Institute of Physics (AIP) through its Division of Education and Manpower, and the Commission on College Physics. In addition, a few of the other professional societies, for example, the Optical Society of America and the Acoustical Society of America, maintain strong educational programs, although smaller in scope.

The AAPT is a voluntary membership organization of over 13,000. Its members come from college and university faculties of all kinds, as well as from high school faculties. Typical of its type, it has confined its programs in the past largely to holding meetings and publishing journals. An array of committees has performed special projects such as the stimulation of laboratory improvement and the production of films (through various competitive events) and the recognition of outstanding achievement (through the bestowal of awards). Currently, circumstances strongly suggest that the AAPT should assume still broader responsibilities. Some of these we recommend in this Report.

The AIP is a federation of professional societies of physicists and astronomers, created as an operating agency to perform functions in the common interest of the Member Societies. Its Division of Education and Manpower has been engaged principally with the collection and dissemination of data and information. It has established the Niels Bohr Library and a History of Physics Program, which over the years has accumulated a unique collection of materials of great value to teachers and scholars. In addition, with funds obtained through grants it has established and maintained a Visiting Scientists Program, which has supplied short-term visitors to smaller or remote institutions seeking help with their curricula and the stimulation of contact with other physicists. Also with grant funds, the Division is undertaking the technical-physics program mentioned earlier in this chapter. With funds derived from its own activities, the AIP supports the Society of Physics Students. Financial pressure is causing the management of AIP to curtail expenditures not supported directly by grants from outside agencies, and the staff of the Division of Education and Manpower is being reduced drastically. The Niels Bohr Library and the History of Physics Project are threatened with an almost complete halt.

Concern with physics education is not confined to those who are members of AAPT, nor to those who are teachers at whatever level. Of the 49,000 individuals represented by AIP, less than one third signify their concern through membership in AAPT. For most of the larger group, programs undertaken by AIP are the only organized expression of their stake in improving physics education. The Member Societies of AIP need to reconsider their interest in preserving, strengthening, and enlarging their commitment to an AIP education program, especially at this time.

In the 1960's, a new way of promoting science education came into being, the College Commission. Each of the basic sciences, mathematics, and engineering established semiautonomous bodies charged with the task of stimulating improvement of college-level education in their fields. All of these bodies were supported by the NSF. The 1970's have brought their demise, apparently as a result of a Foundation policy decision not to be the sole support of any long-lived institution and of wariness toward measures that increase the attractiveness of scientific careers. The Commission on College Physics, therefore, terminated its official existence in 1971. Its decade of existence saw many achievements, including the establishment of independent regional organizations; the stimulation of new curricula, textbooks, and films; and the convening of conferences and dissemination of reports. Perhaps none of its accomplishments was more valuable than the education of hundreds of physicists in modes of action designed to bring about educational reform. The value of an independent, semiautonomous central group charged with responsibility to discover needs and stimulate others to act on them is one of the greatest lessons of the decade in science education. The means by which such a group can be made sufficiently representative of and responsive to the community, yet sufficiently detached to exercise critical judgment, and can be empowered to act promptly remain difficult problems, as they have throughout history.

In its final months, the Commission attempted to maintain a forward-looking stance through organizing two series of studies. One of these was a series of Regional Conferences on Graduate Preparation for Teaching that attempted to stimulate students and their departments to devise and mount programs to give the teaching assistant guidance and practice more befitting his status as an apprentice teacher than was customary. The second activity was a series of preparatory meetings and a final conference, Priorities for Undergraduate Education, held during the summer of 1971. Although operating from a different point of view from that of the Physics Survey Committee, the Priorities Conference advocated many of the same changes in attitude and practice. The final Commission Newsletter * gives a summary of the Conference. The complete report as well as the Final Report of the Commission, which are to be published in 1972,† will be significant documents for anyone concerned with physics and education.

To stimulate educational improvement in physics, AAPT has enlarged the staff and created a new body, the Council on Physics Education. The mandate of the Council directs it to maintain surveillance over the full range of educational levels, recommending and promoting action wherever it is needed. The membership of the Council is more widely representative than was that of the Commission, in recognition of its broader responsibility. More directly responsible to a membership organization than was the Commission, it must find a mode of operation combining this responsibility with prompt and discriminating action. If the Council is to be broadly effective, it should have working relationships with other organizations

---

* *CCP Newsletter,* No. 25 (Nov. 1971).
† See J. M. Fowler, "Progress Report of the Commission on College Physics for 1966–68," *Am. J. Phys., 36,* 1035 (1968); W. V. Johnston, "Pacific Northwest Association for College Physics, a Many-Body Problem," *Am. J. Phys., 38,* 741 (1970); and J. D. Spangler and C. E. Hathaway, "The Consortium for the Advancement of Physics Education," *Am. J. Phys., 39,* 399 (1971).

having interests in physics education. The AIP has been invited to appoint a representative to the Council; other societies might well be included in this formal manner. It was necessary for the AAPT to underwrite the cost of staff increases and Council operation from its own resources, which necessitated a significant increase in dues. The membership of the AAPT has responded favorably to the added amount.

Regional Organizations and Centers

One of the more far-reaching accomplishments of the Commission on College Physics was the conception and nurturing of a number of regional organizations intended to conduct localized efforts to improve physics teaching. The six that have evolved are diverse in constitution and programs, reflecting the quite different circumstances that brought them into being. In some cases, they bring together a fairly homogeneous group of colleges that can easily share instructors and facilities; in others, they encompass the full range of educational institutions, from two-year technical institutes to universities, public and private. Organizations of this kind can apply for grants to support their activities; in at least one case, the approximately 60 member institutions pay dues that partially support a central office and newsletter. Among the explicit aims of such groups, the following are prominent:

1. To increase communication and cooperation among regional physicists (in some parts of the United States a college physics instructor has been known to spend as much as five years without any personal interaction with his colleagues at other institutions);

2. To offer instruction to physics teachers with inadequate or outdated academic training;

3. To offer local opportunities to hear about new instructional materials and developments from those engaged in producing them;

4. To offer auspices and facilities enabling instructors to discuss their problems and ideas with others in similar situations, often guided by more experienced and highly trained physicists.

Experience has shown that, even when funds have been difficult to obtain and programs have had to be curtailed, regional organizations maintain a vitality that could result only from their meeting an important need—the need for a stronger feeling of community than that derived from non-discipline-oriented associations or from the more diffuse and conventional professional organizations. Although aspects of their programs may have a familiar appearance, their role and effectiveness must be judged in a context that differs from the more usual, centralized frame of reference of funding agencies.

A somewhat different means to provide continued and well-focused effort toward improvement is the educational research and development center. Three or four have evolved during the decade just past.* Centers can be established by individual institutions or by consortia and can serve as collection points for teaching materials and information, locations for shop and computer facilities needed to try out laboratory or demonstration experiments, and summer gathering places for those wishing to engage in cooperative projects or to participate in refresher courses in an environment devoted to the improvement of education.

International Activities

Physics education presents problems having a high degree of international commonality despite strong differences in the educational frameworks adopted by the many nations. A comprehensive guide to the variability of physics education can be found in a comparative study, *A Survey of the Teaching of Physics in Universities,*† which also describes preuniversity education. Here one will find diagram and description, examination question and laboratory practice, if he wishes to understand the systems that have produced the colleagues he meets when he attends a conference abroad or an international gathering at home.

The improvement of physics teaching has been explored in a sequence of international conferences sponsored by the International Commission on Physics Education of the International Union of Pure and Applied Physics. These conferences have considered many of the matters discussed in this chapter, including general education in physics, high school physics, the curricula for undergraduates and graduates, and the preparation of secondary schoolteachers of physics.‡ The International Commission has

---

* *CCP Newsletter*, No. 25 (Nov. 1971).

† International Union of Pure and Applied Physics/United Nations Educational, Scientific and Cultural Organization, *A Survey of the Teaching of Physics in Universities* (G. J. Thieme N.V., Konimglijke Drukkerij, The Netherlands, 1966).

‡ S. C. Brown and N. Clarke, *International Education in Physics* (The MIT Press, Cambridge, Mass., 1960); S. C. Brown, N. Clarke and J. Tiommo, *Why Teach Physics?* (The MIT Press, Cambridge, Mass., 1963); S. C. Brown and N. Clarke, *The*

sponsored two seminars as well, one on the Education of Physicists for Work in Industry and another on the Role of the History of Physics in Physics Education.

In addition to the Survey mentioned above, UNESCO and the International Commission on Physics Education have been engaged jointly in publishing a series on *New Trends in the Teaching of Physics* and a *Source Book for the Teaching of Physics in Secondary Schools.*

UNESCO also has undertaken projects in science education in a number of countries needing special help. These have been located in Asia, Africa, the Near East, and Latin America. The last of these included a one-year pilot project on the teaching of physics followed by a series of regional conferences for physics teachers. In many of these areas permanent organizations have been established to carry on developmental work.

A recent review of these activities * expresses the main worry of UNESCO as "the limited participation of physicists and physics teachers from the developing countries," and addresses the question of how to identify and help those with potential for becoming leaders in science education in these countries. Organizations that support promising young scientists from developing nations when they study in the more advanced areas, and the institutions that receive these young people, are asked to pay more attention to this need as they devise programs to accommodate them.

Accreditation?

Accreditation of physics departments is viewed by some as a guarantee that students motivated toward physics will be guided away from institutions offering poor or inadequate education in this subject. We believe that there are more positive ways to accomplish not only the professionally oriented objective mentioned but also improvement of physics education everywhere.

Our reasons are the following: There is a variety of professional societies in physics and an even greater diversity of occupations for physicists; to attempt to formulate standards acceptable to all would be a difficult process and probably fruitless. Rather than being imposed by an external agency, standards should be a concern of the individual educational institution.

Education of a Physicist (The MIT Press, Cambridge, Mass., 1966); S. C. Brown and N. Clarke, *Teaching Physics, An Insoluble Task?* (The MIT Press, Cambridge, Mass., 1966).
* S. C. Brown and N. Clarke, *Teaching Physics, an Insoluble Task?* (The MIT Press, Cambridge, Mass., 1966).

Each institution has its own objectives, its own view of itself, and it has access to outside opinion when it so desires; quality is difficult to define, and high-quality education, no easier. Listings of courses, laboratory apparatus, and facilities, even the citing of textbooks used, give only the façade of education; distinctions won by faculty members provide another dimension but are not likely to be prescribed by an accreditation team. A set of standards once established is difficult to change, and an institution once accredited may find little incentive for further improvement. Those who would innovate need encouragement rather than the reverse. Finally, the students in many accredited institutions would find their instructors performing in ways dictated from the outside rather than from inner convictions.

In the physics community, there is a broad consensus respecting those institutions that mount successful programs of instruction. As we have mentioned earlier, we support the view that periodic publication of the curricula and practices of these institutions will provide a context in which all can view their efforts. If made available to students, compilations of this nature can provide guidance in selecting appropriately oriented undergraduate and graduate programs. The AIP publishes a directory of graduate programs, and in the early 1960's it published information about selected but unidentified public and private undergraduate programs.* We recommend strongly, however, an AAPT program to prepare and publish periodically objective descriptions of the curricula and facilities available for the teaching of physics in all those institutions that offer an undergraduate major in this field. Admittedly a monumental undertaking, it may well require a continuing effort, as the task of completing a survey of 600-odd programs at any one time would require resources beyond those foreseeable. These descriptions, for maximum effectiveness, must be made readily available to high school guidance counselors and students, as well as to undergraduate students. All too frequently students are faced with critical career choices, with woefully inadequate information. The physics community has an obligation to make such information much more readily available.

TRENDS AND OPPORTUNITIES

Education, following the strong trends of social evolution, continually opens new vistas. One that receives public attention from time to time

---

* *Toward Excellence in Physics—Reports from Five Colleges* (American Institute of Physics, New York, 1964); *Toward Excellence in Physics—Reports from Four Public Colleges and Universities* (American Institute of Physics, New York, 1965).

is the potential use of technology to increase the opportunity for individualized instruction. The major effort to develop and use new educational methods with special appeal and effectiveness for the culturally deprived has elicited some thoughts on the place of science in this movement. In an era that is questioning formality in any guise, modes of offering college-level education without requiring attendance at a traditionally structured academy are emerging. Many other new trends and opportunities merit attention. The studies of the Carnegie Commission are an important source of information about them. Space permits us to discuss only a few, including the first attempt to learn systematically on a nationwide basis just what U.S. citizens at a variety of age levels know.

The National Assessment

Critics of U.S. education often cite the achievement of our youth in tests as compared with that of the youth of other nations. Such comparisons, unless the samples of students are carefully matched in respects other than age or school level, can be misleading, because few nations attempt to provide education to as broad a segment of the population as does the United States. Nonetheless, those in this country who are concerned with educational quality seek some objective standard against which to measure U.S. output.

An absolute standard is not yet in sight, but the National Assessment of Educational Progress * offers an objective measure of change. The project was established as an ongoing activity under the Education Commission of the States. It proposes to measure what 100,000 U.S. children (ages 13 and 17) and young adults (26 to 35) actually know in ten areas, including science. Changes in (a) knowledge, (b) skills, (c) understanding, and (d) attitudes are to be assessed by periodic testing of each age group according to a scheme in which some old and some new questions are to be used. The first test including science was held in 1969–1970; the next is scheduled for 1972–1973; a third, in 1975–1976; and a fourth, in 1978–1979. The results are reported by geographic region, size of community, and sex. They also are reported by area of science and objective (1 through 4 above). The Education Commission of the States proposes to issue reports from time to time that, in addition to statistical data, quote examples of performance but neither generalize nor interpret.

The results of the first assessment have been issued as a report accom-

* Material on the National Assessment can be obtained from National Assessment of Educational Progress, 1860 Lincoln Street, Denver, Colorado 80203.

panied by a booklet of observations and commentary by a panel of independent reviewers.* A statement of particular interest in the context of this report is that nine-year-olds were equally knowledgeable in biology, physical sciences, and earth sciences, but that 17-year-olds gave a disappointing performance in physical science. Another reviewer concluded that the performance of all groups was distinctively poor in this area.

The success of the Assessment will depend mainly on the extent to which each question is directed unambiguously toward the objective it is designed to test. The panelists expressed a number of doubts over the trueness of aim. A critical review of the questions by an outside panel would appear more valuable before the test is used, rather than after, with determination to make even extensive modification in response to valid objection.

Much of the content of this chapter has been directed precisely to the objective of the Assessment, and most of the remaining portion is relevant to it. The physics teaching community undoubtedly would have predicted the outcome in physical science suggested by certain of the review panelists and that this outcome will be validated by future, more reliable Assessments. However, "I told you so," is not a fitting response, for this same community has long held the key to a better outcome.

Learning Aids

That scientists should initiate the application of computers to instruction is only reasonable. Some of the earliest efforts to do so were made in the 1960's, one of the first concerted efforts in writing interactive programs in physics having occurred at a conference on new instructional materials held by the Commission on College Physics (CCP) in 1965. During that same year, an additional CCP conference was entirely devoted to computer use in undergraduate instruction. *The Computer in Physics Instruction* † was a result distributed to thousands of educators and scientists. During 1970, a second major convocation was stimulated by CCP but with broader coverage: The Conference on Computers in Undergraduate Science Education, Physics and Mathematics.‡ A conclusion from the Conference is that no general procedure yet exists for the widespread use of computers in education. Many had hoped for, and even predicted, that by this time

---

* *National Assessment of Educational Progress. Report 1, 1969–70 Science: National Results and Illustrations of Group Comparisons* (Education Commission of the States, Denver, Colo., 1970). (See also *Observations and Commentary by a Panel of Reviewers.*)

† Commission on College Physics, University of Maryland, College Park, Md., 1965.

‡ *Computers in Undergraduate Science Education: Conference Proceedings* (Commission on College Physics, College Park, Maryland, 1971).

computers would have found widespread use, but the Conference showed that this goal is not likely to be achieved soon. In various areas of special application, such as the computational mode, computer graphics, the simulation mode, analog computing, and computer-supervised instruction, progress has been made; and, on occasion, a real success, such as the PLATO project, is scored.

Inevitably a younger generation accepts and uses new techniques, while an older one expresses its doubts. The automobile was cited in the 1970 Conference, by A. Oettinger, for having started as a horseless carriage and eventually having transformed society. Not all this transformation is now seen as desirable, or even healthy, but unquestionably society will accept some of it. Oettinger (and the Conference in general) feels the same thrust of inevitability in regard to computers in education. Oettinger believes that, because of the greater flexibility of universities, which facilitates innovation and experimentation, the development and use of computers for education will grow in these institutions before computer-aided instruction becomes widespread in the schools. The federal agencies view computers as a possible means of meeting the crisis that the simultaneously rising costs and standards for education could bring about. Both the NSF and the Office of Education are supporting research and development projects to further the application of computers to education. Physics teaching can play a major part in this great national effort.

Physics and the Disadvantaged

The problems confronting the United States in its effort to engage racial minorities and other disadvantaged citizens with constructive and liberating outlets for personal development touch all facets of life. Physics, which directly as a profession supports only a tiny fraction of the population, would appear at first sight to merit little consideration as a tool for achieving this national goal. The professional activities of physicists do not appear to lead to goods or opportunities having short-range value for this purpose. The identification and encouragement of scientifically talented young disadvantaged people is, of course, an obvious and desirable activity but again does not touch the general problem. However, a growing number of educators, from both minority and majority groups, have seen ways in which physics can serve the broader purpose.

Morris Lerner, a high school teacher and president of the National Science Teachers Association has stated the case in these terms *:

---

* M. R. Lerner, "Physics and the Nation's Racial Problems," *Am. J. Phys. 38,* 126 (1970).

What has physics to offer the black student from the ghetto? As the courses are presently constituted, little or nothing. But I believe that physics has the greatest potential for moving these students into the mainstream of American life. The way out of the ghetto is money, and to get money one must get a good job. These students know this and it is their primary goal. Here is where physics can play a great part. We have a fascinating subject with almost built-in motivation, and we can use it to help these students develop the linguistic and mathematical skills they lack, and at the same time learn basic principles and approaches to problem-solving which can be forever useful to them. *But not in our traditional manner.* In courses for these students the specific content chosen is not particularly important except that it must be relevant to the lives of the students and amenable to development in a laboratory setting in small steps. The purpose being to give the student opportunities to use the methods of science and to apply these methods to solving problems that are of interest and significance to him. For these students physics can be of value as a specific in a technological society, but perhaps its greatest value is as a device to achieve a larger educational end.

With many variations in detailed approach, programs with the objective just described are being tried in various regions and with various disadvantaged groups. Most of these projects are directed toward the high school level and are based on collaboration between the schools and universities or colleges. Many comprise summer programs; such, for example, is a project at the University of Colorado * that has the object of motivating Chicano high school students to go on to college. High school sophomores are selected, who, on completion of the summer program, participate in follow-up activities for two years.

Project Beacon,† in New York City, involves the cooperative efforts of the high schools, York College of the City University of New York, and local industry. Its intent is to stimulate interest in science and engineering among blacks and Spanish-speaking students and to improve the high school curriculum for them. It operates on two interrelated levels, those of students and teachers.

Drawing heavily on two sources of material treated earlier in this Report, PSNS and Project Physics, it takes a flexible approach toward learning, although experience has taught a need for careful structuring at least at the beginning. An explicit aim of the summer phase of the Project is "to demonstrate to the teachers that the inquiry method, combined with an emphasis on student laboratory work, is eminently suited for accomplishing the goals of student motivation and orientation." From this phase, "the teachers saw that students from disadvantaged backgrounds can be

---

* W. R. Chappell, "A College Motivation Program for Minority High School Students," *AAPT Announcer, 1,* 16 (1972).
† F. R. Pomilla and M. S. Spergel, "The Model for Project Beacon," *The Physics Teacher, 9,* 136 (1971).

motivated towards science if they are permitted to handle equipment and if they are encouraged to work independently in a relaxed setting."

The academic-year phase was conducted during six periods per week at the high school and an additional two hours per week at York College. Industry participated by providing career-oriented summer jobs. Teachers in the Project report * on a number of problems that they encountered and techniques that they found useful in coping with these problems. Initially skeptical of the proposal to use Project Physics materials with this population, they now believe that the success of the program in reaching the disadvantaged student consists of a judicious combination of these materials with the standard curriculum, with particular care being taken to accommodate to his abilities and background.

In the belief that inner-city schools need teachers with broader education in the sciences than the average, that their teachers need to have a commitment to teach in this environment together with an understanding of inner-city life and culture, and that inner-city schools, perhaps beyond all others, require the revitalization of academic excellence, a special science teacher training program is under way at Brown University.† The Departments of Physics, Chemistry, Biology, and Education cooperate in this effort with the public school system of Providence. The program prepares students for the AB degree in science education, with an optional fifth year leading to the MA degree. Most of the courses have been developed specifically for the program. They stress the unity of the sciences, emphasize student participation rather than formal lectures, discuss the role of science and technology in modern society, familiarize the student with traditional and new secondary school curricular materials, and provide direct interaction with high school students and teachers. Advanced courses are offered to enable the student to specialize in one of the disciplines.

The study of physical sciences, far more than the humanities and social sciences, appears *a priori* to be "color blind." For this reason there have been no strident demands for racially relevant physics courses. However, precisely because of its freedom from emotionally charged issues, science study can be a useful vehicle for bringing about constructive social change. Programs of recent AAPT meetings give evidence that an increasing number of physics instructors are beginning to use their subject, without demeaning it, in a purposeful way. Today's much-touted quality of relevance is an inherent property of their efforts.

---

* L. Siegel and R. Weinstein, "The High School Experience in Project Beacon," *The Physics Teacher, 9,* 134 (1971).
† W. G. Massey, "Training Science Teachers for the Inner City: A High School and University Cooperative Program," *AAPT Announcer, 1,* 13 (1972).

Open Universities

We view with considerable interest The Open University, which began its teaching programs in England in 1971.* It is intended primarily for adult students in full-time employment or working in their homes. No formal academic qualifications are required. It is "not simply an educational rescue mission" according to its Chancellor, nor is it a rival of the existing universities. It offers six main lines of study: arts, science, mathematics, technology, social sciences, and educational studies. It awards a BA degree and the degrees of Bachelor of Philosophy, Master of Philosophy, and Doctor of Philosophy, as well as Doctor of Letters and Doctor of Science. It operates as a correspondence school with radio or television lectures. Packets of course materials are mailed to students, and the lectures are given at weekly intervals. Students have tutors and examinations. Two-week summer schools are mandatory for some courses. There are plans to establish study centers in locations in which the density of students is sufficient. A recent report on The Open University states that students in a second-level course in electromagnetics and electronics receive packets containing an oscilloscope, a signal generator, a dc supply, and other components. It also relates that the attrition rate for students entering the second year is only half of that expected, so the new first-year intake has been reduced by 20 percent to 20,000 of the 35,000 applicants. Many overseas institutions are purchasing supplies from The Open University to initiate their own versions.

A recent article † provides evidence that in a number of regions of the United States educators have seen in open universities an opportunity to meet the desires of a much more varied component of the population than do the traditional educational structures. Among the ventures planned are national universities, offering enrollment to students anywhere in the nation. One of these would act as a distributor of the British materials, with credit to be granted by individual colleges and universities. Another would grant credit for on-the-job training, internships, courses at local colleges, and courses offered by institutions other than colleges. Degrees would be granted on the recommendation of a council of academic advisors from various fields of study. A number of the ventures are conceived as adjuncts to existing state systems of education, and one is the creation of a group of public and private colleges and universities scattered over the east and midwest.

---

* *The Open University, Prospectus 1971* (The Open University, Bletchley Bucks, Great Britain, 1970).
† P. W. Semas, "Open University Programs Gain Favor in U.S.," *Chronicle of Higher Education, VI* (10), 1 (1971).

Applications for admission to these new programs still are small in number, and, since most of them are in formative stages, enrollments are measured in the hundreds. The apparent success achieved in Great Britain is a strong indication that this mode has great appeal and that their student bodies will grow rapidly.

The physics community may fully expect to be confronted with the need to prepare physics courses of a more self-contained nature than any existing ones. The vital phenomenological component will have to be provided; physicists will have to judge the success of the British venture in mailing kits. In addition, they almost certainly will have to establish well-equipped centers or open existing institutions for laboratory study that cannot be carried out in the home for one reason or another.

12

MANPOWER
IN PHYSICS:
PATTERNS OF
SUPPLY AND USE

INTRODUCTION

This chapter deals with the manpower of physics. It describes patterns in the production and use of physicists, changes in these patterns with time, employment opportunities and problems, and projections of manpower needs in the 1970's.

To provide a background for subsequent discussion, the chapter begins by tracing the development of potential physics manpower resources from the secondary school level through undergraduate, graduate, and post-graduate phases to eventual entry into the national (and international) physics community.* Next, characteristic patterns in the employment of physicists and the ways in which these patterns have changed during the past two decades are examined. Current problems of underemployment and unemployment in physics are then considered and their significance assessed. The chapter concludes with a discussion and synthesis of the projections that have been made in regard to the demand and opportunities for trained physicists during the 1970's and indicates probable further changes in the use of physics manpower in this decade.

In this chapter, natural scientists and engineers are defined as those with some training and experience in one or more of the following disciplines:

---

* A more detailed discussion of physics education appears in Chapter 11, Physics in Education and Education in Physics.

1. Physical sciences (including mathematics, chemistry, earth sciences, and physics and astronomy)
2. Engineering
3. Biological sciences

At present there are approximately 1,000,000 engineers and 300,000 scientists.[1, 2, *] About 10 percent of the scientists are physicists.[4] However, the technical work force does not consist of a large number of people sharply divided into isolated disciplines. Rather, the various parts of the technical enterprise are strongly interdependent, its overall health being determined by that of its many components. Science profits from the new tools emanating from technology and engineering. It also is stimulated by

---

* Numbers pertaining to science, engineering, research, and employing institutions suffer from vagueness in the definition of the entities to which they refer. Different bodies of statistics are collected for different purposes; therefore, appropriately, different definitions and categories are used. A few aspects of this problem that have implications for the data included in this chapter follow.

The American Institute of Physics (AIP), which, under contract with the National Science Foundation (NSF), compiled the physics portion of the National Register of Scientific and Technical Personnel, a major source of data for this chapter, used a fairly rigorous definition of a physicist—one who had two years of experience in physics beyond the BA/BS degree. The definition rested in part on an integrated judgment, based on education, experience, and employment, of whether an individual belonged more appropriately to physics or to some other related discipline.

Less commonly used in this chapter are compilations of the Bureau of Labor Statistics. The Bureau employs a less strict definition of physics; consequently, it reports a larger number of physicists than does the AIP.

Numbers depend also on completeness of coverage. The Committee on Human Resources and Higher Education [3] estimates that the coverage of the 1966 National Register was about 80 percent complete for physicists. The Bureau of Labor Statistics obtains its information through polling employers and probably achieves more nearly complete coverage of this discipline.

The inclusion of astronomers in data on physicists also raises questions. In most of the Physics Survey (although not in the Report of the Panel on Astrophysics and Relativity) astronomy was identified as a separate discipline and deliberately excluded from discussion as a parallel Survey Committee was reporting on it. However, many available statistics, including, for example, the American Science Manpower [4] series published by the NSF, and reports on the Doctorate Records File maintained by the Office of Scientific Personnel of the National Research Council, usually include astronomy as part of physics. Therefore, most of the numbers in this chapter include astronomers in counts of physicists; this does not seriously affect these numbers since astronomers constitute less than five percent of the overall physics population. (The 1970 National Register survey shows the number of astronomers as 3.1 percent of a total physics population of 36,336.)

Another question concerns employment in federally funded research and development centers. Usually in statistical compilations employees of these centers are counted as employees of their managing institutions. Surveys made by the Statistical Data Panel of the Physics Survey Committee treated them as a separate category; however, in this chapter, in which the use of statistics from other sources is necessary, such employees are included in the data on their respective managing institutions.

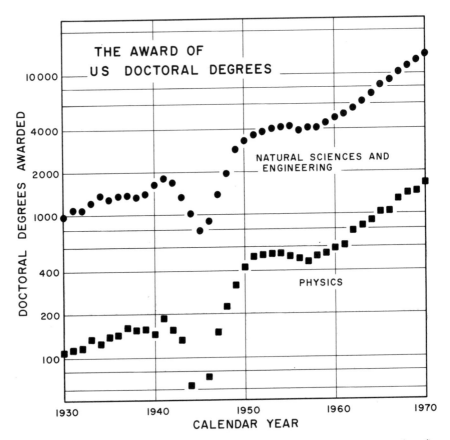

FIGURE 12.1 The production of PhD's in physics and in natural sciences and engineering. The line drawn through PhD production from 1930 to 1957 was determined by the PhD production in 1930 and the total PhD production for the years 1930 through 1957. That it fits the rate of production through the 1930's and in 1957 strongly suggests that the growth of graduate education in physics until 1957 was a natural extrapolation of growth during the 1930's and that the peak of production in the years around 1953 was compensation for the loss of graduate education during the war. The slope of the 1930–1957 line is 5.25 percent per year; the slope of the line after 1958 is doubled to 10.3 percent per year. [Source: Doctorate Records File of the National Research Council Office of Scientific Personnel.]

new problems and discoveries that arise in connection with technical applications. Conversely, engineering and technology need the conceptual guidance and growing base of fundamental knowledge and understanding that science provides.

The core of U.S. science and engineering manpower is the group holding the PhD degree. This degree typically requires six years of education and training beyond the bachelor's degree (or four years beyond the master's degree). The first PhD's were awarded in the United States in 1861 and 1863. Arthur Williams Wright, a physicist, earned the first, and Josiah Willard Gibbs, also a physicist and one of America's most distinguished scientists, was the fourth recipient. Few such degrees were awarded before the first World War, however, and a period of study in Europe typically was part of the education of a prospective scientist in the early part of this century. As awareness of the contributions of science to the achievement of national goals grew after World War I, private foundations, through financial support, began to encourage the development of graduate education. Doctoral training programs grew rapidly in the 1920's, and in the calendar year 1930, for the first time, more than 100 PhD's were awarded in physics in the United States. Figure 12.1, based on the Doctorate Records File, shows the production of PhD's in physics and in the natural sciences and engineering in the United States since 1930. In the early years, the production of physics PhD's increased at a rate slightly greater than 5 percent per year. Academic activity in physics nearly ceased during the war years, with PhD production falling nearly to zero. However, after the war there was a rapid upsurge in the number of PhD's awarded. Many of those who were forced to defer their studies during the war resumed them subsequently, thus leading to a peak in the granting of PhD's in physics around 1953. The deficiency in the normal extrapolation of PhD production was overcome by 1957. In that year, the number of such degrees awarded was about the same as the number obtained by extrapolation of the growth curve for the 1930's.

Figure 12.1 illustrates a rather remarkable characteristic of physics PhD production. Since 1930, the physics PhD's have represented almost precisely 11 percent of the total number of natural science and engineering PhD's produced in any given year. And the constancy of this fraction is not restricted only to the United States; rough statistics from both Canada and the United Kingdom, with rather different educational systems, show this same fraction and constancy.

Science in the 1940's and 1950's differed markedly from science in the 1930's.[5, 6] As Figure 12.2 shows, national expenditures on research and development increased steadily and rapidly following World War II. The scientist and engineer had become part of a major national effort.

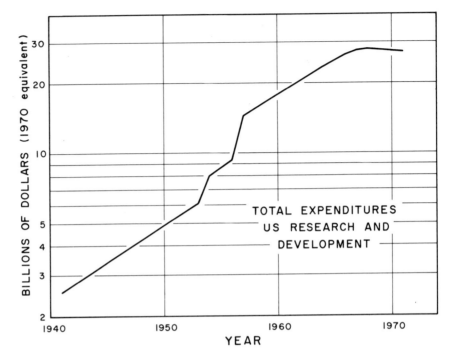

FIGURE 12.2   Total U.S. research and development expenditures.

By 1947, federal support of research and development had reached the unprecedented level of $4 billion,* an amount that was orders of magnitude greater than the level of expenditure for this purpose attained in the 1930's. These vast expenditures created a critical shortage of scientific and technical personnel. Scientists of all disciplines were in great

* All dollar figures used in this chapter are given in equivalent 1970 dollars. The multipliers used to correct the figures for other years are

1940	2.882	1948	1.681	1956	1.486	1964	1.247
1941	2.725	1949	1.695	1957	1.439	1965	1.211
1942	2.577	1950	1.681	1958	1.399	1966	1.178
1943	2.433	1951	1.555	1959	1.368	1967	1.153
1944	2.299	1952	1.524	1960	1.342	1968	1.109
1945	2.247	1953	1.511	1961	1.330	1969	1.043
1946	2.070	1954	1.504	1962	1.325	1970	1.0
1947	1.808	1955	1.508	1963	1.295	1971	0.9653

The figures from 1958–1971 are based on the deflator used by March (University of Chicago Policy Study, 1970); from 1940 to 1958, they are based on the Consumer Price Index given in *Statistical Abstract of the United States 1970* (91st ed.), U.S. Bureau of the Census (U.S. Government Printing Office, Washington, D.C., 1970).

demand, opportunities for the development of new research and training programs in the universities were ripe, and the development of such programs was actively encouraged by government and industry. The new programs that emerged in this period caused a doubling in the rate of increase in the production of PhD's in physics. The continued rapid growth of federal support of research and the economic prosperity of the United States provided ample opportunities for physics and physicists. However, danger signals began to appear in the latter half of the 1960's.

The danger signals and a serious employment problem for physics in 1970 occurred in the three occupational categories in which physicists have traditionally worked[7]: (a) teaching in academic institutions; (b) conducting research and development in federally supported programs; and (c) performing research, development, and related production activities in industry.

That teaching opportunities in academic institutions should decrease after 1969 was predicted by Cartter[8] and Gruner[9] and should have been evident. The year 1947 marked the beginning of the high birth rates following World War II. Eighteen years thereafter children born in 1947 began college and some four years later completed their undergraduate training. Thus, teachers became abundantly available in about 1969; the long-advertised teacher shortage was over at that time.

That research and development in federally supported programs were destined to be curtailed somewhat was also evident because of the exceptionally rapid rate of growth prior to 1965. In real dollars, the rate of growth before 1965 was approximately 10 percent per year. Although this growth matched the production of PhD's in science, it exceeded the growth rates of other national measures such as the gross national product and the size of the industrial and educational enterprises.[1] Figure 12.2 depicts the change in the rate of increase of support for research and development after 1965.

The trend toward level (or, in light of inflation, decreasing) support of federal research and development had particular impact on physics because of the unusually large dependence of physics on federal support as compared with other scientific disciplines, its heavy concentration in academic institutions, and its reliance on large facilities for many of its most advanced and significant experiments. Thus a small change in the level of federal support could produce a sharp change in the availability of funds for salaries to employ physicists.

That research and development in industry were reduced in 1970 is apparent in Figure 12.3, which depicts decreasing expenditures on industrial research and development as a percentage of gross national product.

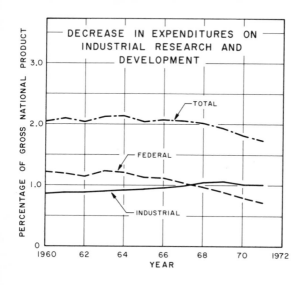

FIGURE 12.3 Decrease in expenditures on industrial research and development as a percentage of gross national product. [Source: National Science Foundation.]

That research, development, production, or any other kinds of jobs that physicists could occupy either in industry or elsewhere were being curtailed sharply in 1970 is evident from the national unemployment percentages that appear in Figure 12.4, as taken from the national press.[10] Unemployment increased dramatically from about 3.5 percent in November 1969 to more than 6 percent in December 1970. A like percentage also was reported in December 1971 and was comprised of three groups of unemployed as follows:

1. Unemployed males, 20 or more years of age	2,141,000
2. Unemployed females, 20 or more years of age	1,710,000
3. Unemployed teenagers, 16–19 years of age	1,365,000
TOTAL	5,216,000

Thus the national unemployment percentage of 6.1 percent (in December 1971) implies a national percentage of 4.4 percent for men 20 or more years of age with which the 3.9 percent unemployment reported within the physics community at that time can be compared.[11] Because the percentage for physicists is comparable with the national average, which became rapidly worse in 1970 and showed no significant recovery in 1971, it is evident that the ability of physicists to move into any kinds of jobs, including those normally occupied by chemists, engineers, mathematicians, and other technically trained personnel, or even into jobs requiring no technical training whatever, was removed suddenly during 1970 when nearly all

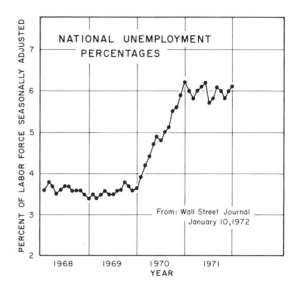

FIGURE 12.4 National unemployment percentages. [Source: *Wall Street Journal,* January 10, 1972; U.S. Department of Labor reports.]

sectors of science and society were experiencing like economic problems.

Obviously, the almost simultaneous disappearance of jobs in the three traditional areas of employment for physicists had a dramatic effect on the physics community. Fortunately, the ability to monitor the experiences almost as they occurred permitted physicists to be informed and to react. Every issue of *Physics Today* from January 1970 to June 1971 contained extensive information and open discussions by physicists about employment questions. Such discussions and self-analyses were made possible by the excellent data collection programs of the National Science Foundation (NSF), the National Research Council (NRC), and the American Institute of Physics (AIP).

A later section of this chapter discusses fully the impact of the worsening job market on physicists.* Briefly, the direct effect was an increase in the percentages of unemployed physicists with bachelor's, master's, and PhD degrees. Not only were the percentages at any one time during the 1970–1971 academic year about double those in 1969–1970, but the percentages were uncomfortably large. For example, in September 1970, 24 percent of the PhD's who had received their degrees in the previous academic year (that is, three or four months before) were still unemployed.

The initial reaction in universities and colleges was to reduce first-year

---

* Chapter 6 of this Report also considers this problem in the broader context of physics as a whole (including the nature and scope of research programs and institutional management policies and practices).

graduate student enrollments and reassign responsibilities in such a way as to assist the postdoctorates and graduate students to whom institutions had commitments. This reaction has had some unexpected and unfortunate consequences that we discuss elsewhere in this Report; in brief, this reaction had as its primary effect not the reduction of the number of PhD's being trained in physics but, rather, a shift in which students denied access to the most highly qualified physics departments continued their work toward the PhD in departments that frequently were much less capable of providing high-quality graduate education.

For two reasons, the future outlook for physics might not be as gloomy as it first had appeared. One reason is that the attitudes of physicists, science planners, and society have changed. The shock of the abruptly worsening job market in 1970 had the following positive effects:

1. Physicists now have an increased awareness that job opportunities require both high-quality work and a competitive attitude. Jobs are no longer freely available and guaranteed.

2. Physicists are emphasizing anew the breadth of a physics education. Physics is not a trade that supplies only specialized training. As the percentages of physicists who were being absorbed by academic institutions in research positions decreased, applications of the expertise of physicists to other disciplines, such as geophysics and biophysics, received increased emphasis. More important, the potential contributions of physics to the solution of many of the problems of greatest national concern and to the needs of industry were newly appreciated and have begun to receive serious study.

3. The need for better data collection and interpretation became more apparent and was generally accepted. Better planning for science and the nation could have minimized the shock of the unemployment experiences during and following 1969.

4. Review of the employment history in physics and other sciences at the beginning of the 1970's has shown less than optimum use of a national resource of trained manpower. As a result there is growing determination to prevent a recurrence of this situation.

A second reason for optimism is that the role of physics in supplying the analysis and measurement capability for science and technology is becoming better understood and accepted. As the economy and technology develop in future years, so also must physics and its associated manpower.

To provide the documentary evidence for the supply and demand picture of physics manpower, this chapter presents detailed descriptions and data in the next five sections. These sections demonstrate that manpower trends and data probably are better understood in physics than in any other science. The final section summarizes the Committee's conclusions and recommendations in this area.

EDUCATION IN PHYSICS

The flow of individuals through the educational system into physics is depicted in Figure 12.5.[12] This figure shows the progress of a group at the same stage in the educational process (a cohort) from contact with physics in secondary school through college and postgraduate study into employment in this discipline. The numbers, of course, change with time and differ for different cohorts, but the example chosen is typical of physics education in the 1960's. (Chapter 11 discusses in detail the broader questions related to physics education; here, only the numbers involved are considered.)

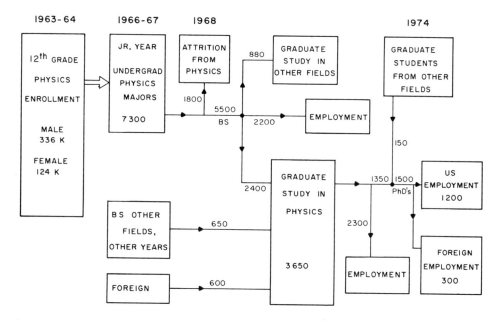

FLOW CHART FOR A SELECTED COHORT OF U.S. PHYSICISTS

FIGURE 12.5 Flow of manpower into physics from secondary school through graduate school and into professional employment in physics.

Secondary School

The data in Table 12.1 show the numbers of physics students at each level of education beginning with the first year of college. Many of these numbers are determined by decisions that students make during their secondary school years in regard to careers and college courses in which they will major. Therefore, a brief discussion of secondary school choices is helpful in interpreting the data on physics education at college and graduate school levels.

Interest in science characteristically develops early. Well over half of those who eventually major in the natural sciences and engineering choose science as a career before leaving high school; however, there is a substantial amount of shifting among major disciplines, in both high school

TABLE 12.1 Consecutive Annual Educational Data for Physics (All Institutions)

Academic Year	Age Group 18 (in Millions)	Total Male Freshmen (Degree Candidates)	Intro. Physics Fresh.	Soph.	Physics Majors Jr.	Sr.[a]	Physics Bachelor Degrees
1953–1954	2.181	342,528	—	—	3000	2600	2439
1954–1955	2.188	383,720	—	—	3200	2700	2420
1955–1956	2.285	415,604	—	—	3700	2950	2883
1956–1957	2.330	442,903	—	—	4300	3400	3293
1957–1958	2.329	441,969	—	—	5301	4000	3891
1958–1959	2.435	465,422	—	—	5903	4800	4669
1959–1960	2.606	487,890	—	—	6504	5172	5042
1960–1961	2.937	539,512	—	—	7161	5759	5293
1961–1962	2.769	591,913	165,200[b]	27,700[c]	7934	6633	5622
1962–1963	2.780	598,099	—	—	7873	6386	5452
1963–1964	2.764	604,282	—	—	7520	6676	5611
1964–1965	3.740	701,524	—	—	7132	6514	5517
1965–1966	3.517	829,215	—	—	7014	6296	5037
1966–1967	3.517	787,000	235,000	43,000	7345	5992	5236
1967–1968	3.495	814,000	—	—	7822	6704	5522
1968–1969	3.577	924,580	—	—	7587	7019	5975
1969–1970	3.671	955,000	210,000	50,000	7480	6700	5782
1970–1971	3.811	1,006,000	—	—	6844	6663	—

[a] These numbers represent 90 percent of the number of juniors.
[b] Plus 24,300 in two-year institutions.
[c] Plus 4200 in two-year institutions.

and college. Some of the relevant facts appear in Table 12.2, which is based on a study of Project TALENT data. In this Project, students were surveyed in 1960 as ninth graders and again in 1964, one year after graduation from high school.[3] About one third of the students preferred careers in natural science and engineering in the ninth grade, but only about one fourth of these persisted in this career choice after high school. However, the natural science and engineering majors were augmented by entries from other disciplines (about 70 percent). A study conducted by the Research Corporation also confirms that two thirds of the science majors are already interested in a science career by the time they leave the ninth grade.[13]

The National Merit Scholarship reports reveal undergraduate major-field choices of the less than 1 percent of the secondary school population

tal ysics aduate roll- nt	1st-Yr. Physics Graduate Enroll- ment	2nd-Yr. Physics Graduate Enrollment	3rd Yr. and above, Physics Graduate Enroll- ment	Physics Master's	Physics PhD's	Post-doc.'s Award-ed to New Gradu-ates	Physics Faculty Full- and Part-Time	Postdoc.'s at Univ.'s
,100	—	—	—	798	515	—	—	—
,200	—	—	—	784	504	—	—	—
,400	—	—	—	796	490	—	—	—
,800	—	—	—	883	446	—	—	—
,700	—	—	—	882	472	—	—	—
,500	—	—	—	958	501	—	—	—
,000	—	—	—	1156	533	—	—	—
,000	—	—	—	1321	615	—	—	—
,308	—	—	5100	1431	699	120	6,600	—
,265	—	—	5500	1850	858	137	6,825	670
,046	4061	3630	5350	1907	792	145	7,070	700
,629	4167	3660	5800	2045	983	186	7,496	750
,876	4358	3800	6690	2050	948	237	8,152	863
,504	4162	3900	7440	2193	1233	304	8,995	950
,305	4010	3500	7800	2077	1325	274	9,962	1150
,475	3669	3100	8700	2223	1355	465	10,575	1054
,372	3918	3550	6900	2268	1545	—	11,120	1140
,327	3494	—	—	—	—	—	—	—

TABLE 12.2   Shifts in Career Choice during High School [a]

Field Choice in Ninth Grade	Number	Field Choice One Year after High School		
		Nat. Sci. and Eng.	Health Professions	Skilled/ Technical
Natural science and engineering	6,370	1550	260	1220
Health professions (MD and DDS)	1,230	100	320	180
Skilled/technical	5,180	470	80	1800
Other	6,590	510	160	1220
TOTAL	19,370	2640	820	4410

[a] Source: Numbers in this table are adapted from Table 6.2 of Reference 3.

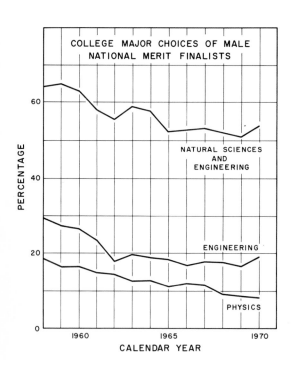

FIGURE   12.6   College   major choices of male National Merit Finalists.[15]

that qualifies for the scholarships. The distribution of choices in recent years appears in Figure 12.6. Among this select group, physics is a popular choice, although the fraction of finalists choosing physics has declined slowly but steadily since 1958. The fraction choosing engineering as their college major has also declined, but the total percentage choosing the natural sciences and engineering has not changed greatly; the principal growth has been in biology and mathematics. "Physics was the choice of 18.8 percent of the boys in 1958, immediately after the first space exploration successes, but steadily declined for a number of years thereafter." [14] The small variation over the years suggests that science will continue to challenge and motivate the brightest students. Further, the early age at which a choice of physics is made and the continued appeal to bright students suggest that aptitude, creativity, and other scores for physicists should be high. Table 12.3, again based on Project TALENT data, shows the scoring on the I.Q. composite variable for those in each of the occupational goal categories; physics undergraduates ranked highest on I.Q. scores. Some academic physicists believe that this situation may be

TABLE 12.3   Mean Quartile Rank [a] on One of Seven Variables Testing Aptitude and Personality Traits (Project TALENT Data Bank)

Occupational Goal Measured at Age 23	Variable, Measured at Age 18 I.Q. Composite [b]
Physics	3.04
Mathematics	2.97
Chemistry	2.90
Geology	2.83
Engineering	2.62
Computer science	2.07
Business administration	1.97
Psychology	2.67
Biological sciences	2.16
Social sciences	2.28
Law	2.64
Medicine	2.84
Arts and humanities	2.07

[a] All scores were based on a 1 to 4 classification in which 1 corresponds to the lowest quartile and 4 to the highest.
[b] Composites are a combination of tests given in one area. The I.Q. composite consists of three tests in reading comprehension, abstract reasoning, and mathematics.

changing; more recent information would be helpful in clarifying this point.

College

Approximately 2,700,000 students, roughly half male and half female, graduated from high school in 1958.[15] Approximately three fourths (73 percent) of the males entered college, as did 55 percent of the females.[15] However, because only a few percent of the physics majors are women,[13] the numbers that follow refer only to male students.

Substantial instability in regard to career choice persists in college. Changes during college lead to a small migration away from the natural sciences and engineering. However, the natural sciences continue to

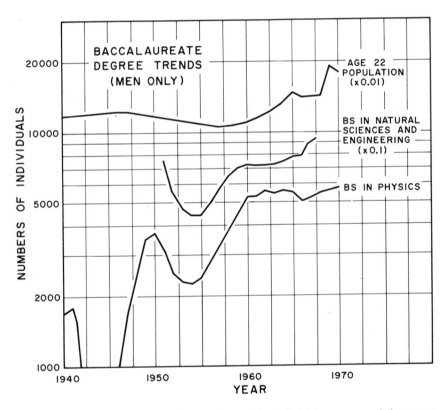

FIGURE 12.7   Comparison of the numbers of bachelor's degrees awarded to men in physics and in all natural science and engineering disciplines with the age 22 male population.

attract the most able students, as the Project TALENT data bank studies show.[3]

At graduation, about 1.4 percent of the bachelor's degrees awarded to men are in physics. This fraction remained roughly constant from 1930 to 1957, rose to a peak of 2.1 percent in 1961 and 1962, then returned to its earlier value.[13, 16–18] The recent decline is a reflection of an increase in the total number of college graduates rather than of a decrease in the number of physics degrees. Figure 12.7 compares the number of bachelor's degrees awarded in physics with the number awarded in all natural sciences and engineering and with the male population of age 22. Physics accounted for approximately 5 percent of the natural science and engineering degrees in the middle 1950's, approximately 7 percent in the early 1960's, and approximately 5 percent in the late 1960's. Before 1960, physics BS degrees represented an increasing fraction of the age 22 population, but the number of physics BS's has not increased in the last decade, although this age group has grown. The number of bachelor's degrees in various disciplines as a percentage of the age 22 population, according to Census Bureau figures, is shown in Figure 12.8.

The implication of a leveling in the relative number of natural science and engineering bachelor's degrees during a period of heavy federal support of graduate study and employment is that the relative number of bachelor's

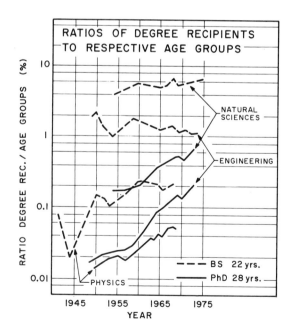

FIGURE 12.8 Ratios of degree recipients to respective age groups.

degrees is largely independent of factors such as graduate school growth and support or, at least, responds slowly to such factors. In addition, the numbers of degrees in the various sciences probably are coupled in such a way that an increase in the number of bachelor's degrees in, for example, mathematics would reflect a decrease in the number in physics or some other discipline. In other words, the relative number of bachelor's degrees in physics and the other natural sciences and engineering in future years appears to be reasonably predictable and largely uninfluenced, on a short-term basis, by Selective Service policies, the influx of foreign scientists, or federal funding. Frederick Terman, in a report to the President's Science Advisory Committee,[19] described the situation in the following way:

The output of baccalaureate degrees in science and engineering depends primarily upon the interests and ambitions of young people and appears to be relatively unresponsive to external manipulation. For example, the fact that salaries for B.S. engineers have in recent years been higher than for any other group of college graduates appears to have had little if any effect on the fraction of college students who choose engineering as a career. Again, there is no lack of opportunity for a young person to study engineering or science; classes in the science and engineering departments of most universities and colleges are underpopulated at the junior and senior levels and could readily handle more students than are available.

Graduate Education

More than half of those receiving a bachelor's degree in physics during the past decade have entered graduate school as full-time students.[20] However, an AIP survey indicates that this fraction has been decreasing recently, as Figure 12.9 shows. Graduate schools also receive a large number of students from abroad. In addition, according to findings of the Physics Survey Data Panel, there is a substantial crossing of disciplines on entrance into graduate school. About half of those who begin graduate studies in physics eventually receive a PhD,[20] spending about six years in this process. The median age of the physics PhD is 28. Seventy-seven percent of those who received a PhD in physics in 1970 possessed a baccalaureate degree in the same discipline, and 81 percent were U.S. citizens.

The history of financial support provided to graduate students is essential to an understanding of the mechanisms involved in the production of master's and doctor's degrees. The first such financial support was provided by fellowships in the last quarter of the nineteenth century; the intent was to induce capable students to undertake graduate study. As the number of students entering college increased, the need for teachers obviously grew. This demand was met early in the twentieth century by the introduction of the second graduate student support mechanism—the

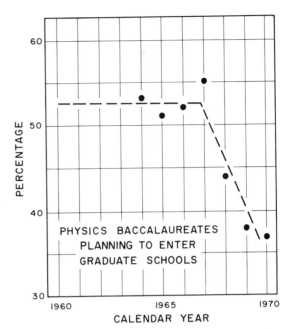

FIGURE 12.9 Physics baccalaureates planning to enter graduate
school in physics.

teaching assistantship. This mechanism was used extensively after World
War II, when the influx of students was large; its use continues, particularly in some of the larger state university systems.

The third major source of support is the research assistantship, which
Chase describes in his report on teaching assistants for the Office of Education[21]:

A third major influence was the initiation by the Federal Government in 1958 (after
Sputnik) of major programs for the advanced training of scarce, specialized manpower. In the fields of perceived national shortages, massive funds were made available in the form of research assistantships, fellowships, and traineeships.

Table 12.4 depicts increasing expenditures in support of graduate education in the natural sciences and in all fields of graduate study from
1954 to 1965.

The success of these various support mechanisms was evaluated by a
panel of the President's Science Advisory Committee in 1962. E. R. Gilliland, the panel chairman, prepared a report titled, *Meeting Manpower
Needs in Science and Technology.*[22] The panel concluded that there was
still a serious shortage of men in engineering and mathematical and physical
sciences, deficiencies in the first two being the most critical. Its recom-

TABLE 12.4   Number and Percentage of Stipends for Graduate Education [a]

Enrollment and Stipends	Year	
	1954	1965
Natural Sciences		
Graduate enrollment	50,864	135,886
Total number of stipends	22,770	91,846
Stipends as a fraction of enrollment	45%	68%
All Graduate Stipends		
Teaching assistantships	9,107 (40.0%)	23,396 (32.0%)
Research assistantships	9,558 (42.0%)	31,115 (33.9%)
Fellowships	4,105 (18.0%)	31,335 (34.1%)

[a] Source: Adapted from Reference 21, pp. 23 and 24.

mendations led to a massive increase in federal support of graduate education, illustrated in Figure 12.10. The number of students supported by traineeships and fellowships rose from about 10,000 in 1960 to a maximum of some 60,000 in 1967 and 1968, as Figure 12.10 shows. A re-examination of the Gilliland projections and goals at that time indicated that the 1970 goals had already been reached. As a result, particularly in the light of the reduced amount of activity in both space and military procurement programs, it was decided to reduce drastically the special training emphasis initiated in 1962 to meet then-perceived national needs. The effect of this decision is also apparent in Figure 12.10.

The consequences of either a rapid and large influx of funds or the sudden withdrawal of support can be far-reaching and can create an unstable situation in any discipline. Although clearly designated as an interim special program to respond to national needs, the Gilliland training expansion was followed rapidly, and perhaps not surprisingly, by a major increase in faculties and the expansion or addition of facilities in the departments of natural sciences and engineering in universities and colleges throughout the United States. Federal support was available for this expansion, which, from the long-range viewpoint, was clearly in the national interest. However, over the short range, it was extremely difficult for the universities and colleges to adapt rapidly and readjust their goals and policies when the Gilliland program was abruptly and sharply reduced in scale.

The nature of the financial support provided to physics graduate students has varied with the number of years of graduate study, as shown in Table 12.5. An AIP Survey of Physics Graduate Students showed that the

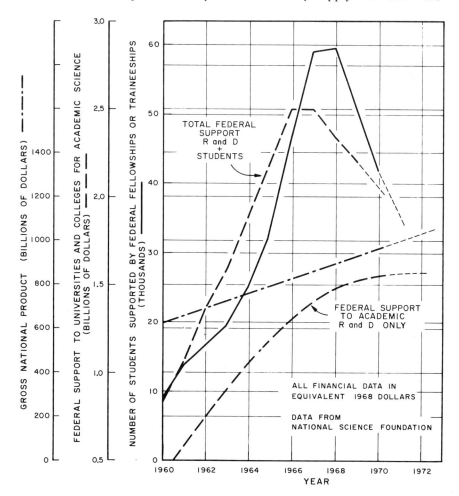

FIGURE 12.10   Federal support of academic science and student stipends compared to the Gross National Product, 1960–1972.

proportion of the total that is composed of research assistantships also has increased with the number of years of graduate study. An additional important characteristic of graduate study is the time lapse between the bachelor's degree and the doctorate. The average total time for physics is 6.4 years, although the NRC Doctorate Records File indicates an average "registered time" of 5.6 years for the doctorate.

Distributions of age, citizenship, and sex of U.S. graduate physics students appear in Table 12.6. This table shows that students from abroad

TABLE 12.5 Financial Support of Graduate Students, 1967–1968 [a]

Source of Support	Number of Years of Graduate Study							Total
	≤ 1	2	3	4	5	6	≥ 7	
Teaching assistantship	1,488	1,085	673	390	182	84	53	3,955
Research assistantship	410	742	1,017	1,145	881	471	250	4,916
Fellowship	928	1,042	881	534	276	68	42	3,771
Employment	841	545	282	166	94	63	23	2,014
Savings	95	58	29	11	12	2	2	209
Other	185	108	56	34	24	16	12	435
TOTAL	3,947	3,580	2,938	2,280	1,469	704	382	15,300

[a] Source: AIP, 1967–1968 Survey of Physics Graduate Students.

constitute 16 percent of the total and that women constitute 3.7 percent of all physics graduate students. Table 12.7 presents data on the number and percentage of doctoral degrees conferred on women in various scientific disciplines.

Progress to the doctorate in physics, and in other disciplines in the natural sciences and engineering, is arduous and demands not only great

TABLE 12.6 Age, Citizenship, and Sex of Graduate Physics Students in the United States, 1967–1968 [a]

Age	U.S. Citizens	Foreigners	Men	Women	Total
≤ 20	26	5	29	2	31
21	93	29	111	11	122
22	439	66	453	52	505
23	1,712	182	1,818	76	1,894
24	1,923	260	2,096	87	2,183
25	1,891	293	2,108	76	2,184
26	1,794	302	2,029	67	2,096
27	1,352	285	1,585	52	1,637
28	993	216	1,179	30	1,209
29	622	220	811	31	842
≥ 30	1,997	600	2,519	78	2,597
TOTAL	12,842	2,458	14,738	562	15,300

[a] Source: AIP, 1967–1968 Survey of Physics Graduate Students.

TABLE 12.7 Fraction of Doctorates Granted to Women [a]

| Year | Doctor's Degrees | | | | | | • |
	Total Con- ferred on Women	Percent- age Con- ferred on Women	Physics (%)	Mathe- matics and Statis- tics (%)	Chem- istry (%)	Bio- logical Sciences (%)	Soci- ology (%)
1960–1961	1112	10.5	1.1	4.9	5.0	11.7	17.4
1961–1962	1245	10.7	1.8	6.1	6.2	11.9	15.0
1962–1963	1374	10.7	1.3	7.3	6.2	12.1	14.9
1963–1964	1535	10.6	1.4	4.9	7.2	11.9	14.6
1964–1965	1775	10.8	2.1	8.7	7.3	11.9	15.7
1965–1966	2116	11.6	2.2	7.3	6.0	15.0	15.0
1966–1967	2454	11.9	2.5	7.1	6.8	15.2	18.0
1967–1968	2906	12.9	2.1	5.5	8.1	15.8	18.5
1968–1969	3436	13.1	2.5	6.2	7.6	15.4	20.0
1969–1970	3800	13.0					

[a] Sources: *Digest of Educational Statistics, 1967, 1968,* U.S. Office of Education, OE-10024-67-68. *Earned Degrees Conferred, 1961, 1962, 1963, 1964, 1967,* U.S. Office of Education, OE-54013-61-62–63-64-67. *Projections of Educational Statistics to 1979–80,* U.S. Office of Education, OE-10030-70.

dedication on the part of a student but also an unusually well-qualified mind. Science and engineering are complex subjects, with many facts and interrelationships that must be perceived, understood, and retained before creative practice of these professions is possible. Consequently, these disciplines are restricted to a relatively small number of the most able people.

Currently, about 40 percent of all undergraduate science majors enter doctoral or medical programs. In physics, the most extreme case, some 70 percent of all bachelor's degree recipients in this discipline enter graduate school, and from one third to one half ultimately earn a PhD. Considering the quality range of undergraduate institutions, one finds it difficult to believe that all these physics PhD's represent the talent usually associated with an advanced research degree, even though the initial pool represents only a little more than 2 percent of the college population. There are indications that the source pool of talent for most fields of science was already beginning to saturate in the late 1960's, and that a deceleration in the growth of graduate enrollment would have been inevitable even in the absence of employment and financial problems. That fewer students were selecting a career in physics even before the impact of the present job shortage was fully perceived is evident in the following statistics. Enrollments of first-year students in physics dropped ˙34 percent in September 1970 in the major universities and 12 percent in all colleges and universities taken together, from the levels of the previous year. The outlook for September 1971 was for a continued downward trend.[23] The supply apparently is beginning to decrease. Although there is clearly an untapped pool of talent among women and minorities, a major shift of these groups into science would have to occur at the undergraduate level, a situation that is not likely to take place rapidly.

POSTDOCTORAL FELLOWSHIPS

A substantial and increasing fraction of physicists continues its training through the acceptance of postdoctoral fellowships or research associate-ships. The role of postdoctoral education in the United States was examined in depth in a report of the National Academy of Sciences, *The Invisible University: Postdoctoral Education in the United States*.[24] This section is based largely on the findings of that report.

The percentage of new PhD's who received postdoctoral appointments was fairly high in the interval 1920–1940; the highest percentage for physics was 18 percent in 1928. Postdoctoral appointments during these years were made possible principally by private funds administered by the National Research Council of the National Academy of Sciences.

Large-scale federal support of postdoctoral training in the natural sciences and engineering began in the 1950's. The result is apparent in Table 12.8, which compares the rate of acceptance of postdoctoral fellow-ships in various scientific disciplines. Such fellowships were far more frequent in some disciplines than in others. For example, they were rarely used in mathematics, engineering, and social science but were frequent in physics, chemistry, and biology. The fraction of doctoral graduates in postdoctoral positions also has been growing rapidly in recent years in those disciplines that typically have large percentages of postdoctoral fellows. The growth in the number of such appointments has been less rapid in the disciplines in which this mechanism is less commonly used. Table 12.8 shows that this difference among disciplines is growing. It also shows a steady increase in the number of postdoctoral fellowships in physics.

The postdoctoral appointment is most often the prelude to an academic career. Three fourths of the postdoctoral fellows in physics anticipate employment in a university; only about 10 percent expect industrial employ-ment.[25] The role of postdoctoral study as a means of entering academic life seems well accepted by university faculties, who strongly encourage

TABLE 12.8 Number of PhD's and Percentages Taking Immediate Postdoctoral Appointment, by Field of Doctorate, 1962–1967 [a]

Field of Doctorate	1962 PhD's N	Taking Post-doct. %	1963 PhD's N	Taking Post-doct. %	1964 PhD's N	Taking Post-doct. %	1965 PhD's N	Taking Post-doct. %	1966 PhD's N	Taking Post-doct. %	1967 PhD's N	Taking Post-doct. %
Mathematics	388	9.2	484	8.4	590	7.0	684	7.0	766	6.6	828	6.9
Physics and astronomy	710	15.8	818	19.0	865	19.9	1,046	21.6	1,049	25.1	1,295	26.1
Earth sciences	249	7.4	322	9.6	312	7.1	374	10.2	399	14.1	419	12.3
Chemistry	1,137	21.9	1,288	30.4	1,351	31.8	1,439	33.2	1,580	33.0	1,764	32.6
Engineering	1,215	3.8	1,357	6.4	1,662	6.1	2,068	6.8	2,283	5.7	2,581	4.8
Agricultural sciences	387	5.8	373	9.7	445	7.3	480	10.6	485	7.0	517	8.1
Biochemistry	286	36.2	300	49.6	371	52.4	391	53.9	446	58.0	495	58.1
Other basic medical sciences	422	25.1	488	29.1	552	30.7	688	34.8	675	36.0	814	35.7
Biology	772	15.2	808	20.5	853	23.4	975	23.6	1,088	23.8	1,114	25.7
Psychology	857	8.9	892	11.1	1,013	10.4	955	14.0	1,133	13.2	1,293	12.5
Social sciences	1,437	2.7	1,575	2.8	1,820	2.3	2,028	2.7	2,178	2.4	2,597	2.4
Arts and humanities	1,196	1.4	1,274	2.2	1,455	1.7	1,718	1.5	1,853	1.1	2,126	1.3
Education	1,898	0.5	2,130	0.6	2,348	0.9	2,727	0.9	3,026	0.5	3,442	1.0
Other fields	553	2.2	611	3.1	687	2.1	729	2.2	904	2.2	1,010	2.6
TOTAL	11,507	8.5	12,720	10.9	14,324	10.8	16,302	11.6	17,865	11.4	20,295	11.6

Number of PhD's and Percentage Taking Postdoctoral by Year of PhD

[a] Source: NRC, Office of Scientific Personnel, Doctorate Records File.

students who plan an academic career to engage in postdoctoral work. They generally do not provide such encouragement for those planning other careers. From the point of view of the postdoctoral fellow, such an appointment is desirable simply because it does help to open the door to a future career in a university. That postdoctoral experience is a gateway to the physics departments of the leading universities is apparent in the finding that 76 percent of the newly appointed junior faculty in physics departments in these universities had held postdoctoral appointments.[26] Often, too, a new PhD needs additional confidence before assuming a faculty position. He may doubt his ability to invent problems and develop techniques or to obtain support for himself and his students in a competitive environment; the postdoctoral appointment allows him to gain this kind of experience.[27]

A few postdoctoral fellows use the appointment to enhance interdisciplinary competence by working in a scientific discipline different from that of their PhD. Another small fraction uses the postdoctoral appointment as an opportunity to extend previous thesis work by remaining at the same institution and working under the same professor. The NAS study showed that those who received postdoctoral appointments reported a shorter than average elapsed time between their baccalaureate and doctoral degrees and that those who changed institutions to accept a postdoctoral fellowship reported an even shorter elapsed time between bachelor's degree and PhD than did those who remained at the same institution.[28] This finding suggests that postdoctoral fellows have either better than average qualifications or exceptionally high motivation and that those who switch institutions are the most highly qualified and motivated.

Universities value the postdoctoral fellowship primarily for its contribution to research. Many university faculties feel that postdoctoral fellows are unusually productive research workers, for, being largely unencumbered by other responsibilities and the pressures of course work, they can allot much more of their time to research than can most other members of a university department.

There is not universal agreement on the desirability of postdoctoral education. Such appointments have been criticized because they isolate the scientist who plans a university career from teaching and other responsibilities related to the welfare of his institution. In addition, this mechanism has been criticized for reinforcing prejudices against industrial employment. However, there is a clear consensus that postdoctoral fellows raise the quality of research at a university.[29]

Postdoctoral appointments also serve another function. About half of the postdoctoral fellows at U.S. universities are citizens of other countries; the appointment offers them an opportunity for research experience in the United States. Most of these postdoctoral fellows plan to return to their native countries after completion of their training. The bonds of interest and shared experience developed during their work and study in the United States provide a basis for continuing communication and cooperation with their U.S. colleagues (see also Chapter 8). Consequently, postdoctoral fellowships have implications for international relations that extend far beyond the confines of the sciences.

EMPLOYMENT

This section deals with patterns in the employment of physicists, especially with changes that have occurred between 1964 and 1970.* It attempts to answer such questions as: Where are physicists employed? What are the characteristics of employment in various types of institutions—governmental, industrial, and academic? What exactly do physicists do?

In this section, PhD and non-PhD physicists are considered separately, since their employment patterns differ markedly: They are employed in different types of institutions, they perform different functions, and their subfield and age characteristics differ greatly.†

Employing Institutions of PhD Physicists

In 1970, half (51 percent) of the PhD physicists worked in colleges and universities. Physicists tend more often than do chemists or agricultural scientists to work in academic institutions; however, the percentages thus employed are less in physics than in mathematics, biology, or social sciences. Nearly one fourth (23 percent) of the PhD physicists are employed in industry. Twelve percent report employment in federally funded research and development centers ‡ and 9 percent in government laboratories.

---

* This section is based on a study conducted by the Statistical Data Panel of the Physics Survey Committee, with the helpful cooperation of the AIP and the NSF. The study deals with the physics portions of the National Register of Scientific and Technical Personnel for the years 1964, 1968, and 1970. The availability of the original Register tapes made it possible to define and subdivide the physics community more clearly and accurately than was the case in other sections of this chapter. Thus, physicists are separated from astronomers in these data, and scientists employed in federally funded research and development centers form a separate category rather than being included in the data on managing institutions. These distinctions cause some of the numbers in this section to differ from those elsewhere in this chapter.

† The introductory section of Chapter 6 includes a number of relevant figures concerning the career distribution of physicists.

‡ These centers, referred to, for convenience, in the tables in this section as research centers, are those recognized and defined as federally funded research and development centers by the National Science Foundation in its series on *Federal Funds for Research, Development, and Other Scientific Activities.*

Other types of institutions, for example, hospitals, medical centers, or secondary schools, employ 5 percent of the PhD physicists.

Table 12.9 shows changes that have occurred in this pattern of employment since 1964. One immediately apparent change is that colleges and universities have attracted an increasing proportion of physics PhD's; from 43 percent in 1964 to 51 percent in 1970. During the same interval, federally funded research and development centers have employed a decreasing fraction of the physics PhD's, the percentage dropping from 19 percent in 1964· to 12 percent in 1970. The percentages in industry, government laboratories, and other types of institutions remained remarkably constant. The implications of these data are discussed in more detail in the following section of this chapter on Supply and Demand.

TABLE 12.9 Changes in Employment Patterns of PhD Physicists, 1964–1970

Physics Subfield	Total Population (excluding nonrespondents)			College or University		
	1964	1968	1970	1964	1968	1970
Astrophysics and relativity	63	123	252	0.83	0.81	0.73
Atomic, molecular, and electron	649	998	1,065	0.46	0.51	0.54
Elementary-particle	889	1,289	1,409	0.54	0.75	0.76
Nuclear	1,277	1,794	1,782	0.41	0.50	0.52
Plasmas and fluids	707	930	1,104	0.31	0.44	0.48
Condensed-matter	2,707	4,064	4,157	0.36	0.43	0.43
Space and planetary	309	567	712	0.31	0.38	0.38
Physics in biology	111	203	274	0.41	0.47	0.46
Optics	477	743	1,078	0.20	0.25	0.27
Acoustics	223	295	324	0.30	0.37	0.31
Other	2,528	2,383	3,457	0.51	0.56	0.58
TOTAL PHYSICS	9,940 [a]	13,389 [b]	15,614 [c]	0.43	0.49	0.50
Astronomy	393 [d]	689 [e]	634 [f]	0.58	0.63	0.59
TOTAL	10,333	14,087	16,248	0.43	0.50	0.51

[a] Nonrespondents = 119.
[b] Nonrespondents = 211.
[c] Nonrespondents = 373.
[d] Nonrespondents = 123.
[e] Nonrespondents = 224.
[f] Nonrespondents = 383.

Employing Institutions of Non-PhD Physicists

Only about one third (30 percent) of the non-PhD physicists were employed in colleges and universities in 1970, in contrast to half of the PhD group. There also were fewer non-PhD's in federally funded research and development centers. Approximately one third of the non-PhD physicists worked in industry in 1970, and about one fifth of them were employed in other types of institutions.

Table 12.10 shows changes in employment for non-PhD physicists from 1964 to 1970. The most obvious change is the increase, from 12 percent to 21 percent, in the number employed in "other" institutions. Analyses of the responses in this category suggest that this change reflects an increase

Industry			Government			Research Center			Other Institutions		
1964	1968	1970	1964	1968	1970	1964	1968	1970	1964	1968	1970
0.03	0.02	0.07	0.02	0.06	0.09	0.05	0.07	0.09	0.08	0.04	0.02
0.22	0.21	0.23	0.11	0.12	0.09	0.17	0.11	0.10	0.03	0.04	0.04
0.02	0.02	0.02	0.05	0.03	0.03	0.24	0.19	0.17	0.05	0.02	0.02
0.08	0.10	0.10	0.07	0.08	0.06	0.39	0.27	0.25	0.05	0.05	0.07
0.22	0.23	0.20	0.08	0.08	0.09	0.35	0.21	0.19	0.04	0.03	0.03
0.38	0.33	0.35	0.08	0.09	0.09	0.14	0.11	0.09	0.04	0.03	0.03
0.22	0.20	0.23	0.22	0.23	0.22	0.15	0.13	0.13	0.10	0.06	0.04
0.13	0.11	0.11	0.17	0.10	0.09	0.04	0.07	0.09	0.25	0.25	0.24
0.51	0.52	0.50	0.09	0.09	0.10	0.15	0.06	0.08	0.05	0.06	0.05
0.37	0.37	0.41	0.13	0.15	0.19	0.11	0.05	0.05	0.09	0.05	0.05
0.23	0.22	0.21	0.08	0.08	0.07	0.12	0.07	0.07	0.06	0.07	0.08
0.24	0.23	0.24	0.09	0.09	0.09	0.19	0.14	0.12	0.05	0.05	0.05
0.05	0.07	0.07	0.17	0.16	0.18	0.12	0.09	0.11	0.08	0.05	0.05
0.24	0.23	0.23	0.09	0.09	0.09	0.19	0.13	0.12	0.05	0.05	0.05

in the number of non-PhD's employed in junior colleges and secondary schools. The percentage of the non-PhD group employed in research centers declined sharply between 1964 and 1968.

A peculiarity in the data on the non-PhD's is the apparent decrease in their total number between 1964 and 1968, followed by an increase to almost the former level in 1970. Table 12.11 presents these data and shows that this pattern was characteristic of both the university and government populations of non-PhD's. The numbers in industry and research centers declined, with only the "other" category showing an increase. Probably these data indicate the use of slightly different and less restrictive criteria for inclusion in the physics section of the National Register in 1964, as well as high mobility of non-PhD's into the PhD category.

TABLE 12.10 Changes in Employment Patterns of Non-PhD Physicists, 1964–1970

Physics Subfield	Total Population (excluding nonrespondents)			College or University		
	1964	1968	1970	1964	1968	1970
Astrophysics and relativity	38	71	113	0.84	0.87	0.76
Atomic, molecular, and electron	682	618	755	0.43	0.47	0.53
Elementary-particle	593	572	682	0.74	0.81	0.80
Nuclear	1,418	1,432	1,414	0.35	0.32	0.38
Plasmas and fluids	745	662	633	0.23	0.24	0.33
Condensed-matter	3,458	3,597	3,126	0.33	0.30	0.37
Space and planetary	591	599	673	0.21	0.32	0.33
Physics in biology	86	120	157	0.40	0.40	0.35
Optics	1,646	2,011	2,029	0.09	0.10	0.11
Acoustics	816	828	745	0.07	0.10	0.10
Other	7,869	5,474	6,925	0.26	0.24	0.23
TOTAL PHYSICS	17,942 [a]	15,983 [b]	17,252 [c]	0.28	0.27	0.30
Astronomy	324	466	427	0.42	0.48	0.44
TOTAL	18,266 [d]	16,449 [e]	17,679 [f]	0.28	0.28	0.30

[a] Nonrespondents = 1356.
[b] Nonrespondents = 1661.
[c] Nonrespondents = 1964.
[d] Nonrespondents = 1416.
[e] Nonrespondents = 1731.
[f] Nonrespondents = 2026.

TABLE 12.11 Non-PhD Employment, 1964, 1968, and 1970

Type of Employer	Year		
	1964	1968	1970
College and university	5,066	4,560	5,309
Industry	6,089	5,432	5,346
Government	2,625	2,496	2,545
Research center	2,293	958	877
Other	2,193	3,003	3,602
TOTAL	18,266	16,449	17,679

dustry			Government			Research Center			Other Institutions		
4	1968	1970	1964	1968	1970	1964	1968	1970	1964	1968	1970
05	0	0.01	0.03	0.04	0.08	0.03	0.03	0.08	0.05	0.06	0.08
23	0.27	0.21	0.15	0.14	0.14	0.15	0.06	0.04	0.05	0.06	0.07
)4	0.03	0.05	0.03	0.03	0.02	0.14	0.12	0.11	0.05	0.02	0.03
14	0.19	0.16	0.12	0.13	0.12	0.29	0.24	0.17	0.11	0.11	0.16
27	0.31	0.29	0.25	0.29	0.24	0.20	0.09	0.09	0.05	0.07	0.05
39	0.46	0.40	0.14	0.14	0.13	0.11	0.04	0.05	0.04	0.06	0.05
32	0.20	0.20	0.30	0.30	0.31	0.09	0.07	0.04	0.09	0.11	0.13
17	0.18	0.20	0.15	0.20	0.14	0.05	0.02	0.04	0.23	0.21	0.27
56	0.63	0.60	0.17	0.18	0.19	0.12	0.03	0.03	0.06	0.06	0.07
45	0.46	0.44	0.28	0.32	0.36	0.12	0.02	0.02	0.08	0.08	0.08
34	0.23	0.25	0.11	0.10	0.10	0.10	0.02	0.03	0.20	0.41	0.39
34	0.34	0.31	0.14	0.15	0.14	0.13	0.06	0.05	0.12	0.19	0.21
.10	0.13	0.16	0.25	0.26	0.27	0.13	0.07	0.08	0.10	0.06	0.06
.33	0.33	0.30	0.14	0.15	0.14	0.13	0.06	0.05	0.12	0.18	0.20

Comparison of Employment Patterns for Physics Subfields

The highest proportion of PhD's in all subfields, except optics and acoustics, was employed in universities in 1970. Elementary-particle physics and astrophysics and relativity had especially high percentages working in universities, 76 percent and 73 percent, respectively, compared with 51 percent of all PhD physicists. The greatest shift toward employment in universities between 1964 and 1970 occurred in plasma physics and the physics of fluids, elementary-particle physics, and nuclear physics, as Table 12.12 shows. The subfields with the highest percentages of PhD's in industry were optics (50 percent), acoustics (41 percent), and condensed matter (35 percent). One fourth of the nuclear-physics PhD's and 26 percent of the plasma physicists (not including physics of fluids) worked in federally funded research centers. And one fourth of the physics in biology PhD's indicated employment in other institutions, principally medical schools and hospitals. Equivalent percentages, 23 percent and 24 percent, of the space and planetary physics PhD group reported employment in industrial and government laboratories, respectively.

Employment patterns of non-PhD's also varied among subfields. The highest percentages of non-PhD's working in universities occurred in elementary-particle physics (80 percent) and astrophysics and relativity (76 percent). About half (53 percent) of the atomic, molecular, and electron physics non-PhD's and half (54 percent) of the plasma-physics group also worked in universities. High percentages working in industry characterized optics (60 percent), acoustics (44 percent), and the physics of condensed matter (40 percent). A number of subfields had substantially higher concentrations of non-PhD's in certain types of institutions than PhD's; for example, one third (36 percent) of the acoustics non-PhD's worked in government laboratories compared with one fifth (19 percent) of the acoustics PhD's. Three times as many non-PhD's in the

TABLE 12.12 Subfields Showing Greatest Increases in University Employment between 1964 and 1970

Subfield	Percentage of PhD's Working in Universities		Percentage Change
	1964	1970	
Plasma physics and physics of fluids	31	48	17
Elementary-particle	64	76	12
Nuclear	41	52	11
ALL SUBFIELDS	43	51	8

physics of fluids worked in government laboratories as did PhD's in this subfield (28 percent compared with 9 percent). Heavier concentrations of non-PhD's tend to occur in institutions conducting a large amount of applied work, which often requires a greater number of assistants than do academic research and teaching. Roughly equivalent percentages of non-PhD's and PhD's in physics in biology worked in "other" institutions. Table 12.13 presents these data and compares PhD and non-PhD employment patterns in various subfields.

Characteristics of Physicists in Different Employing Institutions

The median age of PhD physicists in 1970 was 37.4 years, nearly a year less than the median age in 1964, which was 38.2 years. The median age

TABLE 12.13   Comparison of PhD and Non-PhD Employment Patterns in Various Subfields in 1970

Subfield and Employment	Percentage of Non-PhD's	Percentage of PhD's
Heavily University-Based Subfields		
Astrophysics and relativity	76	73
Elementary-particle	80	76
Atomic, molecular, and electron	53	54
Plasma	54	47
All subfields	30	51
Strongly Industry-Based Subfields		
Optics	60	50
Acoustics	44	41
Condensed-matter	40	35
All subfields	30	23
Relatively High Involvement in Government		
Physics of fluids	28	9
Space and planetary	31	22
Acoustics	36	19
All subfields	14	9
Relatively High Involvement in Research Centers		
Nuclear	17	25
Elementary-particle	11	17
All subfields	5	12
Relatively High Involvement in Other Types of Employment		
Physics in biology	27	24
All subfields	20	5

TABLE 12.14 Comparison of Median Ages of PhD's and Non-PhD's in Various Types of Employment

Employing Institution	PhD Physicists						Non-PhD Physicists					
	1964		1968		1970		1964		1968		1970	
	Mdn.	N	Mdn.	N	Mdn.	N	Mdn.	N	Mdn.	N	Mdn.	N
College and university	38.0	4,478	36.7	7,031	36.7	8,223	28.4	5,066	28.8	4,560	28.5	5,309
Industry	38.5	2,450	38.4	3,194	37.7	3,805	34.6	6,089	34.8	5,432	35.3	5,346
Government	39.4	917	39.6	1,307	39.1	1,468	34.1	2,625	35.5	2,496	35.9	2,545
Research center	37.5	1,930	38.2	1,882	37.8	1,912	33.0	2,293	35.7	958	36.1	877
Other	38.5	558	39.6	673	38.6	840	34.3	2,193	35.0	3,003	34.2	3,602
TOTAL	38.2	10,333	37.7	14,087	37.4	16,248	32.4	18,266	33.2	16,449	32.9	17,679

838

of non-PhD's increased slightly between 1964 and 1970, from 32.4 to 32.9 years. Table 12.14 compares the median ages of PhD and non-PhD physicists in various employing institutions and shows that in both groups the lowest median age was characteristic of college and university employment. The highest median age in the PhD group occurred in the government employment category; the highest median age for non-PhD's was found for those employed in federally funded research and development centers.

The principal work activities of physicists vary substantially among employing institutions. A major difference between PhD's working in colleges and universities and those in other types of employment is, of course, involvement in teaching. More than four fifths (84 percent) of the PhD's in colleges and universities had teaching responsibilities, and 69 percent combined teaching and research. Research was a major work activity in all employment settings. Approximately 85 percent of the PhD's in industry, government, and the universities reported some involvement in research. An even higher percentage, 94 percent, of those working in federally funded research and development centers were engaged in research. Among those working in other types of institutions, the portion having research responsibilities was 70 percent.

Approximately 50 percent of the PhD's working in industry and federally funded research and development centers were engaged only in research, and 40 percent of those in government laboratories had only research responsibilities. The PhD's in universities typically combined other professional responsibilities with research; only 9 percent indicated that they were engaged solely in research.

The research involvement of those in universities was largely basic (90 percent). Basic research also was dominant in federally funded research centers and in government and was the only research activity of one third of the PhD physicists so employed. Sixty percent of the PhD's employed in these institutions who reported any research responsibilities combined both basic and applied research in their work. In strong contrast, only 10 percent of the PhD's working in industry performed only basic research, and less than one third combined basic and applied research in their work. More than half of the PhD's in industry reported some involvement in applied research, and of this number half had responsibilities related to design and development work.

Management responsibilities varied somewhat with employing institution. In universities, only 14 percent of the PhD physicists had some type of management responsibility, usually unrelated to research. In industry, government, research centers, and "other" types of employing institutions, from about one third to two fifths of the PhD's had manage-

TABLE 12.15 Management Responsibilities of PhD Physicists by Employer, 1970

Employing Institution	Total PhD's	Total with Mgmt. Responsibilities	Percentage of PhD's	Basic Research and Mgmt.	Percentage of Mgmt. Group	Applied Research and Mgmt.	Percentage of Mgmt. Group	Nonresearch and Mgmt. or Mgmt. Only	Percentage of Mgmt. Group
College and university	8,032	1,148	14	208	18	61	5	875	76
Industry	3,733	1,492	40	168	11	860	58	464	31
Government	1,432	601	42	227	38	184	31	190	32
Research center	1,876	592	32	233	39	254	43	105	18
Other	823	282	34	50	18	88	31	144	51
TOTAL	15,896	4,115	26	886	21	1,447	35	1,778	43

ment responsibilities that were to a large extent linked with their research activities. In the "other" category, teaching and other nonresearch activities frequently were combined with management. Table 12.15 depicts combinations of management and other responsibilities of PhD's in various work settings.*

Among non-PhD's, applied research was a major activity of two fifths, most of whom were located in government laboratories, industry, and federally funded research and development centers. Those located in the research and development centers also tended to have substantial involvement in basic research; however, those with the greatest involvement in basic research were located in the universities. Design and development work was reported by more than half (52 percent) the non-PhD's working in industry. Teaching was a major work activity of about one third (31 percent) of the non-PhD physicists. Those with teaching responsibilities were located principally in colleges and universities and in "other" institutions, a category that includes junior colleges and secondary schools. Sixty-nine percent of those located at "other" institutions indicated teaching as a major work activity. Management responsibilities, particularly related to research and development, were also reported by relatively high percentages, from one fourth to nearly two fifths, in all types of employing institutions, with the exception of colleges and universities. The work activity pattern of the non-PhD's differed chiefly from that of the PhD's in the relatively greater involvement in applied research and design and development work.

SUPPLY AND DEMAND

In 1940, research and development in the United States was a far smaller enterprise than it is at present. The federal expenditure on research and development was less than 1 percent of all federal expenditures, and the total research and development effort was less than 1 percent of the Gross National Product. World War II, however, was largely a war of advanced technologies, and the federal government found it necessary to devote vast sums to scientific and engineering efforts. The war experience showed that science and technology had enormous potential for changing the nature of man's world and improving his standard of living. Recognition of the potential value of research and development resulted in continued federal support of scientific and technological activities.

---

* Chapter 9 includes several figures that display the employment responsibilities of physicists in both academic and nonacademic environments.

A major factor in fostering awareness that science and scientists are a national resource was a report on basic science, *Science: The Endless Frontier*,[30] submitted to President Truman by Vannevar Bush. Bush proposed the establishment of a National Research Foundation that would have the special responsibility of encouraging and nurturing basic science. He estimated that when this Research Foundation was fully under way its budget should be about $50 million per year! Implementation of Bush's recommendation was slow, but by 1950 all the administrative problems had been worked out and legislation to establish the National Science Foundation was enacted (see also Chapter 10).

A number of federal agencies recognized the relationship of science and technology to their missions and supported both basic and applied scientific research and technological development efforts. The persistently troubled international situation demanded from the Atomic Energy Commission (AEC) and the Department of Defense (DOD) steadily increasing

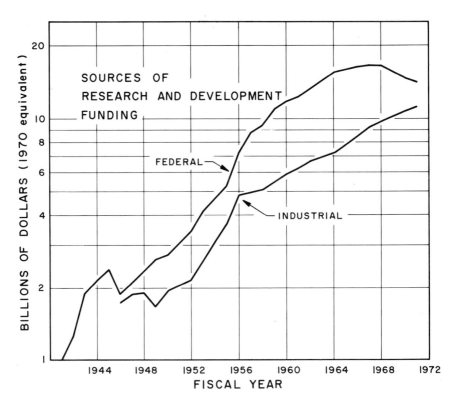

FIGURE 12.11   Sources of funds for research and development.

research expenditures and involvement in science. The promise of atomic energy also provided an impetus to increased investment in science. In July 1958, the National Aeronautics and Space Act brought into being another new agency that assumed a major role in the support of research and development. And in the same year the Advanced Research Project Agency (ARPA) was formed in DOD, with the mission of supporting defense-related science and technology. The scope of ARPA was broad, and it conducted a variety of programs, for example, in materials research and energy conversion. The expansion of programs initiated during and after the war and the steady inauguration of new programs led to a continuing rapid growth in federal support of research and development. Figure 12.11 depicts this expansion.[1, 5, 6] (See also Chapter 10.)

The invention of the transistor in 1948 gave additional impetus to the growing interest in science and technology. It was immediately apparent that the transistor was the beginning of a revolution in electronics and would have enormous impact on industry. This evidence of the industrial impact of research led many industrial corporations to establish new research laboratories and to increase their expenditures on research and development during the 1950's, as shown in Figure 12.11. This growing national scientific effort required more scientists.

Figure 12.12 shows where the growing number of physics PhD's were employed.[1] The trends it depicts are crucial in understanding recent

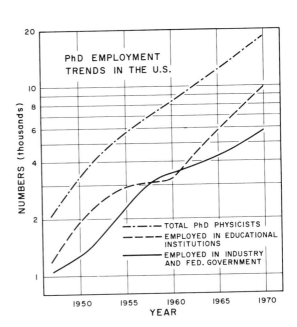

FIGURE 12.12 Employment of PhD physicists in the United States. Government employment includes federally funded research and development centers.

TABLE 12.16 Physicists Employed in Industry and Business

Year	Bureau of Labor Statistics Total No. (Thousands)	National Register Total No. (Thousands)	National Register Total PhD's (Thousands)
1960	12.5	8.6	2.7
1962	14.0	9.8	2.9
1964	15.6	9.0	3.3
1966	16.2	8.3	3.3
1968	17.6	9.4	3.8
1970	18.4	9.8	4.4

developments in physics manpower. In 1950, more PhD physicists were employed in educational institutions than in industry and government taken together. However, during the 1950's new industrial laboratories employing substantial numbers of physicists came into being, and government laboratories were established or expanded. Consequently, although in 1950 employment of PhD physicists in educational institutions substantially exceeded that in industrial and government laboratories, by 1960 this situation had changed, as Figure 12.12 shows, and PhD's in industry and government exceeded the number in universities.

However, increasing numbers of physicists had been employed in universities during this interval, and by 1960 these academic physicists had developed research facilities and programs, including programs in graduate education, that set the stage for an increased rate of production of physics PhD's. But by 1960 the staffing of new government and industrial research establishments, founded during the 1950's, was nearing completion, and the rate of growth of industrial research and development expenditures began to decrease. As a result, the rate of growth of employment of PhD physicists in industry and government slowed. The little growth that still occurred apparently represented an upgrading of the qualifications of industrial physicists * rather than the employment of more physicists. As Table 12.16 shows, the total employment of physicists remained practically constant in industry through the 1960's. This decline in the growth of

---

* A study conducted by the Statistical Data Panel of the Physics Survey Committee confirms the slow growth of the employment of physicists in industrial laboratories. A sample of 17 laboratories who responded to a Data Panel questionnaire reported the relative employment of physicists as follows: 1960, 0.76; 1965, 0.93; 1968, 1.04; 1970, 1.00. Interviews with supervisors in industrial laboratories in 1968 indicated that "no shortage of physicists has existed during the past five years." [31]

employment of physics PhD's in industry resulted from a combination of factors. First, there was a net slowing in the growth of industrial investment in research and development, which was reinforced in the last half of the decade by the decreased federal support of research and development, as shown in Figure 12.11. About half of industrial research and development is federally financed, and a substantial increase in the industrial expenditure would have been necessary to keep the total constant during the 1960's. Second, the confidence in tangible, relatively short-term technological returns on investments in basic research in physics that had characterized the 1940's had not been confirmed by the experience of the 1950's. Thus, there was a steady trend in industry toward greater emphasis on applied work. Third, engineering schools were rapidly increasing their PhD programs and changing the content of these programs to include large sectors of physics. As the applications of physics increased and became more apparent, various aspects that in earlier decades would have been considered physics were incorporated in engineering curricula.* As a result, the number of engineers qualified to perform research was rapidly growing. The engineering PhD's were often in direct competition with physics PhD's for employment.

For these reasons, industry played a relatively minor role in absorbing the increasingly large output of physics PhD's in the 1960's, but continued increases in federal support of basic research in the universities in the first half of the 1960's provided ample opportunity for academic employment of these new PhD's. As they were trained for, and eager to become involved in, research, they enlarged programs of graduate education and developed new ones. In spite of the slow growth of employment in non-academic sectors, the recommendation of the 1962 Gilliland report [22] that the number of PhD's in engineering, mathematics, and the physical sciences should be doubled in eight years was enthusiastically received by industry, the universities, and the professional societies, and the goals set by the report were easily achieved (see Figure 12.10). In fact, the target of 7500 PhD's in these disciplines was attained two years early. Political pressures favoring the geographical spread of graduate education through the development of new graduate programs in science additionally encouraged the growth of PhD programs in the 1960's. Trends in federal

---

* The percentage of the total number of PhD's in the natural sciences and engineering awarded in engineering rose from 16 percent around 1950 to approximately 25 percent in 1970 according to the NRC Doctorate Records File. Interviews with industrial employers [31] contained comments that "the demarcation line between physics and electrical engineering is vanishing" and "the graduate engineering curriculum includes so much physics these days that engineers are qualified for positions physicists used to fill exclusively."

support of basic research, which probably are more descriptive of the impact of such support on physics than on the total research and development effort, are depicted in Figure 12.13.

The new products of the growing graduate education effort, in turn, re-entered academic institutions to continue the cycle. A rather abrupt increase in the rate of production of PhD's, beginning in 1960 and continuing through 1970, resulted. As Figure 12.14 shows, the growth in the number of PhD's awarded in physics paralleled the growth in the employment of physicists in academic institutions. Most of the new PhD's in educational institutions were heavily involved in research; the number of physicists who regarded teaching as their primary work activity increased far more slowly than did the total number working in academic institutions.

In the late 1960's, the rapid growth of federal support of research and development began to slow and level off. Some contributing factors are the following: As the research and development budget became a larger fraction of the total federal budget, it encountered increased scrutiny. From 1964 through 1967, research and development represented about

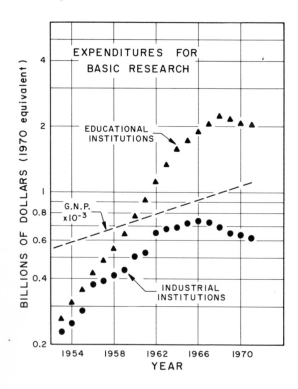

FIGURE 12.13 Funds expended for the support of basic research (in 1970 dollars).

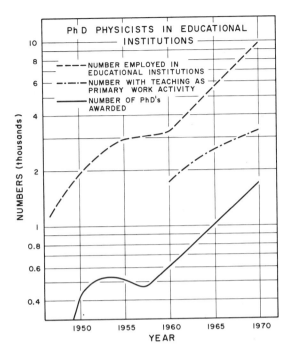

FIGURE 12.14 PhD physicists in educational institutions.

3 percent of the Gross National Product. Federal support of research and development accounted for over 10 percent of all federal expenditures during these years. The sheer size of the enterprise brought into question the necessity for continued rapid growth. Research facilities became larger and more expensive and more concentrated in centers. Consequently, they were more visible. The funding of the National Accelerator Laboratory was a major event that attracted nationwide attention and stirred extended public debate about the desirability of such a large expenditure for fundamental research. Many of the expensive ventures of the space program, with large political and technological motivations, were regarded as science and enhanced the impression of research as being extremely costly.

Much of the increased support of science in the universities had been necessary to meet the need for increased numbers of trained scientists. However, in the late 1960's the manpower needs for research and development in government and industry were easily satisfied, and this argument for research support lost much of its validity.

In addition, the demands on federal funds by other programs became more intense. The goals of the "Great Society," which characterized the years from 1964–1968, shifted attention and resources to programs with little apparent relation to scientific effort. At the same time, the expenses incurred in the war in Indochina also grew rapidly and absorbed a large share of government funds.

Part of the large expenditure on science and technology had been justified by the defense-related inventions of the war and immediate postwar years—the hydrogen and fission bombs, nuclear energy, and the like. However, in the late 1960's, the troubled international situation that had served as a justification for much science with potential relevance to national defense became somewhat more stable. Furthermore, many of the military needs were less obviously research-related than had been the case.

As these various factors combined to cause a decline in the growth of support for science, physics probably was affected more than many other disciplines. Physics had an unusually heavy dependence on large facilities, for example, the great particle accelerators. Physics also was heavily concentrated in the universities; therefore, government support of research and development in industry had less effect on physics than on other sciences. Much of the federal support of physics research was related to defense needs. Since defense-related support declined earlier and more sharply than other forms of federal support, physics was strongly affected. Table 12.17 presents some of these data that illustrate the vulnerability of physicists to declining federal support in comparison with scientists in general. Perhaps the only bright spot in the picture for physics was that physicists were unusually heavily involved in basic research, which fared better than research and development as a whole under growing budgetary constraints, as Table 12.18 shows.

TABLE 12.17   Comparison of Physics PhD's and All Science PhD's in Regard to Research Involvement and Federal Support [a]

Research Involvement and Support	Physics PhD's $N=14,311$ (%)	All Science PhD's $N=111,200$ (%)
Engaged in research and development	54	38
Engaged in research and development in educational institutions	28	19
Engaged in basic research	38	24
Receiving federal support	62	43
Receiving DOD support	17	11

[a] Source: *American Science Manpower 1968* (NSF 69-38).

TABLE 12.18   Federal Funds for Basic Research as a Percentage of All
Federal Funds for Research and Development [a]

Year	Percentage	Year	Percentage
1960	7.9	1966	14.2
1961	9.1	1967	15.1
1962	11.0	1968	15.7
1963	11.7	1969	16.0 [b]
1964	12.7	1970	16.4 [b]
1965	14.0	1971	16.6 [b]

[a] Source: *American Science Manpower 1968* (NSF 69-38).
[b] Estimates.

The long time required to complete graduate education and the tendency
of new PhD's to remain in the academic sector, as a result of federal incentives provided during the early 1960's, made it difficult for universities to
respond to the decline in the growth of support that began around 1965.
Factors related to finances and personnel combined to make the continued
growth of graduate education attractive. In addition, universities could
not accept the waste involved in curtailing the education of the many
students who had invested years in preparation for research careers. Because the educational process is long for a research scientist, there is a
time lag of at least five years in the adjustment of graduate education to
changes in olicy and in the goals and needs of society. Faculties in
general cannot be reduced rapidly because many faculty members attain
tenure, and each tenured faculty member, by long-established tradition,
tends to replicate himself repeatedly with new PhD graduates. Major
facilities, authorized and initiated in the early 1960's during the period of
most rapid growth, were just coming into operation and not only could
not be terminated in good conscience after so massive an investment of
public money but required extensive additional support for their operation.
    A critical factor was the lack of nonacademic employment opportunities
combined with the continuing political pressures toward encouraging the
geographical expansion of graduate education in science. New PhD's
seeking jobs could find them in academic institutions since the support of
basic research in educational institutions continued well into the 1960's,
and colleges and universities throughout the country were attempting to
expand or create graduate programs. The prestige associated with graduate
education had not vanished, and lesser institutions were eager to inaugu-
rate new graduate programs. The decline of alternative employment
opportunities made it easier for them to attract faculty of high quality.

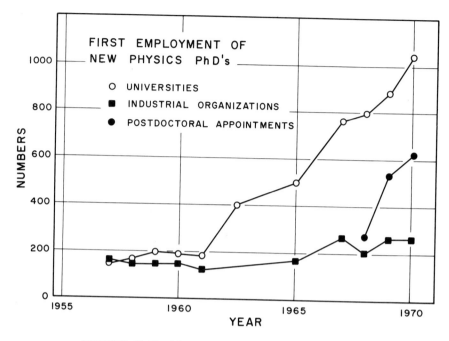

FIGURE 12.15   First employment of new PhD physicists.

Therefore, though the national effort in science and engineering was slowing, many pressures were fostering the continued rapid growth of graduate education.

Figure 12.15 shows the number of new PhD's accepting employment in various kinds of institutions as a function of time. Initial employment in universities and in government and industry were about equal in 1960. As employment of new PhD's in government and industry increased only slowly during the 1960's (see also Figure 12.12), a rapidly growing fraction of the new PhD's found positions in academic institutions. This trend continued throughout the decade, although industrial funding of research and development increased from 30 percent in 1964 to 42 percent in 1971 and federal support declined (see Figure 12.11).

Obviously, the academic institutions could not continue to absorb the rapidly growing number of new PhD's. Difficulties in finding suitable employment became common, and well-known nonacademic employers were able to specify increasingly outstanding qualifications for employment in technical fields.* Awareness of the employment problem began to

---

* Julius P. Molnar, executive vice president of Bell Telephone Laboratories, Inc., is quoted as saying: "The median of those bachelor's [degree holders] we take on is within the top 5% in class standing." (*Business Week,* January 23, 1971, p. 40.)

develop in the physics community in 1967, and concern was particularly apparent after 1969.*

Both the AIP and the National Research Council conducted studies of the employment of new PhD's in 1969. These studies showed that the unemployment percentages of these PhD's in December 1969 were relatively low (about 1.5 percent). However, there were evidences of a deteriorating situation. The AIP interpreted the increasingly frequent appointments of new PhD's to postdoctoral positions as a holding pattern. Subsequent surveys in September 1970 and January 1971 indicated that new PhD's were experiencing greater problems in finding jobs.

The Economic Concerns Committee of the American Physical Society

---

* In June 1966, *Physics Today* reported: "Physicist demand at peak. As our national economy and our inflation surge forward, pressures on all job markets continue to increase, and nowhere is this more evident than in the current demand for physicists [p. 77]." By July 1967, *Physics Today* was reporting: "Physics Job Market a Little Tighter as Economy Wanes [p. 83]."

When the Work Complex Study [31] was funded in 1967, it had become apparent that the problem facing employers was "that of selecting the most appropriate people from a large supply of . . . specialists [p. 7]." [31]

Evidence of the changing climate of employment was present in the early 1960's. For example, the amount of advertising for physicists to fill jobs began to decline sharply after 1960, as Figure 12.16 shows. Further evidence of the decrease in the industrial employment of scientists in general is the decline in advertisements in the *New York Times* for industrial scientists that is shown in Figure 12.17. Concurrently, the number of registrants in the placement service conducted by the AIP at its annual meeting began to increase, as indicated in Figure 12.18. Though registrants increased, the use of the service by employers began to decline in 1968 (the participation of industrial employers had started to decrease three years earlier), leading to an apparent collapse of the placement system in 1971.

Qualitative evidence of the extent of the crisis for physicists appears in letters, an editorial, and articles in *Physics Today* during 1969 (February, pp. 13–16; April, p. 23; May, pp. 12–13; June, pp. 21, 112; July, p. 9; August, pp. 9–10; October, pp. 17, 31; and December, pp. 11–12), in the attention given this subject at meetings of the American Physical Society (*Bulletin of the American Physical Society, 15,* pp. 53, 1319–1320), and in the popular press (*Wall Street Journal,* November 6, 1968; *Scientific Research,* June 9, 1969, p. 15; *U.S. News and World Report,* December 29, 1969, p. 40; and the *New York Times,* January 27, 1970).

The sudden visibility of the employment crisis led the AIP to seek quantitative measures of its extent and seriousness. A questionnaire mailed to physicists who received PhD degrees in the years 1967–1969 showed a disruption of customary patterns of employment. The number of new PhD's taking postdoctoral appointments increased sharply. Presumably this finding shows that persons who cannot find acceptable professional employment are working in postdoctoral positions created for them. If this assumption is correct, these persons are awaiting opportunities to obtain permanent employment and will be appearing on the job market again in the next few years. It also is unlikely that the support of such positions will be continued indefinitely; funds probably were diverted from other uses to create them.

That physicists were having to exert more effort to find jobs was apparent in the finding that from 1967 to 1969 the number of PhD's who submitted more than ten applications to industrial firms increased, while the number submitting three or less decreased; and the number who received no offer as a result of their applications increased. The data on applications to universities were much the same.[32]

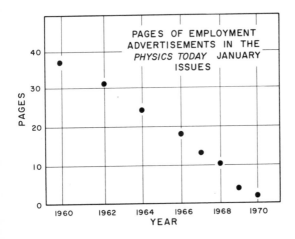

FIGURE 12.16 Pages of employment advertisements in January issues of *Physics Today*.

also undertook a study of employment in 1970. This study emphasized the experiences of older PhD's, a group that other employment studies had not dealt with. The Committee found that only 300 new jobs at the PhD level were available in 1970, although more than 1600 PhD's were granted. This kind of disproportionate relationship probably will hold well into the 1970's.[20] Some details of the flow of physicists into and from various types of employment in 1970 appear in Figure 12.19 (left and right). This figure shows that approximately 1200 PhD physicists did not find employment in physics in the United States in the interval covered by the survey. (Approximately 17 percent of these 1200 went abroad by preference and did not actually attempt to find employment

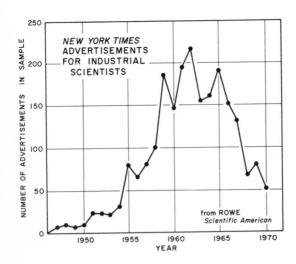

FIGURE 12.17 *New York Times* advertisements for industrial scientists.

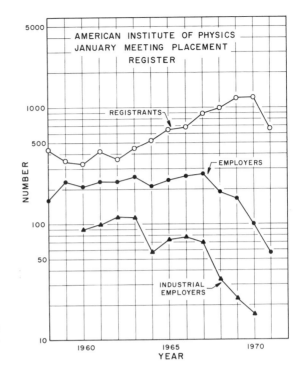

FIGURE 12.18 American Institute of Physics January Meeting Placement Register.

in the United States.) A like diagram for 1967, the last year of adequate employment opportunities, is included in Figure 12.19 (left) for comparison.

In March 1971, the NSF conducted a postcard follow-up survey of respondents to the 1970 National Register. Thus the survey is related to the employment situation of those already established as physicists in January 1970 and does not include data on new PhD's granted in 1970. (Data on approximately 60 percent of the total new PhD's, that is, approximately 900 people, would be included with those on non-PhD respondents.) Ninety percent of the 1970 PhD respondents participated in the March 1971 survey.* It is probable that the nonrespondents to the 1971 survey have significantly different characteristics from the respondents, particularly in view of the substantial dislocation within the physics community that other data suggest. The results so far available (NSF 71-26) from the NSF survey are tabulated only in terms of the numbers actually unemployed at the time of the survey. Taxi driving, for example, would be indicated as "current status: employed part-time;

---

* PhD respondents to the 1970 Register constituted 85 percent of the physics PhD population. The follow-up survey, with a 90 percent response rate, therefore, represents about three fourths of the 1970 PhD population.

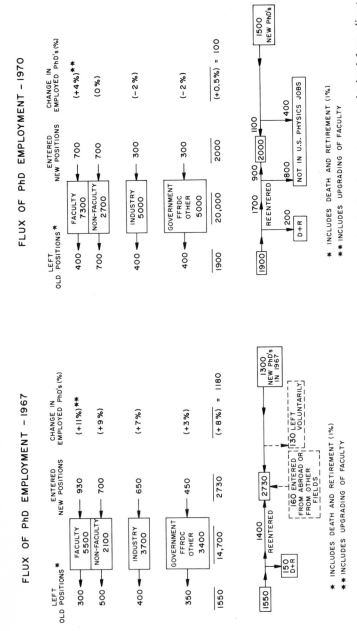

FIGURE 12.19 *Left*: The flux of PhD employment in 1967. The values in the faculty sector were obtained from direct counts of physics faculty directories together with the following assumption regarding PhD's in the various types of faculty. All faculty who entered or left PhD-granting schools were assumed to have a PhD. In the BS- and MS-granting schools, ranks below Assistant Professor were ignored; all who entered or left with the rank above Assistant Professor were assumed to have a PhD; only 25 percent of those who left with the rank of Assistant Professor but all who entered with that rank were assumed to have PhD's. The changes in employment for the other sectors were determined as the averages from 1964 to 1968 according to the American Science Manpower series.
Right: The flux of PhD employment in 1970. The remarks on faculty figures given for the left side of the figure apply here. The changes in employment in the other sectors were estimated from surveys and from information supplied by funding agencies.

nonscience related," as distinct from "unemployed and seeking employment," the response that was examined first in the NSF data. (Data on
part-time employment, nonscience employment, and periods of unemployment will be issued subsequently.) The data show that the unemployment
rate in the established physics community is the highest of the sciences,
and that the rate of unemployment for non-PhD's is higher than that
of PhD's. At the time of the survey, 1126 physicists were unemployed,
258 of whom were PhD's. In contrast to these data, the AIP January 1971
Placement Register listed 608 PhD's and another 343 people who expected
to receive PhD's. The most obvious feature of the NSF data was a sharply
increasing trend in unemployment among relatively well-established PhD's,
depicted in Figure 12.20. The figure shows the unemployment rate per
10,000 PhD nonstudent physics respondents for the 1964, 1968, and
1970 Registers and the 1971 follow-up survey. It should be noted again
that the follow-up survey relates only to the marked increase in unemployment (118 in 1970 to 180 in 1971 per 10,000 respondents) in the 1970
respondent population. (Corresponding figures for the 1964, 1968, and
1970 data would have to be based on a survey of the entire physics PhD
population, including new doctorates and other entrants.)

The parallel development of physics in the academic and the industrial
and government sectors during the 1950's was healthy and productive;
new PhD's were eagerly sought by employers, especially those whose
objectives or mission related to technological development or the solution
of national problems. The 1960's were characterized by a decrease in the
rate of growth of physics in nonacademic sectors and a striking increase

FIGURE 12.20   The number of unemployed per 10,000 PhD respondents to the National Register Survey in 1964, 1968, and 1970 and the
follow-up survey of 1971. The follow-up survey relates only to the
1970 Register respondents and does
not include new PhD's awarded
after January 1970. [Source: National Science Foundation.]

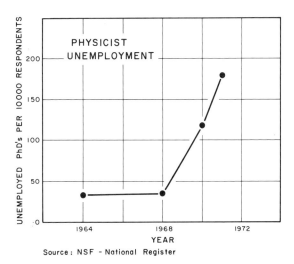

Source: NSF - National Register

in the size of the academic physics enterprise. The 1970's have begun with recognition of a critical employment situation and the realization that the traditional patterns of employment in physics are changing. These changes result in part from a change in the nation's priorities. If physicists understand the nature of this change, are flexible in their attitudes, and will undertake careers not only in physics but in related disciplines and specialties that also offer challenging problems, the outlook for the future could be more encouraging than was first assumed. However, difficult problems lie ahead.

Any career has uncertainties and is dependent on the vagaries of the economy, competitive careers, and the world situation. To examine some of the changes in the employment picture and their implications for the future of physics, we discuss in the following section some analyses and projections.

PROJECTIONS OF EMPLOYMENT FOR PHYSICISTS

The purpose of this section is to estimate the number of jobs of the traditional kinds that will be available for new physics PhD's during this decade. These estimates will show that the number of traditional jobs expected will be far too few to absorb all the PhD's being produced. Therefore, an increasing number of PhD's in physics will have to find employment in nontraditional jobs, for example, research and development in neighboring disciplines or teaching at the college, junior college, and high school level.

The difficulties in making job predictions are generally comparable to those of making economic predictions, particularly over short periods of time. These difficulties arise not so much from the extension of existing patterns into the future but, rather, from the inability to anticipate the effects of attitude changes, changing definitions and boundaries among disciplines and specialties, and the smallness of the numbers involved in physics. A brief review of the employment experiences of physicists from 1969 to 1971 provides a background for future projections.

These two years were unusual in that teaching jobs disappeared simultaneously with the disappearance of both research and development positions supported by the federal government and positions in other disciplines that physicists would have been capable of filling. Such a period of recession, when the job market is lower than at any time since World War II, leads to a cautious approach to the development of projections for future employment.

The initial reaction to the employment problem that was recognized in 1969 was to reduce first-year graduate student enrollments in the larger

PhD-producing universities by about 25 percent on the average. Among the other results were greater attention to manpower data and placement service activities and growing flexibility in the expectations of employment. These attitude changes have helped to reduce unemployment among PhD physicists, because physicists were more willing than ever before to take nontraditional jobs. According to the NSF survey of employment in March 1971, only 258 PhD physicists were unemployed. (The size of the PhD physics population reflected in the 1970 National Register of Scientific and Technical Personnel was 16,631.)

One interpretation of this finding is that physics PhD's, particularly new ones, do not and will not remain unemployed for very long. The educational and research experiences provided by a physics PhD program are sufficiently demanding and broad and of sufficiently high quality to enable physicists to compete successfully on the job market as long as that market is reasonably existent. In addition, the number of physicists produced per year compared to the total population of scientists and engineers is relatively small, and their capabilities are wide-ranging. Therefore, that small number should be absorbable in a reasonably healthy national (and science) economy.

What constitutes employment of a physicist depends on the definition of a physicist. In three of the preceding sections we have attempted to develop a definition of a physicist by delineating his educational back-ground and the nature and loci of his work. (Chapters 3 and 7 offer a more detailed discussion of what a physicist is and does.)

The Committee's estimate of new employment opportunities for physics PhD's in the period 1970–1980, which is shown in Table 12.19, is based on an adaptation to physics of NSF projections for all scientists. (The derivation of the numbers in this table is described in more detail later in this section.) The numbers assume a vigorous economy and a generous allowance for new kinds of employment. We regard them as optimistic but not entirely unrealistic; we estimate approximately 9400 new employment opportunities during the 1970's.

The rate of production of PhD's in physics probably will continue at the present level of approximately 1500 per year for at least the first half of this decade. Sufficient students to sustain the present rate of production are currently at various stages in the graduate education process; few of these will fail to complete an education in which they have already invested a number of years of their lives. Thus, we estimate that 7000 new physics PhD's will be awarded in the next five years. We also estimate that less than half of the 9400 new employment opportunities will develop in the first five years of the decade. Therefore, employment problems will be with the physics community for some years.

The projected employment opportunities should result from the growth

TABLE 12.19   Employment Opportunities for Physics PhD's, 1970–1980[a]

Employment	Number of Jobs (Thousands)		
New employment opportunities	7.2		
Academic		2.7	
Nonacademic, research and development		2.2[b]	
Nonacademic, not research and development		2.3	
Traditional			0.4[b]
New types of jobs			1.9
Replacement employment opportunities	2.2		
TOTAL	9.4		

[a] Based on NSF projections for scientists.[33]
[b] These types of presently existing employment frequently are combined in the discussion in this section and referred to as 2.6 thousand nonacademic research and development positions.

of the universities, the growth of research and development in industry and government, and the replacement of physicists who leave existing positions. Replacement openings will come about not only through death and retirement but also through the transfer of physicists into other kinds of occupations. This trend may be accelerated by the anticipated aging of the overall physics population; the migration of physicists to managerial and administrative jobs not related to research and development increases with age.

The development of institutional means of supporting more physicists by taking turns in an applied science or interdisciplinary adjunct of universities is one possible way to broaden their use. Further, a large program for rotating older university physicists to teaching and research positions in universities in developing countries could offer another means of creating new openings for younger physicists on university faculties. In addition, new large-scale federal research and development programs undoubtedly will develop and will call for the capabilities of physicists and the instrumentation of physics.

With a vigorous economy, some 8000 to 10,000 openings in the traditional types of work performed by physicists could appear by 1980. These estimates are based on more than the simple extrapolation of present growth patterns. Predicted university enrollments [34] and Bureau of Labor Statistics projections for the growth of a post-Vietnam U.S. economy [35] also helped to shape them. They represent an application to physics of projections made by Cartter,[36] the Bureau of Labor Statistics,[37] and the

NSF,[33] and they agree with the lower bounds of estimates made by Grodzins for the Economic Concerns Committee of the American Physical Society.[20]

The essential requirement for physicists in the future is a change in attitude, on the part of both professors and their students, in regard to what constitutes a desirable and challenging employment opportunity for a physicist. Physicists, like most other people, obviously can perform a variety of work tasks other than the ones for which they traditionally have been employed. And, as we have argued throughout this Report, history has shown that physicists traditionally have been characterized by an unusual flexibility and adaptability in this sense. Because projections show that the production of PhD's, at least in the early part of this decade, will substantially exceed the number of new employment opportunities of the traditional type, which are heavily centered on research and development, we anticipate that many new as well as experienced PhD physicists will find other kinds of employment. Probably such employment will not be primarily research- or development-oriented positions in other scientific disciplines, for an overproduction of PhD's in comparison to the number of new employment opportunities currently characterizes virtually all science and engineering disciplines. However, the new PhD's will be better qualified for many types of non-research-and-development positions than many other potential members of the work force. They are not likely to be numbered among the unemployed; probably they will fill positions in various technical services and in secondary school and junior college teaching. These types of employment will not fully exploit the research experience and specialized knowledge that the PhD has devoted six years of his life to acquiring. This situation is currently described as underemployment. In addition to waste of a national resource and of the investment of much student time and educational effort, underemployment also results in disillusionment with science and education on the part of the scientist or engineer in this situation.

Some jobs of the kinds traditionally filled by physics PhD's are held at present by non-PhD's. Because the PhD is in a better position to compete for such opportunities than the non-PhD, the projections include an estimated enrichment of the work force—the movement of PhD's into jobs formerly filled by non-PhD's. This enrichment is a continuing process that has been in progress for many years, as Table 12.16 shows in relation to the industrial sector.

There is a subjective aspect to underemployment. In the 1950's and early 1960's, most new physics PhD's could obtain job offers that afforded a wide choice of geographical location, institutional character, and scientific and technical environment. In recent years, the scope of the choices offered to new PhD's is restricted. Consequently, the new PhD cannot always fully realize his aspirations and frequently considers himself under-employed when an objective observer would have a different view.

Use of Physics PhD's in 1980

Figure 12.21 depicts jobs for physics PhD's in the period 1960–1984—statistical data from the past and projections for the next decade. The university and college sector now employs more than half of the physics community but cannot be expected to maintain so fast a rate of increase in the next decade because of the reduced size and rate of growth of the college age population.[34] Concurrently, the growth rate of the nonuniversity demand for physicists will be declining as the United States shifts to the post-Vietnam, postindustrial era depicted by the Bureau of Labor Statistics[35] in which, however, the supply of physicists by 1980 will again be less than the number needed.

Projection of employment is an analytical exercise in which an assumed distribution of roles is transferred to a quantified description of an assumed future world; the size and composition of the assumed world, weighted by knowledge of what physicists do, yield an estimate for the employment of physicists. The employment projections in this section are based on the following:

1. Existing patterns and established trends that form the basis for the

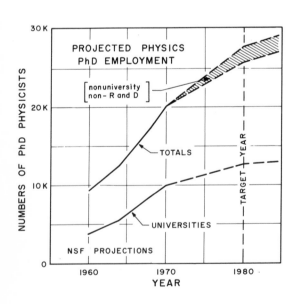

FIGURE 12.21 Projected physics PhD employment. The shaded area, based on National Science Foundation estimates, is nonuniversity, non-research-and-development employment, which is expected to create jobs specifically for PhD's.

relationship between PhD employment and the overall employment requirements for the economy;

2. The existing pattern of university physics involvement scaled according to the size of the student population;

3. The overall Bureau of Labor Statistics projections for the economy, which include the growth of research and development expenditures at a slower rate than the Gross National Product (and thus at a slower rate than during the late 1950's and early 1960's), a national unemployment rate of 3 percent or 4 percent, continuation of high rates of economic growth (4.3 percent in the gross national product), and levels of defense activity that will become approximately like those of 1963—somewhat higher than the levels prior to the buildup of military activity in Vietnam.

The actual pattern of growth between now and 1980 is, of course, uncertain. It is assumed that conditions in 1980 will approximate the underlying assumptions of the Bureau of Labor Statistics model, the target year of which is 1980.

For the target year, we predict 7500 new positions of the traditional types (for example, college and graduate teaching and basic research) for physics PhD's. Less than half of the new jobs, 2700, will be in universities. Nonacademic employment will grow by the remaining 4500 only when substantial changes in the character of nonacademic employment are assumed; that is to say, 2600 openings are physics jobs within the currently normal pattern of work activity, whereas an estimated 1900 of the new jobs will represent an assumed demand for PhD's in non-research-and-development-related scientific positions, a large increase in the number of such positions relative to research-and-development-related employment.

The 7500 jobs include reasonable enrichment of the work force, with PhD's filling jobs that might otherwise have been held by non-PhD's. The projection of the present use pattern at the present concentration of PhD's on the 1980 Bureau of Labor Statistics predicted economy would suggest only 4400 new positions.

In addition to new jobs, openings for new PhD's will result from normal attrition: death, retirement, and transfer from the physics community. These circumstances will produce 2200 openings between 1970 and 1980, an estimate derived from analysis of National Register age distributions and patterns of cohort growth or decline.

New jobs plus attrition can be expected to result in jobs for 9400 PhD physicists between 1970 and 1980, with 7500 of these openings corresponding to today's pattern of teaching and research and development and 1900 more developing as a result of a predicted demand for non-research-

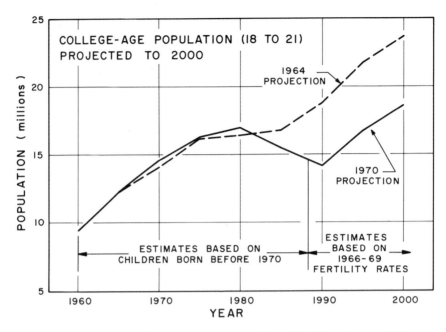

FIGURE 12.22 Projection of college-age population (18–21) to the year 2000.

and-development PhD physicists outside the universities (and probably outside the typical physics work activity pattern of today).

Academic Use Cartter,[36] considering academic institutions generally, assumes employment to be proportional to student populations; in the case of physics, retaining its current share of college enrollment,* the concentration of PhD's among the new faculty that would be needed would then be estimated at about 60 percent of the total. The population data that are the basis of the Cartter model appear in Figure 12.22, and the tabulated population and general enrollment data are shown in Table 12.20. Estimated faculty requirements appear in Tables 12.21 and 12.22; these include physics specifically.

Cartter describes the fundamental limitations of his model as follows [36]:

It is not too difficult to assess the aggregate flows of new teachers, as well as replacement and expansion needs, for errors tend to cancel out. For each 100,000 new students in higher education, about 5,000 new college teachers will commonly be required. But whether those new teachers will be scientists or humanists, specialists or

---

* This is an optimistic assumption; in fact, undergraduate enrollment in physics seems to be declining, as Figure 12.6 shows.

generalists, depends upon a host of factors that are not revealed by the aggregates: factors such as the degree to which the expansion takes place in junior colleges, senior institutions, or in professional or vocational fields; institutional responses to student demands for curricular "relevance"; the extent of the enrollment shift from abstract science to socially applied fields like sociology, psychology, politics; the trend toward eliminating traditional general requirements in liberal arts curricula; and so on. . . . Some very erratic and unpredictable shifts are taking place in undergraduate enrollments in many traditional disciplines today, compounding the difficulty of prediction.

TABLE 12.20   Growth in Age Group, High School Graduates, and College Enrollment, 1965–1990

	Number (Thousands)					
					College Enrollment	
	Age		Annual	High School	First	Total Full-Time
Year	18	18–21	Change	Graduates	Time	Equivalent
Actual						
1965	2771	12,154	659	2665	1442	4564
1966	3550	12,970	816	2672	1380	4936
1967	3543	13,676	706	2679	1440	5380
1968	3514	14,378	702	2702	1630	5810
Projected						
1969	3614	14,218	160	2830	1729	6064
1970	3703	14,371	153	2969	1836	6303
1971	3847	14,680	309	3039	1915	6755
1972	3926	15,119	439	3141	2010	7115
1973	4030	15,532	413	3264	2122	7489
1974	4057	15,884	352	3327	2196	7831
1975	4168	16,206	322	3459	2300	8197
1976	4187	16,465	259	3517	2356	8525
1977	4204	16,641	176	3573	2410	8799
1978	4207	16,790	149	3618	2460	9050
1979	4344	16,967	177	3779	2589	9324
1980	4254	17,033	66	3743	2582	9539
1981	4182	17,011	−22	3722	2585	9705
1982	4120	16,924	−87	3708	2596	9834
1983	3945	16,525	−399	3551	2486	9746
1984	3728	16,000	−525	3355	2348	9514
1985	3625	15,445	−555	3263	2284	9228
1986	3509	14,821	−624	3158	2210	8862
1987	3575	14,442	−379	3218	2252	8639
1988	3564	14,273	−169	3207	2245	8541
1989	3631	14,279	6	3268	2288	8545
1990	3723	14,493	214	3351	2346	8674

TABLE 12.21 Faculty (Thousands) Needed to Maintain Present Quality of Instructional Staff, 1960–1990 [a]

| | Student Enrollments | | | | Faculty Needed | | | |
Year	Full-Time Equiva-lent	Change (2-Year Aver-age)	Change (%)	Total Full-Time Faculty	For re-place-ments	For expan-sion	Total	New (with PhD)
Actual								
1960	2913	145	5.0	169	3.4	7.0	10.4	4.6
1961	3173	117	3.7	177	3.5	8.0	11.5	5.1
1962	3411	249	7.3	190	3.8	13.0	16.8	7.3
1963	3639	233	6.4	202	4.0	12.0	16.0	7.0
1964	4030	310	7.7	220	4.4	18.0	22.4	9.9
1965	4564	463	10.1	233	4.6	13.0	17.6	7.7
1966	4936	453	9.2	245	4.9	12.0	16.9	7.4
1967	5380	428	7.6	261	5.2	16.3	21.5	9.5
1968	5810	437	7.5	279	5.6	17.5	23.1	10.2
Projected								
1969	6064	342	5.7	292	5.8	13.6	19.4	8.5
1970	6303	247	3.9	302	6.0	10.0	16.0	7.0
1971	6755	450	6.7	320	6.3	18.0	24.3	12.0
1972	7115	406	5.7	336	6.6	16.2	22.8	10.0
1973	7489	367	4.9	351	7.0	14.7	21.7	9.4
1974	7831	358	4.6	365	7.2	14.3	21.5	9.3
1975	8197	354	4.3	380	7.5	14.2	21.7	9.4
1976	8525	346	4.1	394	7.8	14.1	21.9	9.5
1977	8799	301	3.4	406	8.0	12.0	20.0	9.5
1978	9050	263	2.9	417	8.2	10.5	18.7	8.2
1979	9324	262	2.8	427	8.4	10.5	18.9	8.3
1980	9537	245	2.5	437	8.6	9.8	18.4	8.1
1981	9705	190	2.0	445	8.8	7.6	16.4	7.2
1982	9834	148	1.5	451	8.9	5.9	14.8	6.5
1983	9746	20	0.2	452	9.0	0.8	9.8	4.3
1984	9514	−160	−1.7	446	8.8	−6.4	2.2	1.0
1985	9228	−259	−2.8	436	8.6	−10.4	−1.8	−1.0
1986	8862	−326	−3.7	423	8.4	−13.0	−4.6	−2.0
1987	8639	−294	−3.4	411	8.1	−11.8	−3.7	−1.6
1988	8541	−161	−1.9	405	8.0	−6.4	1.6	0.7
1989	8545	2	0.0	405	8.0	0.1	8.1	3.2
1990	8674	65	0.8	408	8.1	2.6	10.7	4.7

[a] Source: Science, *172*, 132 (1971).

TABLE 12.22 New Faculty Required to Maintain Quality, and New Doctorates Available: Actual and Projected, 1965–1985

Discipline	Year	New Faculty Needed		C New PhD's			D Ratio C:B		
		A 3-Year Average	B with PhD	NRC[a]	OE[b]	AMC[c]	NRC	OE	AMC
Chemistry	1965	722	(70 percent of A) 505	1439	1439	1439	2.9	2.9	2.9
	1970	703	492	2033	1938	2030	4.1	3.9	4.1
	1975	825	578	2884	2724	2290	5.0	4.7	4.0
	1980	678	475	4273	3153	2888	9.0	6.6	6.1
	1985	−53	−37	n.a.[d]	n.a.	n.a.			
Physics	1965	583	(60.5 percent of A) 353	994	994	994	2.8	2.8	2.8
	1970	568	341	1596	1569	1617	4.7	4.6	4.7
	1975	666	400	2383	2253	1997	6.0	5.6	5.0
	1980	550	330	3708	2608	2680	11.2	7.9	8.1
	1985	−43	−26	n.a.	n.a.	n.a.			
Biology	1965	1535	(74 percent of A) 1136	1928	1928	1928	1.7	1.7	1.7
	1970	1495	1106	3490	2950	3480	3.1	2.7	3.1
	1975	1753	1297	6270	4200	4395	4.8	3.2	3.3
	1980	1446	1070	10719	5020	5774	10.0	4.5	5.4
	1985	−113	−84	n.a.	n.a.	n.a.			
Mathematics	1965	912	(42 percent of A) 379	688	688	688	1.8	1.8	1.8
	1970	888	369	1238	1350	1209	3.3	3.6	3.2
	1975	1042	433	1756	2350	1495	4.1	5.4	3.5
	1980	859	367	2415	3090	2007	6.6	8.4	5.5
	1985	−67	−28	n.a.	n.a.	n.a.			

[a] National Research Council.
[b] Office of Education—projections to 1978 extrapolated to 1980.
[c] A. M. Cartter.
[d] n.a. = not available.

The NSF [33] considers separately graduate and undergraduate faculty needs based again on population figures plus a correction that relates postdoctoral employment to research and development support. Cartter does not go into such analytical detail. The resulting NSF estimate of academic employment agrees with Cartter's overall estimate in the case of physics, which might be expected since the NSF envisions no major changes in the structure of academic physics.

An important consideration in the projected pattern of academic use is that, if faculty size is determined primarily by teaching needs and there is little growth due to a leveling or an actual decline in graduate enrollment in the sciences, then science faculties will tend to grow old together and lose some of their intellectual vitality. The vigor and accomplishments of academic science in the past 25 years have been due in large measure to the continuous infusion of young blood and the opportunity provided, by expansion, for young scientists to achieve intellectual independence at the most productive and innovative periods of their careers. About 40 percent of physics faculties are under 35 years of age, and three fourths, under 45. In ten years, if we assume no migration from the system and only the replacement of deaths and retirements, only 20 percent of these faculties will be under 35, and 50 percent under 45. The picture is, of course, exaggerated, for some migration from the system will occur; its extent will depend on job opportunities in nonacademic research and other settings. However, it is probable that relatively few openings in universities will develop until the 1990's. Thus the future of physics in particular and science in general can be severely affected.

Nonacademic Doctoral Opportunities Nonacademic doctoral opportunities depend for the most part on federal and nonfederal research and development expenditures. In addition, employment outside academic institutions that is not related to research and development must be estimated.

The NSF projects the total research and development effort as between 2.7 percent and 3 percent of the Gross National Product, which, in turn, is projected by the Bureau of Labor Statistics as growing at an average rate of 4.3 percent per year between 1965 and 1980. [Table 12.23 (parts A through D) summarizes the overall economic projections of the Bureau of Labor Statistics, presented in Reference 35.]

The total employment of scientists and engineers in the NSF projection is obtained by combining the projected total of research and development expenditures with the present value and estimated changes in research and development cost per research and development man-year. The number of PhD's required in the total research and development employ-

TABLE 12.23-A Factors Determining Gross National Product, 1957, 1965, 1968, and Projected 1980

Item	1957	1965	1968	Projected 1980 Basic Models 3% Unemployment	Projected 1980 Basic Models 4% Unemployment	Average Annual Growth Rate 1957–1965	Average Annual Growth Rate 1965–1980 Basic Models 3% Unemployment	Average Annual Growth Rate 1965–1980 Basic Models 4% Unemployment
Total labor force (thousands)	69,729	77,177	82,817	100,727	100,727	1.3	1.8	1.8
Unemployed	2,859	3,366	2,817	2,940	3,918	2.1	–.9	1.0
Employed: jobs concept [a] (thousands)	70,953	77,689	84,688	102,896	101,867	1.1	1.9	1.8
Total private	61,197	65,695	70,274	84,396	83,552	.9	1.7	1.6
Annual man-hours (per job) private	2,085	2,052	2,000	1,977	1,977	–.2	–.2	–.2
Total man-hours (millions) private	127.6	134.8	140.5	166.9	165.2	.7	1.4	1.4
GNP per man-hours private [b] (1958 dollars)	3.22	4.21	4.61	6.54	6.54	3.4	3.0	3.0
Total GNP (1958 dollars)	452.6	617.8	707.6	1,168.6 [c]	1,156.9 [c]	4.0	4.3	4.3
Private GNP (1958 dollars)	410.6	567.0	647.9	1,091.9	1,081.0	4.1	4.5	4.4
Total GNP (1968 dollars)	553.8	754.3	865.7	1,427.8	1,415.7	3.9	4.3	4.3

[a] The estimates of 1980 employment start with an estimate of labor force that is a count of people and is converted to a jobs concept that is a count of jobs.

[b] The GNP per man-hour is private since by national income conventions government productivity is set at zero.

[c] This is GNP as was derived; in all other calculations it is rounded to 1,165 and 1,155.

867

ment is then calculated on the assumption that the PhD concentration in 1980 will be 10 percent to 20 percent above the current concentration. Finally, the relative distribution of these PhD's among the different scien-

TABLE 12.23-B   Gross National Product by Major Component Selected Years and Projected 1980 (1958 Dollars)[a]

Component	1957	1965	1967	1968	Projected 1980 Basic Models 3% Unemployment	4% Unemployment
Gross National Product	452.5	617.8	674.6	707.6	1,165.0	1,155.0
Personal consumption expenditures	288.2	397.7	430.3	452.6	758.3	751.9
Gross private domestic investment	68.8	99.2	100.8	105.7	186.3	184.7
Nonresidential	47.4	66.3	73.6	75.8	130.4	129.3
Residential structures	20.2	23.8	20.3	23.3	40.9	40.5
Net inventory change	1.2	9.0	6.9	6.6	15.1	15.0
Net exports	6.2	6.2	3.6	0.9	9.6	9.5
Government	89.3	114.7	140.0	148.4	210.8	208.9
Federal	51.7	57.9	74.8	78.9	85.0	84.3
State and local	37.6	56.8	65.2	69.5	125.8	124.6

Percent Distribution

Component	1957	1965	1967	1968	3% Unemployment	4% Unemployment
Gross National Product	100.0	100.0	100.0	100.0	100.0	100.0
Personal consumption expenditures	63.7	64.4	63.8	64.0	65.1	65.1
Gross private domestic investment	15.2	16.1	14.9	14.9	16.0	16.0
Nonresidential	10.5	10.7	10.9	10.7	11.2	11.2
Residential structures	4.5	3.9	3.0	3.3	3.5	3.5
Net inventory change	.3	1.5	1.0	.9	1.3	1.3
Net exports	1.4	1.0	.5	.1	.8	.8
Government	19.7	18.6	20.8	21.0	18.1	18.1
Federal	11.4	9.4	11.1	11.2	7.3	7.3
State and local	8.3	9.2	9.7	9.8	10.8	10.8

Gross National Product by Major Component
Selected Periods and Projected 1965–80
(Average annual rate of change) [b]

Component	1957–1965	1965–1968	1965–1980 Basic Models 3% Unemployment	4% Unemployment
Gross National Product	4.0	4.6	4.3	4.3
Personal consumption expenditures	4.1	4.4	4.4	4.3
Gross private domestic investment	4.7	2.1	4.3	4.2
Nonresidential	4.3	4.6	4.6	4.6
Residential structures	2.1	−0.7	3.7	3.6
Change in business inventories	—	—	3.5	3.5
Net exports of goods and services	—	—	3.0	2.9
Government purchases of goods and services	3.2	9.0	4.1	4.1
Federal	1.4	10.0	2.6	2.5
State and local	5.3	7.0	5.4	5.4

[a] Source: Historical data are from the Office of Business Economics, U.S. Department of Commerce. The projections are by the Bureau of Labor Statistics.
[b] Compound interest rates between the terminal year.

tific and engineering disciplines is estimated on the basis of the present distribution, past trends in the distribution, and the 1980 projections of the Bureau of Labor Statistics for college-educated scientific workers.

By this method, the NSF estimates that PhD physical scientists doing nonacademic research and development will increase from 25,500 in 1969 to 36,300 in 1980, an average growth rate of 3.4 percent. Of these 10,800 new nonacademic physical science research and development positions, we expect 22 percent to be in physics and 91 percent of the increase to occur between 1970 and 1980. Thus, we find 2200 new positions for physics PhD's.

Non-research-and-development employment of nonacademic scientists also is estimated by the NSF, which identifies the "directors of scientific laboratories in industry and government" as a specific example of this job category. Although an increase of about 11,000 jobs in research-and-development-related work between 1969 and 1980 is expected for PhD's

TABLE 12.23-C Industries Projected To Grow Most Rapidly in Real Output, 1965–1980

		Basic Models						High Durable Models		
			Rate [a]						Rate [a]	
Rank	Sector Number	Industry	3%	4%		Rank	Sector Number	Industry	3%	4%
1	51	Office, computing, and accounting machines	10.3	10.2		1	51	Office, computing, and accounting machines	10.9	10.8
2	63	Optical, ophthalmic, and photographic equipment	8.8	8.8		2	57	Electronic components and accessories	9.3	9.2
3	57	Electronic components and accessories	8.4	8.4		3	63	Optical, ophthalmic, and photographic equipment	9.0	8.9
4	66	Communications, except broadcasting	7.0	6.9		4	56	Radio, television, and communication equipment	7.0	7.0
5	28	Plastics and synthetic materials	6.8	6.7		5–6	66	Communications, except broadcasting	6.9	6.9
6	68	Electric, gas, water, and sanitary services	6.7	6.6		5–6	52	Service industry machines	6.9	6.8
7	52	Service industry machines	6.5	6.4		7	28	Plastics and synthetic materials	6.8	6.8
8–9	32	Rubber and miscellaneous plastics products	6.3	6.2		8	74	Research and development	6.7	6.7
8–9	73	Business services	6.3	6.2		9	68	Electric, gas, water, and sanitary services	6.6	6.5
10	56	Radio, television, and communication equipment	6.2	6.1		10–11	32	Rubber and miscellaneous plastic products	6.4	6.4
11–12–13	10	Chemical and fertilizer mineral mining	6.0	5.9		10–11	73	Business services	6.4	6.3
11–12–13	29	Drugs, cleaning, and toilet preparations	6.0	5.9		12	62	Scientific and controlling instruments	6.1	6.1
11–12–13	74	Research and development	6.0	5.9		13	10	Chemical and fertilizer mineral mining	6.0	5.9

[a] Average annual rate of change in compound interest between terminal years. Output is the gross duplicated value stated in 1958 prices.

in physical sciences, the NSF predicts a greater increase in nonacademic, non-research-and-development-related jobs (10,000 to 15,000) that specifically demand a PhD in physical sciences. In a society becoming ever more problem-oriented rather than scientific discipline-oriented, the Committee views this projection with some doubt.

The relative level of non-research-and-development-related activity in the physics community has been nearly constant during the past six years, specifically 18.6 percent of the nonacademic physicists in 1970. This percentage is included in the Committee's projections. The NSF projection in excess of this steady fraction we regard as outside the present pattern of physics activity.

The Bureau of Labor Statistics has estimated PhD employment in industry by a different method. Rather than linking scientific employment explicitly to a projected level of research and development expenditure, this method estimates the number of scientific workers required to produce the goods and services implied by the Bureau's basic model for the 1980 economy. The current pattern of use of physical scientists in various industries is assumed to be still valid in 1980. However, the growth rates of different industries are estimated separately. The projected annual growth in the employment of physical sciences PhD's is 3.25 percent, a change from 19,500 in 1968 to 29,500 in 1980, including a constant 20 percent non-research-and-development employment factor. Of these 10,000 new jobs, 19.4 percent would be in physics, that being the fraction of physicists among 1968 physical sciences PhD's in industry. Between 1968 and 1970, industry employed about 600 additional PhD physicists; therefore, the growth between 1970 and the target year, 1980, would amount to approximately 1300 jobs.

It is difficult to compare the NSF and Bureau of Labor Statistics estimates, as the NSF estimates total nonacademic research-and-development-related employment—industry, government, and research centers—and the Bureau gives only an estimate of the industrial component. Table 12.24 presents the NSF and Bureau of Labor Statistics projections of new employment opportunities for physics PhD's. Both estimates employ the Bureau of Labor Statistics model of the 1980 economy. The NSF uses the model to define the total level of research and development, and the Bureau follows certain components of the model as a means of developing a picture of industry that includes the demand for PhD physical scientists. The Bureau of Labor Statistics expects stronger economic growth in the industrial component of thet Gross National Product than in the governmental component. Currently, half of the nonacademic physicists work in industry and half in government laboratories and federally funded research and development centers. The growth in employment of physicists in

TABLE 12.23-D Civilian Employment [a] by Major Industry Group, 3 Percent Models, 1965, 1968, and Projected 1980

| Industry Group | 1965 | 1968 | Projected 1980 | | Percent Distribution | | | | | Average Annual Rate of Change [b] | |
| | | | | | | | | 1980 | | 1965–1980 | |
			3% Basic	3% High Durables	1965	1968	3% Basic	3% High Durables			3% Basic	3% High Durables
Total	74,568	80,788	99,600	99,400	100.0	100.0	100.0	100.0	1.9	1.9
Agriculture, forestry and fisheries	4,671	4,154	3,188	3,192	6.2	5.1	3.2	3.2	−2.5	−2.5
Agriculture	4,338	3,811	2,800	2,800	5.8	4.7	2.8	2.8	−2.9	−2.9
Mining	667	646	590	588	.9	.8	.6	.6	−.8	−.8
Construction	3,994	4,050	5,482	5,595	5.4	5.0	5.5	5.6	2.1	2.3
Manufacturing	18,454	20,125	22,358	23,240	24.7	24.8	22.4	23.4	1.3	1.5
Durable goods	10,644	11,854	13,274	14,322	14.3	14.6	13.3	14.4	1.5	2.0
Ordnance and accessories	226	342	250	351	.3	.4	.3	.4	.7	3.0
Lumber and wood products	698	676	685	702	.9	.9	.7	.7	−.1	—
Furniture and fixtures	454	496	640	656	.6	.6	.6	.7	2.3	2.5
Stone, clay, and glass products	646	651	809	830	.9	.8	.8	.8	1.5	1.7
Primary metals	1,308	1,322	1,343	1,413	1.8	1.6	1.3	1.4	.1	.5
Fabricated metal products	1,288	1,417	1,638	1,697	1.8	1.7	1.6	1.7	1.6	1.9
Machinery, except electrical	1,783	2,009	2,495	2,670	2.4	2.5	2.5	2.7	2.3	2.7

Electrical machinery	1,662	1,986	2,334	2,554	1.7	2.4	2.3	2.6	2.3	2.9
Transportation equipment	1,745	2,034	2,014	2,343	2.3	2.5	2.0	2.4	1.0	2.0
Instruments	392	463	553	594	.5	.6	.6	.6	2.3	2.8
Miscellaneous manufacturing	442	458	513	512	.6	.6	.5	.5	1.0	1.0
Nondurable goods	7,810	8,271	9,084	8,918	10.5	10.2	9.1	9.0	1.0	.9
Food and kindred products	1,798	1,811	1,799	1,735	2.4	2.2	1.8	1.7	.0	−.2
Tobacco manufactures	87	84	65	63	.1	.1	.1	.1	−2.0	−2.2
Textiles and apparel	2,311	2,426	2,655	2,590	3.1	3.0	2.7	2.6	.9	.8
Paper and allied products	640	693	801	795	.9	1.1	.8	.8	1.5	1.5
Printing and publishing	1,057	1,128	1,322	1,307	1.4	1.4	1.3	1.3	1.5	1.4
Chemical and chemical products	905	1,024	1,187	1,172	1.2	1.3	1.2	1.2	1.8	1.7
Petroleum and products	183	187	155	152	.2	.2	.2	.2	−1.1	−1.2
Rubber and plastic products	474	560	763	777	.6	.7	.8	.8	3.2	3.3
Leather and leather products	355	358	337	327	.5	.4	.3	.3	−.4	−.5
Transportation, communications, and public utilities	4,250	4,524	4,976	4,961	5.7	5.6	5.0	5.0	1.1	1.0
Wholesale and retail trade	15,352	16,604	20,487	20,501	20.6	20.5	20.6	20.6	1.9	1.9
Finance, insurance, and real estate	3,367	3,726	4,639	4,538	4.5	4.6	4.7	4.6	2.1	2.0
Services	11,118	12,678	18,280	17,785	14.9	15.5	18.4	17.9	3.4	3.2
Government	10,090	11,846	16,800	16,200	13.5	15.0	16.9	16.3	3.5	3.2
Households	2,604	2,435	2,800	2,800	3.5	3.0	2.8	2.8	.5	.5

[a] Civilian employment includes wage and salary employees, self-employed, and unpaid family workers.

[b] Compound interest rates based on terminal years.

TABLE 12.24 Projections of the Number of New Nonacademic Employment Opportunities That Will Develop for PhD Physicists between 1970 and 1980

Nonacademic Employment	NSF Projection	BLS Projection
Research-and-development-related (total)	2600	[a]
Industry	[b]	1300
Government and other	[b]	[a]
Non-research-and-development-related	1900	[c]

[a] Not indicated.

[b] Not separately reported but included in total.

[c] Twenty percent of the 29,500 new industrial employment opportunities predicted by the BLS for physical sciences PhD's are non-research-and-development-related; however, the fraction of these jobs that would be filled specifically by physicists is not indicated.

government since 1964 has been approximately the same as that characterizing industry; continuation of this pattern of parallel growth would result in a 3.25 percent growth or about 600 new jobs in government in 1980. According to the Bureau's model, even an assumed growth parallel to that of industry is optimistic.

The Bureau of Labor Statistics explicitly projects a constant 20 percent non-research-and-development-related employment for PhD's in industry. The additional employment for PhD's that the NSF projects is not envisioned by the Bureau.

College-Educated Workers in Physics

The Bureau of Labor Statistics has estimated the physics component of the college-educated workers on the basis of its basic economic model.[38] The physics labor force used in the Bureau's estimates is much larger and includes less highly qualified people than does the NSF National Register population. On the basis of the Bureau's 1980 model, the physics labor force will grow at a rate of 4.2 percent, changing from 45,000 in 1968 to approximately 75,000 in 1980. The ill-defined character of this occupation group makes these projections essentially irrelevant to the projections of PhD use and only vaguely related to specific physics education. The Bureau's projection of short supply in 1980 rests on estimates of large interfield migration, a further indication of the amorphous character of the occupation group on which it is based. Only half of the physics bachelor's degree group is included in this occupation group. Consequently, about the only conclusions relevant to the Physics Survey that can be drawn from the Bureau's study [38] are (a) that technically trained workers with the equivalent of a bachelor's degree in physics will continue to be in demand, though to a lesser extent than was previously the case; and (b) that occupational opportunities for physics baccalaureates are substantially broader than is the physics occupation group. Table 12.25 depicts employment trends and illustrates the nature of the Bureau's basic model in terms of employment demand for college-educated workers.

Yearly Demand for Physics PhD's, 1970–1980

When expectations for the eventual level of employment of physics doctorates in 1980 are converted into the yearly demand for new PhD's, a clearer perception of economic trends for the decade results. Prospects for the near future are not encouraging, as the following data indicate:

1. Faculty employment is lagging enrollment growth; in 1970 only 182 new faculty jobs were established to deal with a growth in enrollment that normally corresponded to 550 new faculty members. In addition, 21 leading PhD-producing schools [39] had the same number of physics faculty members at the end of fiscal year 1972 as they did at the beginning of fiscal year 1968—four years of zero growth (according to AIP data).
2. The Physics Survey Industrial Questionnaire revealed plans for only an average yearly growth of 1 percent for the next five years.
3. The employment of physicists in the national laboratories has not increased since 1964.

TABLE 12.25 Employment Trends for the College-Educated Work Force [a]

College-Educated Occupation Group	Employment Growth Rate per Year	
	Actual 1960–1968 (%)	BLS Projection 1968–1980 (%)
Physicist	5.9	4.2
Engineer	3.7	2.9
Chemist	3.5	3.8
Life scientist	7.0	2.9
Geologist and geophysicist	6.1	1.6
Mathematician	9.4	4.0
Elementary and secondary schoolteacher	3.5	~0.5

[a] Source: Reference 38.

4. Increases in the employment of physicists by the federal government are not likely in view of the restricted federal budget and the increasing support of other research and development programs than those of defense, space, and atomic energy, which taken together formed 86 percent of the support base for physicists in government employment.

In contrast to this picture of level or lagging employment is the finding of the National Register that new PhD's in physics found jobs between 1968 and 1970. The National Register shows increases of employed PhD's, as Table 12.26 indicates.

These data illustrate an important aspect of physics PhD employment that is characteristic of any kind of employment. That aspect is the considerable elasticity in both supply and demand.

The degree to which oversupply and shortage of manpower in any given field depends on marginal changes in the salary structure is frequently overlooked. Although it is clear that no trained individual will subject himself to the upheaval of moving and job changing *only* on the basis of small incremental changes in salary, it should be emphasized that such small changes *can* be important for those on the margin who need only a slight stimulus to precipitate a change largely brought about by other causes. But the number of such persons is not negligible in any field.

Economists have evolved a rather specific meaning for the terms, "oversupply" and "shortage," that is illustrated [40] by Figure 12.23, which we take as illustrating hypothetical total demand and supply curves for physicists. The supply curve traces the number of physicists available to work in their profession at different salaries. The demand curve traces the total demand, in numbers, for physicists at different salaries. At a salary *P,* in this figure, the market demand is satisfied by and equal to

TABLE 12.26   Employment of New Physics PhD's, 1968–1970

Employment	No. of PhD's		Percentage Increase in Two Years	
University	1195		17	
Faculty		480		8
Nonfaculty		715		23
Industry	610		19	
Government	161		12	
Research centers	30		1	
Other and no response	325		36	
Increased respondents	2312			

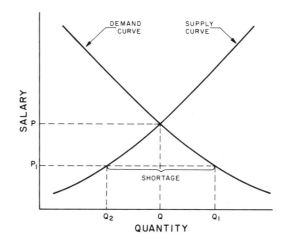

FIGURE 12.23 Hypothetical total demand and supply curves for physicists.

the supply offered. If all employers hire as many physicists at this salary as they wish, they will hire in total the number Q. There will also be Q individuals who wish to work as physicists at the salary P. There will be an equilibrium between supply and demand at this price and numbers of physicists.

Now, suppose for some reason, such as economic recession, particularly affecting employers of physicists, the salary for physicists is forced to remain at a lower value P_1 instead of P. At this lower salary a larger number of physicists, Q_1, will be demanded, but only a smaller number, Q_2, will be interested in working at this lower price. The difference in numbers, Q_1 minus Q_2, is defined as a shortage. It is a situation in which more of a service or commodity is demanded at *the going market price* than is being supplied at that price. An equivalent definition follows for oversupply. It follows that if the physics profession were a typical one, the present apparent oversupply might be ameliorated by a decrease in the average salary expectation of physicists. Indeed, it must be noted that in the past three decades the rate of increase of average salaries for physicists has been one of the highest in the sciences. We believe that some downward adjustment in the average relative salary of physicists is inevitable.

In regard to supply, the number of persons available for different careers after having completed a given level of education is very elastic, particularly those on the borderline between careers who, all else being the same, respond to economic stimuli.[11]

In regard to demand, the number of job openings with given titles, such as physicist or chemist, teacher or counselor, will be influenced by the quality of persons available, the ability to upgrade the job, the willingness to adjust a faculty–student ratio, the availability and acceptability of hiring members of minority groups, and the availability of funds in specific areas such as junior college teaching or oceanographic research.

At the same time, however, it must be noted that, perhaps not surprisingly, we do not believe that physicists represent a typical group in this sense. Among the reasons for this belief are the historical flexibility of physicists in adapting to and making significant contributions in a wide range of fields and the relatively long training period that acts to damp the adjustments discussed above. Cartter [36] has stated this in the following terms in his now classic study of the employment problem for college trained personnel:

If our universities turn out a sufficient number of Ph.D.'s to supply an N% annual expansion in non-academic employment, then there will indeed be an N% growth.

And the Bureau of Labor Statistics [38] has stated:

Supply and demand are not discussed in the usual sense in which wages play a major role in equating supply and demand. The long training period prohibits the immediate adjustments normally associated with the terms supply and demand.

Thus the increase in employed PhD's matches the known production. But what is the nature of this increased employment when the major indications listed previously suggest an essentially static enterprise?

First, according to the Physics Survey Data Panel, the major growth in university employment has been in nonfaculty positions, a manifestation of the much discussed holding pattern that has developed in response to the job shortage. This holding pattern relates to postdoctoral education, which is becoming a central feature of an academic physicist's career. Possibly, in the face of restricted funds and opportunities, the holding pattern does not exploit the breadth of career development opportunities that was formerly an aspect of highly successful postdoctoral education.[24] Rather, it becomes a kind of staying around after the PhD, with the new PhD often turning more or less the same handles as before, his professional growth arrested, like his income, at something closer to the predoctoral level and his expectations of future employment growing dimmer.

The second way in which the employment of PhD physicists has increased while the physics enterprise lagged is apparent in the changing pattern of work activity in industry and government that Figures 12.24 and 12.25 depict. The increase in employment since 1964 has been in development rather than in basic research. The basic research effort has been declining in absolute size. As a result, increased employment in government or industry signifies a changing role for the physicist. Consistent with this picture is the pattern in the national laboratories; there,

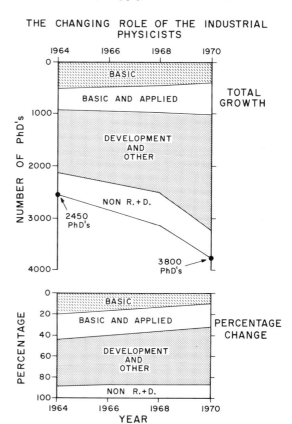

FIGURE 12.24 The changing role of industrial physicists. (Note that the employment of physicists has grown while the emphasis on physics has not, as indicated by the decline of basic research involvement.)

no change in work roles occurred, and there was no change in rate of employment of physicists.

The impact of the current lag in the physics enterprise on the overall employment expectations in this decade is uncertain. Estimates of the duration of the lag and the vigor of the anticipated subsequent expansion are beyond this Committee's expertise. However, we can project the potential employment prospects that would result from various assumptions about the future. We consider specifically the consequences of three possibilities:

1. An indefinite lag;

2. Restoration of the target growth rate (4.3 percent of the Gross National Product), with the rate increasing gradually over the next five years and with target levels deferred beyond 1980;

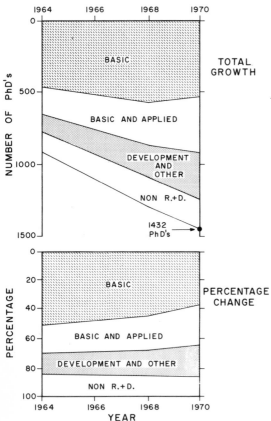

FIGURE 12.25   The changing role of physicists in government laboratories.

3. Rebound of employment opportunities to achieve the actual levels and growth rate of the target year by 1980.

The first possibility is not likely, but it represents a constant base line for the other two growth schemes. The second possibility is one in which the growth rate undergoes a steady increase over the next five years until it reaches the value that is predicted by the basic model for the target year. The overall target level would not be reached until after 1980. The third possibility is considered a reasonable one by the Bureau of Labor Statistics. Their basic model essentially predicts a level of activity in the target year that is not directly dependent on growth patterns. Table 12.27 shows the annual number of openings that we would predict for these three patterns of growth.

TABLE 12.27 Projected Annual Employment Opportunities for PhD Physicists

Year	Continued Lag	Restored Growth	Rebound to Target Level
1971	300	300	300
1972	300	400	500
1973	300	500	700
1974	300	600	900
1975	300	700	1100
Subtotal 1975	1500	2500	3200
1976	300	800	1300
1977	300	900	1300
1978	300	900	1300
1979	300	900	1000
1980	300	900	900
TOTAL 1980	3000	6900	9200
Replacement	2000	2000	2000
New jobs	1000	4900	7200

The current lag in the use of PhD's will either reduce overall employment or set off an oscillatory variation in yearly demand that will be impossible to match in yearly production.

On the other hand, as we have noted above in discussing Figure 12.23, most projections of supply and demand for scientists and engineers do not properly take into account changes in salary in response to supply and changes in demand in response to salary levels. By analyzing historical data, Freeman [41] has shown that the price elasticity of demand is about 0.4, that is, a 1 percent reduction in salary causes an increase in demand of 0.4 percent. Thus one could estimate that a 12 percent reduction in salary level relative to the general level of wages and salaries would be sufficient to eliminate present unemployment of scientists and engineers. Such an adjustment would take place in four years if overall wages continued to rise at 3 percent per year and science and engineering salaries remained level.

Freeman also shows that the long-term elasticity is even greater, about 0.7, largely because salary levels also affect supply after a longer time delay. On the basis of an econometric model, Freeman predicts that both academic and nonacademic demand will be greater by 6 percent in 1980 than estimates by the Bureau of Labor Statistics based on the assumed vigorous growth with full employment. The supply situation caused by relative salary decline compared to Cartter's projections, using physics PhD's as an example, appears in Table 12.28.

Cartter would predict a net increase of the physics PhD manpower pool of 17,700 by 1980, whereas Freeman's econometric model would give an increase of only 6600. The Bureau of Labor Statistics projections, as interpreted by Grodzins, indicate 7200 new job opportunities for physics PhD's by 1980; Freeman's model would yield 7600 jobs.

Although Freeman suggests that the econometric model applies to universities as well as to industry and government, there is some doubt about its validity when so applied. It is possible that a shift away from research in universities to research in other institutional settings may be under way. The decline in growth of faculty, pressure from state legislatures and private boards of trustees to increase productivity in higher education, which means a higher student–faculty ratio, and a possible change in federal policy toward procurement of research from nonacademic institutions better adapted to mission-oriented federal needs—all these as well as a number of academic policy considerations suggest imminent changes and make the development of meaningful projections difficult.

Production of Physics PhD's

The current production of 1500 PhD's in physics annually is at least 50 percent higher than the patterns of use in traditional jobs occupied by physicists and predicted under the best steady growth conditions.

Students already working toward a PhD degree in physics constitute a sizable number. Grodzins has tabulated enrollment data and predicted PhD production on the basis of present experience with the performance of the educational pipeline. Figure 12.26 presents the basic data and future trends.[20] If the number of students earning PhD's is assumed to be about 50 percent of the first-year graduate students, as it has been in the past, then current first-year enrollment data can be used to predict produc-

TABLE 12.28 Comparison of Cartter and Freeman Projections of the Production of PhD's in Physics

| Year | Production of Physics PhD's | |
	Cartter	Freeman
1970	1620	1600
1975	1997	1110
1980	2680	790

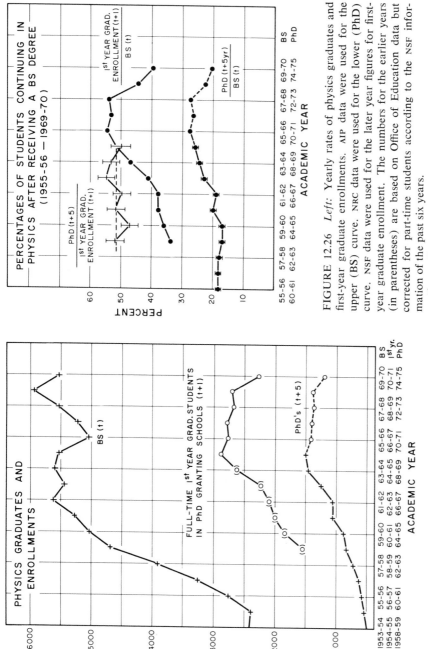

FIGURE 12.26 *Left:* Yearly rates of physics graduates and first-year graduate enrollments. AIP data were used for the upper (BS) curve. NRC data were used for the lower (PhD) curve. NSF data were used for the later year figures for first-year graduate enrollment. The numbers for the earlier years (in parentheses) are based on Office of Education data but corrected for part-time students according to the NSF information of the past six years.

Above: Percentages of students continuing in physics following the BS degree.

[Source: L. Grodzins, American Physical Society, April 1971.]

TABLE 12.29   Estimates of the Production of PhD's in Physics

Fiscal Year	Standard Conditions	Enhanced Attrition
1971	1400	1400
1972	1400	1400
1973	1350	1200
1974	1350	900
1975	1200	600
TOTAL	6700	5500

tion five or more years in the future. This type of prediction suggests an upper limit on future PhD production, as the shortage of jobs can be expected to enhance attrition. Table 12.29 shows estimates of the number of new PhD's under standard conditions and under conditions of enhanced attrition. The lower production alternative in the last column of this table assumes an attrition 50 percent higher than the normal rate for first-year graduates; students closer to receiving their degrees are correspondingly more likely to finish their courses.

Supply versus Use

There is little doubt that a sizable overproduction of physics PhD's will have occurred by 1975 (see Table 12.30), if one assumes employment of physicists only in traditional jobs and ignores the possible effects of a relative decline in science and engineering salaries as compared to overall wages. Beyond 1975, estimates of enrollment are highly uncertain. If the pattern of use projected by the Bureau of Labor Statistics for a robust 1980 economy becomes a reality, approximately 900 PhD's per year would be needed. Although the present production is 1500 annually, downward trends in first-year enrollments in present programs suggest that by 1975 overproduction of PhD's may no longer be a problem. If the economy should achieve the levels of the Bureau of Labor Statistics model, a

Table 12.30   Production and Use of Physics PhD's, 1970–1975

Production		Use	Surplus
High	6700	3200	4200
Low	5500	2500	2300

transient high demand could follow the current lag in employment opportunities. However, it is difficult to estimate the degree to which this trend will be compensated in the post-1975 years by the continued development of new PhD programs. The demand must be evaluated carefully to avoid another cycle of overly rapid growth of graduate education and another large oscillation in the use of science manpower. It is highly probable that the annual demand for physics PhD's in a postindustrial, post-baby boom, post-Vietnam U.S. economy of the 1980's will not resemble the seemingly insatiable demands that characterized the 1950's and early 1960's.

Again, we emphasize that the projections made here for future employment of physicists are based on the present picture of the physicist in U.S. society working in his traditional job pattern—that is, discipline-oriented and heavily committed to academic pursuits and fundamental research. By 1975, perhaps 20 percent of the new physics PhD's will have had to find employment in jobs that currently would be regarded as outside physics. If the physics community can contain these PhD's as a vital part of itself, this employment experience could lead to new definitions of the scope of physics. And if the physics community achieves the broad outlook and flexibility necessary to take an active and effective role in a new partnership of science and society, it can flourish once more as it did during the past two decades.

CONCLUSIONS AND RECOMMENDATIONS

We believe that the long-term health of physics requires a reversal of the trend of the past decade during which a large fraction of new physics PhD's remained in academic institutions. The development of a wider range of employment opportunities for physics PhD's will require substantial changes in the nature of graduate education. Thus, while we reaffirm our belief in the value of graduate education in physics, we recognize that contemporary graduate education contains certain faults, and we recommend specific steps that will increase its value and the attractiveness of the physics PhD as a candidate for nonacademic employment.

This Committee views physics as an essential component of the nation's scientific effort, providing a deep understanding of nature economically reduced to a few broad principles. Consequently, we recommend that physics departments continue to accept the responsibility for providing the best possible graduate education to those who want it and can profit from it. At the same time, we recognize that physics PhD's will have greater difficulties in realizing their aspirations than they have in the past, particularly if they anticipate academic careers. Therefore, we also recommend that physics departments provide information and guidance that ensures that their graduate students are making a well-considered career choice when they undertake graduate study initially. Physics departments also should offer them an education that provides maximum flexibility in adapting to career options. Increased employment opportunities for physics PhD's will have to be developed in nonacademic sectors; thus the nature of graduate education must change. In addition, physics departments are subject to social and economic forces and fluctuations over which they have no control, and they lack the capacity to make an effective short-term response. Therefore, they should be cautious about inaugurating new programs, encouraging the expansion of federal fellowship programs, and increasing the number of tenured faculty. (See also Chapters 2, 9, and 14.)

The federal system of support for science has demonstrated its effectiveness and is to be highly commended. In particular, this Committee favors the project grant system—in which support for research is tied to the scientific excellence of the proposed research—and continued diversity of sources of support. However, we also recognize that in certain circumstances federal support can have some unfortunate results. One is the possible lack of continuity in growth. The nature of physics and of the institutions on which it depends implies a slow response to changes in support. Once physics has adjusted to rapid growth of support, it finds it particularly difficult to adapt rapidly to a different rate of growth; physics is sensitive to the second derivative of the support curve. This situation should be recognized not only by those in government who are responsible for ensuring the wise expenditure of public funds but also by physicists, who must continue to take an active part in determining the future of physics. A second result of the availability of substantial funds under government grants and contracts has been the organization of graduate education into large projects in which a professor directs a group of students, and possibly research associates and postdoctoral fellows as well, in an activity centered around a particular laboratory or facility. Unless great care is exercised to avoid it, the student, rather than having an experience in independent research, can find himself occupying a niche in this organization and may be somewhat removed from the broader life of his department and the university. Thus his training becomes narrow. We believe that greater emphasis should be placed on independent graduate work on small problems of lesser scope, with a student perhaps undertaking more than one such problem, rather than on large team explorations of a subject. The universities should take the initiative in bringing about appropriate modification in the present team-research pattern of graduate work.

We further recommend that physicists in universities devote greater effort to the development of relationships with nonacademic organizations. Strong ties between the universities and industrial and governmental organizations will be increasingly necessary in the future. Universities will depend on industry and government to supply the employment opportunities for their graduates; industry and government will continue to depend on the universities for trained people.

Although in the recommendations that follow we frequently refer only to university–industry relations, the problems involved are shared by government in-house laboratories, nonprofit research institutes, and the national laboratories. All these types of organization should develop such interaction with the university-based physics community for like reasons.

Several general aspects of the university–industry interaction should be recognized. For example, we are not unaware of the difficulties that questions of proprietary information and patent ownership introduce into relations between institutions; however, we are convinced by the many successful examples of the specific kinds of interaction that we recommend that the problems involved can be resolved when the will to do so exists. Further, since the benefits of such interaction accrue to all who are involved —the universities, the industries, and the individual scientists—all should share in the cost of such exchanges. Most examples of satisfactory interchange programs between industry and universities involve research scientists. However, most of the scientists in industry are not in research programs. The maximum beneficial effects of interchange programs will accrue only by extension of university programs to developmental as well as research organizations. In addition, since a substantial amount of new science is generated in Europe, European universities should be included in programs of industry–university interaction. This suggestion applies especially to those industries that have European subsidiaries or close links with European firms.

We recommend that universities recognize the benefit of industrial experience to physicists and that they encourage faculty members to take temporary positions in industrial laboratories.

We further recommend that universities encourage a fraction of their graduate students to perform their thesis research in industrial laboratories. This practice will make available to these students equipment that is not typically found in universities and that could and should be used profitably for educational purposes. In addition, the identification of suitable problems for such student projects will form a useful point of contact between faculty members and industrial scientists.

We also recommend that industries develop strong sabbatical programs allowing their scientists to serve temporarily on university faculties. This step will afford opportunities to industrial scientists to become involved with new people, new disciplines and specialties, and new work environments.

To foster the development of closer relations between industry and the universities, we further recommend that a portion of government funds allocated for the support of research be used to encourage university–industry interaction. In supporting research proposals, the joint participation of universities and industry should be sought, encouraged, and regarded as a positive feature in proposal evaluation. Programs of interchange of personnel between universities and industries also should be fostered by partial government support.

REFERENCES

1. *Statistical Abstracts of the United States* (annual).
2. *Employment of Scientists and Engineers in the United States 1950–66,* NSF 68-30 (National Science Foundation, Washington, D.C., 1968).
3. J. K. Folger, H. S. Astin, and A. E. Bayer, *Human Resources and Higher Education* (Russell Sage Foundation, New York, 1970).
4. *American Science Manpower 1968,* NSF 69-38 (National Science Foundation, Washington, D.C., 1969).
5. *Government and Science,* Hearings before the Subcommittee on Science Research and Development of the Committee on Science and Astronautics, U.S. House of Representatives, 88th Congress (U.S. Government Printing Office, Washington, D.C., 1964).
6. *National Patterns of R and D Resources 1953-71,* NSF 70-46 (National Science Foundation, Washington, D.C., 1970).
7. H. W. Koch, "On Physics and the Employment of Physicists in 1970," Phys. Today *24,* 23 (June 1971).
8. A. M. Cartter, "The Supply and Demand of College Teachers," in *American Statistical Association Social Statistics Proceedings* (American Statistical Association, Washington, D.C., 1965).
9. W. Gruner, Phys. Today *23,* 21 (June 1970).
10. *Wall Street Journal,* January 17, 1972.
11. U.S. Department of Labor, Bureau of Labor Statistics Press, USDL-7310, January 7, 1972.
12. *Physics Manpower 1969,* AIP Rep. R-220 (American Institute of Physics, New York, 1969).
13. W. R. Snelling and R. Boruch, J. Chem. Educ. *47,* 326 (1970).
14. D. J. Watley and R. C. Nichols, "Career Decisions of Talented Youth," National Merit Scholarship Research Rep. *5*(1) (1969) and Eng. Educ., p. 975 (Apr. 1969).
15. U.S. Dept. of Health, Education, and Welfare, *Digest of Educational Statistics 1969* (U.S. Government Printing Office, Washington, D.C., 1969).
16. *Historical Statistics of the United States* (U.S. Government Printing Office, Washington, D.C., 1960).
17. M. W. White, Phys. Today *9,* 32 (Jan. 1956).

18. *Physics Manpower 1966* (American Institute of Physics, New York, 1966).
19. F. E. Terman, Science *173*, 399 (1971).
20. L. Grodzins, Bull. Am. Phys. Soc. *16*, 737 (1971).
21. J. L. Chase, *Graduate Teaching Assistants in American Universities,* Rep. OE-58039 (U.S. Government Printing Office, Washington, D.C., 1970).
22. E. R. Gilliland, *Meeting Manpower Needs in Science and Technology.* A report prepared for the President's Science Advisory Committee (U.S. Government Printing Office, Washington, D.C., 1962).
23. Science Resource Studies, *Highlights,* NSF 71-14 (National Science Foundation, Washington, D.C., 1971).
24. *The Invisible University: Postdoctoral Education in the United States* (National Academy of Sciences, Washington, D.C., 1969).
25. Reference 24, pp. 62–64.
26. Reference 24, p. 68.
27. Reference 24, pp. 64–67.
28. Reference 24, p. 78.
29. Reference 24, pp. 170–173.
30. V. Bush, *Science: The Endless Frontier* (U.S. Government Printing Office, Washington, D.C., 1945); reprinted in *The Politics of Science,* W. R. Nelson, ed. (Oxford U.P., New York, 1968).
31. S. D. Ellis, *Work Complex Study,* Pub. No. R-224 (American Institute of Physics, New York, 1969).
32. S. D. Ellis, invited talk, American Physical Society, Washington Meeting, April 1970.
33. *1969 and 1980 Science and Engineering Doctorate Supply and Utilization,* NSF 71-20 (National Science Foundation, Washington, D.C., 1971).
34. Projections of Educational Statistics to 1979–80.
35. BLS Bulletin 1672 (U.S. Department of Labor, Washington, D.C., 1972).
36. A. M. Cartter, Science *172*, 132 (1971).
37. *Ph.D. Scientists and Engineers in Private Industry 1968–80* (U.S. Department of Labor, Washington, D.C., 1970).
38. *College Educated Workers, 1968–80,* BLS Bulletin 1676 (U.S. Department of Labor, Washington, D.C., 1970).
39. A. M. Cartter, *An Assessment of Quality in Graduate Education* (American Council on Education, Washington, D.C., 1966).
40. J. C. De Haven, *The Relation of Salary to the Supply of Scientists and Engineers,* a Rand Corporation Study, p. 1372-RC (June 1958).
41. R. B. Freeman, *The Market for College-Trained Manpower: A Study in the Economics of Career Choice* (Harvard U.P., Cambridge, Mass., 1971).

13

DISSEMINATION
AND USE
OF THE
INFORMATION
OF PHYSICS

Knowledge is of two kinds. We know a subject
ourselves or we know where to find information
upon it.
    SAMUEL JOHNSON (1709–1784)
    Boswell's *Life of Johnson*, 18 April 1775

THE SCOPE AND IMPORTANCE OF INFORMATION ACTIVITIES

Not only the progress of science but even its very existence depends on the
communication of ideas and knowledge from one person to another. For,
as Ziman has made abundantly clear in his recent book,[1] consensus, without
which science cannot go forward, is made possible only by communication
and discussion in the public record. As he points out, "Objectivity and
logical rationality, the supreme characteristics of the Scientific Attitude, are
meaningless for the isolated individual; they imply a strong social context,
and the sharing of experience and opinion." It is even more obvious, of
course, that the recording and communication of information are essential
for both education and the application of science to practical ends. Thus it
is appropriate that we should devote a chapter of this report to communication in physics.

The mechanisms and patterns of communication are extremely diverse,
and, as many of them have not been fully described before, the Committee
has sponsored a more extensive report on them, of which this chapter is a

890

condensation; we shall refer to this more detailed version from time to time as the Special Panel Report. To facilitate cross-referencing by the reader with a particular interest in any of the many topics to be covered, we have used the same order for sections and subsections in this chapter as for the corresponding ones in the Special Panel Report; figure and table numbers, of course, are different, as there are more of them in the Special Report.

Scale of Values

Although we hope that the reader will be intellectually intrigued, as we have been, by many of the facts and relationships that will emerge in our study of communication, the major reason for undertaking it is a practical one: the value that the effective management of communication activities has to the progress of science and through it to the nation. So it is appropriate that we should start by asking if it is possible to get at least an approximate quantitative measure of the value to society of communication of scientific information or, better, of improvements in this communication. There seem to be two principal sources of information from which such value can be inferred; other possible, but less practical, sources are mentioned in the Special Panel Report.

Decisions of Buyers   Decisions of buyers represent the voice of the marketplace—the amount that individuals, libraries, and sublibraries are willing to pay for journals, books, and secondary services. Of course, to infer what they are willing to pay (always more than what they do pay) requires extensive economic statistics and analysis; also, one must correct for externalities—benefits that society receives but that the buyer does not. Still, a plausible calibration can be made in some cases from this type of information.

Time Spent by Users   Though its significance has not been as widely appreciated as that of free-market prices, the time that users of information services spend in the use of them can be equally valuable for purposes of dollar calibration. Each individual user is perpetually making judgments that balance the value he receives from use of an information service against the value of what he might be doing in the same amount of time devoted to one of his other activities. These judgments are thus a source of information about the value that a large number of knowledgeable individuals place on a service. Here again, the raw input data need to be processed and corrected if we are to extract from them a measure of actual social value.

    Both methods rely on a statistical average of judgments by many individuals. The assumption that has to be made in each case is that these judgments, interpreted as self-interest judgments, are on the average sound. This assumption may not be true, but it usually would be very risky to substitute any other judgments for those of the people who are actually using the information.

Not many applications of either of these approaches to value have been made, of course. Both have, however, been used in a recent study [2] of the economics of primary journals. Both approaches led to the conclusion that, if journal publication of a given small fraction of research were eliminated, the loss to society would have a value of the order of five to ten times the composition and printing costs saved. (As we shall see in the following section, this figure is somewhat reduced when library maintenance costs are included, but it still remains sizable.) We shall make an application of the second (user-time) approach to the full spectrum of communication activities in physics in the discussion of the potential value of improvements in communication services in the following section. Our conclusion there is that any measures that would enable physicists of the United States to communicate as effectively as they do now in λ percent less total time would be worth an amount of the order of ($300 million to $600 million) $(\lambda/100)$ per year.

Though admittedly crude, these are impressive figures and make it clear that all parts of the communication picture deserve serious attention and, where appropriate, support. Another important figure, and one that has not been widely known, is our total national investment of resources in communication activities in physics. Not counting the time of the communicators, or the facilities for the more informal types of communication, we have tried in a subsequent section (Organizations and Resources for the Communication of Information) to make a crude estimate of the total of such resources appearing in identifiable places on budgets. Our estimate comes to about $60 million per year.

We hope that these brief introductory remarks will have given the reader some awareness of the practical importance of efficiency in communication and will have attuned him to keep considerations of value constantly in mind in the reading of the following sections.

Classification of the Subject

We have already alluded to the diversity of communication activities. Because of this diversity, it has not been at all easy to decide how to divide the subject matter into bite-sized pieces. There are at least three orthogonal ways of classifying the subject:

1. We may ask, who is communicating with whom? Who is the giver of information, who is the receiver? Either role may be played by research physicists, by workers in other disciplines, by people here or overseas, and others.

2. We may ask, what kind of information is communicated? There are isolated developments on the research front, there is consolidated knowledge, there is knowledge about where to find other knowledge.

2a. We can ask what function the information is intended to serve. Is it for attention directing or for study and assimilation?

3. We may ask through what medium the communication takes place. There are books, journals, conversations, and the like. Figure 13.1 shows how the important areas of the communication field are distributed over the two dimensions of classifications 2 and 3 (that is, kind of information and medium of communication). The closed contours show where the ground to be covered in each of the third through the seventh sections lies in terms of these coordinates. The eighth section, on communication with other disciplines and the public, which does not fit into the picture of Figure 13.1, is locatable by coordinate 1: It will be devoted to communication between physicists and nonphysicists. Before embarking on detailed discussions of these specific segments of the picture, however, we shall try, in the following section, to develop a little perspective on the uses to which the different components of the communication picture are put and on the

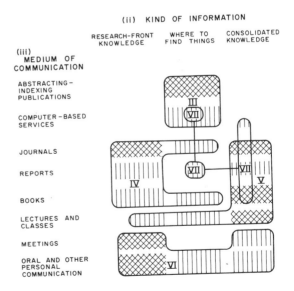

FIGURE 13.1 Organization of the central sections (Roman numerals) of this chapter. In each column, the cross-hatching identifies the rows making the major contributions to this column; the vertical shading, those making secondary but sizable contributions. Key: III, secondary services; IV, primary publications; V, books, reviews, and compilations; VI, oral and interpersonal communication; and VII, information analysis centers.

way in which they relate to each other. We shall conclude this chapter with a brief summary of the roles of the major types of organizations or groups of people concerned in the transfer of information and of our recommendations to them.

The rather unequal lengths of the third through the seventh sections are due more to differences in available information on these areas than to differences in their importance. The same is true of the eighth section; communication between physics and the rest of the world is not only very important, but there is a lot of it. However, there is some further justification for our preoccupation throughout most of this chapter with research-front communication and the consolidation of research-front knowledge—this is where the confusion of the expanding literature causes the most problems.

Having thus sketched the content of the sections to follow, we shall conclude this section with a few words about a use of the hitherto neglected classification 2a, which we shall find very useful in the overview to be given in the next section. In Figure 13.2, the rows again represent the various subdivisions according to criterion 3, now a little more detailed than in Figure 13.1. The columns represent a classification according to 2a, with a further distinction between the roles of giver and receiver. The figure shows principal (black circles) and subsidiary (open circles) ways in which each of the media is used; absence of a circle in any box does not mean

MEDIUM	FUNCTION (USE TO WHICH THE MEDIUM IS PUT)			STUDY
	DIRECTION OF ATTENTION			
	GIVER-INITIATED	RECEIVER-INITIATED		
		BROWSING	SEARCHING	
ABSTRACTING-INDEXING PUBLICATIONS ABSTRACT JOURNALS		○	●	
CURRENT AWARENESS JOURNALS AND PREABSTRACTS		●		
CITATION INDEXES			●	
COMPUTER SERVICES	○ (SDI)		●	
JOURNAL INDEXES			●	
RESEARCH JOURNALS		●	● (Cross refs.)	●
PREPRINTS AND REPORTS	●		○ (Cross refs.)	●
BOOKS AND REVIEWS		○	●	●
LECTURES AND CLASSES		●		●
TALKS AT MEETINGS		●		○
WRITTEN PERSONAL	●		●	○
ORAL PERSONAL	●	●	●	○
OTHER	?	?	?	?

FIGURE 13.2 Modes in which each of the communication media listed in the first column are used by a receiver of information. A full circle indicates major use, an open circle minor use, relative to the total use of the medium.

that such use never occurs but implies that it is currently less important. Three of the boxes contain explanatory notes: SDI refers to selective dissemination of information, a scheme whereby a scientist is automatically sent a small number of titles, abstracts, or reprints chosen by a central agency to match his presumed interests; Cross refs. refers, of course, to the widespread and very fruitful custom of learning about interesting work from its citation in papers one has consulted.

ROLES AND INTERRELATIONSHIPS OF DIFFERENT PARTS OF THE COMMUNICATION PICTURE

Studies of Information-Gathering Habits

We have just classified, in Figure 13.2, the ways in which a scientist can receive information through the various communication media. The simplest types of questions that we can ask about the various combinations in Figure 13.2 are such things as: How do physicists and related scientists distribute their time among these various possible communication activities? What is the yield of each of these activities in terms of useful information? Do physicists prefer some of them over others, and by how much? Many studies have been made on these questions, though few of them have been on physicists, and most of them are unsatisfactory in one way or another. We have tried to put together in Figure 13.3 a composite of some of the principal results that emerge from them; further details are given in the Special Panel Report. Although one should not place too great reliance on any of the figures, the general picture is somewhat confirmed by the results of qualitative preference studies. The following conclusions seem warranted:

1. Physicists, like chemists, spend a sizable proportion of their time, about 15 hours per week, in the reception of scientific information or in give-and-take oral discussion of it.

2. Physicists spend very little time using abstract journals and title listings—much less than chemists. (Nevertheless, they consider them a moderately important medium for current awareness.)

3. Reading of journals, preprints and reports, and books and reviews averages typically 7 or 8 hours per week for research physicists, these three categories of media receiving comparable fractions of reading time. Probably chemists spend a little more time on these. A sizable minor fraction of this reading takes place during other than standard working hours.

4. At least in large institutions, where opportunities for contact with colleagues are good, physicists are apt to spend about as much time in

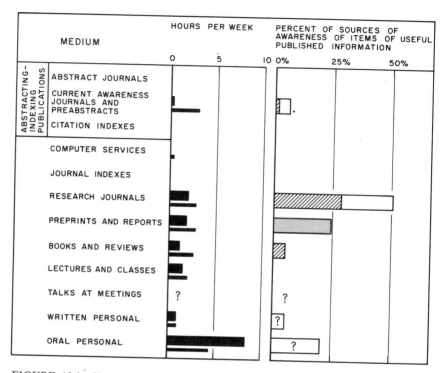

FIGURE 13.3 Use and relative yield of various communication media. The first
column of bars is a composite of estimates of average time spent by physicists (broad
bars) and by chemists (narrow bars) in the use of each of the media listed at the
left. The right-hand column gives, for one population of basic-research physicists,
the distribution of sources of awareness of published items referenced in their papers;
giver-initiated sources are dotted, browsing is open, and searching is diagonally
shaded. Sources and methodology are described in the Special Panel Report.

person-to-person scientific conversations as in all types of reading taken
together.

5. Despite the greater amount of time spent in oral communication, more
leads to published items come from use of journals than from oral sources.

6. Browsing in journals is a major source of useful information, accounting for almost as many leads as the other highly productive category, cross
references.

7. Books and reviews provide a surprisingly small fraction of physicists'
leads (less, for example, than biologists'). There are several further conclusions of some importance from sources not shown in the figure. One,
which will be discussed at greater length in the section on oral communication, is that:

8. Talks and meetings provide an important source of current awareness, probably comparable in average importance to preprints and somewhat below journal browsing and personal conversations. They are less useful as a source of detailed specific information. Many studies of scientists and engineers (for example, Reference 3) have shown that:

9. Easy accessibility is a major determinant of the extent to which any formal or informal information resource is used. Physical proximity to the user's office or laboratory is a very important component of this; an additional 100 ft of distance can retard use quite noticeably.

Individual Differences and the Topology of Communication

The picture we have developed so far is incomplete. It has dealt only with averages over many scientists in a given area and only with one of the functions (columns of Figure 13.2) at a time, most often a direction-of-attention function. Although only very limited studies have been made on differences between different individuals or different classes of physicists, and on the sequences of information-transfer processes that take place in getting information from an originator to an ultimate user, there are a few interesting things that have been discovered.

One finding, a not unexpected one, is that among both physicists and chemists the scientists active in research seem to spend over twice as much time reading journals as nonresearch scientists.[4] Other studies have shown clear positive correlations between amount of journal reading by different individuals and various measures of their productivity.[5,6] An even more striking effect occurs when one goes from science into engineering. Engineers rely very much less than scientists on the reading of journals for the information they use and depend more often on oral sources, catalogues, and the like.[7-10]

The most extensive published studies of the interconnection of different communication channels have been made in industrial laboratories, mostly on engineers and applied scientists. These studies are worth commenting on because they not only have revealed differences between individuals, but have shown how some of these differences play a very important role in the functioning of a scientific or technical group.[3,8,11] These studies have revealed that in typical organizations there exist special individuals—called gatekeepers—who serve as focal points for information originating outside the organizations. Specifically, a person who is chosen by an unusually large number of his colleagues (within the organization) for frequent participation in technical discussions is characterized as a gatekeeper. The studies have shown that the gatekeepers so identified differ considerably from the remainder of their colleagues in that on the average they read many more scientific and professional periodicals and have many more information-producing contacts with friends outside their organizations.

Our best information on communication chains among research physicists has been obtained from a small-scale study of a population of basic-research, condensed-matter physicists. From papers they had written, ideas and facts were selected, at random, which (a) contributed perceptively to the arguments of the paper, (b) did not seem to have originated with the author or authors of the paper, and (c) did not seem to be such common knowledge that they could be expected to be known by the average PhD in the field. Each physicist author was then interviewed and asked:

1. How the specific information he actually used came into his mind, that is, through reading (what?) or listening (to whom? where?).

2. How he was cued to get it. (For example, did someone refer him to a paper? Send him a preprint? Had he undertaken a search for this type of information?)

3. Did he know how the information, or knowledge of its existence, got to the source or cue cited in 1 or 2, and like information?

4. How much time lapsed between the original discovery of the fact or idea and its apprehension by the interviewee? The results are depicted in Figure 13.4. Study of the figure reveals that direct transfer of the information from the originator to the user occurred in about one fifth of the cases. The great majority of the cases involved a single intermediate repository for the information between the originator and the user, most often a journal article but sometimes just the mind of a third person. Oral personal links seem to have been very important, especially for providing leads to printed material; these links were giver-initiated in over half of the cases.

The study of time intervals gave the interesting result that in approximately half of the cases the interval between first availability of the information and its acquisition by the recipient was no more than about a year, but that in most of the remaining cases it was more than five years.

Organizations and Resources for the Communication of Information

A glance at the communication media listed in the first column of Figures 13.2 and 13.3 shows that some involve vast programs of national or international organizations; some involve commercial, industrial, or nonprofit activities of modest scale; and some are at the level of informal everyday activities of individuals. The largest operations are easy to enumerate, but as they are sometimes interdisciplinary and are usually circulated throughout the world, it is not always easy to decide how much of them should be assigned to physics or to the United States. For the activities of commercial

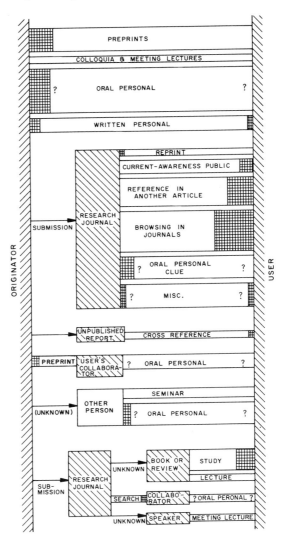

FIGURE 13.4 Distribution of transmission chains for important items of information used by a sample of condensed-matter physicists. Diagonal shading identifies repositories in which the information resided for an appreciable time between transmissions; unshaded rectangles identify different types of clues that led to assimilation of information from use of these repositories. Vertical widths of the latter are proportional to number of cases of each type in the sample; arrows are used in place of rectangles when classification by clues is inappropriate. Rectangular cross hatching at the right or left ends of these rectangles designates the numbers of cases in which the initiative was with the giver (left end) or the receiver (right end) of the information; question marks indicate further cases in which the source of initiative was in doubt. In some cases (for example, lectures), the question of initiative is meaningless and the cross hatching is omitted.

organizations, which may be of large or medium scope, data on costs, circulations, and the like are not easily obtainable. For such small-scale activities as seminar talks, personal correspondence, and other types of interpersonal interaction, not only are data unavailable, but it is often hard to draw a line between communication and other activities. Thus our estimates of U.S. resources invested in communication will have to be very crude.

Still, a rough perspective is better than none. So in Figure 13.5 we have

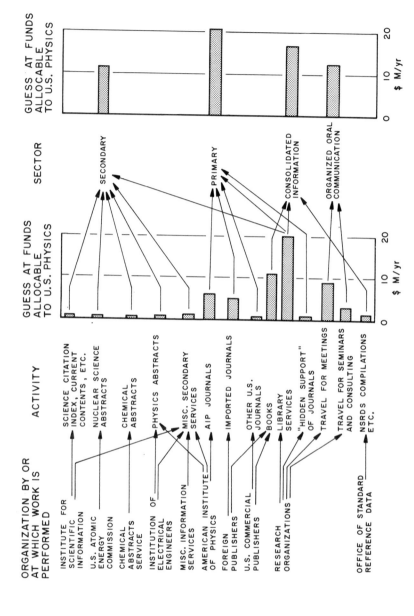

FIGURE 13.5 Organizations and activities relevant to identifiable organized support of communication in physics (excluding education), with rough estimates of the dollar value of expenditures on behalf of U.S. physics.

900

tried to enumerate, in the first column, the leading organizations or groups of organizations that perform work in support of communication in physics. The various information products and services that result are shown in the second column, with arrows from the first column showing which organizations contribute to which products. The bars after the second column give estimates, sometimes quite crude, of the funds expended annually by or on behalf of U.S. physics to produce each product or service, via all the organizations involved. Each of the products or services is classified as pertaining to the primary literature, to the secondary (access) category, to consolidation and reworking of written information, or to organized oral communication, and the total annual expenditures under each of these four headings are shown at the right. (Library services—storing, binding, cataloguing, Xeroxing, and the like—are an exception in'that they contribute to three of the four categories.) Descriptions of the organizations and further details of the dollar estimates are given in the Special Panel Report.

The total of the entries in Figure 13.5 is of the order of $60 million per year. While the primary sector gets the largest allotment, the other three categories in the last column all get more than half as much. The conspicuous feature of the middle column is the large amount estimated as spent on library services (exclusive of acquisition costs). This entry was obtained by a compounding of guesses and so may be very inaccurate; however, independent evidence suggests that the true value could hardly be smaller than half of that shown—still a very sizable sum.

Potential Value of Improvements in Information and Communication Services

The first section of this chapter briefly mentioned that one can use either buyer response or time investment by users to calibrate the dollar value of information services. While we shall not attempt—either here or in the Special Panel Report—to develop either of these approaches in great detail, we do believe it worthwhile to sketch, using the user-time approach, how two effects that are not always appreciated act to augment the real value of information services.

The first effect has to do with the amount of time available for productive work, for example, for research. (For concreteness we shall speak of research henceforth, but the same reasoning could be applied to any other work task whose output one might wish to measure.) The time t that is available for research by a physicist who spends a total time t at work is

$$t_w = t - t_i - \tau, \tag{1}$$

where t_i is the time (of the order of one third t, according to Figure 13.3) devoted to information and communication activities, and τ is the time spent on activities other than communication and research and presumed to be valued lower than the latter. The output of research will generally increase faster than linearly with t_w. Both for this reason and because of the t_i and τ terms in Eq. (1), any time-saving innovation in communication —say, one that enables the same information to be gathered in Δt_i less time —will result in a fraction increase in the output of research that is rather larger than $\Delta t_i/t$.* (If this seems hard to grasp, consider the extreme case $t_i = 0.99t$, $\tau = 0$. Reducing t_i to $0.98t$ will double the output, not merely increase it by 1 percent.)

The second effect has to do with the cooperative interaction of different information channels, which we have seen, for example, in Figure 13.4. Any measure that makes communication via primary journals more efficient, for example, will also significantly increase the effectiveness of personal oral communication, because the people one talks to will be better informed. One can make rough quantitative measures of this effect on the basis of data like those of Figures 13.3 and 13.4.[2]

Crude quantitative estimates in the Special Panel Report lead to the conclusion that *an innovation that would reduce by λ percent the total time physicists need to spend in all types of communication to get a given yield of useful information would be equivalent to augmenting the man-years employed by an amount two or three times λ percent of the time now spent in communication.* Further consideration of the various overhead items attached to the work of physicists shows that some, though not all, of them should be included in estimating the dollar value of the time just mentioned; the appropriate value seems to be of the order of one and one half times what the physicist himself is paid. Taking the number of equivalent full-time physicists affected by a hypothetical improvement in the totality of physics communication channels to be in the range 20,000–27,000, and their mean salary as about $16,000 per year, we arrive at the estimate cited in the first section of this chapter, that is, that a saving of λ percent of the total communication time would be worth approximately ($300–600) ($\lambda/100$) million per year.

---

* It also has been suggested that many scientists probably are prepared to devote approximately one third of their time to communication regardless of its productivity. Beyond this point, time devoted to communication would begin to cut perceptibly into research effort. Therefore, even though scientific communication were made much more efficient, many scientists probably would continue to spend a substantial fraction of their time in this activity, but they would benefit much more from the time so spent, and the effectiveness of their work would be enhanced.

THE SECONDARY SERVICES

There is good reason for starting our survey of communication media with a look at abstracting journals, title listings, citation indexes, and like sources. A great deal of the detailed data we shall present in the sections dealing with the primary literature, the review literature, and meetings and conferences will be extracted from these secondary sources. Thus it is well for us to begin with them so that we can understand their capabilities and limitations.

Number and Variety of Secondary Services

The number of different secondary services in physics is greater than most physicists realize. Thus, in a recent study,[12] some 69 secondary services, covering physics or its subfields or peripheral areas, were identified and discussed. These include four abstracting journals that undertake comprehensive coverage of published articles from all parts of the world in essentially all the subfields of physics, several other publications that undertake a reasonably comprehensive coverage but list titles only, and a number of abstracting and title journals devoted to specific subfields of physics or nonphysics fields that overlap physics. About half of the total are still more specialized: They may have only partial coverage of a particular field; they may cover only work published in a particular geographic region; or they may cover only a particular type of material, such as reports or patents.

So far we have mentioned only those secondary services that supply printed lists of publications in designated subject areas. There are at least three other types of service that physicists recently have begun to use and that have considerable potential utility. One is the citation index, the use of which is discussed later in this section. Another is abstract-index information, and sometimes also citation information, on computer tapes; from these, employers of large numbers of physicists can construct various local services for their employees. A few selective-dissemination services are available that supply individual scientists with current papers that match an individual interest profile.

Bulk and Coverage

Figure 13.6 compares the growth over the years in the numbers of entries in three of the four comprehensive abstracting journals serving physics. All show roughly exponential growth, punctuated by a wartime dip; but

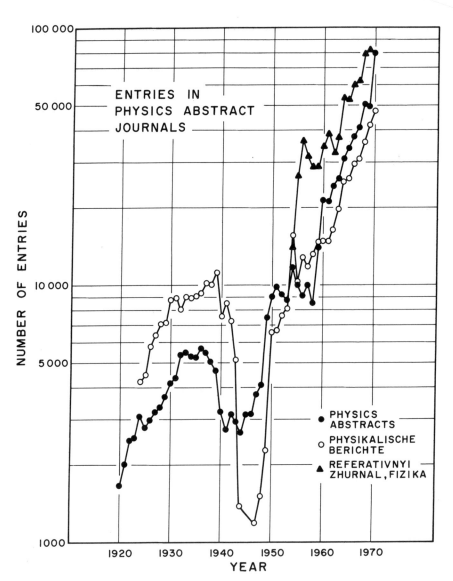

FIGURE 13.6 Growth, in time, of total entries in three abstracting journals covering the literature of physics.

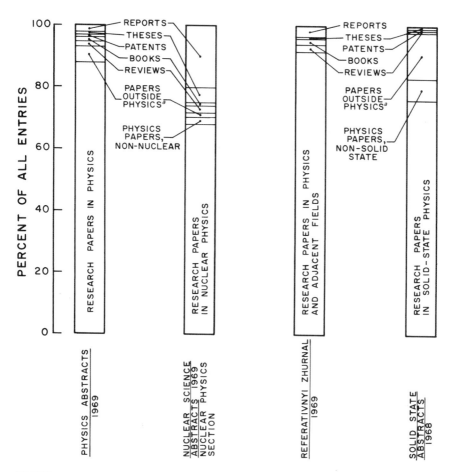

FIGURE 13.7 Rough breakdown of the entries of several abstracting journals by type of material. Since the definitions of different categories of entries overlap, the order of decisions must be specified to clarify the meaning of the numbers. First, each entry was classified according to whether it was published (i.e., available, other than as an abstract, in a regular journal or in a book sold on the open market). Published material was classified into research papers, reviews or compilations, books, popularizations, and patents. Unpublished material was classified into theses and reports, including individual-paper components of reports. Research papers are defined as published (as defined above) articles or letters reporting new research results. *ª* The boundary between physics and nonphysics was chosen rather arbitrarily, perhaps encroaching a little more on chemistry, geophysics, and like sciences than an impartial boundary would; however, it is hoped that the boundary has been drawn consistently for the different abstracting journals.

superimposed on this are some highly individual fluctuations, due presumably to administrative decisions and fluctuating finances. It is interesting to note, however, that despite the fluctuations evident in the figure, the number of journals scanned for *Physics Abstracts* has increased quite linearly with time from about 200 in 1920 to over 800 in 1968.[13]

The variations in Figure 13.6 serve as a reminder that the criteria for inclusion in an abstracting journal of the sort we are discussing are necessarily somewhat arbitrary in each of two dimensions: One must decide what is to be considered physics and what is to be considered published material. Figure 13.7 shows some examples of the way in which the definition of what constitutes published material differs for different abstracting journals; though journal articles predominate, very different mixes of reports, theses, patents, books, and the like can be offered. Figure 13.8 illustrates how arbitrary is the decision on how far afield one should look for articles of interest to the particular community served. Although one finds about 90 percent of the journals covered by *Physics Abstracts* in the 125 leading journals, a few hundred more journals produce some additional articles relevant to physics; no doubt scanning several thousand more journals would still further enlarge the yield, though not by enough to be worthwhile.

Time Lags, Availability, and Other Considerations

The utility of the various secondary services to scientists depends both on the information they offer (discussed in the second and fourth parts of this section) and on their availability in space and time, in other words, their circulation and their promptness.

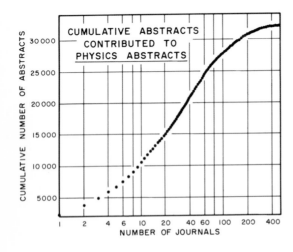

FIGURE 13.8 Cumulative number of abstracts contributed to *Physics Abstracts* 1965 by the n most prolific journal, as a function of n.

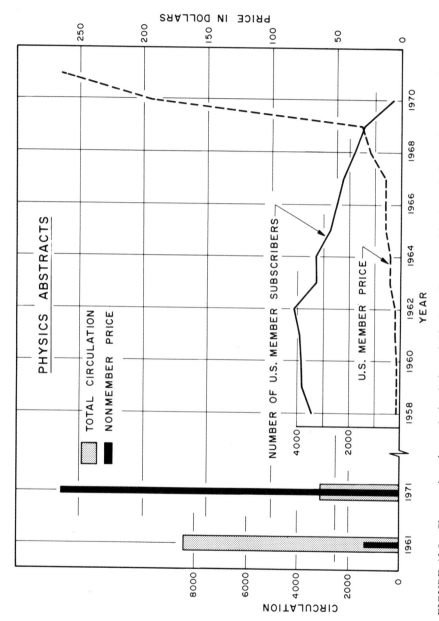

FIGURE 13.9 Changes in price and circulation of *Physics Abstracts* with time. Left portion of the figure shows total circulation (shaded bars, scale at left) and nonmember price (black bars, scale at right) for 1961 and 1971. Right portion of the figure shows number of AIP member subscribers (solid curve, scale at left) and price charged to these (dashed curve, scale at right). Subsidy was removed in 1970.

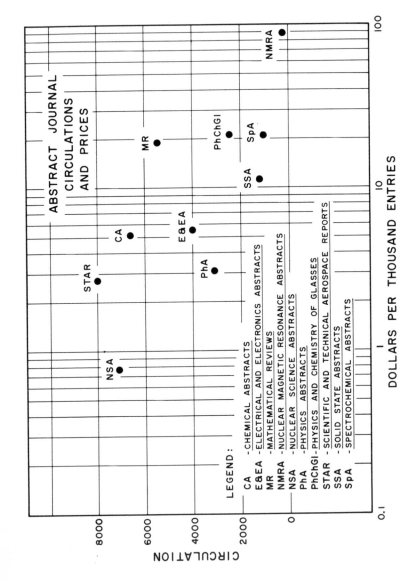

FIGURE 13.10 Circulations and prices per abstract for abstracting journals in physics and related fields. Data come largely from Ulrich's *International Periodicals Directory,* 1968–1969 or 1971–1972.

908

Figure 13.9 shows strikingly how the vast expansion of the literature evident in Figure 13.6 has made *Physics Abstracts* much more costly, and how this cost has reduced its circulation, particularly to private individuals. (Even if the price were low, the sheer bulk of *Physics Abstracts* would doubtless now deter most individuals from subscribing.) That the institutional subscriptions seem to have held up fairly well may be due in part to the transfer of some subscriptions at small institutions from individual to institutional; there may also be a strong inertial hanging-on effect. The partial correlation of circulations of different abstracting journals with their unit price, shown in Figure 13.10, is also illuminating. At the high end of the price range, even institutional subscriptions decrease.

It can hardly be doubted that the decline of widespread individual subscriptions to abstracting journals and the probable decrease in the availability of multiple institutional subscriptions (that is, subscriptions for use in sublibraries and the like) have decreased the frequency with which physicists use these journals. The corresponding loss in utility has been only partly compensated by the growing availability of title listings. Thus we might expect the benefit the physics community receives from abstracting journals to rise if a way could be found to market them at a lower price, so that more copies could be made available to sublibraries and like sources. *Nuclear Science Abstracts,* subsidized by the U.S. Atomic Energy Commission, sets an excellent example. (We shall return to this topic later in this section and in the concluding section.)

The currency of an abstracting or indexing publication is a significant factor in its utility, especially if it is to be used for current awareness. Fortunately, this is an area in which great improvement has occurred in the last decade or so, though there is still wide variation from one publication to another. For example, whereas 10 or 12 years ago the median time lag between appearance of a paper and the appearance of its abstract in *Physics Abstracts* was of the order of five to eight months, and delays of well over a year were not uncommon, today the median time lag of *Physics Abstracts* is only about three and one half months, and delays of as much as a year are very rare. Title listings can be even faster, as they should be, since their use is almost entirely for current awareness.

Content, Organization, and Indexing

Different abstracting journals strive for different standards of quality and relevance in their abstracts. *Mathematical Reviews* represents one extreme, with abstracts written by highly competent mathematicians and designed to be not only informative but in many cases critical as well. While such abstracts were common in the abstracting journals of physics a number of years ago, nearly all of them have now been forced by the swelling bulk of the literature to rely almost entirely on authors' abstracts appearing with each cited publication; *Referativnyi Zhurnal* and *Physikalische Berichte,* however, still have many abstracts prepared by signed abstractors. As for title journals, these sometimes list only the title but occasionally augment this with an indication of whether the paper is experimental, theoretical, or both and with key words or subject index numbers relevant to the content of the paper. Both abstracting and title journals are increasingly adopting the practice of listing authors' institutions. This practice enables a user to make contact with an author whose paper may not be available in the user's library; it can also be extremely useful as a means of searching when the content of the abstracting or title journal is available on computer tape.

Arrangement and indexing are all-important. Arrangement is usually by subject; if this is to be really logical, there have to be a number of duplicate entries or cross references. For example, in *Physics Abstracts,* about 30 percent of the entries are so cross-referenced. Author indexes are reasonably straightforward to construct, but there are often not enough long-term cumulations of them to satisfy the needs of users. Subject indexes (as the discussion of use in the next part of this section shows) have perpetual and probably increasing difficulty in avoiding the Scylla of enormous lists of items under a single subject and the Charybdis of so many subject headings that the user does not know where to look.

Our discussion of the information content of secondary services would be glaringly deficient if it did not point out the extremely rich information content of citation indexes. The author of a scientific paper reveals a great deal about its intellectual content by the articles, books, and other sources that he chooses to cite. Although the set of citations will normally not reveal the nature of the new contributions made by the citing person—as an abstract does—it will usually reveal a great deal about the subject matter and methodology. Thus the great advantage of citations is their convenience as an access tool to tap a rather sizable pool of information based on the expertise of the authors of papers. Citation indexes can be very useful even for simple subject searching.

Patterns of Use of Secondary Services

The use of abstracting services, computer tape searches, and even current-awareness publications by physicists is very different from that by some other types of scientists, especially chemists. Thus, Figure 13.3 showed that industrial research chemists spend over half as much time with secondary services as they do with the primary literature, and about half of this time is spent in retrospective searching. The vastly smaller amount of literature searching by physicists is probably at least partly due to the fact that information needs of the taxonomic variety are much less common in physics than in chemistry. Searches for ideas and conceptual relationships are much more difficult to carry out using subject indexes than are searches for work on specific substances or species. (Some examples are given in the Special Panel Report.) The problem posed for the designers of indexes by the conceptual rather than taxonomic nature of most work in physics is nicely illustrated by an unpublished study of speed of retrieval.[14] A sample of graduate students were given abstracts of *Physical Review* papers, without the authors' names or the journal citation. Some were asked to locate the articles using one subject index, some using another index. It was found that retrieval was significantly slower with the more extensive subject indexes; for example, that of *Chemical Abstracts* required, on the average, nearly twice as much time as that of *Physics Abstracts.*

The utility of citation indexes for retrospective searching is not yet as widely appreciated as it should be. A few quantitative studies have been made, which show that with intelligent strategies one can sometimes do even large-scale searches quite efficiently in terms of both time and retrieval rate. Especially intriguing, though at present expensive, is the computer search of citation tapes to retrieve literature on highly specialized topics.[15,16]

Title journals and other current-awareness services are extensively used by some groups of physicists, but on the whole they seem to be much less used than they might be, and probably much less than they should be. The reasons are obscure, and it will be interesting to see whether the new sectionalized title journals of the American Institute of Physics, *Current Physics Titles,* will have more appeal.

Technology, Economics, and Outlook for the Future

Let us start with a few observations on production costs. Prerun—that is, editorial and composition—costs for an abstracting journal, at 1969 rates, are likely to range from $8 to $12 per abstract, depending on such things as conscientiousness of coverage, depth of indexing, country in which labor is purchased, and the like. Runoff costs—printing and distribution—are never more than a small fraction of the prerun costs, for circulations in the ranges indicated in Figures 13.9 and 13.10. The rise in unit costs due to inflation is being largely compensated by improvements in technology and avoidance of duplication of work by different organizations.

These last points deserve a little enlargement. Photo-offset printing was adopted some time ago by most of the abstracting and title journals of interest to us. But modern techniques of computer-controlled composition —most commonly photocomposition of material for production of offset plates—make it possible for many useful products, with different arrangements or selections of the same material, to be produced with only a single composition (keyboarding) operation. *Physics Abstracts* is already making use of this technology. Although by itself this operation may be a bit more expensive than mere typewriter composition, the same composition can be used for abstracting journals, title lists, indexes, and other media; it is even possible to integrate the composition of the primary journals with the production of these secondary services, so that no additional keyboarding is needed for the latter. This sort of integration of primary and secondary services is one of the goals of the current Information Program of the American Institute of Physics.

Table 13.1 gives details of these new secondary services planned by the American Institute of Physics (AIP) and shows their relation to those of

TABLE 13.1   Secondary Services Offered or about To Be Offered by the American Institute of Physics and the Institution of Electrical Engineers

Service	Description [a]	Prepared at
With Abstracts		
Physics Abstracts	Comprehensive abstract coverage of the world's physics literature (also available in microfiche)	IEE
Current Physics Advance Abstracts	Abstracts of articles in leading journals (initially AIP only), issued prior to publication in these journals	AIP
INSPEC Physics Tapes	Magnetic tapes with the same data as *Physics Abstracts*	IEE
Searchable Physics Information Notices (SPIN)	Abstracts, citations, and subject classifications for articles in about 70 leading journals	AIP
Without Abstracts		
Current Papers in Physics	Comprehensive title listing for the world's physics literature, arranged by subject	IEE
Current Physics Titles	Three journals (*Nuclei and Particles, Atoms and Waves, and Solid State*) listing titles and key words for articles in about 70 leading journals, arranged by subject	AIP
Uncertain		
Current Physics Bibliographies	Specialized bibliographies in relatively narrow area, periodically updated (planned for 1973)	AIP

[a] In addition to the characteristics listed, all the AIP services supply a cartridge and frame number for location of the articles in their primary service *Current Physics Microform,* a film-cartridge form of the full texts of all papers covered in *Current Physics Advance Abstracts.*

the Institution of Electrical Engineers (IEE), with which AIP cooperates. In general, the division of labor is based on the following principles:

1. AIP services are aimed at *selective* coverage of the primary literature; IEE services have comprehensive coverage. Thus the AIP services will cover only articles in several score of the most important journals.

2. AIP aims to integrate the input from its own primary journals with the production of those journals themselves and will supply tapes of the secondary information so obtained to IEE for use in IEE's services, while receiving from IEE taped information from other journals.

It is clear from the table and from what has been said above that a well-planned and efficient set of secondary information services must involve many interrelationships not only among its own elements but also with primary publications, review literature, and other components of the communication picture. (We shall return to this broader picture in the final section.)

Despite the economies we have just referred to, however, the prerun costs of secondary publications will undoubtedly continue to be large enough to price many potential buyers out of the market if they must be recouped from subscription income (see Figure 13.10). This is an unhappy state of affairs. If the value of the product to a buyer, and through him to society's scientific enterprise, is greater than the cost of producing one *additional* copy for him (that is, the runoff cost), society as a whole gains through providing this copy to him; but it will not be provided if the market price is greater than the value to the user, since he will be willing to pay no more than it is worth to him. Because of the large discrepancy between total cost and runoff cost, there will be many potential buyers in this range, and the loss to the scientific enterprise will be considerable. Though we shall not make a specific recommendation, it is clear that there would be advantages to some sort of general social subsidy of input to secondary services, so that they could be marketed at something closer to runoff cost.

Looking farther to the future, it may someday be possible to provide secondary services that use in considerable detail the information contained in citations. As we have mentioned briefly, computer programs to generate clusters of related papers are possible and might form the basis for an extremely useful alerting service, far surpassing the already quite useful services now offered by the Institute for Scientific Information. Ultimately, too, an interactive querying of computer files of the physics literature may become available to physicists generally. Such a capability has already been used experimentally in physics (Project TIP; see, for example, Reference 17). At present, expense bars wide use of such systems; programs to improve their feasibility should be vigorously pursued.

PRIMARY PUBLICATION

We turn now to what is really the heart of communication in any science, primary publication. As is revealed in Figure 13.3, and especially in Figure 13.4, research journals and their informal counterparts, preprints and reports, play the central role in the communication of the details of ideas and data and may well be predominant even for attention-focusing. In this section we shall discuss these two forms of communication and also the communication of new research results in books, for example, conference proceedings or monographs containing previously unpublished material.

Research Journals: Number, Bulk, Price, and Circulation

To get a flavor for the wide diversity of journals in which new research results are published, let us take a look at a few graphical statistics about them. We have already seen, in Figure 13.8, how about two dozen journals account for half the entries in *Physics Abstracts,* whereas several hundred are needed to give all of them. This finding means not only that there are many small journals, but, more importantly, that there are many journals in fields outside of physics that occasionally contain articles of interest to physicists. This latter effect is shown very clearly in Figure 13.11. The number of journals devoted solely to physics is not really so large, but the physicist must keep in touch with a great many interdisciplinary, multi-disciplinary, and nonphysics journals.

The number of journals, of course, has been growing over the years, as one would expect from the vast expansion of activity in physics. For example, in the middle 1930's one could cover half of the entries in *Physics Abstracts* with only about a dozen journals, instead of the two dozen required for the 50 percent level in Figure 13.11. But this growth in numbers has been considerably less than the growth in the total volume of publication, which we can gauge roughly from Figure 13.6. Obviously then, most of the journals have been getting thicker; this is illustrated for a number of the leading journals in Figure 13.12, with some representative growth curves for some of the smaller journals shown for comparison.

Also shown in the figure is the growth of U.S. PhD manpower in recent decades. The slopes of the curves for the larger journals, like those for the totals of material in the leading abstracting journals in Figure 13.6, are greater than for the manpower curve. One suspects that what this means is that the average physicist today is publishing more papers per year than he did some time ago. We have verified this hypothesis by taking random samples of the membership of the American Physical Society (APS) in 1955 and 1969 and noting how many papers published in

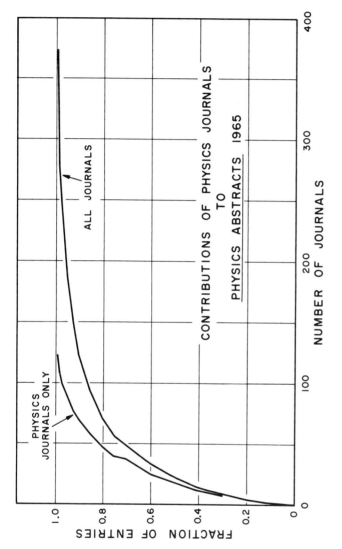

FIGURE 13.11 Contributions of physics journals to the entries in *Physics Abstracts* 1965 compared with that of all types of journals. The ordinate of the lower curve is the fraction of all entries coming from the *n* most prolific journals, where *n* is the abscissa. For each point on the lower curve, the point with the same ordinate on the upper curve gives the number n_p of those n journals that are devoted primarily to physics topics. The differences $(n - n_p)$ consist of journals predominantly devoted to a single nonphysics discipline (about three fifths), multidisciplinary journals (about one fifth), and journals devoted to narrow specialties on the periphery of physics (about one fifth).

915

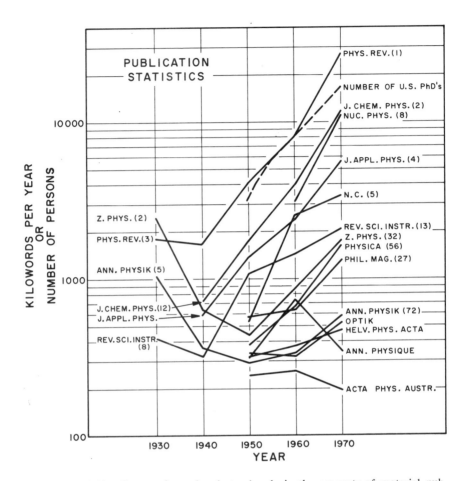

FIGURE 13.12 Changes from decade to decade in the amounts of material published by various physics journals. The numeral appended to each journal at the left is its rank order in number of articles abstracted by *Physics Abstracts* in 1934; the numeral at the right is the corresponding rank order for 1965. If interdisciplinary journals (*Comptes Rendus, Nature,* and the like), letter journals, nonphysics journals (for example, *Journal of the American Chemical Society*), and Soviet journals are excluded, the journals shown include all others in the first ten of the 1934 ranking or in the first 13 of the 1965 ranking. Some smaller journals are also shown for comparison. The dashed curve gives the growth of the number of physics PhD's in the United States.

journals were listed under these names in *Physics Abstracts* for the same year. Even after elimination of those few papers that were in fringe areas in which the coverage of *Physics Abstracts* might have changed between the two years, and after allowance for the fact that the average number of authors per paper has increased slightly during this interval, we found a definite increase in the number of research physics papers contributed to journals by APS members; the numbers were about 0.15 per member in 1955 and 0.23 per member in 1970. This increase could reflect in part an increase in the number of physicists working in academic institutions, in which the incentives for publication are greater than in government and industry; it also could reflect the increase in the number of institutions of higher education that perform research and the consequent increased employment of physicists in research-oriented institutions.

Now let us look briefly at prices and circulations. These variables, and the relation between them, are of interest for two reasons. First, the use of journals depends on their ready availability; as we have already noted, in the discussion of information-gathering habits, any inconvenience in using a journal—even the need to walk an extra hundred feet—detracts significantly from its use. Second, the curve relating price to circulation gives us a measure of the value the scientific community places on the journals. Figure 13.13 shows prices and circulations for a number of physics journals. The following statements, only some of which are apparent from the figure, summarize the situation:

1. Society-run journals in physics tend to be large, especially in the United States, and dominate the publication scene in physics, whereas they do not have so dominant a role in most other areas of science. These society journals have the lowest prices, the largest circulations, and the greatest bulk as compared with other types of journals. Those of the American Institute of Physics (not including translations) head the list in all of these characteristics and account for about one fourth of all the journal papers listed in *Physics Abstracts*.

2. The prices (to institutions) of physics journals vary enormously. In 1968 the most expensive of the commercial journals cost about 21 cents per kiloword as compared with 0.2 cents per kiloword for the *Physical Review*. (The price of the latter is now several times higher than in 1968 but still is at the low end of the price range. The median 1968 price for U.S. society journals was below 1.5 cents per kiloword, that for other Western nonprofit journals was about 3 cents per kiloword, and that for commercial journals about 8 cents per kiloword.

3. In the upper part of the price range, circulations drop significantly with increasing price.

4. Despite buyer resistance, enough libraries will purchase physics

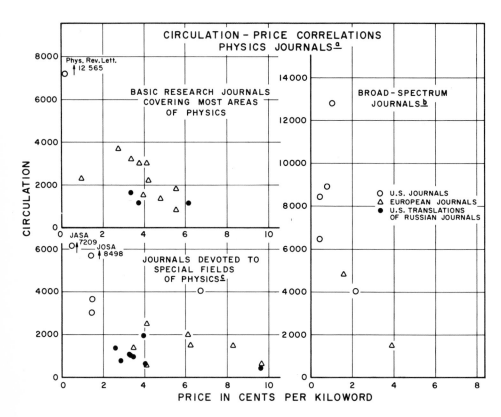

FIGURE 13.13   Circulations of a number of U.S. and Western European physics journals of different types correlated with price (in cents per kiloword) to institutional subscribers. [a] Although it would be better to show institutional and personal subscriptions separately, the separated data were not available for many of the journals. Data on AIP journals suggest that in the 6000–9000 range the subscriptions are divided about half and half between institutional and personal, and that below 2000 they are almost entirely institutional. Availability of data also limited the representation of commercial journals. Although many are shown on the plots, circulation figures for many others, including most of the more expensive ones, could not be obtained. [b] Examples: *Journal of Applied Physics, Journal of Chemical Physics, Review of Scientific Instruments.* [c] Examples: *Physics of Fluids, Applied Optics, Nuclear Physics, Journal of Mathematical Physics.*

journals at quite high prices so that no journal with reasonable content is likely to fail from mere lack of subscribers.

Research Journals: Production and Economics

As one can verify from a glance at Figure 13.5, the production of research journals constitutes a sizable component of the total cost of physics communication, and to this component must be added a comparable amount for the library services that are necessitated by the purchase of journals. These costs have been rising rapidly with the growth of the literature and have caused widespread concern. The problems are, of course, not peculiar to physics, and they have been the object of a recent study encompassing all fields.[2] However, as the journals of the AIP are at the forefront in the use of certain measures to deal with these problems, it is particularly appropriate for us to discuss these issues.

Central to any discussion of the economics of journals is the division of production costs into prerun and runoff items. Prerun items include all operations that are necessary before production of the first copy of the printed work. These costs are independent of the number of copies to be produced, but for a given type of material, they increase with the number of research pages published. Runoff costs, on the other hand, are those that depend on the number of subscribers to the journal, being proportional to this number if it is reasonably large. Thus, if $n(p)$ is the number of subscribers at price p,

$$\text{production cost} = s + rn(p), \tag{2}$$

where s is the total prerun cost and r the runoff cost per copy. For physics journals a typical prerun cost s is about \$60 per kiloword, whereas r, though rather more variable from case to case, may be something like 0.3 to 0.4 cents per kiloword per copy. Thus for physics journals the runoff cost is normally significantly below the prerun cost, and for those journals of lower circulation, it may be only a small fraction.

For journals that try to meet all their costs from subscription income, this dominance of prerun costs has two unfortunate consequences:

1. It necessitates setting a price p much larger than the runoff cost r. This is a deprivation to the class of buyers to whom the value of the journal exceeds the cost r of producing an extra copy but is less than the prorated cost $r + (s/n)$. Society as a whole would gain if extra copies were run off and provided to these buyers, yet the set price excludes them from the market.

2. The finances of the journal become unstable with respect to fluctuations in amount of material submitted or in number of subscriptions.

The physics community can take credit for introducing, in 1930,[18] a method of dealing with these difficulties that has proven so satisfactory that it has come to be adopted by the majority of U.S. society journals in all the other sciences. This mechanism is the so-called page-charge practice, whereby the institution sponsoring a piece of research pays the journal for its publication, at a rate that now is generally sufficient to cover most of the prerun costs. The charge is not compulsory, and payment is not expected if it would constitute a hardship; the decision on acceptance of a paper for publication is made by the editor without knowledge of whether the page charge will be paid. Thanks to diligent education of research institutions and, in particular, to a policy enunciated by the Federal Council for Science and Technology,[19] which allows page-charge payments to be charged to budgets of grants and contracts, the great majority of papers in journals of the AIP do honor the request for page-charge payment.

The high circulation and wide use of page-charge journals attests to the social value of the policy. Moreover, journals with this type of financing can afford to take a cooperative attitude toward Xeroxing, reprinting, and other uses of their product that may be helpful to the scientific community. Nevertheless, there have been, especially in recent years when funds for research institutions have been suddenly constricted, some difficulties and misunderstandings. The issues raised have been reviewed elsewhere [2]; we shall comment here on only one of them. This is the argument that page charges impose a hardship on impecunious institutions. We feel that this view misunderstands, on the one hand, the freedom of those without funds to forego payment and, on the other hand, the fact that a nationwide shift toward publication in the often much more expensive non-page-charge journals could in some cases be more expensive to U.S. colleges and universities (as well as providing poorer dissemination). (We shall elaborate this point further in the final section of this chapter.)

The recent tightening of belts has, understandably but regrettably, led to nonhonoring of page-charge assessments even by some federally funded research projects with a budgeted item for publication. In an effort to counter this trend, many journals have adopted the so-called two-track system, publishing papers that honor the page charge as quickly as possible but delaying the others, so that the rate at which they are published is not greater than can be paid for by available subscription income. In most cases this procedure has been an effective incentive to keep page-charge honoring at a normal level, and only minor delays for the nonhonoring papers have built up; however, delays of over a year developed at one time for the *Journal of Chemical Physics*. Delays of this magnitude are even more harmful to the progress of science than an increase of subscription price would be (Reference 2, Sec. IVB4).

Research Journals: Miscellaneous Characteristics

In this subsection we shall consider a diversity of facts about primary journals as such and the grouping of articles in them; further diverse facts that relate to the research papers, considered one at a time, irrespective of the medium of publication, will be discussed in a subsequent subsection on Characteristics of Research Papers in Physics.

Typography and format, which are discussed in the Special Panel Report but not in this summary, are important both because of their effects on production costs and because of their relation to efficiency of use. Of more concern to most scientists is the time lag in publication. For journals publishing full-length papers, this lag ranges from a minimum of the order of three months or less, imposed by the purely mechanical aspects of editing, composition, and printing, to delays that may occasionally be well over a year. The individual variations are due mainly to the refereeing process and the intellectual aspects of editing but may sometimes be influenced by such factors as transmission overseas for composition or the existence of backlogs. Median lags for some typical physics journals are shown in Table 13.2. Physicists are fortunate in that these lags are, on the average, shorter than for journals in some other fields of science (Reference 2, p. 142).

TABLE 13.2   Median Time Lags[a] for Typical Physics Journals

Journal	Median Lag, Months as of————→ (Epoch)	
Phys. Rev. Lett.	2.6	Oct. 1971
Appl. Phys. Lett.	3.2	Sept.–Oct. 1971
Phys. Rev. A	5.8	Oct. 1971
J. Chem. Phys.	6.8	Oct. 1971
Ann. Phys. (N.Y.)	12.0	July–Sept. 1971
Nucl. Phys. A	4.9	Sept.–Oct. 1971
J. Phys. E	5.4	July–Sept. 1971

[a] Interval from receipt of paper by editor to receipt of published version by a subscriber in the United States. If manuscript was revised, date of receipt of revised manuscript is used. For non-letter journals, only full-length papers are counted.

Although the principal culprit in longer-than-minimum time delays is the practice of submitting papers to outside referees to obtain judgments of their suitability for publication, few physicists would suggest that this practice be abolished. It is used today by practically all journals. The system is sometimes criticized, however, as tending to entrench orthodoxy and to discriminate against authors who are not well known. A recent sociological study of the refereeing process[20] has shown that, at least in the *Physical Review,* refereeing seems to be conducted in a quite impartial manner. The distribution of papers over the various possible outcomes of the refereeing process is shown in Figure 13.14. In any event, it can hardly be doubted that the quality of papers is improved by the authors' knowledge that they will have to undergo refereeing. It should be pointed out, incidentally, that refereeing constitutes an appreciable voluntary, but hidden, subsidy to the production of the journals: the time and effort invested are considerable.

The language distribution in the journal literature of physics is note-

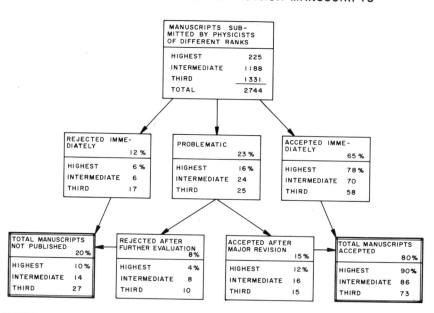

FIGURE 13.14   Fate of manuscripts submitted to the *Physical Review,* 1948–1956, by single authors. Manuscripts are divided into categories according to the estimated level of prestige (rank) of the authors (see Reference 20, p. 28).

worthy in that now, far more than a few decades ago, there is a great predominance of English. English accounts for 68 percent of the entries in *Physics Abstracts;* it is overwhelmingly predominant as a language for the publication of important works in India, The Netherlands, Scandinavia, Japan, and other countries. Russian accounts for 17 percent of the entries in *Physics Abstracts,* French and German for 5 percent to 7 percent each, and all other languages for less than 3 percent.

There has always been a tendency toward segregation of publications originating in a certain nation or region into particular journals. Accompanying this trend is a tendency for physicists of one region to pay more attention to journals of their region than to those originating elsewhere. This sort of provincialism seems to be decreasing, in part because of the birth of increasing numbers of highly specialized but broadly international journals, in part because of the rise of letters journals, and in part because of broad international movements such as the European Physical Society.

Another interesting phenomenon is the intellectual specialization of physics journals. As Figure 13.15 shows, there has been a growth, over recent decades, in the number and proportion of journals that are devoted to relatively highly specialized topics. It is interesting to note, from a table of most productive journals in each of 16 subfields of physics, given in the Special Panel Report, that very few journals occur on as many as four of the 16 lists.[21]

Although the need for some subdivision of the vast physics literature is obvious, one cannot but be distressed by the confusion engendered by the proliferation of so many diverse journals. A possible solution has been receiving serious consideration at AIP.[22] An editor or editorial board oriented toward any one of a large number (perhaps 50–100) of special subfields of physics might select, from articles scheduled for publication in any of several score of key physics journals, those articles likely to be of interest to specialists in this subfield. These articles would then be gathered together and printed in a single issue, with suitable referencing of the archival journals from which they were taken. Such groupings— called user journals—could be circulated inexpensively to individuals interested in each specialty; the same article might well appear simultaneously in the user journals of two or more specialties, if it proved relevant to all of them. Such a system could preserve the role of the original journals as archival repositories, but would transfer the current-awareness role—with a great boost in efficiency—to the new user journals.

The final item of some interest concerns the intellectual relationship of different journals to each other as revealed in their patterns of citation. Certain topics gravitate to certain journals, and mutual citations among

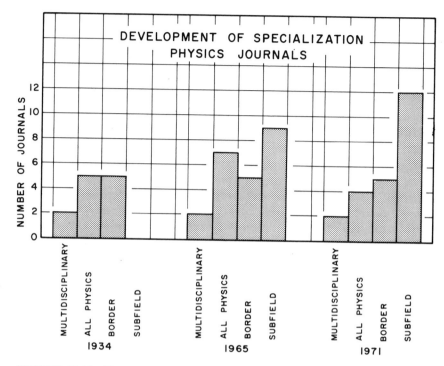

FIGURE 13.15 Time development of the degree of specialization of the leading physics journals. Journals are classified as multidisciplinary (that is, covering all of physics and one or more other disciplines as well), all physics (covering all areas of physics), border (covering areas such as instrumentation or physical chemistry that form interfaces with physics), or subfield (devoted to a subfield entirely within physics). The 1934 data refer to the 12 journals that accounted for half of the entries in *Physics Abstracts* in that year; the 1965 and 1971 data refer to the same set of 23 journals, which accounted for half of the entries in *Physics Abstracts* in 1965, but differ in that physics journals that have subdivided by 1971 have been transferred as units to the subfield column.

these journals are apt to be common; provincialism in reading patterns also affects the distribution of citations, and, of course, large journals are more likely to be cited than small ones. The latter factor can be eliminated by computing the ratio of citations to the bulk of the material published in the cited journal. Figure 13.16 shows a sample citation pattern that emerges when this is done. The ordinate is a composite of the quality of the work in the cited journal, its degree of specialization in the field of the citing journal, and the overlap of readership.

FIGURE 13.16 Relative frequency of citation of different journals in a sample of Section 2 (the first of the solid-state sections) of the *Physical Review* in 1968. For each cited journal, the number of citations to this journal was divided by the current bulk of this journal to give a number roughly proportional to the probability for a given paper in the given journal to be cited. The journal with the highest value for this probability (*Physical Review Letters*) was assigned abscissa 1, that with the second highest probability (*Journal of the Physics and Chemistry of Solids*), abscissa 2, etc. Note that if the two solid-state sections of the *Physical Review,* which can be separately purchased by members though not by nonmembers, were treated as a separate journal, this journal would probably have abscissa 1 or 2 instead of 5.

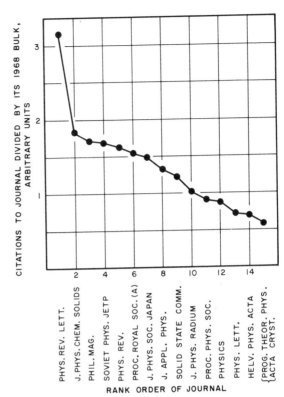

Primary Publication in Books

About one fourth of the articles published in proceedings of conferences appear in books rather than in journals. There are also a few items of original research that are published in collections of the *Festschrift* type. Both types of article are now probably covered reasonably well by *Physics Abstracts,* in which they account for a little less than 2 percent of the entries, or about 2.3 percent of the research papers. In 1969, this output, which amounted to about 1000 papers, appeared in approximately 50 books, priced usually in the range of 5 cents to 12 cents per equivalent kiloword, that is, a range similar to that which obtains for journals of commercial publishers. The circulations of conference proceedings are in one respect superior to those of commercial journals: They are usually purchased—often through registration fees—by many or all of the individual participants in a conference.

The subjective judgment of many physicists is that the average quality of papers in conference proceedings is lower than that of papers in the better journals. Although there is usually an attempt to maintain quality by refereeing, the constraints of rigid deadlines for submission and strict limitations of length undoubtedly have an adverse effect on the quality. Nevertheless, citation statistics are not altogether unfavorable to papers published in conference proceedings, perhaps because the defects of quality are compensated by the fact that the work represented by these papers becomes well known to the community of workers in the field, many of whom will have attended the conference and purchased the conference proceedings. For example, in a sample of all four sections of *Physical Review*, 1971, the ratio of citations to papers in books devoted to conference proceedings, or in *Festschriften*, to the citations of journal articles, was something like 3.25 to 4 percent; this is rather larger than the ratio of the number of such papers in *Physics Abstracts* to the number of journal articles, which we estimated to be less than 2.5 percent. On the other hand, there is fragmentary statistical evidence that conference papers are somewhat less likely to be cited than those in *leading* journals. There is also some intriguing fragmentary evidence that the frequency of citations to conference books varies appreciably from one subfield of physics to another.

An obvious recommendation with which to close is that *wherever possible, the managers of conferences should arrange to publish the proceedings (if publication is deemed necessary) in a journal of wide circulation rather than in a special book.* The proceedings issue should, of course, be available to individual participants by itself at a low price.

Characteristics of Research Papers in Physics

Now we turn to research papers themselves—the intellectual and sociological aspects of the work they describe, their quality, their style. We shall start with a look at the distribution of current physics research papers over subfields of physics, nations, and institutions, as determined in a sampling of entries in *Physics Abstracts* conducted by members of the Statistical Data Panel. The methodology and detailed results are described in the Special Panel Report. Figure 13.17 shows some of the results. We call attention to several striking features:

1. Production of papers in condensed matter is larger by far than that in the other subfields shown, accounting for about 40 percent of the publications in physics. (Condensed matter also accounts for one fourth of the PhD physics manpower in the United States, being the largest physics subfield.)

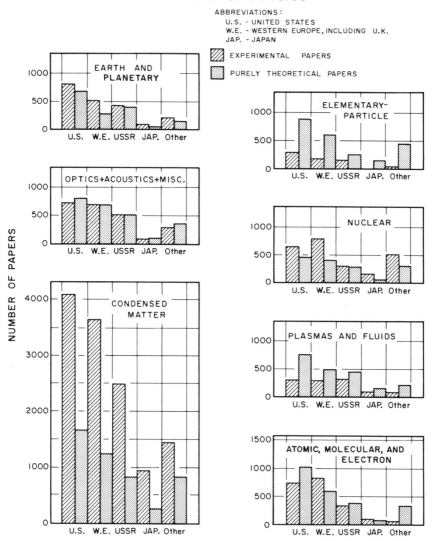

FIGURE 13.17 Production of research papers in 1969 in various subfields of physics by institutions in various countries or regions.

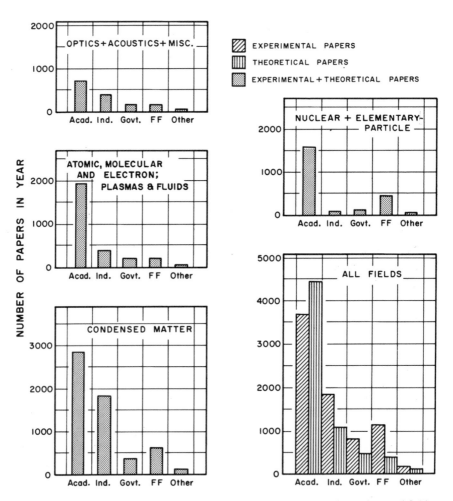

FIGURE 13.18 Distribution of production of research papers in various subfields of physics over different types of U.S. institutions. (Theoretical and experimental papers are grouped together except for the total for all subfields.)

2. The quantity of work in the United States exceeds that in the other countries or regions shown in most, but not all, subfields.

3. Theoretical papers greatly outnumber experimental ones in elementary-particle physics and in relativistic astrophysics (not shown). In most, though not all, other subfields experimental papers predominate; and in biophysics (not shown) there are essentially no purely theoretical papers.

4. Experimental work in fields requiring expensive equipment, most notably elementary-particle physics, is confined to the wealthier countries.

The distributions of published papers from U.S. institutions over subfields and kinds of institutions appears in Figure 13.18. We note:

5. Slightly more of the work in academic institutions is more theoretical than experimental; all the other types of institution produce more experimental than theoretical papers.

6. About 63 percent of the papers from industrial laboratories are in condensed matter; yet, even in this subfield, as in the classical subfields of optics and acoustics (and miscellaneous), academic institutions produce considerably more papers than do industrial ones.

Let us now turn from these gross statistics to some more subtle characteristics of research papers. One such characteristic is length. As Figure 13.19 shows, papers in the *Physical Review* are longer, on the average, than they were one or a few decades ago. This finding suggests, though it does not prove, that papers published in all types of media have been getting longer.

FIGURE 13.19   The increase, over the decades, in the average length of full papers (excluding letters, comments, and the like) in the *Physical Review*. (Equivalent kilowords means the number of words that would occupy the space used by a paper if this space were set in solid text.)

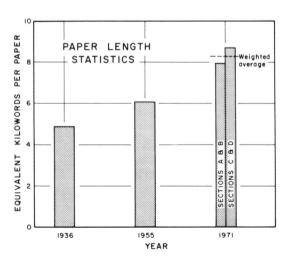

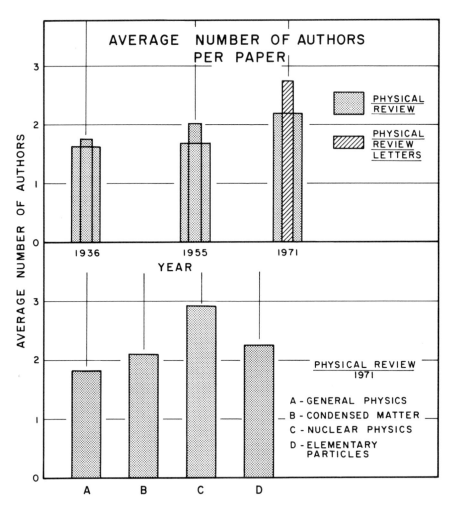

FIGURE 13.20   Average number of authors per paper in the *Physical Review* or *Physical Review Letters*. The top graph shows the growth in this number over the decades (1971 figures for the total of all sections of the *Physical Review*). The bottom graph shows the variation in this number from subfield to subfield in 1971, as exemplified by the comparison of *Physical Review* Sections A (general physics—largely atomic, molecular, and fluid), B (condensed matter), C (nuclear), and D (particles and fields. The relatively low average for papers in Section D probably reflects the large proportion of theoretical papers, which tend to have fewer authors and thus to obscure the high level of multiple authorship associated with the experimental papers.

Another characteristic of papers, which is even more directly related to the changing conditions of research, is the multiplicity of authorship. Some data are shown in Figure 13.20 indicating that the number of authors per paper has been increasing, and that this number varies a good deal from one subfield of physics to another. There are probably two main effects: Experimental work using large machines tends to produce multiauthored papers; theoretical papers tend to have few authors.

Still another measurable characteristic of papers is the number and distribution of the references they cite. According to some counts we have made:

1. The average number of journal articles cited per paper in the *Physical Review* has increased from about 12.5 in 1936 to approximately 16.5 in 1961.* This increase, however, has been less than the increase in length shown in Figure 13.19.

2. Citations of unpublished theses and reports and of research articles in books were much less common in 1936 than today.

Currently, there are, of course, many more productive scientists in each subfield, even when the scope of a subfield is more narrowly defined, than was the case in the 1930's; thus, increasing numbers of citations might be expected.

The distribution of citations in age depends on the rate of obsolescence of the cited papers, the general growth of the literature, and, doubtless, the psychology of the citing authors. Figure 13.21 [23] shows some age distributions for citations in papers of the *Physical Review*. We note:

3. In most subfields the frequency of citation of past literature decreases with increasing age at a rate corresponding to a factor 2 in about 3.5 years.

4. This decrease is compounded in comparable amounts from the exponential growth of the literature capable of being cited (Figure 13.6) and the obsolescence of this literature. The obsolescence half-life appears to be about eight years. (We shall return to this subject in our discussion of Patterns of Use of Primary Publications.)

Preprints, Reports, Theses

Much research work appears initially, and occasionally solely, in various kinds of documents that are unpublished, though often available to anyone

---

* This statistic could be regarded as an index of increased specialization in physics, since the rate of growth in number of citations is much less than the growth in the number of potentially citable articles.

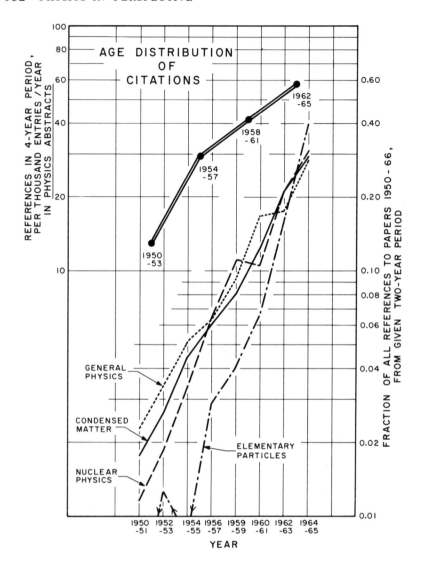

FIGURE 13.21   Age distribution of citations in a sample of papers from the *Physical Review*, 1967 (adapted from Reference 23). The lower curves (see scale at right) depict the fraction of all references to papers dated 1950–1966 that were to papers published in each of eight successive two-year periods for each of four parts of the *Physical Review*. Data for papers published in 1966 are not shown, as most of the citing papers were submitted during 1966. The upper curve (see scale at left) shows total references in all issues of the *Physical Review* divided by number of papers in *Physics Abstracts* for the corresponding year and summed over four-year intervals.

on special order. For example, we have seen in Figure 13.7 that reports and theses account for about 3 percent of *Physics Abstracts* and about 25 percent of the physics part of *Nuclear Science Abstracts*. In addition, many (probably most) papers that are submitted for publication in journals are sent in preprint form by the authors to tens or even hundreds of their colleagues in other institutions. Physicists do more of this than do mathematicians, biologists, or chemists.[24]

The most prolific and best-organized sources of reports are agencies of the federal government such as the National Technical Information Service and the Defense Documentation Center. These offer many tens of thousands of reports a year, most, of course, being of much more interest to engineers and others than to physicists. The number of classified and limited-distribution reports produced by government agencies and contractors is far larger. The information contained in these reports is often publishable and usually does get published, but in a perceptible minority of cases the publishable materal remains unpublished for at least several years. Further details can be found in a recent report.[25]

In one subfield of physics, high-energy theory, there has been a fairly detailed study of the origination, reception, and use of unpublished documents.[26] Among the respondents to an extensive questionnaire there were, of course, wide variations in the number of preprints they sent out. However, the average number of copies distributed was 148; the average number of typewritten pages was 22. The average respondent received about 186 unpublished documents in this subfield per year. From these figures one can estimate that the reproduction and mailing costs for distribution of preprints in high-energy theory have been no more than 5 percent to 10 percent of the total cost (editorial, composition, runoff, profit, and the like) of journal publication of all work in this subfield.

Both in physics and in other fields of science there have been lively controversies in recent years over the desirability of centrally subsidized schemes for large-scale distribution of preprints.[27-29] Proponents cite speed and selectivity of distribution; opponents worry about cost and degradation of quality. Fortunately, data we have presented here and in our discussion of information-gathering habits (in the second section) help one to find a commonsense middle ground. According to Figures 13.3 and 13.4, preprints supplied 10 percent to 20 percent of useful ideas to research physicists; if this is so, the present level of expense seems justified. But, since the average high-energy theorist already receives 186 preprints a year, expansion of preprint distribution by a large factor would not be likely to increase readership correspondingly. What is needed is better aim, not increased quantity. Newcomers to a field should have the same opportunity to receive preprints as should those established in it, yet distribution to those not really interested in the subject should be minimized. These goals might be achieved by a centralized service that circulated lists of available preprints and sent the documents only on postcard request; possibly a small fee should be charged. The service could also send announcements of the appearance in print of items previously distributed, ceasing distribution of these items and enabling journals to ban references to preprints of them.

Quality of Published Research in Different Fields

We come now to a question of great conceptual interest and practical importance: Can one, in any objective way, compare the quality of work in different fields or at different times? Some grounds for hope emerged from an experiment we conducted with the assistance of the Panel on the Physics of Condensed Matter, though the results are only preliminary. The methodology is described in the Special Panel Report. The experiment had two parts, one a study of citations to randomly selected papers in a given subfield, the other an evaluation of the impact of each paper in as objective terms as possible, by experts in the field of the paper. The latter evaluations thus constitute a self-appraisal of the field by its own experts and so avoid bias for or against one field as compared with another. There does seem to be a correlation of grades with citations, and a judicious combination of these two types of information appears to make it possible at times to identify areas of research in which there is a higher-than-average proportion of trivial work or a higher-than-average proportion of exciting work.

Although this approach is only sometimes successful in distinguishing one field from another in regard to quality distribution, the grading experiment gives rather cheering news regarding the health of the condensed-matter subfield in the United States. Thus, of all the papers graded, less than 8 percent were judged worthless by *any* of the experts grading them, and 33 percent were judged by one or more of the experts to add to the body of scientific knowledge at least something of fairly lasting value and conceptual interest. (The remainder, although making identifiable positive contributions, were considered either pedestrian or likely to be soon outdated.) From what few data we have been able to gather, we suspect that many nonphysics fields of science would fare less well under evaluation by their own experts. We feel, therefore, that these results are an indication of good health for the condensed-matter subfield; we have no reason to doubt that other areas of physics would fare comparably well. These data on condensed matter are contrary to many more subjective statements made by prominent physicists about the literature of physics. For example,

Ziman has stated: ". . . the consequences of flabbiness [in science policy] are all too sadly evident in all quarters of the globe—the proliferation of third rate research which is just as expensive in money and materials as the best, but does not really satisfy those who carry it out, and adds nothing at all to the world's stock of useful or useless knowledge." The answer may lie in a well-known phenomenon of questionnaires—that a person's perception of generalities often is quite different from the generalities derived by a researcher from that person's perception of individual instances.

Patterns of Use of Primary Publications

The principal value of primary publications, certainly, comes from their being read. This is true in spite of the often-noted value of publication as a stimulus and morale builder to authors. But publication could not fulfill these roles unless there were, at least on the average, readers. We have seen in Figures 13.3 and 13.4 that the importance of these publications to readers is tremendous. Together, journals, preprints, and like media are by a considerable margin the most important source of detailed information and even of awareness of the existence of information.

One can learn a great deal about the use of primary publications by authors of research papers from studies of citation statistics; we have, in fact, already discussed some of these statistics (see discussion of Figures 13.16 and 13.21). It is interesting to supplement Figure 13.21 with some further data relating to the issue of obsolescence. Thus, Figure 13.22 shows the way in which the number of citations to a sample of papers published in a single year, 1962, varied with time for each of the years 1964–1970. After the initial rise (probably real—a catching-on effect), the citations decrease, at an average rate corresponding to a factor 2 in about four years. An attempt at a self-consistent fit to these data and those of Figures 13.21 and 13.6 and the small changes with the years in the number of references per paper suggest a half-life for citations to a given paper that is closer to five years.

Using the same data, we can also examine the probability for a given paper published in year t to be cited in another paper in the same field, published in year $t + \Delta t$. For a fixed value of the time interval Δt between the two papers, this probability seems to be decreasing exponentially as t increases. This is what may be called a dilution effect: As time advances, all papers, both recent ones and those that have been in existence any given number of years, are less likely to be cited in a particular paper than was the case in former years, because there are so many more papers for them to compete with. This effect almost cancels the rate of growth

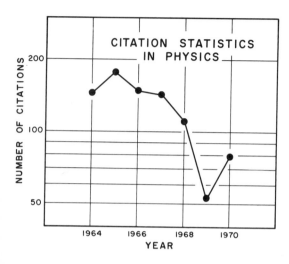

FIGURE 13.22 Number of citations listed in the *Science Citation Index* in each of seven years for a set of 102 articles published in 1962, half in the *Physical Review* and half in *Annals of Physics*. (The largest number of citations for any one article in any year was 20.)

of the literature in its effect on the number of references a given paper cites, but not quite, as evidenced by the slow growth in number of references per paper, which we noted in discussing characteristics of research papers.

We have given special attention to citations because they are measures of that type of use of the literature that is devoted to the work of incorporating new pieces into the structure of science. But this is only one type of use; primary papers may be used by people who are not doing work that is to be published, and even research workers do not cite all the articles they read and benefit from. Obsolescence with regard to this type of use can be inferred from statistics like those shown in Table 13.3. The rate of decline of usage with increasing age of journals is surprisingly slow, being only slightly faster than what one would expect from the general rate of growth of the literature without any obsolescence at all. Despite corrections, discussed in the Special Panel Report, that need to be applied to the raw data, it is hard to escape the conclusion that *the obsolescence half-life of journal articles in physics, as defined by probability of use by any one individual, is probably at least as long as that defined by citations (Figure 13.21, top) and may well be of the order of a decade.*

Although there is an appreciable use by physicists of primary literature in libraries (including the finding of material which is then Xeroxed), studies of reading habits show that literature available only in libraries is consulted much less frequently than that available in more convenient locations. One regrettable consequence of this is a provincialism in reading patterns, enhanced by the reluctance of scientists to read articles in

TABLE 13.3 Age Distribution of Journals Left by Readers on Tables in a Library at Bell Laboratories, 1971

Range of Dates	Observed Number of Physics Journals	Calculated [a] Number
1970–1971	13	0
1965–1969	73	63.2
1960–1964	36	36.0
1950–1959	32	32.7
1940–1949	4	10.7
1940	5	5.3

[a] Based on the assumption

Probability of consultation of a journal from year $\tau \propto \exp{(\tau/9 \text{ yr})}$

and fitted to the observed number for the years 1960–1964.

foreign languages. There is some evidence that exciting items appearing in the literature are significantly more frequently and more quickly picked up by American physicists if they appear in a certain few of the most popular journals than if they appear elsewhere. Moreover, there is further evidence that even material appearing in the most popular journals will often fail to come to the attention of a physicist highly interested in it unless he systematically scans these journals or some current-awareness service. In other words, preprints and the highly effective oral grapevine cannot be relied on to communicate all the more interesting new results to people who might be excited by them.

THE REVIEW AND CONSOLIDATION OF INFORMATION

The Need for Consolidation

The sifting, evaluation, and consolidation of new knowledge have always been essential to the advance of science. In the old days, many of the greatest minds in physics constructed magnificent anchor points for its further progress by gathering together, in one mighty treatise, all that was known about some area and presenting it from a critical and unified point of view. Books or *Handbuch* articles by Rayleigh, Lamb, Sommerfeld, Born, Bethe, and many others were the Bibles of their day, and sometimes remain so today. But now things have changed for the worse; although

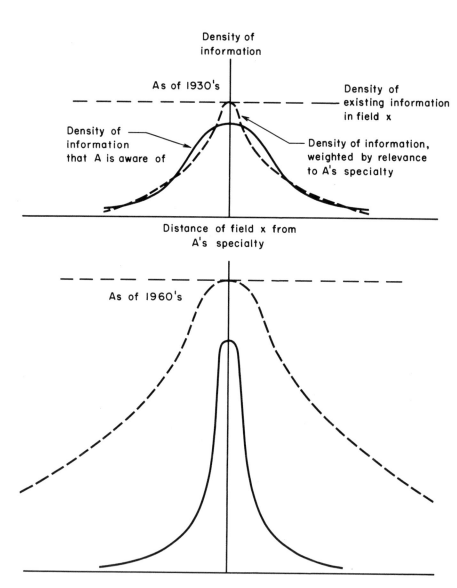

FIGURE 13.23 A schematic representation of awareness versus relevance for fields neighboring the specialty of a typical physicist, Mr. A.

the need for such syntheses, or even a weak approximation to them, has increased, the vast expansion of the literature has made it practically impossible to assemble them.

The way times have changed is illustrated schematically in Figure 13.23.[30] The full curves represent the level of awareness of a typical research worker, Mr. A, in various fields x, plotted as a function of the intellectual remoteness of x from A's specialty. Because of the vastly greater rate of production of new information now than in the 1930's, the lower full curve has to be much narrower than the upper, the area under it being little, if any, greater. The dashed curves, on the other hand, represent the amount of information in different fields that is reasonably relevant to A's specialty. Advancing knowledge is continually establishing relations between fields that were previously unrelated. So the dashed curve for the 1960's is broader than that for the 1930's; the full and dashed curves have changed in opposite directions, and their relation to each other is now, in most fields, qualitatively different from what it was then.

The situation will surely be worse in the future, not only because of further expansion and developing interrelatedness of the literature, but because interpersonal contacts, which play such an important role in information transfer today, will probably become less adequate. For one thing, although they may be relied on even more in the wings of the dashed curve of Figure 13.23 than at the center, they are undoubtedly less effective there; as the relative importance of the wings will grow with time, this shortcoming will be felt more. A second consideration is that it is likely for economic reasons that the current exponential growth of information will slow down considerably in the next generation. As the slowdown proceeds, there will be an increase in the proportion of the total store of information, relevant to a given piece of current work, that is more than a few years old. This older information is less likely, as compared with fresh information, to emerge from casual contacts with one's colleagues.

As we have seen in the discussion of technology and economics in the section on Secondary Services, much thought is currently being given to improved retrieval and indexing schemes. But although these promise to be very useful, they are not by themselves the answer. Not only are there grave difficulties (see Patterns of Use of Secondary Services) in making them adaptable to the wide variety of user needs; more seriously, even if they work perfectly, they may only choke the customer with an indigestible surfeit. Herring [30] has cited as an example a theoretical specialty in which the literature consists of about 150 very abstruse papers, some wrong, some overlapping, but most containing some morsel of value. A physicist, interested in this topic but not planning to spend years in the field, would be poorly served by having the 150 papers dropped in his lap, or even any subset of them. An orderly assemblage of all the valid morsels (which could be given in a tenth the total space) would be vastly useful.

The same sort of thing holds true for experimental data: A worker who wants a particular datum will be shabbily served by being given a bibliography of some dozen or so conflicting measurements. To name but one major field, a recent study by a panel of the National Research Council [31] has pointed out that compilation of nuclear data has fallen dangerously far behind the accumulation of experimental results. Only compilations accompanied by critical evaluation will rescue the situation.

As these examples show, there is no substitute for evaluating and compacting information and presenting the result—the hard core of scientific knowledge—in an easily accessible form.

Amount and Types of Syntheses Now Being Produced

Two of the most obvious questions to ask about the existing review literature are: How much of it is there? How is this bulk distributed among books, articles, compilations, and the like? These are not at all

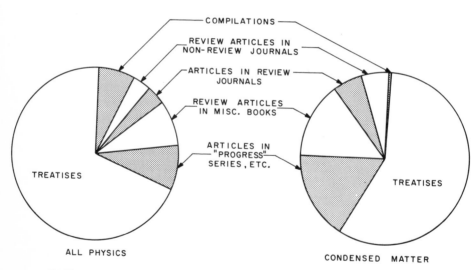

FIGURE 13.24   Two kinds of estimate of the relative page bulk of different kinds of review literature in physics as a whole and in the physics of condensed matter. Estimates for all physics in 1966 (see Reference 31) were obtained from an analysis of *Physics Abstracts,* book lists, and the like. Estimates for the physics of condensed matter, 1946–1967, were obtained by querying a sample of authors of papers in the physics of condensed matter.

easy questions to answer; some of the difficulties are discussed in the Special Panel Report. With some rather arbitrary decisions on the boundaries of physics and of review literature, the conclusion reached there is that the total page bulk of review literature produced annually in physics—the literature of nonnegligible interest to research workers—is currently of the order of 80,000 to 100,000 pages per year, or perhaps one fourth of the bulk of the primary literature. (This figure may make the reader wonder: If there is so much of it, why do we need more? The answer, as we shall see presently, is that this literature is highly redundant because it is—properly—aimed at many different audiences; its coverage of recent research is still not adequate.)

The distribution of this bulk among the various types of review literature is shown, for all physics and for one of its subfields, condensed matter, in Figure 13.24. It will be seen that pagewise the book material enormously outweighs that in journals, and that the material in single-topic books—treatises—greatly outweighs that in books made up of separate articles.

Coverage and Obsolescence

How much of the journal literature do these treatises and review articles cover? Counts of references show that in condensed matter the treatises typically have rather less than one reference per page, the review articles rather more than two. As one goes from theoretical and logically structural fields to more empirical or taxonomic ones, the density of references in treatises and reviews gets higher. Even in condensed matter, where the density of references is fairly low, the total number of references in a year's production of review literature is half again as large as the total number of new condensed-matter research papers published in a year. But of course, by no means all the papers in the bibliographies are papers in condensed matter, and there is great overlap.

A much more significant question, but unfortunately quite difficult to answer, has to do with how well the really significant knowledge is covered. The only study seems to be one in the physics of condensed matter, briefly mentioned by Herring,[32] which concluded that a paper whose significance would entitle it to a place in the top tenth or fifth of all papers published would have a chance of the order of 0.8 of having at least its main results quoted somewhere in the review literature within five years. But in the most comprehensive and respected single source of review literature—the Seitz-Turnbull-Ehrenreich series *Solid State Physics*—the probability would be only about $\frac{1}{6}$.

Because of the relatively slow rate at which review literature is written, it is usually some years before a given item of new research is picked up for incorporation in a treatise or review. And even after the author of the review has completed his work, there is a delay before it appears in print. Thus the review literature can never be fully up to date. Figure 13.25 illustrates this and also shows that, as one might expect, treatises are less current than review articles in journals. Even review articles in journals seem typically to achieve their optimal coverage of the literature only for published papers two or more years old. Books containing separate contributions from different authors (not shown in the figure) are apt to be slowest of all.

We see, then, that even a good treatise or review article is to some

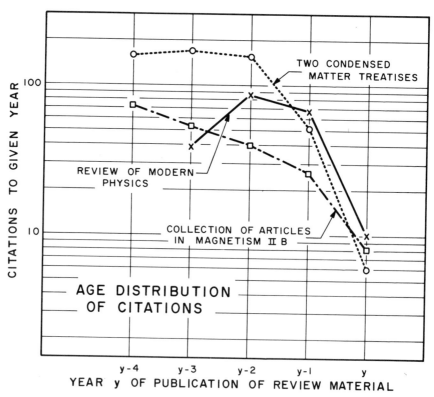

FIGURE 13.25 Age distribution of citations in the bibliographies of three samples of review literature: a sampling of review articles from *Reviews of Modern Physics,* 1964–1967; a sampling from the collection of articles in *Magnetism II B* (G. T. Rado and H. Suhl, eds., Academic Press, New York, 1966); and a sampling of references in two condensed-matter treatises.

extent obsolete when it appears. Subsequent obsolescence seems, however, to be surprisingly slow on the average. Studies of citations of books and reviews have shown, as described in the Special Panel Report, that the probability for a particular one of them to be cited in a research paper decays much more slowly with increasing time interval between the latter and the former than does the corresponding probability for a research paper to be cited. The half-life for obsolescence seems to be at least of the order of ten years in the cases that have been studied.

Authors and the Writing Process

Now let us take a look at the people who write reviews and other consolidated material. Who are they, how much time do they spend, what are their problems and rewards?

As studies have shown that Fellows of the American Physical Society typify a majority of the American authors of treatises and reviews, a natural experiment [32] was to query a random sample of Fellows in condensed-matter physics to find out how many books and reviews had been written by them and of what type. The results, which were consistent with other evidence (see especially Reference 24), suggested that the men in this group had been writing review material (including advanced textbooks) at a rate of the order of 15 pages per year per man. Although the investment of time per page written varies greatly from case to case, data supplied by a few authors suggest that an average value of the order of four hours per page would not be unreasonable. Thus in the writing of books and reviews for the use of other research workers, the average expenditure of time by condensed-matter physicists of the level of APS Fellows may be something like 60 hours per year.

These statistics having suggested that review writing is not in general at a level that cuts seriously into the time of most of the competent physicists, it was natural that more detailed studies of the motivations and problems of authors should be undertaken to determine what measures would be effective in increasing this activity. Several such studies were conducted in 1970–1971 by the AIP and the Subcommittee on Reviews and Compilations of the Advisory Committee on its information program. These studies uncovered a good deal of information about authors' motivations, rewards, mode of operation, and difficulties. Some of these results are listed in the Special Panel Report. Here it will suffice to note just one of them: Although assistance by graduate students or postdoctoral associates was only rarely employed, nearly two thirds of the authors said that they would have used such assistance if funds to pay for it had been available. An independent study by the APS Division of Nuclear Physics revealed a similar result: Many leading workers in the field expressed a willingness to supervise the preparation of nuclear-data compilations if adequate postdoctoral assistance could be provided. (In the subsection on Goals and Suggestions, we shall mention recent actions by the AIP and by the NRC that endeavor to capitalize on these findings.)

Distribution and Use

Remembering that the amount of use made of any information resource is strongly correlated with its accessibility, we ought now to look at the circulations of books, review journals, and the like and their market prices, which undoubtedly influence the circulation.

Let us look first at books. Prices vary widely, from below 1 cent to over 8.5 cents per page; paperback editions are apt to be not much more than half as costly as hard-cover editions. The median price is around 4.5 cents per page, or 9 cents per kiloword. Although information on circulation usually is not publicly available, our impression, based on information supplied by a handful of authors, is that books of the type used by research workers typically have total sales in the range of 2000–3000 copies. Of course, a very successful graduate text may run considerably higher; an outstandingly successful "Progress in . . ." series may attain a circulation of 5000.

As for review journals, the commercially produced ones seem to have prices per kiloword similar to those of books or of commercially produced research journals but circulations a little larger than for primary journals of similar unit price. Subsidized review journals, sold at a low price, can have vastly larger circulation; *Reviews of Modern Physics* seems to be the largest, with a circulation of around 12,000. The low circulation of the translation journal *Soviet Physics—Uspekhi* (approximately 1100 a year or two ago) is typical of translation journals, yet the quality and price of the material offered seem to justify a higher circulation. Does the thinking of Soviet and American schools mesh poorly? Or is it merely that the subscription patterns of libraries and sublibraries have a great deal of inertia?

Now let us turn to the actual use of books, reviews, and compilations. In spite of the fact that questionnaires have repeatedly shown that physicists desire more reviews, the evidence of Figures 13.3 and 13.4 is that physicists spend only a small fraction of their time in the use of these sources of information, and that only a small proportion of the useful information they assimilate in their daily work comes via them. However, it is not hard to think of two or three factors that act to depress usage. One is the mismatch between their typical lack of currency, illustrated in

Figure 13.25, and the comparative recency of so much of the information currently used by physicists. Another thing that retards use is the distribution of review material over such a large number of books, review journals, and other types of journal that many users may not be aware of the existence of material that could help them. Still another cause may be the difficulty of finding and extracting useful material, even when it is known that it is in a certain book or review article.

These considerations suggest that when we look at citations of review literature, we do so with an eye to finding out what types of such literature are most used. One type of data that can be obtained is illustrated by Figure 13.26,[32] which shows the distribution of citations in the *Physical Review* over the various kinds of review literature and contrasts this distribution with that previously presented in Figure 13.24 for the bulk of the review literature produced in the various categories. It will be seen that treatises, although still the most cited category, are less cited in proportion to their bulk than the shorter items; it is not difficult to think of reasons for this. More thought-provoking, however, is the fact that the review articles appearing in nonreview journals and conference reports, summer school notes, and the like seem to be less often cited in proportion

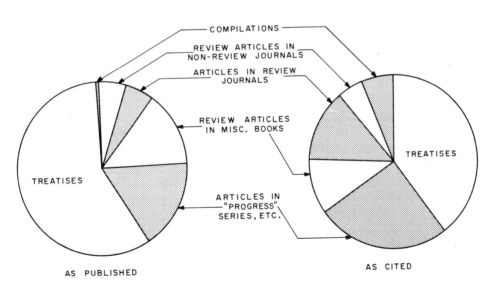

FIGURE 13.26 Comparison of the relative page bulk of different kinds of review literature in the physics of condensed matter (1956) with the relative numbers of citations to these same kinds of review literature in part A of the *Physical Review* (1964).

to their bulk than the longer ones appearing in progress books or review journals. This fact, which has been confirmed in several other studies, probably means that *the review articles that are the most used are the ones that are reasonably thorough and are published in fairly obvious and accessible places.* An author serves himself and his colleagues poorly if, after making the effort to write an acceptable review, he publishes it in an out-of-the-way book or a journal one does not normally think of as a repository for review material.

Because one factor in the accessibility of review literature is its circulation, which depends in large degree on its price, it is desirable, just as in the case of primary and secondary publications, that as much as possible of it be so supported that it can be marketed near runoff cost.

Goals and Suggestions

A few obvious desiderata are suggested by what we have learned about the use of review material. There is obviously a need for many different types and levels of books and reviews, corresponding to the different interests and different levels of prior preparation of the various types of user. The matching of the review literature to its users can be facilitated if each author identifies, at the outset of his work, the audience to which it is addressed and the level of the treatment. If his treatment is other than exhaustive, he should give reference to and his critical opinion of available treatments of the subject that go into greater depth. Access tools—for example, annotated bibliographies of reviews—are badly needed to guide the user to review literature that is appropriate for his particular need. Authors of longer reviews and treatises should so locate definitions of symbols and concepts as to facilitate piecemeal use.

Measures to enhance the production of review literature and guide it toward these other goals should be taken by sponsors of research, by academic and other organizations, and by the physics community. The need for enhancing the prestige of this kind of work is obvious and has often been commented on. A variety of more specific measures can and should be tried. One, which we shall describe in the section on information analysis centers, is the setting up of what may be called "review writing centers." Another measure is suggested by a fact noted above as an important practical deterrent to the preparation of reviews and compilations: At present they rarely have a place in the responsibilities for which scientists and their assistants are explicitly funded. *It should be just as easy for a scientist to obtain grants for review preparation as for new research, and such grants should allow for participation, when appropriate, of graduate students and postdoctorals and for travel, computer time, over-*

head, and the like. A specific proposal for an experimental program of grants of just this type has recently been made by the AIP[33]; even further advanced is a two-year program, administered by the National Research Council with National Science Foundation support, for compilation of nuclear data by teams of senior physicists and postdoctoral fellows.[31]

ORAL AND INTERPERSONAL COMMUNICATION

In this section we shall discuss the channels of communication shown in the lower part of Figure 13.1, that is, oral communication, both formal (lectures and meetings) and informal (conversation), and person-to-person written communication. As we have seen in Figures 13.3 and 13.4, these channels play a role in the physics community comparable in importance to that of the journal literature and certainly consume more time.

Meetings

Let us start with some rough statistics about the number, diversity, and size of meetings attended by physicists. Data assembled in the Special Panel Report give the following picture:

1. Total attendance by U.S. physicists at domestic meetings probably came to something of the order of 40,000 to 50,000 man-meetings in 1969.

2. Total attendance by U.S. physicists at conferences abroad is harder to estimate, but was probably within ± 50 percent of 5000 man-meetings.

3. The regular periodic meetings of scientific societies, which are often rather large, account for close to half of the domestic man-meetings mentioned in 1. But miscellaneous meetings, most of them aperiodic, are far more numerous.

4. Similarly, the periodic international meetings, especially the ones sponsored by the various Commissions of the International Union of Pure and Applied Physics, probably account for over half of the man-meetings involving foreign travel by U.S. physicists.

5. The total cost of meetings (nearly all of it travel and living expenses) is something of the order of $8 million to $10 million per year for U.S. physicists. About three quarters of this goes for domestic meetings, one quarter for foreign. The sizes of society meetings seem to grow in time approximately proportionally to society membership. Figure 13.27 shows some examples and also suggests that current shortages of funds have either caused attendance to drop off or led to extensive evasion of registration fees.

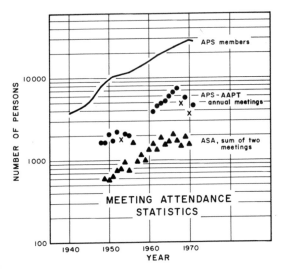

FIGURE 13.27 Comparison of attendance at meetings with society membership over the last two to three decades. The upper curve depicts the membership of the American Physical Society (APS). The circles indicate registered attendance at the Annual Meetings of the APS and the American Association of Physics Teachers (AAPT) when held in New York. (Unregistered attendance is a small but unknown fraction of the figures shown.) The crosses indicate registered attendance at APS and AAPT Annual Meetings not held in New York. The triangles show the sum of registered attendance at two yearly meetings of the Acoustical Society of America (membership was 4403 in 1968).

Many qualitatively familiar facts have been roughly quantified in a study of meetings of various scientific societies by the Johns Hopkins University Center for Research in Scientific Communication. In its study of a large meeting of the Optical Society of America [34] this group found that the contributions of the meeting to scientific communication were made in comparable measure through three channels: receipt by authors of comments and inquiries regarding their presentations; information acquired by listeners from presentations or from contact with authors that was stimulated by these presentations; informal contacts unrelated to formal presentations. (The role of the latter type of communication may have been abnormally low at this meeting, because of lack of favorable opportunities.) It must be remembered, too, that contacts initiated at meetings are often responsible for other types of informal communication between workers at different institutions (whose important role we shall examine later in this section).

In considering the findings on communication at meetings, it must be remembered that those who attend meetings are a rather select group. For example, the average member of the APS goes to 0.40 or 0.45 meetings of the Society a year, whereas there are many members who go to two or more.

A plausible guess is that, except for the regular national and regional meetings of societies, about half of all conferences publish their proceedings.

A little less than half the proceedings, but rather less than a third of the total of papers, appear in books, the rest in journals. Journal publication probably gives better accessibility to nonparticipants.

Colloquia and Seminars

Most physicists probably spend only a small fraction of their time attending seminars at their own institutions. In small institutions there are few to attend; in large institutions, as Figure 13.3 shows, shortage of time seems to cause them to restrict their attendance to the few that are of most outstanding interest to them. Nevertheless, the total time the average physicist spends in seminars is probably over five times as large as the total time he spends in meetings away from home. Now, in most universities and probably in many other organizations, a majority of speakers at seminars are visitors from outside the host institution. This type of communication between scientists attached to different institutions has some advantages over communication at meetings; presentations are more unhurried, and there is more opportunity for discussion. Travel expenses are incurred by the speaker only, not by the listeners. Thus this type of communication probably surpasses meetings in the total volume and utility of communication by it—and certainly surpasses them in the value received per dollar expended. However, it cannot do all the things that meetings do, nor can it be indefinitely expanded.

Personal Oral Communication

As we have seen in Figures 13.3 and 13.4, personal oral communication is by far the most extensively used medium of communication for physicists and has a place second only to primary literature in terms of the amount and value of the information it communicates. It is very difficult to study satisfactorily, and very few of the many studies that have been attempted have involved physicists. Nevertheless, when one puts all the studies together, a few facts seem to emerge:

1. For many areas and institutions, at least, conversations with colleagues from other institutions are quite frequent, and their value in the transmission of useful information is of the same order as that of conversations with colleagues in the home institution. One can guess that conversations with invited seminar speakers constitute an important part of the total.

2. The relative use of conversations with external colleagues increases

with increasing educational level and decreases as one passes from research to development to operation. It is positively correlated with attendance at meetings.

3. Much of the information received through personal contacts is information that the receiver would not have thought of seeking but that nevertheless turns out to be very useful.

The low average use of extraorganizational sources by engineers and applied scientists may not mean that such sources are unimportant in the total chain of communication of information to them, just as their low average use of literature does not mean that literature is unimportant. As we have briefly noted in the discussion of individual differences in communication in the second section, information from outside an industrial organization typically penetrates it through special individuals, called technological gatekeepers.[3,8,11] We have seen in Figure 13.4 that information flows in a network of channels, connected now in series, now in parallel. The gatekeepers happen to be in series with many of the channels.

The general importance of physical proximity and convenience in the use of information sources, which we noted above in the second section, applies strongly to informal oral communication.[3] As we shall see, this factor is especially important for interdisciplinary communication.

Personal Letters

As Figures 13.3 and 13.4 show, personal correspondence has a small but far from negligible role in communication of physics information. As this channel has been little studied, we will make only one comment on it: It is especially inportant for communication with colleagues abroad, and this aspect of its use no doubt depends strongly on personal contacts initiated at international meetings.

INFORMATION ANALYSIS CENTERS

An earlier section of this chapter was devoted to the consolidation of information and discussed several kinds of consolidation. But one important kind was omitted, and this we shall take up now: the information analysis center. This is an organization, large or small, whose primary purpose is to sift useful information (usually, though not necessarily, numerical data) from the primary literature, evaluate it, order it, and repackage it. Clearly, if enough people, and enough at the highest level of scientific

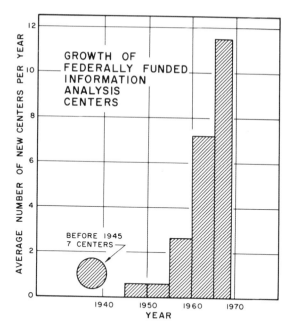

FIGURE 13.28 Growth of federally funded information analysis centers (in all fields). Heights of bars are the mean numbers of centers existing in 1968 that were started per year over the interval spanned by the widths of the bars; thus areas of the bars are contributions to the total number of centers (see Reference 25).

competence, were employed at such centers, the problem of consolidating information could be solved in a very orderly and satisfactory way. The Weinberg report [35] in fact envisioned this as the hope of the future, with the words, "Ultimately we believe the specialized center will become the accepted retailer of information, switching, interpreting, and otherwise processing information from the large wholesale depositories and archival journals to the individual user."

How far have we come toward this goal, and what are the prospects for the future? Let us first look at a few facts; perhaps then we can make a modified prospectus. The number of information analysis centers has been growing rapidly: In 1968 there were some 113 such centers supported by the federal government; the distribution of the dates at which these were started is shown in Figure 13.28. This rapid growth no doubt influenced, and was in turn stimulated by, the commendation of such centers in the Weinberg report. Although these statistics refer to centers in all fields, many of them in the domain of the social sciences, a large proportion of them pertain either to physics or to sciences closely related to it. Table 13.4 lists some 29 centers that might reasonably be regarded as pertaining more closely to physics than to any other major discipline. Two characteristics of these physics-related centers are worth noting:

TABLE 13.4 Some of the Federally Financed Information Analysis Centers Most Closely Related to Physics [25]

Name of Center	Location [a]	Sponsor [a]	Scope (Physics Part)
Alloy Data Center	NBS	NBS	Structure-insensitive properties of metals and alloys
Applied Science Data Group	LBL	LBL	Neutron and photon cross-section data
Atomic and Molecular Processes Information Center	ORNL	NBS, AEC	Atomic collisions, atom–surface collisions
Atomic Energy Levels Data and Information Center	NBS	NBS	Atomic spectra
Atomic Transition Probabilities Data Center	NBS	NBS, ARPA	Atomic transition probabilities
Berkeley Particle Data Center	LBL	LBL, CERN	Properties of elementary particles and resonant states
Center for Diffusion in Gases	U. of Md.	NASA	Gaseous transport data
Charged-Particle Cross-Section Information Center	ORNL	AEC	Charged-particle-induced reaction cross sections
Cryogenic Data Center	NBS, Boulder	NBS, NASA	Low-temperature properties of materials
Crystal Data Center	NBS	NBS, ACA	Crystal structure data
Diatomic Molecule Spectra and Energy Levels	NBS	NBS	Spectra of diatomic molecules
Diffusion in Metals and Alloys Data Center	NBS	NBS	Diffusion in metals and alloys
Electronic Properties Information Center	Hughes	USAF	Electronic, magnetic, and optical properties of solids
High Pressure Data Center	BYU	NBS	High-pressure properties of materials
Infrared Information and Analysis Center	U. of Mich.	ONR	Infrared technology
Isotopes Information Center	ORNL	AEC	Production and use of radioisotopes

Center	Institution	Agency	Description
Soviet Institute for Laboratory Astrophysics Information Center	U. of Colo.	NBS, ARPA	Collisions between electrons, photons, ions, atoms, and molecules
Light Scattering Data Center	Clarkson Coll.	NBS	Light scattering in gases and liquids
Low-Temperature Specific Heats	NBS	NBS	Heat capacities 0–300 K
Microwave Spectra	NBS	NBS	Microwave absorption spectra of molecules
Nuclear Data Project	ORNL	AEC	Nuclear levels
Photonuclear Data Center	NBS	NBS	Photonuclear cross sections, etc.
Radiation Effects Information Center	Battelle	NASA, DASA	Effects of radiation on materials
Radiation Shielding Information Center	ORNL	AEC, DASA	Shielding of reactors, accelerators, radioisotopes
Research Materials Information Center	ORNL	AEC	Availability, preparation, and electronic properties of high-purity inorganic solids
Shock Wave Data Center	LBL	LBL	Hugonoit curves for chemically classifiable materials
Superconductive Materials Data Center	GE	NBS	Properties of superconducting materials
Thermophysical Properties Research Center	Purdue	NBS, USAF NASA, NSF, ONR, and industry	Thermodynamic and thermal transport properties of materials and surfaces
X-Ray Attenuation Coefficient Information Center	NBS	NBS	X-ray attenuation data useful in shielding

[a] Abbreviations: ACA, American Crystallographic Association; AEC, U.S. Atomic Energy Commission; ARPA, Advanced Research Projects Agency; Battelle, Battelle Memorial Institute; BYU, Brigham Young University; CERN, European Organization for Nuclear Research; DASA, Defense Atomic Support Agency; GE, General Electric Co.; Hughes, Hughes Aircraft Co.; LBL, Lawrence Berkeley Laboratory; NASA, National Aeronautics and Space Administration; NBS, National Bureau of Standards; ORNL, Oak Ridge National Laboratory; USAF, United States Air Force.

1. Some 11 of the 29 centers are operated by the National Bureau of Standards (NBS), and six more are located elsewhere with NBS support. These are components of the National Standard Reference Data System.

2. Most of the centers are small groups of a few physicists with a little clerical, bibliographic, and other support. However, they are usually located in large institutions. Two or three centers seem to have more than ten equivalent full-time scientists.

The Office of Standard Reference Data, which, as we have just seen, supports or administers a majority of the centers in physics subfields, was set up in the NBS in 1963 on the initiative of the Federal Council for Science and Technology; its powers and responsibilities were later spelled out by Congress in the National Standard Reference Data Act of 1968. Since that time it has had a direct congressional appropriation; its current budget is approximately $2.4 million. Besides supporting many in-house and external information analysis centers (about half in physics), it also stimulates and funds a limited number of short-term data-compilation and bibliographic projects.[36]

What will be the future role of information-analysis centers? In regard to the gathering, evaluation, and compilation of numerical data—the most important activity for the present centers—their role ought to grow and become quite predominant. But in physics, more than in most other fields of science, theoretical structures and networks of conceptual relations between experimental facts are especially important. The synthesis and compaction of information of these types is, we feel, rarely appropriate for a center with a small staff, although it might occasionally be done by one or two extremely brilliant people located at a large institution. At least two new types of center, however, might be worth considering. One would be a large center, like a national laboratory, with a sizable staff in many areas of physics but especially concentrated in one of them; the staff would carry on some original research, but its principal task would be the preparation of reviews in the central area of interest of the laboratory. The other possible type of center, of more modest scale, would be what might be called a Review Writing Center.[37] Such a center, which could be fairly small in terms of permanent staff, would be located at a large institution doing leading work in a certain field. Although it would not undertake to review this field comprehensively, it would provide assistance to the staff of this or any other institution in the preparation of reviews in this field, by providing them with intellectual backup, including professional-level assistance, bibliographic and clerical services, and assistance in writing and editorial work.

THE COMMUNICATION OF PHYSICS WITH OTHER DISCIPLINES AND WITH THE PUBLIC

Our preoccupation thus far with the channels used by physicists in communicating with each other has been necessary; without these channels, physics would be of no use to anyone, for it would not be a science. But neither can physics be of much value to this country unless it communicates information and ideas to other sciences, to technology, and to culture. Thus the subject matter of this section is fully as important as that of all the others. The comparative brevity of our treatment here results from two things. First, by far the most important channel for communication between physics and all these other areas is that of formal education. As this topic is discussed at length in Chapter 11 of this Report and in the Report of the Panel on Physics in Education, we shall omit it from our discussion. (Interdisciplinary education is also touched on in the reports of several of the subfield Panels.) The second reason for our brevity is that what remains of interdisciplinary communication has been very little studied.

Communication with Other Scientific and Technological Disciplines

Despite the difficulty that specialists in one discipline often have in adapting themselves to the jargon of another, there is a sizable transfer of information across disciplinary lines through the regular primary journals. Physics journals are particularly likely to be cited in papers of neighboring disciplines. Journals that could reasonably be called interdisciplinary (between physics and another major area) seem to account for over one third of the entries in *Physics Abstracts*. Although such journals are obviously very important for interdisciplinary communication, it is noteworthy that the *Physical Review,* whose domain is the very center of physics, is quite often cited in journals of the neighboring disciplines.

Many books, too, are addressed to an interdisciplinary readership. For example, in a random sample that we have taken from the physics shelves (Dewey decimal numbers in the 530's) of a research library devoted to physical science and technology, we found the distribution of primary intended readership, as inferred from statements in the Preface, to be 34 percent research workers in physics; 11 percent to 17 percent students who have committed themselves to physics; 39 percent to 45 percent students or workers in other fields of science or technology; and 10 percent people who do not necessarily have a career interest in science or engineering.

Another very useful medium for interdisciplinary communication is the short review article. Such articles, aimed at scientists or engineers generally, may appear in interdisciplinary journals like *Science;* alternatively, they may be written for research workers in a particular discipline and appear in journals of that discipline.

One would expect oral communication to be even more useful for the crossing of disciplinary lines than within a discipline, at least if the right people can be brought together. Interdisciplinary conferences, of which there are a great many, are particularly useful for this purpose, as they bring together people whose paths might not ordinarily cross. Probably the most effective such conferences, though unfortunately only a small number of people can participate in them, are the Gordon Research Conferences. With unhurried review talks and ample opportunity for person-to-person communication, these have proved very effective at surmounting interdisciplinary barriers; however, they sometimes have difficulty maintaining a proper balance on the interdisciplinary fence.

Within organizations, much interdisciplinary communication occurs in seminars and day-to-day personal contact. Whether these are effective or not depends very much on organizational and geographic factors. On both counts the interdisciplinary communication in industrial laboratories is often much better than it is in universities; interdisciplinary work assignments and transfers are more common, and, perhaps most important of all, workers in different disciplines are more likely to be in the same building or even quite close together.

Communication with Leaders in Culture and Public Affairs and with the General Public

This very important type of communication relies more completely than any other on the educational process, which we are not discussing here. Personal contacts are rather few, and transmission of information about physics in the formal channels, the press and television, is rather meager.

Useful communication in the popular press is largely dependent on professional science writers. Fortunately, the number and esprit de corps of these are growing, though slowly. There is a national organization, the Council for the Advancement of Science Writing, which assists working journalists, offers training programs, and conducts seminars for science writers. Some schools of journalism offer training in the field. The AIP has a Public Relations Division that acts as middleman between reporters and physicists who present interesting new work in talks and articles. Yet despite these favorable trends, the reporting of physics developments in even the leading newspapers and magazines is very spotty. A few statistics illustrating this point have been assembled in Figure 13.29. The figure

SCIENTIFIC PUBLICITY

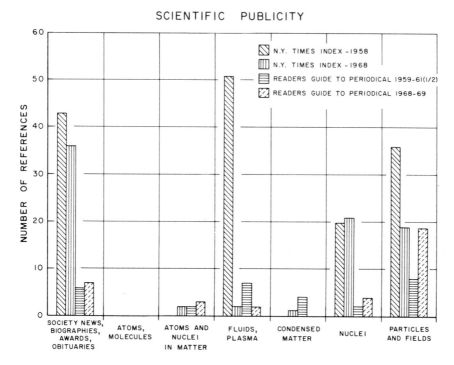

FIGURE 13.29 Coverage of physics in the *New York Times* and leading periodicals.

indicates a marked decrease in physics in the news in the past decade, an era during which the percentage of the population with a college education increased dramatically, thus producing a more highly educated readership. However, the declining references to physics and physicists in the *New York Times* (Figure 13.29) during the last decade may not result only from lack of communication on the part of physicists but from the emergence of other public issues demanding attention and space.

Magazines addressed to the intelligentsia are especially important when they contain material about physics, because some of their readers may be capable of a fairly deep level of comprehension, and some may be in positions of leadership where their knowledge can be put to good use in cultural or practical concerns of the nation.

Lectures in a hall do not seem to be popular in the United States. By contrast, we have been told that in Kharkov, U.S.S.R., a public lecture series on the special theory of relativity about a decade ago drew an audience of about 500. Television is another matter. There have been some very successful physics programs for children, and in those areas where educational stations exist, there should be real possibilities for communicating to adults exciting and thought-provoking material about physics as well as education of a more routine sort.

ROLES AND RESPONSIBILITIES

In this final section we shall focus, one at a time, on the various major groups that have roles relating to communication in physics * in the United States, and for each group shall review what has been said or implied in the previous sections about its role and its responsibilities for future progress. (Most of the groups we shall consider can be identified in Figure 13.5.)

1. *Government agencies.* These directly provide information services central to physics, at a level corresponding to perhaps a few million dollars a year, and in many cases have large mission-oriented information programs peripheral to physics. More importantly for our present interests, however, they fund many of the communication-related activities carried out by other groups, whose dollar values are listed below.

2. *Scientific societies.* These are concerned with primary publication, secondary services, meetings, and the like, the total of budgets for these purposes (in physics) being of the order $8 million to $9 million per year.

3. *Commercial publishers and commercial information services.* Remembering that most of the commercial journals in physics are foreign, as are some of the books, we can estimate that the portion of the work in Figure 13.5 done by U.S. commercial organizations amounts to about $7 million to $8 million per year.

4. *Research organizations.* With their management of travel and their provision of library services, these oversee the lion's share of the resource commitment that goes into physics communication, probably over $30 million per year.

The other large component in the breakdown of activities in Figure 13.5 is, of course, foreign organizations, whose work affecting the entries in the figure is worth approximately $10 million per year. We shall not discuss this component further here. There is another group, however, that we

---

* These groups, of course, provide communication services and media employed in other scientific, engineering, and nontechnical fields. However, this discussion is concerned only with their roles in relation to the communication of physics information.

should include as having roles and responsibilities affecting communication in physics, though it does not appear in Figure 13.5, namely,

5. *Individual physicists.* Though individuals do not control large-scale activities, they make decisions that in the aggregate have a great influence on the effectiveness of the communication media operated by the larger groups.

Government Agencies

In several places in the previous sections we have encountered a very important principle of economics that pertains to most kinds of information services and that practically requires a partial government subsidy of these services if the public interest is truly to be served. The reader will recall that in discussing primary journals we argued that society is poorly served by pricing journals to recoup both the prerun and the runoff costs. For there will then be cases in which provision of a copy of the journal to an additional user will benefit his work, and through it society, by an amount greater than the cost of providing one more copy for him (the runoff cost) but less than the sum of runoff and prorated prerun costs. The positive net benefit cannot be realized if the journal is priced at the latter figure. The same general principle (which is well recognized by economists) applies to many other information services, for example, abstracting services, data compilations, and treatises. In each case there is a product that is costly to produce initially but inexpensive to replicate. In each case different prospective purchasers place widely different values on the product, but in each case the purchaser's valuation is a plausible estimate of its value to society. Although a fully adequate treatment of what society gets for its money has to take account of such things as producer motivation and interactions between different users—these are discussed for the primary-journal case in Reference 2—the essence of the argument as we have just sketched it remains valid.

As we have seen in the discussion of the production and economics of research journals, this incremental-pricing rule provides the major argument for the page-charge system of financing, as long as the funds for payment of page charges can be obtained from a source that has a long-range concern for the advancement of science. The Federal Council's policy [19] of encouraging the inclusion of page charges in research budgets reflects an intelligent concern in this direction, described in the statements that publication is an integral part of doing research and that those who support research have an obligation to support its publication. We commend continuation of this policy and urge further measures to ensure that the governmental contribution to page-charge support operates in the public interest. (See, for example, the discussion in Reference 2.)

Much the same is true for the basic abstracting and indexing services, whose role is just like that of a primary publication, namely, to make possible effective public access to the results of research. Although, for reasons mentioned in the discussion of these services in an earlier section, no techniques for subsidy of extragovernmental abstracting and indexing services have been worked out that are comparable in practicality to the page-charge system, we feel that agencies that support research should be aware that their work is incomplete as long as the basic secondary services that provide access to it have to recoup their entire costs from subscriptions. The mission agencies that support in-house secondary services—most notably the Atomic Energy Commission with *Nuclear Science Abstracts*—are to be commended for their recognition of this principle. We agree, however, with the SATCOM report [38] that basic secondary services for all physics should be managed by scientific societies, rather than by mission agencies.

In the realm of compaction of information, the principle is again just as valid, and, though far easier to apply, it seems to have been even less generally appreciated than for secondary services. We urge that all agencies supporting research accept a proportionate responsibility for consolidation of its results, that is, for funding the production of critical data compilations and reviews. As we have noted in the sections dealing with books and reviews and information analysis centers, there are several measures that may be appropriate, most notably making available grants to individuals or small groups, similar to research grants but for review or compilation work, and the funding of review-writing centers. Where the need exists, support of ongoing information analysis centers is recommended. Charges for the products produced by these should be close to distribution cost, at least in the case of products with a wide range of users.

Our general principle of providing support adequate to allow marketing of information services at incremental or distribution cost does not especially imply the desirability of subsidy of preprint distribution. As we mentioned in our discussion of the distribution of reports, preprints, and the like, a centralized preprint distribution of small volume may be worth considering in some cases but should be virtually self-supporting in its steady state.

Insofar as it is possible to plan an orderly response to the fluctuations in the funding of science that national economic conditions may require, it would be desirable to put a larger proportion of the budget into consolidation activities in lean periods, a smaller proportion in periods of ample funding. For it takes fewer dollars to support a given number of scientists in consolidation work than in research.

Before closing this section we should mention the broad issue of copyright, particularly as it applies to replication and to computerized information systems. Though this issue is far broader than science and technology, its great importance for these areas, including, of course, physics, has been described at length in the SATCOM report [38]; the legislative and judicial branches of the government have a major responsibility to lay down guidelines that will serve the public interest by recognizing the valid points that have been made on both sides of the issue. In the meantime, intelligent use of direct and indirect subsidies to encourage pricing at output cost can significantly ease dislocations resulting from copying. For example, page-charge support for commercial journals, though admittedly very difficult to implement practically, might be further explored.[2]

Scientific Societies

As we have seen in Figure 13.5, the largest role of the societies, in economic terms, has been the management and production of primary journals. In the case of the AIP and its member societies, this work has been performed quite effectively, and the low prices and high circulations of its journals are a commendable example to the other sciences. Improvements in the technology of journal production are continually being made. We might offer a few cautions, however. The rapid growth in bulk of some of the journals implies the need for vigilance and imagination in making decisions about subdivision. The use of the two-track system to encourage honoring of page charges (see section on Research Journals: Production and Economics), though generally more beneficial than harmful (see Reference 2, Sec. IVB4 for a quantitative balance of pros and cons), can become contrary to the best interests of society if the nonhonoring papers are subjected to long delays.

Prompt repackaged journals—the user journals described in the section on Research Journals: Miscellaneous Characteristics [22]—could conceivably be not only the answer to the loss in usefulness that journals have suffered due to the decline in personal subscriptions but also an unprecedentedly effective current-awareness tool. We encourage experimentation with them.

Although swamped by costs resulting from primary-journal publications, the secondary services have in recent years occupied a major place in the thinking and planning of the AIP. Their projected National Information System for Physics and Astronomy [33] actually envisions a highly integrated structure of primary literature, access services, and the consolidation of information; the health of this structure, and of many other aspects of the nation's physics effort, is to be promoted by an ongoing program to provide information about the functioning of the physics and astronomy community. Economy and speed can be served by the integration of several different secondary services with each other and with primary journal composition; these possibilities are being exploited, as are those of cooperative agreements with other organizations that generate inputs for secondary services.

The consolidation portion of this AIP program has so far two clearly enunciated components, both of which seem promising. One is a new journal for publication of data compilations, which, it is to be hoped, will make many of them much more accessible than they have been in the past. The other is a proposal for a government-funded system of all-expense grants to authors of consolidation services or compilations (Reference 33, final part of the section on consolidation). It is hoped that the AIP or its Member Societies could take the initiative in implementing some of the other measures that we have suggested in our discussion of review and consolidation needs.

In conclusion, we urge most especially that the physics community support an adequate ongoing operational-research group at the AIP to monitor data and perform studies relating to the functioning of the physics community, including problems of information and communication.

U.S. Commercial Publishers and Information Services

The vital role of commercial publishers as producers of books is not due merely to the fact that other groups produce so few of them. It is also due to their unique capability for recognizing the needs of special groups of users. Most of the advanced books produced on physics subjects are addressed to one kind or another of interdisciplinary readership. The interaction of commercially motivated managers with diverse would-be authors is a very effective mechanism for discovering when an important class of users with special needs exists and getting the right sort of information package prepared for them.

In regard to the advanced review literature of pure physics, we would like to see a little more stress on the thorough and comprehensive types of review articles and treatises; however, we recognize that achieving this emphasis will depend more on authors than on publishers. At the other end of the range of types of consolidations, collections of reprints of important papers in a special field can be extremely useful. An intermediate type of consolidation that often can be helpful is the reprinting of individual review articles from review journals when they have a broad readership and are much in demand. We would like to encourage the collaboration of journal publishers with publishers of books in ventures of these last two types.

We also would like to see more emphasis on a small number of "Progress in . . ." series of review-article collections and less emphasis on the publication of conference proceedings in isolated books. We feel that these are usually more accessible if published in journals, with simultaneous availability separately bound.

Serious concern has been expressed over the squeeze on publishers of monographs due to rising costs and the decline of sales resulting from increasing specialization.[39] Fortunately, some publishers are reducing costs by typewriter composition and offset printing; moreover, as we noted earlier, paperback editions, usually fairly satisfactory for the individual purchaser, are becoming increasingly available at prices of the order of half those of hardcover editions. (In general, *for a given level of profit,* the welfare of society is enhanced by any scheme of differential pricing that can create a correlation between the prices paid by different users and the value of an information product to them.)

Research Organizations

Our remarks in regard to the responsibilities of scientific societies strongly endorsed the page-charge system of support for primary journals. To be successful, however, this system requires the cooperation of the organizations at which research takes place. We would like to stress that it is precisely these organizations' own interests that are at stake; the object of the page-charge system is to make possible the easy accessibility of journals in as many libraries, sublibraries, group collections, and individual holdings as possible. This accessibility is surely desired by these organizations as users of information and by their staff members in their roles as authors. We would like to reiterate the point, briefly mentioned in the discussion of the production and economics of research journals, that a decision of a U.S. author to publish a paper in an expensive non-page-charge journal can sometimes actually cost more money, from the budgets of U.S. research institutions, than publication, with payment, in a page-charge journal. Use of typical numbers in Eq. (2) (in the section on Primary Publication) shows, for example, that this can be the case when the price is above approximately 10 cents per kiloword, unless the U.S. circulation of the journal in question is below 750. This arithmetic, moreover, is based solely on the purchase cost of the journals and does not take account of the fact that library administration, binding, and storage costs, which typically are comparable with acquisition costs, are appreciably higher per kiloword for the smaller journals. Thus an organization that encourages its authors to publish in expensive journals is really asking its sister institutions to support its publication, and at a rate that may be more expensive than that in a page-charge journal.

In our discussion of the responsibilities of government agencies, we stressed that those who foster research have an obligation to accept a proportionate responsibility for its compaction. This obligation applies as much to universities and to industrial basic-research groups as it does to government. Actually, the interests of both types of organizations would be advanced if they would more actively encourage their staffs to prepare books, reviews, and compilations. The educational value of such work is obvious. Further, since a primary motivation for the performance of basic research in industrial laboratories is to place the laboratory in touch with research-front activity in the rest of the world, what could be better than an occasional comprehensive review? Faculty and research administrators must eliminate the notion that three research papers are a better basis for promotion than two research papers and a review. Similarly, we urge research organizations of all types to be hospitable to information analysis and review-writing centers and to encourage their staffs to take leaves to work at such centers.

Research organizations have a special responsibility to foster informal communication and to facilitate the use of formal media by their staffs. We have commented briefly in the sections on oral communication and communication with other disciplines on organizational and administrative measures to foster oral communication, especially among workers in different disciplines. Both in this realm and in the realm of access to books, journals, and secondary services we would like to stress again the importance of spatial proximity of the user to the resource. Large buildings housing many workers in related fields, and containing an adequate library in the middle, are much to be preferred to a scattering of isolated buildings, some without libraries. Even within large buildings, subcollections of books and periodicals useful to a particular group can be very helpful. Research departments and libraries should cooperate closely and be aware of each other's problems.

Individual Physicists

Effective communication depends on the freedom of each individual to choose his degree of involvement with each of a wide range of information resources. It is only he who can know what he needs and how much. Unfortunately, several parts of our study have shown that the physicist often does not know these things as well as he should, and that when he does know his needs he often does not know how to go about satisfying them. The basic facts of the communication picture should be much more widely known. There should be less provincialism in journal-scanning habits. The sluggish response to new information tools—citation indexes, current-awareness services, translation journals—should be accelerated. Education to foster awareness and effective use of information media and services should be a part of normal graduate training.

It is individuals, too, who undertake the preparation of books and reviews. We have made a general exhortation for more and better work in this direction and have sketched specific desiderata for improving quality and readability. We reiterate, too, our recommendation, based on the citation statistics in the section on Books and Reviews, that authors not dissipate their review-writing energies in too many sketchy reviews and that they publish their reviews in media of high visibility and wide circulation.

Our remarks regarding the folly of publishing papers in expensive journals of low circulation should be taken to heart by authors as well as by administrators, as it is usually the authors who decide where to submit their work.

Last but far from least among the roles of individual physicists is the collective one of an informed and dedicated citizenry. It is their preferences that determine the demand for marketed products, their votes and letters that determine the policies of the societies; their aggregate pressure has a large though not necessarily controlling influence on the policies of governmental agencies and other organizations. We hope that in examining these responsibilities they will give serious attention to problems of information and communication.

References

1. J. M. Ziman, *Public Knowledge* (Cambridge U.P., New York, 1968).
2. Committee on Scientific and Technical Communication, *Report of the Task Group on the Economics of Primary Publication* (National Academy of Sciences, Washington, D.C., 1970).
3. T. J. Allen, in *Communication among Scientists and Technologists,* C. E. Nelson and D. K. Pollack, eds. (Heath, Lexington, Mass., 1970).
4. M. W. Martin, Jr., IRE Trans. Eng. Management *EM-9,* 66 (1962).
5. R. von Zelst and W. A. Kerr, J. Abnormal Soc. Psych. *46,* 470 (1951).
6. R. E. Maizell, Am. Doc. *11,* 9 (1960).
7. D. G. Marquis and T. J. Allen, Am. Psychol. *21,* 1052 (1966).
8. T. J. Allen, "Managing the Flow of Scientific and Technological Information," thesis, MIT (1966).
9. T. J. Allen, Ind. Management Rev. *87* (1966).
10. W. M. Carlson, Trans. N.Y. Acad. Sci. *31,* 803 (1969).
11. T. J. Allen and S. I. Cohen, Admin. Sci. Quart. *14,* 12 (1969).
12. M. Cooper and E. Terry, *Secondary Services in Physics,* Rep. ID 69–2 (American Institute of Physics, New York, 1969).
13. L. J. Anthony, H. East, and M. J. Slater, Repts. Progr. Phys. *32,* 709 (1969).
14. R. E. Maizell, unpublished work (1960).

15. N. Price and S. Schiminovich, Info. Storage Retrieval *4*, 271 (1968).
16. S. Schiminovich, Info. Storage Retrieval *6*, 417 (1971).
17. M. M. Kessler, Phys. Today *18*, 28 (Mar. 1965).
18. H. A. Barton, Phys. Today *16*, 45 (June 1963).
19. Federal Council for Science and Technology, Scientific Information Notes *3*(5), 1 (1961).
20. H. Zuckerman and R. K. Merton, Minerva *66* (1971); see also Phys. Today *24*, 28 (July 1971).
21. "World Literature of Physics as Seen through Physics Abstracts 1964 Issues" (Abstracting Board of the International Council of Scientific Unions, 1967).
22. H. W. Koch, *The Role of the Primary Journal in Physics,* AIP Rep. ID 70–1 (American Institute of Physics, New York, 1970).
23. A. L. Schawlow, unpublished work (1967).
24. W. O. Hagstrom, in *Communication among Scientists and Technologists,* C. E. Nelson and D. K. Pollack, eds. (Heath, Lexington, Mass., 1970).
25. *Panel Report on the Role of the Technical Report in Scientific and Technological Communication* (Federal Council for Science and Technology, 1968 (available from the National Technical Information Service, Springfield, Va. PB 180 944).
26. M. A. Libbey and G. Zaltman, *The Role and Distribution of Written Informal Communication in Theoretical High Energy Physics,* Rep. AIP/SDD-1 (American Institute of Physics, New York, 1967) or USAEC Rep. No. NYO-3732–1.
27. M. J. Moravcsik, Phys. Today *18*, 23 (Mar. 1965).
28. "A Debate on Preprint Exchange," remarks by M. J. Moravcsik and S. Pasternack, Phys. Today *19* (June 1966).
29. D. Green, Intern. Sci. Technol. p. 82 (May 1967).
30. C. Herring, J. Chem. Doc. *8*, 232 (1968).
31. Ad Hoc Panel on Nuclear Data Compilations, *Nuclear Data Compilations: The Lifeblood of the Nuclear Sciences and Their Applications* (National Academy of Sciences, Washington, D.C., 1971).
32. C. Herring, Phys. Today *21*, 27 (Sept. 1968).
33. The AIP Program for Physics Information (American Institute of Physics, New York, 1971).
34. *The Dissemination of Scientific Information, Informal Interaction, and the Impact of Information Received from Two Meetings of the Optical Society of America,* Report #3 of the Johns Hopkins University Center for Research in Scientific Communication (1967).
35. *Science, Government, and Information,* Report of the President's Science Advisory Committee (U.S. Government Printing Office, Washington, D.C., 1963).
36. *Critical Evaluation of Data in the Physical Sciences—a Status Report on the National Standard Reference Data System,* NBS Tech. Note 553 (U.S. Government Printing Office, Washington, D.C., 1970).
37. S. A. Goudsmit, Phys. Today *19*, 52 (Sept. 1966).
38. Committee on Scientific and Technical Communication, *Scientific and Technical Communication: A Pressing National Problem and Recommendations for Its Solution.* (National Academy of Sciences, Washington, D.C., 1969).
39. C. Benjamin, Sci. Res. p. 32 (Sept. 16, 1968).

14

POLICY CONSIDERATIONS: CONCLUSIONS AND FINDINGS

The present science establishment of the United States is unique. Whatever criticisms can be leveled, it is still the most productive, most innovative, most forward looking of all such efforts in the world.

> Research Management Advisory Panel
> Subcommittee on Science, Research and Development
> Committee on Science and Astronautics
> U.S. House of Representatives—1970

First—the continuing development of science and technology at an optimum rate is vital to the nation.

Second—our Federal science policy has been— to support science and technology where and when it appears promising.

Third—Federal science policy has thus far been based on the principle that control of the support for science and technology should not be centralized.

Basic philosophy

The Federal Government should formally recognize its debt to and dependence on science and technology and establish herewith a national policy for their support and furtherance.

967

In this regard we are recommending fundamentally three things.

• That a National Science Policy be stated
and maintained as public law.

• Such policy be incorporated into the
operations of every department or agency of
the U.S. Government which utilizes science and
technology in its mission.

• Such policy be flexible and subjected to
continual review and re-evaluation in light of
changing national goals and priorities.

> Toward a Science Policy for the United
> States
> Subcommittee on Science, Research and
> Development
> Committee on Science and Astronautics
> U.S. House of Representatives—1970

The United States will strive to remain competitive at or near the forefront of each of the
major areas of science and, to this end, will
continue to identify and support excellence.

The Federal Government has a responsibility
to ensure that new scientific knowledge is
utilized as rapidly and effectively as possible in
support of national goals and for the welfare of
the world's peoples.

> Basic Tenets for U.S. Science Policy
> National Science Board—1970
> Report to the Congress

INTRODUCTION

The Physics Survey Committee strongly endorses the above statements,
both generally and more particularly as they relate to physics. It is our
belief that continued development of physics as part of science and technology is necessary to ensure continuing economic growth, improvement
in the education of the citizenry, and the maintenance of strong national
defense. Physics can be expected to play a major role in the use of science
and technology to contribute to the amelioration of environmental problems
and the alleviation of undesirable consequences of increasing population.
Improvement in the standard of living and the quality of life can be
maintained only by the fullest utilization of science and technology, of
which physics forms an essential part. Physics, as a principal contributor
to science and technology, has made, and will continue to make, major
contributions to culture and to present new opportunities for human
achievement. Quite apart from this, and of great importance, is the role
that physics plays not only in establishing fundamental guidelines for

science but also in evolving and developing the concepts of public knowledge, of verifiable fact and evidence, and, ultimately, of truth itself, which underlie all human endeavor.

Basic Premise

Underlying the discussions and recommendations throughout this Report is the Committee's conviction that only by maintaining U.S. physics research at or near the forefront of international activity in each of its major subfields can this science realize its potential for service to society and to the nation. This is the basic premise that the Committee has attempted to document and support herein.

This Report outlines the present status of U.S. physics and indicates the massive contributions to the intellectual and material well-being of society that physics and physicists have been able to make in the past. These derive from the steady, devoted pursuit of basic research over many decades, with generous public support, yet at a cost minuscule compared with other public expenditures—and minuscule compared with the documented yield of technology based on this research.

Since 1967, federal support of physics has dropped by some 8 percent in terms of real purchasing power. At the same time, the cost of doing research has grown more costly as ever more sophisticated questions need to be answered; this sophistication escalation is very real and has been variously documented by the National Science Foundation (NSF) and others as falling between 3 and 5 percent per year on an average basis.

Effective support of U.S. physics has been reduced by about one fourth since 1967. It is difficult to assess the real loss to society, but there are tangible losses in underutilized capital equipment and underemployment of highly skilled manpower. Even such major national installations as the Stanford Linear Accelerator and the National Magnet Laboratory, in the face of heavy pressure from the scientific community and with vitally important problems awaiting study, have been forced to suspend research operations for several weeks in recent years in order to remain within available budget; the situation at smaller facilities is even more serious.

In 1967, 1300 new physics PhD's were graduated and there was a net increase of 1180 in the number of physics PhD's employed in the physics community; in 1970, 1500 new physics PhD's were graduated, but the net increase in the number of PhD's employed in physics was only 100. Projections of the demand for scientific manpower in the 1970's suggest that the employment picture for physicists is bleak indeed. This situation is examined in some detail in Chapter 12. If present trends continue, a rather reliable estimate indicates that a total of 6700 new physics PhD's will be produced between 1970 and 1975, inasmuch as they are already in the educational pipeline. They will have to compete for only some 1500 employment opportunities in physics. Even the most optimistic projection raises the number of employment opportunities to only 3200.

Although it is clear that there are very real benefits that can accrue to society through utilization of trained PhD physicists in employment other than in physics, the level of mismatch represented by these estimates represents a waste of an important national resource and one that the nation can ill afford. Unknown is the effect of new discoveries not yet made and the loss of momentum, which changes the whole tone of scientific pursuit from the exciting thing it was, with new discoveries emerging, often daily, to a more pedantic and cautious business of trying to keep the system moving lest it stop altogether.

It may be true that individual scientists, like individual artists, can flourish under adversity, but a whole community does not. Like the great ages of art, a great age of science requires generous public support, support in material resources and in interest, enthusiasm, and involvement. Only in this way is it possible to maintain the eager spirit and the tumultous excitement of new discovery of the world around us and the concomitant increase of control of that world for human benefit. Continuation of present trends can strangle much of U.S. physics.

The Committee in its discussions and interactions with its subfield panels developed a variety of conclusions and findings, both general and specific, that are presented herein. As is emphasized in Chapter 5 of this Report, it is clearly recognized that these conclusions must be viewed as part of a much larger fabric representing the entire U.S. scientific and technological enterprise; even within the restricted field of physics it would be unrealistic for this Committee to attempt to evolve any detailed national program. Physics serves the nation in a multiplicity of roles, and many of the important inputs are beyond the purview of the present survey. At the same time we believe that it is important to attempt to establish certain policy principles that can underlie the administration of national programs for physics in this decade; we believe that many of these principles have broader validity within all of science.

The Committee's conclusions and fiindings are directed to the attention of several different audiences: the federal government in both its policy and operational agencies; the Congress; the U.S. scientific, and specifically the physics, community, including all its institutional aspects; the U.S. educational enterprise as it concerns both education in physics and physics in education; industrial organizations; state governments; private foundations; and the general public. These conclusions and findings support and augment the specific recommendations presented in Chapter 2.

GENERAL POLICY CONSIDERATIONS

The Importance of Stability

Physics, and all of science, is particularly vulnerable to discontinuity in its support patterns because of the long time constants involved in both research and the graduate education process (see Chapters 6 and 12). Thus it should be recognized that as a consequence of recent economic conditions, especially those resulting from decreased emphasis on military and space research and development, serious dislocation has occurred in the research and development community, with significant unemployment of scientific and technical manpower and underutilization of research facilities. (See Chapter 12.)

Physics is indeed a frontier, and the opportunities that it presents for new exploration—for new forays into the unknown—are virtually unlimited. In general, and understandably, the discussions in this Report and those throughout the community have focused on the value of a healthy scientific establishment and on the consequences of rapidly decreasing effective funding. At the same time, it should be recognized that funding that increases too rapidly can also carry severe penalties. One of the important elements of a long-range national scientific policy is the recognition that, although exploitation of new ideas and opportunities will from time to time demand rapid expansion of the nation's scientific capability, such expansion must be carefully planned to avoid the necessity for later wasteful retrenchments. (Specific recommendations directed toward achievement of increased stability are presented in Chapter 2.)

The Support of Excellence

Inasmuch as the institutional base for graduate research in the United States has just undergone a period of rapid growth, we strongly believe that, as long as present budget restrictions remain, available funding should be used primarily to improve the quality of existing institutions and research groups of demonstrated competence rather than to increase the number of institutions and groups having the capability of providing graduate education and research in physics. This conclusion is based on the conviction that any other course of action, under current and forseeable budgetary limitations, would lead to a deterioration of the overall quality of U.S. research and education in physics.

We fully recognize that this policy would mark a most significant change from the thrust of federal support of higher education in recent years. During this time, a major effort has been devoted to the broadening of the institutional base for graduate education and research in physics, both in the total number of universities and colleges involved and in their geographic distribution. (These questions are discussed at greater length below and in Chapter 5.)

Multiplicity of Support Channels

One of the important bases for the present strength of the U.S. science and physics communities, and one of the important reasons why these communities have been able to make significant contributions in times of both war and peace, lies in the multiplicity of support channels available to them.

A single source of support can too easily become captive to the views, prejudices, or idiosyncracies of a particularly strong individual or a clique of advisers or a Congressional committee with a special axe to grind. Such a situation can result in the stultification of a whole field of science. A lively competition among supporting agencies as well as among research groups is best for the health of science, even though occasionally this results in apparent duplication of effort. Pluralism is further enhanced when some fraction of support, and hence of scientific choice, is delegated through coherent area and similar types of broad grants to local scientific management within a university.

In particular, *all* the federal mission-oriented agencies should be encouraged to support a level of basic scientific activity commensurate with the demands implied by their stated short- *and* long-range mission objectives (see Chapter 10).

The Support of Industrial Research

Industrial research activity in the basic sciences is an important resource for both industrial innovation and the health of the national economy. The Committee is distressed by the deterioration that is taking place in the health of the overall U.S. industrial enterprise and in its competitive posture in the international market (as discussed in Chapter 7). Modification of federal policies and practices to ensure the encouragement and growth of industrial research, both basic and applied, throughout the nation is of increasing importance in reversing the present trend.

Physics in Major Problems of Society

Because we view physics as a profession, not a trade, and because we believe that an adequate physics education instills a breadth of view and a flexibility of approach that are not limited to the problems of physics but are broadly applicable, we conclude that a larger proportion of physicists should become involved on either a part-time or full-time basis in both direct and interdisciplinary approaches to major societal problems, such as those in the areas of health, energy generation and transmission, environmental monitoring and management, law enforcement, and transportation. Such involvement is important because of the contributions that physicists can make and the feedback to physics of new basic problems that must be solved.

In many of the social-problem areas, the program requirements for basic physics as such may be relatively small, but in the systems approach to many of these rather unstructured problems the physicist's training, research experience, and style of attack can be a major input. There is need to broaden the horizons, especially of the younger physicists trained in recent years or now in training, to include more of the opportunities and challenges inherent in these interdisciplinary or societal programs and to foster a greater willingness on the part of physicists in general to attack these important applied problems. These needs place a heavy burden of responsibility not only on the universities but also on the federal agencies or industrial organizations most concerned with societal problems. (These questions are discussed in greater detail in Chapters 6, 7, 9, and 11.)

General Science Education

The quality of general science education in the United States is distressing, and a renewed national commitment to the improvement of this education, clearly stated and implemented at both federal and state levels, is essential.

A vital part of such a commitment would be the strengthening of the scientific backgrounds and qualifications of teachers in both elementary and secondary schools. Repeated surveys have demonstrated that the qualifications of these teachers in general science are usually very weak and in the physical sciences they are frequently entirely lacking.

By the time they have completed their secondary education, the great majority of students have made general career choices. These choices should be as well informed as possible. No less important is the vital development of a degree of scientific and technological literacy, now lacking in much of the U.S. public but increasingly necessary if the rising demand for broader participation in decisions involving technological development and application is to be realized without disruption of many aspects of the nation's well-being.

An important part of any program to improve science education is the direct participation of active scientists in educational policy making and practice, both federal and local (see Chapter 11).

Dissemination and Consolidation of Scientific Findings

In many scientific fields, and particularly in physics, the mechanisms available for the dissemination and consolidation of research results have not kept pace with the rate of production of new knowledge, so that, increasingly, research findings are available only with difficulty, if at all, to other scientists and technologists most needing them. These problems are considered in Chapter 13.

We strongly commend the approach developed in the major governmental agencies since the 1961 policy statement of the Federal Council for Science and Technology, whereby applicants for research grants and contracts, as well as management of governmental and national laboratories, are encouraged to include explicit budget items for the support of publication, including the payment, where appropriate, of page charges in nonprofit scientific journals. A similar approach for university and industrial laboratories should be adopted.

Much has been achieved in the abstracting and indexing of the physics literature in the last few years, and there is extensive publication of review articles, monographs, and nonspecialized summaries of existing knowledge. Nevertheless, these activities have not kept pace with the needs of all the audiences served by the physics literature, especially technologists and policy makers, as well as physicists. The physics profession, employers of physicists, and most government agencies have done too little to encourage, support, and reward such activity. Studies made in this survey (see Chapter 13) show that investments to improve the effectiveness and reduce the time required for scientific communication among physicists and between physicists and technologists can often be more productive per dollar expended than the corresponding additional research support. These questions are considered again in this chapter when conclusions relevant to future planning in both federal agencies and university and college departments are listed.

Increasing Public Awareness of Science

Physicists much more generally should recognize and accept the obligation of making their work and its possible implications widely available and understandable to an interested public. Unfortunately, there is an unhappy tradition of poor communication between large components of the physics community and the public. In part, this situation has reflected implicit assumptions that the public generally was uninterested and that much of modern physics could not be made comprehensible to a public audience. The Committee rejects both assumptions. At the same time we recognize that the burden of responsibility falls on physicists to develop more effective methods of communication, to devote a larger fraction of their time and resources to such activity, and to make themselves more available to public forums and public media. We believe that the mutual benefits to physics and to the public through such increased awareness can be impressive.

Conclusions and Findings

Having listed the above general policy considerations, we turn to more specific areas in which the Committee wishes to bring its findings to the attention of one or more of the audiences that we have listed above. In doing so, we shall have frequent occasion to emphasize our support for certain recommendations contained in the Report of the Subcommittee on Science, Research and Development (October 15, 1970) to the Committee on Science and Astronautics of the Ninety-First Congress. For convenience, we group these findings under the heading of the appropriate agency or agencies.

Office of Management and Budget (OMB) and Office of Science and Technology (OST)

We strongly support the conclusion in the above-mentioned report that the OST should take the lead in formulating basic federal science policy within a continuing framework, in addition to evaluating and reporting, on a regular basis, on overall federal scientific and technological activities. At the same time, it is extremely important to retain the diversity of objectives and missions characteristic of the many federal agencies to provide the all-important flexibility and breadth of implementation of any such basic federal policy.

The Committee also strongly supports the following specific recommendations of the Subcommittee on Science, Research and Development, which stem from the foregoing conclusion:

The Office of Management and Budget should develop a "stable funding" procedure with regard to basic research which will avoid seriously disruptive funding fluctuations.

As a crucial aid to the Nation's general welfare and security, the Office of Management and Budget should make 5-year projections of scientific and technological trends, probable national needs for scientific resources during such times, plus indications of probable levels of Federal support for meeting the needs.

OST should submit an annual report to the President and the Congress setting forth

(1) a comprehensive review of the status of research and development in the United States and (2) a recommended program of scientific research and development for the coming year.

The Office of Science and Technology should develop criteria for the support of basic research by the mission-oriented agencies.

Concerning Continuity   With regard to the first of these recommendations, we would find it difficult to add to the comments given in the Subcommittee Report, which we quote:

> One of the most serious problems encountered by Federal agencies in their support of research is the extreme vulnerability of the research dollar when it is forced into headlong competition with, say, the weapons dollar, or the crop support dollar, or the welfare dollar, or the veterans dollar, or the postal dollar, or the hot-lunch dollar— even the foreign aid dollar. Anything with a humanitarian, security, service or other popular-appeal tag on it is apt to take priority over scientific research.
>
> We do not argue that this is invariably wrong. On the contrary, for the short term it is often right. Our point is that effective science and technology, the critical nature of which we have tried to show in this report, cannot be turned on and off like a water spigot.
>
> It does not make sense, in fact it is enormously expensive and dangerous, to dissolve good research teams, dismantle first-rate facilities, fire research professors, discourage graduate students and "turn off" prospective science education majors through sudden budgetary downturns—just as it is wasteful and inefficient to increase research budgets too rapidly.
>
> It is recommended, therefore, that the Budget Office study ways and means of considering the science research and education phase of the Federal budget as an entity— that this be regarded as an integral program of prime importance—and that policies be adopted which will assure a minimum of fluctuation over any consecutive 5-year period. Perhaps anything beyond a 5 percent change per year (adjusted for inflation) during this period should be avoided even if other, apparently more "critical" programs must absorb the difference. It is emphasized that any percentage change should be viewed as a relative one which takes into account growth factors, cost-of-living trends and other inflationary components.

As a first step in such a program, as recommended in Chapter 2, the possibility should be explored of establishing forward funding for perhaps one third of the basic physics research projects receiving federal support funds. Few projects can be completed in a year, and the length of the training period of graduate students is measured in several years. It is desirable to invent mechanisms for providing longer-term support that are compatible with the need to maintain quality and the constraints of the annual appropriations process. At the same time, it must be emphasized that such mechanisms should retain the benefits to both support agency and the investigator that derive from the requirement of an annual review of the program involved.

This recommendation is directed toward basic physics research; one third of the total research support has been rather arbitrarily chosen as an appropriate initial fraction to avoid loss of flexibility and the ability to respond to totally new opportunities and to permit an essential degree of experimentation in such a significant change in funding patterns.

But the problem is a much more general one. Similar problems of stability are encountered in both applied research and development. However, in these cases the demands and responsibilities external to physics work to establish effective bounds on the support fluctuations to a greater extent than in basic research.

Indeed a case can be made for budgeting all federal research (not development) on a manpower basis, ensuring a certain level of effort in man-years over broad areas of research. Translation of manpower levels into dollars can be made on the basis of unit cost indices, which change with time and field. Changes in science policy would then be conceived in terms of movement of manpower resources from one area to another. This is essentially the way U.S. agricultural research has been planned over the years. It ensures a measure of stability, hence more efficient use of resources. What would be needed in such a program is a continuously updated five-year manpower budget for all scientific research, based on a projected utilization of a certain percentage of all scientific and technical manpower within defined upper and lower limits. The program might be divided into several support categories:

1. Projects that are selected for support on the basis of national competition among proposals judged by peers, with allocation among fields being determined by evaluated proposal pressure;

2. Selective support of work related to major facilities such as accelerators, telescopes, magnet facilities, and reactors;

3. Infrastructure support in which program choices are made by local management (this category would include institutional and departmental grants, traineeships, training grants, general research support grants, and the like);

4. Selective emphasis support, in which certain broad areas of science are chosen for special attention, either because of special scientific opportunities or because of social relevance or both, but in which selection of individual projects is still made according to some competitive basis, local or national (the approach to selection of program emphases developed in Chapter 5 would be directly applicable to the selection of these support areas).

Part of the current discontinuity in scientific research support reflects major changes in the national mission-oriented activities in space and defense, to give two examples. We believe that a balancing mechanism should be made a central part of a national policy for continuing support of science. Such a mechanism would make it possible to expand and contract mission-related programs without the overall dislocations that are now experienced. Thus the Committee supports the 1970 Subcommittee on Science, Research and Development Report recommendation:

> . . . responsibility for basic research should center in the National Science Foundation which should provide approximately a third of all Federal support in this area.

We believe that the stated fraction, one third, as opposed to the present approximately 18 percent, would permit the balancing function.

We also support strongly the following statement from the Report of this same Subcommittee, concerning the NSF, that was addressed to the Eighty-Ninth Congress in 1966:

> At a time when NSF testifies that it cannot support all the worthwhile research proposed to it, and that the projects it does select often are approved at budgets and terms less than requested, the Foundation nonetheless must assume a responsibility that will impose great future financial strains upon it. This responsibility is that of becoming the Federal balance wheel, for scientific research. That is to say, the Foundation, having identified national needs for science and the intentions of other agencies, has a responsibility to compensate for any disparity in level of support or allocations among different fields of science.

The involvement of the NSF in other programs of national significance should not be permitted to interfere with, or degrade, its primary responsibility for overall health of the national basic research enterprise, including academic research and related graduate education. It is also essential to the continuing health of U.S. science that the mission-oriented agencies, in the orderly pursuit of their missions, support both basic and applied physics, each of which contributes to these missions in the long term. It must be recognized, however, that the aggregate program, which comprises the summation of the science programs of the mission-oriented agencies, does not necessarily constitute a balanced national scientific effort—balanced, that is, in terms of the evolution of the conceptual structure of science. In this connection, the NSF has a particular responsibility to plan its own programs to create a balanced national effort, taking fully into account the plans and programs of all other federal agencies and also industry. By balance we do not necessarily mean equitable treatment of all disciplines or fields of science but rather an adequate and internationally competitive effort in those fields that are currently at the forefront of the advancing intellectual cutting edge of contemporary science. The judgment of what these fields are, and what is required to stay at the cutting edge, must be a prime responsibility of the Foundation, working in close collaboration with the scientific community.

Emphasis on Excellence   Stimulation of rapid general growth in graduate education in physics seems undesirable in the immediate future. As a consequence, in the use of those forms of aid that have as one of their primary results an increase in the output of graduates (for example, institutional development awards, traineeships, and certain formula grants), the objective must be quality rather than quantity. Under present conditions, the quality of the proposed research and the quality of the students to be involved in it increasingly should be a paramount consideration in the awarding of research grants or contracts.

As noted in Chapters 11 and 12, however, retention and modest expansion of the federal program of competitive national fellowships is important as a means of identifying and encouraging the most able students. On the other hand, a program must be balanced by devices that achieve a broader institutional spread than experience shows to be achievable with the fellowship approach alone. Nationally competitive fellowship holders tend to concentrate in a very limited number of institutions. Their interaction under such circumstances is extremely important, and experience has shown that the student interactions are among the most effective in stimulating and developing scientific potential. A balanced national program, including both fellowships and traineeships, in perhaps a one-third to two-third ratio, would foster both the special excellence achieved when the numbers of first-rank students in a few institutions exceed a critical minimum and the more general excellence that demands a distribution of the most able students much more broadly throughout the educational enterprise.

Concerning Federal Scientific Status Reports   The annual reports that the 1970 Subcommittee on Science, Research and Development Report recommends from the Office of Management and Budget and the Office of Science and Technology, can and should be highly complementary. On the one hand, they direct attention to the short- and long-range national and societal needs as they may act to determine scientific policies and strategies; on the other, they assess the health of the national scientific and technological enterprise and the opportunities that the internal logic and development of each field present for exploitation. Such reports can provide the opportunity for any fine adjustments of the scientific support structure. These may be required to avoid the extremely unproductive oscillations in level of scientific and technological activity and in utilization of trained manpower that have occurred in recent years. We urge that mechanisms be developed and publicized that will facilitate direct input from the various components of the scientific and technological communities in the preparation of these reports.

Mission Agency Support   As discussed elsewhere in this Report, a number of the older mission agencies such as the Atomic Energy Commission (AEC) and Department of Defense (DOD) have played a central role in the development of mechanisms for broad federal support of basic research. The excellence of U.S. science reflects a long-standing debt to the wisdom and skill of a few far-sighted officials in these agencies. However, in recent years certain problems have developed. We quote again from the 1970 Report of the Subcommittee on Science, Research and Development:

. . . Some confusion has resulted, within the agencies as well as without, as to whether this basic research is "relevant to the mission of the agency." Senator Mansfield has done an important service in calling this situation to the attention of the Congress and the Nation with respect to one mission agency, the Department of Defense.

Mission agencies should be involved in the support of basic research. This is important not only to help assure the generation of knowledge to carry out their missions, but in order that the applied research and development programs in the agencies have adequate access to highly innovative scientists and ideas operating throughout the basic research spectrum.

Admittedly there is great difficulty in ascertaining the relevance of basic research to an agency's mission. This is almost impossible to do with respect to a specific research project of an individual researcher. At the same time, it does seem possible to identify broad scientific areas of basic research which do have some relation to some technological requirements important to the mission agency funding the inquiry.

OST should develop general criteria which the Federal departments might reasonably use as a guide for their decisions as to which areas of basic research should be supported.

To this rationale for the support of basic research by the mission agencies we add and emphasize that, within areas of broad general relevance to its mission, support should be based on the recognized excellence of the principal investigator. Another factor to be considered is support of the education and training of personnel required for effective pursuit of the agency missions.

In the case of physics, the interaction with the DOD and with the AEC is one of long standing. Apart from recent difficulties such as those noted above and others relating to a growing disaffection with some of the perceived policies of the DOD on the part of some members of the scientific community, the interaction has been a remarkably effective and smoothly functioning one. Physics and physicists also can make significant contributions to some of the newer mission agencies such as the Department of

Transportation (DOT) and the Department of Housing and Urban Development (HUD). To that end, we urge the introduction of imaginative programs in areas such as new forms of energy generation, transportation, housing, and biomedical engineering that will enable the physics community to make contributions in areas of great national need. It is of particular importance in the case of the newer mission agencies for OST to exert leadership in the development of appropriate criteria for their support of basic research.

As emphasized elsewhere in this Report, the U.S. scientific community and the nation have been exceedingly well served by that small band of dedicated scientific administrators in the different agencies. A basic strength in the U.S. approach to federal support of science has been the attempt to identify areas and people of excellence and to support these on the implicit assumption that the people involved were most qualified to make the necessary scientific choices and decisions. On the other hand, agencies have the responsibility of ensuring that their funds are being used in the most effective and efficient fashion in the pursuit of agreed-upon goals. Under certain pressures, the exercise of this responsibility is exaggerated. Hence, in any overview of federal support of science, due attention must be devoted to the dangers of possible overmanagement of the detailed scientific program activities by agency personnel.

The administrative guidelines set forth by the Office of Management and Budget (Circular No. A-101, Attachment A, paragraphs IB1a and IIC) deal with this matter in an equitable fashion. They direct granting agencies to allow grantees considerable freedom in selecting and altering their research procedures.

Concerning a Continuing Data Base for U.S. Science   In formulating a more comprehensive federal plan for science, it is essential to have available a more complete data base concerning all the sciences than has previously existed. For example, in the present survey it has been necessary to devote extensive effort to the establishment of a more extensive data base in the three major areas of funding, manpower, and publications than has previously been available in physics. Similar data bases have been established recently in several of the other sciences. We consider that, once established, it is of particular importance that these be maintained and updated. Such a procedure is enormously more efficient than are the sporadic and less-effective crash efforts that have characterized the preparation of this and previous scientific survey activities. In a continuing program, the completeness and quality of the data base can also be greatly improved with time and experience. We believe that a natural home for these activities falls within the major professional societies or organizations such as the American Institute of Physics and the American Chemical Society. At the same time, all these separate society activities must be coordinated to ensure standardization of definitions and intercomparability of data throughout the scientific community. There has already been discussion within the National Academy of Sciences concerning this coordinating activity. We, therefore, urge that the OST convene a standing panel, drawing on the Smithsonian Institution, the National Academy of Sciences and the National Academy of Engineering, the National Bureau of Standards, and any other appropriate agency for the purpose of coordinating the establishment and maintenance of a national data base for all science and technology, and that, wherever possible, these activities be subcontracted to the major professional organizations and societies involved.

Concerning Education The questions concerning education in physics and physics in education are discussed in some detail in Chapter 11, and a number of detailed recommendations are included. The Committee is aware of impending changes in the overall federal structure for the support of education in the nation and, lacking detailed information on these changes, does not presume to make detailed recommendations or comment on them. At the same time, however, we are particularly sensitive to the importance of maintaining close contact between working scientists and education in science at all levels including that of broad policy formulation. We have been somewhat discouraged by the scientific content of the existing programs of the Office of Education and we urge that, in any restructuring of federal support mechanisms for both general and more specialized scientific education, due recognition be given to the importance of providing and maintaining a viable communication channel to the scientific community and to the working scientists as contrasted to the professional educational specialists. We believe that active involvement of working scientists having a broad knowledge of both science and the current problems in scientific education is essential to improvement of the present situation. Basic to all these considerations is much more effective cooperation, at all levels, between working scientists and working teachers and educators.

The Support Agencies

The conclusions and findings recorded in this section are directed primarily to the support agencies, although it should be noted that all the recommendations in the previous section, if implemented, would also affect these same agencies.

Concerning Communication with the Scientific Community   Despite the success of the national physics program and its great achievements, significant problems have arisen—frequently as a result of misunderstandings by the scientific community and the public. On occasion, these have reflected the rapidly changing character of our times. In the case of the DOD, as discussed above, marked tensions have developed in some areas, on the one hand, reflecting disenchantment with our national defense establishment and policies, particularly with our current involvement in Southeast Asia and, on the other, reflecting problems resulting from the implementation of the Mansfield Amendment. In the case of the AEC, the problems have involved segments of both the public and the scientific community dissatisfied with alleged insensitivity of the AEC to environmental issues. In the case of the NSF, almost the reverse situation has occurred, with large segments of the scientific community registering alarm at what appeared as overconcern on the part of the agency with environmental and societal problems at the expense of its role as the primary federal sponsor of basic research. In turn, the National Aeronautics and Space Administration (NASA) has been severely criticized for the proportion of its resources devoted to manned spaceflight activities as opposed to other areas of scientific activity with lower unit costs.

In all these examples, a more coherent program designed to further communication between the individual agencies and the scientific community could significantly reduce tensions. More effort should be devoted by these agencies to the presentation of their program objectives and plans to the scientific community and to soliciting specific suggestions and views than has previously been the case. The American Association for the Advancement of Science is to be commended for the initiative it has taken through its official publication, *Science,* toward making knowledge of federal agency activities and goals available to a broad scientific audience.

Concerning the Importance of Multiplicity of Support Channels   Extensive comments have already been made on the importance of the NSF exercising its role as a balance wheel in the federal funding of research. The recent increases in the Foundation budget to this end are applauded.

The importance, to science and to the nation, of maintaining a multiplicity of research-support channels at the federal level also has been discussed. The apparently limited success of more centralized support systems where they have been utilized—in the United Kingdom, France, and the Soviet Union, to name only three instances—serves to strengthen our conviction that this multiplicity of support, enjoyed in the United States during the past two decades, has played a decisive role in the establishment of present scientific strength. The competition that is involved here, on the part of the investigator for support of his program and on the part of the agency, not only for its proper share of the available federal support but also for the most able investigators, provides a continuing filter mechanism that acts to eliminate weak or obsolescent programs and activities. It also improves greatly the possibility that the very novel or unusual proposal that is outside or between traditional subfields or missions—often with relatively small chance of success but with enormous potential should it succeed—can find support somewhere.

For these reasons among many others discussed elsewhere in this Report, including our conviction that physics and physicists have significant contributions to make to the solution of the problems of U.S. society, we urge all mission-oriented agencies not only to strengthen their current programs of basic research support but also to act, as soon as possible, to initiate support of innovative and imaginative programs of interdisciplinary character directed toward the solution to those societal problems falling within their formal purview or to which their special expertise might, even in the long range, be expected to make significant contributions. Mechanisms by means of which the scientific community may be able to help in identifying and formulating such programs are discussed below.

Concerning Coordination between Support Agencies   Chapter 10 describes some of the current decision-making mechanisms now used by the NSF and the mission-oriented agencies in arriving at their current program emphases. The extent to which internal coordination among the programs of the different agencies is now undertaken is commendable; such coordination should be strengthened wherever possible. This is of particular importance in times such as the present in which major reorientation of support for relatively large facilities or laboratories is involved. Without adequate coordination there is the hazard that a high-quality activity—and one readily adjudged so by the agencies involved—may simply wither and die in an administrative or fiscal hiatus during a transfer of support. At the same time, the importance of maintaining the independence of the different agency programs within such a cooperative framework must be re-emphasized; otherwise, the many advantages that derive from the multiplicity of available support sources can be lost.

Concerning the Balance between In-House and Contract Research   The mission agencies in general are deserving of high commendation for maintaining a reasonable balance between their in-house and their grant and contract activities in research. Obviously the different missions have required quite different fractions of the two classes of activity. With a general decrease in available funding, however, there appears to be the beginning of an understandable tendency to protect in-house activities at the expense of the external programs. In the long run this practice can have unfortunate consequences for the health of U.S. science. Therefore, we urge mission agencies to exercise special care in administering budgetary reductions to maintain a balance between in-house and external research activities. Explicit criteria and procedures by which choices between the two may be more objectively made need to be evolved and applied. A recommendation in Chapter 2 suggests a possible guideline toward which these agencies might work in allocating their resources to production, development, and research activities. Wherever they have not already done so, mission agencies should initiate explicit studies directed toward the identification of new roles for their in-house facilities in attacking problems of major national importance. Certain of the present national laboratories deserve much credit for the initiative and imagination they have already demonstrated in this area.

We urge that some of the older agencies, such as the National Institutes of Health (NIH), which, as noted earlier, do not at this time support any significant level of physics research, should initiate immediate action to support a level of such physics activity commensurate with the demands the mission of the agency makes on physics. In the case of NIH and the newer agencies such as DOT and HUD, which do not as yet provide significant direct support of physics research, the development of physics advisory groups drawn in representative fashion from the physics community could not only assist the agencies in the identification of particularly relevant research areas but also, with developing mutual insight and confidence, constitute an effective communication channel between the physics community and their particular agency administration and staff.

Utilizing advisory groups and other appropriate mechanisms, the newer agencies should develop, as quickly as possible, general statements concerning their scientific and technological goals, both short- and long-range, and where possible *ab initio,* those areas of research both basic and applied that appear most relevant. Broad dissemination of this information throughout the scientific community could then serve to attract first-class minds in all the sciences to the solution of some of the pressing problems being addressed by these agencies. It is our current impression that there exists a widespread interest and enthusiasm throughout the physics community for participation in this work; what is lacking is a focus or mechanism for involvement.

The newer agencies, in particular, need to be exposed to physics, and physicists need to be exposed to their problems on an intermittent but continuing basis. Considerable progress has already been made. The chief scientific officers of several of the science agencies are physicists, or are highly physics-oriented engineers, and in several cases effective inter-action with the physics community has already been established. A rep-resentative of DOT is currently making a systematic tour of U.S. universities to identify their basic research capabilities potentially relevant to DOT's interests. It must be emphasized that, understandably, these agencies are looking for areas where significant payoff might be anticipated in a reason-able time; 10 to 15 years might still be considered as reasonable, whereas 25 years probably would not.

Concerning Communication in Physics In Chapter 13, and in the Special Panel Report on Dissemination and Use of the Information of Physics, which appears in Volume II of the Survey Report, a detailed examination of the many channels of scientific communication employed in the physics community and between physicists and other scientists, technologists, and the general public is presented. A number of conclusions follow from this examination; several are not at all in accord with much of the traditional wisdom in the physics community and thus merit particular emphasis. Among these conclusions are the following:

1. Primary scientific journals are the most important single source of physics information, with informal oral communication a close second.

2. Any improvement in physics communication that would produce a saving of P percent in the total time devoted by physicists to communica-tion and information activities to get a given yield of useful information would result in overall savings that would be worth between $3P$ and $6P$ percent of the total cost of the physics enterprise per year.

3. The obsolescence half-life of primary journal articles is much longer than commonly believed and almost certainly lies between five and ten years, with eight years being the most probable value.

4. For both primary journals and secondary abstracting and indexing services, the product is expensive to produce intially but inexpensive to replicate for additional users. There will be a very large number of users for whom the benefits will justify their paying the runoff costs but not the prerun costs. However, the resulting benefit to their work, and hence to society, is almost certainly greater than the cost of providing an additional copy for their use. Thus from an economic point of view society as a whole, or the research enterprise as a whole, should support most of the prerun costs of scientific publication. This principle applies to primary publication, secondary information services, and information-analysis centers. Physics has been a pioneer in the application of this principle, which should be even more generally recognized than it is, especially by those organizations and agencies that support research.

5. It should be just as easy for a scientist to obtain a grant for the preparation of a review article as for the conduct of new research, and such grants should allow for participation of graduate students and post-doctorals and for travel, computer time, information-retrieval services, and overhead. The resulting reviews should be thorough and should be published in media that ensure wide accessibility and dissemination. If this is done, an important by-product will be greater recognition and prestige for critical reviewing and information compaction as an activity of value comparable to original research in the overall physics enterprise.

6. The sluggish response of physicists to the availability of new information tools should be accelerated; one way is the more explicit inclusion of education in the use of information tools in the graduate training of physicists.

7. Empirical research indicates that, contrary to commonly held belief, most of the primary literature of physics is useful and used. A study in the condensed-matter subfield indicated that 33 percent of the papers probably add something of lasting value and conceptual interest; only 8 percent of the papers were judged worthless by any of the sample of experts rating them.

The Congress

In the past, the direct interaction between the scientific community and the federal government, with few outstanding exceptions, has involved the executive branch. The recent increasingly direct interaction with the legislative branch is a most welcome development.

Congress appears increasingly to be faced with acting upon issues having a significant scientific or technological content. An indication of congressional concern with research and development is provided by Table 14.1.

The Joint Committee on Atomic Energy (JCAE), over the years, has developed a record of understanding and sound judgment concerning those aspects of the nation's scientific enterprise over which it has cognizance. It has developed a rare insight into the scientific underlay of the field of atomic energy and has in many instances provided leadership. The recent change in the Atomic Energy Act, taken at the initiative of the JCAE, which broadened the responsibility of the AEC to include other than nuclear-energy sources implies a continuing and expanding role for the JCAE in the energy field. We applaud and welcome this step and suggest that the physics community stands ready to assist the JCAE and the AEC in implementing this new role.

Some of the questions that arise are extremely difficult ones, involving complex relationships to the nation's physical, social, and economic

TABLE 14.1 Committee Activities by Topic[a]

Senate

Committee	Agriculture	Atomic Energy	Economic Development (Private)	Economic Development (Public)	Education	Energy and Communication	Environmental Pollution	Environmental Sciences	Foreign Affairs	National Security	Natural Resources	Public Health	R&D Management	Science Resources	Space	Transportation	Urban Affairs	Water Resources
Aeronautical and Space Sciences	X								X						X			
Agriculture and Forestry					X		X		X									
Appropriations																		
Armed Services										X					X			
Banking and Currency			X	X														
Commerce			X	X	X	X	X	X	X		X	X	X		X	X	X	X
District of Columbia	X															X		
Foreign Relations									X									
Government Operations			X	X			X	X	X	X		X	X	X		X	X	
Interior and Insular Affairs			X	X		X					X							
Judiciary			X	X							X		X					
Labor and Public Welfare					X							X		X				
Public Works				X			X							X		X	X	X
Rules and Administration													X					
Select Committee on Small Business			X	X								X						
Special Committee on Aging																		X

House

Committee												
Agriculture	X					X		X		X		
Appropriations												
Armed Services									X			
Banking and Currency		X				X					X	X
District of Columbia											X	X
Education and Labor		X	X	X					X			X
Foreign Affairs	X											
Government Operations		X	X	X	X	X	X	X	X	X	X	X
House Administration												
Interior and Insular Affairs		X	X	X	X	X	X		X			X
Interstate and Foreign	X		X	X	X	X	X	X	X	X	X	
Judiciary								X				
Merchant Marine and Fishery				X	X		X		X		X	
Post Office and Civil Service				X	X				X	X		
Public Works		X	X	X		X					X	X
Rules												
Science and Astronautics	X	X	X						X	X	X	
Select Committee on Government Research		X	X						X	X		
Select Committee on Small Business											X	
Veterans Affairs												

Joint

Committee												
Atomic Energy	X	X		X		X					X	X
Economics		X				X						

989

ª Source: *An Inventory of Congressional Concern with Research and Development (88th and 89th Congresses), A Bibliography,* prepared for the Subcommittee on Government Research of the Committee on Government Operations, United States Senate, December 15, 1968.

environment; many involve highly specialized knowledge, and, even then, unambiguous answers may not be possible. Ascertaining the facts may be only a first step. Although they can claim no special expertise in political, economic, or social aspects of these issues, most scientists consider it a privilege to make their technical knowledge available to the appropriate bodies in Congress. However, they do not always know how to go about doing so.

If, to assist it in dealing with specific legislative problems, Congress wishes more fully and effectively to avail itself of the resources represented by the scientific community, the development of a well-identified, more formal and centralized mechanism for bringing this about would be desirable.

The Survey Committee supports the 1970 Subcommittee on Science, Research and Development Report recommendation:

> In the legislative branch, the Congress should seek a centralized Senate jurisdiction over science and technology, and establish an Office of Technology Assessment as recommended previously by the Committee.
> In the House of Representatives responsibilities for the Nation's scientific and technological activities in a general way and for overview of Federal scientific research and development in particular are, by House rule, centered in the Committee on Science and Astronautics. The Senate has no such counterpart, no science focal point.

Properly organized, such an Office of Technology Assessment could serve a critically important function in identifying broad areas of research in which new knowledge will be needed in the future to understand and monitor the overall effects of technology and manage them more intelligently. It could provide an effective input to the Congress in regard to the relationship of current basic research and effective future decision making on technology.

In addition to this Office of Technology Assessment, it is hoped that both the House and Senate will establish standing scientific advisory groups. These should be regarded as complementary to, and in no sense supplanting, the OST and other advisory groups reporting specifically to the Executive Branch. The existence of such groups would provide a much-needed focus for the interaction of scientists with the Congress.

We also recommend that the House and Senate committees hold more frequent hearings than is now the case on matters of national scientific importance as distinct from the more customary hearings specifically relating to the budget presentations of the different agencies. Of particular importance here would be hearings on the status and needs of specific disciplines, with emphasis on their potential relevance to national missions —quite apart from their problems, opportunities, and achievements.

Finally, and most important, we urge that members of the scientific community, and specifically the physics community, make themselves available to members of the Congress, and in particular to their own representatives and senators, on the supposition that some significant work will be expected of them. We suggest that briefing of congressional staff when appropriate, the preparation of position papers on matters of scientific interest, and the drafting of public statements are among those ways in which scientists could make effective contributions to congressional activities while at the same time building a much broader base of scientific awareness and interest than that which the nation currently enjoys.

Universities and Colleges

In Chapter 11 of this Report questions of education—at all levels—are discussed in considerable detail. Included here are only certain conclusions that in particular should be brought to the attention of the administration and faculties of universities and colleges.

Concerning the Character of Contemporary Education in Physics Traditionally an education in physics has been characterized by a deep interest in first principles, an economy of thought, and a style and flexibility that opened a wide range of career opportunities to one so educated. Frequently it has been the physicist's style, his approach to physical problems, or his familiarity with mathematical techniques, rather than his physics expertise, that have represented his major contribution to a given problem or activity.

During the past decade, a significant change in this situation appears to have taken place, with a general shift of the traditional physicist's approach to the engineer. At the same time, in part reflecting an essentially open employment market, an unfortunate tendency toward overspecialization in the education of the physicist can be noted, with a corresponding shrinking of his intellectual horizons and his acceptable employment opportunities. The gradual elimination of much of the more classical physics—including optics, acoustics, hydrodynamics, and even thermodynamics from physics curricula at all levels—has been a part of the change.

More important than this, however, is a change in motivation that may be sensed in many physicists. In some academic communities an often unconscious arrogance has developed; a career in academic science has become equated with first-class citizenship; other careers, in industry, government, and the like, have been rather subtly reduced to lower status. In part this situation has occurred because academic careers were so readily available, and industry felt that it had to offer high salaries or pseudo-academic careers or both, at least initially, to attract scientists of the desired quality. Although neither the academic nor the industrial communities are blameless, the Committee considers this situation totally unacceptable. Therefore, we urge that all universities and colleges immediately initiate steps directed toward regaining something of the older breadth of training and approach traditional in physics. This effort will normally involve the reinstitution of certain courses in more classical physics; it will certainly involve specific emphasis—in examples, problems, seminars, and the like—on the opportunities and challenges that lie outside the normal academic channels.

In reaching this conclusion we are, of course, mindful of the current employment problems facing young physicists. The major motivation here is the conviction that this broader education is essential to full enjoyment and realization of the physicist's potential as a scholar and as a member of society.

At the same time, industrial organizations are urged to be as explicit as possible in presenting to potential employees the opportunities and responsibilities available to them. These are often inherently different from those available in an academic community; industrial organizations should not feel it necessary to compete with academic communities on their grounds but should emphasize their own strengths.

Concerning Interdisciplinary Programs   There has been much discussion in recent years of the desirability of initiating broadly interdisciplinary majors at undergraduate levels. Many of these, involving physics, mathematics, chemistry, and like fields, have been available for many years; recently the emphasis has shifted to environmental-science areas. Although broadly based interdisciplinary attacks on major societal problems are obviously deserving of support, the undergraduate college years are not the appropriate milieu. Optimum progress toward the solution of many national problems will be effected by bringing together individuals, each deeply knowledgeable in his own discipline. On this basis, interdisciplinary activities as such might better be deferred to graduate training periods, while in undergraduate training programs specific efforts be made—in the selection of examples and problems, visiting speakers, seminar topics, and the like—to acquaint undergraduate students with the opportunities and challenges that can be theirs outside their disciplinary boundaries. At the graduate education level, the development of interdisciplinary groups, the granting of advanced degrees on the basis of work of an interdisciplinary or interdepartmental character, and other similar mechanisms are strongly supported.

Concerning the PhD   Although there is widespread dissatisfaction with certain aspects of the PhD as traditionally awarded, and in particular with its specific original research requirements, in a fundamentally experimental science such as physics the problems may be much less severe than in many other fields. Without doubt, experience in original research is essential to the education of any practicing physicist. At the same time, even though (as detailed in Chapter 11) PhD programs in the physical sciences are significantly shorter in both registered and calendar time than is the case in many other major areas, it is still a matter of concern that the time required to attain the PhD is increasing to between five and six years beyond the baccalaureate degree. This trend suggests that universities should carefully examine their current PhD programs with the view to making possible the degree award to well-qualified students after, on the average, three to four years of full-time graduate study. In general, the average initial formal educational period is simply too long, and emphasis must be given to arrangements by means of which students can conveniently spend one to several years following the baccalaureate degree in professional employment prior to embarking on graduate study *or* an equivalent period following the award of an intermediate graduate degree such as the Master of Philosophy (M. Phil.), as recently instituted at a number of universities, before returning for the PhD.

Concerning Midcareer Education The rapid pace of contemporary science and technology—and particularly of physics—suggests that universities in particular should investigate the possibility of establishing special fellowships to enable physicists who have been in the profession for a number of years to return for at least six months and preferably a year for an intensive period of refresher education. To be effective it may well be necessary to evolve special seminars and the like specifically for this purpose; the benefit to the institution from the presence of these nonacademic scientists and their direct participation with students and faculties can be very large. Indeed, this could be one of the important channels for achievement of the broader student motivations, the desirability of which has been noted above. A crucial aspect of this program, of course, will be the willingness of the employers involved to release their people for the time required for this further education.

The development of such nonacademic fellowships by the federal support agencies is recommended. We deplore the apparent termination of the NSF Senior Postdoctoral Fellowship Program. Although this program was dominated by academic participants, some few industrial scientists benefited from it, and this fraction might have been substantially increased. Because of the disproportionately large benefits that accrue to the entire U.S. scientific community relative to the required funding for such programs, we have recommended, in Chapter 2, that the NSF Senior Postdoctoral Fellowship Program be retained and expanded and that an effort be made to develop greater participation in it by industrial scientists.

Important benefits can also be derived from the kinds of intensive summer program activity that have already been initiated by a number of universities and that frequently are more accessible than are the longer sabbatical or fellowship programs to senior industrial scientists who may have administrative or other ongoing commitments at their own institutions. Universities and colleges should examine the extent to which they might make available such intensive summer programs in areas in which their faculties or facilities might make them particularly well qualified.

The Senior Review that has been operated for the past 14 years by the Department of State should be considered as a model for much broader utilization, in both government and industry. In this program, some 40 senior governmental officials are freed from their normal responsibilities for a 10-month period during which they are exposed to, and participate in, an extremely concentrated and high-level educational experience. This involves not only lecture series by recognized leaders in most areas of the nation's activity but also field trips and individual research projects for which extensive logistics support and facilities are made available. Because the participants in such programs are highly capable and highly motivated, a remarkable amount of material can be included in the 10-month program. The Committee urges that other federal agencies and major industries consider the benefits that could accrue to them through making similar midcareer educational opportunities available to their staffs.

It may be that in individual instances it would be more effective and feasible to consider the practice of awarding sabbatical leaves of absence, with full salary and other support, to selected agency and department staff so that they might devote themselves for a year to uninterrupted work in the communities that they represent or with which they interact during the normal course of their duties. Such interaction in depth can renew and restore the all-important sense of perspective and commitment that is essential to the most effective and efficient discharge of their duties in government or in industry.

Finally, it should be noted that, particularly in engineering areas, the evening extension course program has frequently provided an effective and convenient means through which members of a greater university community could take advantage of the available educational opportunities on a continuing basis. Physics departments, which have traditionally played rather minor roles in these extension activities, are urged to examine the extent to which they might make such activities more accessible to their communities.

Concerning Postdoctoral Opportunities in Physics During the past decades, in almost all areas of physics, the postdoctoral appointment has become a standard component of the educational process—a period following the award of the PhD but before taking on completely independent research activity—in which the individual works closely with one or more senior physicists and their students. This period can be an enormously profitable time inasmuch as it is one in which every attention can be directed toward research activity free from the all-too-common pressures of other teaching or administrative responsibilities. Except in isolated cases involving use of very large facilities, the typical research group involves significantly more students than postdoctoral fellows. However, at least until the present financial stringencies are resolved, it is recommended that these traditional ratios be changed in favor of a greater number of postdoctoral fellows. Recent graduates who have already committed themselves to physics merit such consideration, and it is much more humane and effective to apply throttling in the graduate training pipeline at the point of input rather than output; throttling has all too frequently been applied at the latter point in recent years. This recommendation is already being implemented at some of the major universities; it should be considered much more widely.

In consequence of the foregoing, the termination of the NSF Postdoctoral Fellowship Program, without the compensating increase in research support that could provide alternate postdoctoral opportunities, is a situation viewed with dismay. The NSF, at least for the next few years, should increase markedly the number of competitively available postdoctoral fellowships, and program officers in support agencies, especially the NSF, should view with more sympathy than in the past requests for support of postdoctorals rather than graduate students on research grants.

Concerning Graduate Education in Physics The recent graduate admission statistics presented in Chapter 9 are most disturbing. They suggest that in responding to current financial pressures, and to apparent overproduction of graduates in some subfields of physics, the most distinguished universities are those that have cut most deeply into the size of their entering classes—some 50 percent on the average between 1969 and 1971. At the same time, other institutions, in responding to increasing undergraduate enrollments and reduced finances—and in recognizing that typically several graduate teaching assistants can be supported at the same cost as one junior faculty member—are, in fact, increasing their graduate enrollments. In some cases, too, this situation is exacerbated by the fact that, unable to gain admission to the most distinguished graduate programs because of their marked reductions in class size, well-qualified students increasingly have to seek enrollment in the less-distinguished or emerging graduate programs. Over a long period, such a situation will tend to reduce the overall quality of the nation's scientific manpower resources.

For these reasons, in each of the major scientific and engineering fields, including physics in particular, it is urgent that a joint committee be assembled by the appropriate National Academy (NAS, NAE) and professional societies (the American Association of Physics Teachers and the American Physical Society in the case of physics) to develop and disseminate a set of criteria for graduate programs. It is further urged that, in fairness to prospective students who may otherwise lack the necessary information, such a committee as we have suggested above should develop new and more stringent national criteria for graduate programs in science and engineering and make the results available on a national basis.

Concerning the Structure of Physics Departments Because any given university (or college) should not unnecessarily be precluded from participation in graduate-level activities, it is essential to address one of the prevalent concepts that has developed in physics education during the past decade. For various reasons, in an expansionary climate, it became a common belief that to be of highest quality a physics department must simultaneously have research activity in *all* the recognized major subfields of physics. Particularly with constricting resources, this belief has often led to coverage at the expense of quality. We suggest that each physics department examine its internal structure critically with the view to selecting those areas in which it has the faculty and facilities to achieve excellence. Although breadth is of course desirable, it is not essential. It is further urged that, particularly but not exclusively in the case of smaller institutions, attempts be made to identify groups of geographically related institutions that could enter into joint arrangements such that each might select a particular field, or subfield, for specialization at a depth and quality not otherwise possible, with the students of all institutions having access to any of these regional activities. As discussed elsewhere in this Report, this concept of regional educational activity may well be the only means by which maintenance of quality in the face of rapidly increasing costs is possible. It is noted that the NSF Science Development Program has actively encouraged, and in a number of cases assisted, the formation of such cooperating regional groups.

Concerning Elementary and Secondary Education in Physics As discussed at length in Chapters 11 and 12, physicists as a group make their career decisions at an unusually early stage of their academic careers; it is thus of particular importance to the physics community that the teaching and counseling in the elementary and secondary schools be of as high quality as possible. Furthermore, and in the long run perhaps even more important than this professional consideration, involved and enthusiastic science teachers in the elementary and secondary schools have the opportunity to present something of the meaning, the adventure, and the goals of science to students at a receptive and formative period in their lives. This is clearly the time at which a golden opportunity to engender scientific literacy in the general public—as a vital part of our ongoing culture—can either be exploited or almost irretrievably lost.

In the past, the physics community—both academic and industrial—has had far too little contact with the forces of elementary and secondary education in this country. All evidence indicates that by far the largest fraction of physics teachers in secondary schools are woefully ill-prepared in their subject matter; hence the recommendation that secondary school systems emphasize subject-matter requirements for science teachers and that universities offer, in both graduate and undergraduate programs, opportunities for training of prospective science teachers, with a view to meeting formal qualifications for secondary school positions.

Furthermore, physicists, acting as individuals or through the state and local branches of their professional societies, should make contact with their state and local departments and colleges of education, with specific offers of assistance in planning, staffing, and even executing some of their scientific educational activities. The area of curriculum reform is a particularly important but sensitive one. The nation has much to gain from such activity; it is already in progress in isolated instances.

Concerning the Change to User-Group Activity in Physics Returning finally to the major university departments themselves, we have recommended certain changes in the structure of doctoral programs and the like, but perhaps the most far-reaching recommendation in terms of internal structure, and one of particular relevance to physics, relates to the changing mode of frontier research in many of the physics subfields.

For some years, astrophysicists and high-energy physicists working in the universities have found that successful work in these subfields necessitates the use of major facilities at regional or national laboratories. This type of user activity at a place remote from the home campus is now spreading to other subfields of physics, as the frontiers that are ripe for exploitation in them begin to require the use of facilities too large to be acquired or supported by any single university. For example, some of the best opportunities for frontier research in nuclear physics will soon be offered by meson factories and heavy-ion accelerators; high-magnetic-field work in condensed-matter physics may require the facilities of the National

Magnet Laboratory. In Chapters 6 and 9 some of the problems that this shift of activity from the campus to a central or national facility entails are considered.

As previously indicated, universities must be prepared to be flexible in their demands on faculty and students, who, because of their field of research, find it essential to spend significant periods of time at remote sites, using major facilities not otherwise accessible. Universities must recognize that increasingly this situation will be the price required to retain personnel active at the frontiers of their fields, and that the returns to the university in contagious excitement and intellectual vitality will repay manyfold whatever inconvenience and additional expense may be entailed. In much the same spirit, it should be noted that in certain subfields of physics—and most particularly in more applied areas—the research frontiers are not at the universities but rather at industrial or more specialized governmental laboratories. In these cases, procedures should be developed that will permit the pursuit of dissertation research in these frontier facilities under joint supervision of a university faculty member (full-time or adjunct), who must protect the student's interest and maintain educational standards, and a representative of the facility, who must be responsible for the relevance of the project to its overall mission. It is recognized, however, that the *in absentia* thesis presents problems. If too large a fraction of students choose this route, the whole character of a graduate department can be altered. Students lose the value of interaction with each other, which is half of their education. Mechanisms must be developed to provide an equitable balance within any given department. One basic criterion that should be included is that of equitable student compensation; the university should retain the right to establish the compensation of its students working *in absentia* in order that the choice between the normal departmental and the *in absentia* routes is not made on a purely financial basis.

Concerning Communication in Physics Earlier in this chapter (and in Chapter 13) conclusions based on an examination of communication mechanisms relating to physics were presented. Many of these conclusions are broadly applicable throughout all science. They include the findings that the primary journals remain the most important single source of physics communication, that the obsolescence half-life of articles in such journals is about eight years, thus much longer than has frequently been estimated, and that at least in the area of condensed-matter physics, in which a sampling of published articles was examined, most of this primary literature is both useful and used—again at variance with common belief.

To the emphasis previously placed on the importance of high-quality review articles in disseminating and consolidating research results, a further conclusion should be added because of its special relevance to university and college communities:

1. In hiring and promotion procedures, high-quality review articles should be given weight comparable with original research papers of equal quality.
2. Graduate curricula for physicists should contain explicit training and experience in the use of modern information tools.

National Laboratories

As indicated above, the national laboratories have made, and continue to make, important contributions to the national life through their research and development activities. The function of the national laboratories in the case of a major piece of scientific instrumentation too large to fit comfortably, either physically or administratively, on a university campus— such as a super-accelerator—is obvious. There are, however, many other unique characteristics of these institutions. Some areas of science, particularly those of an interdisciplinary character, can be pursued effectively only within such a framework. In nuclear physics, for example, the coming activity in supertransuranic species requires a unique combination of people and facilities in physics, chemistry, large-scale instrumentation, and health physics. Similarly, major programs applying the concepts and technologies of physics to the central problems of biology, such as the MAN program at the Oak Ridge National Laboratory, would be difficult to mount in another institutional framework. However, there are other types of major problem of importance to society in which the universities also can present a unique competence; an example would be one requiring not only scientists but also economists, political scientists, lawyers, and historians. It would be difficult to find such a mix of recognized experts in other than a university context.

Although many national laboratories were established specifically to attack major physics-related problems, their management must continue to search for broad areas of national concern to which their personnel and facilities might be particularly appropriate. Several of the laboratories are to be commended for the initiatives they have already demonstrated in these areas.

Concerning University–National Laboratory Interactions Some tension has developed over the years between the university and the national laboratory communities. This reflects a feeling on the part of some university faculty that, in research areas in which both communities were active, they were in a competitive situation and one not favoring them in terms of support and facilities. As a necessary part of their program to attract and hold very competent personnel in their mission programs, as well as to establish a core group to maintain the intellectual vigor of an enterprise and uphold its scientific standards, the national laboratories quite properly have established and maintained basic research programs of high quality. We suggest, however, that their managers should examine their basic physics research programs carefully and, when activities are found that do not contribute in a significant way to the mission of the laboratory or do not require the special support or facilities characteristic of the laboratory, should consider transferring them to a suitable university or industrial milieu.

Concerning the Role of Younger Physicists　There is a concern that in responding to short-term financial pressures the national laboratories, just as universities, may have sacrificed young personnel in favor of older and more established members of the laboratory staffs. Chapters 6 and 9 discuss many of the problems inherent in any such reaction. When they have not already done so, national laboratories should develop and implement an employment policy for professional staff that ensures an adequate mix of both young and old, temporary and permanent staff. We recognize that the policies usually in force at most universities may well be inappropriate to a given national laboratory in view of the degree of continuity required in certain applied programs.

Concerning Review Mechanisms　A description of some of the operational mechanisms of a typical national laboratory will be found in Chapter 9. Unfortunately, there appears to be a rather widespread misunderstanding of these mechanisms, particularly in academic communities and with reference to mechanisms for peer group or other review of the quality of the detailed research activity in progress. This review is, of course, ultimately the responsibility of the laboratory director, and a variety of different mechanisms exist by means of which this responsibility is discharged. Any misunderstanding might be minimized if the individual laboratories publicized more widely the ongoing review mechanisms that they already have.

Industrial Organizations

Many of the recommendations noted earlier in this chapter have related to the importance of furthering communication between industrial research activities and those in both universities and national laboratories. In a time of increasing international competition, particularly in high technology and other science-related industries, it is especially important from a national viewpoint to optimize this communication and expedite the transfer of new concepts, techniques, and technologies among these institutions.

Concerning University–Industry Interactions   There are two basic reasons why stronger university–industry interactions are desirable. One is that the rapid growth of research that characterized the past two decades has decreased substantially. This means that fewer new research people will be produced, and the average age of the research people presently employed will increase. Their skills and knowledge will tend to become outmoded. The flux of new people into most research institutions will greatly diminish or entirely halt. To maintain vitality and substitute another kind of flux of people through research organizations, new kinds of mechanisms must be developed. This is a problem not only for industry but for universities and for government and national laboratories. The most stimulating kinds of interactions are those that take place among people in different kinds of institutions. Because most of those involved are in either universities or industry, it follows that university–industry interactions are particularly important.

The second aspect of the scientific and research community that demands stronger industry–university interactions is education. Industry will continue to depend on the universities for people it needs and universities will depend on industry to supply the jobs that their graduates will fill and to provide new problems, both basic and applied, to attack.

Several general aspects of university–industry interaction should be recognized. First, because the benefits of such interaction accrue to everyone involved—the universities, the industries, and the individual scientist—all should share in the associated costs. Further, various examples of satisfactory relations between research people in industry and university faculties, such as personnel interchange programs, now exist. Most of the scientists in industry are not in basic research programs, however, and the maximum beneficial effects to industry will accrue only if universities provide for interaction with developmental as well as research organizations. A substantial amount of new science is generated in Europe, and European universities should be included in industrial interaction. This is especially appropriate for industries that have European subsidiaries or business. It should be re-emphasized, too, that the problems of university laboratories and departments in many respects are shared by government in-house laboratories and national laboratories and that these institutions also should be encouraged to develop similar interaction with industry for like reasons.

As a mechanism for increasing the flexibility of physics baccalaureates and their awareness of the nonacademic employment opportunities open to them, more industrial research organizations should give serious consideration to the possibility of accepting young men at the BS level, with the understanding that superior performance could result in return for graduate study at a later date under sponsorship of the parent company. This practice could do much to change the motivation of the individuals involved, and a flow of such students would have a salutary leavening effect in the graduate schools that they subsequently would attend. A number of the larger industrial organizations already have well-developed programs of this type, and their initiative is commendable. Smaller companies would face special problems in implementing such a program, and direct federal agency assistance could be particularly important and beneficial.

Concerning the Expansion of Industrial Research Activity Concern regarding the foreseen competition between the scientifically based industries of this nation and those abroad has led to a recommendation that tax policies and other federal practices be examined and perhaps modified with the aim of strengthening the national industrial research base. In many of the medium-sized industrial research organizations, there appears to be a lack of conviction on the part of senior management regarding the extent to which the long-term continuity of their research activity represents one of their especially important company resources. A heightened vulnerability of the research groups in time of dwindling profits or contracting budgets is a reflection of this lack of conviction. The Committee believes that an important component of any improved competitive position of U.S. high-technology industry in the world market must involve extension of the strong research activity now fostered in a few of the major industrial organizations to a much larger fraction of U.S. industry.

Recognizing that there is a certain critical mass phenomenon in any research activity, we understand that many smaller U.S. industries have thus far refrained from attempting to establish other than minimal in-house research capabilities. Just as in the case of the smaller universities and colleges, smaller industries should be encouraged to establish joint research activities, perhaps with joint federal participation as well, as has been the case in the United Kingdom and the Scandinavian countries for some time. Clearly there are serious questions of proprietary rights, sharing of research results, and the like, but there appears to be no fundamental reason why these questions cannot be solved. This approach could significantly strengthen the U.S. economic posture in international competition.

The Scientific Societies

As the only truly grass-roots organizations—at least in principle—in U.S. science, the scientific societies can, and should, play a much larger part in national scientific affairs than they have in the past. At the same time, it is important for them to concern themselves primarily with furtherance of the internal goals of their respective fields.

To play a more effective role in responding to the needs of their memberships in such areas as employment, public information, education, and maintenance of a statistical data base, in addition to the current major activities in conjunction with scientific meetings and scientific publication, scientific societies should move aggressively to improve their financial positions through development of increased federal, industrial, and foundation support and particularly through substantial increases in their membership fees or assessments, reflecting the increased services that they could provide to their members.

One of the areas of greatest concern in many of the sciences at this time is the vast expansion of literature that already bids fair to outrace the ability of even the most avid specialist to keep pace. A crisis point has already been reached, and it is now necessary to evolve readily accessible and convenient mechanisms through which both active scientists and others outside the sciences obtain the scientific data that they require. The present archival journals simply do not suffice. (These serious problems are discussed at length in Chapter 13.)

The scientific societies provide an appropriate and natural milieu for creative action in addressing this problem. The recommendation regarding markedly increased membership fees is partially directed toward supporting their efforts to improve communication.

Each of the scientific societies working with the appropriate federal coordinating groups should appoint a standing committee charged with the development of recommendations for coping with the publication problems in their respective fields. It is hoped that whatever mechanism emerges may offer an acceptable solution in many fields. To wait until some giant computer takes over is not feasible; innovative ideas are needed more than anything else.

Quite apart from the demanding task of compiling, collecting, and evaluating the vast mass of scientific data now produced, there is an overriding need in almost all fields for critical reviews and condensations. An added stimulus would be provided if each scientific society established, as soon as possible, a rather generous series of prestigious prizes to be awarded each year for the most outstanding review articles in their field.

National Academy of Sciences (NAS), *National Academy of Engineering* (NAE), *and National Research Council* (NRC)

The 1970 Subcommittee on Science, Research and Development Report recommended that:

> The role which science and technology have to play in promoting solutions for the major problems of the day should be explained forcefully to the public with particular help from the National Academies of Sciences and Engineering.

We concur. The Academies have a unique role to play, as they represent the entire science and engineering communities. Therefore, it is extremely important for the appropriate committees (NAS Committee on Science and Public Policy and NAE Committee on Public Engineering Policy, for example) to continue the work started in this and similar survey reports that they have commissioned and to develop a report that not only presents the case for science and technology, as noted in the recommendation quoted above, but also faces the questions (analogous to those we considered for physics) of priorities and contingency alternatives for all the sciences. This is a fearsome but vitally important task.

We have already commented on the part that we hope the NAS-NAE-NRC will undertake to coordinate the maintenance of an adequate statistical data base for science. In the same spirit, we recommend that the NAS-NAE-NRC establish a library devoted to the collection and handling of information about science and scientists.

Following our earlier recommendation to the scientific societies and with the same rationale, we recommend that the NAS-NAE act to establish annual prizes to be awarded for the best review articles in the different areas of science in any given year. We believe it to be absolutely essential that the practice of devoting extensive effort to the preparation of high-quality critical reviews and syntheses be returned to the position of high esteem it once occupied. Our recommendations in regard to prizes are directed to this end.

President Nixon has recently announced programs for a National Institute of Education and for a National Foundation for Higher Education. We urge that the NAS-NAE, at the earliest opportunity, appoint a committee of senior scientists and engineers who can provide input relevant to the founding of the new Institute and Foundation and to the proper inclusion of science and technology in their programs from the outset. We have already commented on the importance of direct scientific input to federal education groups; we believe that this channel through the NAS-NAE to the new Institute and Foundation can be a most important one.

Finally, we urge that the NAS-NAE-NRC, as a recognized central voice of the U.S. science and engineering community, increasingly take the initiative to issue well-documented, reasoned position papers on the scientific facts underlying public issues in which science and engineering play an important role—it is to be hoped at an early enough stage of public discussions to preclude or minimize some of the questionable and almost hysterical scientific discourse that has marked similar public discussions in the recent past.

Private Foundations

In the early days of physics in the United States, and indeed until the late 1940's, a major component of the research activity was made possible only through the generosity of a small number of private foundations. In the postwar period, with the development of big science and the general identification of physics with it, most of the foundations adopted the not unreasonable view that the scale of operations in the new physics was such that only federal sources could provide the necessary resources and that, as a consequence, they might better devote their more limited resources to the support of other fields. This trend has continued.

There are certain features of the present situation, however, that might commend physics support to these foundations again. First is that science and physics play an ever-growing role as a foundation for contemporary education in all fields; they are essential to effective participation in a technological civilization. Hence very limited awards, often to individuals in smaller institutions, can pay remarkable dividends in improved teaching and understanding. Also, for many reasons, alternate support is frequently lacking.

In big science, as already mentioned in this Report, large leverage factors apply when a substantial percentage (say, 85 percent) of the total funds available to a major facility or laboratory are committed to keeping its doors open, quite apart from any specific scientific research. Under such circumstances, a relatively small sum—at most a few percent of the total running cost—of a facility can make the difference between being able to do a fundamental experiment and having to discard or seriously delay it. Certain of these high-leverage opportunities should also be attractive to the private foundations on occasion.

Therefore, we urge that immediate past traditions should not be allowed to prevent opportunities such as those just discussed from being brought to the attention of private foundations; and foundations are urged to view these opportunities in terms of what might be accomplished for a given expenditure even though the sums involved may seem small in comparison with the total required by the field. The Sloane Foundation program of providing unencumbered funding to highly selected young scientists is to be commended. Other foundations should find similar programs attractive. The Guggenheim Foundation has a long tradition of support of older scientists. Over the years, the Sloane and Guggenheim Fellows have achieved a deserved and enviable reputation for excellence and achievement.

State Governments

In recent years, many of the states have established research councils or committees on a continuing basis and have made available to them funds for the initiation of support of research and development activities. These actions are to be applauded and, despite a less favorable national economic climate than that which we enjoyed in the mid-1960's when much of this activity took place initially, all state governments that have not yet done so should consider the creation of a research council specifically charged with an overview responsibility and the furtherance of science and technology in their respective states.

The strong ties between science and industrial and economic growth have already been discussed. Route 128 in Massachusetts, despite recent vicissitudes, remains a remarkable demonstration of this linkage. Hence, the Committee concurs in the following recommendation of the 1970 Subcommittee on Science, Research and Development:

The scientific method and technological research should be increasingly utilized by regional, state, and local organizations in seeking solutions to societal problems.

For some years it has been recognized that the most effective application of science and technology to certain problems such as crime and pollution abatement can be made on a regional or local basis. Progress in implementing this concept has been slow, and the State Technical Services program in the Department of Commerce has been dropped. We must not cease our attempts, however, to provide the necessary support for this important area, and at the same time increase the ability to utilize such support by State and local officials.

For many years surveys have shown that secondary school science teachers tend to be inadequately prepared in their subjects, and that, in particular, high school physics teachers are less well-prepared than other science teachers. We believe this circumstance has had an adverse influence on student interest in physics, whether for cultural or for career reasons. It has been fostered by the willingness of school authorities to employ teachers who are inadequately prepared in physics. To correct this situation, physics faculties in colleges and universities are urged to participate in improved and realistic teacher-preparation programs. In turn, states should adopt and enforce improved standards of subject-matter qualification in physics.

The science background of elementary schoolteachers is known to be weak and in physical science almost nonexistent. This circumstance may have contributed eventually to indifference and even antipathy toward science on the part of much of the general public. In Chapter 4, physics faculties in the colleges and universities are asked to participate actively in the development and implementation of new, realistic programs to improve the physical science backgrounds of preservice and in-service elementary schoolteachers. In turn, states are urged to support these moves and adopt and enforce improved standards for the science backgrounds of elementary schoolteachers.

The teaching of science cannot be adequate at any level, from kindergarten through graduate school, without continued refresher courses for teachers and without a laboratory component. Science teaching places demands on the teacher well beyond classroom hours. For laboratory instruction, these demands are often beyond the capacity of the teacher himself; he needs assistance. Hence states in their budgeting processes should give adequate recognition to the monetary and time requirements for effective science teaching.

APPENDIX A:
PANELISTS AND
CONTRIBUTORS*

Acoustics

Robert T. Beyer, Brown University, *Chairman*
Andrew V. Granato, University of Illinois
Theodore Litovitz, Catholic University of America
Herman Medwin, Naval Postgraduate School
Wayne Rudmose, TRACOR, Inc.
Jozef Zwislocki, Syracuse University

CONTRIBUTORS

David T. Blackstock, University of Texas
Floyd Dunn, University of Illinois
Cyril M. Harris, Columbia University
Harvey Hubbard, National Aeronautics and Space
    Administration
Lewis Larmore, Douglas Advanced Research
    Laboratories
J. J. G. McCue, Massachusetts Institute of Technology
Kenneth N. Stevens, Massachusetts Institute of
    Technology
J. E. White, Globe Universal Sciences, Inc.
Charles R. Wilson, University of Alaska

*The Chairman of the Physics Survey Committee is an *ex-officio* member of all panels.

In general, no agency was represented on the Committee or any of its panels by more than one person at a given time.

LIAISON REPRESENTATIVES

Edward I. Garrick, National Aeronautics and Space Administration

Rolf M. Sinclair, National Science Foundation

READERS

Leo L. Beranek, Bolt Beranek & Newman
James H. Botsford, Bethlehem Steel Corporation
John V. Bouyoucos, General Dynamics
Richard K. Cook, National Bureau of Standards
Ira J. Hirsch, Washington University
Karl D. Kryter, Stanford Research Institute
R. Bruce Lindsay, Brown University
Isadore Rudnick, University of California, Los Angeles
John C. Snowden, Ordnance Research Laboratory
Henning E. Von Gierke, Wright-Patterson Air Force Base

Astrophysics and Relativity (Joint Panel with the Astronomy Survey)

George B. Field, University of California, Berkeley, *Chairman*

Geoffrey Burbidge, University of California, San Diego

George W. Clark, Massachusetts Institute of Technology

Donald D. Clayton, Rice University

Robert H. Dicke, Princeton University

Kenneth Kellermann, National Radio Astronomy Observatory

Charles Misner, University of Maryland

Eugene N. Parker, University of Chicago

Edwin E. Salpeter, Cornell University

Maarten Schmidt, California Institute of Technology

Steven Weinberg, Massachusetts Institute of Technology

CONSULTANT

David D. Cudaback, University of California, Berkeley

LIAISON REPRESENTATIVES

Nancy Roman, National Aeronautics and Space Administration

A. W. Schardt, National Aeronautics and Space
Administration
Harold S. Zapolsky, National Science Foundation

READERS

Elihu A. Boldt, National Aeronautics and Space
Administration
Giovanni G. Fazio, Smithsonian Institution
Astrophysical Observatory
Hugh M. Johnson, Lockheed Palo Alto Research
Laboratory
D. P. McNutt, Naval Research Laboratory
Alan T. Moffet, California Institute of Technology
J. A. Simpson, University of Chicago
James W. Truran, Yeshiva University

Atomic, Molecular, and Electron Physics

N. Bloembergen, Harvard University, *Chairman*
P. L. Bender, Joint Institute for Laboratory
Astrophysics
R. Bernstein, University of Wisconsin
Ronald Geballe, University of Washington
J. A. Giordmaine, Bell Telephone Laboratories, Inc.
D. A. Kleppner, Massachusetts Institute of Technology
M. O. Scully, University of Arizona
R. W. Terhune, Ford Motor Company Research
Laboratory

LIAISON REPRESENTATIVES

Harold Glaser, National Aeronautics and Space
Administration (alternate)
Harry Harrison, National Aeronautics and Space
Administration
Richard J. Kandel, Atomic Energy Commission
Goetz Oertel, National Aeronautics and Space
Administration
Rolf M. Sinclair, National Science Foundation

READERS

Benjamin Bederson, New York University
R. Stephen Berry, University of Chicago
Wade L. Fite, University of Pittsburgh
Henry M. Foley, Columbia University
Keith B. Jefferts, Bell Telephone Laboratories, Inc.

J. V. Phelps, Joint Institute for Laboratory
Astrophysics
Francis Pichanick, University of Massachusetts
Francis M. Pipkin, Harvard University
T. N. Rhodin, Cornell University
Felix T. Smith, Stanford Research Institute
G. King Walters, National Bureau of Standards

Earth and Planetary Physics

Richard M. Goody, Harvard University, *Chairman*
Donald Anderson, California Institute of Technology
W. I. Axford, University of California, San Diego
Henry G. Booker, University of California, San Diego
A. G. W. Cameron, Yeshiva University
Joseph W. Chamberlain, The Lunar Science Institute
William Gordon, Rice University
Robert Phinney, Princeton University
William Richardson, Nova University

CONSULTANTS

Robert Fleagle, University of Washington
Marvin Goldberger, Princeton University
Walter Munk, Scripps College
Frederik Zachariasen, California Institute of
Technology

LIAISON REPRESENTATIVES

David B. Beard, University of Kansas, Consultant to
the Atomic Energy Commission
Albert E. Belon, National Science Foundation
Neil Brice, National Science Foundation
Robert F. Fellows, National Aeronautics and Space
Administration
Albert Opp, National Aeronautics and Space
Administration
Erwin R. Schmerling, National Aeronautics and Space
Administration

Education

Ronald Geballe, University of Washington, *Chairman*
E. Leonard Jossem, The Ohio State University
John H. Manley, Los Alamos Scientific Laboratory
Eugen Merzbacher, University of North Carolina
Philip Morrison, Massachusetts Institute of
Technology

Lyman G. Parratt, Cornell University
George T. Reynolds, Princeton University
Robert H. Romer, Amherst College

CONSULTANTS AND CONTRIBUTORS

J. Hamilton Andrews, National Science Foundation
Arnold B. Arons, University of Washington
Wilbur V. Johnson, American Association of Physics Teachers
Justin C. Lewis, National Science Foundation
Morris Lerner, Barringer High School, Newark, New Jersey
Bernard E. Schrautemeier, Meremac Community College, St. Louis, Missouri
W. Michael Templeton, University of Washington

LIAISON REPRESENTATIVES

Marcel Bardon, National Science Foundation
Charles H. Carter, National Aeronautics and Space Administration
Earle W. Cook, Atomic Energy Commission
Paul F. Donovan, National Science Foundation

Elementary-Particle Physics

Robert G. Sachs, University of Chicago, *Chairman*
James D. Bjorken, Stanford Linear Accelerator Center
James W. Cronin, University of Chicago
Thomas Fields, Argonne National Laboratory
Louis N. Hand, Cornell University
Donald H. Miller, Northwestern University
William J. Willis, Yale University

LIAISON REPRESENTATIVES

Marcel Bardon, National Science Foundation
Albert Opp, National Aeronautics and Space Administration
Erwin R. Schmerling, National Aeronautics and Space Administration
William A. Wallenmeyer, Atomic Energy Commission

READERS

Hans A. Bethe, Cornell University
Robert Hofstadter, Stanford University
J. David Jackson, University of California, Berkeley
W. K. H. Panofsky, Stanford University

Oreste Piccioni, University of California, San Diego
D. D. Reeder, University of Wisconsin
Karl Strauch, Harvard University
Robert L. Walker, California Institute of Technology

Instrumentation

Nathan Cohn, Leeds & Northrup Company,
  Chairman
Ernest Ambler, National Bureau of Standards
W. A. Anderson, Varian Associates
J. G. Atwood, The Perkin-Elmer Corporation
T. G. Berlincourt, Colorado State University
George Downing, Merck & Company
E. C. Dunlop, E. I. du Pont de Nemours & Co., Inc.
Erwin Eichen, Ford Motor Company
J. W. Frazer, Lawrence Berkeley Laboratory
R. S. Gordon, formerly with The Monsanto Company
F. W. Luerssen, Inland Steel Company
Emery Rogers, Hewlett-Packard Company
D. B. Sinclair, General Radio Company
W. E. Vannah, The Foxboro Company

LIAISON REPRESENTATIVES

Marcel Bardon, National Science Foundation
Robert L. Butenhoff, Atomic Energy Commission
Paul F. Donovan, National Science Foundation
James H. Trainor, National Aeronautics and Space
  Administration

Nuclear Physics

Joseph Weneser, Brookhaven National Laboratory,
  Chairman
Edward Creutz, Gulf General Atomic, Inc. (until
  June 1970)
Herman Feshbach, Massachusetts Institute of
  Technology
Bernard G. Harvey, Lawrence Berkeley Laboratory
Robert Hofstadter, Stanford University
O. Lewin Keller, Jr., Oak Ridge National Laboratory
Thomas Lauritsen, California Institute of Technology
Leon Lederman, Columbia University
Malcolm H. Macfarlane, Argonne National
  Laboratory
Philip Morrison, Massachusetts Institute of
  Technology

Louis Rosen, Los Alamos Scientific Laboratory
Martin Walt, Lockheed Missile & Space Company
G. C. Wick, Columbia University
Eugene Paul Wigner, Princeton University
C. S. Wu, Columbia University
Alexander Zucker, National Academy of Sciences

Charles K. Reed, *Executive Secretary*

CONSULTANTS

Hans A. Bethe, Cornell University
G. E. Brown, State University of New York at Stony
  Brook
Keith Brueckner, University of California, San Diego
Maurice Goldhaber, Brookhaven National Laboratory
Arthur Kerman, Massachusetts Institute of Technology
Robert S. Livingston, Oak Ridge National Laboratory
Ben Mottelson, Nordisk Institute for Teorisk
  Atomfysik

LIAISON REPRESENTATIVES

Marcel Bardon, National Science Foundation
C. Neil Ammerman, Department of Defense
Paul F. Donovan, National Science Foundation
Harry Harrison, National Aeronautics and Space
  Administration
William Metscher, Department of Defense
William S. Rodney, National Science Foundation
George L. Rogosa, Atomic Energy Commission

READERS

John D. Anderson, Lawrence Livermore Laboratory
Peter Axel, University of Illinois
Thomas A. Cahill, University of California, Davis
Joseph Cerny, Lawrence Berkeley Laboratory
J. L. Fowler, Oak Ridge National Laboratory
Gerald T. Garvey, Princeton University
Lee Grodzins, Massachusetts Institute of Technology
Ernest M. Henley, University of Washington
John R. Huizenga, University of Rochester
Edwin Kashy, Michigan State University
James E. Leiss, National Bureau of Standards
Robert J. Macek, Los Alamos Scientific Laboratory
Edwin M. McMillan, Lawrence Berkeley Laboratory
G. C. Phillips, Rice University
John O. Rasmussen, Yale University
M. E. Rickey, Indiana University

D. Robson, Florida State University
John P. Schiffer, Argonne National Laboratory
Fay Ajzenberg-Selove, University of Pennsylvania
T. A. Tombrello, California Institute of Technology

Optics

W. Lewis Hyde, New York University, *Chairman*
J. G. Atwood, The Perkin-Elmer Corporation
Herbert P. Broida, University of California, Santa
  Barbara
Peter Franken, University of Michigan
Arthur Schawlow, Stanford University
Elias Snitzer, American Optical Corporation

CONTRIBUTOR

Jarus W. Quinn, Optical Society of America

LIAISON REPRESENTATIVES

John D. Magnus, National Aeronautics and Space
  Administration
Rolf M. Sinclair, National Science Foundation

READERS

John A. Armstrong, International Business Machines
  Corporation
Joseph W. Goodman, Stanford University
John Nelson Howard, Air Force Cambridge Research
  Laboratories
Charles J. Koester, American Optical Corporation
Robert V. Pole, International Business Machines
  Corporation
Roderic M. Scott, The Perkin-Elmer Corporation
F. Dow Smith, Itek Corporation

Physics in Biology

Robert Shulman, Bell Telephone Laboratories, Inc.,
  Chairman
Victor P. Bond, Brookhaven National Laboratory
Franklin Hutchinson, Yale University
Seymour H. Koenig, International Business Machines
  Corporation
Robert Langridge, Princeton University
Alexander Mauro, Rockefeller University

LIAISON REPRESENTATIVES

Eugene L. Hess, National Science Foundation
Robert W. Wood, Atomic Energy Commission
Richard S. Young, National Aeronautics and Space
Administration

Physics in Chemistry

R. Stephen Berry, University of Chicago, *Chairman*
William Goddard, California Institute of Technology
Richard E. Merrifield, E. I. du Pont de Nemours &
Co., Inc.
John Ross, Massachusetts Institute of Technology
Jerome D. Swalen, International Business Machines
Corporation

LIAISON REPRESENTATIVES

Bernard G. Achhammer, National Aeronautics and
Space Administration
Richard J. Kandel, Atomic Energy Commission
Rolf M. Sinclair, National Science Foundation
Donald K. Stevens, Atomic Energy Commission

Physics of Condensed Matter

George H. Vineyard, Brookhaven National
Laboratory, *Chairman*
Frederick C. Brown, University of Illinois
Albert M. Clogston, Sandia Corporation
Morrel H. Cohen, University of Chicago
Theodore H. Geballe, Stanford University
Hillard B. Huntington, Rensselaer Polytechnic Institute
Seymour P. Keller, International Business Machines
Corporation
Arthur S. Nowick, Columbia University
Roland W. Schmitt, General Electric Research
Laboratories

LIAISON REPRESENTATIVES

Howard W. Etzel, National Science Foundation
Harry Harrison, National Aeronautics and Space
Administration
Donald K. Stevens, Atomic Energy Commission

READERS

John Bardeen, University of Illinois
Robert J. Birgeneau, Bell Telephone Laboratories, Inc.
David H. Douglass, Jr., University of Rochester
Walter A. Harrison, Stanford University
John J. Hopfield, Princeton University
Walter Kohn, University of California, San Diego
George T. Rado, Naval Research Laboratories
Paul L. Richards, University of California, Berkeley
Ralph O. Simmons, University of Illinois
George D. Watkins, General Electric Research and
   Development Center

Plasma Physics and the Physics of Fluids

Stirling A. Colgate, New Mexico Institute of Mining
   & Technology, *Chairman*
Stanley Corrsin, The Johns Hopkins University
T. K. Fowler, Lawrence Berkeley Laboratory
Harold P. Furth, Princeton University
Uno Ingard, Massachusetts Institute of Technology
Richard J. Rosa, AVCO-Everett Research Laboratory
Ascher H. Shapiro, Massachusetts Institute of
   Technology

CONSULTANTS

Robert H. Kraichnan
Cecil Leith, National Center for Atmospheric
   Research
Douglas K. Lilly, National Center for Atmospheric
   Research

LIAISON REPRESENTATIVES

Harry Harrison, National Aeronautics and Space
   Administration
Robert L. Hirsch, Atomic Energy Commission
Erwin R. Schmerling, National Aeronautics and Space
   Administration
Rolf M. Sinclair, National Science Foundation

READERS

S. J. Buchsbaum, Bell Telephone Laboratories, Inc.
George F. Carrier, Harvard University
John M. Dawson, Princeton University
Melvin B. Gottlieb, Princeton University
Harold Grad, New York University
Robert A. Gross, Columbia University

Garrett Guest, Oak Ridge National Laboratory
Francis H. Harlow, Jr., Los Alamos Scientific
   Laboratory
Wallace D. Hayes, Princeton University
Robert L. Hirsch, Atomic Energy Commission
Russell G. Meyerand, Jr., United Aircraft Corporation
Richard Morse, Los Alamos Scientific Laboratory
Harry E. Petschek, AVCO-Everett Research Laboratory
Richard F. Post, Lawrence Berkeley Laboratory
E. L. Resler, Jr., Cornell University
Norman Rostoker, Cornell University
William R. Sears, Cornell University
R. A. Shanny, Naval Research Laboratory
A. W. Trivelpiece, University of Maryland
James L. Tuck, Los Alamos Scientific Laboratory
Peter P. Wegener, Yale University

Statistical Data

Conyers Herring, Bell Telephone Laboratories, Inc.,
   Chairman
Thomas Lauritsen, California Institute of Technology,
   Vice Chairman (until May 1970)
Fay Ajzenberg-Selove, University of Pennsylvania
Lee Grodzins, Massachusetts Institute of Technology
   (from January 1971)
Robert Herman, General Motors Corporation
Geoffrey Keller, The Ohio State University
Robert W. Keyes, International Business Machines
   Corporation
Philip M. Morse, Massachusetts Institute of
   Technology
Arthur H. Rosenfeld, Lawrence Berkeley Laboratory
Louis B. Slichter, University of California, Los
   Angeles

Ex officio

Lewis Slack, American Institute of Physics

STAFF SCIENTIST

Dale T. Teaney, International Business Machines
   Corporation

LIAISON REPRESENTATIVES

Walter C. Christensen, Department of Defense
Wayne R. Gruner, National Science Foundation
Walter E. Hughes, Atomic Energy Commission
A. W. Schardt, National Aeronautics and Space
   Administration

APPENDIX B:
PHYSICS SURVEY—
A CHARGE TO THE
SURVEY PANELS

The following are topics on which the Survey Committee requests input information from the Panels:

THE NATURE OF THE FIELD

It is vitally important that we communicate to our audiences some coherent presentation of what we believe physics is all about. Please help us in this by considering how best to present your field to (a) other physicists, (b) other scientists, (c) nonscientists. Particularly in the latter case it will be helpful to provide the Committee with what the Panel may well consider an oversimplified and overpopularized view—previous panels have erred in the opposite sense. Examples, illustrations, case history—and indeed some historical perspective generally—will be most helpful.

THE STATUS OF THE FIELD

(a) What have been the major developments (both in theory and experiment) during the past five years? If possible, put these into context with reference to the status statements in the Pake Survey and Panel reports.

(b) What are the implications of these developments for the growth of the field during the next five years?

1020

(c) What are specific examples of major changes or advances that these new developments afford? Can we do things now that were simply impossible before? Are there examples that could provide striking graphic treatment in our report?

(d) What are the present frontier areas of the field? How are these defined?

(e) Is the balance between experimental and theoretical activity in the field at a desirable level? If not, what are the Panel recommendations concerning an optimum balance and how might it be achieved?

INSTITUTIONS OF THE FIELD

(a) How is activity in this field now divided among the various types of research institutions, i.e., academic, national laboratory (e.g., Brookhaven), government laboratory (e.g., NRL), industrial laboratory, etc.?

(b) What recommendations does the Panel have concerning this balance and its possible modification in the next five years? The next decade?

(c) In this field, what are the characteristic features of activity in the different institutions?

(d) What are the interactions between these institutions? Are there areas where this interaction could or should be improved? What are the effective barriers, if any, that may prevent ready communication between, or direct exchange of, personnel for example?

INTERACTION WITH OTHER AREAS OF PHYSICS

(a) Illustrating with specific examples wherever possible, what have been the outstanding examples of interaction between this and other fields of physics recognizing that this is almost always a two-way interaction?

(b) What are specific examples of techniques—either experimental or theoretical—that cut across field boundaries? Detailed studies of selected examples would be particularly useful.

INTERACTION WITH OTHER AREAS OF SCIENCE

Questions identical to those above seem appropriate again with stress on the desirability of specific examples and possible illustrative material. The most important interactions will, of course, vary with the field; areas such as chemistry, medicine, biological sciences, ecology are obvious candidates for consideration.

INTERACTION WITH TECHNOLOGY

Research and technology have long advanced through mutual stimulation. In this field, what are the outstanding examples of such interaction in recent years? Case studies are particularly useful here. Purely as an example that has been suggested, it might be useful to consider an essay covering a tour through a modern hospital, a chemical processing plant, a paper mill, or the like, noting in passing those techniques and instruments that have arisen from work in the field. Cooperative efforts with other Panels would seem profitable. The inverse should not be neglected; some emphasis on the great dependence of research progress on technological progress is clearly indicated.

The Data Panel will attempt to arrive at methods of quantifying some of the available information in the area—both within and outside of this country. Close collaboration with the Data Panel in identifying areas of particular importance and interest would be most helpful.

INTERACTION WITH INDUSTRY

(a) Illustrating, wherever possible, with *specific* examples, what have been the outstanding interactions between this field of physics and the industrial sector in the past five years?

(b) Can any of the recent developments in the field be extrapolated, at this time, as having such interaction in the near, or distant, future?

(c) What is the inverse situation? What impact have techniques, products, or people in the industrial sector had on this field?

(d) How can the interaction between this field and the industrial sector be made more effective?

(e) It has been suggested that the development of biotechnology represents the conversion of the last of the guilds into an industry. What contributions has this field made to this conversion?

(f) Succinct case studies would be very valuable here.

INTERACTION WITH SOCIETY

(a) In what areas is the field already having major impact on questions of direct social importance?

(b) What other areas are candidates for such interaction?

(c) What aspects of training in this field are of particular importance for utilization in problems of broader social implication—which of these latter in particular?

(d) What would be the Panel recommendations concerning broader utilization of present personnel and facilities on such problems? Examples of possible situations would be most helpful.

(e) A few groups have already decided to devote some selected fraction of their effort to such activities. A discussion of such approaches would be helpful.

(f) One of the major questions facing physics (and science generally) is that of educating the nonscientific public to its very real relevance— however defined—in a technological civilization. The Survey Committee would welcome suggestions, case histories, examples, and any material that would assist in its consideration of this question for physics generally, as well as more specifically within the context of the Panel's subfield.

RELATIONSHIP TO OTHER AREAS OF SOCIAL CONCERN

Traditionally, physics has been recognized as being relevant to national defense, atomic energy, space, etc., and has enjoyed support from the corresponding federal agencies. Today our society is moving its center of concern to areas for which, at first sight, physics is less relevant: health, pollution, racial tension, etc. The new federal agencies organized to deal with such questions, such as NIH, HUD, DOT, accept much less, or no, responsibility for physics. How strong a case can be made for the relevance of your subfield to the achievement of the missions of these other agencies? In general, this will come through the help your field can give to technologies that will further these social ends: for example, the role of computers (and therefore solid-state physics) in automating hospital care. However, there may be other more direct inputs to your field that do not go through technology.

CULTURAL ASPECTS OF PHYSICS

Knowledge of the physical universe has more than utilitarian value. Each advance in fundamental understanding becomes an indestructible asset of all educated men. It is not suggested that each Panel should provide an essay on the contributions of its field to human culture, but it would be helpful in developing a broad exposition of this aspect of physics to have suggestions or compelling examples related to your field. A rather obvious concrete example: we know how old the earth is; that knowledge came through physics. Examples less obvious, and especially examples of important questions that may be answered in the foreseeable future, would be welcome.

We would welcome assistance from the Panel in answering such questions as (a) How best do we bring out the cultural relevance of physics?

(b) To what extent should our report develop the cultural arguments as a basic justification for continuing support of physics? (c) How can we best address ourselves to the resurgence of mysticism and of anti-intellectual and antiscience attitudes among students? Among the citizenry generally? (d) What is the role of physics in countering these developments?

RELATIONSHIP TO NATIONAL SECURITY ACTIVITIES

(a) What role has the field played in national defense activities?

(b) What future role is envisaged? How important is the field to these activities? Disarmament activities should be carefully considered in this context.

(c) What have been the respective roles of the different institutions of physics in this area?

(d) Again the Committee would welcome the assistance of the Panel in addressing the general questions relating to the overall interaction of physics in national security activities.

TRAINING IN THE FIELD

(a) It is often implied that contemporary graduate and postdoctoral training is becoming so narrow that students have lost the traditional breadth of outlook and flexibility expected of a physicist. Is this situation true in this field? What can be done to improve the situation? What recommendations does the Panel have for modification of contemporary training programs?

(b) In what ways is this field of particular importance for physics education?

(c) Although clearly the question relates to all of physics, can the Panel provide relevant input to the Committee concerning (i) the adequacy of current secondary school training in physics and mathematics; (ii) the effectiveness of some of the more modern secondary school curricula, e.g., PSSC; (iii) the relative intellectual standing, at the secondary school level, of those students who choose to major in undergraduate physics? (There is a widespread element of folklore that suggests that physics no longer attracts the most intellectually gifted secondary students. Can this be supported or refuted? What is the significance of this statistic in whichever case emerges?)

(d) Again, although relating to all of physics rather than to this Panel specifically, the Committee would welcome input concerning such topics as (i) what has been accomplished in bridging the gap between physics and other disciplines at the undergraduate level? How successful have general science or interdisciplinary courses been for entering—for advanced—students? How can we better illustrate the fundamental impact of physics as an underlying discipline in many areas of undergraduate education? (ii) Are teaching materials adequate? Do presently used textbooks adequately reflect the contemporary structure of physics? (iii) How important a demand for trained physicists will teaching requirements represent at established university centers—at newer campuses—at the colleges?

(e) To what extent has obsolescence of training overtaken members of the field? What can be done about it?

(f) What effective mid-career training opportunities now exist in the field? What are the Panel recommendations in this area?

(g) How effective are existing summer school programs in meeting the need for continuing training and education in the field?

(h) How effective are conferences and symposia in the field as training mechanisms?

(i) What are the Panel recommendations concerning the number and character of such conferences and symposia now available in the field?

Postdoctoral Training

(a) What is the role of the postdoctoral appointment in the field? This will, of course, be different in the different institutions.

(b) What is the average duration of the postdoctoral appointment? How has this changed with time?

(c) What has been the distribution, by nationality, of postdoctoral people in the field, and what fraction of these have remained in the United States following their postdoctoral training? How has this changed with time?

(d) How has the leveling of funding affected the availability of post-doctoral appointments in the different institutions (e.g., industrial laboratories, national laboratories, government laboratories, universities)?

Training in Applied Areas of the Field

(a) What are the applied areas that draw most heavily on this field?

(b) Does the supply of physicists in this field suffice to meet the demand in these areas?

(c) Is the current training adequate? Would modification of current training patterns be expected to open up significant new employment opportunities?

(d) It might be argued that there has been a significant failure in communication between prospective applied physics employers and the academic groups involved in the applied training. Is this true in this field? If so, how can it be improved?

(e) How is the applied work distributed with regard to the type of institutions involved?

MANPOWER PROJECTIONS

(a) What is the current population in the field, and how has this population developed since 1965 (as covered in the Pake reports) in (i) academic research, (ii) industrial research, (iii) government laboratory research, (iv) postdoctoral training, (v) graduate student training, (vi) other?

(b) During the same period what migration has occurred into—and out of—the field? What have been the major sources and recipients of this migration?

(c) In the light of current challenges in the field and/or new or anticipated facilities, what projected manpower needs can be expected in each of the above area in the next five years—the next ten (recognizing that this latter is an extreme extrapolation at best and closely related to available funding)?

(d) The argument is often advanced that the shortage of jobs requires additional funding in the field. This is more frequently reversed in Washington to imply simply that there are too many physicists being trained. What is the situation in this field?

(e) To what extent is the claim of inadequate employment opportunities legitimate (i.e., to what extent does this simply reflect the fact that for perhaps the first time physicists are not able to obtain the job that they would find most attractive)? What fraction of current PhD graduates were unsuccessful in finding employment where they were in a position to utilize their broad physics training if not their immediate specialty training?

(f) Will adequate manpower be available to staff emerging institutions in the field? How can qualified staff be attracted to and retained by such institutions?

(g) Does this field have unique or special characteristics that recommend it for consideration by an emerging institution?

(h) With leveling funding it may well be impossible for new (and indeed old) institutions to span as broad a spectrum of fields of physics as has been traditional, and while regrettable from a training viewpoint further specialization may be required in any given institution. How feasible are joint activities in this field as compared to others in physics? What recommendations would the Panel have in this difficult area?

FACILITIES

(a) Existing Facilities

(i) What are the major facilities in the field, and how are they distributed as to type?

(ii) Are the existing facilities now being utilized to full capacity? If not, explain.

(iii) How are present facilities being utilized, i.e., are they shared by more than a single group, how are decisions made regarding the research scheduling?

(iv) What are the outstanding problems now faced in the use of existing facilities?

(v) Is the distribution of existing facilities adequate?

(vi) Is modernization of the existing facilities feasible? What is the estimated effective lifetime of typical existing facilities in the field?

(vii) What criteria should be applied in reaching decisions to close down existing facilities?

(viii) To what extent do such criteria differ in different institutions (e.g., a facility might have training potential in an academic environment when it has reached a stage of unacceptable obsolescence elsewhere)? Is relocation of facilities a viable suggestion under such conditions? There are clearly pitfalls of which the receiving institution should be aware. What are they in this field?

(b) New Facilities

(i) What new facilities will be required to exploit the potential of the field? What is the priority ordering of these facilities? Please support with detailed discussion.

(ii) To what extent could existing facilities now used by other areas of physics be adapted for frontier use in this field?

(iii) What are the Panel recommendations regarding siting and operation of new facilities?

(iv) Within this field, what is an optimum balance between large centralized facilities and smaller more widely distributed ones? Please discuss.

(v) What new developments now on the horizon show promise of evolution as major facilities in the field? Is an estimate of the probable gestation period and possible cost now possible for each?

THE IMPACT OF COMPUTER TECHNIQUES ON THE FIELD

(a) What have been the outstanding impacts of computer technology in this field?

(b) Would larger and/or faster computers be of significant value? What would be the relative priority assigned to the higher costs that would be involved here as compared to other major capital needs of the field?

(c) Has any particular scheme of utilization, i.e., small local computers, institutional computer centers, regional computer centers emerged as preferable in this field?

(d) Do existing software and languages pose significant limitations in the field?

(e) What estimate does the Panel have for the present and projected utilization of computers in the field? Can a dollar level be attached to this?

(f) What impact has the field had on computer technology?

(g) Are there outstanding examples of studies that would simply have been impossible without sophisticated computer utilization? Specific examples would be most useful.

Cost Increases

(a) Selecting, say, ten instruments much used in the field spanning the cost range involved—how have the individual costs varied with time in the last decade?

(b) How has the average (very crudely defined) overall cost of an experiment, typical of those at the frontier of the field at the time, varied with time in the last decade?

(c) How have average postdoctoral and student training costs varied over the same interval? It would be advantageous to consider experimental and theoretical situations separately in this instance.

(d) Illustrating with specific examples, what would be a reasonable annual estimate of the cost escalation in the field reflecting increasing sophistication of the studies themselves? Reflecting aging of the institutional staff?

(e) To what extent is progress in the field *really* dependent upon the availability of the most modern instrumentation? It has been suggested that in some fields the instrumentation has become over-sophisticated, over-flossy and that in at least some instances the Ferrari could be replaced by a Ford without undue restriction of the research quality and productivity. To what extent is this suggestion true in this field? To what extent can (and should) it be countered? Specific illustrations and examples would be extremely helpful here.

Funding Levels

(a) What have been the actual funding levels *and* expenditure levels annually in the field since 1965? Compare these with the Pake Report projections. Insofar as possible, separate academic, industrial, and governmental laboratory operations for consideration. In some instances the leveling off of federal funding has been counteracted, for a time at least, by infusions of institutional funds, so that actual expenditure levels have not tracked funding limitations. What information is available on such phenomena in this field?

FUNDING MECHANISMS

(a) How has the available funding been distributed among these sources: federal (AEC, DOD, NSF, NASA, others), state, industrial, local (university contributions, etc.), foundations, and other sources?

(b) How does the funding process actually work for each of the above sources? What are the relative distributions, advantages, disadvantages, etc., of grants and of contracts? What are the effective differences between these two approaches? What improvements might be suggested?

(c) What is the relative importance of project and of institutional grants in this field?

(d) How are decisions reached concerning grant and contract applications? Please comment on the decision-making processes at the national level—for example, by administrators in the various federal agencies and by advisory committees to these agencies. Is the present practice satisfactory or would change be desirable? What are the Panel recommendations?

THE IMPACT OF LEVELING FUNDING

(a) Discuss in some detail, with specific illustrations, the overall impact of leveling funding on the field. The following subtopics might prove useful:

    (i) Utilization of current facilities

    (ii) Exploitation of new discoveries

    (iii) Employment of physicists

    (iv) Support of the young researcher

    (v) Alienation of young physicists

    (vi) Possible new approaches to training in the field

    (vii) The support of offbeat proposals. There is always a tendency, under limited funding conditions, to eschew risk or adventure, to bet on the sure thing.

    (viii) Long-range implications for the field generally.

(b) It is clear that level funding is not synonymous with level productivity. The Committee will welcome case histories, etc., to illustrate this general point.

(c) What are the relative advantages of expanding (or contracting) activities in this field by expanding (or contracting) the size of existing groups active in the field as opposed to proliferating (or reducing) the number of such groups?

FUNDING PROJECTIONS

In the past, survey reports have generally made specific projections and recommendations which have very often been negated by large departures of the total budgets available from those on which the recommendations were based. To be responsive, our Report must provide for a spectrum of possible situations; in doing so it must carefully spell out, in as detailed fashion as possible, both the short- and long-range consequences of funding at levels below those necessary for both orderly growth and exploitation of new developments in each of the fields of physics. Specific examples and case histories will be particularly effective in illustrating these consequences. With these points in mind,

(a) What level of funding, quite apart from any current estimate of future funding, would be required to enable this field to realize its full potential during the next five years? The next ten years? How would it be distributed broadly over the subareas of the field—recognizing that detailed projections are, in many cases, impossible?

(b) Consider a spectrum of possibilities ranging downward in 10% increments from that developed above to a level some 10% below that currently in effect. At each step indicate as clearly as possible,

(i) What opportunities would be missed—what developments would not be exploited?

(ii) What new facilities would necessarily be postponed or eliminated entirely from consideration?

(iii) What programs or facilities would necessarily be phased out or closed down?

(iv) What would the impact be on the manpower and employment situation?

(c) A detailed discussion of the basic issues that underlie the Panel's assignment of priorities within the field would be an essential component of the Panel report. It is essential that long-range implications be developed realistically; it is essential that we not predict greater catastrophic impact than can be clearly justified.

(d) Separate discussion of major new facilities—in order of priority—with careful discussion of the bases for the priority ordering and of the relative justifications will be particularly important.

(e) The question of laboratories, as distinct from facilities, will be appropriate in some fields. The need for and justification of such laboratories will require careful consideration. What are the recommended criteria for closing down an existing laboratory in this field? To what extent are the laboratories in the field adaptable to broader use and to alternate modes of support during periods of fiscal stringency?

(f) To what extent can the Panel assist in developing a balanced presentation of the overall impact on the continuity of physics (i.e., the faucet effect—it is not generally appreciated that the re-emergence of funding after an indeterminate drought will not guarantee re-emergence of a healthy physics—or science—community)? Can this be quantified in this field? Are there relevant examples or case histories?

(g) A clear statement of the basic fiscal assumptions underlying the Panel projections is essential. The Data Panel will provide basic information concerning inflation rates, etc., which should be used systematically by all Panels to permit later direct comparisons by the Committee.

PHYSICS DATA IN THE FIELD

(a) How effective is communication of scientific information in the field generally? Are there adequate review articles—conferences and conference proceedings? Are there too many of the latter?

(b) What is the role of the preprint in this field? Is the present system effective?

(c) How adequate are the present data compilation and dissemination mechanisms in this field?

(d) What are the Panel recommendations in this area? Are new approaches or mechanisms required? How can manpower, adequate both in quantity and quality, be integrated into the data compilation activities?

(e) What is the estimated cost involved?

(f) Quite apart from data communication and compilation within the field, (i) how effective is communication with related fields that may have need of your data, and (ii) how effective are your data formats and presentations for their use?

INTERNATIONAL ASPECTS

(a) Where does this field in the United States at the present time stand with respect to the same field abroad?

(b) How does U.S. activity in the field compare on a manpower or funding basis with that in the most active foreign countries? What are the relative growth rates? What are the major points of similarity or dissimilarity in the overall programs? What has been the significance of the different funding techniques and levels?

(c) What international cooperation now exists? What would be the

direct and indirect benefits to the United States in expanding such cooperation in this field? Are there particular facilities that should be considered
in this light?

(d) What problems now exist with regard to the implementation of
foreign cooperation and exchanges? Have problems been encountered in
the obtaining of requisite visas—of permission to travel freely across international boundaries—of access to national or governmental laboratories
in this country or abroad?

(e) What is the situation vis-à-vis international cooperation in physics
in the industrial sector? Are there outstanding difficulties in this area?
How important is fostering of such cooperation in this field?

(f) To what extent does this field encompass well-defined national
schools of thought (e.g., the Copenhagen School in quantum mechanics
and nuclear physics)?

(g) What has been the impact of foreign work and foreign research
centers on activity in this field in this country?

(h) How do developing countries attain critical mass in this field? Are
there specific mechanisms in this area? Should there be?

(i) What international laboratories *should* be developed in this field?
Upon what criteria should the establishment of such laboratories be based?

ILLUSTRATIVE MATERIAL FOR THE SURVEY REPORT

It will be particularly important that the Committee receive from each Panel
a selection of illustrations and photographs carefully selected to highlight
progress or particularly interesting vignettes in each field. It would be helpful if the Panels would address themselves to this request at an early stage
of their deliberations. The members of the Data Panel will devote considerable effort to the development of new techniques for the presentation of
statistical data and will cooperate closely with each of the subfield Panels.

INDEX